Advances in Intelligent and Soft Computing

169

For further volumes:
http://www.springer.com/series/4240

David Jin and Sally Lin (Eds.)

Advances in Computer Science and Information Engineering

Volume 2

Editors
David Jin
Wuhan Section of ISER Association
Wuhan
China

Sally Lin
Wuhan Section of ISER Association
Wuhan
China

ISSN 1867-5662 e-ISSN 1867-5670
ISBN 978-3-642-30222-0 e-ISBN 978-3-642-30223-7
DOI 10.1007/978-3-642-30223-7
Springer Heidelberg New York Dordrecht London

Library of Congress Control Number: 2012937219

Printed on acid-free paper

Springer is part of Springer Science+Business Media (www.springer.com)

Preface

In the proceeding of CSIE2012, you can learn much more knowledge about Computer Science and Information Engineering all around the world. The main role of the proceeding is to be used as an exchange pillar for researchers who are working in the mentioned field. In order to meet high standard of Springer, the organization committee has made their efforts to do the following things. Firstly, poor quality paper has been refused after reviewing course by anonymous referee experts. Secondly, periodically review meetings have been held around the reviewers about five times for exchanging reviewing suggestions. Finally, the conference organization had several preliminary sessions before the conference. Through efforts of different people and departments, the conference will be successful and fruitful.

During the organization course, we have got help from different people, different departments, different institutions. Here, we would like to show our first sincere thanks to publishers of Springer, AISC series for their kind and enthusiastic help and best support for our conference.

In a word, it is the different team efforts that they make our conference be successful on May 19–20, Zhengzhou, China. We hope that all of participants can give us good suggestions to improve our working efficiency and service in the future. And we also hope to get your supporting all the way. Next year, In 2013, we look forward to seeing all of you at CSIE2013.

March 2012 CSIE2012 Committee

Committee

Honor Chairs

Prof. Chen Bin	Beijing Normal University, China
Prof. Hu Chen	Peking University, China
Chunhua Tan	Beijing Normal University, China
Helen Zhang	University of Munich, China

Program Committee Chairs

Xiong Huang	International Science & Education Researcher Association, China
Li Ding	International Science & Education Researcher Association, China
Zhihua Xu	International Science & Education Researcher Association, China

Organizing Chair

ZongMing Tu	Beijing Gireida Education Co. Ltd, China
Jijun Wang	Beijing Spon Technology Research Institution, China
Quanxiang	Beijing Prophet Science and Education Research Center, China

Publication Chair

Song Lin	International Science & Education Researcher Association, China
Xiong Huang	International Science & Education Researcher Association, China

International Committees

Co-sponsored by

International Science & Education Researcher Association, China
VIP Information Conference Center, China
Beijing Gireda Research Center, China

Reviewers of CSIE2012

Z.P. Lv	Huazhong University of Science and Technology
Q. Huang	Huazhong University of Science and Technology
Helen Li	Yangtze University
Sara He	Wuhan Textile University
Jack Ma	Wuhan Textile University
George Liu	Huaxia College Wuhan Polytechnic University
Hanley Wang	Wuchang University of Technology
Diana Yu	Huazhong University of Science and Technology
Anna Tian	Wuchang University of Technology
Fitch Chen	Zhongshan University
David Bai	Nanjing University of Technology
Y. Li	South China Normal University
Harry Song	Guangzhou Univeristy
Lida Cai	Jinan University
Kelly Huang	Jinan University
Zelle Guo	Guangzhou Medical College
Gelen Huang	Guangzhou University
David Miao	Tongji University
Charles Wei	Nanjing University of Technology
Carl Wu	Jiangsu University of Science and Technology
Senon Gao	Jiangsu University of Science and Technology
X.H. Zhan	Nanjing University of Aeronautics
Tab Li	Dalian University of Technology (City College)
J.G. Cao	Beijing University of Science and Technology
Gabriel Liu	Southwest University
Garry Li	Zhengzhou University
Aaron Ma	North China Electric Power University
Torry Yu	Shenyang Polytechnic University
Navy Hu	Qingdao University of Science and Technology
Jacob Shen	Hebei University of Engineering

Contents

Vehicle Style Recognition Based on Image Processing and Neural Network

ZhengWei Zhu and YuYing Guo

College of Information Engineering,
Southwest University of Science and Technology,
Mianyang, 621010, China
zhuzwin@163.com

Abstract. A vehicle style recognition method using computer vision, image processing and RBF neural network is presented in the paper. First, a vehicle side image is acquired by using a high-speed vidicon. Then a vehicle edge outline image was obtained by a series of image processing and the vehicle features are extracted from the edge outline image. Finally, the vehicle is recognized and classified using a RBF neural network. Experimental results show that the proposed method has a good classification effect in the practical application of vehicle style recognition at vehicle toll stations.

Keywords: vehicle, style recognition, feature extraction, image processing, RBF neural network.

1 Introduction

At present, the traffic and transport of China is very busy. In order to improve the efficiency of transport, it is significant to introduce a modern intelligent vehicle style recognition technology into the traffic control system. Aiming at this problem, a vehicle style recognition method using computer vision, image processing and RBF neural network is presented in the paper.

2 Flow of Vehicle Type Recognition

To recognize the type of a vehicle, it is necessary to select and extract some efficient features about the vehicle. We shoot a side vehicle image with a vidicon installed on the side of a highway first. Then a series of image processing is executed for the vehicle image and some efficient vehicle features are extracted. Finally the type of the vehicle is recognized using a RBF neural network. The block diagram of the system is shown in Fig.1.

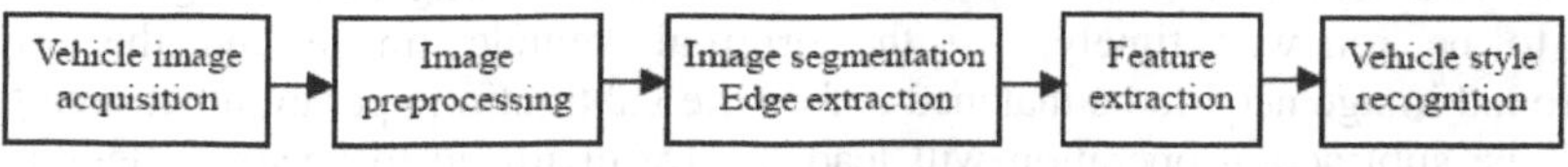

Fig. 1. Flow of vehicle type recognition

D. Jin and S. Lin (Eds.): Advances in CSIE, Vol. 2, AISC 169, pp. 1–8.
springerlink.com

3 Vehicle Image Acquisition and Processing

3.1 Acquisition and Preprocessing of Original Vehicle and Background Image

The side image of the vehicle is dynamically acquired by a high-speed vidicon installed on the side of a road. An original background image and an original side vehicle image acquired by the vidicon are shown in Fig.2(a) and (b). In the process of the vehicle image acquisition, transmission and transformation, the image is often stained by various noises. So to improve the recognition rate of vehicle types, the original background and side vehicle image must usually be preprocessed before the feature extraction. The preprocessing includes noise filtering and edge enhancement, which can remove the disturbing noise and outstand the vehicle features.

In our system, median filtering is used for removing the image noise. To improve the speed of filtering, we presented a fast algorithm of median filtering: When the algorithm seeks the median of current window, it takes only the impact of the pixels moving in and out of front window over current window into account, so it can reduce the comparison times effectively and quicken the speed of median filtering greatly.

Filtering can remove or weaken the noise, but it simultaneously makes the image outline illegible, which is harmful for subsequent vehicle type recognition. To solve the problem, we use Sobel operator to generate an outline-enhanced image according to the below-mentioned method:

Supposing that $f(x, y)$ represents the original image, G represents the sharp operation such as Sobel operator and $g(x, y)$ represents the outline-enhanced image, so we have:

$$g(x,y) = \begin{cases} G[f(x,y)], & G[f(x,y)] \geq T \\ f(x,y), & others \end{cases} \tag{1}$$

Where T is a threshold value, which is nonnegative. If only we select T appropriately, it can make the image outline outstanding and don't destroy the background, in which the gray change of the pixels is mild.

3.2 Procurement of Pure Vehicle Image

By taking a subtraction operation between the original vehicle image and the original background image, we can get a pure vehicle image. Supposing that $g_1(x,y)$ and $g_2(x,y)$ represent the vehicle image and the background image respectively, $h(x, y)$ represents the result image of the subtraction operation, *Zero* represents the gray value of dark pixels, so we have:

$$h(x,y) = \begin{cases} g_1(x,y) - g_2(x,y), & g_1(x,y) > g_2(x,y) \\ Zero, & g_1(x,y) \leq g_2(x,y) \end{cases} \tag{2}$$

After the subtraction operation, we obtain a *pure* vehicle image, shown in Fig.2(c).

Where two aspects need to be pointed out: (1) if the background image changes, it needs to be renewed timely; (2) the original vehicle image and the original background image need to be matched before the subtraction operation between them, or else the subtraction operation will lead the result image to become illegible and affect subsequent vehicle style recognition.

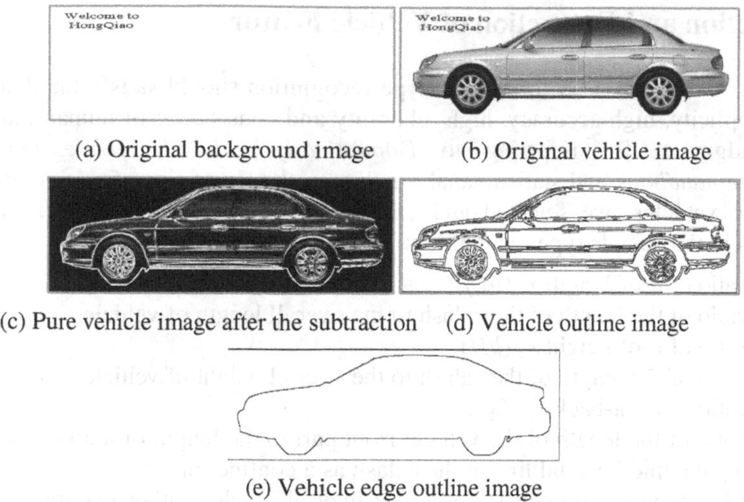

(a) Original background image (b) Original vehicle image

(c) Pure vehicle image after the subtraction (d) Vehicle outline image

(e) Vehicle edge outline image

Fig. 2. Vehicle image processing

3.3 Extraction of Vehicle Edge Outline Image

To implement automatic vehicle style recognition and classification, the *pure* vehicle image obtained by the subtraction operation needs to be processed as follows:

(1) Transform the *pure* vehicle image into a binary image by using a threshold image segmentation operation

(2) Remove isolated points and isolated lines

(3) Filling processing

First we take some transverse filling processing for every horizontal scanning line, that is, first find from left to right a bright dot whose gray value is 255 and note down its location (i_1, j_1), then find from right to left a bright dot whose gray value is also 255 in the same scanning line and note down its location (i_2, j_2), finally the gray value of all the pixels between the two points is transformed into 255. After transverse filling processing for every horizontal line, we take longitudinal filling processing for every vertical row, its processing method is similar to the above transverse filling processing, where not to repeat.

(4) Pruning

First set four parameters d_1, d_2, l_1 and l_2, and then judge the length size of every horizontal bright line or every vertical bright row in the image from four directions namely top, bottom, left and right respectively, if its value is smaller than the above given parameter, then the gray value of all the pixels in the line or row is set as 0, or else the gray value of the bright line or row is reserved.

(5) Edge extraction

By extracting the edge of the filled and pruned vehicle image, we can get a continuous expected vehicle edge outline image ultimately, shown in Fig.2 (e).

4 Selection and Extraction of Vehicle Features

The features which are used for vehicle type recognition should satisfy the demands such as simplicity, high accuracy, high reliability and consistence of human-machine in the judgment. By referring to *Domestically-produced vehicle technical performance handbook* and statistic analysis data to the shape size of a large amount of domestic vehicles, we selected and extracted the following parameters as the eigenvalues in the vehicle type recognition:

(1) The ratio of roof-length x_1 (l/L)

It is the ratio of the length of the calash to the overall length of vehicle.

(2) The ratio of roof-height x_2 (l/H)

It is the ratio of the length of the calash to the overall height of vehicle.

(3) The ratio of front-back x_3 (L_1/L_2)

It is the ratio of the length of the vehicle front part to the length of the vehicle back part by taking the middle axial line of the calash as a confine line.

Besides the above-mentioned features, we may also select other features such as the relative area of the vehicle side image to the entire image, the quantity of the axles, the wheeltrack and the distance between the axles, etc.

In the following part, we will adopt three above-mentioned eigenvalues to recognize the vehicle type.

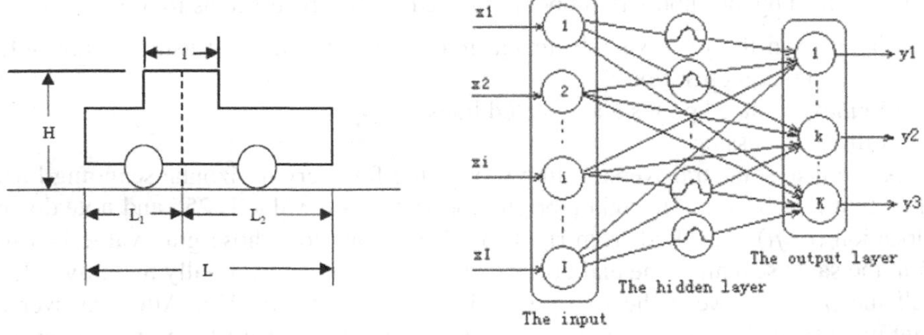

Fig. 3. Feature parameters of a vehicle

Fig. 4. Structure of RBF network

5 Design of Vehicle Type Classifier

5.1 Network Model and Network Structure

We adopt a RBF network to classify the vehicle type in the system. The RBF network takes the vehicle eigenvector $X = (x_1, x_2, x_3)$ as the input, the vehicle type code (where we supposed that (1 0 0) represents the passenger car, (0 1 0) represents the carriage truck and (0 0 1) represents the saloon car) as the expected network output $T = (t_1, t_2, t_3)$, and the actual network output is expressed as $Y = (y_1, y_2, y_3)$. The structure of the

RBF network is shown in Fig.4. It is composed of three layers, where $I = 3$ and $K = 3$. In the network, the input layer only transfers the input signal to the hidden layer, the nodes of the hidden layer are composed of some radial basis functions and the nodes of the output layer are usually some simple linear functions.

The network output may be expressed as follows:

$$y_k(x) = \sum_{j=1}^{J} \omega_{kj}\phi_j(x - c_j),\ j = 1,2,\cdots,J \tag{3}$$

Where J is the quantity of the neurons in the hidden layer, ω_{kj} is the weight between the neuron j in the hidden layer and the neuron k in the output layer, $\phi_j(\ )$ is the RBF kernel function, x is the input vector and c_j is the centre vector of the jth RBF. $\phi_j(\ )$ exists various formations to be selected, where we adopt the following Gauss-model kernel function:

$$\phi_j(x - c_j) = \exp\left[-\frac{(x - c_j)^T(x - c_j)}{2\sigma_j^{\,2}}\right] \tag{4}$$

5.2 Learning Algorithm

A two-stage learning algorithm presented by J. Moody and C. J. Darken is used in the RBF network. In the first stage of the algorithm, the centre location c_j and the width σ_j of radial basis function of every node are decided according to the overall input samples, which usually adopt a clustering learning method with no teacher such as the K-means method (in which the clustering quantity is fixed). After the parameters of the hidden layer are decided, the second stage will be taken. The weight ω_{kj} between the hidden and output layer is decided by taking the least square criterion and a learning method with teacher such as the LMS algorithm in the stage. After the two stages of learning, the parameters still need to be tinily readjusted by a learning method with teacher. If J represents the quantity of clustering and I represents the quantity of data, then the steps of the algorithm are as follows:

1) Initialize the clustering centre vector

First to ascertain J initial clustering centre vectors c_{10}, c_{20}, $\cdots$, c_{J0}. Usually the initial J data in the set of the data are used as the initial clustering centre vector.

2) Input the sample data and select the corresponding clustering

Input the sample data x_i and select the nearest clustering j^*,

$$c_{j^*} = \arg\min_j \|x_i - c_j\| \tag{5}$$

The operation is executed for all the sample data (i=1, 2, …, I).

3) Update the clustering centre vector

The clustering centre vector is updated according to the following method:

$$c_j^{new} = \frac{1}{N_j} \sum_{x \in cluster(j)} x \tag{6}$$

Where N_j is the quantity of the data in the J clustering. The operation is executed for all the data (i=1, 2, …, I) as well.

4) Decision-making

If all the clustering centre vectors have no changes, then go to carry out the second stage, or else return to the above step 2).

In the second stage, the weight ω_{kj} between the hidden and output layer is decided by using a learning method with teacher. In the paper we take the LMS algorithm, whose steps are as follows:

5) Initialize the weight ω_{kj} using a comparatively small random datum.

6) Calculate the network output

Input the sample data and calculate the network output according to the formulas (3) and (4). We adopt the average distance between the clustering centre and the datum which belongs to the clustering as the RBF width σ_j in the formula (4), the formula of calculating the clustering centre is:

$$\sigma_j^2 = \frac{1}{N_j} \sum_{x \in cluster(j)} (x - c_j)^T (x - c_j) \tag{7}$$

7) Update the weights

First calculate the output error of every neuron in the output layer:

$$e_k = d_k - y_k(x) \tag{8}$$

Where d_k is the expected output of the output neuron k. Then update the weight by using the following formula:

$$\omega_{kj}^{new} = \omega_{kj}^{old} + \eta e_k \phi_j (x - c_j) \tag{9}$$

Where η is a learning constant.

8) Decision-making

If the given termination condition is satisfied, then end the overall process, or else return to the above step 6).

5.3 Establishment and Testing of RBF Network

In order to train and test the presented network, we acquire 60 samples including three vehicle types, in which every vehicle type (in which includes various styles) includes 20 samples. We take randomly eight sample data in every vehicle type to be used for network learning (training), other twelve sample data to be used for testing. The training data are shown in Table 1. The partial testing data and testing results are shown in Table 2. We can see from the testing results that the presented method in the paper can recognize effectively various vehicle types.

Table 1. Training data

Vehicle style	Sample number	Vehicle sample eigenvector			Sample number	Vehicle sample eigenvector			Vehicle style code		
		x_1 (l/L)	x_2 (l/H)	x_3 (L_1/L_2)		x_1 (l/L)	x_2 (l/H)	x_3 (L_1/L_2)	t_1	t_2	t_3
Passenger car	1	0.8645	2.3053	0.9288	5	0.7797	1.5103	0.9315	1	0	0
	2	0.8329	1.9963	0.9625	6	0.8805	1.5903	0.9466			
	3	0.9498	2.4015	0.9466	7	0.7453	1.3771	0.9543			
	4	0.8735	2.4342	0.8956	8	0.8456	1.7366	0.9721			
Carriage truck	9	0.1473	0.4372	0.3926	13	0.2311	0.9459	0.1306	0	1	0
	10	0.1526	0.4592	0.4588	14	0.1534	0.5280	0.5172			
	11	0.3421	1.0006	0.5166	15	0.3005	0.8357	0.1874			
	12	0.3164	0.9250	0.1875	16	0.3926	1.0277	0.1875			
Saloon car	17	0.3962	0.7500	1.5238	21	0.3319	1.0754	1.4878	0	0	1
	18	0.3687	0.8138	2.8823	22	0.3745	1.3827	1.3600			
	19	0.4444	1.0012	2.1818	23	0.2747	0.9566	1.4878			
	20	0.2921	0.9845	2.3000	24	0.3331	1.1811	1.5578			

Table 2. Partial testing data and testing results

Vehicle style	Sample number	Vehicle sample eigenvector (Input sample vector)			Expected output			Actual network output		
		x_1 (l/L)	x_2 (l/H)	x_3 (L_1/L_2)	t_1	t_2	t_3	y_1	y_2	y_3
Passenger car	25	0.8819	2.3588	0.9156	1	0	0	1.005	0.0106	-0.0160
	26	0.9112	2.4393	0.9547				1.078	-0.0805	0.0021
	27	0.9340	2.8712	0.9912				0.8401	-0.3508	0.5107
	28	0.9740	2.5451	0.9702				1.0435	-0.1064	0.0629
Carriage truck	29	0.1874	0.8576	0.2628	0	1	0	-0.0202	1.0801	-0.0600
	30	0.4334	1.0677	0.5166				0.1647	0.8965	-0.0612
	31	0.1473	0.6312	0.5666				-0.0446	1.005	0.0397
	32	0.3227	0.7238	0.1875				0.0800	0.9757	0.1043
Saloon car	33	0.3095	1.0331	1.3225	0	0	1	-0.0401	0.1198	0.9203
	34	0.2675	0.8105	1.4148				0.0221	0.0216	0.9563
	35	0.3282	1.1258	1.4528				-0.0128	0.0269	0.9858
	36	0.4578	1.4347	1.8372				-0.0320	0.0247	1.007

6 Conclusions

We can see from Table 2, the actual network output is consistent with the expected output generally. The system can accurately recognize and classify 58 samples among 60 samples by using the above well-trained RBF network. The experimental results indicate that when we select the ratio of roof-length, the ratio of roof-height and the ratio of front-back as the vehicle features and adopt the presented method to classify the passenger car, the carriage truck and the saloon car, the system has a few computation, a quick recognition speed and a high recognition accuracy. The accurate rate of the vehicle type recognition can reach 96.7%. So the system has a good recognition capability of vehicle types.

Acknowledgment. The work is supported by National Natural Science Foundation of China (Grant No.61105020).

References

1. Petrovic, V.S., Cootes, T.F.: Vehicle type recognition with match refinement. In: Proceedings of the 17th International Conference on Pattern Recognition, UK, pp. 956–961 (2004)

2. Zheng, M.X., Gotoh, T., Shiohara, M.: A hierarchical algorithm for vehicle model type recognition on time-sequence road images. In: Proceedings of the IEEE ITSC 2006, Canada, pp. 542–547 (2006)
3. Gupte, S., Masoud, O., Martin, R.F.K., Papanikolopoulos, N.P.: Detection and classification of vehicles. IEEE Trans. Intell. Transport Syst. 3(1), 37–47 (2002)
4. Tang, K.W., Kak, S.: A new corner classification approach to neural network training. Circuits Systems and Signal Processing 17(3), 459–469 (1998)
5. Zhou, H.X.: Application of neural network in automatic recognition of vehicle style. Microcomputer Applications 24(3), 161–164 (2003)
6. Zhang, X.J., Feng, H.W.: Automatic vehicle classification based on radial basis function neural network. Science Journal of Northwest University Online 4(2), 1–7 (2006)
7. Liu, S.F., Liu, Y.B.: Research of Vehicle Type Recognition Technology. Computer and Digital Engineering 33(1), 71–76 (2005)
8. Zhang, Y.X., Fan, D.Q., et al.: Study on Automatic Vehicle Recognition System. Traffic and Transport 33(1), 45–48 (2006)
9. Shen, Q., Wang, T.: A Learning Algorithms for Optimizing RBF Neural Network Structure. Microelectronics and Computer 23(4), 14–18 (2000)
10. Feisi R&D centre of technological product. Neural network theory and MATLAB 7 realization. The electronic industry publishing house, Beijing (2005)

Feature Model Application in No-Mold-Drawing Control

ShuRong Ning, ShengLi Gan, ChaoEn Xiao, and LinJie Miao

School of Computer and Communication Engineering
University of Science and Technology Beijing
Beijing, China
fancyning@163.com

Abstract. Model in temperature field analytical approach is very complex. The relationship between parameters is not binding with model of high dimension, not conducive to actual control. Based on the experimental data, this work gives a relational model of feature-based model drawing ratio and diameter difference under a specific power. It extracts the temperature parameters from the model and dealt with separately. In addition, it reduces the model dimensions and provides a new way of thinking for the research of shape model of no mold drawing deformation zone. It also verifies the superiority of this model through experimental equipment.

Keywords: no-mold-drawing, temperature field analysis, feature model control.

1 Introduction

No mold drawing forming technology [1] is a flexible near-net shape technology with high-efficiency, high-precision, low-power, non-polluting, non-friction, lubrication-free and little or no cutting. During the process of no mold drawing forming, some phenomenon exists such as the difficult of precise temperature control, poor uniformity of product size (For example, metal wire along the axial surface is easy to produce wavy or bamboo-like defects), unstable product quality, and prone to breakage. For the phenomenon, this paper proposes a solution.

2 Feature Model of No Mold Drawing

The so-called feature model combines dynamic model of controlled object. Environmental characteristics and control performance request that we must establish a model which is easy-to-controller designed and simple and practical [2]. The basic idea of feature modeling is regarding the object as a black box system [2][3].

When analyzing the basic equation with no mold drawing temperature of deformation zone, for all kinds of the same nature of the material, we assume that the physical performance parameter is constant. That is the thermal conductivity, heat capacity and density of materials and other parameters have nothing to do with the temperature. So in the cylindrical coordinate system of the thermal conductivity equation is:

$$\frac{\partial^2 T}{\partial r^2}+\frac{1}{r}\frac{\partial T}{\partial r}+\frac{1}{r^2}\frac{\partial^2 T}{\partial \theta^2}+\frac{\partial^2 T}{\partial z^2}+\frac{qv}{k}=\frac{pc}{k}\frac{\partial T}{\partial r}. \tag{1}$$

D. Jin and S. Lin (Eds.): Advances in CSIE, Vol. 2, AISC 169, pp. 9–14.
springerlink.com

Where: T is the temperature, ℃, qv is the internal heat source strength, k is the thermal conductivity, ρ is the material density.

No mold drawing process is the use of induction heating coil with Partial depth of the heating the workpiece material, and then drawing deformation. Because cooling water only needs to calculate heat transfer coefficient and then apply to the wire on the drawing, so the model is not required. Therefore, in the simulation we should mainly select the part of the heating section as the basis for modeling. In the early on the basis of exploration of a large number of numerical simulation, with the actual experimental device, in this simulation study, modeling the length of wire drawing is taken as 70mm (Including the deformation zone and part of the undeformed zone).

Use the current-density to load the induction heating coil. In the process of modeling, we should use a rectangular represent of the induction heating coil cross-section. The width of the rectangle is equal to the diameter of induction heating coil. Then you can determine the length according to the different number of turns of induction heating coil, and establish calculation model. As the induction heating coil and pull wire are axially symmetric, obviously it is an axisymmetric problem for the wire of no mold drawing. For the problem of axis of symmetry, equation (1) can be simplified as:

$$\frac{1}{r}\frac{\partial}{\partial r}\left(rk\frac{\partial T}{\partial z}\right)+\frac{1}{\partial r}\left(k\frac{\partial T}{\partial z}\right)+q=\rho c\frac{\partial T}{\partial t}\,. \tag{2}$$

Assuming a constant ambient temperature, consider the nickel-titanium shape memory alloy wire contact with the surrounding air for heat transfer and the convection, radiation heat transfer, etc. The equation of feature model of no mold drawing wire reduces to:

$$[C]\{U\}+[K]\{T\}=\{Q\}\,. \tag{3}$$

Where: [C], [K] are the overall heat capacity matrix and the overall heat transfer matrix. {U} is the temperature rate vector. {T} is the temperature vector. {Q} is the heat flux vector node. As can be seen by the above formula, heat flux vector node has a relation with temperature rate vector and temperature vector.

In the no mold drawing process of forming, temperature distribution and changes will directly affect the stress field and strain field distribution and changes. Therefore, analysis of no mode drawing temperature field is very important to the actual process of development [4]. As inside the no mold drawing deformation zone, influence the final shape of the wire is strong coupling between factors and non-linear, and also has a large time lag. It is difficult to establish accurate mathematical model between various factors and drawing the final shape of deformation zone [5]. Therefore, the model which is established on the basis of temperature field analysis is very complex and not conducive to effective control. In view of this paper, we consider the temperature extracted from the model and for separately controlling. This goal can be achieved that reduce the model dimensions to improve the controllability.

3 Relationship Model between Drawing Ratio and Diameter Difference

In the process of forming test to no mold drawing, from detecting a change in diameter difference to adjustment process to the target diameter difference, there is a lag links which means there is a time delay. For this reason, from the engineering point of view combined with the characteristics of the test, the relational model of drawing ratio and diameter difference in line with the following characteristics model:

$$y(k+1) = f_1(k)y(k) + f_2(k)y(k-1) + g_0(k)u(k-k_0) + g_1(k)u(k-k_0-1)\,. \quad (4)$$

Where, k0 means delay time corresponding to the sampling cycles. In this experiment, the sampling period is 0.5s.

The target diameter has relation with the speed of the feed and drawing speed ratio. Under a specific power, the changes of diameter differences comply with this rule: $y=f(v1/v2,\Delta t)+g0$. Here y -- Tolerance, mm; v1 -- drawing speed, mm/s; v2 -- the feed rate, mm/s; Δ t -- delay time, s; g0 -- a constant.

3.1 Data Collection

During the simulation, We make the following assumptions on the actual situation: (a) material of the heating, cooling and deformation is axisymmetric; (b) material is isotropic; (c) Physical properties of the material parameters (such as thermal conductivity, heat capacity and density) is independent of temperature; (d) The heat of deformation is small compared to the heat of heating and absorption, so it can be ignored; (e) Radial deformation rate had no effect on the temperature.

In the course of the experiment, under a specific heating power (Laboratory set to 30KW), feed speed is set to 0.5mm/s. The drawing speed from 0.54-1.20mm/s is divided into 20 groups. Do a drawing experiment each group. Each experiment collects a data to 0.5s as a time domain. Total 1000 data were collected. After finishing the experimental, data is shown in Table1.

The initial diameter of test bars is 6mm. According to the experimental data and the definition of 1, we can fit a feature model of drawing ratio and diameter difference. Taking into account the use of simulation systems to do comparison tests, time lag can be reduced significantly, which mean CPU computation model time is only a subtle level. Therefore, the delay time Δt can be extracted from the model dealt with separately. So, we can achieve the purpose of reducing the model dimension and improving the availability of the model.

Table 1. The experimental data of drawing ratio and diameter difference

No.	Drawing ratio	Diameter difference (mm)
1	1.083	0.302
2	1.181	0.465
3	1.231	0.587
4	1.251	0.625
……	……	……
19	2.215	1.927
20	2.298	2.099

3.2 The Establishment of This Model

After repeating fit comparison by MATLAB tools, when x is taken to the fifth order, it is the closest to the actual path difference. So we take the model order for the fifth-order. Drawing ratio satisfies the following relationship with the diameter difference:

$$Y(x) = p1\times x^5 + p2\times x^4 + p3\times x^3 + p4\times x^2 + p5\times x + p6. \tag{5}$$

Where Y (x) -- Diameter difference; x -- drawing ratio; $p1$ = -0.5468; $p2$ = 5.774; $p3$ = -22.64; $p4$ = 41.35; $p5$ = -33.72; $p6$ = 9.949. Feature model diagram as follows:

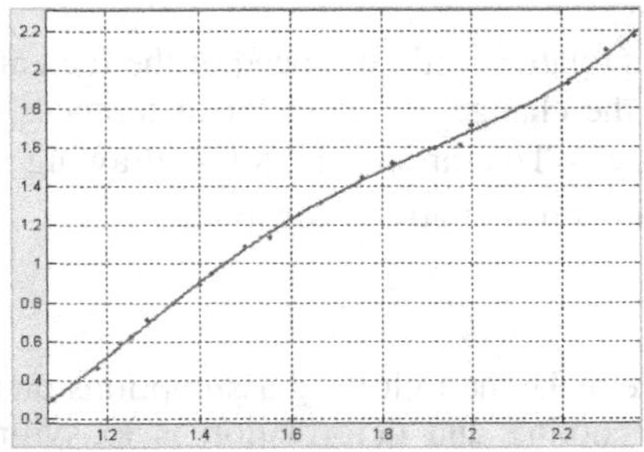

Fig. 1. Relational model between drawing ratio and diameter difference

4 Comparative Experiments between Feature Model Control and Temperature Field Analysis

Firstly, let the feed motor and pulling motor to drive four rollers run with 0.50 mm/s speed. It makes no tension transmission wire to achieve. Then open the heating and cooling. After wire temperature is stability to 1050 ± 2 ℃, gradually increase the speed of drawing roller. In accelerating remember to pay attention to changes in temperature. If large fluctuations happen, immediately stop accelerating. You should wait until the temperature stable again before continuing to accelerate. When the drawing roller speed reach at 0.75 mm/s, stop accelerating. It's time of stable drawing stage.

The experiment based on temperature field analysis is carried out with the initial diameter of 6.00mm, feed rate of 0.50mm/s, drawing speed of 0.75mm/s, heating power of 30KW of process parameters. After the experiment the wire is shown in Figure 2.

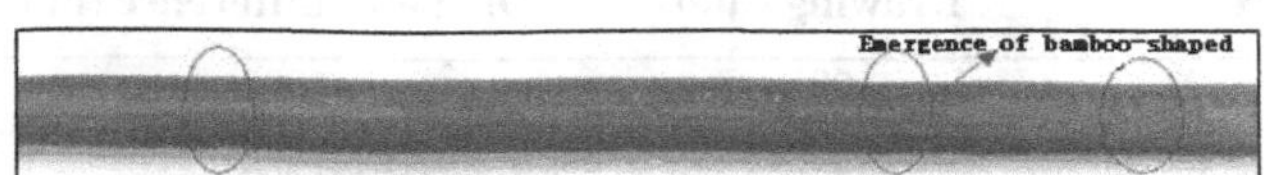

Fig. 2. Drawing results of temperature field parsing

As shown in Figure2, we use the finite element method for drawing control. After drawing the wire is still the case of bamboo. Although a lot has been improved compared to the previous, drawing results are still not good. Bamboo-like and uniform diameter at the diameter error remained at 0.4-0.6mm, after the measurement.

Under the same conditions, that the experiment of feature model control is with the initial process parameters with diameter of 6.00mm, feed rate of 0.50 mm/s, drawing speed of 0.75 mm/s, the initial induction heating power of 30KW. After the experiment the wire is shown in Figure3.

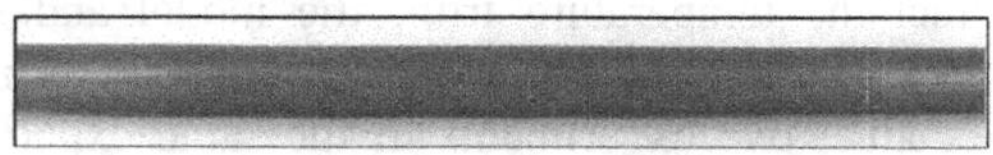

Fig. 3. Drawing results of Characteristics of the model control under the same conditions

We can see no obvious bamboo rods for drawing through characteristics of the model control. Error between the diameter where bamboo appears and the diameter which is uniform is maintained at 0.15-0.2mm.After measure the results of two control methods of drawing, initial data as shown in Table 2.

Table 2. The diameter of collected data

No.	Diameter (mm) after using temperature field analysis	Diameter (mm) after using feature model control
1	3.63	3.93
2	3.84	4.01
3	3.77	4.03
4	3.81	3.92
......	......	......
999	3.92	3.90
1000	3.79	3.89

Calculating on the initial data, Comparing indicators include average diameter, the frequency of bamboo, diameter difference range, the maximum diameter difference of wire diameter, average diameter, and mean square diameter and so on, calculated data is shown in Table 3.

Table 3. Comparison of measured values and various indicators

	Average diameter (mm)	Frequency of bamboo (times/m)	Diameter difference range (mm)	Maximum diameter difference(mm)	Mean square Diameter (mm)
Temperature field analysis	3.79	0.35	0.4-0.6	0.76	0.122
Feature model control	3.93	0.16	0.15-0.2	0.33	0.053

We can see that it is more effective for improving the quality of drawing using feature of the model control than temperature field parsing. It also verified the feasibility of feature model control and greatly improved the quality of drawing.

5 Conclusions

Through analysis of no mold drawing control process, the model which is based on temperature field parsing is very complex and is not conducive to effective control. In view of this, we extract the temperature from the model and separately treat it. Establish a relationship model based on drawing ratio of feature model and diameter difference. Then we conduct two experiments in the same experimental conditions. Laboratory equipment, laboratory environment and the parameters of experiment are all the same. Execute the control through temperature field analysis. Keep the error in diameter between 0.4-0.6mm, while keep the error in diameter between 0.15-0.2mm through characteristics of the model control. The frequency of bamboo is down to 0.16 from 0.35. This shows that it greatly improve the quality of drawing through characteristics of the model control than through temperature field analysis.

References

1. Huang, Z.Y., Wang, P., Kong, W.B.: Technics and development of dieless drawing. Journal of East China University of Metallargy, 118–120 (2000)
2. Yang, Y.Q., Xie, J.X.: The Research of On-line Control System for Wire Diameter in Dieless Drawing Forming. University of Science and Techology, Beijing (2007)
3. Wu, H.X., Liu, Y., Liu, Z.: Control of feature modeling and flexible structure. Scientia Sinica Techologica, 137–149 (2001)
4. Zhang, H.Q., Xia, H.Y., Chen, A.: Study on forming test and numerical analysis on temperature field in drawing without die. Metalforming Machinery, 38–42 (1999)
5. Furushima, T., Sakai, T.: Finite Element Modeling of Dieless Tube Drawing of Strain Rate Sensitive Material with Coupled Themo-Mechanical Analysis. In: AIP Conference Proceedings, pp. 522–527 (2004)

Research of E-Commerce Based on Cloud Computing

Chunling Sun

School of Information Science and Technology Heilongjiang University
chunling6666@163.com

Abstract. Cloud computing is a new Internet-based computer techno logy. The paper analyzed the current actuality of the application for enterprise E-commerce, and pointed the main issue of that application. The paper put forward that cloud computing has a wide perspective in the application of E-commerce by describing the conception and characteristic of cloud computing, and special analyzed the main aspect of improving E-commerce by cloud computing.

Keywords: Cloud computing, E-commerce, Characteristics, Influence.

1 Introduction

Internet has changed the world, it is an indisputable fact. The 21st century, the trend of the gate to the social normalization brings the huge impact, based on the Internet and the rise of the Internet technology one at tremendous speed changing people's existence and study way, Cloud Computing so arises at the historic moment. Cloud Computing is a new technology, is a hi-tech product, has broad prospects for development. A good structure, reliable and extensible security model structure of the Cloud Computing development plays a very important role, it is the grid and cloud can actually used in the powerful guarantee of real world.

E-commerce is online commerce verses real-world commerce. E-commerce includes retail shopping, banking, stocks and bonds trading, auctions, real estate transactions, airline booking, movie rentals—nearly anything you can imagine in the real world. Even personal services such as hair and nail salons can benefit from e-commerce by providing a website for the sale of related health and beauty products, normally available to local customers exclusively.

While e-commerce once required an expensive interface and personal security certificate, this is no longer the case. Virtual storefronts are offered by a variety of hosting services and large Internet presences such as eBay and Yahoo!, which offer turnkey solutions to vendors with little or no online experience. Tools for running successful e-commerce websites are built into the hosting servers, eliminating the need for the individual merchant to redesign the wheel. These tools include benefits like shopping carts, inventory and sales logs, and the ability to accept a variety of payment options including secure credit card transactions.

Cloud computing is seen as the next revolution of science and technology industry, it will radically change the way of working and business model. Combining with e-commerce, cloud computing is to have great influence on the enterprise all aspects.

D. Jin and S. Lin (Eds.): Advances in CSIE, Vol. 2, AISC 169, pp. 15–20.
springerlink.com

2 What Is Cloud Computing?

Everyone is talking about "the cloud." But what does it mean? Business applications are moving to the cloud. It's not just a fad—the shift from traditional software models to the Internet has steadily gained momentum over the last 10 years. Traditional business applications have always been very complicated and expensive. The amount and variety of hardware and software required to run them are daunting. You need a whole team of experts to install, configure, test, run, secure, and update them. When you multiply this effort across dozens or hundreds of apps, it's easy to see why the biggest companies with the best IT departments aren't getting the apps they need. Small and mid-sized businesses don't stand a chance.

With cloud computing, you eliminate those headaches because you're not managing hardware and software—that's the responsibility of an experienced vendor like salesforce.com. The shared infrastructure means it works like a utility: You only pay for what you need, upgrades are automatic, and scaling up or down is easy.

Cloud-based apps can be up and running in days or weeks, and they cost less. With a cloud app, you just open a browser, log in, customize the app, and start using it.

Businesses are running all kinds of apps in the cloud, like customer relationship management (CRM), HR, accounting, and much more. Some of the world's largest companies moved their applications to the cloud with salesforce.com after rigorously testing the security and reliability of our infrastructure.

As cloud computing grows in popularity, thousands of companies are simply regrinding their non-cloud products and services as "cloud computing." Always dig deeper when evaluating cloud offerings and keep in mind that if you have to buy and manage hardware and software, what you're looking at isn't really cloud computing but a false cloud.

The latest innovations in cloud computing are making our business applications even more mobile and collaborative, similar to popular consumer apps like Face book and Twitter . As consumers, we now expect that the information we care about will be pushed to us in real time, and business applications in the cloud are heading in that direction as well. With Cloud 2, keeping up with your work is as easy as keeping up with your personal life on Face book.

Cloud computing is a way of using computers where the computer resources (software and hardware) are provided as a service over the internet and are dynamically scalable and often virtual (i.e. not necessarily in one known place). What this means to users is that the information they use is stored on computers somewhere else (other than there local PC) and can be accessed where, when and how they want it.

Cloud computing customers don't generally own the physical infrastructure on which the applications run and store the data. Instead, they rent usage from a third-party provider and then use the system as they need it, much as people use gas or electricity. The more resources they use (such as more users having access to an application or using more disk space for storing data) the more they pay.

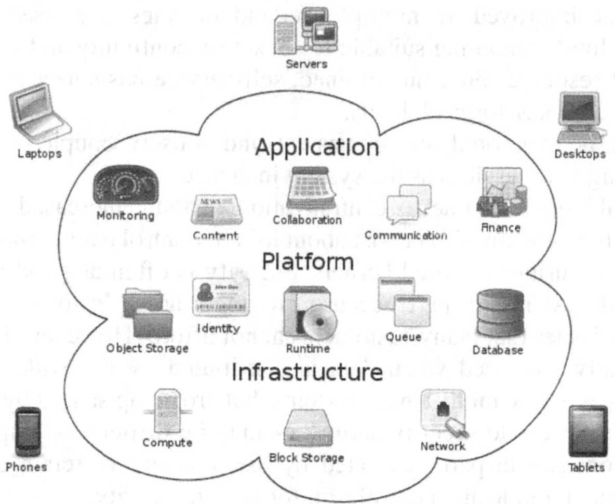

Fig. 1. Cloud computing logical diagram

3 Cloud Computing Exhibits the Following Key Characteristics

Empowerment of end-users of computing resources by putting the provisioning of those resources in their own control, as opposed to the control of a centralized IT service. Agility improves with users' ability to re-provision technological infrastructure resources.

Application programming interface (API) accessibility to software that enables machines to interact with cloud software in the same way the user interface facilitates interaction between humans and computers. Cloud computing systems typically use REST-based APIs.

Cost is claimed to be reduced and in a public cloud delivery model capital expenditure is converted to operational expenditure. This is purported to lower barriers to entry, as infrastructure is typically provided by a third-party and does not need to be purchased for one-time or infrequent intensive computing tasks. Pricing on a utility computing basis is fine-grained with usage-based options and fewer IT skills are required for implementation (in-house).

Device and location independence enable users to access systems using a web browser regardless of their location or what device they are using . As infrastructure is off-site (typically provided by a third-party) and accessed via the Internet, users can connect from anywhere.

Multi-tenancy enables sharing of resources and costs across a large pool of users thus allowing for:

(1) Centralization of infrastructure in locations with lower costs (such as real estate, electricity, etc.)

(2)Peak-load capacity increases (users need not engineer for highest possible load-levels)

(3)Utilization and efficiency improvements for systems that are often only 10–20% utilized.

Reliability is improved if multiple redundant sites are used, which makes well-designed cloud computing suitable for business continuity and disaster recovery. Provisioning of resources on a fine-grained, self-service basis near real-time, without users having to engineer for peak loads.

Performance is monitored and consistent and loosely coupled architectures are constructed using web services as the system interface.

Security could improve due to centralization of data, increased security-focused resources, etc., but concerns can persist about loss of control over certain sensitive data, and the lack of security for stored kernels. Security is often as good as or better than other traditional systems, in part because providers are able to devote resources to solving security issues that many customers cannot afford. However, the complexity of security is greatly increased when data is distributed over a wider area or greater number of devices and in multi-tenant systems that are being shared by unrelated users. In addition, user access to security audit logs may be difficult or impossible. Private cloud installations are in part motivated by users' desire to retain control over the infrastructure and avoid losing control of information security.

Maintenance of cloud computing applications is easier, because they do not need to be installed on each user's computer.

4 The Influence of the Cloud Computing upon the E-Commerce

4.1 Cloud Computing Can Improve the Safety of Business Enterprise E-Commerce Application

The business enterprise scale is more and more big, business enterprise the backlog more information resources. Along with the rapid development of network, the business enterprise data gets effectively and savagely to also lead the attack of coming a lot of viruses and black guest at the same time and then makes the safety that the business enterprise data saves be subjected to serious threat and made also more and more big in the devotion on the information safety. Apply cloud in the business enterprise calculation, can is saving the data in the high in the clouds, is computed service by cloud to provide ascend provide profession, efficiently and safety of data saving, thus the business enterprise need not worry again because various safe problem causes the data throw to lose. Therefore, cloud computing can provide the data of credibility and safety saving center for business enterprise.

4.2 Cloud Computing Can Improve the Vivid and Profession of Business Enterprise E-Commerce Application

Cloud computing can provide economic dependable E-commerce system to make to order service for business enterprise, software's namely serve (SaaS) is a kind of service type that cloud computing provides, it software Be a sow in line service to provide. Compute a technical electronics to outside wrap according to cloud is the importance of business enterprise application E-commerce service to apply of a. Business enterprise while using network frame and application procedure makes use of cloud computing can make it more of convenience, the electronics outside wraps actually be with need but change a kind of form of E-commerce. The business enterprise is adopting cloud computing service, to the E-commerce system carry on a development and get stripe have already no longer needed to cost a great deal of funds and manpower, don't need singly trap software and procedure of investing the establishment inner part. The business enterprise being a customer to carry can more expediently use various service that cloud computing provides and needs to install a network browser then and makes the business enterprise be getting less for supporting and getting stripe E-commerce system but throwing in of expenses like this just at this time.

4.3 Cloud Computing Can Carry Out Common Calculation Environment Hard the Data Attaining Handle Ability

Cloud computing passes to definitely adjust one degree strategy, can pass logarithms ten thousand is carry on consociation's providing super strong calculation ability for customer to the of 1,000,000 common calculators, using the door can complete to use list set calculator hard completion of task. In "cloud", cloud computing mode will be according to needing to be adjusted to provide strong calculation ability with numerous calculation resources in cloud while being to hand in one to compute a claim. Compute mode in cloud in, business enterprise no longer from own calculation on board, is not from a certain appointed server either, but passed various equipments(such as move terminal etc.) on the net to acquire from the Internet need of information, therefore the speed got leaping of quality.

4.4 Cloud Computing Can Provide Good Economic Efficiency for the E-Commerce Application

A great deal of calculator and network equipments are what business enterprise sets up E-commerce system to provide with, remarkable BE, business enterprise for satisfying more and more business needs, have to also periodically carry on replacing to the calculator and the network equipments. E-commerce system establishment of cost more big, and develop and the empress supporting expect to need higher expenses, for funds opposite limited small and medium enterprises to speak, is is hard to undertake, and apply with network service and business of rapid growth to request to be hard also to match. Cloud computing can reduce the establishment cost of business enterprise E-commerce system in the application in the E-commerce. The business enterprise passes cloud computing any further the dissimilarity continue to purchase an expensive hardware equipments, also the dissimilarity bears a large amount of maintenance fee, this is mainly cloud computing to provide IT foundation structure, at this time the equipments that needs to rent high in the clouds is all right. From this, cloud computing can provide good economic efficiency for the application of business enterprise E-commerce.

Today, cloud computing is getting all the rage. Cloud computing is still a very young technology and we still having more room for improvement. Although the meaning of cloud computing may be differ from one point of view of a person to another, it still all boils down to sharing one meaning which is – delivering information over the internet. It is very similar to: autonomic computing, client server model, grid computing, mainframe computer, utility computing, peer to pee and service oriented computing.

5 Conclusion

(1) All of your gadgets will be synchronized at all times.
(2) You can access your stored information any time you like.
(3) Organizing mine data online.
(4) Sharing of data instantly regardless of type and size.

So with the given benefits of using cloud computing, you are probably thinking now that you could have been using it already, but you are just not aware. To give you an example, of what is cloud computing, have you ever used gamily? If yes, then maybe you are aware of how they share files with their Google docs, right? With Google docs, you get to share information whether written document or spread sheet file. Anyone with permission to view and edit the file can do so. That is a classic meaning of cloud computing, we just don't use the term for it more often. We might be expecting something like: our gadget knowing what we like and what we want to do even without our command. It may sound weird today, but the possibility of it happening in the near future is crystal clear.

References

1. Qiao, W.-B.: The Lacks of Cloud Computing Applications and Recommendations for Improvement. Computer Knowledge and Technology 16 (2010)
2. Yao, T.-X., Xu, Y.-H., Liu, S.-X.: Research of Cloud Computing in Small Medium Enterprises Application. Computer Knowledge and Technology 14 (2011)
3. Sun, X.: Cloud Computing Research and Development. Computer
4. Li, J.-X., Chen, H.-G.: Design of a Geological Cloud System Based on Cloud Computing. Computer Engineering & Science 06 (2011)
5. Zhao, W., Geng, Q.: Application of Cloud Computing to GIS Model. Geospatial Information 06 (2010)

Study on Chaos-Based Weak Signal Detection Method with Duffing Oscillator

Fengli Wang, Hui Xing, Shulin Duan, and Hongliang Yu

College of Marine Engineering, Dalian Maritime University, Dalian 116026, P.R. China
wangfl@dlmu.edu.cn, xingcage@163.com, duanshulin66@sina.com, yhl1202@dl.cn

Abstract. Aiming at the feature extraction of weak periodic signal in early fault of machinery, a novel weak periodic signal detection with Duffing oscillator based on empirical mode decomposition (EMD) was presented. The chaotic character of Duffing oscillator was analyzed, and the Melnikov method of determining chaotic threshold of Duffing oscillators was discussed. The principle of weak signal detection based on the change of phase trace was described. In practical engineering measurement, the influence of noise to the system status in the chaos detection process was studied. EMD was proposed to avoid component interference, and weak characteristic signal can be separated from background signal and noise. The analysis results show that the weak periodic signal can be detected efficiently.

Keywords: Duffing oscillator, chaos, empirical mode decomposition, weak signal detection.

1 Introduction

In the early stage of machinery faults, the interested characteristic signal is usually submerged in heavy noise. To realize the early-stage diagnosis of machinery faults, it is necessary to conduct weak signal detection in the background of strong noises. The chaotic methods on weak signal processing is a new field of signal processing, and have wide engineering application on weak signal processing[1]. Because a tiny perturbation of a parameter might cause an essential change of the state in a non-linear system, a great many researchers have developed various methods to detect weak periodic signals by using the sensitivity of the non-linear system to its parameters [2]. Of them, the chaotic oscillator has been extensively studied over the past few years[3]. In practical machinery fault diagnosis, the chaotic oscillator is adequate for detecting a single and fixed periodic signal in a narrow frequency band and has obtained satisfactory results. But when the weak periodic signal is in a wide frequency band, it is difficult to distinguish the state change of the chaotic oscillator.

Empirical mode decomposition(EMD) could decompose the complicated signal into a number of intrinsic mode functions (IMFs) [4]. Frequency component contained in each IMF relates to sampling frequency and most importantly changes with the signal itself. So, after the misalignment vibration signal is decomposed by EMD, the interested characteristic signal can be separated from background signal and noise signal, and detected by Duffing oscillator efficiently.

D. Jin and S. Lin (Eds.): Advances in CSIE, Vol. 2, AISC 169, pp. 21–26.
springerlink.com

2 Principle of Weak Signal Detection with Chaotic Oscillator

As a chaotic oscillator described by the Holmes Duffing equation, a tiny perturbation of a parameter might cause an essential change of its state. Its performance has also been studied extensively [5,6].

The Holmes Duffing equation is chosen for modeling,

$$\ddot{x} + c\dot{x} - x + x^3 = F\cos\omega t . \tag{1}$$

where $(-x + x^3)$ is the non-linear recovery force; c is the damping constant of the Duffing oscillator; F is the amplitude of the periodic driving force in the Duffing oscillator. When $\omega = 1rad/s$, the corresponding state equation is

$$\begin{cases} \dot{x} = y \\ \dot{y} = -cy + x - x^3 + F\cos\omega t \end{cases} \tag{2}$$

where x, y are the outputs of the Duffing oscillator.

After making a time scale conversion of Eq. (2), the corresponding state equation by which it can detect arbitrary frequency component ω is

$$\begin{cases} \dot{x} = \omega y \\ \dot{y} = \omega(-cy + x - x^3 + F\cos\omega t) \end{cases} \tag{3}$$

If we fix c, let F increase gradually from 0 to more than a certain threshold F_a and keep increasing to above another threshold F_b, then the rule of the state change of the Duffing oscillator in the phase space will be: small-scale periodic state $\rightarrow$ chaotic state $\rightarrow$ large-scale periodic state. Eq.(3) is to be solved by discretizing the equation and using the fourth-order Runge–Kutta algorithm. When $c = 0.5$, $\omega = 1rad/s$, and calculation step $h = 0.01$, $F_b = 0.8245$. To facilitate the observation of the change of the phase trajectories of the oscillator, we take the initial $x_0 = 0$, $y_0 = 0$, and the computing time equal to 50s. When external to-be-detected periodic signals and noise are introduced into the oscillator, Eq.(3) changes to

$$\begin{cases} \dot{x} = \omega y \\ \dot{y} = \omega(-cy + x - x^3 + F\cos\omega t + F_1\cos(\omega_1 t + \varphi) + N(t) \end{cases} \tag{4}$$

where $F_1\cos(\omega_1 t + \varphi) + N(t))$ is called external perturbation or externally applied signal; F_1 is the amplitude of the to-be-detected periodic signal; ω_1 is the angular frequency of the to-be-detected periodic signal; φ is the initial phase of the to-be-detected periodic signal; $N(t)$ is the Gaussian white noise.

Let the frequency ω of the internal periodic driving force of the oscillator is supposed to be equal to the frequency ω_1 of the external to-be-detected weak periodic signal, F is a little less than F_b, that is, when the external signal is not introduced, the oscillator is in the chaotic state. When the external periodic signal with frequency ω_1 and amplitude F_1 is introduced into the oscillator, the oscillator will be transformed from the chaotic state to the large-scale periodic state so long as $F + F_1 > F_b$.

However, it is difficult to identify the state change of the Duffing oscillator when the background signal is too heavy. So, we should decompose the measured vibration signal into IMFs, the interested characteristic signal can be separated from background signal and noise. It becomes monocomponent in a narrow frequency band and can be detected by Duffing oscillator efficiently.

3 Empirical Mode Decomposition(EMD)

By using EMD [4], any signal could be decomposed into a number of IMFs $c_1(t)$, $c_2(t)$, . . . , $c_n(t)$, and a residue $r_n(t)$,

$$x(t) = \sum_{i=1}^{n} c_i(t) + r_n(t) \tag{5}$$

Thus, one can achieve a decomposition of the signal into n-empirical modes, and a residue $r_n(t)$, which is the mean trend of $x(t)$. The IMFs $c_1(t)$, $c_2(t)$, . . . , $c_n(t)$ include different frequency bands ranging from high to low. The frequency components contained in each frequency band are different and change with the variation of signal $x(t)$. So, EMD is a self-adaptive signal decomposition method.

4 Weak Signal Detection with Duffing Oscillator Based on EMD

The weak periodic signal detection follows two operations:

(1) Decompose the original signal into several IMF components $c_j(t)$, $j = 1,2,3 \cdots n$. After decomposition, the interested characteristic signal can be separated from background signal and noise.

(2) Introduce the IMF components $c_j(t)$ of the to-be-detected signal into the oscillator, when the oscillator is transformed from the chaotic state to the large-scale periodic state, it indicates that the to-be-detected signal includes the characteristic frequency component and complete the weak signal detection.

5 Results and Discussions

5.1 Numerical Simulation

we consider the function $0.025\sin t + 0.3\sin(5.5t)$, the frequency of to-be-detected component is $\omega = 1rad/s$. Thc simulated signal is introduced into the oscillator, the output phase trajectory of the Duffing oscillator is shown in Fig. 1. As a result, we cannot recognize the weak signal correctly owing to the interference of the strong background signal.

The simulated signal is decomposed by EMD and the decomposition result is shown in Fig.2. The second IMF $c_2(t)$ in Fig.2 include the to-be-detected component with frequency $\omega = 1rad/s$ is introduced into the oscillator in, the output phase trajectory of the Duffing oscillator is shown in Fig. 3. From Fig. 3, it is easy to find that the chaotic oscillator is transformed from the chaotic state to the large-scale periodic state. It is obvious that the $\omega = 1rad/s$ frequency component exists.

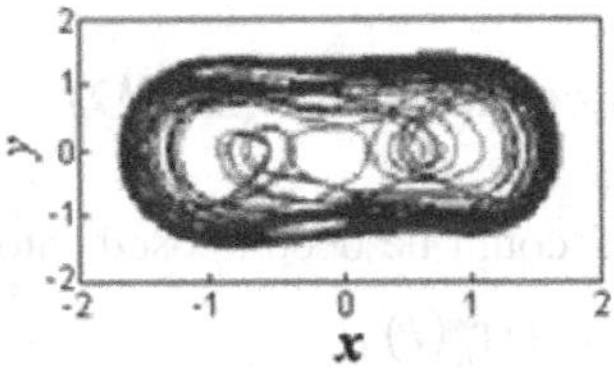

Fig. 1. The output phase trajectories of simulated signal

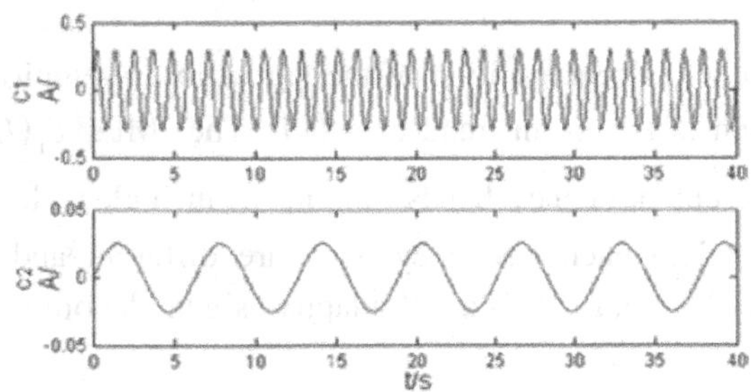

Fig. 2. The decomposition result of simulated signal

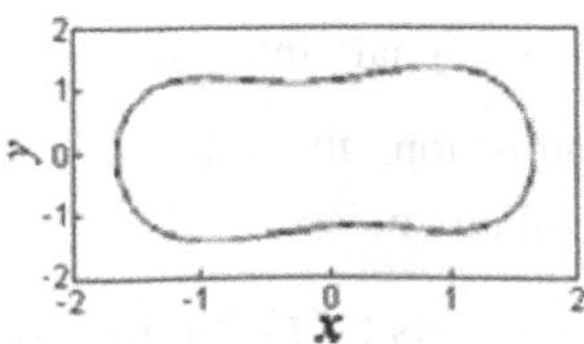

Fig. 3. The output phase trajectory of IMF $c_2(t)$

5.2 Experiment

For the rotating machinery, the 2X harmonic vibration response indicates that there is misalignment fault in the rotating machinery. To verify the validity of the proposed method, the misalignment was conducted on the rotor test rig. The radial displacement vibration signal with misalignment fault picked up by the displacement sensor is shown in Fig. 4. The rotating frequency is 24 Hz and the sample frequency is 2560 Hz.

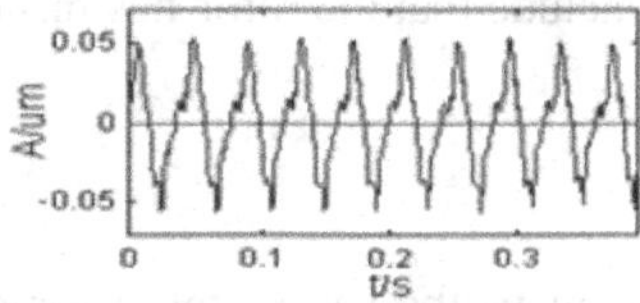

Fig. 4. The waveform and spectra of misalignment signal

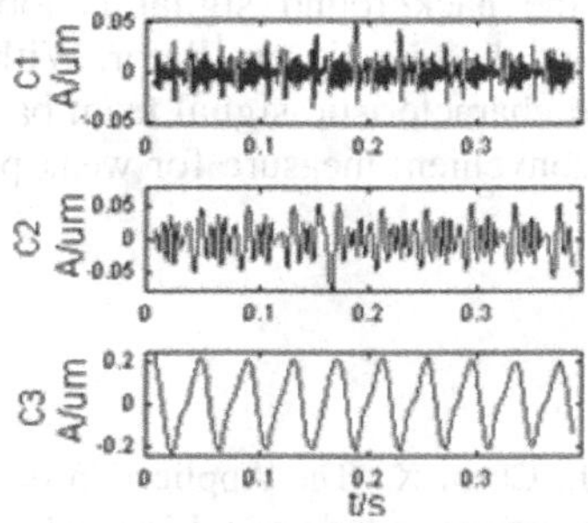

Fig. 5. The decomposition result of misalignment signal

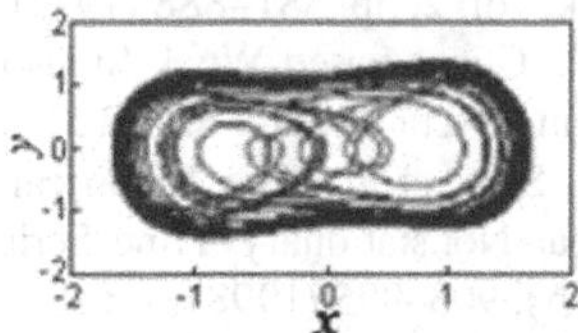

(a) Original state of Duffing oscillator

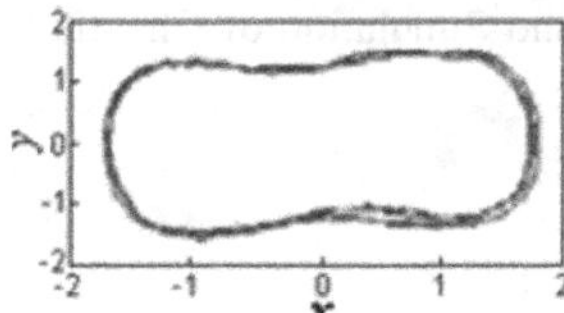

(b) Output phase trajectory under IMF $c_2(t)$

Fig. 6. The output phase trajectory of Duffing oscillator

Fig. 5 shows the decomposition results of the misalignment vibration signal obtained by using EMD. In Eq. (4), let ($F_1 \cos(\omega_1 t+\varphi)+N(t)$) be replaced by the second IMF $c_2(t)$ of misalignment signal. When the value of frequency of the periodic driving force in Duffing oscillator is 2X rotating frequency, $\omega = 2\pi \times 48 rad/s$, and the corresponding output phase trajectory of chaotic oscillators is shown that in Fig. 6. From Fig.6, it is easy to find that the chaotic oscillator is transformed from the chaotic state to the large-scale periodic state. It is obvious that the $\omega = 2\pi \times 48 rad/s$ frequency component exists. According to the above analysis, it is easy to conclude that the rotor has misalignment defect.

6 Summary

The chaotic oscillators detecting method is suitable for detecting a single periodic signal in a given narrow band. It requires the background signal and noise to be not great enough to affect the oscillator state. But when the weak periodic signal is in a wide frequency band, and the background signal is too strong, it is difficult to distinguish the state change of the chaotic oscillator. With the help of EMD, we are able to separate the interested characteristic signal from background signal and noise, and complete a reliable and convenient measure for weak periodic signal detection by Duffing oscillator efficiently.

References

1. Wang, G., Chen, D., Lin, J., Chen, X.: The Application of Chaotic Oscillators to Weak Signal Detection. IEEE Transactions on Industrial Electronics 46(2), 440–444 (1999)
2. Birx, D.L.: Chaotic Oscillators and Complex Mapping Feed Forward Networks (CMFFNS) for Signal Detection in Noisy Environments. In: Proceeding of the IEEE International Joint Conference on Neural Network, vol. 2, pp. 881–888 (1992)
3. Yue, L., Baojun, Y., Wu, S.Y.: Chaos-based Weak Sinusoidal Signal Detection Approach under Colored Noise Background. Acta Physica Sinica 52(3), 526–530 (2003)
4. Huang, N.E., Shen, Z., Long, S.R., et al.: The Empirical Mode Decomposition and The Hilbert Spectrum for Nonlinear Nonstationary Time Series Analysis. Proceedings of the Royal Society of London 454(A), 903–995 (1998)
5. Guckenheimer, J., Holmcs, P.: Nonlinear Oscillations, Dynamical Systems, and Bifurcations of Vector Fields. Springer, New York (1983)
6. Hoppensteadt, F.C.: Analysis and Simulation of Chaotic Systems, 2nd edn. Springer, New York (2000)

The Application of Association Rule Mining in CRM Based on Formal Concept Analysis

HongSheng Xu* and Lan Wang

College of Information Technology, Luoyang Normal University, Luoyang, 471022, China
xhs_ls@sina.com

Abstract. CRM (Customer Relationship Management) is to select and manage valuable customer relationships and a business strategy, CRM requires a customer-centric corporate culture to support effective marketing, sales and service processes. As a branch of applied mathematics, FCA (formal concept analysis) comes of the understanding of concept in philosophical domain. This paper presents the application of association rule mining in CRM based on formal concept analysis. Experiments show that the proposed algorithm in the CRM more effective than the traditional algorithm.

Keywords: Formal concept analysis, CRM, association rule mining.

1 Introduction

Customer Relationship Management of the implication is that through the details of clients' in-depth analysis, to improve customer satisfaction, thereby enhancing the competitiveness of enterprises as a means of it [1]. CRM (Customer Relationship Management) that is customer relationship management. Literally, is an enterprise with the CRM to manage customer relationships? In different contexts, CRM may be a management academic language, may be a software system, and are usually referred to CRM, is a computer automated analysis of sales, marketing, customer service and application support processes of the software system. Its goal is to reduce the sales cycle and marketing costs, increase revenue, expand their business needs and new markets channels and enhance customer value, satisfaction, profitability and loyalty. CRM is to select and manage valuable customer relationships and a business strategy, CRM requires a customer-centric corporate culture to support effective marketing, sales and service processes.

As a branch of applied mathematics, FCA (formal concept analysis) comes of the understanding of concept in philosophical domain. It is to describe the concept in formalization of symbol from extent and intent, and then realize the semantic information which can be understood by computer. It is to extract all connotative concepts and connections between them from formal context according to the binary relationship so as to form a hierarchical structure of concept. Concept lattice in formal concept analysis as the core data structures, in essence, describes the link between

* Author Introduce: Xu HongSheng(1979-), Male, lecturer, Master, College of Information Technology, Luoyang Normal University, Research area: Data mining, CRM.

D. Jin and S. Lin (Eds.): Advances in CSIE, Vol. 2, AISC 169, pp. 27–32.
springerlink.com

objects and features, indicating generalization between the concepts of relationship with patients, the corresponding Hasse diagram is the visual realization of the data of it.

In the data mining study, found that the rule has become a central issue. This paper first introduces the concept of a classical lattice-based algorithm for extracting implication rules, the incremental algorithm to build grid, and update the rule set, we needed on the grid structure modified accordingly, so you can get frequent incremental itemsets. This article describes the basic theory of association rule mining knowledge. This paper proposes the association rule mining in CRM using formal concept analysis. The algorithm is based on two concepts removed conjunct implies the rule set as input, the rule set according to their content to be divided, the division will focus on a rule one by one into the other rule set, and thus get the final merged result.

2 The Research of Association Rule Mining Based on Formal Concept Analysis

Concept lattice model is the product of introduction and lattice theory combined with practical application, here is some of the basic definitions of introduction and lattice theory.

Formal Concept Analysis is a philosophical concept of a mathematical process in which people organize and analyze data in a way, is the data and its structure, nature and visualize dependencies for a description. In formal concept analysis, the concept is to understand the grounds of extension and intension of two parts. Refers to the concept of extension of this concept is the set of all objects, meaning it refers to all characteristics common to these objects (or attributes) set [2]. Concept lattice in formal concept analysis as the core data structures, in essence, describes the link between objects and features, indicating generalization between the concepts of relationship with patients, the corresponding Hasse diagram is the visual realization of the data, vivid and concise reflection of the generalization relationship between these concepts. Therefore, the concept lattice is considered to be a powerful tool for data analysis. Order theory and lattice theory as a practical application combined with a product concept lattice model study has important theoretical significance.

A formal context (formal context) is a triple K = (G, M, I), where G is a collection of objects, M is the set of attributes, I G and M is a binary relation between, the $I \subseteq G \times M$. gIm that $g \in G$ and $m \in M$ there is a relationship between I, read as an object g has attribute m, is shown by equation 1.

$$\begin{aligned} &\forall O_1 \subseteq G : f(O_1) = \{m \mid \forall x \in O_1 (xIm)\} \\ &\forall M_1 \subseteq M : g(M_1) = \{x \mid \forall d \in M_1 (xId)\} \end{aligned} \tag{1}$$

If $(M, \leq)$ is a partial order set, a, b, c and d are the elements of M and $b < c$. Then set[b, c] : = $\{x \in M \mid b \leq x \leq c\}$ called interval (interval), collection (a] : = $\{x \in M \mid x \leq a\}$ called principal ideal (principal ideal), set [d) : = $\{x \in M \mid x \geq d\}$ called principal sub filter (principal filter). Besides, $a \prec b \Leftrightarrow a<b$ and $[a, b]=\{a, b\}$, is shown by equation 2.

$$\left(\bigcup_{t\in T} A_t\right)' = \bigcap_{t\in T} A_t' \tag{2}$$

Set (A,≤) is a partial order set, if for any the unempty set $S \subseteq A$, there exists $\vee S$, $(A,\leq)$ is called a full merger half lattice. Similarly, if for any the unempty set $S \subseteq A$, there exists $\wedge S$, $(A,\leq)$ is called a full cross half lattice. If $(A, \leq)$ is a full merger half lattice and also a full cross half lattice, it is a full lattice.

The two mapping is called Galois connection between the power set of A and the power set of B. binary group (A1, B1) ∈ P (A) x P (B), if meet the A1 = g (B1) and B1 = f (A1), then is called a formal concept of formal context C, A1 called denotation, B1 called connotation, all the formal concept sets of C writes down as F(C), as is shown by equation 3.

$$\psi x := \left(\{g \in G | \tilde{\gamma} g \leq x\}, \{m \in M | x \leq \tilde{\mu} m\}\right), \tag{3}$$

The progressive construction concept lattice is under the given original formal context K = (X, D, R) corresponding to the original concept lattice L and new object X * situation, solving formal context K * = (X∪{X *}, D, R) corresponding to the concept lattice L *.

Given formal context K = (G, M, I), if formal context $K_1 = (G_1, M_1, I_1)$ and $K_2 = (G_2, M_2, I_2)$ meet the $G_1 \subseteq G$, $G_2 \subseteq G$, $M_1 \subseteq M$, $M_2 \subseteq M$, then says K_1 and K_2 is the same domain formal context, they are all the son formal contexts of K, also says the concept lattice L (K_1) of formal context K_1 and the concept lattice L (K_2) of formal K_2 are the same domain concept lattice. The similarity is calculated as follows equation 4.

$$\mathfrak{B}(G,M,I) \cong \mathfrak{B}\left(G, M \setminus \{M\}, I \cap \left(G \times \left(M \setminus \{M\}\right)\right)\right). \tag{4}$$

For the formal contexts K1 = (G, M_1, I_1) and $K_2 = (G, M_2, I_2)$ of the same object domain, if $M_1 \subseteq M$, $M_2 \subseteq M$, $M_1 \cap M_2 = \varnothing$, then says $K_1$and K_2, L (K_1) and L (K_2) were connotation independent; If $M_1 \subseteq M$, $M_2 \subseteq M$, $M_1 \cap M_2 \neq \varnothing$, for any $g \in G$ and arbitrary $m \in M_1 \cap M_2$ meet $gI_1m = gI_2m$, it says K_1 and K_2, L (K_1) and L (K_2) are respectively connotation consistent [3].

Given formal context K = (G, M, I), if formal context $K_1 = (G_1, M_1, I_1)$ and $K_2 = (G_2, M_2, I_2)$, if $G_1 \subseteq G$, $G_2 \subseteq G$, $M_1 \subseteq M$, $M_2 \subseteq M$, $G_1 \neq G_2$, $M_1 \neq M_2$, $G_1 \cap G_2 \neq \varnothing$, $M_1 \cap M_2 \neq \varnothing$, then the same domain formal context fold set to $(G_1 \cup G_2, M_1 \cup M_2, I_1 \cup I_2)$, as is shown by equation 5.

$$g \swarrow m \Leftrightarrow \gamma g \wedge \mu m = (\gamma g)_* \neq \gamma g \tag{5}$$

Let I = {i1, i2, i3, ..., im} for the entry space; a collection of items called itemsets, with k-item set of items is called k-itemsets. Transaction database TD of each transaction Tr has a unique identifier TID, and contains a term set ⊆T I. Association rule is of the form B⇒A implicate, in which I⊂A, B and B∩A =∅ Rule is an objective measure of support. Support the rule that the percentage of samples to meet the rules. Support is the probability P (A∪B), of which, A∪B that contains both A and B services, that is, itemsets A and B and. Another association rule is an objective measure of confidence. Confidence is the conditional probability P (B|A), which includes the A's work also includes the probability of B.

Algorithm 1. association rule mining based on formal concept analysis

Input: Concept lattice L.

Output: Lattice L and after inserting the updated rule set R1 obtained from the algorithm array Rules [1, ..., | | L | |], Rules [N] represents the grid node N set of rules related to output.

(1) IF f*({x*})⊄Intent(inf(L)) THEN;
(2) Intent (inf (L)):=Intent (inf (L)) ∪f*({x*})
(3) IF C=0 OR ||D||≤1 THEN
(4) Rules[N] := GenerateRulesForNode(N);
(5) R: = R∪Rules [N];
(6) FOR each parent node of N, DO ∅ ≠ D ∩(C, D) (LHS∩D≠∅)
(7) Adding new lattice node H: (Φ, Intent (inf (L)) ∪ f * ({x*})), making H becomes inf (L);
(8) IF Q≠∅ THEN;
(9) Notes for B [I]: = {C: | | Intent (C) | | = I};
(10) return R: = R∪Reduce (N);

A given concept lattice L and inserted into the grid to the new transaction T, T, after inserting a new record for the lattice L'. And then compared to the original lattice L, after you insert the new transaction T, L ', there are three types of nodes. Way to keep the same. The other will change, but only change the extension, which are updated grid nodes. Another is the new grid node, which is to be inserted by the transaction and pay grid nodes generated in the original format does not exist in the content of the composition. The following diagram, as is shown by figure1.

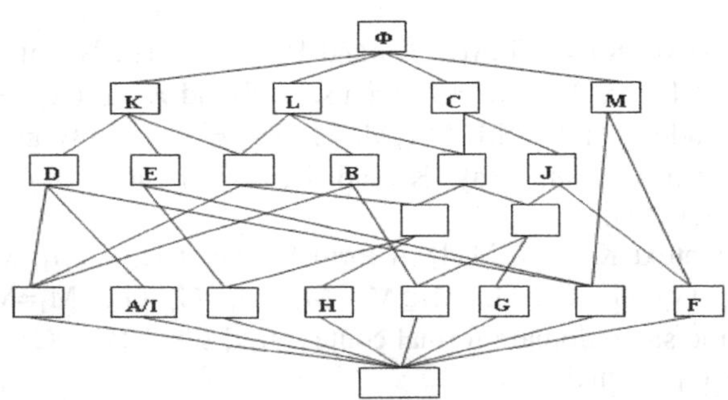

Fig. 1. The result of association rule data mining based on formal concept analysis

Marked with a simplified approach to the representation of the concept of property, the principle is: the same word in the graph node label attribute appears only once, the top node is unique, its meaning for the empty set, and the extension contains all object file; the bottom node is unique, contains all the terms of its content properties, and the extension of the empty set.

3 Application of Formal Concept Analysis in the Association Rule Mining of CRM

CRM is an enterprise business strategy, which according to customer segmentation and effective organization of corporate resources, develop a customer-centric business practices and the implementation of customer-centric business processes, and as a means to improve profitability capacity, revenue and customer satisfaction. Customer is an important asset, customer care center of the CRM, customer care and aim to establish long-term customers with the selection and effective business relationship with each customer "touch points" are closer to the customer, to understand customers, maximize profits and profit share [4].

Concept lattice, in addition to the classification and definition from the data concepts, it can be used to find objects, properties, dependencies between. This has two meanings: (1) part or all of the scanning grid structure can be used to generate the rule set in the future; (2) browse lattice structure, to test a certain given rules established.

New grid node: set transaction with item set T to be inserted into the lattice L Tr, if a grid node N1 = (C1, D1) satisfy: (1) T,∩ Intersection = D1 while for L in any of the node N2 have Intersection; (2)≠Intent (N2) for L in any meet N3> N1 node N3, Intersection; T ≠ ∩Intent (N3) the N1 is generated as a sub-grid nodes, the N1 can produce a new grid node (C=C1+1, Z=D1∩T).

Algorithm 2. Application of association rule mining in CRM based on formal concept analysis

Input: Conjunct implies the rule set R : P⇒Q as well as an array Rules[1...||L||], Rules[N] represents the grid node N set of rules related to L.

Output: $R=R_1 \cup R_2$. Rule sets R1 and R2 set the rules of the division R2 (Di), R1 and R2 are the same domain and two independent sets of rules.

(1) R1 will be divided according to their content, build into the relationship between father and son;

(2) **IF** inf(*L*) = (ϕ, ϕ) **THEN**

(3) **FOR** k := 0 **TO** size **DO**

(4) Intent(inf(*L*)) ← Intent(inf(*L*)) ∪ *f**({*x**});

(5) Notes for B [I] : = {C: | | Intent (C) | | = I};

(6) **IF** $D_k=D_i$ **THEN** exit algorithm **ENDIF**

(7) Int: = Intent(C)∩ f ({x*}));

(8) *GenerateRulesForPartition*(G_k, $Count_{1k}$, D_k);

(9) R:= R∪Reduce(N);

(10) **IF** Intent(*C*) = *f*({*x**})) **THEN** exit algorithm;

The paper is using WindowsXP operating system, and using Visual C + +6.0 to achieve the above rule sets and computing algorithms. For randomly generated data sets, 80% probability of their relationship, the number of attributes is 50, we do scalability testing, each increase in the number of objects 582, recorded by the child form the background to generate the corresponding concept lattice implication and operation of the rule set time spent Apriori at the same time a direct comparison of the original form of the background corresponding to the generated concept lattice implication rules set the time, comparing the results shown in Figure 2.

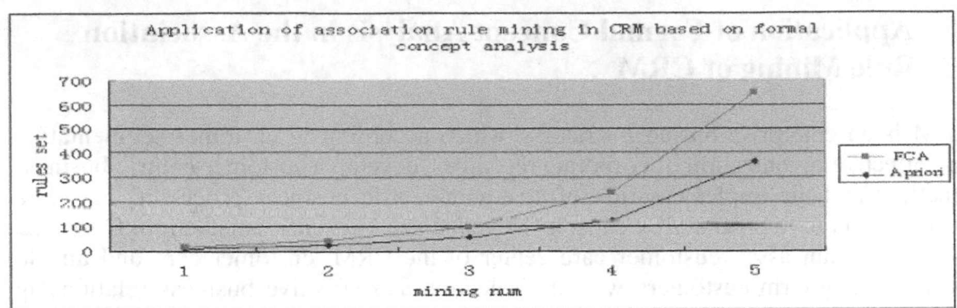

Fig. 2. The compare of association rule mining in CRM based on FCA with Apriori

4 Summary

As a branch of applied mathematics, FCA (formal concept analysis) comes of the understanding of concept in philosophical domain. It is to describe the concept in formalization of symbol from extent and intent, and then realize the semantic information which can be understood by computer. This paper presents the application of association rule mining in CRM based on formal concept analysis. Experiments show that the proposed algorithm in the CRM more effective than the traditional algorithm.

Acknowledgement. This paper is supported by not only Henan Science and Technology Agency science and technology research in 2010 (Key Project) under Grant no. 102102210472, but also Education Department of Henan Province Natural Science Research Program (2010A520030).

References

1. Lin, S.-C., Tung, C.-H., Jan, N.-Y., Chiang, D.-A.: Evaluating Churn Model in CRM: A Case Study in Telecom. JCIT 6(11), 192–200 (2011)
2. Burusco, A., Fuentes-González: Concept lattices defined from implication operators. Fuzzy Sets and Systems 114, 431–436 (2000)
3. Wille, R.: Concept Graphs and Formal Concept Analysis. In: Delugach, H.S., Keeler, M.A., Searle, L., Lukose, D., Sowa, J.F. (eds.) ICCS 1997. LNCS (LNAI), vol. 1257, pp. 290–303. Springer, Heidelberg (1997)
4. Liu, B., Hsu, W., Ma, Y.: Mining association rules with multiple minimum supports. In: Proc. KDD 1999, San Diego, CA, USA, pp. 337–341 (1999)

The Design of Data Collection Methods in Wireless Sensor Networks Based on Formal Concept Analysis

Yi Wang*, Jian Zhang, and HongSheng Xu

Department of Information Technology, Guangdong Textile Polytechnic, Foshan, 528041, China
xhs_ls@sina.com

Abstract. Data collection is one of the main research topics of wireless sensor networks in recent years, and, data collection research outside users through wireless sensor networks to collect perception data from the monitoring area. Formal concept analysis is a data analysis tool, especially for investigation and treatment can be given information to discover important information hidden in the data behind. This paper proposes the data collection methods in Wireless Sensor Networks based on formal concept analysis. Experiments show that the proposed FCA-based data collection algorithm in WSN more effective than the traditional algorithm.

Keywords: Formal concept analysis, Wireless Sensor Networks, data collection.

1 Introduction

The wireless sensor network is highly cross emerging multidisciplinary cutting-edge research, which combines the embedded computing technology, sensor technology, network and wireless communication technology, distributed information processing technology and other cutting-edge technology. Wireless sensor networks with the development of research models [1]. It combines the advanced microelectronic technology, system-on-chip SOC design technology, modern information and communication technology, computer network technology to achieve multi-functional, miniaturized, systematic, integrated and networked.

Formal Concept Analysis was born in about the 1980s, is a data analysis tool or method, especially for investigation and treatment can be given information. The data should be the unit of meaningful and understandable from a human thinking - the concept of extraction and formalization of the unit. Formal indicate that data processing is a formal mathematical entities do not have exactly the same and the concept in human thinking, it also pointed out that the basic data in the form of the formal concept analysis is a formal context, formal background in the human background knowledge a small part.

* Author Introduce: Wang Yi(1979-), Male, lecturer, Master, Department of Information Technology, Guangdong Textile Polytechnic, Research area: Data mining, WSN.

D. Jin and S. Lin (Eds.): Advances in CSIE, Vol. 2, AISC 169, pp. 33–38.
springerlink.com

The sensor network is an integrated monitoring, control, and wireless communication network system, a much larger number of nodes (thousands or even tens of thousands), more densely distributed nodes;, the node is more prone to failure due to the environmental impact and energy depletion; environmental interference easily lead to changes in network topology and node failures; under normal circumstances, the majority of sensor nodes is fixed. WSN data collection study included many aspects, such as: data collection protocol, data transmission routing protocols and sensor node scheduling. This paper mainly focuses on data collection protocol and data collection in the node scheduling mechanism study. This paper proposes the data collection methods in Wireless Sensor Networks based on formal concept analysis.

2 The Research of Data Collection in Wireless Sensor Networks

The primary design goals of traditional wireless networks is to provide high service quality and efficient bandwidth utilization, followed by considering energy conservation; the primary design goals of sensor networks is the efficient use of energy, which is the sensor networks and traditional networks one of the most important distinction.

People only care about the value of an observation of an area sensor networks do not care what the observational data of a specific node, which is the characteristics of the sensor network data-centric. Traditional network is the IP address of the network, data transmitted in the traditional network and the physical address of the node linked to data-centric characteristics of sensor networks can out of the traditional network addressing, fast and effective organization from the information of each node and aggregation to extract useful information delivered directly to the user.

The data collection protocol is the routing layer protocols, research and sensor nodes how the data transmitted to the base station. Sensor node's energy consumption is mainly concentrated in the communications equipment, including transmission of data, receive data, and the listener; the routing nodes consume more energy than in non-routing node [2]. Effective data collection protocol can make the data collection process used to transmit data, receive data, and listen for the energy minimization, and energy consumption of each node to reach consensus, thus extending the system life cycle, is shown by equation 1.

$$H[c_j(t)] = \frac{1}{\pi} p \int_{-\infty}^{\infty} \frac{c_j(t')}{t - t'} dt' \tag{1}$$

Network data aggregation play an important role as an important mechanism in the data collection process in WSN, the WSN data collection, performance data as follows: each node in the network monitoring area are collected in each data collection cycle 32network data aggregation from different nodes sensing data integrated as a single, and then aggregates the data transfer to a distant base station, is shown by equation 2.

$$PSNR = 10 \times \log_{10} \frac{MN\,[\max(F(x,y)) - \min(F(x,y))]}{\sum_{i=1}^{M}\sum_{j=1}^{N}[R(x_i,y_j) - F(x_i,y_j)]^2} \tag{2}$$

Lindsey and Raghavendra noted that at a distance within the energy consumed by the node transceiver circuit is greater than the energy consumed by the amplifier circuit, they to propose PEGASIS agreement in order to reduce the energy loss of the sensor nodes, the idea is that all sensor nodes in the system with greed algorithms constitute an edge of chain length and close to the minimum, in each time cycle, any node receiving data from an adjacent node located on the chain, and aggregate the received data with the data of this node, the aggregate data the chain sent to the other adjacent node to send and aggregate from the end node in the chain along the chain until the specified node.

In WSN, many related to the methods of data collection study describes three typical data collection from the sensor nodes in the monitoring area to distribute data to the Sink node topology in Figure1.

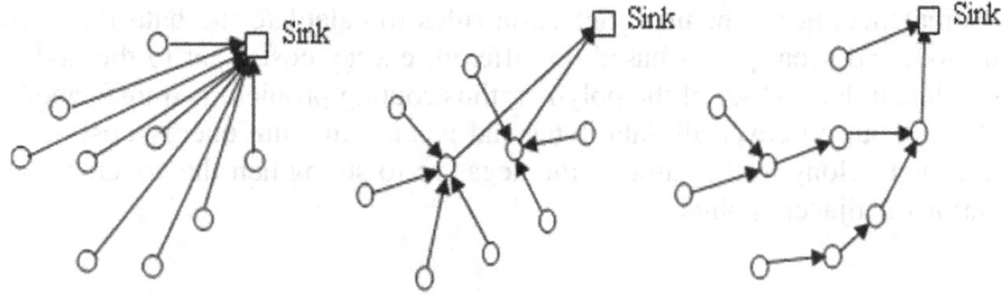

Fig. 1. The result of WSN data collection topology

Set composed of a WSN system identified as a sensor node, and identified as a focal point, the sensor nodes are randomly distributed in a monitoring area, the position of the nodes and base stations for fixed predictable, each node monitoring a certain amount of information collected, pass a packet to the base station assumes that each node in each time unit time unit for one cycle of direct communication between nodes in the network can also be directly transmitted to the base station, each node has a certain energy, the base station has infinite energy supply, as is shown by equation 3.

$$N_D \le (k-1)X + Min(d_i) \tag{3}$$

Stage a complete node scheduling point in the monitoring area selection, the excitation of the Sink node, k nodes are assigned r equal scheduling time slice t1, t2, ..., ti, the value of ti is relatively small and successively continuous monitoring of the regional random the introduction of the back off algorithm to solve the same layer of redundancy neighboring nodes choose the same scheduling time slice. Nodes in the network focal point Sink incentive redundant adjacent nodes exchange information to determine the assigned scheduling time slice, if satisfied: (1) the same layer adjacent nodes select the same scheduling time slice. (2) Whether the adjacent nodes of different layers to select a different scheduling time slice. The similarity is calculated as follows equation 4.

$$x(t-\tau)=\sum_{i=1}^{n} c_i(t-\tau)+r_n(t-\tau) \tag{4}$$

Preliminary scheduling point selection (ti value is relatively small) node after the stage of a network monitoring area scheduling node monitoring in the monitoring probability and energy information to complete the scheduling time to adjust. The set of nodes in network collected data, data collection, and data collection and acquisition time; there are certain changes with the application of linear and nonlinear relations, as is shown by equation 5.

$$E_{jA}^{\xi}(m,n)=\sum_{m'\in J,n'\in K} w_A^{\xi}(m',n')[D_{jA}^{\xi}(m+m',n+n')]^2 \tag{5}$$

Select a routing rule based on energy efficiency in data collection, elimination or weakening of invalid data transmission paths, invalid path to the introduction of negative strengthening mechanism. Selection rules to calculate the data items cover the set of nodes adjacent points based on different energy costs sent to the node that contains different data subset of the polymer, thus routing problem is transformed into looking for a group to cover all data items and has a minimum energy cost adjacent points, does not belong to the same in the negative to strengthen the coverage of the concentration of adjacent points.

3 The Design of Data Collection Methods in Wireless Sensor Networks Based on Formal Concept Analysis

A formal context K: = (G, M, I) by a collection of G, M, and the relationships between them composed of elements of G are called objects (objects), the elements of M are called attributes (attributes), to describe an object binary relation between g and an attribute m to write gIm or (g, m) ∈ I, said that "the object g has attribute m".

Assumes that given a form of the background of a formal context K: = (G, M, I), where G is a collection of objects, M is the collection of properties, I is a binary relation between them, then there exists a partially ordered set with which correspondence, and this collection of partial order a lattice structure, which induced by the background of a formal context (G, M, I) lattice L is called a concept lattice. Lattice L in each node is a pair (concept) is denoted by (X, X), where XG is called the extension of the concept, and X-M called the connotation of the concept [3]. An ordered pair (X, X) of relation R is complete, that is the nature of equation 6.

$$\begin{aligned} X' &= \left\{ x' \in M \;\middle|\; \forall\ x \in X\ ,xR\ x' \right\} \\ X &= \left\{ x \in G \;\middle|\; \forall\ x' \in X'\ ,xR\ x' \right\} \end{aligned} \tag{6}$$

To establish a partial order relation between the concept of grid nodes, given C1 = (X), C2 = (X2, X2), then C1, <C2, X2 <X2, we can understand this partial order relations for Asia concept - super concepts.

Let K = (G, M, I) is a formal context, B (K) = ((G, M, I), ≤) is the concept lattice of formal context K, then B (K) is a complete lattice, any nonempty subset of (G, M, I) least upper sector sup (B (K)) and the greatest lower bound inf (B (K)) ,as is shown by equation7.

$$\begin{aligned} \bigvee_{i\in I}(X_i,g(X_i)) &= (g(f(\bigcup_{i\in I}X_i)),\bigcap_{i\in I}g(X_i)) \\ \bigwedge_{i\in I}(X_i,g(X_i)) &= (\bigcap_{i\in I}X_i,f(g(\bigcup_{i\in I}g(X_i)))) \end{aligned} \tag{7}$$

Routing in sensor networks, the routing structure of the data collection does not change over time, that the survival time of the entire network [4], there is only one routing tree is used for data collection, called network mode is static, otherwise, in WSN data collection time, data routing structure with a series of routing tree routing mode for data collection, known as dynamic, a series of data collection tree for each data collection cycle, be used as a data collect routing tree.

(1) Determine the basic re-election time is Δt set.

(2) IF inf(L) = (ϕ, ϕ) THEN

(3) Define the parameters k, k = Min[reselect – times, m] set, so.

(4) If $J > L$ the set threshold (for applications) was established, the monitoring area is scheduled to enter the active state of the node to select a relatively large time slice, otherwise the node to maintain the scheduling phase of a set state;

(5) Randomly retrieve a value from a discrete $\left[0,1,\ldots,(2^k-1)\right]$ set of integers, denoted by r, the re-election after the time-piece.

(6) the threshold is reached when the re-election time slice m times can not be completed to show that the network topology and the time slice is set wrong, the focal point Sink send an error report prompts

(7) GenerateRulesForPartition(Gk, Count1k, Dk);

(8) Incidence graph of Super League side that the data correlation of sensor nodes to the associated collection of data of a group of sensor nodes;

(9) IF Intent(C) = f({x*})) THEN exit algorithm;

Figure 2 shows the minimum energy and data link building process, the initial stage of data link the round0 round1 with the nearest neighbor of the same build process, the current chain from node1 and node3 time round2, join node2 node1 and node4link the increase in the minimum energy, delete the link between node1 and node4, the same process is repeated until round5 chain 4-5-3-2-7-8 contains all nodes of the network, relative, neighbor data collection chain constructed the method, based on the minimum energy of the distance between nodes and the formation of the energy consumption of data link reduced from 20 units to 54 units.

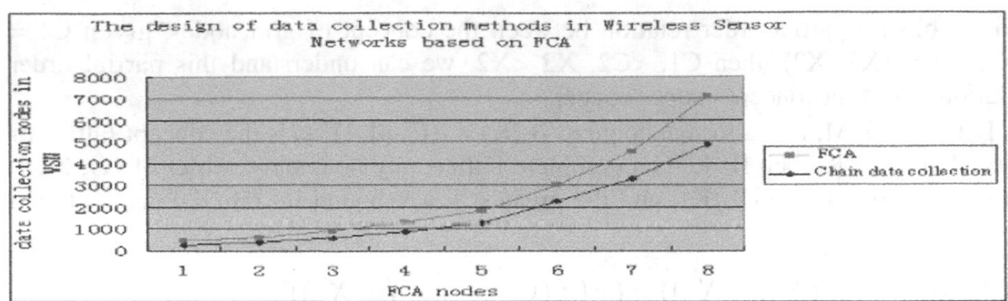

Fig. 2. The compare of data collection in WSN based on FCA with Chain

4 Summary

The concept lattice theory is the core data structure of the formal concept analysis theory, knowledge discovery and data analysis, a powerful mathematical tool. The concept lattice has good mathematical properties and suitable for batch processing, the concept lattice data for solving the parallel distributed data mining, distributed storage and parallel processing, we can say is the ideal tool. Therefore, the concept lattice has a very important theoretical significance. Most sensor network data collection applications have in common: a large number of static sensor nodes are densely arranged in a broad area, the base station is responsible for connection to a wireless sensor network and the outside world, the sensor nodes to the local perception to analyze the data transfer to the outside of the system life cycle requirements higher. In addition to these commonalities, each specific application, there are some differences, such as the requirement of real-time, fault tolerance, data acquisition frequency. This paper proposes the data collection methods in Wireless Sensor Networks based on formal concept analysis. Experiments show that the proposed FCA-based data collection algorithm in WSN more effective than the traditional algorithm.

References

1. Rickenbach, P.V., Wattenhofer, R.: Gathering correlated data in sensor networks. In: Proceedings of the 2004 Joint Workshop on Foundations of Mobile Computing, Philadelphia, PA, USA, pp. 60–66 (2004)
2. Krishnamachari, B., Estrin, D., Wicker, S.: The impact of data aggregation in wireless sensor networks. In: Proc. of the 22nd Int'l Conf. on Distributed Computing System Workshop, pp. 575–578. IEEE Computer Society, Vienna (2002)
3. Wille, R.: Concept Graphs and Formal Concept Analysis. In: Delugach, H.S., Keeler, M.A., Searle, L., Lukose, D., Sowa, J.F. (eds.) ICCS 1997. LNCS (LNAI), vol. 1257, pp. 290–303. Springer, Heidelberg (1997)
4. Tan, H.O., Korpeoglu, I.: Power efficient data gathering and aggregation in wireless sensor networks. SIGMOD Record 32(4), 66–71 (2003)

The Research of Network Management System Based on Mobile Agent and SNMP Technology

Shiwei Lin[1] and Yuwen Zhai[2]

[1] College of Information and Control Engineering, Jilin Institute of Chemical Technology, Jilin, China
[2] Mechanical & Electrical Engineering College, Jiaxing University, Jiaxing, China
shiwei_lin512@126.com

Abstract. Aimed at the shortcoming of traditional SNMP network management, a hybrid network management model is presented based on mobile Agent and SNMP technology in the paper. Nodes in each management domain will be divided into two categories: Mobile Agent can migrate directly to the nodes that are installed MAS and implement management functions by starting the SNMP agent in the local; the approach of management is used based on MA / SNMP gateway for the nodes that are not installed MAS. Application analysis showed that hybrid network management is more prominent in performance than the SNMP.

Keywords: Mobile Agent, SNMP protocol, network management, MA / SNMP gateway.

1 Introduction

Most of the current network management model is based on protocol SNMP. There are some shortcomings, most manufacturers of network devices are directly supported because of a wide range of applications. Mobile Agent as a unified interface is not feasible because of existing network equipment does not allow third-party software loaded, such as routers, switches, etc.. In this paper, the Mobile Agent technology combined with SNMP, SNMP network management technology at the same time to retain full use of the flexibility of Mobile Agent technology and intelligence, the Mobile Agent as complement traditional network management systems to improve the traditional C / S model network distribution management and flexibility.

2 The Design of the MA/SNMP Reference Model

The application conversion of MA and SNMP is required for combining Mobile Agent technology and the traditional SNMP protocol. This component is designed to convert an intermediate. It is located in the managed devices and network management centers, known as MA / SNMP gateway. MA / SNMP reference model shown as Figure 1.

D. Jin and S. Lin (Eds.): Advances in CSIE, Vol. 2, AISC 169, pp. 39–44.
springerlink.com

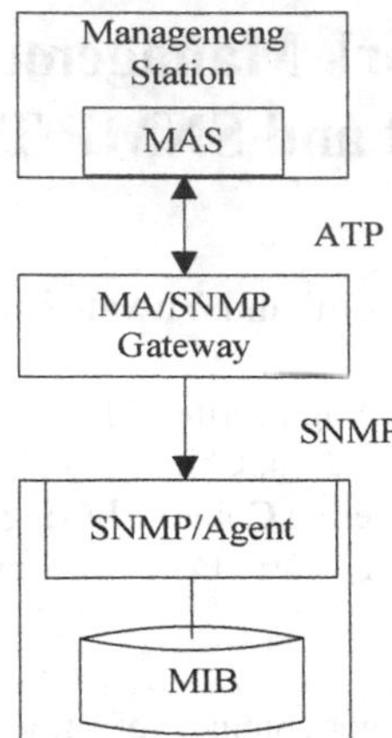

Fig. 1. The Reference Model Based on MA/SNM

3 The Design of the System Frame

The architecture of hybrid network management system based on Mobile Agent is shown as Figure 2.

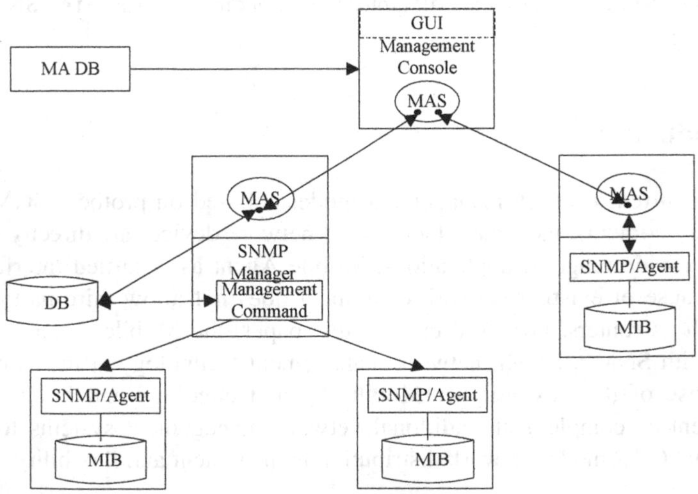

Fig. 2. The Hybrid Network Management Model Based on Mobile Agent

The model consists of network management console network management (NMS), Mobile Agent (MA), Mobile Agent Server (MAS), the managed node (NE), MA / SNMP gateway five elements.

4 The Work Flow Chart of System

The work flow chart of system is shown as Figure 3.

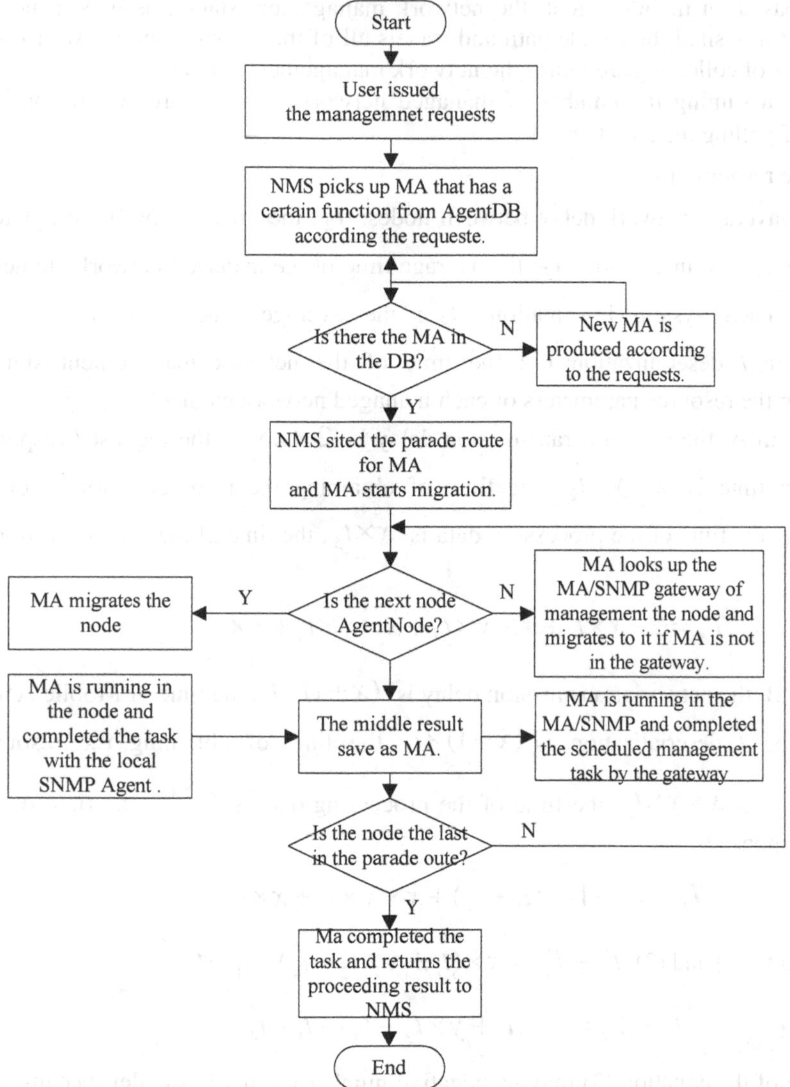

Fig. 3. The work flow chart of system based on mobile Agent

5 The Analysis and Comparison of the System Performance

The system response time and data traffic are two very important indicators that evacuee a network management system performance. In this paper, we analyze one importance.

The network management system based on SNMP (System A) compared with the network management system based on mobile Agent (System B).The network management stations collect the data through polling each network element devices in the network management systems based on SNMP. In the network management systems based on mobile Agent, the network management station assigns a mobile Agent that it is sited the parade path and travels all of the network nodes. At last, the data results of collecting are return the network management station.

In this, assuming the number of managed network elements are x, the objects number of polling the target are y.

① The response time

t_1: the average network delay between nodes; t_2: the time of SNMP complete a request / response interaction; t_3: the average time of the managed network elements obtaining each system information. t_4 : the average time of Mobile Agent serialization / deserialization; t_5 : the time of the network management station processing the resource parameters of each managed network element.

In system A, the network transmission delay is $2\times x\times t_1$, the request / response interaction time is $x\times y\times t_2$, the time of obtaining the resource information is $x\times y\times t_3$, the time of the processing data is $x\times t_5$, the time of the system response is:

$$T_A = 2\times x\times t_1 + x\times y\times t_2 + x\times y\times t_3 + x\times t_5 \tag{1}$$

In system B, the network transmission delay is $(x+1)\times t_1$, the time of Mobile Agent serialization / deserialization is $(x+1)\times t_4$,the time of obtaining the resource information is $x\times y\times t_3$, the time of the processing data is $x\times t_5$, the time of the system response is:

$$T_B = (x+1)\times(t_1 + t_4) + x\times y\times t_3 + x\times t_5 \tag{2}$$

According to (1) and (2): $T_A - T_B = x\times(t_1 + y\times t_2 - t_4) - t_1 - t_4$

$$T_A - T_B = x\times(t_1 + y\times t_2 - t_4) - t_1 - t_4 \tag{3}$$

The value of the equation (3) may be negative number when x is smaller, because the time of Mobile Agent serialization / deserialization is lager than the request / response interaction time. As the x value increases, the number of managed network elements increases, to a certain extent, t_1 and t_4 can be ignored in the formula (3).

The value of y is the factors of affecting the results when the value of x unchanged. When the value of y is smaller, the value of the formula (4) may be negative because the value t_4 may be larger. However, the value t_1、 t_2 and t_4 are not changing, the value of will be positive when the value y increases (acquisition of the target number of MIB objects) to a certain extent and $(t_1 + y\times t_2 - t_4)$ will be positive and the value of the formula (4) will continue to increase large.

From the analysis result, NMS for the management of the managed network element complexity (in this article reflect the number of MIB objects in the collection) are two different systems in the critical response time due to move to the Mobile Agent on a managed network element local management, so the complexity of managing little change, reflecting a better side than system A, and with the number of managed network elements increases more and more prominent.

② The data flow

M: the average data flow that the information is obtained from managed network elements; N: the data flow that the mobile Agent is migrated; so the total data flow system A is $L_A = 2\times x\times y\times M$, the total data flow system B is $L_B = (x+1)\times N$. The data flow of system B is about $x\times N$ when the value of x is enough lager. Then $L_A - L_B = x\times(2\times y\times M - N)$.

According to the analysis, when the value of y is small, the data flow of the system B is higher than system A, because mobile Agent migration data is higher than the average flow of the SNMP polling. However, the data flow of the system B remains basically unchanged because it is not effect by y. So, with the number of polling target MIB object is increasing, system B has a more prominent advantage.

6 Conclusion

Most of the current network management model is based on protocol SNMP. There are some shortcomings, most manufacturers of network devices are directly supported because of a wide range of applications. Mobile Agent as a unified interface is not feasible because of existing network equipment does not allow third-party software loaded, such as routers, switches, etc. A hybrid network management model is presented based on mobile Agent and SNMP technology in the paper. Application analysis showed that hybrid network management is more prominent in performance than the SNMP.

References

1. Chen, J.-S., Shi, H.-D., Chen, C.-M., Hong, Z.-W., Zhong, P.-L.: An efficient forward and backward fault-tolerant mobile agent system. In: Proceedings - 8th International Conference on Intelligent Systems Design and Applications, vol. 2, pp. 61–66 (2008)
2. Tsaur, W.-J., Yeh, L.-Y.: A novel mobile agent authentication scheme for multi-host environments using self-certified pairing-based public key cryptosystem. International Journal of Innovative Computing, Information and Control 7(5A), 2389–2404 (2011)

3. Lee, Y.-C., Rahimi, S., Gupta, B.: Resource constraint mobile agent framework for ambient intelligence. Proceedings of World Academy of Science, Engineering and Technology 64, 582–589 (2010)
4. Zhang, Q.: Location-transparent mobile agent communication. Advanced Materials Research 225-226, 1054–1058 (2011)
5. Lee, J., Choi, S., Lim, J., Suh, T., Gil, J., Yu, H.: Mobile Grid System Based on Mobile Agent. In: Kim, T.-H., Yau, S.S., Gervasi, O., Kang, B.-H., Stoica, A., Ślęzak, D. (eds.) GDC and CA 2010. CCIS, vol. 121, pp. 117–126. Springer, Heidelberg (2010)
6. Liu, S.-P., Ding, Y.-S.: A scalable policy and SNMP based network management framework. Journal of Donghua University (English Edition) 26(2), 143–146 (2009)
7. Janeiro, J., Liebing, C.: Developing management applications based on SNMP. In: Modern Problems of Radio Engineering, Telecommunications and Computer Science - Proceedings of the 10th International Conference, TCSET 2010, p. 151 (2010)
8. Zeng, X., Cheng, C.: TT-ERCR: A flexible SNMP management method. In: 2009 International Symposium on Intelligent Ubiquitous Computing and Education, IUCE 2009, pp. 476–479 (2009)

A Research of PLD Training for IT Applications

Wei Zhang and Limin Liu

Institute of Embedded Systems
IT School, Huzhou University
Huzhou, Zhejiang, 313000, China
zhangwei@hutc.zj.cn, liulimin@ieee.org

Abstract. PLD, Programmable Logic Device, is a common hardware of IT applications now. A popular PLD is FPGA, Field Programmable Gate Array. Its design is based on VHDL, a hardware description language. Since the development of PLD is concerned to hardware and software, its study is helpful for most IT specialists. In this paper, a research of PLD training for IT applications is discussed.

Keywords: PLD, IT application, FPGA, computer study.

1 Introduction

PLD, Programmable Logic Device, is an essential hardware for many applications of IT, especially embedded system applications, and is used popularly[1].

FPGA, Field Programmable Gate Arrays, and CPLD, Complex Programmable Logic Device, are common PLD devices. PLD presents a relatively new development in the field of VLSI, Very Large Scale Integrated Circuit, circuits and is an ideal target technology for VHDL, a hardware description language, based designs for both prototyping and production volumes. It is different from firmware or micro-controllers that are widely used nowadays. PLD has more advantages in speed, development time and future modification[2-3]. There are many applications of PLD in various fields, such as custom reconfigurable processors, telecommunication systems and networks, fault diagnosis, signal processing and so on[4-6].

In this paper, the concept and training of PLD and its applications will be discussed. The concept of PLD mainly is concerned to FPGA and CPLD for RAM and ROM semiconductor devices. The training of PLD includes theory and practices.

Actually PLD applications are based hardware, software and platform. The common hardware is FPGA and CPLD. The software has to involve HDL, Hardware Description Language, program and other application software. The VHDL, Very high speed integrated circuit HDL is a popular HDL language. The platform of PLD development normally is an EDA, Electronic Design Automation, platform.

2 Concept of PLD

The PLD is an electronic logic device. Its hardware can be designed through software programming. The PLD is a newer device. Before PLD, FLD, Fixable Logic Device,

D. Jin and S. Lin (Eds.): Advances in CSIE, Vol. 2, AISC 169, pp. 45–48.
springerlink.com

devices are common to be used in circuit design as IC devices. FLD is an electronic logic device. Its logic circuit is fixable with one or one set function, can not be updated after to be produced.

With the development of semiconductor technology, PLD is produced on PROM, Programmable Read Only Memory, PAL, Programmable Logic Array, PLA, Programmable Array Logic, GAL, Generic Array Logic. Resent years, FPGA and CPLD are most popular PLD devices. The devices based on PROM, PAL, PLA and GAL, the less than 500 logic gates, are simple PLD devices. FPGA and CPLD are complex PLDs.

In programming, the PLD can be classified as Fig. 1.

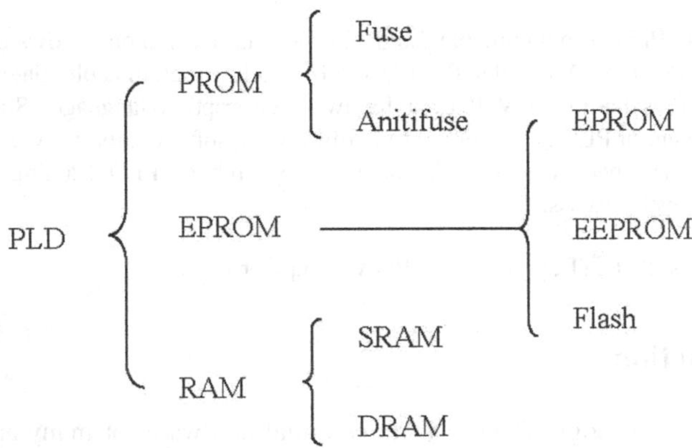

Fig. 1. The PLD classification in programming

The PROM PLD devices are built on fuse or antifuse procedures. Since the programming process of fuse or antifuse is not recovered, the PLD is just programmed one time and in manufactories normally.

The EPROM PLD means that some program on the device can be erased and rewrite again. EPROM, Erasable Programmable Read Only Memory, EEPROM, Electric Erasable Programmable Read Only Memory, Flash, Flash memory are EPROM devices. The program in EPROM is erased by UV rays. To clear codes in EEPROM and Flash memory is with electric current on their PCD board. Obviously, the latter is used more easily.

RAM PLD normally involves SRAM, Static Random Access Memory, and DRAM, Dynamic Random Access Memory, chips. The SRAM PLD is composed of 6 MOS transistors, but the DRAM with 1-4 transistors. So, the cost of DRAM for one memory unit is the cheaper. But the DRAM devices are required refreshing information in a short time.

The main difference between ROM and RAM is that ROM keeps information without power, but RAM not.

3 Training of PLD

In order to study the design skill of PLD applications, a student should learn some basic knowledge, such as hardware, software and platforms.

3.1 The Basic Knowledge of PLD Design

The knowledge frame of PLD application design is shown as Fig,2. Here, hardware, software and platform are basic knowledge for PLD development.

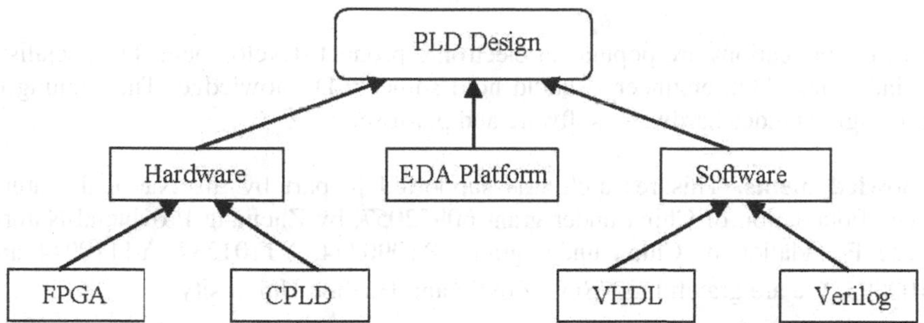

Fig. 2. The knowledge frame of PLD design

For the hardware, FPGA and CPLD should be known. FPGA is based on look-up table and SRAM normally. CPLD is a PLD device based on product term and EEPROM. The product term is a programming structure with programmable AND array and fixed OR array. Actually, for different manufactures the definition FPGA or CPLD is various.

The software for PLD is HDL, a special language to design hardware by software. The common HDL languages are VHDL and Verilog.

Platform is based on EDA technology. It is a design or programming environment as some tool and kids.

3.2 The Software and Platform

The VHDL language is not very difficult to be learned, if the student is familiar with C or Assembly language. The programming essential elements of VHDL are Entity, used to define external view of a model, such as symbol, and Architecture, used to define the function of the model, such as schematic.

Similar to C, there are some libraries in VHDL. It can supply some common code modules and let program more easily. A VHDL routine is as follow.

```
Library
Entity use_fct is
        Port( );
End use_fct;
Architecture struct of use_fct is
 Begin
   Dsp: process;
        Data handle;
   End process Dsp;
End struct;
```

The PLD development platforms are mainly designed by professional suppliers or manufactures, such as Cadence, Xilinx or Atera.

4 Conclusions

The PLD applications are popular in electronic product development. IT specialists, especially embedded engineers, should hold some PLD knowledge. The training of PLD design includes hardware, software and platforms.

Acknowledgments. This research was supported in part by the National Natural Science Foundation of China under grant 60872057, by Zhejiang Provincial Natural Science Foundation of China under grants R1090244, Y1101237, Y1110944 and Y1100095. We are grateful to NSFC, ZJNSF and Huzhou University.

References

1. Kilts, S.: Advanced FPGA Design: Architecture, Implementation, and Optimization. Wiley, New Jersey (2007)
2. Alluri, V.B., Heath, J.R., et al.: A New Multichannel, Coherent Amplitude Modulated, Time-Division Multiplexed, Software-Defined Radio Receiver Architecture, and Field-Programmable-Gate-Array Technology Implementation. IEEE Transaction on Signal Processing 58(10), 5369–5384 (2010)
3. Ostua, E., Viejo, J., et al.: Digital Data Processing Peripheral Design for an Embedded Application based on the Microblaze Soft Core. In: Proc. 4th South. Conf. Programmable Logic, San Carlos de Bariloche, Argentina, vol. 3, pp. 197–200 (2008)
4. Dimond, R.G., Mencer, O., Luk, W.: Combining Instruction Coding and Scheduling to Optimize Energy in System-on-FPGA. In: Proceedings of 14th Annual IEEE Symposium on Field-Programmable Custom Computing Machines, Napa, CA, USA, vol. 4, pp. 175–184 (2006)
5. Sifakis, J.: Embedded systems design - Scientific challenges and work directions. In: Proceedings of DATE 2009, Nice, France, vol. 4, p. 2. IEEE Press (2009)
6. Liu, L., Luo, X.: The Reconfigurable IP Modules and Design. In: Proc. of EMEIT 2011, Harbin, China, vol. 8, pp. 1324–1327 (2011)

Terminal Sliding Mode Control for STT Missile Based on RBF Neural Network

Xianjun Shi, Hongchao Zhao, and Jie Chen

Naval Aeronautical and Astronautical University, Erma Road 188, 264001 Yantai, Shandong, China
sxjaa@sohu.com, navyzhc@163.com, hjhychenjie@tom.com

Abstract. The output-redefinition approach is applied in the overload control system, in order to remove the nonminimum phase characteristic of overload output. It utilizes the combination of overload and angle rate to redefine the new output for the sake of easy measurement. In order to improve the tracking speed and to estimate the lumped uncertainty, the Terminal sliding mode (TSM) controller based on RBF neural network is designed in the overload control system. A simulation example is illustrated to compare the TSM controller with the conventional SMC, and the simulation results show that the former has faster tracking speed than the later.

Keywords: output-redefinition, overload control, TSM, RBF neural network.

1 Introduction

Recently, the overload control method has been widely used to design the flight control system of STT (skid-to-turn) missile, for it can improve the maneuverability of the STT missile [1, 2]. Since the dynamic characteristic from control fin deflection to missile overload output is of nonminimum phase for tail-controlled configuration, the output-redefinition approach is often applied to remove it, and thus the overload control system is transformed into a minimum phase system [3, 4]. The former researchers usually utilize the combination of overload, angle-of-attack and angle acceleration to redefine a new output. But it is very difficult to exactly measure angle-of-attack and angle acceleration. In this paper we utilize the combination of overload and angle rate to redefine the new output for the sake of easy measurement.

Sliding mode control (SMC) approach is well-known for its robustness to parameter perturbations and external disturbances, so it is successfully applied in many fields. The switching surface of conventional SMC is chosen as a linear hyper-plane. Its representative property is that the system state converges to the equilibrium point asymptotically but not in finite time. Terminal sliding mode (TSM) control approach can offer some superior properties such as fast finite time convergence and less steady state errors [5, 6]. Therefore, in the paper we adopt the TSM control approach to design a controller so as to make the new output to track its desired signal fastly. In addition, we consider the uncertainties of missile parameters and the external disturbance. The RBF neural network (RBFNN) is used to estimate them.

D. Jin and S. Lin (Eds.): Advances in CSIE, Vol. 2, AISC 169, pp. 49–54.
springerlink.com

2 Missile Dynamics

The following assumptions are adopted for the STT missile.

Assumption 1: The missile has pitch and yaw symmetry.

Assumption 2: The missile is roll-position stabilized.

From above assumptions, pitch dynamics and yaw dynamics of the STT missile are very similar, and the structure of pitch control system can be converted to that of yaw control system with ease. Hence, only the pitch control system is studied here. For small angle-of-attack and sideslip angle, the pitch dynamics of the STT missile with actuator dynamics is described as follows,

$$\begin{cases} \dot{\alpha} = \omega_z - a_{34}\alpha - a_{35}\delta_z \\ \dot{\omega}_z = a_{24}\alpha + a_{22}\omega_z + a_{25}\delta_z \\ \dot{\delta}_z = -\dfrac{1}{\tau}\delta_z + \dfrac{k_\delta}{\tau}u_\delta \\ n_y = \dfrac{v}{g}a_{34}\alpha + \dfrac{v}{g}a_{35}\delta_z \end{cases}, \tag{1}$$

where α is the angle-of-attack, ω_z is the pitch rate, δ_z is the fin deflection angle, τ and k_δ are the actuator's time constant and gain, respectively, u_c is the control input command, n_y is the normal overload, a_{22}, a_{24}, a_{25}, a_{34} and a_{35} are the missile's aerodynamic coefficients.

In order to help the state variables represent the missile's performances directly, we take a coordinate transformation on the system (1) as $[\alpha\ \ \omega_z\ \ \delta_z]^T \to [\alpha\ \ \omega_z\ \ n_y]^T$. Note that the relative degree of the system is 1, so there is no need to transform it into the normal form, i.e. n_y is the output and $[\alpha\ \ \omega_z]^T$ constitutes the zero dynamics.

$$\begin{cases} \dot{n}_y = Q_1\eta + P_1 n_y + bu_\delta \\ \dot{\eta} = Q_2\eta + P_2 n_y \end{cases}, \tag{2}$$

where $Q_1 = \begin{bmatrix} \dfrac{v}{g\tau}a_{34} & \dfrac{v}{g}a_{34} \end{bmatrix}$, $\eta = [\alpha\ \ \omega_z]^T$, $P_1 = -a_{34} - \dfrac{1}{\tau}$, $b = \dfrac{vk_\delta}{g\tau}a_{35}$, $P_2 = \begin{bmatrix} -g/v \\ ga_{25}/(va_{35}) \end{bmatrix}$, $Q_2 = \begin{bmatrix} 0 & 1 \\ a_{24} - a_{25}a_{34}/a_{35} & a_{22} \end{bmatrix}$. It is easy to see that the eigenvalues of Q_2 coincide with the open-loop transmission zeros. For the missile overload output is of nonminimum phase, it has unstable zero dynamics. We will stabilize the zero dynamics by output-redefinition approach.

The new output is redefined as, according to the measured signals n_y and ω_z,

$$\xi_m = n_y + c\omega_z = n_y + K\eta , \tag{3}$$

where $K = [0 \quad c]$, thus the system (2) is transformed into

$$\begin{cases} \dot{\xi}_m = \bar{P}_1 \xi_m + \bar{Q}_1 \eta + b u_\delta \\ \dot{\eta} = (Q_2 - P_2 K)\eta + P_2 \xi_m \end{cases} , \tag{4}$$

where $\bar{P}_1 = P_1 + KP_2$, $\bar{Q}_1 = Q_1 + KQ_2 - P_1K - KP_2K$. Here the zero dynamics is transformed into the second equation of (4). For (Q_2, P_2) is a controllable pair, we can assign arbitrary eigenvalues of $(Q_2 - P_2K)$ in the left-half phase plane by choosing *K* (i.e. *c*). *K* plays the role in only stabilization of zero dynamics by pole placement. Omitting the detailed deduction, we give the stabilization condition $c < -v/g$.

For the output tracking problem, the following tracking errors are defined,

$$\tilde{\xi}_m = \xi_m - \xi_m^d , \ \tilde{n}_y = n_y - n_y^d , \ \tilde{\eta} = \eta - \eta^d , \tag{5}$$

where ξ_m^d, n_y^d and η^d denote the desired signals of ξ_m, n_y and η, respectively.

From (3), ξ_m^d can be expressed as

$$\xi_m^d = \xi^d + c\omega_z^d = n_y^d + K\eta^d . \tag{6}$$

We design a controller to make $\tilde{\xi}_m$ to converge to zero in a finite time *T*, i.e. $\tilde{\xi}_m(t) = 0$, $\forall t \ge T$, then the $\tilde{\eta}$-dynamics changes to be

$$\dot{\tilde{\eta}} = (Q_2 - P_2K)\tilde{\eta} , \ \forall t \ge T . \tag{7}$$

Because ($Q_2 - P_2K$) is a Hurwitz matrix, we get $\tilde{\eta} \to 0$ as $t \to \infty$ and

$$\tilde{n}_y = \tilde{\xi}_m - K\tilde{\eta} \to 0 \text{ as } t \to \infty , \tag{8}$$

which represents the tracking error of original output also converges to zero.

3 Terminal Sliding Mode Controller Design Based on RBFNN

Form above analysis, in order to ensure the tracking of n_y to its desired signal n_y^d, we need only to design a controller to force $\tilde{\xi}_m \to 0$ in a finite time.

In actual application the missile's aerodynamic coefficients can't be get exactly for all flight conditions. They have some uncertainties and may be expressed as:

$$a_{22} = \hat{a}_{22} + \tilde{a}_{22} , \ a_{24} = \hat{a}_{24} + \tilde{a}_{24} , \ a_{25} = \hat{a}_{25} + \tilde{a}_{25} , \ a_{34} = \hat{a}_{34} + \tilde{a}_{34} , \ a_{35} = \hat{a}_{35} + \tilde{a}_{35} , \tag{9}$$

where the sign "^" represents a measured value and "~" an uncertain value. Accordingly the parameters metrics of (4) are expressed as:

$$\bar{P}_1 = \hat{\bar{P}}_1 + \tilde{\bar{P}}_1,\ \bar{Q}_1 = \hat{\bar{Q}}_1 + \tilde{\bar{Q}}_1,\ b = \hat{b} + \tilde{b},\ P_2 = \hat{P}_2 + \tilde{P}_2,\ Q_2 = \hat{Q}_2 + \tilde{Q}_2, \tag{10}$$

Remark 1: Suffering from the uncertainties, the open-loop transmission zeros may depart from their original places. However, as long as K is chosen to place the zeros far from the origin of the phase plane in the left, the stability of the zero dynamics can be always guaranteed.

The aim of the controller design is to force $\xi_m \to 0$ in a finite time T. The ξ_m subsystem of (4) is expressed as, considering the uncertainties and external disturbance,

$$\dot{\xi}_m = (\hat{\bar{P}}_1 + \tilde{\bar{P}}_1)\xi_m + (\hat{\bar{Q}}_1 + \tilde{\bar{Q}}_1)\eta + (\hat{b} + \tilde{b})u_\delta + d\,. \tag{11}$$

where d is the external disturbance. A TSM hyperplane is designed as

$$S = \tilde{\xi}_m + \rho \int_0^t \tilde{\xi}_m^{q/p}(\tau)d\tau, \tag{12}$$

where $\rho > 0$, q and p are odd positive integers with $q < p$. If the system states are trapped in the TSM hyperplane, then $S(t) = \dot{S}(t) = 0$, and the equivalent dynamics of the system (11) is

$$\dot{\tilde{\xi}}_m = -\rho \tilde{\xi}_m^{q/p}. \tag{13}$$

Thus the time of $\tilde{\xi}_m$ converging to zero is given as

$$T = \frac{p}{\rho(p-q)}\left[\tilde{\xi}_m(0)\right]^{(p-q)/p}. \tag{14}$$

The time derivative of (12) is

$$\dot{S} = \hat{\bar{P}}_1\xi_m + \hat{\bar{Q}}_1\eta + \hat{b}u_\delta + (\tilde{\bar{P}}_1\xi_m + \tilde{\bar{Q}}_1\eta + \tilde{b}u_\delta + d) - \dot{\xi}_m^d + \rho\tilde{\xi}_m^{q/p}. \tag{15}$$

Let $\Delta(t) = \tilde{\bar{P}}\xi_m + \tilde{\bar{Q}}_1\eta + \tilde{b}u_\delta + d$, which is called the lumped uncertainty and is estimated by an RBFNN. The RBFNN is a three-layer feedforward neural network. For only one uncertain function $\Delta(t)$ is required to be approximated, the RBFNN has one output layer node. The network output is

$$y = W^T\Phi(X), \tag{16}$$

where $W = [w_1, \cdots, w_J]^T$ is the weight vector of RBFNN, $X = [\tilde{\xi}_m, \tilde{n}_y]^T$ is the input vector to the network, $\Phi(X) = [\phi_1(X), \cdots, \phi_J(X)]^T$, $\phi_i(X)$ is a Gaussian basis function, and let $\phi_i(X) = 0$ if $\phi_i(X) < 0.03$.

The following assumptions are introduced according to the universal approximation of RBFNN.

Assumption 3: $\Delta(t)$ is an unknown bounded continuous function.

Assumption 4: For $\Delta(t)$ and arbitrary $\varepsilon > 0$, there always exists an optimal weight vector W^*, such that

$$\Delta(t) = W^{*T}\Phi(X) + \varepsilon, \tag{17}$$

where ε is approximation error of RBFNN. But W^* can't be obtained beforehand, and it is estimated by $\hat{W}$. The estimate error is $\tilde{W} = \hat{W} - W^*$.

The controller design is stated as the following theorem.

Theorem 1: For the uncertain system (11) with the TSM hyperplane (12), the TSM controller based on RBFNN is designed as

$$u_\delta = -\frac{1}{\hat{b}}\left\{\hat{\bar{P}}_1\xi_m + \hat{\bar{Q}}_1\eta - \dot{\xi}_m^d + \rho\tilde{\xi}_m^{q/p} + \hat{W}^T\Phi(X) + \lambda_1 S + \lambda_2 sign(S)\right\}, \tag{18}$$

where λ_1, $\lambda_2 > 0$, and the weight vector is updated online by

$$\dot{\hat{W}} = \Gamma\Phi(X)S, \tag{19}$$

where $\Gamma = diag(\eta_1, \cdots, \eta_J)$, $\eta_i > 0$ is the update rate; then the tracking error $\tilde{\xi}_m$ will converge to zero in a finite time.

The theorem is easily proved by the Lyapunov stability theory.

In the application, the sign function $sign(S)$ may cause chattering effect, so it is usually replaced by a continuous function $S/(|S|+\sigma)$, where σ is a little positive number.

4 Simulation Example

In the section, a supersonic STT missile is illustrated, its pitch dynamics is of (1) with the aerodynamic coefficients uncertainties (9) and the disturbance *d*. The overload control method is used to design its flight control system, where the controller is designed by TSM control technique based on RBFNN according to (18) along with (3), (12) and (19). In the simulation the uncertainties are assumed to be $\tilde{a}_{22} = 0.32\cos(3t)$, $\tilde{a}_{24} = 1.95\sin(6t)$, $\tilde{a}_{25} = 1.6\cos(5t)$, $\tilde{a}_{34} = 0.22\sin(5t)$, $\tilde{a}_{35} = 0.07\sin(4t)$, and the disturbance $d = 2.5\sin(6t)$. The desired signal of missile overload output is chosen as $n_y^d = 1.0$. For the TSM controller we choose $q/p = 5/7$. In order to compare the TSM controller with the conventional SMC, we also perform the simulation with $q = p$. The simulation result is shown in figure 1. It is obviously that the TSM controller has faster tracking speed than the conventional SMC.

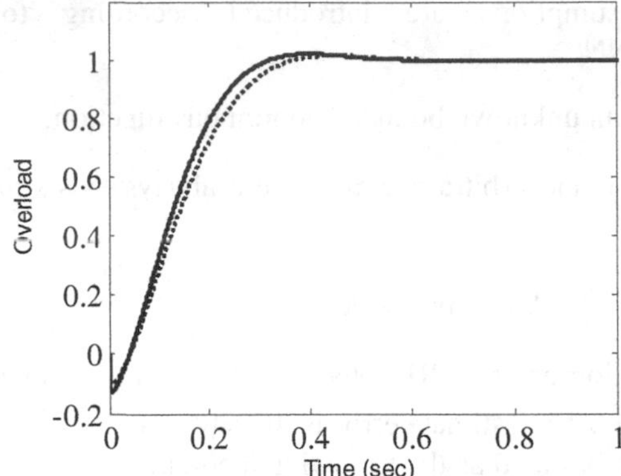

Fig. 1. Responses of missile overload output with the TSM controllers (*solid line*) and the conventional SMC (*dotted line*)

5 Conclusion

A new overload control method is studied in the paper. It uses the combination of overload and angle rate to redefine a new output for the sake of easy measurement. The output-redefinition approach solves the nonminimum phase problem. In order to improve the tracking speed and to estimate the lumped uncertainty, the TSM controller based on RBFNN is designed in the overload control system. A simulation example is illustrated to test the performance of the controller. The simulation result shows the effectiveness of the designed controller.

Acknowledgments. This work is supported by National Science Foundation of China (No. 61174031).

References

1. Jackson, P.B.: Overview of Missile Flight Control Systems. Johns Hopkins APL Technical Digest 29(1), 9–24 (2010)
2. Adounkpe, I.J., Godoy, E., Harcaut, J.P.: Robustness Analysis of a Nonlinear Missile Autopilot: an Integral Quadratic Constraint Approach (2004), `http://ascc2004.ee.mu.oz.au/proceedings/papers/P266.pdf`
3. Gu, W., Zhao, H.: Fuzzy Sliding Mode Control for Slid-to-Turn Missile Based on Output-Redefinition. Dynamics of Continuous Discrete and Impulsive System (12), 162–171 (2005)
4. Yu, J., Gu, W., Zhang, Y.: Adaptive Fuzzy Sliding-Mode Control for Nonminimum Phase Overload System of Missile. Joumal of Jilin University (Engineering and Technology Edition) 34(3), 412–417 (2004)
5. Feng, Y., Yu, X., Man, Z.: Adaptive Fast Terminal Sliding Mode Tracking Control of Robotic Manipulator. In: 40th IEEE International Conference on Decision and Control, pp. 4021–4026. IEEE Press, New York (2001)
6. Man, Z., Shen, W., Yu, X.: A new Terminal Sliding Mode Tracking Control for a Class of Nonminimum Phase Systems with Uncertain Dynamics. In: International Workshop on Variable Structure Systems, pp. 147–152. IEEE Press, New York (2008)

Design of Broadband Receiver in Battlefield Electromagnetic Spectrum Monitoring System

Wenjun Lu, Libiao Tong, Lianfei Duan, Fei Ye, and Hongjun Zhang

New Star Research Institute of Applied Technology, Hefei 230031, China
{lwj2963,nuptzhj}@sina.com, libtong@sohu.com

Abstract. By the increasing complexity of future battlefield actual electromagnetic environment, a radio receiver based on spectrum monitoring was designed, and the overall design based on triple conversion mechanism was presented. The key designs were focused on the synthesized local oscillator, digital IF, and electromagnetic compatibility. The receiver implements the reception of signal covered 3GHz.

Keywords: Electromagnetic Environment, Battlefield Electromagnetic Environment, Spectrum Monitoring, Receiver, Triple Conversion Mechanism, Electromagnetic Compatibility.

1 Introduction

Electromagnetic environment (EME, electromagnetic environment) generally is present in a given place the sum of all electromagnetic phenomena. Battlefield electromagnetic environment is the environment in a particular military operation, system or platform to carry out its mandate may encounter in various frequency range by radiation or conduction of the electromagnetic emission (horizontal) and temporal distribution of power results. Military operations in increasingly are implemented in complex electromagnetic environment. As war battlefield electromagnetic environment by geographic distribution of equipment, operating frequency, radiated power, radiation pattern, in which geographic features in the environment, weather conditions and other effects, and thus, the battlefield electromagnetic environment by time domain, frequency domain, the domain can and on the distribution of dense airspace, the vast amount of style complex, dynamic random signal overlap from a variety of complex electromagnetic environment. Complex electromagnetic environment has become different from the traditional battlefield, the battlefield information the most prominent signs.

Battlefield environment, the radio frequency spectrum monitoring and management system for battlefield communication and command staff to provide high-level, intelligent frequency monitoring and management, is the frequency of communication and command personnel in planning an effective and dynamic management tool and an important tool. War, according to battlefield communications network topology, transmission equipment and battlefield performance information such as geographical features, spectrum monitoring and management system for the development of

D. Jin and S. Lin (Eds.): Advances in CSIE, Vol. 2, AISC 169, pp. 55–60.
springerlink.com

communication and command staff the best frequency assignment program, battlefield communications network with high electromagnetic compatibility nature, and part of the network band interference in the case, adjust the frequency of use of the program.

Based on the above background and needs of military applications, battlefield environment, further study of the radio spectrum monitoring and management system, a systems integration approach, preparation and management of war-time spectrum monitoring equipment is necessary.

2 Theory of Radio Spectrum Monitoring

2.1 The Concept and Characteristics of Radio Spectrum

Radio spectrum is a band from 9 kHz to 3000GHz, spread without physical connection to the space of oscillation power and magnetic energy. Due to technical limitations and radio equipment, ITU (International Telecommunication Union) defined 9 kHz to 400GHz range, in fact, only dozens of now-GHz, and cellular operations in to 3GHz. The radio spectrum is available for people to use in the production and life, with the object of ownership and use of natural substances, is a resource, which has limited, non-expendable, objective laws, four features easy to interference, so the actual application in order to improve the spectrum efficiency, only through the development of new systems, develop new band, reducing the transmitter power and optimize system parameters such as means to achieve.

2.2 The Concept of Radio Spectrum Monitoring

Is the use of radio spectrum monitoring techniques and some equipment for the basic parameters of radio emission and spectral characteristics (frequency, frequency error, RF level, emission bandwidth, modulation) to measure, monitor analog signals, digital signals spectral characterization, and utilization of the frequency band occupies a degree of statistical analysis to test, and illegal radio stations and to locate sources of interference measurements to investigate.

3 Receiver Design Based on Radio Frequency Spectrum Monitoring

3.1 Machine Design

RF receiver design in the selection of commonly used traditional high-frequency mode, this program easy to deal with image rejection, but first require a higher vibration. In order to ensure coverage of a wide tuning range and low noise sidebands of the requirements, often using multi-segment modular voltage-controlled YIG oscillator (VCO) using broadband microwave devices are more expensive machine design and development. The design of this machine will give up the traditional wideband receiver design, the use of interactive multi-frequency tuning mode, which

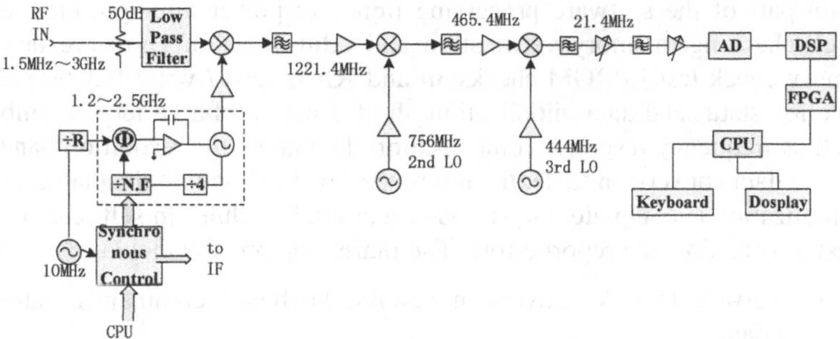

Fig. 1. Block Diagram of RF Receiver Design

reduces the bandwidth of the oscillator and the broadband requirements of microwave devices. The first local oscillator design uses a single-loop fractional-N and digital sweep mode, with all-digital IF processing be achieved. Machine design diagram is shown in Figure 1.

RF input signal through the attenuator and 3GHz low-pass filter, which is fed into the first mixer. In the first mixer, the input signal and the 1.2 ~ 2.5GHz local oscillator signal of the first mixer, first IF signal to be 1221.4MHz, after the mixer is a low noise amplifier to compensate for first-level mixed frequency conversion loss, then the signal through a 3dB bandwidth of 400MHz first IF filter, this requires the first local oscillator signal is provided by the broadband VCO. The first IF signal in the second mixer, second local oscillator and mixer 756MHz, 465.4MHz to get the second IF signal, after the mixer is also a low-noise amplifier, followed by the second intermediate frequency through a 3dB bandwidth of 20MHz IF filter, the signal after the third mixer, and 444MHz to be the third local oscillator mixing the third intermediate frequency signal 21.4MHz, 21.4MHz signal through two controllable-gain amplifier and a 3dB bandwidth of 2MHz filter, the signal is fed into the AD converter, is converted to a digital IF signal. Digital IF signal is sent to the digital board, the first in the FPGA are down converted to base band signals and digital RBW filters used to filter, then the signal is sent to digital detectors, video filtering and video detector and converted to the several formats to give the main CPU, the last show on the display.

The receiver enables a fully digital IF processing, all of its IF filter bandwidths, the video bandwidth filter and video detector by software. Microwave part of the third-down conversion to IF of 21.4MHz, anti-alias filter into the AD, follow-up digital IF processing. The digital IF receiver hardware block diagram shown in Figure 2.

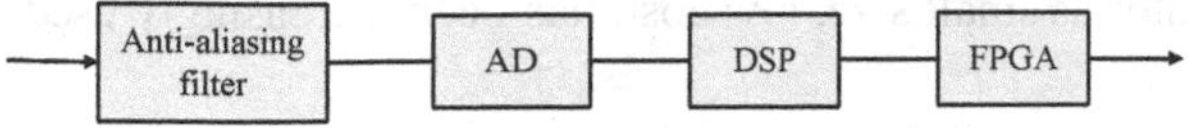

Fig. 2. Digital IF Receiver Hardware Block Diagram

Of these, DSP and FPGA chips to implement functions. DSP main digital IF filter, FPGA implementation of digital mixer and all major digital video filters and video detector.

The main part of the software processing unit are: power after power, the CPU features self-checking, memory verification and adjustment of hardware detection, main memory check test EPROM checksum and RAM read / write function testing, and completion status and data initialization, the formation of a variety of calibration forms, such as frequency response compensation data table, the resolution bandwidth compensation data conversion table, the reference level calibration data tables, and so on. All initialization is complete, the various functional modules in self-test, and save the self-test information and report errors. The main loop process includes:

(1) Data received: D / A conversion results, keyboard commands, interfaces, commands and data.

(2) Data processing: to quantify A / D conversion result and the corresponding data processing, measurement data processing according to display data and the need to generate a variety of forms, data form the final display, display control and display refresh.

(3) Measurement and control: the first, second and third local oscillator frequency control, frequency resolution bandwidth, video bandwidth and IF gain control.

(4) Calibration: reference level, resolution bandwidth filter calibration.

(5) Error detection and diagnosis: exception handling, fault location.

3.2 Digital Design Swept Local Oscillator

With the increasing test requirements, the receiver's local oscillator technology is gradually evolving, and has become the instrument to measure the performance level of the main signs. Early use of open-ended LO YIG oscillator frequency stability is poor, and the remaining FM to 1 kHz; after the development of frequency-locking type local oscillator, the remaining FM reached 30Hz; and the current high-performance receiver, the most used PLL type synthesized local oscillator, the output signal is locked to a stable reference base, the residual FM of less than 1Hz. This LO is generally formed by a number of phase-locked loops, each loop of the oscillator's contribution requirements vary. The sweep control is usually constituted by a ramp generator and a digital-scan circuit, resulting in the slope and rise time control of the scanning ramp voltage, and current drive YIG oscillator to convert the main coil. In order to improve the accuracy of sweep, the oscillator phase-locked-type synthesis using "lock a roll" approach to control, that is the starting point of each scan loop lock, then open-loop scanning, scanning the end of the count and set the frequency and comparison to be swept error, and then the next scan of the ramp voltage to be amended. In this way, the analog scan sweep oscillator has tuning voltage and high linearity requirements, if the primary broadband VCO oscillator device more difficult to guarantee. In the modern design of the RF receiver's local oscillator, the voltage-controlled oscillator as small size, low cost, has been increasingly used.

3.3 IF Filter

Design of IF processing is a core technology of the receiver. In general super heterodyne receiver, RF signal frequency after repeated after the completion of most of the intermediate frequency signal processing functions. These functions include the resolution bandwidth (RBW) filter, video bandwidth (VBW) filter and detector type selection, etc.; will eventually be converted into a continuous spectrum of information displayed on the screen points. For a long time, the receiver IF frequency processing have been using analog technology, digital signal processing technology with the recent rapid development of digital IF technology has been successfully applied to a new generation of receivers. In the receiver design, we use the large bandwidth of the analog IF processing technology, the processing of large bandwidth while using the latest digital IF technology.

Test signal into the RF front-end by three after the output 21.4MHz IF frequency analog signals with the LO mixer is 25kHz, where the frequency back again to help reduce the A/D sampling process, but also help to improve the device performance. Mixer output low-pass filter to remove clutter. Stepping through the programmable gain amplifier signal the main purpose is to optimize the IF signal level to meet the A/D converter's input level required to achieve the maximum dynamic range. In the A/D conversion before the anti-aliasing filter to eliminate possible signal generated by the sampling aliasing frequency components. To improve data accuracy and increased signal processing dynamic range, high-precision 18 A/D, the A/D converted data into the DSP processing, and then displays the video detector output. Here, we use the generic Motorola DSP56002 digital signal processing chip to complete the digital intermediate frequency processing.

Linear phase digital IF frequency than analog RBW filters allow faster scan speed. Digital implementation makes it easier read data frequency and amplitude compensation, typically allow the scan rate is twice to four times the old receiver. Number of ways to achieve with digital zoom is very accurate; the whole of the typical error is small.

IF bandwidth of the precision set by the ability to filter the digital part and analog pre-filter to determine the uncertainty of calibration, and analog frequency analyzer 10 to 20% compared to the majority of RBW in their specified bandwidth of 2%.

3.4 Video Detector

Signal from the IF part of the resolution bandwidth filter and logarithmic detector, the detector video signal obtained by low-pass filter after the video detector processing. The purpose of the video detector is based on the different test needs to be synchronized with the vibration amplitude value, and display the screen corresponding to the position. Conventional receiver, the signal detected by the detector range, the most used in the video detector in a large number of discrete components to achieve different detection methods, the A/D conversion data is later read by the CPU, so the circuit is complex and bulky. We are in the design of large-scale FPGA implementation using a variety of video detection methods, only to conduct A/D conversion, data processing of the detector inside the FPGA implementation; CPU read the data directly, this circuit design, small size, easy to retrofit high reliability and easy adjustment.

References

1. Ronald, B.: The Fourier Transform and Its Applications. McGraw-Hill College (2005)
2. Li, L.: Modern electronic test technology. National defense industry (2008)
3. Meihua, L.Z.G.: The Discussion on the Design Technique for Tracking Pre-selector of The Spectrum Analyzer. Journal of Electronic Measurement and Instrument 14(1), 37–40 (2000)

4. Chen, C.-T.: Digital Filter Design. Oxford University Press (2002)
5. Oppenheim, A.V.: Discrete-Time Signal Processing. Prentice Hall (1989)
6. Zhang, B., Kiasaleh, K.: Broadband receiver design on FPGA. In: IEEE International Symposium on Broadband Multimedia Systems and Broadcasting, pp. 1–5 (2008)
7. Pamarti, S., Galton, I.: Phase-Noise Cancellation Design Tradeoffs in Delta-Sigma Fractional-N PLLs. IEEE Transactions on Circuits and Systems II: Analog and Digital Signal Processing 50(11), 829–838 (2003)
8. Qi, J.-G., Cao, G.-P.: Design of a RF Wideband Receiver. Telecommunication Engineering 47(2), 88–91 (2007)

Discussion of "Sensors and Detection Technology" Curriculum Construction

Shenhui Du and Zheng Li

School of Electrical Engineering and Information Science,
Hebei University of Science and Technology, Shijiazhuang 050018, China
hbdsh999@sina.com

Abstract. As a professional foundational course, the theory of Sensors and Detection Technology is playing a guiding role in the field of industrial manufacturing and scientific research. Considering it has a strong combination of a diverse of subjects as well as the problems arise in traditional teaching model, the article poses new teaching methods mainly on class-teaching model and how to put into practice and good teaching results have been obtained in practical teaching work.

Keywords: Electrical and automation Sensors and detection courses Teaching experience Teaching research.

1 Introduction

Sensors are located in the interface of the subject and test system. Sensors are used to obtain information from all scientific research and automatic manufacture process and transform them into electric signal that are easy to transport and deal with. Hence Sensors and Detection Technology is an important professional basic course for engineering majors like monitoring and control, automation, electrical engineering and automation etc. This course serves as a connecting link between the preceding and the following between the establishments of specialties [1, 2]. On one hand it is a combination of Analog electronics, digital electronics, circuits, automatic control theory etc. One the other hand it makes up SCM Principles, Process Control Engineering etc into smart meters, forming a complete intelligent control unit. This does have a big impact on students' professional development. In the article, we take Hebei University of Science and Technology as an example to discuss the study of "Sensors and Detection Technology" teaching.

2 Professional Development Requirements

Sensors and Detection Technology is a basic course for automation, electrical engineering and automation, Monitoring and control.40 classes for automation, electrical engineering and automation including 8 classes of experiments. 48 classes for monitoring and control with 8 classes of experiments are included. It is a required course for all three majors. Through the study of the course, the students should master the basic principles and applications of typical types of the sensors, grasp the

D. Jin and S. Lin (Eds.): Advances in CSIE, Vol. 2, AISC 169, pp. 61–66.
springerlink.com

basic working principles and performances of the measuring instruments and they can select the electronic equipments reasonably and measure the circuits. After the course we also demand them to design the measurement systems according to the variously measuring requirements, analysis the errors of the measuring results and process the data, all of which help them to achieve the unity of theory and practice.

3 Course Features and the Lack of Traditional Teaching Model

Sensors and Detection Technology contains a large amount of information and involves a wide range of professionals but each chapter is relatively independent and is lack of continuity and systematic. Thus arises the problem "hard to teach and difficult to learn". The traditional teaching model is in center of teachers and students are in a passive state to accept. The lack of flexibility and initiative is not conducive to mobilize students in active learning. Students after school are lack of enthusiasm for the extension of curriculum only limited to materials. Their ideas are narrow and they do not understand the related relationship between related courses so they do not form a whole concept of the major. Performances as follows: 1.Teaching arrangement ideas as "many principles and little practice". Assessment methods are "theory instead of experiment". Teaching is focused on the principles of teaching. Generally verification experiments leave students little space to unleash initiative, not even. Student's curiosity and exploration are not met, so they gradually lose their interests. 2. Amendments of teaching materials are delayed to the development of the subject and new sensors are timely introduced into materials. 3. Teaching methods use mainly classroom teaching, with emphasis on learning of the sensor work principles, practical application for the sensor described rarely. Students lack an intuitive understanding, thus thinking activity restricted, over the time students tend to lose the initiative to learn [3, 4].

4 Methods of Teaching Reform and How to Achieve

Sensor and Detection Technology is a very applied course which not only possesses distinctive feature of theory but also distinctive characteristics of the actual operation on which the overall teaching direction is settled as follow: Theory serves practice by mode that theory study and teaching practice go neck and neck and by theory teaching and practice combined, teaching effectiveness will be improved.

4.1 Curriculum Instruction and Daily Life Link Ingeniously

In the learning process of Sensor and Detection Technology, give some related examples in real life so that students understand the application of sensor is at hand. For instance: When introducing contact and non-contact measurement, list stations , hospitals and other public places using infrared temperature detector in the special period of SARS is a typical example of non-contact measurement. When introducing strain gauge sensor, list common large vehicles weighing device , that is a detection unit using strain gauge which can form a bridge. When introducing piezoelectric sensor, list why sparks appear when we press the lighter. Understand that the force on the piezoelectric generated charge then ignited gasoline and then spark. Instance by side, so that students truly feel the sensor and our lives are close connected, to stimulate students' interest and enthusiasm.

4.2 Introduction of Multi-media Teaching and Complementary to Traditional Blackboard

The advantage of multimedia teaching is vivid, compared to a large section of text description; the screen is more intuitive, clear results. Especially the need to explain the contents of the site for graphics, multimedia teaching can save time, graphics is more standard. When teaching variable-gap-type differential inductive pressure sensor, multimedia animation indicates that when armature moves up and down the output voltage will be constantly changing. Hand-drawing effect is laborious as well as not good [5].

But for engineering courses, only by using multimedia to teach may easily fall into the embarrassing situation that what the teacher taught and what the student thought are not synchronized. Multimedia demonstrates quickly, the amount of information increases teacher lectures invisibly, the time for students to reflect becomes shorter. Complex problems require complex calculations and derivation, the formula currently displayed and the text reflects the limited amount of information, related content can not display on the same page, resulting in lectures difficult to understand. Taught using blackboard can solve the problem that the derivation of complex equations, difficulty explanation or the problem of slow response from the students when introducing the basic principles of the sensor. Writing while teaching, it can make teaching and listening to the synchronization. In view of two teaching methods have their own strengths, we should make appropriate arrangements and give full play to their strengths in actual class teaching according to their strengths and the different characteristics of the teaching contents.

4.3 A combination of Sensors and Detection Technology and Microcontroller Courses-Production Practice

Under normal circumstances, curriculum design as engineering courses serves as an important mean to combine theoretical knowledge and practical .While the schools use existing laboratory experiments platform to set up a two-week school production internship link. The link combines with the Sensor and Detection Technology and Microcontroller course. Teachers are supposed to draw up the subject while students to choose their topics. It can also be submitted by the students themselves according to interests while teachers review the feasibility of the subject. In the production practice process, the students design their subjects according to teachers' arrangement by going to the library or surf the Internet for materials, the specific content of the designer can be simulated by students in the study and the instructor should check the correct before soldering circuit boards in the laboratory and then use the laboratory burner for software debugging. In the final inspection process, teachers raise questions at site according to design requirements. The students are required to answer questions according to their specific design or required to make temporary changes on the design of software solutions to meet the teachers' satisfaction. With a variety of assessment methods, we can ask questions rely on students' actual design, thereby avoiding the shortcomings that some of the students do not attach importance to production practice. For example, the environment temperature measurement devices based on Pt100 design is mainly used in low ambient temperature test occasions such as weather monitoring, digital thermometer, food production transportation, storage temperature detection and monitoring, etc. When designing their own signal amplification circuit, students can use bridge or three-wire connection and finally transform the temperature signal into a standard voltage signal output. The voltage signals enter the input controller from the A/D converter chip, with the corresponding temperature displayed on LED. Changes in a combination of corresponding buttons decide the set of alarm value. In the beginning of the design, students make work distribution for the program design, forming a complete solution after colligating various parts. Through this production practice, this group of students work together closely, not only improved ability of practice, but also strengthened the collaboration among students which does lay a good foundation for future work.

4.4 Knowledge Unit of the Course and Knowledge Points Requirements

This course aims to equip students with a variety of sensor work principles, the basic structure, measurement conversion circuit and the actual application, the basic knowledge and basic skills for Sensor and Detection Technology. Students are also required to have a systematic study of sensor selection, debugging, analysis of measurement data. While lay a necessary foundation for the follow-up control courses. Knowledge unit of this course, knowledge points and learning requires are shown in Table 1.

Table 1. Knowledge unit, knowledge points and requires of "Sensor and Detection Technology"

Knowledge Unit		Knowledge Points	Proficiency
Basic concepts of information acquisition and processing	Detection methods and principles	Structure of detection system and the basic types	Master
		Direct and indirect measurement	Master
		Contact and non-contact measurement	Master
		Static and dynamic measurement	Master
	Sensor	Definition and constitution of the sensor	Master
		Classification of sensors	Master
Measuring uncertainty and regression analysis	Basic concept	True value, the definition of measurement accuracy	Grasp
		Sources of error, classification and express	Grasp
	Error handling and the estimating of the true value	Random error estimation and correction	Grasp
		Propagation algorithm of indirect measurement error , error synthesis and basic method of distribution	Grasp
		Best estimate of the true value and uncertainty	Grasp
	Regression analysis	Least-squares method	Grasp
		One linear fitting	Grasp
		Multiple linear fitting	Understand
		Curve fitting	Understand

Table 1. (*continued*)

Static and dynamic characteristics of detection system	Detection system features	Concept of static and dynamic characteristics	Grasp
		General mathematical model	Grasp
	Indicators of static and dynamic characteristi cs	Static characteristics of the basic parameters and indicators	Grasp
		Indicators and analysis of dynamic response characteristics	Grasp
		Indicators and analysis of frequency response characteristics	Grasp
	Static calibration and dynamic calibration	Basic method of static calibration and calibration	Understand
		Basic methods of dynamic calibration and calibration	Understand
Detection transform theory and the sensor	Sensor working principle and application	Resistive sensor	Grasp
		Inductive sensor	Grasp
		Capacitive sensor	Grasp
		Photoelectric sensor	Grasp
		Magnetic sensor	Grasp
		Thermoelectric sensor	Grasp
		Piezoelectric sensor	Grasp
		Wave sensor	Understand
		Ray sensor	Understand
		Chemical sensor	Understand
		Biosensors	Understand
Parameter detection	Process parameters	Measure of temperature, pressure, flow, level and other parameters	Grasp
	Mechanical amount parameters	Measure of is placement, speed, speed, vibration and other parameters	Grasp
The initial design of automatic detection system	Automatic detection system	Composition and basic design methods	Grasp
		Sensor selection	Grasp
		Microprocessor, A/D Converter Selection	Grasp
		Determination of the sampling period	Grasp
		The concept of scale transformations	Grasp
	New technologies in detection area	Soft-sensing technology	Understand
		Multi-sensor data fusion	Understand
		Fuzzy Sensor	Understand
		Smart sensor	Understand
		Network sensor	Understand

4.5 Mobilize the Initiative of Students—The Students Themselves Participate in Scheduling Examination Papers

Teachers are no longer independent of the content of the examination papers; students are required to participate in. The students are divided into several groups together. One the one hand, teachers grasp students' situation from the quality of the examination papers designed by students, the other hand teachers make appropriate choice of 10 scores of the contents as a recognition of the students' work. Students of selected group can get a full score of the usual results in order to mobilize the enthusiasm of students.

5 Conclusions

Teaching reforms of Sensor and Detection Technology have been carried out for 2 years and achieved good teaching results. Particular the join of production practice, not only the enthusiasm of students have been greatly improved, but also lays a good foundation for graduation and future employment. In the future, this course continues to introduce advanced teaching philosophy and teaching methods and play a role in teaching demonstration fully in order to play its important role of reaching to a new level for professional system.

References

1. Jiang, Q., Lv, J., Yun, F., et al.: Detection teaching reform for the training of application-oriented technical persons. Hefei University of Technology (Social Science Edition) 24(3), 145–147 (2010)
2. Zheng, C.: Several reform measures of methods of teaching and examination on "Sensor and Detection Technology". Liaoning University of Technology (Social Science Edition) 13(3), 128–130 (2011)
3. Zhu, L., Li, Y., et al.: Innovation discusses of the teaching mode of "Sensor and Detection Technology". East China Institute of Technology Journal (Social Science Edition) 12(1), 77–80 (2011)
4. Zhang, X.: Teaching reform research of "sensor". Jilin Teachers' institute of Engineering and Technology Journal 10(2), 43–44 (2009)
5. Xia, A., Tan, Z.: Thermocouple sensor teaching discusses. Changzhou College of Information Technology Journal 16(3), 61–64 (2006)

Development of Interactive and Virtual Algorithm Animation of C Programming Language

YangNa Su

Department of Education, HanShan Normal University, China
suyangna@hstc.edu.cn

Abstract. By analyzing the characteristics such as abstract, complication and baldness of algorithm of C Programming Language and the problems existing in algorithm teaching, aiming at the shortcomings of algorithm animation at present such as single format, incomplete function, static data and lack of amusement, a gaming and instructional virtual algorithm animation is developed with strong interaction and dynamics. It focuses on the cases of select sort algorithm and implements the dynamic data acquisition and freedom of input, and deduces the sorting process actively. Take the case of recursion algorithm for example, it realizes the interaction and amusement of virtual algorithm. the purpose is to turn the complexity and baldness of algorithm study into direct, vivid, and interesting gaming study.

Keywords: C Programming Language, algorithm, animation, sorting, recursion.

1 Problems Identification

1.1 Problems Exist in C Programming Language Teaching

Algorithm is the foundation and spirit of C Programming Language, but the concept of algorithm is very complex with various rules, abstract content which makes study boring. For instance, sorting algorithm is varied and the sorting rules are highly similar except for some differences. Part of algorithm such as Eight Queen Problem, Tower of Hanoi, the idea and the rule of algorithm are different. There is no link between knowledge points to make knowledge transfer possible. Moreover, the running system only provides the input and output data and the process of algorithm running is abstract and invisible, so learners are not able to see the deduce process of algorithm directly. It is easy to make the novice students feel confused and confusing various algorithms, arising boredom even fear, thus impede the Programming Language learning.

1.2 Limitation of Existing Algorithm Animation

Although a small amount of algorithm animation on the web can make a simple demonstration of the process of algorithm, mostly the format is simple, and the function is not completely, and lacks interest and entertainment. These algorithm animations are

D. Jin and S. Lin (Eds.): Advances in CSIE, Vol. 2, AISC 169, pp. 67–72.
springerlink.com

independent existence and are in lack of connection and comparison between systems. The distributed knowledge points make the learning lake continuity and systematic. More important, the input data is static not dynamic. That is, the data for sorting is initialized by the system default and couldn't be input dynamically and refreshed by the learner. And the system couldn't interact with and make feedback to the user, though some realize the dynamic data input, but with limitation on the number and length of input data, thus limit the freedom of data input.

2 Design Idea of Virtual Algorithm

Virtual Algorithm is designed for the beginners of C Programming Language, with simulation covering as many algorithm as possible, making knowledge point comprehensive, showing interactive and entertaining. The system is outstanding of interactive, dynamic and freedom by the sorting algorithm, and is vivid and entertaining with the animation game of *Tower of Hanoi*, *Eight Queen Problem* and *recursion algorithm.* It aims to turn the complex and boring algorithm learning into directly, interesting and entertaining learning.

2.1 Dynamic

Dynamic of Virtual Algorithm not only refers to the vividly and moving of algorithm animation, but also indicates that the accurate, variable and changeable of input data. First, it inspires students by scene animation to introduce the concept, and then input the data by the learner dynamically and variably. The system turns the data into figure demonstration; the learner can manipulate the process of algorithm, and the system can provide interaction and feedback function, and make automatic process and evaluation of operating results.

2.2 Edutainment

Edutainment theory promotes the design idea that education must be unified with entertainment. The basic idea of edutainment theory is that acknowledges and respects the current value of life of learner; promote learner's true life experience and fun integrated with purpose and means of learning.; The boring algorithm learning is turned into amusement learning by the design of the complex algorithm into vivid and interesting entertainment game.

3 Implementation Process of Virtual Algorithm Animation

3.1 Development Tools

Currently most of virtual algorithm animations on the web are appeared in JAVA or Flash format. However, operation of JAVA needs to be supported by virtual machine and most of Flash file player is supported by common operating environment. Based on Vector Graphics and streaming technology, and with the advantages of small capacity, high quality, quick transmission speed, simple and easy to learn, Flash Animation becomes the tool of choice of this system. ActionScript is an object-oriented programming language, it's strong interaction can implement data input and variable transmit; It can turn the abstract algorithm idea into direct and interactive, virtual algorithm animation.

3.2 Implementation of Virtual Select Sort Algorithm Animation

3.2.1 Idea of Select Sort Algorithm

To find a least element (compare by traversal) from an array that contains n element,, exchange it's position with the first element, then find the least from the remaining array, finally exchange it's position with the second element.

3.2.2 Design of Virtual Sort Algorithm

① Dynamic data: the input data is not initialized by the system but by the learners. ② Freedom of input: the number of input data can be changed. ③ The system will demonstrate the sort deduce process step by step by the input data, and can look over or go back to previous step. ④ The users can participate the manual sort by themselves, the system will give some instruction, and produce feedback information by operate steps, then process and evaluate the operate results automatically.

3.2.3 Technology Implementation

The process of virtual sort algorithm involves four steps: data input, data traversal, data comparison and data exchange. Each trip must involve the four steps; the number of trips is determined by the input data.

Data input: The system prompt the user to input a set of data, each data element is not bigger than 999, the number of element is not bigger than 10, the data must be separated by comma. With the input textfield to input data, give a variable name to the input textfield, use *myarray = new Array();* to create array to storage data, make myarray = txt.split(",") function to separate the values of data, and use *n = myarray.length()* to compute the length of array, shown as Figure1.

Data traversal: The process of data traversal is to use i,j,k as the traversal pointer, use i point to the begin position of disorder region, use j as scan pointer to select the mini record R[k] in the disorder region. The movement of traversal pointer is realized by control the coordinate property of pointer movieclip by *setPorperty()* function, shown as Figure2.

Data compare: For the process of data compare is a repeated process, so we can turn the process of comparison into a function of *compare()*, through the pairwise comparison of *R[i]* and *R[j]*, use mathematical operators ">,=,<"to compare data. Because the data input get by input textfield is character, we must use *Number()* function to transform the character data into number data, shown as Figure3.

④ **Data exchange:** After data comparison, if *R[i]* is bigger than *R[j]*, exchange the two data, if *R[i]* is smaller than or equal to *R[j]*, then don't make any process of data. We can make the process of *data compare* to independent function of *change()*, when it needs to change, the system call the function automatically. We use the function of *duplicateMovieClip()* the duplicate movieclip to make the animation of data exchange process. Exchange of data is using *t* to temporary storage variable, using statement of *t=R[i]; R[i] =R[j]; R[j]=t;* to exchange data, shown as Figure4.

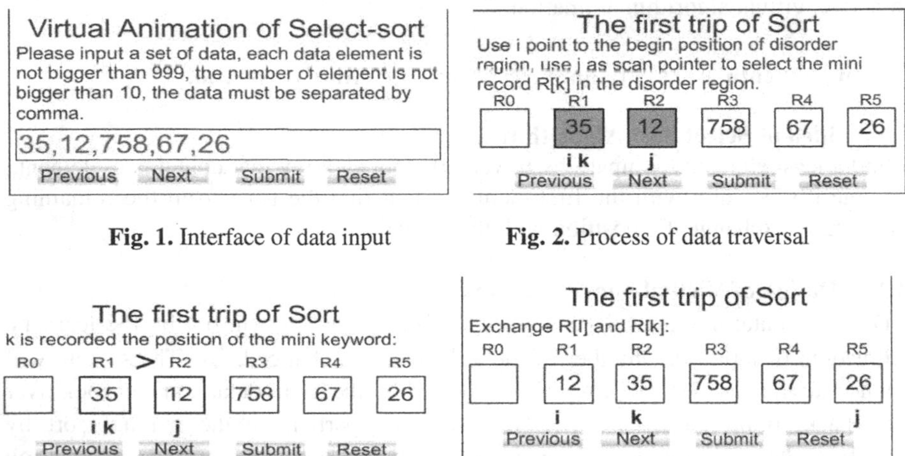

Fig. 1. Interface of data input

Fig. 2. Process of data traversal

Fig. 3. Data compare

Fig. 4. Data exchange

3.3 Recursive Algorithm Game

3.3.1 Introduction of Recursive Algorithm

Recursive algorithm is a directly or indirectly self-call algorithm. Generally, recursive process is achieved by a function or sub-process by calling its own algorithm directly or indirectly in the internal of function or subroutine. In order to strengthen the learners' understanding of the recursive algorithm, strengthen the practical application, and enrich the interest of algorithm learning, recursive algorithm can be shown in the form of games. The most common recursive algorithm on the network is the Tower of Hanoi Game; the following is the realization of the algorithm of recursive games.

3.3.2 Designation of Animation Game of the Tower of Hanoi Algorithm

(1) Idea of algorithm and Designation of games. Tower of Hanoi algorithm is to move the sample sizes of plate from the first column to the third column with the help of the second column, you can only move a plate each time, and abide the rule that the bigger plate could not cover the smaller plate, shown as Figure5. The learner can use mouse to drag the plate of the column, if the drag position is correct, then the plate is stop in the target position, but if the drag position is wrong, the plate go back to the original position. The idea of algorithm is through the way of game animation, helping the learner to master the rules of game and the algorithm idea step by step from easy to difficult. Only after learner finish the round of operation, can he conduct the next advanced operation. The system will increase the difficulty and level of game gradually. The characteristic of interaction, simulation, entertainment and competitive of the game ensure the learner to study in entertaining.

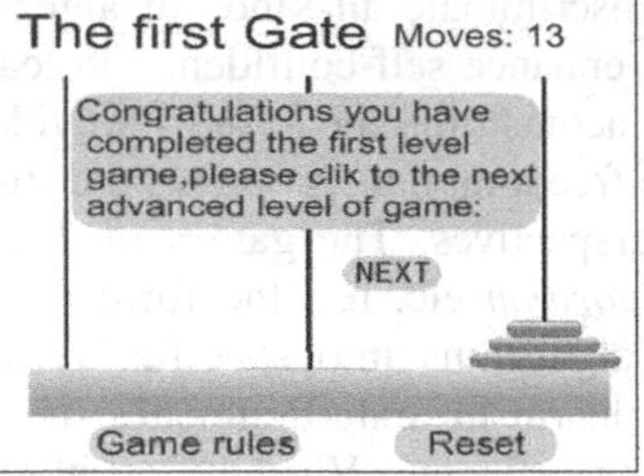

Fig. 5. First gate of *Hanoi* algorithm game

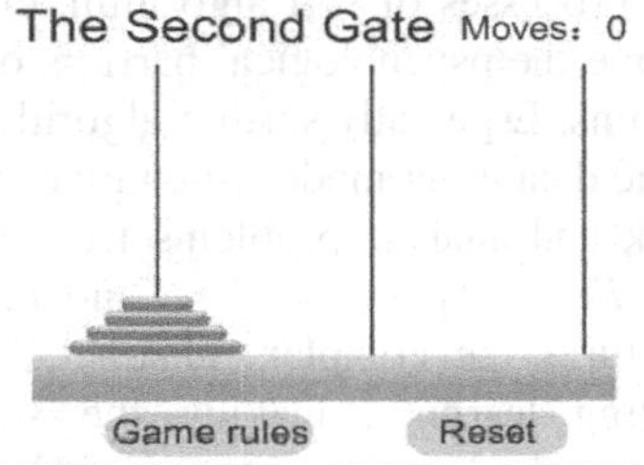

Fig. 6. Second gate of *Hanoi* algorithm game

(2) Implement of function. The function of drag the plate is realized by using the *startDrag()* and *stopDrag()* of actionScript, using the collision detection function *hitTest()* and *dropTarget* property to detect the target position. when the plate is drag to the target position, identify the order of the plate on the column, make sure that all plates on the column are in the order of small to large, shown as Figure5. The key code is show as Code 1.

Code1 : Game algorithm of Tower of Hanoi

```
for (i=1; i<=pnumb; i++) {
_root["pan"+i].onPress = function() { //the operation
executed when drag the plate
          n = this._name.slice(3,4);
          _root["kpan"+n]._visible = true; //show the drag
target of the plate
          pres=true; };
_root["pan"+i]. onReleaseOutside=function () {// the
operation executed when release the plate
          _root["kpan"+n]._visible = false;// Hide the drag
target of the plate
          pres=false;  };}
if (md&pres) {
if (Math.abs(_xmouse-zh2._x)<=50) {// Judge the position
and order by the coordinate of the plate
drag(n,zh); //the function of drag
chek(n,zh2); }// check the order of arrange
else if(shift) {
movepan(n,zh); //call the recursion function
if(_root.zh3.a.length==pnumb){//all plates is stacked and
order right
gotoAndStop("end"); } }
```

4 Conclusions

Development of virtual algorithm animation transforms the abstract algorithm running process into direct and vivid animation demonstration. It turns the algorithm idea into entertainment games and enriches the pleasure of learning algorithm. The teaching applications of virtual sort algorithm help students to understand and master all kinds of sort processes of sort algorithm, clarify and discriminate all kinds of algorithms, overcome the psychological barriers of learning, enhance self-confidence in learning algorithms. Especially sorting algorithm for data acquisition, the system provides the dynamic data input mode which gives users more freedom, thus promotes the students to think and analyse problems from multiple perspectives. The games of *Tower of Hanoi, Eight Queen Problem,* and *recursion algorithm* etc. use the form of game, which turns the complex, abstract and boring algorithms into specific, vivid and interesting learning, making the students to learn in entertainment, therefore, improving the learning interest of C programming language. Virtual algorithms can stimulate students' learning interest and thirst for knowledge. It cultures students' creative thinking and imagination. It can adjust teaching content according to different learners' individual learning.

The interactive virtual algorithm animation creates a new, self-study and free exploring learning space; it is one effective way to solute the teaching difficulties of C programming language. It not only assists the teachers in teaching algorithm in classroom, but also helps the students in their self-study. The extending application not only suits for C program language study, but also for other language study, such as VB,VF or C++ etc. algorithm teaching of similar courses.

References

1. Kehoe, C., Stasko, J.T., Tayloe, A.: Rethinking the evaluation of algorithm animations as learning aids: An observational study. International Journal of Human Computer Studies 54(2), 265–284 (2011)
2. Kerren, A., Stasko, J.T.: Algorithm Animation - Introduction. In: Diehl, S. (ed.) Dagstuhl Seminar 2001. LNCS, vol. 2269, pp. 1–15. Springer, Heidelberg (2002)
3. Stasko, J.T.: Tango: A framework and system for algorithm animation, vol. 11, pp. 32–35. IEEE Computer Society Press (2008)

Clustering on Multiple Data Streams

Li Tu[1,2]

[1] Jiangsu Engineering R&D Center for Information Fusion Software, 214405 Jiangsu Jiangyin, China
[2] Department of Computer Science, Jiangyin Polytechnic College, 214405 Jiangsu Jiangyin, China
yzutuli@yahoo.com.cn

Abstract. This paper presents a multiple data streams clustering algorithm CA-cluster based on Kendall correlation analysis. CA-cluster instantaneously adjusts the number of clusters and detects the development of the data stream. Since the high velocity and the large number of the data stream, it is not feasible to retain the raw data to calculate the correlation coefficient. We propose a compression mechanism based on AU statistics to support the only once-scanned algorithm to calculate the Kendall correlation coefficient. Experimental results show that our algorithms are more superior to other methods in the aspect of clustering quality, speed and scalability.

Keywords: Clustering, Multiple data streams, Kendall correlation analysis.

1 Introduction

In recent years, there have been a lot of studies on data streams clustering [1-3]. However, sometimes it is necessary to cluster the streams themselves rather than single data items. There is few research on it so far. Yang [4] used the weighted aggregation of snapshot deviations as the distance measure between two streams, which could observe the similarity of data values but ignore the trends of streams. Beringer et al. [5] proposed a method which used a discrete Fourier transforms (DFT) approximation of the original data.

The important trends in the information contained in the data streams are completely abandoned, because the data streams with similar trends may be not closed to each other in the Euclidean distance. Therefore, an algorithm for multiple data streams clustering based on Kendall correlation analysis is proposed which can well reflect evolving changes in the data stream.

2 Similarity Measurement between Data Streams

We use Kendall correlation coefficient[6] as a similarity measurement for clustering multiple data streams. For $X=(x_1,...,x_n)$, $Y=(y_1,...,y_n)$, the Kendall correlation

D. Jin and S. Lin (Eds.): Advances in CSIE, Vol. 2, AISC 169, pp. 73–78.
springerlink.com

coefficient between X and Y is $\tau_{XY} = 1 - 2U$, $U = \frac{1}{n(n-1)} \sum_{i \le i < j \le n} h(i, j)$. Here, $h(i, j) = I(x_i < x_j, y_i > y_j) + I(x_i > x_j, y_i < y_j)$ and $I(a,b) = \begin{cases} 1 & a = b \\ 0 & a \ne b \end{cases}$.

2.1 U Statistics

To calculate the Kendall correlation coefficient, we record U-statistics[7] at each time fragment. Suppose that $h(x_1,\ldots,x_m)$ is a measurable function defined on $\Re^m$ and satisfies $\vartheta = E[h(x_1,\ldots,x_m)] < \infty$. Then an unbiased estimator of ϑ is given by $U_n = \binom{n}{m}^{-1} \sum_{1 \le i_1 < \ldots < i_m \le n} h(x_{i_1},\ldots,x_{i_m})$. Here, U_n is called a U-statistic. When calculating the Kendall correlation coefficient, we set m=2. For $k = 1,\ldots,m$, we define $h_k(x_1,\ldots,x_k) = E[h(x_1,\ldots,x_k, x_{k+1},\ldots,x_m)]$. Then the projection of U_n on $(x_1,\ldots,x_n)$ is defined as $\breve{U}_n = E(U_n) + \sum_{i=1}^{n} [E(U_n | x_i) - E(U_n)] = \theta + \frac{m}{n} \sum_{i=1}^{n} \tilde{h}_1(x_i)$, where, $\tilde{h}_1(x) = h_1(x) - E[h(x_1,\ldots,x_m)]$.

2.2 AU Statistics

It is obviously impossible for calculating the U statistic of entire data stream. Therefore, we calculate the U-statistic by the use of AU statistics. Let $\{x_1,\ldots,x_n\}$ be a random sample from an unknown distribution P. Firstly, partitioning the random sample into K subsets with observations in the kth subset denoted by $\{x_{k1},\ldots,x_{kn_k}\}$ and the U-statistic based on them as U_{kn_k} . The AU statistic $\tilde{U}_n = \frac{1}{n} \sum_{k=1}^{K} n_k U_{kn_k}$.

In stream processing, we use sliding window to update the AU statistics where using fixed-length fragments as the basic unit of time. we should remove the old piece of statistic from AU statistics and add new one after each time fragment. The kth subset is defined as $\{x_{k1},\ldots,x_{kn_k}\}$, $k = 1,\ldots K$. Obviously, $\sum_{k=1}^{K} n_k = L$, $\tilde{U}_L = \frac{1}{L} \sum_{k=1}^{K} n_k U_{kn_k}$. Assuming the U-statistic of the 1st subset $\{x_{11},\ldots,x_{1n_1}\}$ is U_{1n_1} , $\tilde{U}_{L-n_1}$ reducing U_{1n_1} is $\tilde{U}_{L-n_1} = \frac{L}{L-n_1} \tilde{U}_L - \frac{1}{L-n_1} n_1 U_{1n_1}$. $\tilde{U}_L$ adding the U-statistic of U_{kn_k} is $\tilde{U}_L = \frac{L-n_K}{L} \tilde{U}_{L-n_K} + \frac{1}{L} n_K U_{Kn_K}$.

3 The Process of the Multiple Data Stream Clustering

In this paper, the algorithm CA-Cluster keeps detecting and reporting the clusters for the most recently data streams of fixed length *L*. For every *l* time steps, it first computes the compressed term.

The number of the time fragments in the sliding window is $K = L/l$. Assuming the present time is *t*, the sequence in present sliding window is $X_i = \{x_{i(t-L+1)},\ldots,x_{it}\}$. When the most recent time fragment *k* ($k \in [1, K]$) arrives, the algorithm should save the compressive term (CTerm) defined as CTerm(*t*)=$\left\{(\vec{U}(t,l), \vec{\tilde{U}}(t,l), \vec{\tau}(t,l)\right\}$. Where, matrix $\vec{U}(t,l) = \left[U_{kn_k}(i,j)\right]$, $\vec{\tilde{U}}(t,l) = \left[\tilde{U}_L(i,j)\right]$, $\vec{\tau}(t,l) = [\tau(i,j)]$, $1 \le i < j \le N$. $U_{kn_k}(i,j)$ denotes the U-statistic of X_i and X_j at the *k*th time fragment. $\tilde{U}_L(i,j)$ denotes the AU statistics of X_i and X_j at present sliding window L. $\tau(i,j) = 1 - 2\tilde{U}_L(i,j)$ denotes the Kendall correlation coefficient of X_i and X_j.

```
CA-Cluster algorithm:
  begin
      t = 0;
      while data stream is not terminated do
           t = t + 1;
           read new data items x_k(t), k=1…n, one from each
           of the n data streams;
           if (t mod l = 0) then
               calculate the Cterm for[t-l+1,t] ;
                    remove the expired one and update CTerm;
                if t = L then calculate initial U(t,l) for
                [t-L+1,t], update CTerm;
               else incrementally update CTerm;
                       call dynamic -k-means();
                       call adjust_k();
               output the clustering results;
            end if
        end while
  end.
```

The number of the clusters is adjusted by splitting or emerging some current clusters, and the clustering quality is measured by an objective function *G*.

$$G = \frac{1}{n}\sum_{i=1}^{k}\sum_{j=1}^{n_i}\left(1/\tau_{X_{ij}c_i}\right) - \frac{1}{k(k-1)}\sum_{i=1}^{k}\sum_{j=i+1}^{k}\left(1/\tau_{c_i c_j}\right). \tag{1}$$

```
dynamic -k-means(k, Center_k, R_k) algorithm:
  begin
     repeat
       for i = 1 to N
         calculate the correlation distances between X_i
         and centers of k clusters and assign X_i to the
         cluster with the shortest distance;
      end for
      compute the new center of each cluster, update
      the set of centers Center_k;
    until no change of clustering result
    calculate the objective function G_k.
  end.

adjust_k(k, Centerk, R_k) algorithm:
  begin
    Calculation of R_k+1:
         Among all the clusters, choose X which is the
         farthest from its cluster center, set a new
         cluster with X as its center;
         Center_k+1= Center_k ∪ {X};
         Call dynamic-k-means(k+1, Center_k+1, R_k+1);
    Calculation of R_k-1:
         Choose two closest clusters, suppose their
         centers are C_1 and C_2, combine these two
         clusters into a new cluster, compute C_3 of the
         new cluster Center_k-1= Center_k ∪ {C_3} {C_1, C_2};
         Call dynamic-k-means(k-1, Center_k-1, R_k-1);
    Choose the one with best G form R_k-1,R_k,R_k+1, set k'
    and R_k' accordingly;
  end.
```

4 Experimental Results and Analysis

We evaluate our algorithms on both synthetic data sets and real data sets on Visual C++ 6.0, PC with 1.7GHz CPU and 512 MB memory running Window XP.

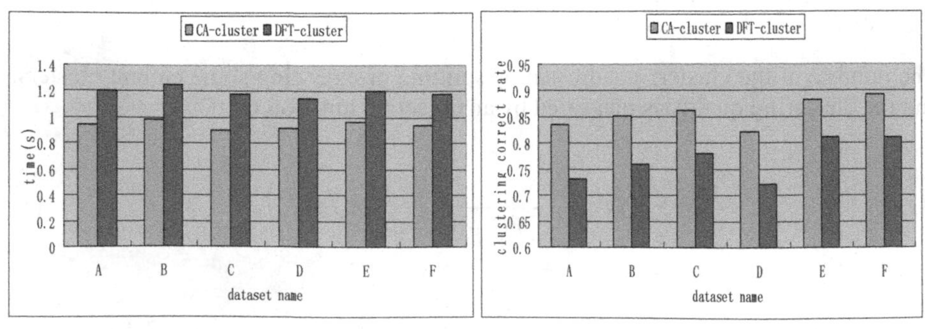

Fig. 1. Computation time and clustering quality of two algorithms on synthetic data.

The synthetic data are generated in the same way as in [5]. We test the processing speed and clustering correct rate of CA-cluster compared with DFT-cluster [5]. We generate 6 data sets each containing 100 streams. Each stream has 65,536 data items. As shown in Fig. 1, CA-cluster is faster than DFT-cluster. Reducing the number of DFT coefficients can save time but will lead to worse quality.

We ran CA-cluster on the real data set as shown in Fig. 2(a) downloaded from the website http://www.engr.udayton.edu/weather/default.htm which contains average daily temperatures of 169 cities around the world. Each city is regarded as a stream which has 3,416 points. We set L = 360 and l = 30. CA-cluster gave five clusters as shown in Fig. 2(b)-(f).

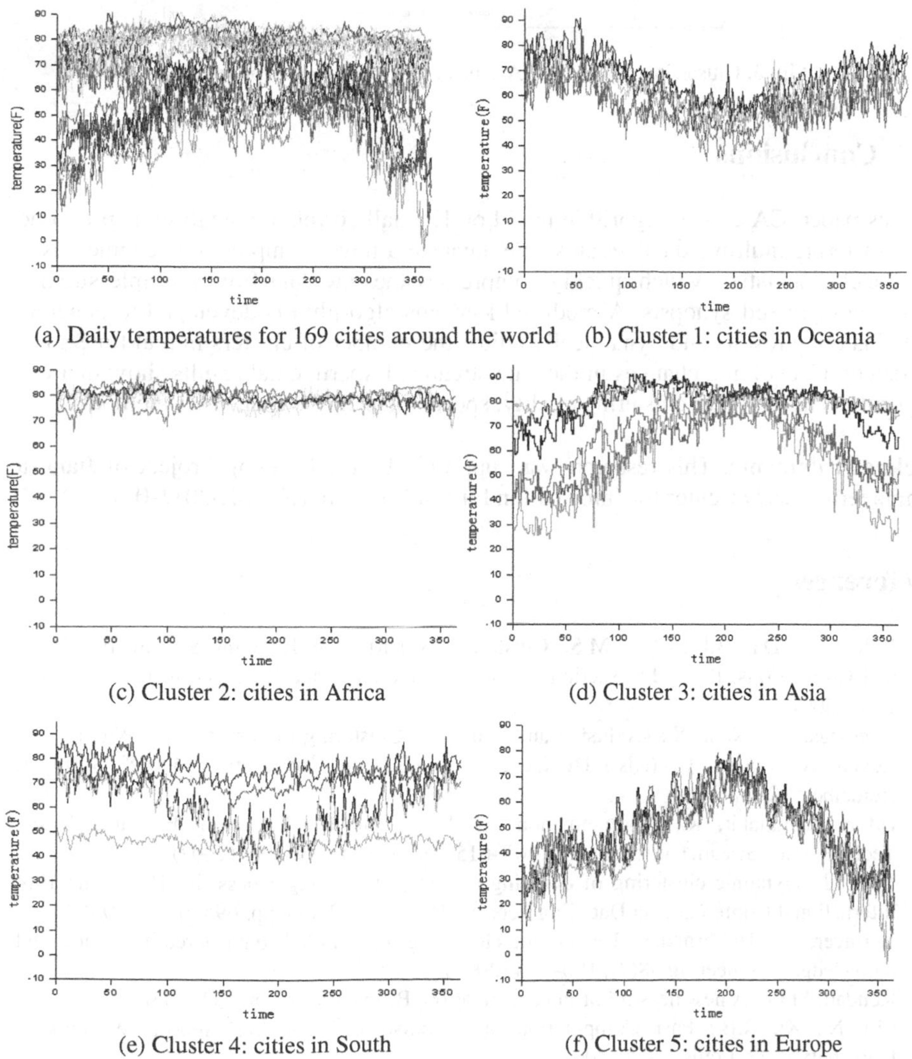

(a) Daily temperatures for 169 cities around the world (b) Cluster 1: cities in Oceania

(c) Cluster 2: cities in Africa (d) Cluster 3: cities in Asia

(e) Cluster 4: cities in South (f) Cluster 5: cities in Europe

Fig. 2. Clustering on daily temperatures of cities around the world.

As shown in Fig. 3, CA-cluster always has a better quality than DFT-cluster. The performance of DFT-cluster largely depends on the number of DFT coefficients. The clustering process does not take into account trends in the data, affecting the quality of clustering. While in CA-cluster, there is no such constraint.

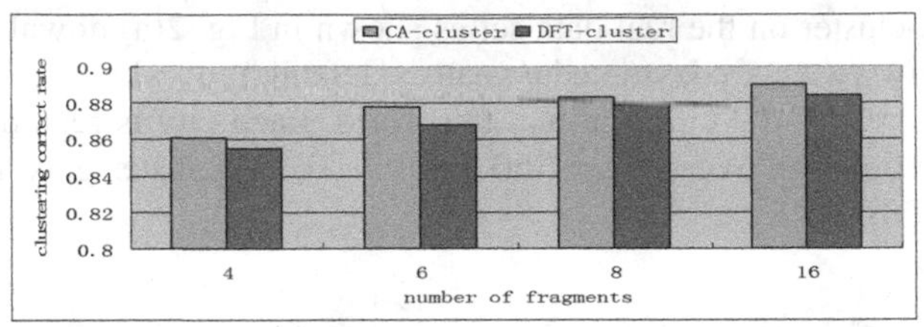

Fig. 3. Clustering quality of CA-cluster and DFT-cluster on real data.

5 Conclusions

In this paper, CA-cluster algorithm based on Kendall correlation analysis is proposed for clustering multiple data streams. We advance a novel compression scheme based on the AU statistics, which quickly compresses the raw data from multiple streams into a compressed synopsis. A modified k-means algorithm is developed to generate the clustering results and dynamically adjust the number of clusters in real time so as to detect the evolving changes in the data streams. Experimental results show that our algorithms have higher clustering quality, speed and stability than other methods.

Acknowledgments. This research was supported by the Opening Project of Jiangsu Engineering R&D Center for Information Fusion Software (No. SR-2011-05).

References

1. Yeh, M.Y., Dai, B.R., Chen, M.S.: Clustering over Multiple Evolving Streams by Events and Correlations. IEEE Transactions on Knowledge and Data Engineering 19(10), 1349–1362 (2007)
2. Bhatnagar, V., Kaur, S.: Exclusive and Complete Clustering of Streams. In: Wagner, R., Revell, N., Pernul, G. (eds.) DEXA 2007. LNCS, vol. 4653, pp. 629–638. Springer, Heidelberg (2007)
3. Udommanetanakit, K., Rakthanmanon, T., Waiyamai, K.: E-Stream: Evolution-Based Technique for Stream Clustering, pp. 605–615. Springer, Heidelberg (2007)
4. Yang, J.: Dynamic clustering of evolving streams with a single pass. In: Proc. The 19th International Conference on Data Engineering, Bangalore, India, pp. 695–697 (2003)
5. Beringer, J., Hüllermeier, E.: Online-clustering of parallel data streams. Data and Knowledge Engineering 58(2), 180–204 (2006)
6. Kendall, M.G.: A new neasure of rank correlation. Biometrika 30, 81–93 (1938)
7. Lin, N., Xi, R.B.: Fast Computation of U-statistics. Technology report. Washington University in St. Louis, USA (2008)

Research on Automatic Scoring for Java Programming Questions

ShanPing Qiao and XueSong Sun

Shandong Provincial Key Laboratory of Network based Intelligent Computing, Jinan, China, 250022
{ise_qiaosp,ise_sunxs}@ujn.edu.cn

Abstract. Automatic scoring for Java programming questions is a very important and hard problem. Currently automatic scoring technology is mainly based on whether the program runs properly. This way is not always feasible because many programs can not be executed normally only for a few errors. In this paper, an integrated method based on weight, credibility and tree structure is proposed. In this method, source file, class file, method information, output result and source code are all checked and scored with different weights. Especially for the source code, a tree-based thought is put forward. This method constructs a reasonable flow to score a Java program question. As a result, a final score is calculated successfully. The test results show this method is feasible and effective.

Keywords: automatic scoring, Java programming question, credibility, tree.

1 Introduction

Paperless examination is very popular in nowadays and a lot of computer examination systems are developed [1-3]. This kind of examination has brought many benefits to us. But to design a feasible examination system is a very difficult problem, especially the subjective questions, for example, programming questions, included.

Java [4] is a kind of Object Oriented Programming (OOP) [5] language and becoming more and more important. Java Programming is a required course in many universities. Automatic scoring by computer for Java programming questions is a very important and hard problem. The current test systems mainly focus on questions of objectivity, including the question form, multiple-choice questions and judgments. However, programming questions are necessary for the examination of this course. How to develop a feasible examination system to test and mark the programming questions becomes an urgent thing.

Some scoring methods have been proposed [6-9]. But current automatic scoring technologies for Java programming test are based mainly on whether the program runs properly. This is not feasible because many programs can not be executed only for a few errors. A candidate will get a zero under this condition. Obviously, this is neither fair nor reasonable.

D. Jin and S. Lin (Eds.): Advances in CSIE, Vol. 2, AISC 169, pp. 79–84.
Springerlink.com

In this paper, an integrated method based on weight and credibility is proposed. In this method, source file, class file, method information, output result and source code are all checked and scored with different weights. Especially for the source code, a tree-based thought is put forward. This method constructs a reasonable flow to mark a Java program question. As a result, a reliable score is calculated successfully.

2 Java Programming Question

The teaching aim of Java Programming course is to let the students learn how to program effectively in Java. A teacher should give his students a big project to design and develop in normal time. But in an examination, allow for the limit of time, the test questions can not be very difficult and big. As a rule, a teacher often requires his students to design a class with some functions to be implemented during an examination. On this condition, the scoring strategy becomes the way that how to measure a class which a student designed in the test.

3 Scoring Method for Java Programming Questions

There are countless ways to implement a class. How to score a Java programming question automatically? In this proposed method, from source file, class file, method declaration and output result to source code are all checked and scored with different weights. Especially for the source code, a tree-based technology is put forward. In this technology, credibility is used to measure the whole code.

3.1 Scoring Flow Chart

In order to measure a Java class comprehensively, a clear flow is designed. This chart is depicted in Fig. 1. The first step is to set some weights and initialize several variables. In the following each step, a corresponding score will be marked based on the corresponding algorithm. There are 5 kinds of score in all. A total score can be calculated after these 6 steps are all executed successfully. The details are in the following sections.

3.2 Set Weights and Initialize Variables

A java class is measured from 5 aspects: java, class, method, output and code. Java is to check the code exists or not. Class is to check the class file can be compiled successfully or not. Method is to check the method can be executed successfully or not. Output is to measure the method with the test data. Code is to measure the code. For each aspect, a weight is set. These weights are named java_w, class_w, method_w, output_w and code_w. The first 4 weights are satisfied with the equation: java_w+class_w+method_w+output_w=1. The last weight, coed_w, is less or equal to 0.5. In order to memorize the score of each aspect, 5 variables are defined. They are java_s, class_s, method_s, output_s and code_s. These variables are initialized to 0 respectively. The score of this question is stored to the variable which is named score.

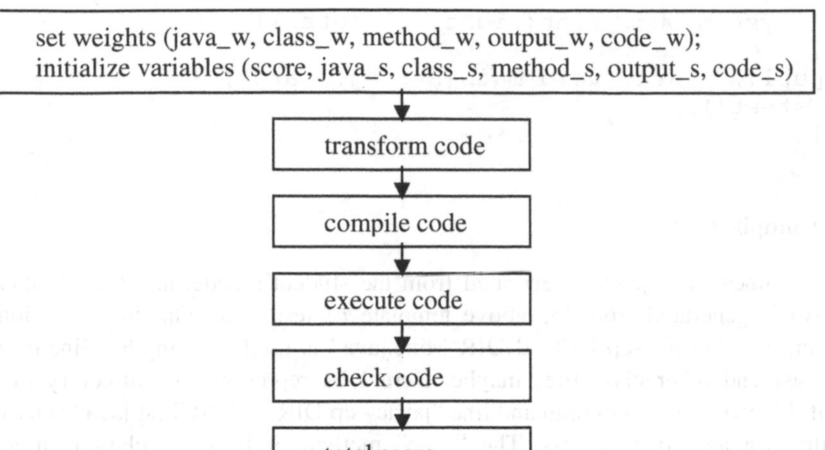

Fig. 1. Scoring flow chart

3.3 Transform Code

If the student's code is not null then java_s is set to score*java_w. The next, this code and the standard code will be combined and transformed to a corresponding test code. The package information in the student's code and the method information in the standard code all need to be resolved. Added with the test data, the test code can be generated. In this step, how to generate the parameters list is difficulty. Here we limit the type scope in the set {boolean, byte, char, short, int, long, float, double, String} and use the "parseXxx" method provided by JDK to generate the parameters list. The template of test code is shown below. The generated test code embeds the standard method code and the student class code. It also calls the student's method and the standard method to measure the correctness of the student's method. Finally, a ratio of right in testing is output.

```
public class Test {
  METHOD_DECLARATION METHOD_BODY
  private static void test() {
    String indata = IN_DATA;
    String[] params = indata.split(">/");
    int total = params.length;
    int right = 0;
    Test test = new Test();
    PACKAGE_NAMEMine mine = new PACKAGE_NAMEMine();
    for (int i = 0; i < total; i++) {
      String[] param = params[i].split("\");
      String stdout = "" + test.method(PARAM_LIST);
      String stuout = "" + mine.method(PARAM_LIST);
      if (stdout.equals(stuout)) right++;
    }
```

```
    System.out.print(right+" "+total));
  }
  public static void main(String[] args) {
    test();
  }
}
```

3.4 Compile Code

The file named Mine.java is generated from the student's code, and the file named Test.java is generated from the above template of test code. On this condition, a command line "javac –cp DIR –d DIR Mine.java" is used to compile Mine.java to Mine.class and other class files maybe. Here DIR represents the directory of the student. Likewise, another command line "javac –cp DIR –d DIR Test.java" is used to compile Test.java to Test.class. The "exec" method of Runtime class is used to execute the two command lines. If the file Mine.class is generated successfully then class_s is set to score*class_w. If the file Test.class is generated successfully then method_s is set to score*method_w.

3.5 Execute Code

After the two class files was generated successfully, Test.class can be executed followed. A command line "java –cp DIR Test" is used to execute the Test class. We also use the "exec" method of Runtime class to execute this command line. If this class is executed successfully the right number in the test data will be generated. Then the right ratio named output_ratio is calculated according to the function (1). Finally, output_s is set to score*output_w*output_ratio.

$$output\_ratio = (e^{right/total} - 1)/(e - 1) \tag{1}$$

3.6 Check Code

If the Test class was not executed successfully or the left score is large than 0 the student's source code will be checked. How to measure this code is a very hard problem. There are some methods have been developed [10-14]. In this paper, a tree-based measure strategy is proposed. First, a tree is constructed according to the standard code. Second, another tree is constructed according to the student's code. Third, the structures of these two trees and the content in each node are compared with some weights. Finally, a credibility of similarity named code_ratio is calculated. And then code_s is set to score*code_w*code_ratio.

3.7 Total Score

The final score of this question is set to java_s+class_s+method_s+output_s+code_s. This is a comprehensive value. In most cases, this value is good.

4 Analysis of Testing Results

Now a programming question is described as follows. Please design a class. In this class, there is a method. The function of this method is to judge an integer is a prime or not. The format of this method has been given. You don not change it, but you can add any code in it and this class. Please submit you java file of this class finally.

The provided standard code of this method and 3 students' codes are shown below. The test data are 2, 3, 5, 7, 11, 13, 17, 19, 23, 50 and 100. The weights are 0.1, 0.1, 0.1, 0.7 and 0.5. The score of this question is 20.

```
public boolean method(int n) {
  for(int i=2;i<=(int)Math.sqrt(n);i++){
    if(n%i==0) return false;
  }
  return true;
} //This is the standard code
```

```
public boolean method(int n) {
  for(int i=2;i<=n;i++){
    if(n%i==0) return false;
  }
  return true;
} //This is the first student's code
```

```
public boolean method(int n) {
  for(int i=2;i<=n/2;i++){
    if(n%i==0) return true;
  }
  return false;
} //This is the second student's code
```

```
public boolean method(int n) {
} //This is the Third student's code
```

The final score of the first one is: java_s=2, class_s=2, method_s=2, output_s=14, code_s=0, final score=2+2+2+14+0=20.

The final score of the second one is: java_s=2, class_s=2, method_s=2, output_s=0, code_s=6, final score=2+2+2+0+6=12.

The final score of the third one is: java_s=2, class_s=0, method_s=0, output_s=0, code_s=2.5, final score=2+0+0+0+2.5=4.5.

5 Conclusion

The test results show that this method is feasible in most cases. In fact, the output and code are the bottlenecks of scoring programming questions. Tough in some complex and particular cases the results of this method are not very reasonable; we can improve its accuracy by designing the test data more well-rounded and developing the better code comparing algorithm.

Acknowledgments. This paper is supported by the Key Teaching Reform Project of University of Jinan (Grant No. JZ1012).

References

1. Li, Z., Hao, P.: Research and Implementation of Agent-based Network General Examination System. Journal of Zhejiang University of Technology 37, 610–613 (2009)
2. Yuan, B., Yang, C.: Design of Distributed Examination System Based on Task. Computer Engineering and Design 32, 3530–3533 (2011)
3. Li, Y.: Distance Education Online Exam Marking System Design and Security. Journal of Northwest University (Natural Science Edition) 40, 239–242 (2010)
4. Oracle Technology Network for Java Developers, `http://www.oracle.com/technetwork/java`
5. Wikipedia, `http://en.wikipedia.org/wiki/Object-oriented_ programming`
6. Lou, B.: Design of Auto-marking Software Framework for Java Programming Questions. Computer Engineering and Design 31, 5343–5346, 5358 (2010)
7. Ma, P., Wang, T., Su, X.: Automatic Grading of Student Programs Based on Program Understanding. Journal of Computer Research and Development 46, 1136–1142 (2009)
8. Liu, P., Li, Z.: Research on Methods of Automatic Checking over Subjective Examination Based on Fuzzy Conceptual Graphs with Weight. Application Research of Computers 26, 4565–4567, 4584 (2009)
9. Luo, Y., Liu, J.: Design and Implementation of National Computer Rank Examination System Process Evaluation Based on Feature Table. Computer Applications and Software 28, 45–47 (2011)
10. Wang, Q., Su, X., Ma, P.: Automatic Grading Method for Program with Syntax Error—Via Local Syntax Analysis and Key Point Matching. Computer Engineering and Applications 46, 239–242 (2010)
11. Zhou, H.: Study on Application of Levenshtein Distance in Programming Test Automatic Scoring. Computer Applications and Software 28, 209–212 (2011)
12. Chang, Q., Ma, Y.: Online Examination System Based on Short Word Fuzzy Matching. Coal Technology 30, 243–244 (2011)
13. She, S., Zhou, S.: Application of Regex in Auto-Checking Paper of Programs. Computer Technology and Development 17, 244–246 (2007)
14. Tian, T., Zhang, Z.: Research on Automated Assessment Technology for Subjective Tests. Computer Engineering and Design 31, 3697–3699, 3704 (2010)

Study on Blind Source Separation of Single-Channel Signal with EEMD

Fengli Wang, Shulin Duan, and Hongliang Yu

College of Marine Engineering, Dalian Maritime University, Dalian 116026, P.R. China
wangfl@dlmu.edu.cn, duanshulin66@sina.com, and yhl1202@dl.cn

Abstract. A new blind source separation method is proposed to solve the single-channel mechanical signal separation. The new approach consists of ensemble ensemble empirical mode decomposition (EEMD) and blind source separation. Firstly the single-channel signal was decomposed into a set of proper intrinsic mode functions(IMF) by EEMD. A multi-dimensional signal was obtained by the combination of the denoised single-channel signal and its IMFs. Then mechanical sources number was estimated by a singular value decomposition and a Bayesian criterion. The multi-dimensional mixed signal was recombined according to estimated sources number and mechanical source was estimated. Simulation results indicate that the single-channel source signals are separated correctly by the proposed method.

Keywords: blind source separation, ICA, EEMD, singular value decomposition.

1 Introduction

Blind source separation (BSS) has recently been flourishing in numerous fields. However, most of the algorithms in BSS are applicable in overdetermined or determined BSS. In some fields like wireless communication, radar and mechanical engineering, there exist some single-channel signals that are superposed in time–frequency–spatiality domains [1,2].

The EMD can be directly used to separate the nonstantionary single-channel signal when the source signals in the mixture satisfy the conditions of intrinsic mode function(IMF) [3]. But the EMD algorithm will suffer from the problem of mode mixing, and a number of phantom sources will appear in the decomposed signals. The ensemble empirical mode decomposition (EEMD) defines the true IMF components as the mean of an ensemble of trials, can overcome the mode mixing [3]. A novel BSS method is proposed to solve the ill-conditioned problem in blind source separation of single-channel signal.

2 Ensemble Empirical Mode Decomposition

Different to almost all previous methods of data analysis, in the EMD[3], any signal $x(t)$ can be decomposed into IMFs $c_i(t)$, and a residue $r_n(t)$,

D. Jin and S. Lin (Eds.): Advances in CSIE, Vol. 2, AISC 169, pp. 85–90.
springerlink.com

$$x(t)=\sum_{i=1}^{n}c_i(t)+r_n(t) \tag{1}$$

However the EMD algorithm will suffer from the problem of mode mixing. To solve the problems, the EEMD is introduced, which consists of sifting an ensemble of white noise-added signal (data) and treats the mean as the final true result. The proposed EEMD is developed as follows [3]:

1) add a white noise $N(t)$ series to the targeted data $x(t)$;

$$X(t)=x(t)+N(t) \tag{2}$$

2) decompose the signal $X(t)$ with added white noise into IMFs;

$$X(t)=\sum_{j=1}^{n}c_j(t)+r_n(t) \tag{3}$$

3)repeat the step 1 and step 2 m times;

$$X_i(t)=x(t)+N_i(t) \tag{4}$$

$$X_i(t)=\sum_{j=1}^{n}c_{ij}(t)+r_{in}(t) \tag{5}$$

4) obtain the (ensemble) means of corresponding IMFs of the decompositions as the final result.

$$c_j(t)=\frac{1}{m}\sum_{i=1}^{m}c_{ij}(t) \tag{6}$$

3 Blind Source Separation

Blind source separation consists in the recovering of the various independent sources exciting a system given only the measurements of the outputs of that system. One of its most representative types is independent component analysis (ICA), the main purpose of ICA is to find a linear representation of non-Gaussian so that the components are statistically independent. Consider n sources s_1, s_2,..., s_n, which are statistically independent, m measurements from sensors x_1, x_2,..., x_m, which are represented as a linear combination of sources s_i as follows [4]:

$$x=As. \tag{7}$$

Where A and s are unknown and x is known. A is the mixing matrix.

Our aim is to seek a demixing matrix which recovers the source vector s from the observed vector x. The elements of y are estimates of the observed vector x which can be used to represent the observed vector x as follows:

$$y = Wx = WAs. \tag{8}$$

Where W is the demixing matrix.

4 Blind Source Separation of Single-Channel Signal Based on EEMD

In order to apply blind source separation to the single-channel signal, the dimension of the single-channel signal should be ascended to satisfy the conditions of that the number of the observed mixtures is more than or equal to the number of sources. EEMD method can decompose the observed data into a series of IMFs, and the denoised single-channel signal $xs_1(t)$ can be obtained by the corresponding IMFs reconstruction. Then both $xs_1(t)$ and the IMFs are combined into multidimensional signal, which is treated as the observed mixtures.

4.1 Estimation of the Number of Sources

ICA method for high performance depend on the determination of the number of sources. In order to obtain good separation performance of ICA, the unknown mixing matrix A in Eq.(7) should be square. A novel estimation method is proposed to estimate the number of sources based on EEMD. Generalizing Eq. (7) to the case with additive Gaussian noise, the observed vector x can be estimated

$$x = As + \mathrm{v}. \tag{9}$$

Where $A \in R^{m\times n}$ is a row full rank matrix and has the size of $m \times n$, and $m > n$. $s \in R^n$ is the zero mean vector. v is the zero mean Gauss noise vector, the covariance matrix of which is $R_{vv} = \sigma_v^2 I_m$, and σ_v^2 is noise power.

The covariance matrix of the observed vector can be estimated

$$R_{xx} = E[xx^H]. \tag{10}$$

According to the theorem of the SVD, for the covariance matrix R_{xx}, it always satisfies

$$R_{xx} = V_s \Lambda_s V_s^{\ T} + V_N \Lambda_N V_N^{\ T}. \tag{11}$$

where the matrix Λ_s is a diagonal matrix. Λ_N is the feature vector to the noise characteristic value.

The valuation of real dimension effective criterion is presented based on Bayesian model, which is called Bayesian model selection [5],

$$BIC(k) = \left(\prod_{j=1}^{k} \lambda_j\right)^{-N/2} \sigma_k^{-N(\tilde{l}-k)/2} N^{-(d_k+k)/2}. \tag{12}$$

4.2 Blind Source Separation of Single-Channel Signal Based on EEMD

The blind source separation of the single-channel signal based on EEMD is summarized into the following procedure:

1) The single-channel signal $x_1(t)$ is decomposed in terms of IMFs without mode mixing by EEMD. The denoised single-channel signal $xs_1(t)$ can be obtained by the corresponding IMFs reconstruction.

2) Combine $xs_1(t)$ and the IMFs into multidimensional signal $xs=[xs_1,c_1,...,c_n,r_n]^T$, which is then treated as the observed mixtures. The covariance matrix of a multi-dimensional signal was obtained by the combination of the denoised single-channel signal $xs_1(t)$ and its IMFs. Then the number of source signal is estimated by singular value decomposition and Bayesian criterion.

3) The multi-dimensional mixed signals $x=[xs_1,c_1,c_2,...]^T$ is recombined according to estimated sources number. The source signals can be obtained by performing FastICA algorithm on the multi-dimensional mixed signals.

5 Numerical Simulation

We consider the simulation mixing signals with two sine signals and one amplitude modulated signal, $s_1(t)=\sin(2\pi f_1 t)$, $s_2(t)=0.8\sin(2\pi f_2 t)$, $s_3(t)=\sin(2\pi f_c t)(1+1.2\sin(2\pi f_z t))$, and $f_1=20Hz$, $f_c=80Hz$, $f_z=8Hz$. The simulation source signals are shown in Fig.1. In order to simulate the measured signal with noise, choose a random matrix A with the size of 3×3, three mixing signals can be composed in terms of instantaneous mixed signal model $x = As$. Add Gaussian noise to the three mixing signals respectively, which are shown in Fig.2. Considering case of receiving in single channel in practical engineering, choose the first signal shown in Fig2 as the single-channel observed signal.

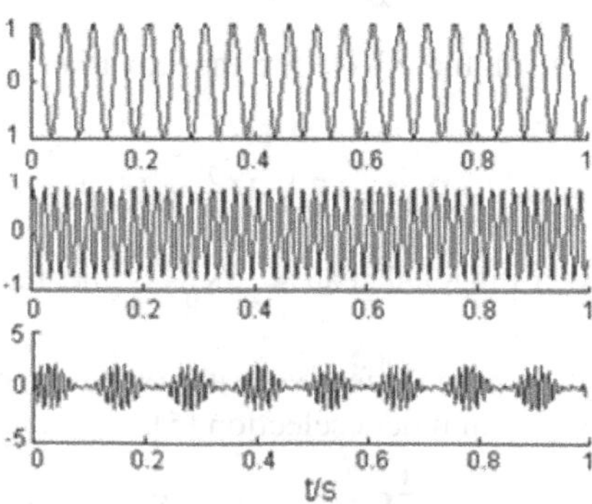

Fig. 1. Simulation original signals

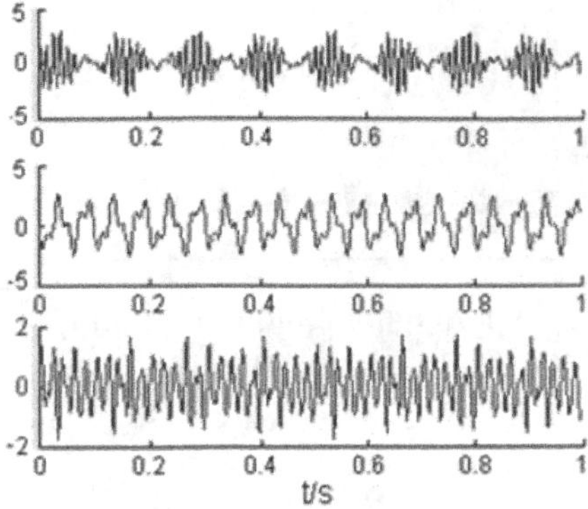

Fig. 2. Simulation mixture signals

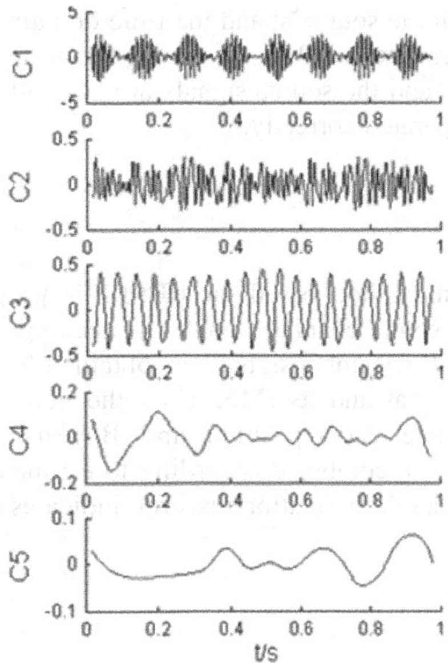

Fig. 3. The decomposition result of simulation signal

Firstly, the single-channel observed signal is decomposed into five IMFs shown in Fig.3 by EEMD. Regard the sum of the first three IMFs as the denoised single-channel signal, A multi-dimensional signal was obtained by the combination of the denoised single-channel signal and its IMFs. Then the estimation of source number is 3 by singular value decomposition and Bayesian criterion, which is equal to the number of simulation source signals. In term of the estimated sources number 3, the 3 dimensional mixed signal is recombined with the denoised single-channel signal and IMFs c1 and c2. In the end, the estimated source signals can be obtained by performing FastICA on the three dimensional mixed signal x shown in Fig.4.

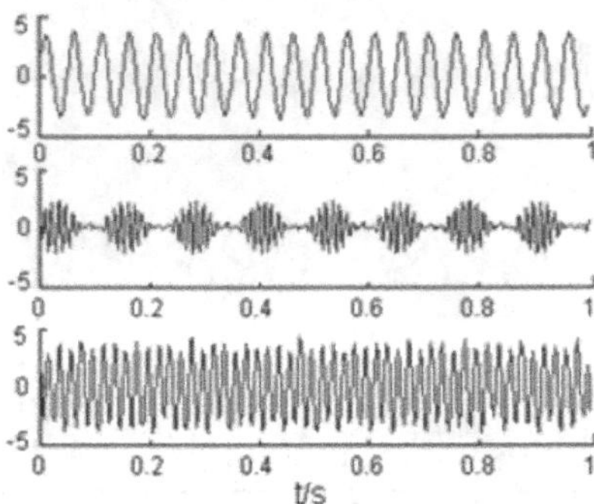

Fig. 4. The estimated source signals of single-channel signal

Compare Fig.3 with Fig.4, the sequence of the three estimated signals shown in Fig.4 is not consistent with the source's, and the time domain amplitude also change. This reflects the two uncertainty of BSS. However, the time domain oscillogram of the three estimated signals and the source signals are basically same, it indicates that the source signals were separated correctly.

6 Summary

A blind source separation method based on EEMD is proposed to solve the ill-conditioned problem. The single-channel signal was decomposed into a set of proper IMFs by EEMD. A multi-dimensional signal was obtained by the combination of the denoised single-channel signal and its IMF. Then the number of source signals is estimated by singular value decomposition and Bayesian criterion. The multi-dimensional mixed signal is recombined according to estimated sources number and mechanical source is estimated. Simulation research indicates the effectiveness of the proposed method.

References

1. James, R.H., Peter, J.R.: Single Channel Nonstationary Stochastic Signal Separation Using Linear Time-varying Filters. IEEE Transactions on Signal Processing 51(7), 1739–1752 (2003)
2. Jang, G.J., Lee, T.W.: A Maximum Likelihood Approach to Single-Channel Source Separation. Journal of Machine Learning Research 28, 1365–1392 (2004)
3. Wu, Z.H., Huang, N.E.: Ensemble Empirical Mode Decomposition: A Noise Assisted Data Analysis Method. Advances in Adaptive Data Analysis 1, 1–41 (2009)
4. Hyvarinen, A.: Fast and Robust Fixed-Point Algorithms for Independent Component Analysis. IEEE Transactions on Neural Networks 8, 622–634 (1999)
5. Minka, T.P.: Automatic Choice of Dimensionality for PCA. In: Advances in Neural Information Processing Systems, vol. 13, pp. 598–604 (2001)

Computer Multimedia Technology Status and Development Prospects

Yan Huang

Jiangxi University of Finance and Economics, Nanchang, Jiangxi, 330032, China
hhyy199@sohu.com

Abstract. The paper discussed the two aspects of the application of multimedia data processing and computer multimedia communications technology about the status of computer multimedia technology. There is a prospect of multimedia development on the two aspects of multimedia technology integration from the computer multimedia and the development of multimedia three aspects of computer vision technology.

Keywords: computer multimedia technology, application status, prospects.

1 Introduction

Computer multimedia technology is the one with the fastest growing, the most widely used, fastest changing in the field of modern information technology, and which focus the hot of electronic technology development and competition. Multimedia technology is a synthesis with various functions constructed by the smart, voice, data, images, video and communications, and which is made by the global computer networking and sharing of information resources, with the growing popularity of high-speed information network. So it is changing our lives, with widely used in industry, agriculture, services, education, communications, military, financial, medical and other industries.

2 A Computer Multimedia Technology Related Content

Computer interactive multimedia technology is the use of computer technology and integrated digital communications network technologies that deal with a variety of media - data, text, graphics, images, video, sound, variety of perception, measurement and other information technology. Use a variety of hardware and software, to establish a logical connection to a variety of information, integration of an interactive system. It mainly involves the following parts: multimedia data processing techniques, including interface design, audio technology, video technology, imaging technology, compression and encoding, and virtual reality; multimedia communication technology, refers to the data, voice, video, image transmission; artificial intelligence, including intelligence technology and software and hardware.

D. Jin and S. Lin (Eds.): Advances in CSIE, Vol. 2, AISC 169, pp. 91–97.
springerlink.com

3 Application of Computer Multimedia Technology

In recent years, rapid development of multimedia technology, multimedia applications but also to a strong penetration into all areas of human life. Application of existing technology, multimedia technology is a hot area, new tools, new phenomena occur every day, brought a new feeling, new experience is not ever imagine. Human aspects of work and life have felt the changes it brings.

3.1 A Multimedia Data Processing Technology

Computer multimedia technology to deal with text, data, images, audio, video and all kinds of perception, measurement and other information technology. It is the transformation of the data collection, storage and transmission. A variety of digital information is a very large number of multimedia data processing currently heavily dependent on the ability of the processor, memory storage capacity, communications transmission capacity and processing efficiency of these systems. Multimedia data processing techniques involving audio technology, video technology, imaging technology, compression and encoding, and virtual reality.

3.1.1 Video Technology

Video technology, including digital video and video coding technology aspects. Video is digitized by the analog video signal analog-digital conversion and color space conversion into digital signals your computer can handle, so the computer can display and process the video signal. Video coding technology is the digital video signal is encoded into the video signal, which can record or play. For different applications there are different techniques can be used, for example, television stations broadcast coding techniques.

3.1.2 Audio Technology

Audio technology mainly includes four aspects: digital audio, speech processing, speech synthesis and speech recognition. Audio technology developed earlier, some of the technology has matured and product-oriented, have poured into the family. More and more audiovisual information storage and transmission in digital form, which is more flexible for people to provide the possibility of using this information. Speech recognition has long been one of the beautiful dream, let the computer understand human speech is to develop a new generation of intelligent man-machine voice communication and the main objective of the computer. With the proliferation of computers, more and more people using the computer, how to people who are not familiar with computers to provide a friendly human-computer interaction means is that people are interested in the problem, and voice recognition technology is one of the most natural one means of communication. Currently, speech recognition research in the field is in the ascendant. In this new algorithm, new ideas and new applications are emerging. Meanwhile, the voice recognition is also in a very critical period, the world's researchers are to the highest level of speech recognition applications --- non-specific, large vocabulary, continuous speech dictation systems research and practical system for sprint, can be optimistic that people expect practical speech recognition technology will soon become a reality the dream. Currently, the world has developed a Chinese, English, Japanese, French, German and other languages, text to speech conversion system, and in many areas has been widely used.

Stitution of using software in the process of Physical education teacher making use of courseware

authoring tools						
Powerpoint	authorware	director	flash	coursewareMasters	FangzhengAosi	others
Percentage %						
23.81	21.43	11.90	2.38	14.29	2.38	23.81

3.1.3 Data Compression Technology

Data compression techniques, including images, video and audio signal compression, file storage and use. Image compression technology has been one of the hot, the computer processing images and video as well as an important foundation for network transmission, the current ISO standard that developed two compressed JPEG and MPEG, real-time processing while the computer audio, video information, to ensure that the playback of high quality video and audio.

3.1.4 Virtual Reality

Multimedia computer and simulation technologies can produce a people as if you were in a virtual world, its true no doubt, this technique usually called "virtual reality" (Virtual Reality, referred to as VR). In other words, virtual reality and multimedia technology is a combination of simulation technology to generate an interactive artificial world, in this artificial world, you can create a completely immersive real feeling. VR technology is currently mainly used in a small number of difficult and a number of military and medical research, but education and training in the field of VR technology can not be replaced very encouraging prospect, so this trend should cause us Note.

3.2 Computer Multimedia Communications Technology

Computer multimedia communications is to provide multi-media information --- sounds, images, graphics, data, text and other new means of communication. It is the communications and computer technology product of the combination.

Current physical form of communication are only two ways: First, wired communication; second wireless communication. These two means of communication are currently widely used, but in the middle into the computer multimedia technology to form a computer multimedia communications. And telephone, telex, fax, computer, communications and other traditional means of communication compared to a single media, the use of computer multimedia communications, the user can not only thousands of miles apart illustrated audio-visual exchange of information, distribution of multimedia information in different locations, but also in unison presented as a complete information to the user, and the whole process of communication with the user complete interactive control. This is the distribution of multimedia communication, synchronization and interactive features. Today, with multi-media computer technology and communication technology, a combination of both the formation of multi-media communication and distributed multimedia information systems to better address the issue. Multimedia communication technology involves multimedia data compression, multimedia communication networks, etc.

The courseware overall design process

Steps: determine the topics → Scripting →courseware module function → material collection → courseware → courseware integrated debugging →courseware evaluation.

Overall flow chart:

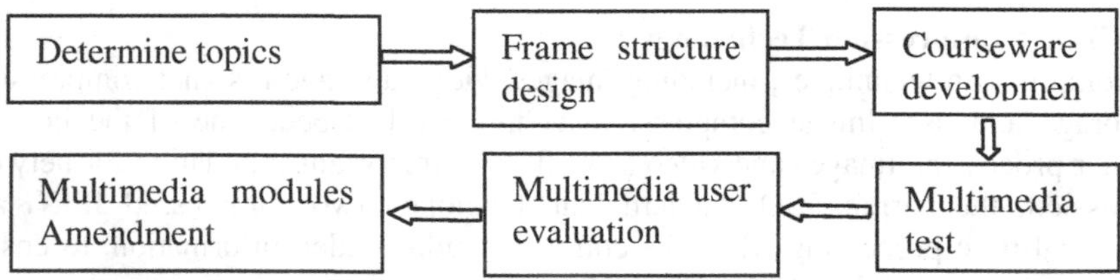

3.2.1 Multimedia Data Compression

Multimedia system for multimedia data need to capture, storage, transmission and playback processing, data compression technology is the core of multimedia communication technology issues. Advanced data compression technology, especially video compression technology to achieve low latency and high compression ratio, to achieve better image quality, which is widely accepted by multi-media services can be a major factor. International Organization for Standardization (ISO), International Electro technical Commission IEC), International Telecommunication Union (ITU) to develop a series of video compression coding standard. With the continuous development of multimedia services, the new video and audio compression standards continue to emerge, gradually expanding the scope of application, compression efficiency is greatly improved, so that the gradual application of multimedia communication in all areas of daily life.

3.2.3 Multimedia Communications Network

Communication network is the transmission of multimedia applications environment, the transmission of multimedia communication and exchange of information have raised new and higher requirements, network bandwidth, exchange and communication protocols will have a direct impact on the availability of multimedia communication services and multimedia communications quality.

3.3 Multimedia Applications of Artificial Intelligence

Artificial Intelligence is the late 1950s the rise of the use of computer simulation research field of human intellectual activity, that is, from the machine instead of human intelligence to do some work. It combines some of the results and multimedia in industrial, commercial, military, medical and cultural education, applied. Multimedia technology involving computer-aided system of artificial intelligence, intelligent information systems, etc.

3.3.1 Computer-Aided System

According to certain objectives, the preparation of a series of computer programs, design and control process, the user through the use of the program, to complete the task, this series of computer programs known as computer-aided software (Computer Assist, computer-aided).

3.3.2 Intelligent Information Systems

Intelligent information systems acquisition, processing, application information, the main application field of information. Image and voice processing technology compared to the higher starting point of intelligent software, the application of professional level is also relatively high, with the development of computer technology, intelligent information technology gradually developed into a new direction.

4 Prospects for the Development of Computer Multimedia Technology

Prospects for the development of multimedia computer technology has several aspects: integration of multimedia technology, network development etc.

4.1 Computer Multimedia Technology Integration

In traditional computer applications, most have adopted the text media, so information is limited to the expression "display."In the future multimedia environment, the coexistence of a variety of media, visual, hearing, touch, taste and smell the integration and synthesis of media information, you can not just use the "Show" to complete the performance of the media. A variety of media time and space arrangements and effect, between the synchronization and synthesis effects, the interpretation and description of such interactions are expressed information. Film sound technology is widely used to make synthetic space-time multimedia, synchronization results, visualization, auralization and flexible interaction methods is a multimedia development. Interactive multimedia technology, the multimedia technology in pattern recognition, holographic images, natural language understanding (speech recognition and synthesis) and the new sensor technology, based on the use of human sensory channels and action channels (such as voice, writing, facial expressions, posture, vision, movement and sense of smell, etc.), by data transmission and special forms of expression, such as the perception of facial characteristics, synthetic facial movements and expressions to parallel and non-precision approach to interact with the computer system. Can improve the natural and human-computer interaction efficiency, to achieve realistic output as a symbol of virtual reality. Virtual reality is the people through the computer visualization of complex data operations with a whole new way to interact with the traditional man-machine interface and the popular Windows operating compared to virtual reality technology has been a qualitative leap in thinking. Virtual reality in "reality" generally refers to functions in the physical sense or meaning there anything in the world or the environment, it can be actually be achieved, it can be difficult to achieve or simply impossible to achieve. Virtual reality is a multi-technology multi-disciplinary mutual penetration and integration of technology, research very difficult..Interactivity is the substantive features of virtual reality, the reality of the concept of space-time environment (that is inspired by ideas, access to information process) is the ultimate goal of virtual reality.

4.2 Computer Multimedia Technology Network Development

Computer multimedia technology depends primarily on the development of network communication technology, network communications technology with the development and integration with each other, so that multimedia technology into life, technology, production, business management, office automation, education, health, transportation, military, culture and entertainment, control and other fields. The rapid development of modern communications technology, satellite communication, optical fiber communications, the world has entered the digital, networked, global integration of the information age. Information technology penetrated into every aspect of people's lives, which is the network technology and multimedia technology to promote the full realization of the world's key information technology. Development and application of Bluetooth technology, the wireless multimedia network technology, miniaturization. It can be near the end to form a small number of networks, digital information appliances, personal area networks, wireless local area network, a new generation of wireless, Internet communication protocols and standards for next-generation network multimedia software development, integrated original range of multimedia services will make the sudden emergence of computer multimedia technology wireless network, set off a new wave of the Internet age, making multimedia everywhere, all kinds of things are interesting in the interaction, depicting the Computer multimedia technology network can be described as a decisive (key) technologies, these technologies through access to global networks and devices to achieve the use of multimedia resources, be sure the subject of future development.

5 Conclusion

All in all, the application of computer and multimedia technology development is in process of rapid development, with a variety of ideas, technology continues to evolve and innovate, and integrate into multimedia technology, the future will appear colorful, fresh multimedia phenomenon, it destined to change human lifestyles and values. Multimedia technology in pattern recognition, holographic images, natural language understanding (speech recognition and synthesis) and the new sensor technology, based on the use of human speech, writing, facial expressions, posture, vision, movement and sense of smell and other sensory channels and action channel, through the expression of data transmission and special way, is interact with the computer system with the most extensive prospect in the future.

References

1. Zhuo, Z., Cai, L.-H., Li, S., et al.: Multimedia Computer Technology and Applications. Higher Education Press, Beijing (1999)
2. Liu, M.: Multimedia Application, 2nd edn. Higher Education Press, Beijing (2000)
3. Zhong, Y., Shen, H.: Multimedia technology infrastructure and applications. Tsinghua University Press, Beijing (2005)
4. Zeng, C.: Multimedia technology in teaching the application. Ningbo Radio & TV University (2006)

The Algorithm for 3D Lotus Construction Based on Fractal

Bo Cong

Shenyang Sport University, Shenyang 110102, China
congbo_2002@163.com

Abstract. This paper gave the flower structure patterns on the observation and statistic. It provided computer simulation based on plant morphology and the principles of fractal graphics technology。 The longitudinal curve of the petals is self-similarity. According to the results of observation and a lot of mathematical experiments, the author gave the 3D IFS code of lotus. The method generates the 3D petals and 3D flowers based on a second curve. The simulation result of the flowers is more realistic.

Keywords: fractal, plant morphology, IFS(Iterated Function System).

1 Introduction

Flower variety, form the myriads of changes. Flowers by shape classification include: single flowers, such as peach and apricot; double petals, such as Amaryllis and camellia; multilayer petals, such as rose and chrysanthemum. Some flowers, such as the cherry blossoms and Prunus triloba has monolayer, bilayer and multilayer cent. Flower petals are arranged according to a mode, between the layers of self-similar, surface has a wealth of detail and random variation of irregular shape. It is difficult to use the conventional Euclidean geometry description.

Nineteen seventies, B·Mandelbrot created the fractal geometry, used to describe the irregular and Euclidean geometry to describe the geometric phenomena and objects, known as "nature's own geometry", make natural scenery depicted as possible, this is also the fractal geometry was highly attention to one of the reasons. Fractal description of nature is the best language, are widely used in natural landscape modeling. Natural landscape modeling is mainly directed against the trees, mountains, grassland, and other big sky scene, less involved in flower modeling. This paper will combine the plant morphology and the principles of fractal graphics technology, realize the lotus, chrysanthemum and rose 3D modeling, and from different points of view can browse, enhance the reality of virtual scene effect.

2 Principle and Method

Self similarity fractal is the most important feature. Considering the fractal self similarity and contraction mapping fixed point principle, Hutchinson in 1981 first

D. Jin and S. Lin (Eds.): Advances in CSIE, Vol. 2, AISC 169, pp. 99–104.
springerlink.com

proposes the use of compression mapping method for generating fractal, and then the United States of America Georgia Polytechnic mathematics professor M·Barnsley in 1985 invented the iterated function system IFS. IFS is not only a powerful mathematical method to describe the fractal and fractal image compression, the basic tool, the mathematical theory and application have been studied.

The basic IFS idea: in the affine transformation of sense, geometric object in whole or in part with self similar structure. It is this self similarity, makes it possible to use the iterative method for generating image. From a point or a simple geometry, according to certain rules iteratively, until it generated a complex image. Finally get the target point set with the initial set of unrelated, and depends only on the iterative rules, i.e. a set of affine transform coefficients, also known as the IFS code, the image generation system called iterative function system.

Definition 1. Transformation $W : R^3 \to R^3$ has the form

$$W\begin{bmatrix} x \\ y \\ z \end{bmatrix} = \begin{bmatrix} a & b & c \\ d & e & f \\ g & h & k \end{bmatrix}\begin{bmatrix} x \\ y \\ z \end{bmatrix} + \begin{bmatrix} u \\ v \\ r \end{bmatrix}$$

a,b,c,d,e,f,g,h,k,u,v,r are real number, W is called an affine transformation. When $X \in R^3$, the type can be written as $W(X) = AX + t$, where $A = \begin{bmatrix} a & b & c \\ d & e & f \\ g & h & k \end{bmatrix}, t = \begin{bmatrix} u \\ v \\ r \end{bmatrix}$。 A is the four kind of affine transformation of the compound. $\begin{bmatrix} sx_n & 0 & 0 \\ 0 & sy_n & 0 \\ 0 & 0 & sz_n \end{bmatrix}$ is to control the size scaling transformation, $\begin{bmatrix} 1 & 0 & 0 \\ 0 & \cos\gamma_n & -\sin\gamma_n \\ 0 & \sin\gamma_n & \cos\gamma_n \end{bmatrix}$, $\begin{bmatrix} \cos\alpha_n & 0 & \sin\alpha_n \\ 0 & 1 & 0 \\ -\sin\alpha_n & 0 & \cos\alpha_n \end{bmatrix}$ and $\begin{bmatrix} \cos\beta_n & -\sin\beta_n & 0 \\ \sin\beta_n & \cos\beta_n & 0 \\ 0 & 0 & 1 \end{bmatrix}$ are separately controlled rotation transformation around X axis, the Y and Z axis. t is translation transform.

Theorem 1. A iterated function system is composed by a complete metric space (X,d) and a limited compression mapping sets $W_n : X \to X$, $n = 1,2,...,N$. To express with IFS $\{X; W_n, n = 1,2,...,N\}$, canonical transformation of W is defined as:

$$W(B) = \bigcup_{n=1}^{N} W_n(B)$$

$\forall B \in \chi(X)$ is a compression mapping in complete space $(\chi(X), h(d))$ with compression factor, i.e. $h(W(A), W(B)) \le sh(A,B)$, $\forall A, B \in \chi(X)$, and $P = \lim_{n\to\infty} W^n(B), P \in \varphi(X)$ is called IFS attractor, the attractor is a fractal.

Theorem 2. Set the affine transform $W : R^3 \to R^3$ shaped like $W(X) = AX + t$, only if the spectral radius of $r_\sigma(A) < 1$, transform W is compressed.

Theorem 3. let (X,d) be a metric space, $\{W_n, n = 1,2\cdots, N\}$ is also a compression mapping on $(\chi(X), h(d))$, corresponding to the W_n compression ratio of s_n, defined by $W : \chi(X) \to \chi(X)$

$$W(B) = W_1(B) \cup W_2(B) \cdots \cup W_n(B) = \bigcup_{i=1}^{n} W_i(B), \forall B \in \chi(X)$$

W has a compression ratio of $s = \max\{s_1, s_2 \cdots s_n\}$compression mapping.

3 The Algorithm Realization

3.1 Lotus Algorithm

3.1.1 Lotus Petals Affine Transformation Matrix

Lotus is a perennial aquatic plant. Flowers highly stand above the surface of the water. It has a single, double, complex flap and heavy units such as the flower. Colors include white, pink, red, purple or color change. Lotus petals longitudinal lines are self-similar characteristics. According to the characteristics of lotus petals, with a two curve as iteration start, taking into account the speed and accuracy of 1, according to the definition and theorem 1 and theorem 2 gives lotus petals iterated function system is composed of 7 affine transformation matrix. Lotus petals natural feature selection as shown in Table 1 of the IFS codes iterative.

3.1.2 The Algorithm Description of the Lotus Petals

(1) Set the initial parameters by the limited user (such as the iteration number of levels of lotus petals). The number of iterations to control the size of the petals and undivided shares;

(2) Structure the initial region AR of the construct iteration of the lotus petals of the initial parameter values according to user settings;

(3) Determine the value of $level$, if $level == 0$, to the end, otherwise the implementation of (4);

(4) $W(AR)=\bigcup_{n=1}^{7}W_n(AR)$, $level=level-1$, $AR=W(AR)$, and draw a new AR, recursive lotus petals;

Among them, $W_1, W_2, \ldots W_n$ met the conditions $r_\sigma(A_n)<1, n=1,2,\ldots,7$. Then constructed of the IFS attractor is to get three-dimensional fractal lotus petals. Based on IFS code in Table 1, iteration generated lotus petals shown in Figure 1.

Table 1. 3D Lotus petals IFS code

n	sx	sy	sz	α	β	γ	u	v	r
1	1.00	0.50	0.30	0	0	0	0	0	0
2	1.00	0.50	0.30	15	0	0	0	0	0
3	1.00	0.50	0.30	30	0	0	0	0	0
4	1.00	0.50	0.30	45	0	0	0	0	0
5	1.00	0.50	0.30	60	0	0	0	0	0
6	1.00	0.50	0.30	75	0	0	0	0	0
7	1.00	0.50	0.30	90	0	0	0	0	0

3.1.3 Lotus Affine Transformation Matrix

Shows the morphological structure of the lotus, the general Lotus 4-5 layers, each of the five petals between the layers at an angle of dislocation, belong to the tweezers together like petals arrangement, so each petal are receiving to full light, is the result of natural selection. By definition and Theorem 1 and Theorem 2 gives the Lotus iterated function system consists of five affine transformation composition. Selected as shown in Table 2 of the IFS code to iterate based on the natural characteristics of the Lotus.

Table 2. 3D Lotus IFS code

n	sx	sy	sz	α	β	γ	u	v	r
1	0.80	0.80	0.80	30	0	0	0	0	0
2	0.80	0.80	0.80	30	60	0	0	0	0
3	0.80	0.80	0.80	30	120	0	0	0	0
4	0.80	0.80	0.80	30	180	0	0	0	0
5	0.80	0.80	0.80	30	240	0	0	0	0

3.1.4 The Algorithm Description of the Lotus

(1) Lotus initial parameters (such as the number of iteration level, etc.) set by the user is limited. The number of iterations to control the size and stratification of the Lotus number;

(2) Structure the initial region AR of the construct iteration of the Lotus initial parameter values according to user settings;

(3) Judge the value of $level$, if $level == 0$, to the end, otherwise the implementation of (4);

(4) $W(AR) = \bigcup_{n=1}^{5} W_n(AR)$, $level = level - 1$, $AR = W(AR)$, Draw a new AR , recursive Lotus;

Among them, $W_1, W_2, \ldots W_n$ met the condition $r_\sigma(A_n) < 1, n = 1,2,\ldots,5$. Then constructed of the IFS attractor is to get the 3D fractal Lotus. Based on IFS code in Table 2, iteration generated lotus shown in Figure 2.

Fig. 1. Lotus petals

Fig. 2. Lotus

4 Conclusion

Through the statistical observations on the lotus, the laws of their shape can be drawn: the petals of the longitudinal section curve is approximated by a quadratic curve, the front-end midpoint and at the bottom of the midpoint of the longitudinal and again on both sides followed by longitudinal a set of longitudinal line curve has self-similarity; petals and between multi-layer structure of petals between the layers of self-similar figures.

In this paper, combined with statistical data, on the basis of computer graphics, it constructs the algorithm model and algorithm model of the flowers of the petals. The petals of the algorithm model to a quadratic curve as the initial iteration element to generate three-dimensional petals, the flowers of the algorithm model-generated three-dimensional petals as the initial iteration to generate three-dimensional flowers. Painted flowers form rounded appearance, realistic higher. In OpenGL environment using a variety of mature light, and texture to enhance the realism of the petals and flowers, can be applied to the structure of flowers in the virtual scene.

Using fractal methods to draw the petals, to avoid the triangle faces as the initial generation flower authenticity, but also to avoid the complexity of the constructed curve equation of three-dimensional surface petals. This method is lower complex. You can quickly and flexibly generate various forms of the petals. However, the method generates petal shape is too smooth, the surface of the lack of change.

References

1. Wang, L., Zeng, L.: Chrysanthemum modeling algorithm. Journal of Liaoning Technical University 26(3), 0410–0411 (2007)
2. Barnsley, M.F.: Fractals Everywhere. Academic Press, New York (1993)
3. Wang, Y., Zeng, L.: Fractal based 3 dimensional bamboo modeling algorithm. China Journal of Image and Graphics 12(1), 0177–0182 (2007)
4. Zeng, W., Wang, X.: Fractal and fractal computer simulation. Northeastern University press, Shenyang (2001)
5. Li, S., Wu, T.: The fractal and wavelet. Science Press, Beijing (2003)

Mobile Agent Routing Algorithm in Wireless Sensor Networks

Sheng Zhang[1,2], Zhang He[1], and Huili Yang[1]

[1] College of Information Engineering, Nanchang Hangkong University, Nanchang, Jiangxi Province, China
[2] College of Computer Science and Technology, Zhejiang University, Hangzhou, Zhejiang Province, China
zwxzs168@126.com, heqq_92@163.com, 459541002@qq.com

Abstract. Wireless sensor network is resources-restricted network and similar to traditional mobile ad-hoc networks in the sense that both involve multi-hop communications. An improved ant colony algorithm based on ant colony system is put forward which to find the initial optimal migration path for mobile agent in wireless sensor networks environment. This improved algorithm selects a part of optimal routes from a large of initialization routes with leaving pheromone which consider the resident energy of nodes. A mutation operator is introduced to avoid invalid path since the limitation of communication capability of wireless sensor nodes. The simulation results indicate that the improved ant colony algorithm can enhance the global search capability significantly and solve the migration path problem of mobile agent effectively.

Keywords: wireless sensor networks, mobile agent, migration path, ant colony algorithm, mutation.

1 Introduction

Because of wireless sensor nodes hardware defects, including limit in node energy, communication ability, computing and storage capacity, etc. wireless sensor network (WSN) is a resources-restricted network [1]. In order to get information from the monitored area by wireless networks, studying of WSN routing algorithm is very useful for improving the reliability and adaptability of the network, also, prolonging the network life time extremely.

Early, the sensor networks data transmission methods are mainly based on traditional Client/Server wireless sensor networks model [2-4]. The data processed is routed towards the server based on some opportune routing algorithm. However, there are plenty of defects associated with client/server model, such as resource-wasting, imbalance of loading, lack in fault tolerance and security, etc. Hairong Qi proposed Mobile Agent model [5] to change information gathering method in wireless sensor networks against traditional Client/Server model. It is showed in Fig.1.

In contrast to traditional Client/Server mode, Mobile Agent model has efficient data gathering ability in WSN. Mobile agent is a type of computing entity which can collect circumstances information sensed, operate independently and achieve a series

D. Jin and S. Lin (Eds.): Advances in CSIE, Vol. 2, AISC 169, pp. 105–113.
springerlink.com

of goals on behalf of users. It performs data processing autonomously while migrating from node to node and has a number of features, including reactivity, autonomy, target-oriented, else including mobility, adaptability, communication skills (including consultation and cooperation, etc), etc. The mobile agent visits the network either periodically or on demand (when the application requires) and carry back data.

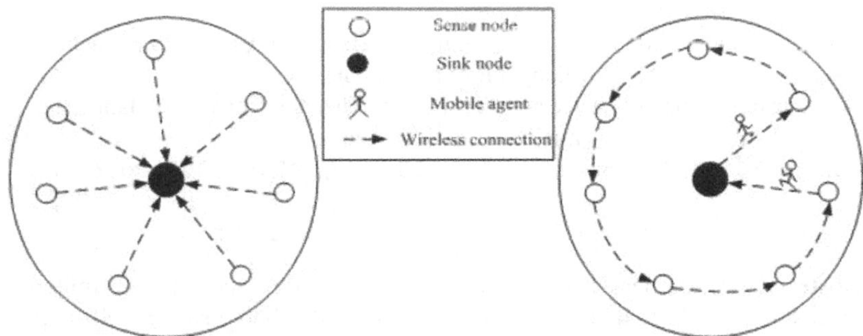

Fig. 1. The comparison between Client/Server model and Mobile Agent model

Mobile agent routing problem is a kind of complex combinatorial optimization problem which solves the optimal path according to sequence of visited nodes and energy efficiency in WSN. Ant colony optimization (ACO) is put forward to solve migration of mobile agent. And some improved mobile agent route algorithms based on ACO are proposed [6-9]. But the function of algorithm reach optimum difficultly because its parameters are usually set by experiments that lead the optimization performance of algorithm is closely related to people's experience. Furthermore, ACO Algorithm has some weakness, such as easy to bring premature convergence and take long computing time.

As mentioned above, this paper presents a novel improved ACO in order to strengthen convergence speed and avoid partial optimal solutions. Meanwhile, in order to avoid invalid path, a mutation operator will be imported.

2 Problem Description

Firstly, mobile agent is sent to monitored areas for information gathering by sink node. Mobile agent exchanges data with cluster-head nodes one by one and carries data back to sink node. Another mobile agent is responsible for visiting all source sensor nodes in a cluster and carrying information to the cluster-head node.

The power consumption should be proportional to data transmission distance in WSN. The shortest migration route means minimal energy consumption in the migration path for mobile agent. Therefore, solving the mobile agent can be described to seek one shortest path that is an open curve including all N cluster-head nodes. The curve can be formulated in the following way.

Set S of cluster-head nodes $N(x_1, x_2, ...x_N)$, objective function f_d is described by

$$f_d = \sum_{i=1}^{N-1} d(x_{\Pi(i)}, x_{\Pi(j)}) \tag{1}$$

In (1), $d(x_i, x_j)$ denotes distance of two nodes, then $1 \le i, j \le N$. The goal of this paper is to seek nodes sequence that achieves the minimum function value. Let S_d is nodes sequence. Then, S_d is equal to S_d ={ $x_{\Pi(1)}$, $x_{\Pi(2)}$,..., $x_{\Pi(N)}$ }. Number sequence of $\Pi(1), \Pi(2), ..., \Pi(N)$ is one permutation of 1 to N.

3 Related Algorithm and Description

3.1 The Primitive Ant Colony Algorithm—Ant System

In this paper, we discuss application of ant colony optimization (ACO) in WSN environment. Given a set of n nodes, an ant visits all the n different nodes. A data structure called $tabulist_k$ is provided for memorizing the nodes already visited by the time t and forbidding the ant to visit them again before a travel completed. When an iteration is completed, the $tabulist_k$ is used to compute the ant's current solution (i.e., the distance of the path visited by the ant). At beginning of next circle, the list is emptied and the ant is free again to choose nodes.

In ant system model, initially, pheromone intensity of trail should be set to arbitrarily the same value. Ants use the so-called random proportional rule, in which ants migration directions depend on pheromone intensity on the process of traveling. We define the transition probability from node to node for the $k-th$ ant as

$$p_{ij}^k(t) = \begin{cases} \dfrac{\tau_{ij}^\alpha(t)\eta_{ij}^\beta(t)}{\sum\limits_{s \in allowed_k} \tau_{is}^\alpha(t)\eta_{is}^\beta(t)}, & j \in allowed_k \\ 0, & otherwise \end{cases} \tag{2}$$

Where $allowed_k$ is equal to $N - tabulist_k$, and where α and β are the parameters that control the relative importance of the pheromone versus the heuristic information η_{ij}, is equal to $1/d_{ij}$. Let d_{ij} is the length of path between nodes i and j (i.e., $d_{ij} = [(x_i - x_j)^2 + (y_i - y_j)^2]^{1/2}$).

Let $\tau_{ij}(t)$ be the intensity of trail on edge (i, j) at time t and τ_{ij}^k be the quantity per unit of length of trail substance laid on path (i, j) by the $k-th$ ant. Ant-cycle algorithm model is most-used model in which τ_{ij}^k is not computed at every step, but after a complete iteration. The value of the trail is updated every $n-1$ steps according to the following formula

$$\tau_{ij}(t+n)=\rho\cdot\tau_{ij}(t)+\tau_{ij}(t,t+n) \tag{3}$$

Where $\tau_{ij}(t,t+n)=\sum_{k=1}^{m}\tau_{ij}^{k}(t,t+n)$, and ρ is pheromone evaporation coefficient, $0<\rho<1$. Let $m=\sum_{i=1}^{n}b_i(t)$ be the total number of ants and let $b_i(t)$ be the number of ants in node i at time t. The value of $\tau_{ij}^{k}(t,t+n)$ is given by

$$\Delta\tau_{ij}^{k}(t,t+n)=\begin{cases}\dfrac{Q}{L_k} & \text{if } k-th \text{ ant uses path(i,j) in its tour}\\ 0 & \text{otherwise}\end{cases} \tag{4}$$

Where L_k is the tour length of the $k-th$ ant and Q is a constant left on path (i,j) after $n-1$ steps. In ant-cycle model, pheromone update occurs at the end of a whole cycle.

3.2 The Improved Ant Colony Algorithm: Ant Colony System

Ants have memory function to achieve communication by sensing some heuristic information. In the algorithm initial time, concentration of the pheromone is equal for every edge (i,j). It is difficulty for ants to find out an optimal path from mass paths. Traditional ACO has drawbacks, such as slow convergence speed and stagnation. In our experiment, we adopt a strategy of selecting a part of optimal routes from a large of initialization paths and leaving pheromone at the initial time, also, considering the resident energy of nodes to lead ants choose different routes. Ant Colony System (ACS) is one of improvement ant colony algorithm.

(1) Pseudo random proportional of state transition rule

In AS model, ants are entirely dependent on probability to search new paths from local short paths. In contrast, the ACS pseudo random proportional state transition rule provides a direct way to balance between exploitation of a priori and accumulated knowledge and exploration of new states tendentiously.

In ACS model, the probability for an ant to choose next node j depends on a random variable q uniformly distributed over $[0,1]$, and the parameter q_0.

$$j=\begin{cases}\arg\max\limits_{j\in allowed_k}\left\{\tau_{ij}^{\alpha}(t)\bullet\eta_{ij}^{\beta}(t)\right\} & ,if\,(q\le q_0)\\ S & \text{else formula (2) is used}\end{cases} \tag{5}$$

Where q_0 is given for leading ants choose next node j in order to avoid selecting path depending on probability absolutely. If $q\le q_0$, then

$j = \arg \max_{j \in allowed_k} \{\tau_{ij}^{\alpha}(t) \cdot \eta_{ij}^{\beta}(t)\}$, otherwise depending on probability formula. The heuristic information η_{ij} is equal to $\frac{E_i + E_j}{d_{ij}}$, and E_i, E_j are residual energy of nodes i and j respectively.

(2) Local Pheromone Updating

The local pheromone update is performed by all the ants after each step from node i to node j. The principle of pheromone update is given by

$$\tau_{ij}(t+1) = (1-\rho)\tau_{ij}(t) + \rho\tau_{ij} \tag{6}$$

Where $\tau_{ij} = \sum_{k=1}^{m} \tau_{ij}^{k}$, and $\tau_{ij}^{k} = \frac{E_i + E_j}{d_{ij}}$. In (6), $\rho \in (0,1)$ denotes the pheromone decay coefficient, and $1-\rho$ is the pheromone residual parameter. Let τ_{ij} be pheromone increment leaved by all ants passing the pass from node i to node j. The local updating rule of pheromone is modified to avoid local convergence and increase retrieval efficiency.

(3) Global Pheromone Updating

In ACS model, global pheromone updating does not process until all ants have computed their tour. Only global optimal ant is permitted to release pheromone that leads ants to perform an instructional search for next node. So ant search scope is mainly concentrated in the current cycle of the best path field in order to improve searching speed. After a complete iteration, the value of the trail is updated according to the following formula,

$$\tau_{ij}(t+1) = (1-\rho)\tau_{ij}(t) + \rho\tau_{ij} \tag{7}$$

$$\Delta\tau_{ij} = \begin{cases} 1/L_{gb} & if \quad (i,j) \in L_{gb} \\ 0 & otherwise \end{cases} \tag{8}$$

Where τ_{ij} is equal to $1/L_{gb}$ if node i and node j belong to optimum path. Let L_{gb} be global optimum path of each circle. Pheromone which only belongs to the optimal path is enhanced during global pheromone updating.

3.3 Mutation Operator

Using ACO for TSP, the path between cities was effective. By contrast, to solve migration of mobile agent in WSN, the communication capability of nodes should be considered as WSN is resources-restricted network in which nodes have limited energy. In order to avoid invalid path, a kind of mutation operator will be imported. The specific idea is to find out common neighbor node as relay node from terminal nodes of invalid path as every node has a list of neighbor nodes. Mutation model is shown in Fig. 2 and Fig. 3.

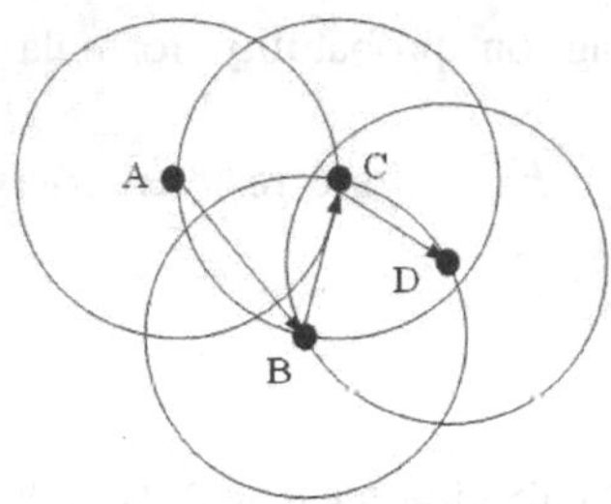

Fig. 2. Before mutation

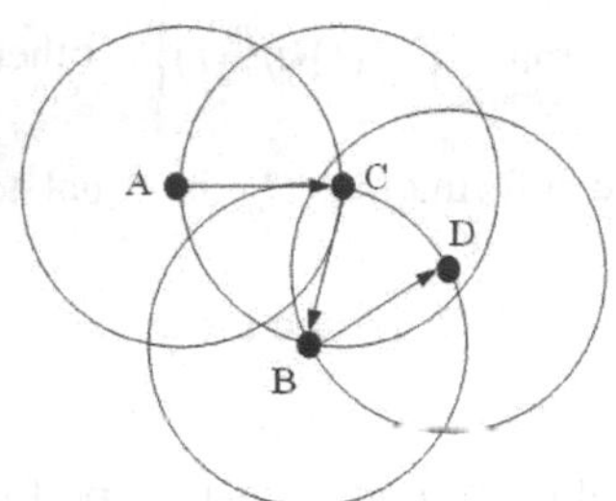

Fig. 3. After mutation

As shown in Fig.2, let r be communication radius. Computed by ACO, the nodes sequence visited by mobile agent is A-B-C-D. Obviously, the path is invalid and unreachable in the actual network because that distance between A and B is greater than radium of communication ($d_{AB} > r$). The mutation idea is to change the nodes sequence visited by mobile agent by choosing common neighbor node C as relay node. After mutation, the sequence of nodes is A-C-B-D, and the tour is effective.

3.4 Algorithm Description

Improved ACO algorithm description as follows:

Step 1: Initialize N paths randomly, choose $l(1 \le l \le N)$ near-optimal paths and release pheromone that leads ants to choose different nodes, due to different pheromone as well as different length of routes.

Step 2: Initialize, set t=0, Initial iteration $Nc = 0$, MAX denotes max iteration; set α, β, ρ and q_0 value. Place n ants on every node i. Begin to iterate according to improved ACO and insert starting node in $tabulist_k$ of the $k-th$ ant.

Step 3: Compute the value of p_{ij}^k according to formula (2) and formula (5). Ant $k(1 \le k \le n)$ chooses next node j and place node j in $tabulist_k$. For every edge (i, j), compute local pheromone updating according to formula (6).

Step 4: If ant k has not visit all nodes and $tabulist_k$ is not full, return step 3. Otherwise, continue step 5.

Step 5: Judge n routes generated after one iteration. If invalid paths exist, mutation operator was imported. Then, record optimal result.

Step 6: According to equation (7), compute global pheromone updating.

Step 7: If the number of iteration was not reach MAX, empty $tabulist_k$ and return to step 3.

Step 8: While the number of iteration reach MAX, compare the optimal result of each circle and print out final optimal result.

4 Experimental Results and Analysis

In order to verify availability of improved ant colony algorithm for migration of mobile agent, simulation experiments are carried out on Microsoft Visual C++ 6.0. Experimental parameters are shown in Table 1.

Table 1. Experimental parameters

Parameters	Value
Coverage of network	1000×800
Nodes numbers	30
Communication radius	300
Ants numbers	30
Max iteration	100
Q	100
α	1
β	3
ρ	0.5
q_0	0.15

To compare the three models, we ran each algorithm 20 times using the parameters given in Table 1 and analyzed their relative advantages and drawbacks by experimentation. The comparisons among three algorithms are shown in Table 2.

Table 2. The comparison of three algorithms

Item	AS	ACS	Improved ACO
Optimal path	3984.16	3973.28	3963.35
Average length	3988.6590	3976.1665	3964.6475
Invalid path	7	3	0

As shown in Table 2, three algorithms (AS, ACS and Improved ACO) generated 7, 3 and 0 invalid paths respectively. Both AS and ACS algorithms have given worse results than those obtained with Improved ACO. The improved ACO algorithm received good effect and avoided generation of invalid path effectively in wireless sensor network.

Observing Fig.5, the results generated by improved ACO are shorter than two other algorithms. In WSN, power consumption should be proportional to data transmission distance, so improved ACO can provide less network energy consumption and prolong network lifetime. In addition, the trend chart of improve ACO is smoother than other algorithms that indicates performance of Improved ACO is stable and practical. Experiments demonstrate that the improved ACO has better performance of effectiveness and robustness on migration of mobile agent.

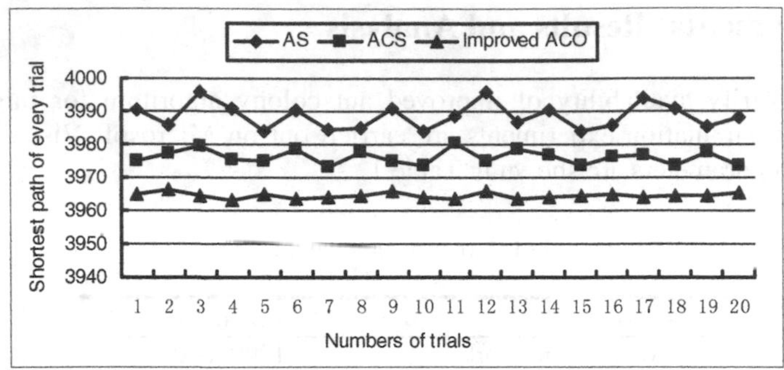

Fig. 5. The comparison of shortest path generated by three algorithms

5 Conclusion

For mobile agent routing in the wireless sensor network, an improved ACO algorithm was proposed. In order to avoid invalid path, a mutation operator is presented since the limitation of communication capability of wireless sensor nodes. At last, we implemented the three algorithms (AS, ACS, Improved ACO) and investigated their relative advantage and disadvantage by experimentation. The results of contrastive experiment show that the proposed algorithm is superior to two other algorithms both on convergence speed and global search capability. How to improve ACO and make it suit to WSN environment are the focus of research next step.

Acknowledgments. This work is partially supported by the National Natural Science Foundation of China (NSFC) Grant #61161002 to Sheng Zhang, and the Aviation Science Funds Grant #2009ZD56009 to Sheng Zhang.

References

1. Krishnamachari, B., Estrin, D., Wicker, S.: The Impact of Data Aggregation in Wireless Sensor Networks. In: International Workshop on Distributed Event-Based Systems (2002)
2. Ganesan, D., Estrin, D., Heidemann, J.: Why do we need a new Data Handling architecture for Sensor Networks? SIGCOMM Computer Communication Review 33(1), 143–148 (2003)
3. Choi, W., Das, S.K.: A Novel Framework for Energy-Conserving Data Gathering in Wireless Sensor Networks. In: Conference of the IEEE Computer and Communications Societies (INFOCOM), vol. 3(1), pp. 1985–1996 (2005)
4. Chen, M., Kwon, T., Yuan, Y., Choi, Y., Leung, V.C.M.: Mobile Agent-Based Directed Diffusion in Wireless Sensor Networks. EURASIP Journal on Advances in Signal Processing 2007, Article ID 36871, 13 pages (2007)
5. Qi, H., Xu, Y., Wang, X.: Mobile-agent-based Collaborative Signal and Information Processing in Sensor Networks. Proceeding of the IEEE 91(8), 1172–1183 (2003)
6. Du, R., Yao, G., Wu, Q.: Application of an Ant Colony Algorithm in Migration of Mobile Agent. Journal of Computer Research and Development 44(2), 282–287 (2007)

7. Ma, J., Zhang, J., Yang, J., Cheng, L.: Improved Ant Colony Algorithm to Solve Travelling Agent. Journal of Beijing University of Posts and Telecommunications 31(6), 46–49 (2008)
8. Xu, Y., Peng, P., Guo, A., Zhang, G.: Study of mobile agent route algorithm based on improved ant colony algorithm in wireless sensor networks. Computer Engineering and Application 45(4), 126–129 (2009)
9. Zheng, H., Zhang, W., Teng, S.: Improved ACO-based wireless sensor networks routing. Application Research of Computers 27(1), 99–100 (2010)

The Design of Electronic Transformer Sophisticated Calibration System Based on FFT Algorithm

Lujun Cui[1,*], Hui-chao Shang[1], Gang Zhang[2], You-ping Chen[2], Yong Li[1], and Ze-xiang Zhao[1]

[1] Zhongyuan University of Technology, School of Mechanical & Electronic Engineering, 450007 Zhengzhou, China
[2] Huazhong University of Science and Technology, School of Mechanical Science & Engineering, 430074 Wuhan, China
cuilujun@126.com

Abstract. The present work investigated a sophisticated calibration system based on the FFT algorithm for electronic transformers. The structure and the characteristics of LabVIEW were introduced in detail, at the same time, and the detailed demonstration of the process of harmonic analysis was given via FFT based on LabVIEW. Through these methods, the disturbance of the frequency leakage and the noise could be reduced and the detecting accuracy of the harmonic analysis could be improved. Serials of the experiments and error analysis were given, and the simulation results have verified the effectiveness and practicability of the algorithm for the electronic transformer calibration system.

Keywords: electronic transformer, Calibration System, FFT.

1 Introduction

Great interest has concentrated on research and development of the technology in digital substation since the generation of the novel electronic transformers [1-3], Being of excellent insulation and no saturation, electronic transformer (ET) meets the needs of the electric power system development towards to large volume, high- level voltage and automation, Due to the such the advantages compared with the traditional electromagnetic transformers, so the electronic transformers become the substitute of the electromagnetic transformers. The measurement and calibration of electronic transformers have been a challenging technological problem and attracted much research and development efforts in recent years. Experience has shown that it is technically and economically advisable to carry out a strict check for current and voltage transformers directly before as market products [4], to meet the market requirements, many types of checking system are currently studied. The cost was one of the key factors for such sophisticated electronic transformers calibration system.

The objectives of work were to study the fabrication of the electronic transformer sophisticated calibration system based on Fast Fourier Transform (FFT). The paper is

[*] Corresponding author.

D. Jin and S. Lin (Eds.): Advances in CSIE, Vol. 2, AISC 169, pp. 115–120.
springerlink.com

expected to design efficient algorithm to enhance the measurement accuracy, according to international electronic transformer standards, the checking accuracy experiment for ET is described based on USB data acquisition cared and the virtual instruments experiment platforms, through the technology of virtual instruments, the calibration system could analysis the amplitude and phase of signal in the electronic transformers accurately.

2 Fabrication of ET Calibration System

Fig.1 illustrates the fabrication of electronic transformer calibration system. The performances of calibration system were made up of the following several parts: standard electromagnetic transformer, standard current/voltage signal conversion device, voltage/current regulator, secondary signal conversion device, USB data acquisition card. The precision class of conventional electromagnetic transformer was two-grades higher than the electronic transformer. The function of standard signal conversion device was to transform the current/voltage signal to the acceptable voltage signal. A high speed USB synchronous data acquisition card with 24-bite precision is used for the calibration system, its acquisition accuracy and sampling velocity affect the calibration accuracy for electronic transformer. At the same time, through the algorithm of FFT, the model of error analysis in computer could measure fundamental component and harmonic component of signal in electronic transformer.

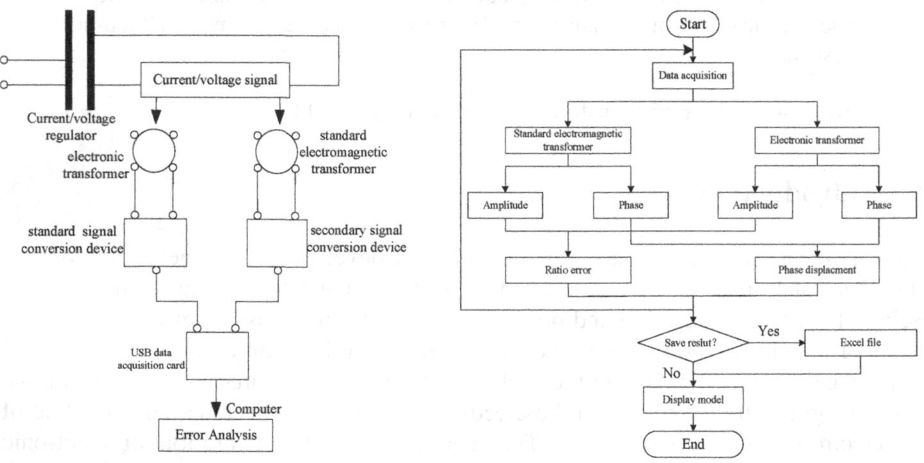

Fig. 1. The sketch of optical path for optical fiber hydrogen sensor

Fig. 2. Software flow pattern of error analysis in calibration system

As shown in Fig. 2, the two-way output voltage signal were send into the computer, the model of error analysis calculated the amplitude and phase of signal separately, according to the definition of ratio error and phase displacement, the error analysis could calculated the differences of amplitude and phase compared with the standard electromagnetic transformer, at the same time, the error analysis could save the data for further processing. The implementation of virtual instrument and the USB data acquisition card for the calibration system could reduce cost, and convenient for programming.

3 Principle

The existing of harmonic results in the distortion of the supply network waveform, the harmonic analysis is the an important index for the supply power network, the increasing the harmonic content maybe lead to local over-heat, overvoltage and so on, so it's necessary to measure and monitor the harmonic for electronic transformer and the others electric power unit.

In an ideal power network, the current and voltage are sine wave, so the current and voltage are also sine wave in the linear load, so it was expressed by the

$$u(t) = U\sin(\Omega t + \theta)\,. \tag{1}$$

However in real power supply system, as the nonlinear load distort in the sine wave, and results in the generation of nonstandard periodic sine wave. Nevertheless, the current or voltage in power supply system satisfy Dirichlet, so the *u(t)* could expressed by the equation (2)

$$u(t) = a_0 + \sum_{n=1}^{\infty}(a_n\cos(n\omega t) + b_n\sin(n\omega t)) = a_0 + \sum_{n=1}^{\infty}c_n\sin(n\omega t + \varphi_n)\,. \tag{2}$$

The parameters a_0 stands for direct-current component, $c_n = \sqrt{a_n^2 + b_n^2}$, $c_1\sin(\omega t + \varphi_1)$ and $c_n\sin(n\omega t + \varphi_n)$ stands for fundamental component and harmonic component respectively. It's concluded that the current/voltage signal in power supply system can be expressed the sum of the fundamental component and harmonic component. So the emergence of higher-order harmonic has seriously affected the sine wave in electric system.

In recent the harmonic measuring based on FFT in power supply system, the time domain waveform could be transformed into time domain waveform frequency domain waveform through the Discrete Fourier Transform (DFT). After sampling in data acquisition card, $U(t)$ was changed into the $U(nT)$,

$$u(k) = \sum_{n=0}^{N-1}u(n)e^{-j\frac{2\pi}{N}nk} \qquad k = 0,1,2\cdots,N-1\,. \tag{3}$$

In equation (3), T stand for an sampling period.

So in an sampling period, $\{u(k)\} = (u(1),u(2),\cdots,u(N-1))$. When the discrete time $t = kT/N = kdt$ (dt is sampling inter-arrival time, $dt = T/N$), the sampling value equals to $u(k)$, so the coefficient of Nth-degree harmonic[5] could be calculated by the equation (4).

$$a_n = \frac{2}{N}\sum_{k=0}^{N-1}u(k)\cos\frac{2\pi}{N}kn, b_n = \frac{2}{N}\sum_{k=0}^{N-1}u(k)\sin\frac{2\pi}{N}kn \qquad n = 1,2,\cdots,N-1\,. \tag{4}$$

4 Experiments and Discussion

In calibration system for electronic transformers, its core algorithm is FFT. With undergoing the frequency domain analysis in LabVIEW, the two signals were decomposed into phase information and amplitude information, and during the processing of FFT for the signals, it may cause the long range spectrum leakage and short range spectrum leakage, in order to avoid the negative effect of the spectrum leakage, therefore, it is absolutely necessary to add the window function before the FFT calculation in the calibration system. The typical windows function including the rectangular window, triangular window, hanning window, and hamming window. Different window function before the FFT for the electronic transformer calibration system affected the measuring accuracy. So in order to verify the rationality of soft algorithm in the calibration system, so a series of the simulated experiments carried out in the paper.

$$u(n) = \sum_{i=1}^{8} A_i \sin(2\pi i \frac{f_1}{f_s} n + \varphi_i) \,. \tag{5}$$

In equation (5) f_1 stands for fundamental frequency, and equals to 49.5Hz, the sampling frequency f_s equals to1KHz, the coefficient of N equals to 1024. So the experiment data was listed in table1.

Table 1. Simulated electronic transformer signal with harmonics

	Simulated signal							
harmonic order	1	2	3	4	5	6	7	8
Amplitude A_i (Kv)	150	1.2	0.1	0.01	0.05	—	0.02	—
Phase φ_i (o)	162	30	150	70	50	—	30	—

The experiments results of calibration system under the window function mentioned above were listed in table 2. in the electronic transformer calibration system based on the FFT, it's concluded that the lower sidelobe was beneficial to the inhibition of leakage, a serials of experiments based on the different fundamental frequency or sampling rate, were also carried on, the same experiment conclusion were obtained.

Table 2. The measurement results and error using the rectangular window in calibration system

rectangular window	measurement results and error							
harmonic order	1	2	3	4	5	6	7	8
Amplitude A_i (Kv)	101.384	0.753	0.060	0.006	0.030	—	0.013	—
Error (%)	30	37	40	40	40	—	35	—
Phase φ_i (°)	115.673	28.66	145.72	74.71	45.93	—	28.34	—
Error (%)	28.5	4.46	2.85	6.73	8.14	—	5.53	—

Table 3. The measurement results and error using the triangular window in calibration system

triangular window	measurement results and error							
harmonic order	1	2	3	4	5	6	7	8
Amplitude A_i *(Kv)*	142.317	1.096	0.089	0.008	0.044	—	0.018	—
Error (%)	5.12	8.67	11	20	30	—	10	—
Phase φ_i *(°)*	148.56	29.03	146.77	67.88	46.84	—	28.77	—
Error (%)	8.30	3.23	2.15	3.02	6.32	—	4.1	—

Table 4. The measurement results and error using the hanning window in calibration system

hanning window	measurement results and error							
harmonic order	1	2	3	4	5	6	7	8
Amplitude A_i *(Kv)*	150.684	1.183	0.099	0.01	0.049	—	0.020	—
Error (%)	0.45	1.41	2	0	2	—	0	—
Phase φ_i *(°)*	162.154	29.66	151.08	69.48	50.422	—	29.901	—
Error (%)	0.09	1.13	0.72	0.74	0.84	—	3.27	—

Table 5. The measurement results and error using the hamming window in calibration system

hamming window	measurement results and error							
harmonic order	1	2	3	4	5	6	7	8
Amplitude A_i *(Kv)*	137.864	1.158	0.096	0.008	0.043	—	0.018	—
Error (%)	8	3.5	4	20	14	—	10	—
Phase φ_i *(°)*	167.83	31.26	148.31	72.09	48.891	—	28.76	—
Error (%)	3.60	4.21	1.13	2.99	2.22	—	4.13	—

Comparison for the measurement results and errors in the tables for the calibration system, Conclusions were drawn that the hanning window was fit for the electronic transformer calibration system and it enhanced the measuring precision and suppress spectrum leakage effectively in the electronic transformer calibration system.

5 Conclusions

In the present work, the sophisticated calibration system based on FFT has been estimated for electronic transformers. It was found that the detecting accuracy was affected by spectrum leakage during processing the signals. Through the simulation results and contrast experiments, it's concluded that the FFT algorithm and hanning window could enhance the detecting accuracy for the power quality in electrical power system, the LabVIEW provide a preferable means for the harmonic analysis in real power system.

Acknowledgments. This work was supported in part by the National Natural Science Foundation under Grant 50974062, Basic and Frontier Technology Research Project of Henan Province under Grant 082300410320, and Zhengzhou Technology Innovation Team Project under Grant 10CXTD153.

References

1. Emerging Technologies Working Group: Fiber Optic Sensors Working Group. Optical Current Transducers for Power Systems: A Review. IEEE Transactions on Power Delivery 9, 1778–1787 (1994)
2. Lee, B.: Review of the Present Status of Optical Fiber Sensors. Optical Fiber Technology 9, 57–79 (2003)
3. Satioh, M., Kimura, T., Minami, Y.: Electronic Instrument Transformers for Integrated Substation Systems. Transmission and Distribution Conference and Exhibition 1, 459–464 (2002)
4. Nagornyi, P.D., Ryabchuk, S.V.: A mobile laboratory for checking current and voltage transformers. Measurement Techniques 50, 97–99 (2007)
5. Ding, Y., Gao, X.: Numerical signal processing, Xi'an, pp. 28–43 (2000)

Application of Data Mining Technology Based on RVM for Power Transformer Fault Diagnosis

Lin Niu, Jian-guo Zhao, and Ke-jun Li

State Grid of China Technology College, Jinan, China
niulin2008@126.com

Abstract. One of the most challenging problems in real-time operation of power system is the state monitoring and fault diagnosis for large-scale transformer. Fast and accurate techniques are imperative to achieve on-line state assessment. To counter the shortcoming of common machine learning methods, a novel machine learning technique, i.e. 'relevance vector machine' (RVM), for on-line state assessment is presented in this paper. The proposed method is tested and compared with 'support vector machine' (SVM) classifier. It is demonstrated that the RVM classifier can yield a decision function that is much sparser than the SVM classifier while providing higher classification accuracy. Consequently, the RVM classifier greatly reduces the computational complexity, making it more suitable for real-time implementation.

Keywords: Relevance vector machine, support vector machine, data mining, fault diagnosis, power transformer.

1 Introduction

In the power system, power transformer is in a pivotal position, which is one of the most important and crucial electrical equipment, its reliability and security is directly related to the security and stability of the power system. Therefore, it is desirable to instantly detect incipient faults of power transformers. Several methods have been commonly employed for monitoring the conditions of transformers: the dissolved gas in oil analysis (DGA), the recovery voltage measurement and the frequency response analysis (FRA). Among the various diagnosis methods, DGA methods have been broadly used to detect incipient faults of power transformers.

Dissolved Gas Analysis (DGA) is one of the most convenient and effective means for the technology of transformer oil fault diagnosis. It can more accurately and reliably find the potential faults in the progressive development, which is conducive to promote the transit from regular maintenance to the state maintenance. But the actual regulations or rules based on DGA can only give determinant boundaries, which cannot represent the relation between the faults and the exterior character. Therefore, many artificial intelligence technologies based on DGA, such as neural network [1], gray clustering [2], Petri network [3], information fusion [4] have been applied to transformer diagnosis and produced some results. However, from the perspective of statistics, most of these methods adopt the principle of empirical risk minimization, not expected risk minimization. Therefore the generalization ability of

D. Jin and S. Lin (Eds.): Advances in CSIE, Vol. 2, AISC 169, pp. 121–127.
springerlink.com

these methods is not very good. From the above-mentioned methods, we can conclude that data mining methods have the potential to be applied to real-time on-line fault diagnosis practically with the improved machine learning algorithms. In [5], the state-of-the-art SVM based on the statistical learning theory has been used to fault diagnosis with good generalization performance. However, the SVM has a number of the significant and practical limitations. In the SVM, the predictions are not probabilistic and the kernel function $K(x, x_i)$ must satisfy Mercer's condition.

In this paper, we propose a new power transformer fault diagnosis model, which performs system optimization and generalization simultaneously using relevance vector machine (RVM) with a probabilistic Bayesian learning framework that does not suffer from above disadvantages [6]. This paper takes full advantages of RVM to solve the problem, such as probabilistic predictions, automatic estimation of 'nuisance' parameters, and the facility to utilize fewer arbitrary basis functions. The proposed model was tested on actual power system, it was shown that the prediction model could generalize well and provide accurate fault diagnosis results at low computational cost, thereby making it more suitable for real-time applications.

2 Relevance Vector Machine

In this paper, power transformer fault diagnosis is formulated as a binary classification problem. Especially, for the real-time fault diagnosis, we can apply a classifier to determine whether a fault state will be present or not in advance. To begin, let vector $\mathbf{x} \in R^n$ denote a pattern to be classified, and let scalar t denote its class label (i.e. $t \in R$).

2.1 RVM Classifier Model

For an input vector $\mathbf{x}$, an RVM classifier models the probability distribution of its class label $t \in R$ using a sigmoid logistic function σ as:

$$p(t = 1|\mathbf{x}) = \frac{1}{1 + \exp(-f_{RVM}(\mathbf{x}))} \tag{1}$$

where $f_{RVM}(\mathbf{x})$, the classifier function, is given by:

$$f_{RVM}(\mathbf{x}) = y(\mathbf{x}; \mathbf{w}) = \sum_{i=1}^{N} w_i K(\mathbf{x}, \mathbf{x}_i) + w_0 = \mathbf{\Phi w} \tag{2}$$

where N is the length of the data, weight vector $\mathbf{w} = [w_0, \cdots, w_N]^{\mathrm{T}}$ and $\mathbf{\Phi}$ is the $N \times (N+1)$ design matrix with $\mathbf{\Phi} = [\phi(\mathbf{x}_1), \phi(\mathbf{x}_2), \cdots, \phi(\mathbf{x}_N)]^{\mathrm{T}}$.

Adopting the Bernoulli distribution for $p(t|\mathbf{x})$, we write the likehood as:

$$p(\mathbf{t}|\mathbf{w}) = \prod_{n=1}^{N} \sigma\{y(\mathbf{x}_n; \mathbf{w})\}^{t_n} \left[1 - \sigma\{y(\mathbf{x}_n; \mathbf{w})\}\right]^{1-t_n} \tag{3}$$

where target vector $\mathbf{t} = [t_1, \cdots, t_N]^{\mathrm{T}}$ with the targets $t_n \in R$. To avoid this, a zero-mean Gaussian prior distribution over $\mathbf{w}$ with variance $\boldsymbol{\alpha}^{-1}$ is added as:

$$p(\mathbf{w}|\boldsymbol{\alpha}) = \prod_{i=0}^{N} N(w_i|0,\alpha_i^{-1}) = \prod_{i=0}^{N} \sqrt{\frac{\alpha_i}{2\pi}} \exp(-\frac{\alpha_i}{2} w_i^2) \tag{4}$$

where hyperparameter $\boldsymbol{\alpha} = [\alpha_0, \alpha_1, \cdots, \alpha_N]^{\mathrm{T}}$.

The posterior distribution over the weight from Bayes rule is thus given by:

$$p(\mathbf{w}|\mathbf{t},\boldsymbol{\alpha}) = \frac{\text{Likelihood}\times\text{Prior}}{\text{Normalizing factor}} = \frac{p(\mathbf{t}|\mathbf{w})p(\mathbf{w}|\boldsymbol{\alpha})}{p(\mathbf{t}|\boldsymbol{\alpha})} \tag{5}$$

Unlike the regression case, however, the marginal likehood $p(\mathbf{t}|\boldsymbol{\alpha})$ can no longer be obtained analytically by integrating the weights because of the discontinuity of the likelihood $p(\mathbf{t}|\mathbf{w})$, and an iterative procedure has to be used.

Let α_i^* denotes the maximum a posteriori (MAP) estimate of the hyperparameter α_i. This is equivalent to maximizing the following objective function:

$$J(w_1, w_2, \cdots, w_N) = \sum_{i=1}^{N} \log p(t_i|w_i) + \sum_{i=1}^{N} \log p(w_i|\alpha_i^*) \tag{6}$$

where the first summation term corresponds to the likehood of the class labels, and the second term corresponds to the prior on the parameters w_i. In the resulting solution, only those samples associated with nonzero coefficients w_i (called relevance vectors) will contribute to the decision function.

The gradient of the objective function J with respect to $\mathbf{w}$ is:

$$\nabla J = -\mathbf{A}\mathbf{w}\text{-}\boldsymbol{\Phi}^{\mathrm{T}}(\mathbf{f}\text{-}\mathbf{t}) \tag{7}$$

where $\mathbf{A} = \mathrm{diag}(\alpha_0, \alpha_1, \cdots, \alpha_N)$, $\mathbf{f} = \left[\sigma(y(x_1)), \cdots, \sigma(y(x_N))\right]^{\mathrm{T}}$, matrix $\boldsymbol{\Phi}$ has elements $\phi_{i,j} = K(x_i, x_j)$. The Hessian of J is:

$$\mathbf{H} = \nabla^2(J) = -(\boldsymbol{\Phi}^{\mathrm{T}}\mathbf{B}\boldsymbol{\Phi} + \mathbf{A}) \tag{8}$$

where $\mathbf{B} = diag(\beta_1, ..., \beta_N)$ is a diagonal matrix with $\beta_i = \sigma(y(x_i))\left[1 - \sigma(y(x_i))\right]$.

The posterior is approximated around $\mathbf{w}_{MAP}$ by a Gaussian approximation with covariance:

$$\boldsymbol{\Sigma} = -(\mathbf{H}|\mathbf{w}_{MAP})^{-1} \tag{9}$$

and mean:

$$\mu = \mathbf{w}_{MAP} = \boldsymbol{\Sigma}\boldsymbol{\Phi}^{\mathrm{T}}\mathbf{B}\mathbf{t} \tag{10}$$

Using the new $\mathbf{w}_{MAP}$, the new target t^* is then obtained through:

$$t^* = \mathbf{\Phi}\mathbf{w}_{MAP} + \mathbf{B}^{-1}(t - \sigma(y(\mathbf{x}_n;\mathbf{w}))) \tag{11}$$

Using $\mathbf{\Sigma}$ and $\mathbf{w}_{MAP}$, hyperparameter α_i can be updated by:

$$\alpha_i = \frac{\gamma_i}{\mathbf{w}^2_{MAP}}, \quad \gamma_i = 1 - \alpha_i \mathbf{\Sigma}_{ii} \tag{12}$$

2.2 Comparison to SVM Learning Model

SVM is a constructive learning procedure rooted in statistical learning theory. It is based on the principle of structural risk minimization, which aims at minimizing the bound on the generalization error rather than the minimizing the mean square error over the data set. As a result, this lead to a good generalization and an SVM tends to perform well when applied to data outside the training set.

An SVM classifier in concept first maps the input data vector $\mathbf{x}$ into a higher dimensional space F through an underlying nonlinear mapping $\mathbf{\Phi}(\mathbf{x})$, and then applies linear classification in this mapped space. Introducing a kernel function $K(\mathrm{x,y}) = \mathbf{\Phi}(\mathbf{x})^{\mathrm{T}}\mathbf{\Phi}(\mathbf{y})$, we can write an SVM classifier $f_{SVM}(\mathbf{x})$ as following:

$$f_{SVM}(\mathbf{x}) = \sum_{i=1}^{N} w_i K(\mathbf{x}, \mathbf{s}_i) + b \tag{13}$$

where $\mathbf{s}_i, i = 1, 2, ..., N_s$, is a subset of the training samples $\{\mathbf{x}_i, i = 1, 2, ..., N\}$ (which are called support vectors).

The SVM classifier in (13) is noted to resemble in form the RVM classifier in (2), yet the two classifiers are derived from different principles. As will be illuminated later by the simulation results, for SVM the support vectors (SVs) are typically formed by "borderline", difficult-to-classify samples in the training set, which are located near the decision boundary of the classifier; for RVM the relevance vectors (RVs) are formed by samples appearing to be more representative of the two classed, which are located away from the decision boundary of the classifier.

3 Fault Diagnosis Model of Power Transformer

Data mining is defined as the efficient discovery of valuable, non-obvious information embedded in a large collection of data. The working object of data mining is data warehouse or original data. The general data mining scheme is shown briefly as follow: data selection, data transformation, data mining and knowledge representation.

3.1 Preparation of Training Data Set

The analytical base of diagnosis is some diagnostic gas obtained by DGA. The content information reflects the states of transformer. The diagnostic gases include hydrogen (H_2), methane (CH_4), ethane (C_2H_6), ethylene (C_2H_4), acetylene (C_2H_2), et al. So in this paper we chose these 5 gases as input features.

Based on analyzing the characteristic of transformer fault, the transformer fault diagnosis which is identified relying on the effective information provided by DGA is divided into six types of transformer state. The six types as follows: partial discharge (PD), discharges of low energy (D_1), discharges of high energy (D_2), thermal fault (T_1), thermal fault (T_2), thermal fault (T_3). In this paper, the collected sample set has 960 samples of the power transformer fault with known fault kinds, and it was divided into two parts. Among them, 770 samples are used to as the relevance vector training set to obtain the relevance vector, while the others are used as the verifying set.

3.2 Machine Training for RVM

As power transformers are planned to operate most of the time under stable conditions, operational studies usually generate highly unbalanced classes. This unbalance between the normal and fault classes can be troublesome for some classifiers. Aimed to improving the classification performance, several strategies exist for dealing with unbalanced data sets for classifier training, including resizing the training data sets, adjusting misclassification costs, and recognition-based learning (learning from the minority class)[7]. The approach of resizing training sets is through over-sampling the minority class and/or under-sampling the majority class. It has been concluded that such an approach for adjusting the balance of the samples between the two classes in training can be beneficial for training of classifiers using the data set.To determine the fine-turning parameters of the RVM classifier model for optimal performance, we apply a leave-one-out cross validation [8] procedure in the training set. The parameter setting with the smallest generalization error is chosen. The resulting classifier is then tested using the testing data sets for performance.

We summarize the training results in Table 1, where the generalization error obtained by the trained classifier is listed under different parametric settings. From these results, we see that the best error level (4.89%) was obtained by an order-2 polynomial kernel; a similar error level (5.62%) was obtained by the RBF kernel with $\sigma = 2.5$. Consequently, the order-2 polynomial kernel was chosen for the RVM classifier, which was subsequently retrained with all the samples in the training set. The resulting classifier was then used for performance evaluation.

Table 1. Generalization errors obtained by RVM Classifier under different parametric settings

Polynomial kernels	Order1	Order2	Order3	Order4
Error rate	0.0877	0.0489	0.0512	0.0537
RBF kernels	σ =1	σ =2.5	σ =5	σ =10
Error rate	0.0706	0.0526	0.0603	0.0621

For comparison, we also trained the SVM classifier in (13) using the same training procedure as for the RVM. The best error level (5.11%) was achieved when an order-3 polynomial kernel was used and the regularization parameter C was set to 9. As for RVM, the SVM classifier was retrained using these tuned parameters with all the samples in the training set.

For the RVM classifier, the number of RVs (produced during training) was found to be only 17(1.72% of the number of training samples); for SVM, the number of SVs was found to be 130(13.1% of the number of training samples). Indeed, the RVM classifier is much sparser than the SVM. The trained classifiers were used subsequently for performance evaluation.

3.3 Performance Evaluation Results

The results of power transformer fault assessment are evaluated by successful classification rate (SCR) and false discriminate rate (FDR). SCR is the ratio of correct classified samples over all samples. The higher the SCR is, the better the performance is. FDR is the ratio of the fault samples classifier as normal samples over all samples. In real world, these false discriminate samples are unacceptable and can be considered as one of the indices for classification performance. The lower the value of FDR is, the better the performance is. It is shown In Table 2, that RVM classifier can yield a decision function that is much sparser than the SVM classifier while providing more higher classification accuracy. This can lead to significant reduction in the computational complexity of the decision function and faster decision time, thereby making it more suitable for real-time application. In Table II, the training time is the average amount of time taken for each sample of training in the cross-validation procedure during the training phase; the testing time is the average amount of time taken by the trained classier for classifying each sample during the testing phase. Table III shows a sample of the results for the testing data, each line representing a case. The reader will notice some cases where the RVM made a correct identification while the SVM failed and one case of failure of the RVM where also the SVM table was not useful.

Table 2. The performance comparison of SVM and RVM for Fault Diagnosis

Method	SCR		FDR		Time		Number of RV or SV
	Train	Test	Train	Test	Train	Test	
SVM	0.971	0.952	0.01	0.012	46.7s	32s	130
RVM	0.992	0.986	0.002	0.004	52.8s	9.2s	17

Table 3. Test Results (T: Thermal Fault; PD: Partial Discharges; D: Discharge; NI: Not Identified)

H_2	CH_4	C_2H_2	C_2H_4	C_2H_6	RVM	SVM	TEST
13	17	0.3	4.7	4.2	T	T	T
39	1.7	0.1	0.1	0.6	PD	PD	PD
1600	3600	0	14	670	PD	NI	PD
95	10	39	11	0	D	NI	D
1570	1110	1830	1780	175	D	D	D
41	112	4536	254	0	D	T	D
835	76	16	10	29	T	NI	D
10	13	0.1	25	7	T	T	T
33046	619	0	2	58	PD	PD	PD

4 Conclusions

In this paper, we have introduced a new approach for power transformer fault diagnosis using the relevance vector learning mechanism based on a probabilistic Bayesian learning framework. The number of rules and the parameter values of membership functions can be found as optimizing posterior distribution of the RVM in the proposed model. Because the RVM is not necessary to satisfy Mercer's condition, selection of kernel function beyond the limit of the positive definite continuous symmetric function of SVM. The relaxed condition of kernel function can satisfy various types of membership functions in prediction model. The RVM, which was compared with support vector learning mechanism in simulations, had the small model capacity and described good generalization. Simulation results showed the effectiveness of the proposed method for modeling of nonlinear dynamic systems with noise, and the practicability for real-time fault diagnosis.

References

1. Zhang, Y.: An artificial neural network approach to transformer fault diagnosis. IEEE Transaction on Power Delivery (1996)
2. Sun, X.: Study on fault diagnose method of transformer DGA with fuzzy model hierarchy classification. Elec. Eng. (2001)
3. Wang, J.: Application of fuzzy Petri nets knowledge representation in electric power transformer fault diagnosis. Proceedings of the CSEE (2003)
4. Shang, Y.: Synthetic insulation fault diagnose model of oil-immersed power transformers utilizing information fusion. Elec. Eng. (2002)
5. Moulin, L.S.: Support vector machines for transient stability analysis of large-scale power systems. IEEE Transactions on Power Systems 19(2), 818–825 (2004)
6. Tipping, M.E.: Sparse Bayesian learning and the relevance vector machine. Mach. Learning 1, 211–244 (2001)
7. Kaplowicz, N.: Learning from imbalanced data sets: a comparison of various strategies. In: Proceeding of Learning from Imbalanced Data Sets, pp. 10–15. AAAI Press, Menlo Park (2000); Tech. Rep. WS-00-05
8. Browne, M.W.: Cross-validation methods. Journal of the Mathematical Psychology 44, 108–132 (2000)

Application of MATLAB in Physical Experiment Processing

Jianguo Lu and Ming Hu

Jiaozuo Teachers College, Jiaozuo, Henan Province, China
zcactus03@126.com

Abstract. Take the data processing of the volt-ampere characteristic of diode experiments and the single slit Fresnel diffraction for example, this paper gives an introduction to Matlab in physical experiment processing. Compared with traditional experiment methods, the application of Matlab in experiment data processing can avoid the error effectively which caused by manual handling, and the Matlab method is simple to operate. While the simulation to single slit Fresnel diffraction gives a more precise study on the diffraction phenomenon of light.

Keywords: Matlab, volt-ampere characteristic of diode, single slit diffraction.

1 Introduction

Matlab is a high-level technical numerical calculation and visualization soft of Mathworks Company launched in 1982. Matlab is a software platform for science and engineering computing integrated with numerical analysis, symbolic computation, graphics processing, system simulation and other functions. With friendly interface, powerful functions, Matlab is regarded as the indispensable software in applied science computer aid analysis, design, simulation and teaching. Using this software in the experiment data processing can reduce the computation and obtain the accurate fitting curve, and the graphical display results can determine the accuracy of the calculation directly. At present, Matlab has become the most popular software of engineering, and it can be used to solve the questions on specific subjects. Matlab has the features of expandability, easy usage and high efficiency etc. In Europe and America, Matlab has become the basic teaching tools in application in linear algebra, automatic control theory, mathematical statistics, digital signal processing, time series analysis, dynamic system simulation and so on. In industry, Matlab is widely used to study and solve engineering problems. The application of Matlab in calculations, graphics, animation and other functions to simulate physical phenomenon and demonstrate the movement process can free the users from tedious low-level programs, so they have more time and energy to explore the scientific problems themselves. This paper studies the application of Matlab in physical experiments from two aspects: data processing and experiments simulation.

D. Jin and S. Lin (Eds.): Advances in CSIE, Vol. 2, AISC 169, pp. 129–133.
springerlink.com

2 Application of Matlab in Experiment Data Processing

Adopting Matlab in experiment data processing can reduce the compute capacity, obtain an accurate fitting curve. Moreover we can judge the accuracy of the calculation directly by the graphical display results. Next we will take the experiment data processing of the volt-ampere characteristics of diode for instance to illustrate the application of Matlab in data processing.

The following table is the voltage and current data segment measured in experiments, saved as an Excel “data.xls”.

	Data
Voltage	-5.03 -5.02 -5.01 -5.0 -4.99 -4.98 -4.97 -4.95 -4.5 -4 -3.5 -3 -2.5 -2 -1.5 -1 -0.5 0.1 0.4 0.74 0.75 0.76 0.77 0.78 0.79 0.80
Current	-18.1 -16.6 -15.0 -12.3 -9.6 -7.5 -5.9 -3.0 -1.8 -1.4 -1.30 -1.2 -1.11 -1.1 -0.8 -0.5 -0.1 0.5 0.7 1.4 2.4 3.3 4.9 6.1 6.9 7.9

Write the m files as follows:

```
clear, clf, clc
S=xlsread('data.xls') ;
v=S(2, 2:end) ;
A=S(3, 2:end) ;
vInter =- 5.03:0.01:0.80;
t =interp1(v, A, vInter, 'pchip')  ;
plot(v, A, '.', 'markersize', 20) ;
hold on
plot(vInter, t, 'r') ;
hold off
grid on
xlabel ' voltage(v)', ylabel ' current(mA)' ;
```

The display after operation is as follows:

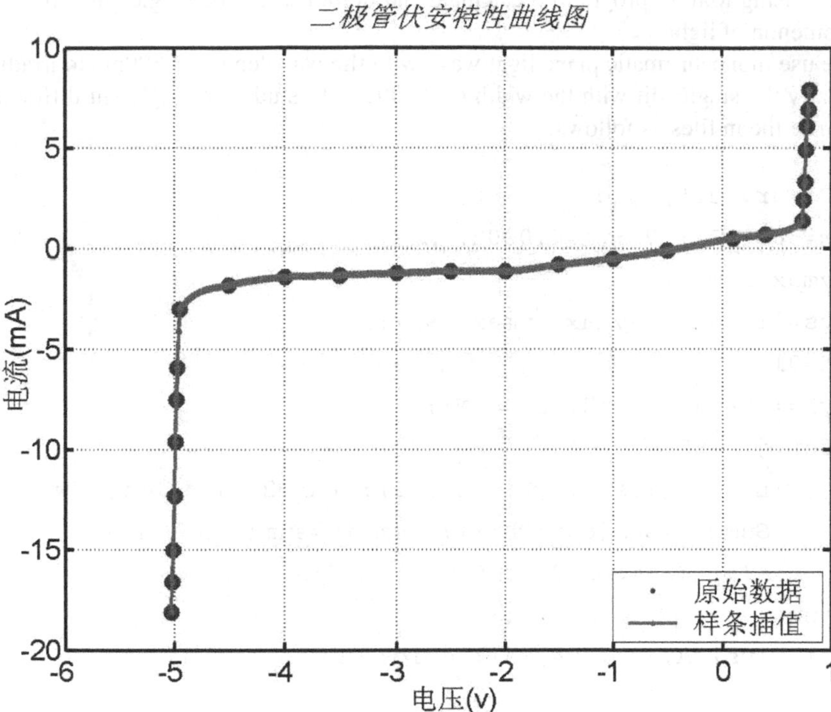

According to the experimental requirements, the figure above gets a smooth curve with using one-dimensional interpolation method. The red line is the correct curve of the volt-ampere characteristic of diode.

It is simple and practical to apply Matlab in physical experiment data processing. The accuracy of adopting fitting curve is higher than human drawing.

3 Application of Matlab in Calculating and Demonstrating of the Single Slit Diffraction of Light

Diffraction is a basic property when a light wave spreads in space. During the spread process of light, as long as the surface of the light wave is constrained such as the mutation of amplitude or phase, the phenomenon diffraction will occur. In optical system, any light in the spread process actually is the diffraction process under the relevant optics modulation. According to the scalar diffraction theory, the diffraction process can be described with the Fresnel diffraction integral. The computer simulation with its advantages of good controllability, non destructiveness, easy observability and low cost etc which provide a good tool for digital simulation of the modern optical experiments. The study of various forms diffraction screen under different experiment conditions is of great significance not only for the classical physical optics but also the modern optics.

Next using Maltab program to calculate and demonstrate the single slit diffraction phenomenon of light.

We use monochromatic plane light wave with the wavelength of 500nm to irradiate vertically the single slit with the width of 0.0002m, to study the single slit diffraction and write the m files as follows:

```
Clear, clf, clc
w=500*10^(-9); a=0.0002; z=1;
ymax=3*w*z/a;
ys=linspace(-ymax, ymax, 91);
N=91;
yP=linspace(-a/2, a/2, N);
for j=1:91
    L=sqrt((ys(j)-yP).^2+z^2); Phi=2*pi.*(L-z)./w;
    SumCos=sum(cos(Phi)); SumSin=sum(sin(Phi));
    A(j)=(SumCos^2+SumSin^2)/N^2;
end
plot(ys, A,'o', ys, A); grid;
```

The display after operation is as follows:

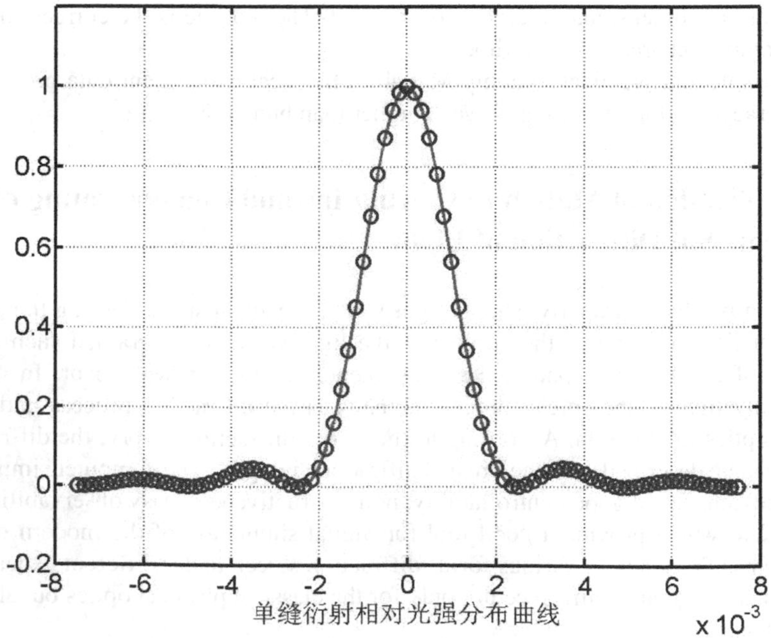

The figure shows that, for single slit diffraction of light, the highlight in the middle concentrates of most of the brightness and energy of the whole single slit diffraction. With Matlab simulation, the single slit diffraction of light can be performed accurately.

4 Conclusion and Prospect

From the above two experiments we can see, Matlab has extremely advantageous conditions in experiment data processing and optic simulation, and it can save a lot of time in computing and has high accuracy. Matlab is an indispensable tool in this field. Actually, Matlab has a wide range of applications in other aspects of physics. For example, in engineering design, people build model by study objects, achieve the operation process of system and work out the results by computer program, and find out the optimal solution for physical implementation. That is computer simulation. Computer simulation has many advantages in good controllability, non destructiveness, reproducibility and so on.

Expansion: take into account the advantages of Matlab on programming efficiency, portability and scalability, Matlab can be used to design a program interface with powerful functions, stable performance and high quality. In this interface, Matlab function can achieve more functions without restriction such as the rotation of three-dimensional image, the flash play of images and etc.

References

1. Zhang, Z.: Proficient in MATLABS 5.3, pp. 244–263. Beihang University Press, Beijing (2000)
2. Zhang, Y.: Proficient in MATLABS, pp. 175–181. Tsinghua University Press, Beijing (2000)
3. Yang, G.: Mordern Physic Experiment, pp. 22–26. Zhejiang University Press, Hangzhou (1997)
4. Gao, J.: The Computer Simulation of the Coherent Optics System. Chinese Journal of Lasers 24(10), 957–960
5. Yu, M.: Optical Holography and Information Processing. National Defence Industry Press, Beijing

Video Traffic Detection Method for Deep Packet Inspection*

Jun Huang, Botao Zhu, and Zujue Chen

School of Computer Science and Telecommunication Engineering
Jiangsu University
Zhenjiang 212013 P.R. China
340031804@qq.com

Abstract. The basic concept of deep packet inspection over wireless or wired networks and the main challenges from deployment are described. One of the new traffic (video for gaming) detection issue encountered in the trials is reported, fast Hurst calculation approach to solve the particular issue is proposed. Experimental measurements are carried out to verify the special traffic behavior based approach. Section I is the introduction of the challenges, section II is the briefing for the customized issue and solution, section III is the detail measurement analysis and the last section is the summary and future work. The major challenge in this area is providing fast detection algorithm with quick turn around time, under limited computation, memory and time constrains. This paper offers an alternative complimentary methodology that is practical when traffic volume especially for video applications become heavy beyond normal deep packet inspection capability.

Keywords: Deep packet inspection, Video Pattern, Fast Hurst.

1 Introduction

DPI (Deep packet inspection) is an advanced method of packet filtering those functions up to the Application layer of the OSI (Open Systems Interconnection) reference model. The use of DPI makes it possible to find, identify, classify, reroute or block packets with specific data or code payloads that conventional packet filtering, which examines only packet headers, cannot detect [1].

Using DPI, communications service providers can allocate available resources to streamline traffic flow. For example, a message tagged as high priority can be routed to its destination ahead of less important or low-priority messages or packets involved in casual Internet browsing. DPI can also be used for throttled data transfer to prevent P2P (peer-to-peer) abuse [2], improving network performance for most subscribers. The security implications of DPI are widespread because the technology makes it possible to identify the originator or recipient of content containing specific packets, a capability that has sparked concern among advocates of online privacy. For this reason, we are proposing a non-intrusive method, which can identify potential abuse without breaking the privacy law, as a complimentary solution.

* This work was partially supported by Jiangsu University Distinguished Expert Grant #10JDG071, and the NJLECO and MPX Technology Collaboration Project #561957.

D. Jin and S. Lin (Eds.): Advances in CSIE, Vol. 2, AISC 169, pp. 135–140.
springerlink.com

The current DPI method has at least three significant limitations. First, it can create new vulnerabilities as well as protect against existing ones. While effective against buffer overflow attacks, denial of service attacks and certain types of malware, DPI can also be exploited to facilitate attacks in those same categories. Second, the existing DPI method adds to the complexity and unwieldy nature of existing firewalls and other security-related software. All DPI methods require it own periodic updates and revisions to remain optimally effective. Third, DPI can reduce computer speed because it increases the burden on the processor. Despite these limitations, many network administrators have embraced DPI technology in an attempt to cope with a perceived increase in the complexity and widespread nature of Internet-related perils. Numerous companies, including such major players as Alcatel-Lucent, Cisco, Huawei, Ericsson, IBM, Microsoft, Nokia and Symantec have begun to aggressively market DPI technology as components of hardware and software firewalls. Our new fast computation method is trying to partially overcome the above-mentioned pitfalls of the existing DPI methods.

Deep packet inspection is the process of examining the non-header content of a packet by a system that is not an endpoint in the communication. It is a successful by-product out of a failed concept and work called layer 7 switching, due to the amount of 7 layers processing is beyond physical limitations [3]. Most DPI systems reconstruct communication streams and maintain state information for large numbers of concurrent connections. When a packet arrives each layer is fully parsed and inspected [4]. State information associated with the communication stream at each layer is updated and stored [5]. All of this work requires many CPU cycles. Fortunately, the work is highly parallelizable; therefore, multi-core architectures can be used to increase the available CPU cycles. So, that raises the question of how many packets can be scheduled on a multi-core platform. In this paper we explore the traditional problem with the advanced Hurst based pre-processing method to assist and off-load the existing PAM (Protocol Analysis Module), when the network load is too high to be handled by PAM alone.

PAM is a software library that uses DPI to detect and respond to malicious network traffic. It is a proprietary module originally used by IBM Internet Security Systems in all products that implement intrusion detection and prevention. PAM fully parses all protocol layers and emulates the state transitions of both the client and server at each layer. PAM detects when a client or server is in a vulnerable state and watches for any attempts to exploit it. PAM is constantly updated. New protocol parsers are added and old ones are updated so they can detect the latest threat. This continuous addition of new code causes the average number of instructions needed to inspect a packet to slowly increase with each security update. In addition, network speeds continue to increase.

No matter which parallel methods are used to catch up the traffic growing speed, there is a significant cost impact associated with it; our approach is trying to balance the cost versus the depth of the inspection, during the current economy down turn we are experiencing. Our main contributions in this paper are summarized as follows:

(1) Designed and simulated non-intrusive fast new video traffic detection algorithms.

(2) Demonstrated that Hurst based detection has the advantages of encryption-agnostic, especially useful over wireless link.

(3) Compared our detection algorithms against a packet detection algorithm commonly used in network applications, in terms of complexity and cost.

(4) Compared with Periodogram, Whittle, Wavelet, our IDC algorithm has the characteristics of simple, efficient, accurate and reliable. It is a heuristic method and calculates Hurst quickly.

2 Research Method and Measurement Analysis

2.1 Hurst Based Approach

During a week long trial in Nanjing for Hong Kong customer, we found that the new traffic detection cycle could be as long as a week, by the time it is detected, the network Quality of Service is already affected. There are a few reasons behind the slow detection process:

(1) Some video traffic is encrypted, and this prevent the existing protocol analyze to decode it; obtaining password is a lengthy approval process.

(2) Part of video traffic is using some proprietary algorithms, deviate from the standard video format.

(3) The other video traffic is complete disguised format, trying to sneak through firewall.

(4) The large volume of traffic forced DPI filter switched into sampling mode, by pass potential harmful traffic.

To solve above issues one by one is not an easy task, as such, we looked at the problem from a complete different angle. Based on the observations that encryption or change of the data format does not affect video stream's traffic activity behavior or more precisely the Hurst parameter [6], and the Hurst parameter of single video stream will not diminish, when mingled with a large volume of average traffic. The stream with high Hurst (more active) hurt the QoS of entire network the most, and thus need to be watch out for.

For these reasons, we propose two detection schemes, with the first is easy to implement, and the second is more accurate.

(1) We collect the packet size only, skip the collection of time interval, to speed up calculation and lower the cost of implementation, and calculate the Hurst based on the size.

(2) Analysis based on the autocorrelation curve of size, so that the relative of time relationship is not skipped completely [7].

IDC (Index of Dispersion for Counts) can be used to evaluate Hurst parameters with minimum calculations. From the nature of slow variance attenuation of self-similar process $X_1^{(m)}$, when the sample size $m \to \infty$, the variance of arithmetic average decay slower than its bottom of the sample size [8]. Therefore, when $m \to \infty$, we have the following formula:

$$\sigma^2\left(X_1^{(m)}\right) \sim \alpha m^{-\beta} \tag{1}$$

α is limited normal number and independent of m, $0 \langle \beta \langle 1$.

X={ X_K ,1≤k≤L, k∈ N}, X is time series, its length is L. It is divided into d data blocks; the size of one data blocks is m. So, d=L/m.

Calculate the sample mean of each data block:

$$E_n^{(m)} = \frac{1}{m} \sum_{i=(n-1)m+1}^{nm} X_i, 1 \le n \le d \tag{2}$$

Calculate the sample variance to different m:

$$\sigma^2_{(m)} = \frac{1}{d}\sum_{n=1}^{d}(E_n^{(m)})^2 - (\frac{1}{d}\sum_{n=1}^{d}E_n^{(m)})^2 \tag{3}$$

Take logarithm to (1):

$$\log\left|\sigma^2\left(x_1^{(m)}\right)\right| \sim \beta \log m + \log \alpha \tag{4}$$

Draw points in the logarithm coordinates: $p\left(\log m, \log\left(\sigma_m^2\right)\right)$.

According to the generalized least squares, make linear fitting lines, and find out the slope of the line, $-\beta$. The slope of the IDC curve gives the Hurst parameter from equation (5).

$$H = 1 - \beta, 0 < \beta < 1 \tag{5}$$

Autocorrelation function means relevance of one sequence at different times.

$$R_x(m) = \frac{1}{N-m}\sum_{n=1}^{N-m} X_n X_{n+m}, m = 0,1,2..., M \tag{6}$$

$$Xmix(t) = Xaudio(t) + Xvideo(t), t > 0 \tag{7}$$

Plug equation (7) in equation (6).

$$R_{X_m}(m) = \frac{1}{N-m}\sum_{n=1}^{N-m}(X_a(n) + X_v(n))(X_a(n+m) + X_v(n+m))$$

$$= \frac{1}{N-m}\sum_{n=1}^{N-m}(X_a(n)X_a(n+m) + X_v(n)X_v(n+m) + X_a(n)X_v(n+m)$$
$$+X_v(n)X_a(n+m))$$
$$= R_{X_a}(m) + R_{X_v}(m) + R_{X_aX_v}(m) + R_{X_vX_a}(m) \tag{8}$$

Assume that audio and video are independent, $R_{X_aX_v}(m) = R_{X_vX_a}(m) = 0$, we have that the cross correlations are zero, and we have

$$R_{X_m}(m) = R_{X_a}(m) + R_{X_v}(m) \tag{9}$$

Since $PSD = FFT(Rx)$, PSD means power spectrum density, according to the definition, we further have

$$PSD(mix) = PSD(audio) + PSD(video) \tag{10}$$

2.2 Measurement Analysis

Most DPI products support H.232 and SIP protocols, dynamic opening and closing ports of transmission voice and video. Voice over Broadband is the use of broadband access networks and terminals to provide services, connected with the traditional PSTN. Users on the Internet to realize the voice communication through the PSTN gateway to access networks. We have collected 50 audio traces (MagicJack) and 1 video gaming trace (RuneScape) via WireShark, Matlab processing the collected data. According to IDC curve, we get Hurst parameters. We choose 12 typical traces from them and merged together, video's Hurst parameter is 0.741, as show below. Ha means Hurst of merged audio, Hm means Hurst of mixed audio and video.

Table 1. Hurst parameters

			H			Ha	Hm
0.536	0.573	0.577	0.578	0.582	0.583	0.524	0.785
0.655	0.657	0.658	0.671	0.678	0.679	0.611	0.791

The table can be seen, H became large when adding video. In PSTN network, fractional T1 can be sold in nearly any fractional of a 24 channel, 1.54Mbps circuit. For example, out of the 24 channels in a full T1 line, 12 channels might be used in a fractional T1 to provide a 768 k connection.

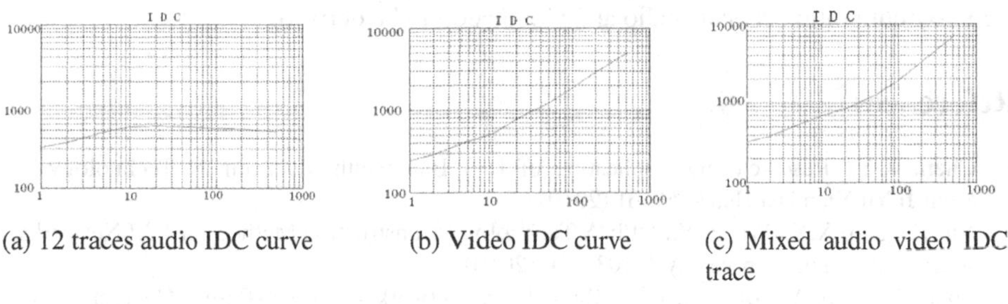

(a) 12 traces audio IDC curve (b) Video IDC curve (c) Mixed audio video IDC trace

Fig. 1. IDC curves

Figure (a), shows the Hurst is 0.536 when video is off, Figure (c) shows the Hurst is 0.745 when video is on.

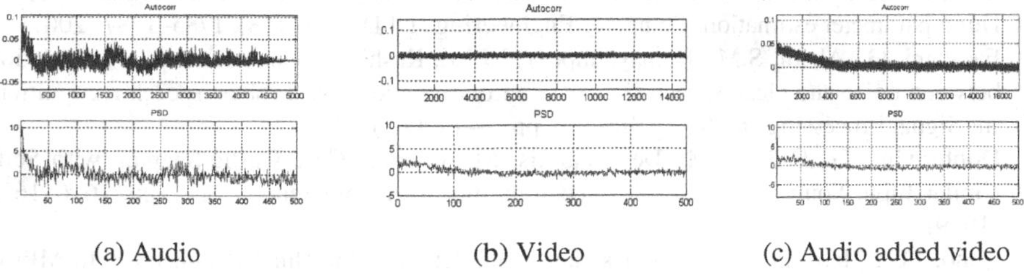

(a) Audio (b) Video (c) Audio added video

Fig. 2. Correlation and PSD curves

The correlation curve is shown in Figure 3. Correlations show a predictive relationship in a sequence of data. Fig.3 shows that video frames are strongly correlated. When adding even one video into 12 audios, the strong correlation shows up immediately. This way, we are surely to identify the presents of the video traffic. We can further calculate the power spectrum density of the correlation length, and use the power density near the origin as a signature of the concerned video. In our case, PSD (Audio) =1.5, PSD (Video) =10, and PSD (Audio + Video) =11.5; from which we can understand the amount of the activities that could adversely affecting the network Quality of Service. In another words, how "bad" is the traffic behavior itself.

There are three fundamental massages been transmitted over the network audio, video and text. Video is the only one has strong correlation; the main reason is that the current video frame is always similar to the previous video frame, it can not be completely different, and otherwise human eyes can not see anything meaningful [9]. However, audio frame or text frame can make sense alone; although added correlation may help on better understanding of the whole story.

4 Conclusion and Future Work

The challenges for the new unknown video traffic detection has been discussed, a brand new calculation method is proposed, the measurement is conducted to verify the effectiveness, during an international link tests. The theoretic signatures are derived, the flow chart based on the derivation is offered. More measurement works need to be done on the production network. More rigorous theoretic bound in terms of the maximum ratio between audio and video need to be derived.

References

1. Chen, X.Y.: Research and Application of DPI Bandwidth Management Technology. Ji Suan Ji Yu Xian Dai Hua 9, 59–61 (2010)
2. Yu, Z., Yao, X.X., Wang, Y.: DPI: A Technology Construction Method for P2P Networks. Journal of Sichuan University 4, 103–110 (2010)
3. Zhao, L., Luo, Y., Bhuyan, L.N., Ravi, I.: A Network Processor-Based, Content-Aware Switch. IEEE Micro 26, 72–84 (2006)
4. Lowen, S.B., Teich, M.C.: Fractal-Based Point Process. WILEY (2005)
5. Cascarano, N., Este, A., Gringoli, F., Risso, F., Salgarelli, L.: An Experimental Evaluation of the Computational Cost of a DPI Traffic Classifier. In: Global Telecommunications Conference, pp. 1–8 (2009)
6. Mao, L., Lin, Y., Ma, S.N.: Research on method of network abnormal detection based on Hurst parameter estimation. Computer Engineering and Design 28(8), 1785–1787 (2007)
7. Farahani, G., Ahadi, S.M., Homayounpoor, M.M., Kashi, A.: Consideration of correlation between noise and clean speech signals in autocorrelation-based robust speech recognition. In: Signal Processing and Its Application, pp. 1–4 (2007)
8. Doubrovina, G., Falkner, M., Devetsikiotis, M.: Optimal Cost Traffic Shaping with Self-similar Input Sources. In: Global Telecommunications Conference, vol. 2, pp. 1616–1622 (1999)
9. Cano, J.C., Manzoni, P.: On the Use and Calculation of the Hurst Parameter with MPEG Video Data Traffic. In: Euromicro Conference, pp. 448–455 (2000)

Automatic Knot Adjustment for Reverse Engineering by Immune Genetic Algorithm

Li Xu, Xiuyang Zhao*, and Bo Yang

School of Information Science and Engineering, University of Jinan, Jinan 250022 P.R. China
{nic_xul,zhaoxy,yangbo}@ujn.edu.cn

Abstract. B-spline surface reconstruction is a challenging problem in CAD design, data visualization, computer animation, virtual reality, especially in reverse engineering. Fairing B-spline surface is regarded as one major precondition for reverse engineering. There are two main stages in B-spline surface reconstruction: 1) knot vector selection, 2) Surface Parameterization. Stage one plays a decisive role between the two stages. To obtain a good surface approximation, knot vector is treated as variable and optimized by the immune genetic algorithm (IGA). The experimental results show that this method performs well in terms of accuracy and flexibility.

Keywords: B-spline surface, Immune Genetic Algorithm, Knot adjustment, Reverse engineering.

1 Introduction

In reverse CAD/CAM, laser scanner systems can yield a cloud of 3D data points from a 3D model easily. Although there are numbers of outstanding achievements in the field of the research on the surface reconstruction, a lot of key issues and difficult problems remain to be solved and perfected [1].

The paper [2], [3], [4] describe that the distribution of knots is a multivariate and multimodal nonlinear optimization problem. Zhao el. [5] proposes a B-spline curve approximate that optimize the knots by the GMM-based continuous optimization algorithm. Akemi el. [6] optimizes the knots by GA, but does not talk much about random points. Meanwhile, traditional GA is easy to fall into the local optimal solution [7].

In this paper, in order to obtain a better fairing B-spline fitting surface, we introduce a new method for surface reconstruction from a cloud of random 3D data points. Firstly, knots are optimized by IGA [8]. Then the fitting surface can be achieved by solving least squares equations through LU factorization method.

2 Surface Reconstruction

A order $k \times l$ B-spline surface $p(u,v)$ [9] is parametric surface as following equation:

$$p(u,v)=\sum_{i=0}^{m}\sum_{j=0}^{n} d_{i,j}N_{i,k}(u)N_{j,l}(v),\quad u_k \le u \le u_{m+1}, v_l \le v \le v_{n+1} \tag{1}$$

* Corresponding author.

D. Jin and S. Lin (Eds.): Advances in CSIE, Vol. 2, AISC 169, pp. 141–146.
springerlink.com

Where, $d_{i,j}(i=0,1,\cdots,m; j=0,1,\cdots,n)$ are control points determine the shape of the surface, $N_{i,k}(u)$ and $N_{j,l}(v)$ are the B-spline basis functions of order k and l . Respectively, two knots are $U=[u_0,u_1,\cdots,u_{m+k+1}]$ and $V=[v_0,v_1,\cdots,v_{n+l+1}]$.

Step1. Data points parameterization. Firstly, the initial surface, usually called based surface, is constructed to be a bicubic coons surface using the given four boundary curves $\{\Gamma_i, i=0,\cdots,3\}$ of the original points [10]. Then the random points project onto base surface and parameterize to be $(u_r,v_r), r=0,\cdots,M$ [11].

Step2. Computed control points. The control points can be computed by performing the least-squares solution fitting of equation F .

$$F=\alpha F_{lsq}+\beta F_{fair}, \tag{2}$$

where F_{lsq} is least-squares term, F_{fair} is fairing term.

$$F_{lsq}=\sum_{r=0}^{M}[p(u_r,v_r)-q_r]^2 , \tag{3}$$

$$F_{fair}=\sum_{r=0}^{M}\left[p_{uu}(u_r,v_r)^2+p_{vv}(u_r,v_r)^2+2p_{uv}(u_r,v_r)^2\right] \tag{4}$$

α is approximate weight, β is fairing weigh, $\alpha+\beta=1$, the value of α is bigger the approximation is better, the value of β is bigger the surfaces is more fairing. Order $\partial F/\partial P_{ij}=0$ $(i=1,\cdots,m-1; j=1,\cdots,n-1)$. Basing on least-squares solution, we can deduce a nonsingular linear systems of equation as $AX=B$ [12]. Then solve the equations by means of LU factorization method to achieve the inner control points.

3 Automatic Knot Adjustment by the Immune Genetic Algorithm

Knots playing a crucial role in surface reconstruction, so knots are adjusted by IGA. Let $U=[u_0,u_1,\cdots,u_{m+k+1}]$ and $V=[v_0,v_1,\cdots,v_{n+l+1}]$ as the initial knots can be gained by chord length parameterization method. Antigens represent the target surface. Respectively, antibodies represent knots. Note that antibodies represent every potential solution of the problem.

Step1. Create the initial random population.

Taking B-spline surfaces bidirectional knots as a random variable vector, antibody can be defined as $Anti_l=\{u_0^l,u_1^l,\ldots,u_s^l;v_0^l,v_1^l,\ldots,v_t^l\}$, where $s=m+k+1$ and $t=n+l+1$. Initially, for each $Anti_l=\{u_0^l,u_1^l,\ldots,u_s^l;v_0^l,v_1^l,\ldots,v_t^l\}$, its u_i^l and v_j^l are chosen randomly from the initial knots $[(u_i+u_{i-1})/2,\ (u_i+u_{i+1})/2]$ and $[(v_j+v_{j-1})/2,(v_j+$

$v_{j+1})/2]$. Then $AntiPop = \{Anti_1, Anti_2, \ldots, Anti_N\}$ can be achieved as the initial anti-body population, where N is the number of population.

Step2. Calculate the fitness between antibody and antigen, and select the better ones as vaccines.

The fitness function is defined by the approximate error of the measured data points $q_r (r = 0,1,\cdots,M)$ from the surface. In this paper, we consider the sum error value as fitness function, given by:

$$f = \sum_{r=0}^{M} \left| p(u_r, v_r) - q_r \right|, r = 0,1,\cdots,M \tag{5}$$

Note that the good knots is the one with the minimum fitness.

Step3. If the current population reached to iteration limit, then the course halts, or else, continue to iterate.

If the whole error value is smaller than the θ(a smaller constant is threshold).Or the number of iteration is bigger than the MAXGEN (a constant given by prior knowledge) the course halts, or else, continue to iterate.

Step4. Adjusted the probability of antibody based on its concentration.

Giving two antibodies $Anti_s = \{u_0^s, u_1^s \ldots, u_e^s; v_0^s, v_1^s, \ldots, v_f^s\}$ and $Anti_t = \{u_0^t, u_1^t \ldots, u_e^t;, v_0^t, v_1^t, \ldots, v_f^t\}$ if they satisfy the requirement

$$\sum_{i=0}^{e} \left| u_i^s - u_j^t \right| + \sum_{j=0}^{f} \left| v_i^s - v_j^t \right| < \varepsilon \tag{6}$$

where ε is threshold, then $Anti_s$ and $Anti_t$ are considered to be same or similar.

The antibody's concentration p_d represent the same or similar antibodies' proportion; The antibody's fitness probabilities p_f is defined as its fitness value proportion in the whole fitness.

Step5. Perform selection, crossover and mutation.

Firstly, update antibodies' probabilities of selection by using

$$p = p_f (1 - \alpha p_d) \tag{7}$$

Where, $0 < \alpha < 1 \;\; 0 < p_f, p_d < 1$.

Then perform selection on *Kth* parent *AntiPop* obtain the next *tempAntiPop* by the selection operator which is implemented as the classical biased roulette wheel with weighted in p. Finally, perform crossover and mutation on *tempAntiPop* and obtain the new *tempAntiPop'*.

Step6. Perform vaccination. Generate the new population and go to step2. Perform vaccination on *tempAntiPop′* and obtain $K+1th$ parent *AntiPop* . Vaccination means that the antibodies with smaller fitness are replaced by vaccine.

4 Experimental Results

In this section, we report the experimental results of reconstructing B-spline surfaces with knots adjustment by the IGA. In this paper, we validate our algorithm on a set of points from a tooth model. It should be noted that the number of knots affects the fitting result. More knots lead to a better fitting result.

The set of parameters used for IGA is presented in Tab.1. Fig.1 is original points. Fig.2 shows its B-spline surface without knots adjustment. Fig.3 is the result with knots adjustment. A comparison with some previous approaches (shows in Tab.2, Fig.2 and Fig.3) express that our method can get a better fitting result.

Table 1. The set of parameters

Parameter	IGA
Population	100
Crossover rate	0.6
Mutation rate	0.4
Concentration regulate ratio	0.2
The maximum generation	100
3D data points	1717
Approximate weight α	1
Fairing weigh β	0
Number of knots	15×15

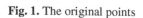

Fig. 1. The original points

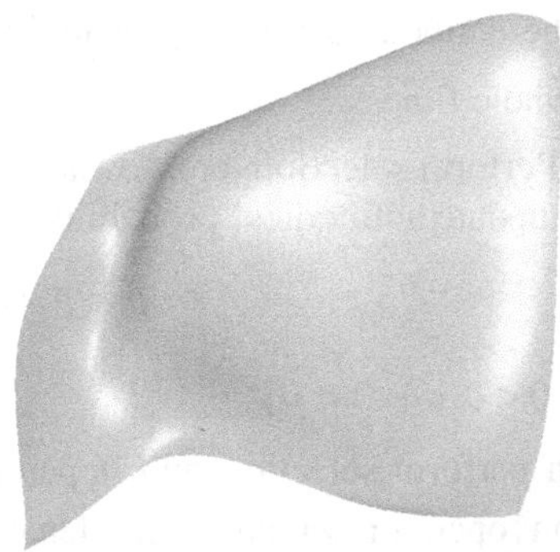

Fig. 2. The result without knots adjustment

Fig. 3. The result with knots adjustment

Table 2. Fitness of the IGA

Generation	Fitness	Average norm distance	Maximum norm distance
1	3.945666	0.002298	0.029988
10	3.653776	0.002128	0.029891
20	3.653776	0.002128	0.029891
30	3.591964	0.002092	0.029911
34	3.586813	0.002089	0.029959
53	3.549039	0.002067	0.030134
100	3.425415	0.001995	0.029743

5 Summaries

This paper introduce a new IGA-based B-spline surface reconstruction. The knots of B-spline surface are treated as antibodies, they are optimized by the IGA. Then, the fitting surface is calculated by least squares system of equations through LU methods. The experiment shows that the method yields a good result, and the method can obtain more precise surface reconstruction.

Acknowledgment. This research was supported by National Natural Science Foundation of China under contract No.60873089, China Postdoctoral Science Fund under contract No.20080431210, Postdoctoral Innovation Fund of Shandong Province under grant No.200802026, Doctor Foundation of Shandong Province under grant No.2007BS04018, Technology Plan Project of Department of Education of Shandong Province under grant No.J07YJ23, Science Foundation of Shandong province under contract No.ZR2010FM047 and Postgraduate Innovation Fund under contract No.YCX10007.

References

1. Liu, L.: Research on Surface Reconstruction in Reverse Engineering. Shandong University, Shandong (2007)
2. Yoshimoto, F., Moriyama, M., Harada, T.: Automatic knot placement by a genetic algorithm for data fitting with a spline. In: Proceedings of the International Conference on Shape Modeling and Applications, pp. 162–169. IEEE Computer Society Press (1999)
3. Dierckx, P.: Curve and surface fitting with splines. Oxford University Press, Oxford (1993)
4. Yoshimoto, F., Harada, T., Yoshimoto, Y.: Data fitting with a spline using a real-coded genetic algorithm. Compute. Aid. Design, 751–760 (2003)
5. Zhao, X., Zhang, C., Yang, B., Li, P.: Adaptive knot placement using a GMM-based continuous optimization algorithm in B-spline curve approximation. Computer Aided Design 44(6), 598–604 (2011)
6. Galvez, A., Iglesias, A., Puig-Pey, J.: Iterative two-step genetic-algorithm-based method for efficient polynomial B-spline surface reconstruction. J. 182, 56–76 (2012)
7. Mitchell: An Introduction to Genetic Algorithms. MIT Press, Cambridge (1996)
8. Jiao, L., Wang, L.: IEEE 30(5), 552–561 (2000)
9. Shi, F.: Computer aided geometric design with non-uniform rational b-spline. LNCS. Higher Education Press, Beijing (2001)
10. Floater, M.S.: Parameterization and smooth approximation of surface triangulations. Computer Aided Geometric Design 14, 231–250 (1997)
11. Piegl, L., Tiller, W.: Parameterization for surface fitting in reverse engineering. Computer Aided Design 33(8), 593–603 (2001)
12. Weiss, V., Andor, L., Renner, G., et al.: Advanced surface fitting techniques. Computer Aided Geometric Design 19(1), 19–42 (2002)

Research on the Key Technology of Computer Multimedia Terminal Software

JingMin Li

Jilin Business and Technology College, 130062 Changchun, Jilin, China
ljmaww@163.com

Abstract. The computer multimedia mobile terminal software research and development of key technologies were analyzed, and the completion of the basic module of the graft, core image processing algorithms, as well as computer mobile email and mobile terminal of the two computer game core software realization. Mobile multimedia terminal and the computer core software development have certain academic significance and application value.

Keywords: Multimedia terminal, e-mail, game, transplantation.

1 Introduction

With the rapid development of computer technology, the computer functions of the mobile terminal are increasingly high requirements. In a few short years, computer mobile terminal mobile phone products emerge in an endless stream of new function. Major manufacturers in order to occupy this Chinese huge market, undertake the technology innovates ceaselessly, make the function of the product is ceaseless and rich. Computer software of mobile terminal core technology research is imminent, it realize some function of the terminal software, developed computer software key technology of mobile multimedia terminal. Further multimedia and intelligent, will be the main tendency in the development of mobile communication terminal.

Motion estimation algorithm is the core technology of the multimedia terminal software, and motion estimation algorithm can affect the coding bit rate, the reconstructed image quality and speed. If the prediction error as the criterion, the search is the best, the simple matching algorithm, but due to its computation amount is too high, not suitable for real-time application, it presents many improved fast algorithms such as the three steps, three steps, four steps and the new MPEG4 calibration model of rhombus method. Because they have a common characteristic is that they search range are determined, which will inevitably bring a certain redundancy. These algorithms for those running vector deviation smaller blocks also brought search effect, and for those running vector deviation of the larger block, is more likely to cause loss of image quality. Because of the use of the image motion correlation to find the initial search center position, and dynamically determined for each search block area size, thus it can according to the frequency and amplitude of the motion of different, different level land to reduce computational complexity. At the same time, this algorithm can also be associated with any other fast motion estimation algorithm in combination, and can basically not loss of signal to noise ratio of the premise, improving the operation speed of this search algorithm.

D. Jin and S. Lin (Eds.): Advances in CSIE, Vol. 2, AISC 169, pp. 147–151.
springerlink.com

The return to basic concise design, do not have many options, get much better than H.263 compression performance, enhance the ability to adapt to the various channels, using network friendly structure and grammar, is conducive to the bit error and packet loss processing, application of a wide range of objectives, to meet the needs of different speed, different resolution and different transmission requirements. Can be widely used in digital broadcast, video conferencing, broadband TV, network stream media, digital image storage, digital film and other fields. In addition, the digital audio video coding standard and foreign standard is not the same, China standard by foreign limitation, it is domestic manufacturers save a lot of the patent fees.

2 Adaptive Fast Motion Estimation Search Algorithm

A fast motion estimation algorithm based on process include the prediction of initial search center position, to determine whether the stationary macroblock, dynamic adjustment of the search area size and according to the set value ahead of the end of search steps. On current macro block motion estimation, firstly, image motion correlation, find the initial search center position, and then according to some criterion for rapid determination of whether it is stationary macroblock, directly on the next block motion estimation, otherwise, according to the adjacent macroblock motion vector to dynamically determine the current block search area size, so as to reduce the number and reduce the computation complexity.

Motion estimation algorithm commonly used in the matching criteria are of three types: the minimum absolute difference, minimum mean square error and the normalized cross correlation function.

Minimum absolute difference MAD (I, J) = 1 / m * n

(I, J) is the displacement vector, m * n is for macro block size.

Minimum mean square error of the MSE = 1 / M * H,

M * H which are Fang Hong block size, when the value of MES minimum time is the best matching point.

Normalized cross correlation function

NCCF (I, J) = MH / NT, when the NCCF value maximum points for the best matching point.

In motion estimation and matching criterion for matching accuracy is not great, because the MAD guidelines do not require multiplication implementation is simple and convenient use so long, usually use SAD instead of MAD.

Among the many measure optimal matching criteria, absolute difference and the smallest computation, also used the most, this paper uses SAD to best match measure.

If the motion vector of the adjacent plates with large absolute value, the predictions of the motion vector of the current block and finally calculates motion vectors between the deviation of the domain of convergence will be relatively large, then the need to enlarge the search to find the best matching block. When the motion vectors of adjacent blocks are more hours, then need to enlarge the search to find the best matching block, according to the search area is different, also can be related to specific fast motion estimation algorithm combining, in order to ensure the image quality of the premise, achieve the purpose of reducing the operation complexity.

The corresponding movement is smaller, because the image in the vast majority of macroblock motion vector are relatively small, many part of macro block search region may be defined in a smaller range, greatly reducing the matching points, reduces the computational complexity, but also does not affect the quality of the reconstructed image. Because most of the images of macroblock motion vector is still relatively small, so the matching points will be greatly reduced, but not high frequency, movement amplitude under the condition of a small decrease much, the image quality is also a certain loss, is also in the performance decline. According to the characteristics of sports area, especially on the same object, because it is highly associated with the macroblock motion vector, so if the use of adjacent macroblock motion vector to predict the current macroblock of the initial motion vector, then find the initial search center position, can make the final of the motion vector is closer to the global optimal value.

3 Simulation Result

In the experiment, the coded reference model to implement fast motion estimation algorithm, while the test environment is formulated according to encoder testing environment is extended to. In order to verify the effect of formation estimation algorithm, select broadly representative standard test sequence for the simulation test. Set the frequency, the quantization step size, the search range. After the coding sequence in addition to the first frame, the frame into a 5 frame. In order to compare, achieve full search method, three steps, three steps, four steps of new diamond and method five Classical algorithm. And the other 10 sets of data as the test samples, the number of training can set to 500, choose a different number of hidden layer nodes of the experiment conclusion is consistent.

Based on the analysis of neurons in the hidden layer, choose the number of hidden layer neurons to (6, 8, 10) experiment, and with the other 10 groups of data as the test samples, the number of training can be set to 500, choose a different number of hidden layer nodes of the experimental conclusion is consistent. Sample data for D, F, A, R.

Sample data

D	F	A	R
1	**7**	**9**	**12.27**
1	**5**	**9**	**24.74**
1	**4**	**7**	**12.89**
2	**3**	**4**	**9.52**
6	**3**	**4**	**4.59**
3	**2**	**2**	**8.11**

MARS Algorithm for training after completion of the samples

D	F	A	R
1.00	**3.22**	**3.20**	**10.32**
0.99	**4.55**	**4.50**	**20.45**
0.99	**7.88**	**7.79**	**61.35**
0.99	**6.55**	**6.47**	**42.39**
0.99	**4.58**	**4.53**	**20.72**
0.99	**8.32**	**8.22**	**68.39**
0.99	**7.22**	**7.12**	**51.40**
0.99	**4.87**	**4.80**	**23.36**
0.99	**4.87**	**4.81**	**23.43**

PNSR Algorithm for training after completion of the samples

D	F	A	R
1.20	14.88	15.86	48.58
0.99	24.55	14.50	20.45
0.99	27.88	27.79	61.35
0.99	6.55	6.47	42.39
0.99	24.58	4.53	20.72
0.99	8.32	8.22	68.39
0.99	7.22	7.12	51.40
0.99	4.87	4.80	23.36
0.99	4.87	4.81	23.43
0.98	7.87	7.74	60.95
1.20	14.88	15.86	48.58

To sum up, the MARS algorithm the image and PNSR algorithm is the most close to the. At the same time, when the test sequence of movements more hours, using MARS algorithm the SNR loss less. When the test sequence of movements is large, then the algorithm the image SNR loss will be small amplitude is increased, the main reason is because when the motion vector is compared, because of the use of SD algorithm, and introduce a little bit of quality loss.

The average search points of comparison, the MARS algorithm requires the least number of comparison, because the motion vector of the current block are mostly close to the prediction of initial search center location. When the search range is equal to 1, MARS algorithm need only at most 9 search points, when the search range is equal to 2, need only at most 25 search points. Due to the introduction of static block method, a search will continue to decrease; the search is significantly smaller than the other types of search algorithms. This algorithm in matching accuracy, although with the full-search equivalent, but the calculation is greatly reduced. It can be in accordance with the adjacent motion correlation to predict the motion vector of the current block, and through some calculation for company to dynamically determine block search range. At the same time, because it can accord the specific conditions and the types of fast algorithm combining, therefore is very wide.

With the rapid development of computer technology, the computer functions of the mobile terminal are increasingly high requirements. In a few short years, computer mobile terminal mobile phone products emerge in an endless stream of new function. Major manufacturers in order to occupy this Chinese huge market, constant technology innovation, make the function of the product is ceaseless and rich. Computer software of mobile terminal core technology research is imminent, it realize some function of the terminal software, developed computer software key technology of mobile multimedia terminal. Further multimedia and intelligent, will be the main tendency in the development of mobile communication terminal.

4 Conclusion and Prospect

Motion estimation algorithm is the core technology of the multimedia terminal software, and motion estimation algorithm can affect the coding bit rate, the reconstructed image quality and speed. Aiming at the key technology of multimedia terminal core software in motion estimation algorithm, extensive access to a large number of domestic and foreign literature, and thus to establish computer mobile multimedia terminal basic software development model and development process, some core computer multimedia mobile terminal application software and game software development and implementation has a deeper understanding of.

References

1. Fang, W.Y.: Based on the characteristics of the personalized recommendation system research and J2MR. Northeast Normal University, Changchun (2006)
2. Zhi, Z.: When the soldiers. Improved personalized recommendation algorithm J2MR. Journal of Changchun University (2005)
3. Zhang, D., Wang, Y.: Rough neural network in image fusion for leachate application research. Hunan University press (2011)

Effect of Turbocharging on Exhaust Brake Performance in an Automobile

Chengye Liu[1] and Jianming Shen[2]

[1] School of Mechanical & Automotive Engineering, Jiangsu Teachers University of Technology 213001, Changzhou, China
lccyyyy@163.com
[2] Changzhou Institute of Light Industry Technology, 213164, Changzhou, China
sjmczjs68@163.com

Abstract. More and more diesel engines had been used as power source in an automobile when power was increasing and exhaust regulations for an automobile became more and more strict. Because compression ratio of diesel engines was higher than that of gasoline vehicles its performance of exhaust brake was prior to that of gasoline vehicles. Turbocharger was commonly used to improve power and reduce fuel consumption ratio. Study on effect of turbocharger on exhaust brake for diesel engines had practical significance. In this paper the effect of turbocharger on exhaust brake had been studied by theory method.

Keywords: Automobile, Turbocharging, Exhaust brake, Diesel engine.

1 Introduction

With the development of highways, logistics and the pace of life weight and velocity of vehicles become more and larger, which make driving safety of vehicles lowered. Braking load of vehicles increases quickly so that primary brake system is easy to be overloaded and damage for overheating, then traffic accident will take place. In addition, for vehicles in the mountains, hills and city, the driver have to use the main braking system frequently for security reasons, which leads to the average speed lowed, and it would affect the operating cost. In Europe, UN-ECE R13 regulatory requirements that auxiliary braking system must be installed as compliant brake set for heavy vehicles [1]. Currently, more types of auxiliary brake, such as an eddy current retarder, hydraulic retarder, engine braking, etc., and they have their own characteristics, and have better braking effect only in a certain speed range [2]. The more popular method is that more than one retarders are installed to compensate for each other in their lack of braking, so brake load of primary braking system can be reduced to improve traffic safety [3].

The initial exhaust brake appeared in Alps mountain area during World War I, and it was used for a long history [4], then there was significant development in research aspect. Working principle of the exhaust brake is that a butterfly valve or similar body is built in the engine exhaust manifold, and when the valve closes wind tunnel the brake force of the vehicle starts to increase, so the engine's exhaust resistance

D. Jin and S. Lin (Eds.): Advances in CSIE, Vol. 2, AISC 169, pp. 153–158.
springerlink.com

increases accordingly, and it makes the engine that takes as the source of automotive power becomes a set that consumes vehicle kinetic energy, in fact, the engine is transferred into the air compressor. The exhaust brake can prevent vehicle accidents due to rapid deceleration and skidding. Exhaust brake has simple structure, reliable performance and easy operation, and its braking power can reach about 60% rated power of the engine [5]. Many diesel vehicles are equipped with this device, such as the Steyr series, Dongfeng EQ Series, Isuzu EHD. Series of diesel vehicles are equipped with a structure similar to the exhaust brake. Especially in Japan, diesel exhaust brake usage of the car has more than 70% [6]. As the diesel engine compression is relatively larger than gasoline engine, when an engine works as the air compressor its retarder resistance is better than gasoline (because of its higher absorption of energy), so this set is mainly used for diesel engine. Turbocharged diesel engine unit is commonly used to improve the power per quality of the engine and to reduce fuel consumption rate, so effect of the exhaust gas turbocharger on braking performance has certain significance.

1.1 The Structure of the Exhaust Brake

Exhaust brake is generally divided into two kinds of pneumatic and vacuum form, and pneumatic controlled exhaust brake is used more. Figure 1 is electronic-controlled valve exhaust brake structure diagram. It consists of actuator, control organization and break-oil organization. The actuator includes brake cylinder and the butterfly valve, and control organization includes the brake switch, pedal and clutch switch, work lights, electromagnetic valve and air tubes, and break-oil organization includes the oil tank and other institutions and so on.

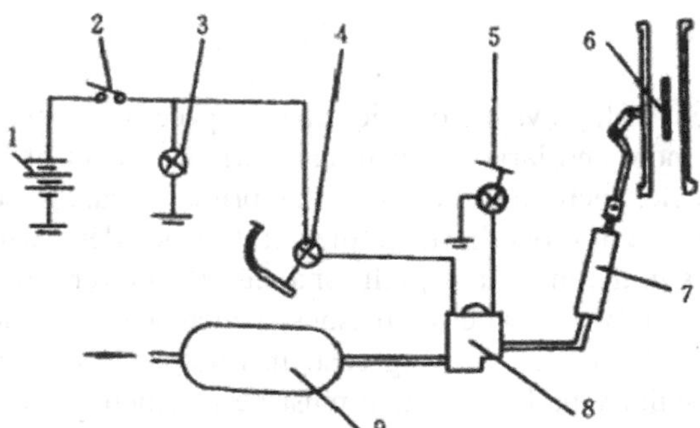

Fig. 1. Electronic-controlled valve exhaust brake structure
1. Battery 2. Brake Switch 3. Work Lights 4. Clutch Switch 5. Accelerator Pedal Switch 6. Brake Valve 7. Brake Cylinder 8. Electromagnetic Valve 9. Gas Tank

1.2 Working Principle of the Exhaust Brake

When a four-stroke diesel engine works, as shown in Figure 2 heat energy that fuel releases is transferred into mechanical energy and transmitted by the camshaft out. Exhaust brake is the set that a special butterfly valve is installed in the engine exhaust manifold as shown in Figure 3. In the normal working conditions the butterfly valve is opened, and this condition is similar to the status that exhaust brake is not installed, and flow area that exhaust manifold is cut is slightly reduced. When the exhaust brake retarder need work, firstly the fuel supply system is cut off, and this will make the engine flameout, secondly pressure cylinder and rocker promote the butterfly valve closed, and at this time exhaust gas emission is limited, and back pressure in the engine exhaust manifold start to increase, in the exhaust stroke gas in the cylinder and manifold is compressed again, and it will make the diesel engine become a air compressor that absorbs kinetic power of thc vehicle, and movement of the engine piston is hampered by the compression of the gas, so this resistance is like a big spring stiffness (air springs), acting directly on the engine piston, and the equivalent retarder device provides a vehicle through the transmission gear wheel and makes the vehicle speed reduction without the use of or less the case with the primary braking system, and the speed of the vehicle is easy under control, meanwhile this affect can reduce the primary braking system maintenance chance.

As there is cushion effect between the piston and cylinder it is benefit for extending the life of the engine after exhaust brake is used. And it can be overcome bigger inertial force when the vehicle is drove at the over speed, and in addition because the engine is used as brake set kinetic energy of vehicle is transfer into pressure energy and heat energy, at this time brake the engine temperature can keep constant, so engine thermal fatigue can be reduced in the mountain area [7,8].

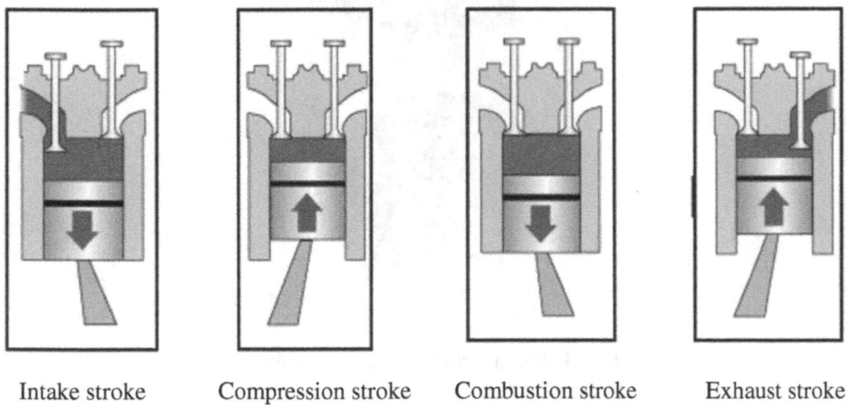

Fig. 2. Working process of four-stroke diesel engine

1.3 Turbocharging

Turbocharger driven by exhaust gas can add inlet air pressure to increase the amount of fresh air for diesel engine. So quality per power of diesel engine can be reduced, meanwhile the dimensions also is reduced, and raw materials and fuel consumption of the diesel engine is decreased. Above advantage can satisfy lightweight of a vehicle.

Work principle of turbocharging is as followed. Exhaust manifold is connected to the turbine housing of the turbocharger, and diesel exhaust has a certain pressure and high-temperature, and exhaust gas pass through the turbine housing into the nozzle ring, because the channel area of the nozzle ring changes from large to small, thus pressure and temperature of exhaust gas has a very large drop, while the rotation speed of the turbocharger is rapidly increasing, and high-speed exhaust gas flows and has a certain impact on the direction of the turbine, such as the turbine rotation speed, exhaust gas temperature. Pressure and temperature is higher, the higher the rotation speed of the turbine. Finally, exhaust gas go into the atmosphere through the turbine. And the shaft of turbine is connected with the shaft of a compressor, so their shafts are the same speed of rotation, and fresh air through the filter is inhaled into the compressor shell. High-speed rotation of the impeller presses fresh air through edge of the compressor impeller to increase the pressure and velocity, and fresh air flows into the diffuser to slow speed and increase pressure further. The pressurized air flows through the engine intake manifold into the combustion cylinder and more fuel can be injected to ensure a greater issue of diesel power.

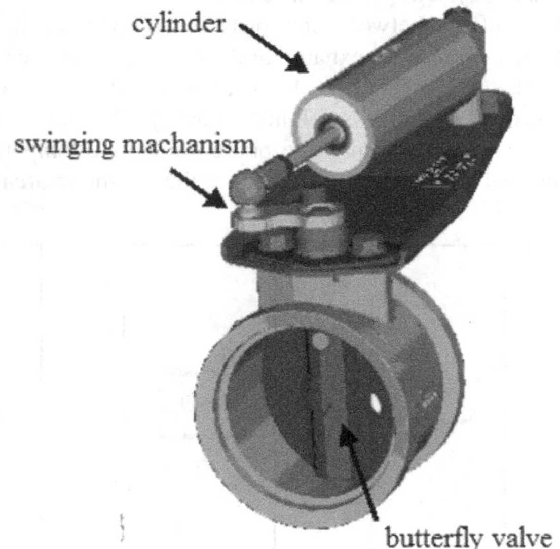

Fig. 3. The structure of exhaust brake

2 Effect of Turbocharging on Exhaust Brake

Without using turbocharger of exhaust gas the working process of engine is 1′→ 2 → 3 → 4 → 5′, in a complete cycle negative power of the diesel is made up of two parts: A and B as Fig.4 shows. A part indicates consumption of power of the compression and expansion stroke for diesel engine, in this part the negative work is a relatively small; however in the B part exhaust stroke in the normal working hours (intake valve is closed at this time) is changed into the compression stroke (equivalent to the air compressor), so negative power caused by the compression and expansion stroke has an much larger area than the negative power in the normal working. Negative power can increase the engine power loss to improve the exhaust brake retarder effect. When the intake valve is opened the air inside the cylinder with high pressure and temperature will cause air to the intake pipe, so air will flow back in the intake at this time.

After installation of turbocharger diesel engine at this time does not work, and butterfly valve is closed, and in exhaust manifold there is no exhaust emissions so the turbine blades can not work, but because turbine blades has the blocking effect, and this effect can make lower inlet pressure of the diesel engine than it with no turbocharger, so air put is relatively small. At the end of the compression stroke or the piston reaches TDC the peak pressure will decrease a little, so the PV curve will have a slightly decreased (this condition is not shown in order to show clear). Above condition shows the installation of turbocharger of exhaust gas diesel engine will have some negative impact on exhaust brake. Diesel engine with turbocharger will consume more air and fuel and produce more gas, and in the same circumstances exhaust gas has higher air pressure and temperature, so at the beginning of the implementation of exhaust braking system a greater dynamic torque will generate, after the intake valve opens brake torque will decline, and at this time brake torque is lower than that no turbo exhaust brake system at same rotation speed.

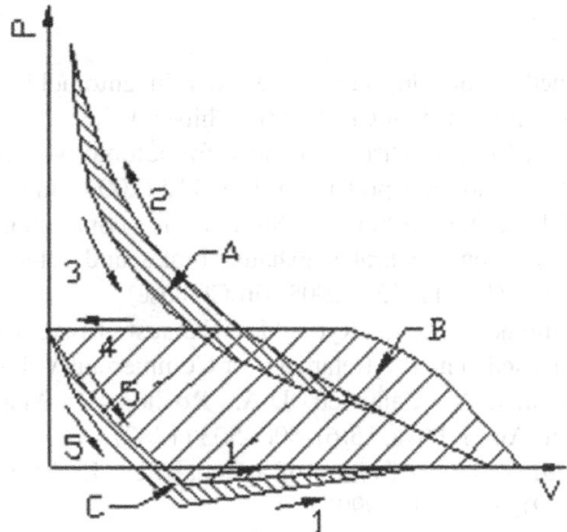

Fig. 4. P-V diagram of the diesel with supercharged and no turbocharged

We can see from the characteristics of the exhaust brake that exhaust brake energy is closely related to engine speed. At lower rotation speed the braking energy is smaller, and the higher speeds, greater braking energy, and this performance is in line with vehicle braking requirements, so in order to maximize the effect of exhaust brake some certain measures such as exhaust valve is mounted near the exit of the turbocharger shell to minimum compression volume in order to improve braking performance; meanwhile brake body of the exhaust valve can be added, and its effect is that the exhaust valve can be opened instantly when piston reaches at the end of the TDC at compression stroke , so that air pressure inside the cylinder declines quickly, and in next stroke compression air do not form a thrust piston, then energy absorption in compression stroke plays a role in the effective retarder, thereby the sum of both brake function can enhance the braking effect.

3 Conclusion

Followed conclusion can be obtained through PV diagram of diesel engine with exhaust gas turbo:

① There are greater braking force for exhaust gas turbocharged diesel than that no turbocharged diesel at the beginning, then brake torque start to reduce and has approximately equal to torque with on turbocharged diesel, and it is said that effect of turbocharger on exhaust brake is very little.

② Some certain measures can be taken to improve the braking performance such as the exhaust brake (butterfly valve) is mounted near the exit of the turbocharger shell to minimum compression volume, or the exhaust valve can be opened instantly when piston reaches at the end of the TDC at compression stroke.

References

1. He, J.: The design method of eddy current retarder in automobile and its performance analysis. Jiangsu University, Zhenjiang (2003) (in Chinese)
2. Ma, J., Chen, Y., Yu, Q., Guo, R.: Distance control for automotive's stopping with retarder. China Journal of Highway and Transport 16(1), 108–112 (2003) (in Chinese)
3. Yu, Q., Chen, Y., Ma, J., Guo, R., Zhang, Q.: Study of non continuous linear control system of combining action with engine brake, exhaust brake and retarder. China Journal of Highway and Transport 18(1), 117–121 (2005) (in Chinese)
4. von der Bie, H., Summerauer, J., Ryti, M.: Exhaust Brake-The viewpoint of the manufactures of engines and vehicles. Retarders for Commercial Vehicles (August 1980)
5. Wiederick, H.D., Gauthier, N., Campbell, D.A., Rochon, P.: Magnetic Brake: Simple Theory and Experiment. Am. J. Phys. 55(6), 500–503 (1987)
6. Lee, K., Paek, K.: Optimal Robust Control of a Contactless Brake System Using an Eddy current. Mechatronics (9), 615–631 (1999)
7. Loehner, K., Stabl, G.: Die Schleppleistung der Viertkt-Dieselmotor bei Talfahrt. ATZ (11) (1956)
8. Smith, H.B.: Function analysis on exhaust brake. Retarders for Commercial Vehicles (August 1980)

A Study on Classification Method of Discrete Data Basic on Improved Association Rules

YueQin Cao

Computer Department of Wenzhou Vocational and Technical College
WenzhouCity, Zhejiang Province China
416637928@qq.com

Abstract. A large amount of data is involved in the processing of discrete data. As there is a large quantity of redundancy discrete data in mass data, it causes the error of discrete data association and reduced classification effect. To solve this problem, we brought forward the classification method of discrete data to improve associated mining algorithm. Before establishing association rule, we performed secondary validation of information which may contain redundancy data to lower down the probability of error in discrete data association, thus the possibility of false classification of data, and remove the defect of traditional method. As proved by simulation experiment, this improved algorithm is able to largely raise the preciseness of discrete data association and achieve better effect.

Keywords: data classification, association rule, improved algorithm.

1 Introduction

With the explosive development of data capacity and processing technology in recent years, mass data becomes the effective carrier of various new-typed information and a great progress has been made in processing capacity. With increasing popularity of informatization worldwide, there is growing demand for data processing capacity of computer [1]. The world also raises higher demand for data processing, for example, rapid growth of applications, like websites, as well as required data processing scale and complexity between discrete data volume and information. Data classification encounters great challenge [2]. Current mainstream classification algorithms include association rule, cluster analysis and so on. As the demand of data process becomes widespread, more and more scholars participate in wider range of study on related data process algorithms and make it a hot science [3-4].

Facing the increasingly huge information volume to be processed, traditional treatments meet some problems in processing of discrete data. Firstly, more and more redundant data exists in the large volume of information. While this kind of data makes no contribution to classification, efforts should be made to perform effective identification to obtain contents of candidate set in large range. Traditional methods work by the feature of discrete data. However, the widespread existence of redundant data may weaken the uniqueness of data characteristic, which turns out that redundant data is combined in classification reducing the rationality of data classification and lessening its effectiveness. Therefore, the precise classification of mass data,

D. Jin and S. Lin (Eds.): Advances in CSIE, Vol. 2, AISC 169, pp. 159–164.
springerlink.com

especially the one with large volume of useless redundant discrete data, becomes a problem to be solved [5].

To clear the above defect, we brought forward the classification method basic on improved association mining algorithm. Before establishing association rule for large volume of discrete data, we should firstly make secondly validation of potential data containing large amount of redundant data to lower down the probability of error in association of discrete data and thus decrease the possibility of false classification and set up strong bedding for data association in later stage. Thc improved algorithm removes the defect of traditional method. Simulation experiment shows that this improvement can greatly increase the rate of inquiry and bring a good effect.

2 Principle of Classification of Discrete Data

An important premise of classification method basic on association rule is the establishment of precise original feature cluster. By judgment of original characteristics, we can enrich this original cluster with association rule and evaluate the association similarity of discrete data under the rule, and then complete the prior association of characteristics of association type. Algorithm of association rule compares to obtain new cluster result and thereby completes the partition of data type. The principle is as follow:

Firstly, supposing the effective type of data to be calculated in database is N, set up the model of data vector to be equipped with association rule, assuming $y=\{y_1,y_2,\ldots,y_n\}$. In this set, most elements are randomly selected without specific association. Randomness of data is great in the original data to be associated. If the randomness is represented by ω, we can have the following rule:

ω_k fluctuates between 0 and 1 representing that the compactness of data association is often related with the characteristic of data and the relevance of data varies with the connection between data. The maximum relevance of data is 1 and minimum value being 0; among different types, there may be data of greater similarity as well as more non-associated data. Optimization aims to make precise classification. Among the establishing methods of original association rule, it is a must to establish the relevance of data and then make the judgment of data dependency. The formula of dependency is shown as follow:

$$D(w\ ,\ e) = \sqrt{\sum_{k=2} \omega_k \,/\, e_n} \tag{1}$$

Thereinto, P is the quantitative evaluation of data to be associated by association rule; ωk represents the characteristic of dynamic association between data and is the variable related with characteristic; through formula 1, while meeting the premise of $1\leq p\leq\infty$, $\omega_k>0, 0<k<N$, we can figure out the relevance of discrete data and then perform data classification.

As known from the principle of algorithm, traditional classification algorithm of data association rule is basic on the characteristics of data. With the increasing volume of data, there is a large amount of redundant data in mass data, which is somewhat similar to the characteristic of classification, as a result, the ω value of formula 1 may deviate and the result, which is gained by W, is mixed with great amount of redundant data. For that, the result of formula 1 may have a great deviation and the association rule may end up as a fault in later stage. The effect of classification is lowered down. Therefore, sorting of redundant discrete data becomes the challenge of research.

To solve this problem, we brought forward the discrete data classification method to improve association mining algorithm and then verified the classified data by data verification method to decrease the possibility of false data classification. Simulation experiment shows that this improved algorithm can greatly accelerate the access to data and achieve a very good effect.

3 Classification Method of Discrete Data Basic on Improved Association Rule

As known from traditional classification method of discrete data, the processing of discrete data is greatly affected by large volume of redundant data. When there is excessive redundant data of same characteristics in the large volume of discrete data, it may cause large amount redundant data in discrete data during characteristic collection. As traditional algorithm doesn't have a mechanism to detect redundant data, once redundant data is excessive, the effect of classification of discrete data is weakened. We will describe the principle of improved classification algorithm of discrete data:

3.1 Collection of Characteristics of Discrete Data

To make a classification, we need to have effective detection of data feature. The principle of discrete data detection is to complete data setting and characteristic collection by related calculation in order to stabilize the fluctuation of data feature within a certain scale.

1) Firstly, we need to collect data characteristics to obtain the feature of discrete data. The method is as follow:

$$t = \sum_{i=1}^{n} t_i / n \tag{2}$$

t_i represents a single-item characteristic of data; n is the number of characteristics; t is the final result of characteristic of discrete data.

$$\nabla e(t) = t - t_0 \tag{3}$$

2) Perform characteristic regulation for collected feature of discrete data, t. The deviation of data characteristic $\nabla e(t)$; $\nabla e(t) = t - t_0$

3) For collected characteristics, we can make fine adjustment according to characteristic derivation $\nabla e(t)$. The method is as follow: $u(t) = k_p \nabla e(t) + \beta k_i \sum_{j=0}^{k} \nabla e(t) T$ $+ k_d (\nabla e(t) - \nabla e(t-1)) / T$

Thereinto, k_p is the scale factor of transverse characteristic adjustment; k_i being coefficient of quantitative adjustment; k_d the scale factor of longitudinal characteristic adjustment. These 3 parameters can dynamically modify the collected characteristic u (*t*) of discrete data. By parameter adjustment, we can make it fall in a precise interval.

3.2 Verification Calculation of Redundant Data

After we obtain the characteristic of discrete data, we need to remove the redundant data through an efficient data verification mechanism to guarantee the correctness of data classification. Though redundant data is with the similar characteristic of discrete data, the real discrete data to be classified is with certain homogeneity. Repeated comparison of data works better to discover redundant data.

Firstly, we assume a verification data area, $\Delta A_k B_k C_K$. A large number of discrete data points spread around this area. The characteristic points of the discrete data can be described by Q. This characteristic point of discrete data contains the average weighting value of discrete characteristic. Verification area can be described by the following formula: $(\underline{P}-\underline{A})\bullet \underline{n}=0$

Thereinto, *P* is a random characteristic vector in verification area of discrete data. It meets the following equation in the 3D space: $P=\underline{P}=x\underline{i}+y\underline{j}+z\underline{k}$

$\underline{A}=x_A\underline{i}+y_A\underline{j}+z_A\underline{k}$ is the vector expression of random discrete data A; $\underline{n}=a\underline{i}+b\underline{j}+c\underline{k}$ is the vector expression of data verification area. Through the above formula, we can have: $a(x-x_A)+b(y-y_A)+c(z-z_A)=0$

By applying the 2D characteristic of data characteristic point *F* in the area and another discrete data point *D*, we can have the linear equation for data verification as follow:

$$\begin{cases} \underline{r}=l\underline{i}+m\underline{j}+n\underline{k} \\ l=x_D-x_F, m=y_D-y_F \end{cases} \tag{4}$$

To facilitate the operation, it can be turned to parameter method: $x=lt+x_F$, $y=mt+y_F$

In the above formula, parameter t represents the coefficient of variation of characteristic which tends to be empirical coefficient. The calculation is as follow:

$$t_Q=-\frac{p}{s} \qquad s=al+bm+cn\text{。}$$

After calculation, we have the expression of a 2D coordinate of a new characteristic point.

$$x_Q=lt_Q+x_F, y_Q=mt_Q+y_F \tag{5}$$

In the result of verification, we can have the expression method (x,y) of 2D coordinate of redundant data. After obtaining the expression method of 2D characteristic of redundant data, we need to verify it again to confirm the accuracy of result. Through 2D plane, we can confirm whether it fits the request of deletion. Supposing that $\triangle A_kB_kC_K$ contains the standard characteristic of data, so that the judgment of redundant data is turned into the matter of verifying the inside of area.

4 Experimental Result and Analysis

To validate the effectiveness of algorithm mentioned in this paper, we carried out contrast experiment to complete the validation of algorithm. The flow diagram of experiment is as follow:

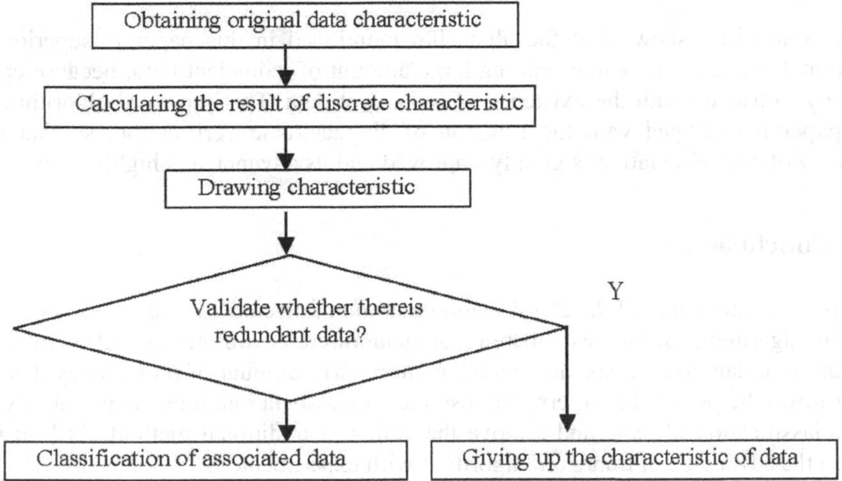

Fig. 1. Test Flow

We classified the 200,000 data in database with traditional and improved algorithm. To construct experimental environment, a large amount of redundant data of same characteristics was added to database. The trend of results is shown in fig 2.

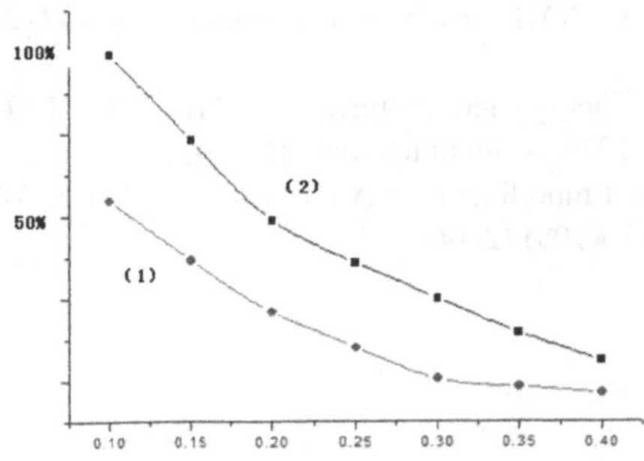

Fig. 2. Contrast Trend of Classification Veracity of Different Algorithms

Fig 2 shows that X axis represents the magnitude of data; Y axis represents the accuracy of classification; 1# curve is the trend of result of traditional operation; 2# curve is the trend of algorithm mentioned in this paper. Through comparison, we can see that improved algorithm is superior to traditional one.

Statistical result is shown in table 1

Table 1. Experimental Data

	Data Magnitude	Accuracy
Traditional Algorithm	0.4	67%
Algorithm in this paper	0.4	79%

A comparison shows that the algorithm mentioned in this paper is superior to traditional one, especially after adding large amount of redundant data, because error is likely to happen with the existence of redundant data. The improved algorithm of this paper is equipped with the function of characteristic verification, so that the accuracy of data association is greatly improved and its advantage is highlighted.

5 Conclusion

This paper brings forward the classification method of discrete to improve association mining algorithm. Before establishing association rule, we performed secondary validation of data to be associated which contain large amount of redundancy data to lower down the probability of error in discrete data association, thus the possibility of false classification of data, and remove the defect of traditional method. At last, we proved the advantage of improved algorithm with experiment.

References

[1] Bray, T., Paoli, J., Sperberg-McQueen, C.M.: Extensible Markup Language(XML)1.0, 2nd edn., pp. 21–24. W3C Recommendation (October 2000)

[2] Abiteboul, S., Bnueman, P., Suciu, D.: Data on the Web, pp. 89–93. Morgan Kaufmann Publishers (2000)

[3] Xuan, W.C., Ping, L.X.: XML Database Technology, pp. 137–200. Tsinghua University Press, Beijing (2008)

[4] Zhong, S.Y.: XML Theory and Application Basic, pp. 130–134. Press of Beijing University of Posts and Telecommunications, Beijing (2000)

[5] Wen, T.J.: Design and Embodiment of A Kind of Distributive Web Log Mining System. Computer Simulation 10(109) (2006)

The Research of Collaborative Learning Based on Network Environment

DaoJun Tan

The Department of Information Technology and Education, Hunan University of Science and Engeering, Yongzhou, Hunan, China
tandaojun2012@126.com

Abstract. Collaborative learning based on network environment builds collaborative learning environment by using computer networks and multimedia technology. At first place, this paper introduced the present situation of the research. Then, this paper analyzed the advantages and characteristics of collaborative learning based on network environment. In the middle of this paper, the describe of collaborative learning based on network environment including :the design of learning objectives, the design of learning resource, the design of collaboration tools, the creation of learning community, the design of learning evaluation. The collaborative learning based on network environment requires careful organization. Each link has to be carefully weighed and system planning. Of course, this research need more practice to test and developing.

Keywords: collaborative learning, network, research.

1 Introduction

With the development of the network technology, learning style has changed a lot. The new methods and means of learning are appearing constantly. Collaborative learning based on network environment is a new learning method. Collaborative learning is cooperation and mutual assistance of all relevant acts. Learners learn the knowledge as a team. They have the same learning goals. In order to achieve the goals, they formed a team. They can maximize the learning effect by cooperation. The power of individuals is limited. The learning effect is better by cooperation as a team.

With the rapid development of computer-mediated communication and Internet technology, collaborative learning has broken the traditional boundaries of classroom walls, and the network environment, collaborative learning, began to be widely used. Collaborative learning based on network environment builds collaborative learning environment by using computer networks and multimedia technology.

2 The Advantages and Characteristics of Collaborative Learning Based on Network Environment

2.1 The Advantages of Collaborative Learning Based on Network Environment

Breakthrough time and space constraints. Teachers and students are no longer subject to the classroom and geographical constraints, collaboration can range expansion

D. Jin and S. Lin (Eds.): Advances in CSIE, Vol. 2, AISC 169, pp. 165–170.
springerlink.com

from their own class to class, grade even collaboration between the schools. Partners may also be learners of another region or another country. In this way, the various forms of interaction are likely in this collaborative circle, so that students get to varying degrees, a variety of different forms of experience.

Online collaboration platform can effectively control the collaborative learning process. Collaborative learning based on network environment is to build collaboration platform set up by computer technology, collaborative activities and participants can not be divorced from the collaboration platform to achieve the ultimate learning goals. Therefore, the controllability of the collaborative process has been a strong guarantee [1].

The web collaboration platform is facilitating the collaboration of student communication. The collaborative learning based on network environment can take advantage of online collaboration platform, or a variety of exchange communication tools, such as the QQ, E-mail, video conferencing systems, etc., to overcome the communication barriers between the study partners. Due to the instant network messaging, the speed of collaboration and communication between students is very fast.

2.2 The Characteristics of Collaborative Learning Based on Network Environment

Compared to the collaborative learning in traditional classroom teaching, the collaborative learning based on network environment has the following characteristics:

Access to learning resources is diversified. To some extent, as the carrier of the transmission of information, network is a large resource library. In the collaborative learning based on network environment, students can query through a search engine to the information he needs, they can also send E-mail to the teacher or the relevant experts to obtain information and knowledge they need. With dynamic web technology gradually moving towards maturity, the interactive features of the network information is also more powerful. Learners can get a lot of learning resources from the network, and the form of learning resources is becoming more and more vivid, emotional.

Student grouping method is more flexible and diverse. In collaborative learning, the grouping of students is a flexible mechanism may include the individual units, for group as a unit and all the activities. This grouping can change. You must use the different groups in order to improve the learners' learning efficiency especially in the different periods of learning. With the aid of computer network, the grouping in collaborative learning is more convenient, flexible, and diverse. Students not only can be combined freely within the class group, but also can communicate and collaborate with other learners on the network when it necessary.

A wide range of ability of students can be improved. It can develop students 'information skills, learning ability and social skills, help students adjust to campus learning outside the classroom teaching and distance learning, and help develop students' lifelong learning abilities. Teacher-led role can be into full play. In collaborative learning by means of computer networks, the role of the teacher is the wizard and monitoring, and students can be fully autonomous learning, students' learning attitude is dynamic [2].

The interaction between the learners is better. The communication efficiency is improved. Through the network, the learners can achieve a broader and higher quality of collaborative learning. It is helpful to communicate the information effectively between the learners.

3 The Learning Objectives

An important principle of collaborative learning in the network environment is to have appropriate learning goal orientation to ensure that the "lively" collaborative learning activities are not only formal, but able to achieve substantial learning effect. To this end, we must ensure that learners engage in meaningful problem solving activities, namely to solve the problem with understanding. It has the appropriate learning goal orientation. It collaborative learning activities with their domain knowledge behind (such as basic principles, concepts, methods, etc.), prompting the learner to focus sexual, reflective inquiry, guide students to the understanding of the basic relations, rather than just blindly try. We must be combined with the discipline-specific content when the analysis and design of learning objectives.

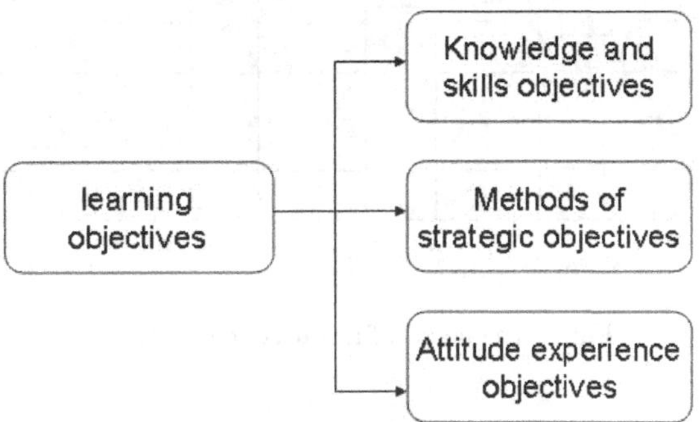

Fig. 1. The learning objectives

Knowledge and skills objectives: we hope that learners can be integrated application of what knowledge points, or intellectual framework, or to find out what new knowledge or the development of what new skills.

Methods of strategic objectives: we hope that learners able to develop what problem-solving strategies, learning strategies, cooperation, communication strategies and information literacy.

Attitude experience objectives: we hope that learners get what attitude of emotional experience through collaborative learning activities. These objectives are the banner of the guidelines for the design of collaborative learning activities.

4 The Learning Resource

In order to solve the problem, students need to master new knowledge and additional information. To this end, it must be through the network to build a diverse library of knowledge resources and related cases. Students search and locate information in accordance with their own list of issues. So that students can learn strategies for solving problems in broad learning resources [3]. In the design of resources should follow the following principles:

The amount of information is adequate, and is closely related with the learning content.

Resource structure should have a certain breadth, depth, and hierarchy in order to meet the needs of students at different levels.

Resources manifestations (text, graphics, animation, video, case, etc.) should be diversified, easy to search and find.

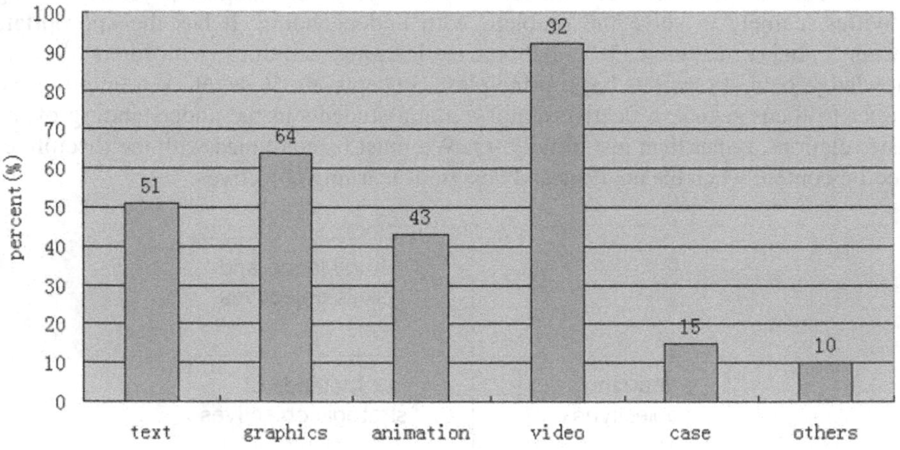

Fig. 2. The needs of resource in the class

5 The Research of Collaboration Tools

In the problem-solving process, team members, teems, students and teachers need to discuss and communicate the information of problem. Therefore, need to design a collaborative communication space, and collaboration tools in this space include the following aspects:

Communication tool: This is an essential tool for cooperation and exchange between student and student, students and teachers, as well as teachers and teachers in the network environment. Through synchronous and asynchronous communication, the team members can communicate timely for problem solving suggestions and share the distribution of resources and awareness. Commonly used network communication tools have: QQ, UC, MSN, ICQ, Instant Messenger, E-mail, Internet telephony and BBS forums.

Upload tool: Used to upload the student's observational data, the researchers report form, the mutual evaluation between the groups and to develop solutions, and many more in the form of the E-mail and FTP transfer.

Expert advice: According to each learning activities and learning tasks, we should hire experts in the field as a consultant. Students can directly ask the relevant questions to the experts or regularly organize the problems and the answers placed on the website for users to browse and learn.

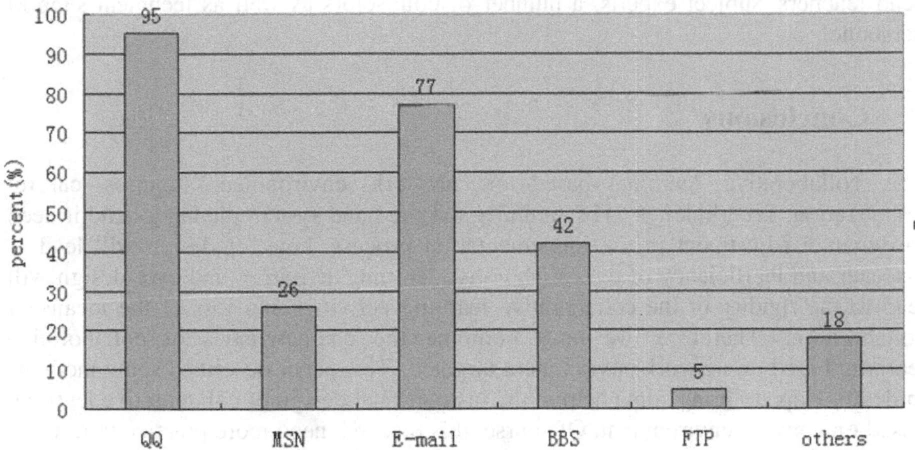

Fig. 3. The choice of collaboration tools in the class

6 The Learning Community

Collaborative learning is going in a Learning Community. Teaching organizers should take effective measures to create web-based active learning community. The so-called learning community is a group constituted by the learner and students (including teachers, specialists, counselors, etc.). They often communicate each other in the process of learning, they share the learning resource, and they complete a same learning mission. So they can influence and promote each other. Learning communities based on the network are often not just a class, but the joint multiple classes of different schools and stakeholders to participate [4]. These classes may come from different regions or even different countries, geographical and cultural differences are often better reflecting the significance of collaborative interaction.

In order to create the learning community, we need to do something as flowing:

Determine the range of the Learning Community. We should consider to organize which students in where and invite which experts or practitioners as a consultant.

Enhance the sense of community. We should make learners aware of that they are learning in a group, but also to feel the value and significance of the groups. In the beginning of learning, allows each learner to write a self-introduction, to all members of the mailing list. Each learner can also create a personal home page, and attach a photo and personal information. Moreover, in the case of possible can also arrange some face-to-face activities; enhance the realism of the exchange. In addition, to encourage mutual assistance and to assist learners in the learning process, not just the issue sent to the teacher, so that it enables the learners to feel the value of the Community.

Determine the organization contact. According to the characteristics of learning tasks and learners to choose a certain organizational contact, for example, group cooperation, and each team to arrange a leader, at the site level arrangements subject head teachers, subject experts, a number of counselors as well as technical support personnel.

7 Conclusions

The collaborative learning based on network environment requires careful organization. Each link has to be carefully weighed and system planning. And it need to dynamic adjustment in the implementation process. Lack of design will lead to disorder and inefficiency of the collaborative learning activities, and over-design will lead to the rigidity of the collaborative learning activities and loss of the means of collaboration. Therefore, we must combine the characteristics of collaborative learning based on network environment targeted. This paper described some methods in detail. Hope to bring some help to the instructional design of collaborative learning based on network environment. Of course, this research need more practice to test and developing.

References

1. Sakawa, M., Kato, K.: Cooling load prediction in a district heating and cooling system through simplified robust filter and multilayered neural network. Applied Artificial Intelligence (15), 633 (2001)
2. Al-Tabtabai, H.: A framework for developing an expert analysis and forecasting system for construction projects. Expert System With Application (14), 259 (1998)
3. De Raedt, L., Kersting, K.: Probabilistic logic learning [EB/OL] (2002), http://www.acm.org/sigs/sigkdd/explorations
4. Venkata, R.: Taxonomies, categorization and organizational agility[EB/OL] (2002), http://www.kmworld.com/publications/whitepapers/kmworldpaper02.pdf
5. Turban, E., Aronson, J.E.: Decision Support Systems and Intelligent Systems, pp. 754–756. Tsinghua University Press, BeiJing (2000)
6. Trifonova, Knapp, J., Ronchetti, M.: E-learning versus M-learning: Experiences, a prototype and first experimental results. In: Proc. ED-Media 2005, Montreal, Canada, June 27-July 2, pp. 4751–4758 (2005)

The Investment Decision-Making Index System and the Grey Comprehensive Evaluation Method under Hybrid Cloud

Donglin Chen[1], Min Fu[1], Xuerong Jiang[2], and Dawei Song[1]

[1] School of Economics, WHUT, Wuhan, China
[2] School of Management, WHU, Wuhan, China

Abstract. Aiming at the problems existing in the current study on the uncertain evaluation and neglected quantitative analysis when the enterprises making investment decisions under hybrid cloud environment, this paper concludes five indicators that impact their decisions. Then, it introduces the Grey Comprehensive Evaluation System and organizes experts in grading, so that the qualitative indicators will be more quantified and the analysis more intuitive. Finally select the most suitable program when making investment decisions.

Keywords: hybrid cloud, evaluation index, grey comprehensive evaluation method, investment decision-making.

1 Introduction

As a new IT delivery model at this stage and with its advantage of low cost, large scale and virtualization, cloud computing has become the fifth utility, together with water, electricity, gas and telecommunications [1]. According to its operational mode, Cloud computing can be divided into three types: Public Cloud, Private Cloud and Hybrid Cloud. While most customers do not want to migrate the system to all business on the public cloud because of the lack of security in public cloud [2], so that the " hybrid cloud "model which contains part of "private cloud" and part of "public cloud "has become the ideal solution for customers. Hybrid clouds combine the advantage of both public and private clouds, so that enterprises can not only retain their private cloud core systems, but also some of the services.

To make IT investment decisions is the way to describe the process of identification and solution of environmental choice behavior of the decision-makers in the process of investment. At this stage, the research of hybrid cloud is mainly based on systems and technology, such as Lifeng using random keys genetic algorithm to solve the problem of the lowest-cost mix of customers under the cloud resource allocation and so on. But in the current study under hybrid cloud there weren't an index system which affects the integrity of enterprises to make investment decisions, or the current impacts about the service indicators and the relative lack of data have caused the existence of some difficulties. However, in contrast to the qualitative indicators, quantitative indicators are more intuitive and convincing. To address the issue, this paper summarizes five indicators that impact the IT investment decisions, and introduces the Grey Evaluation System to comprehensively evaluated related index, presenting a quantitative model of these qualitative indicators.

D. Jin and S. Lin (Eds.): Advances in CSIE, Vol. 2, AISC 169, pp. 171–176.
springerlink.com

2 Combination Method of Investment Decision-Making Index under Hybrid Cloud

Under hybrid cloud, the factors that affect enterprises in investment decision-making include cost and Cloud QoS (quality of service). Costs mainly represent the cost of cloud computing and cloud storage. And the latter includes data storage, retrieval and transmission costs, while CQoS including execution time, security, availability, reliability and other factors.

Taking example by the computing model of grid and web services, we can define the computing method of cost and CQoS.

1) Define the expression of total cost of customers under hybrid cloud while using Cc to show the cost of cloud computing and Cs to show the cost of cloud storage.

$$\mathrm{TCO} = \mathrm{Cc} + \mathrm{Cs} \tag{1}$$

2) Define the expression of cost of cloud storage while using CS1 to show the cost of data storage and Ct to show the cost of Cloud transmission.

$$\mathrm{Cs} = \mathrm{Cs1} + \mathrm{Ct} + \mathrm{Cs2} \tag{2}$$

3) The total cost can be also defined as :

$$\mathrm{TCO} = \mathrm{Cc} + \mathrm{Cs1} + \mathrm{Ct} + \mathrm{Cs2} \tag{3}$$

3 The Investment Decision-Making Index System of Corporations under Hybrid Cloud

Under hybrid cloud, a combination of customer-oriented quality of service HCQoS largely depends on the security, availability, reliability and execution time of service platform.

1) Security. Since their data are stored in the cloud, the control of customer's data totally depends on their suppliers. Safety has always been the key indicator of customers to make investment decisions.

2) Availability. It mainly refers to customers' accession to cloud service in a certain time and under certain conditions.

3) Reliability. It means the capabilities for corporations to provide customers with stable and accurate service.

4) Execution time. It also means waiting time, referring to the time required for the system from receiving commands to finishing data analysis.

Therefore, based on the characteristics of enterprises in investment decision-making under hybrid cloud environment, the index system is designed to include cost and quality of service, while quality of service also includes security, availability, and execution time.

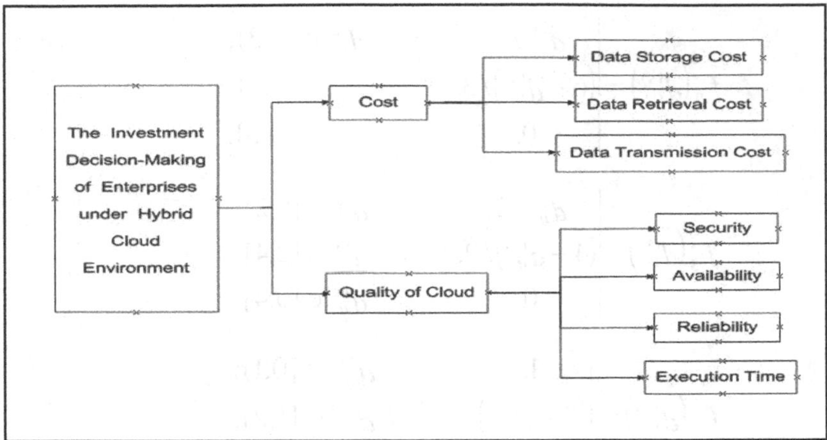

Fig. 1. Investment Decision-making system of Enterprises under Hybrid Cloud

4 Index Analysis of Investment Decisions Based on Gray Evaluation Method

Four representative options are selected as evaluation objects, and the gray evaluation method is used to rate special steps as follows:

Case I: rent all IaaS and SaaS public cloud.

Case II: rent all IaaS public cloud, including virtual servers and storage, build private SaaS.

Case III: rent virtual servers of IaaS public cloud; build the part of storage of IaaS cloud and SaaS.

Case IV: build all IaaS and SaaS cloud.

(Considering the depreciation of hardware and software elements, all the indicators in the four cases are observed in 5 years.)

4.1 Build Evaluation Matrix

Evaluation of the factors of the grade is set at "highest, high, medium and low", assigned to the scores of four, three, two and one, and the scores of 3.5, 2.5, and 1.5 are assigned between them. Then organize experts to score the factors in each case, constructing an evaluation sample matrix. Assume that the number of the gray class is e (e=1,2,…,g). Since there are four evaluation gray class in this case, e=1, 2, 3, 4.The corresponding four whitening functions are:

$$f_1\left(d_{ij}^{(s)}\right)=\begin{cases} d_{ij}^{(s)}/4, & d_{ij}^{(s)}\in[0,4); \\ 1, & d_{ij}^{(s)}\in[4,\infty); \\ 0, & d_{ij}^{(s)}\notin[0,\infty). \end{cases} \tag{4}$$

$$f_2\left(d_{ij}^{(s)}\right)=\begin{cases} d_{ij}^{(s)}/3, & d_{ij}^{(s)}\in[0,3); \\ \left(6-d_{ij}^{(s)}\right)/3, & d_{ij}^{(s)}\in[3,6]; \\ 0, & d_{ij}^{(s)}\notin[0,6]. \end{cases} \tag{5}$$

$$f_3\left(d_{ij}^{(s)}\right)=\begin{cases} d_{ij}^{(s)}/2, & d_{ij}^{(s)}\in[0,2); \\ \left(4-d_{ij}^{(s)}\right)/2, & d_{ij}^{(s)}\in[2,4]; \\ 0, & d_{ij}^{(s)}\notin[0,4]. \end{cases} \tag{6}$$

$$f_4\left(d_{ij}^{(s)}\right)=\begin{cases} 1, & d_{ij}^{(s)}\in[0,1); \\ \left(2-d_{ij}^{(s)}\right), & d_{ij}^{(s)}\in[1,2]; \\ 0, & d_{ij}^{(s)}\notin[0,2]. \end{cases} \tag{7}$$

If Vie represents the eth gray class of the factor Vi, the evaluation-weight of the eth gray class of the Vi evaluation index in case S is:

$$r_{ie}^{(s)}=\frac{Vie}{Vi} \tag{8}$$

The gray evaluation matrix in case S is:

$$R^{(s)}=\left(r_{ie}^{(s)}\right)_{p\times4} \tag{9}$$

Table 1. The table of evaluation quantization in investment program

	Cost(ten thousand yuan)	Security	Availability	Reliability	Execution time
Case I	1000	0.4	0.39	0.4	0.4
Case II	1500	0.52	0.49	0.49	0.5
Case III	2000	0.6	0.6	0.6	0.6
Case IV	4000	0.8	0.8	0.8	0.8

Then the weight vector is calculated as: λ= (0.30, 0.20, 0.10, 0.25, 0.15).

4.2 Determine the Gray Correlation

Select the relatively optimal value of every Vi in the evaluation matrix, then compose the optimal reference vector. The worst reference vector is selected in the same way. Finally initialize the matrix composed by these two vectors. Define

$$\xi_i(k)=\frac{\min\min\left|f_g(k)-f_i(k)\right|+\rho\max\max\left|f_g(k)-f_i(k)\right|}{\left|f_g(k)-f_i(k)\right|+\rho\max\max\left|f_g(k)-f_i(k)\right|} \tag{10}$$

Then calculate the gray relational grade vector.

In this case, the gray correlation factor related to the optimal reference vector is:

$$\gamma(x, x_b) = (0.568, 0.4487, 0.581, 0.7061)$$

The gray correlation factor related to the worst reference vector is:

$$\gamma(x, x_b) = (0.568, 0.4487, 0.581, 0.7061)$$

4.3 Comprehensive Evaluation of the Four Cases

Make an order according to the two gray correlation factors

$$v_i = \frac{1}{\left[1 + \gamma_{ib} / \gamma_{ig}\right]^2} \tag{11}$$

The gray comprehensive evaluation value of the four cases are: V=(0.5124,0.4384,0.3043,0.1408), the maximum value of vi is in Case I, so the best investment program is case I.

5 Conclusion

This paper firstly summarized the indicators of enterprises to make investment decisions under hybrid cloud environment. Then evaluate each indicator by the gray evaluation method, taking full advantage of the ambiguity of experts to quantify these qualitative indicators. Finally prove the method's practically in actual project.

Acknowledgement. This research was supported by the National Natural Science Fund of China under Grant71172043, 71072077 and 70972094; the National Technology R&D Program 2011BAH16B02, the Fundamental Research Funds for the Central Universities and Teaching and research projects of WHUT.

References

1. Buyya, R., Yeo, C.S.: Cloud computing and emerging IT platforms: Vision, hype, and reality for delivering computing as the 5th utility. Future Generation Computer Systems (25), 599–616 (2009)
2. Qi, Z., Lu, C.: Cloud Computing: State-of-the-Art and Research Challenges. Journal Internet Services and Application (1), 7–18 (2010)
3. Bi, X., Zhang, H.: Study on the Influencing Factors and Model of IT Investment Decision-making Behavior. Journal of Intelligence (3), 5–9 (2007)
4. Li, X., Yang, H., Qiu, N.: On the Theory of Semi-structured Decision-making Fuzzy Optimization Method of Investment Program. Statistics and Decision (14), 21–22 (2010)

5. Dong, X., Lv, T.: View of the Cloud Computing and Its Future Development. Journal of Beijing University of Posts and Telecommunications (Social Sciences Edition) (5), 76–81 (2010)
6. Wei, G.: Grey relational analysis model for dynamic hybrid multiple attribute decision making. Knowledge-Based Systems (24), 672–679 (2011)
7. Kuo, Y., Yang, T., Huang, G.-W.: The use of grey relation analysis in solving multiple attribute decision-making problems. Computer & Industrial Engineering (55), 80–93 (2008)
8. Zhang, W.-G., Mei, Q., Lu, Q., Xiao, W.-L.: Evaluating methods of investment project and optimizing models of portfolio selection in fuzzy uncertainty. Computer & Industrial Engineering (61), 721–728 (2011)
9. Zhao, T., Han, K., Yang, Z.: Risk Evaluation of Enterprise Resource Planning Projects Using Multilevel Grey Evaluation Method. Industrial Engineering Journal (5), 101–105 (2008)
10. Xu, Y., Tang, W., Wu, B.: The Design Principles and Application of Technology Evaluation Index System. China Soft Science Magazine (2), 48–51 (2000)

A Novel Convergence DRM System Based on a Trusted Third Service Center

Wenxue Jia, Yang Song, and Yanhui Xia

Shijiazhuang Tiedao University, 050043 Shijiazhuang, China
Songyang1981@stdu.edu.cn

Abstract. With the development of information technology, digital works such as software, video, music etc. have been the hottest resources in cyberspace. But the values conflicts severely between the intellectual property protection and public access to the knowledge. This paper tries to use a new improved system to manage the digital copyrights with a convergence service center. With the new system we can achieve the purpose that license to the clients automatically, meanwhile the copyrights owner will get the adequate economic interests, too.

Keywords: Digital copyrights protection, Convergence managements, Center of services.

1 Introduction

21st century is the period of knowledge boom, and the internet is the most active economic growth point. Profit from the character of the internet, the mode of the works transmission has been revolutionarily changed. Paper has been the transmission media for more than 20 centuries, in these centuries, the content and the tool of works is always at peace. The copyrights owners and the disseminators didn't bother each other. But the technology changed everything, digital disk substituted the paper as the most important media in a certain extent, this situation ends the honeymoon between the content and tools. The dispute occurs repeatedly because the tacit privities which deemed as constitution have been broken.

Firstly, from the theory aspect, the traditional mode of license was criticized by many reputed scholars. For example, profess Samuelson [1], the winner of Nobel Prize in economics, argued that the copyrights act 1976 of United States has been outdated severely. And another expert of copyright, Littman [2] holds the opinion it is unfair that the Copyrights Act made too much comprise to the enterprises which are copyrights owner. If set the limitation to the Individual users, the copyrights will become the obstacle of the knowledge communication. This phenomenon deviates from the original intention of the copyright system. Viewing the individual download as commercial privacy is extremely unacceptable. For this reason the new copyrights system should attach importance to the custom and practical need, instead of waving the club at the public communication. And she also hold the idea that the allowance of

D. Jin and S. Lin (Eds.): Advances in CSIE, Vol. 2, AISC 169, pp. 177–183.
springerlink.com

personal digital copying behavior's benefits may outweigh the harm. If we have to regulate the harm, he also proposes to levy the special fee as well.

From the technical aspect, Kalker Ton et al. [3] proposed that use of Digital Rights Management (DRM) technologies for the enforcement of digital media usage models is currently subject of a heated debate. Content creators, owners and distributors argue that DRM technologies are needed to protect their Intellectual Property (IP) from unauthorized access.

Zhiyong Zhang et al. [4] insisted that effective usage control technologies can guarantee that end consumers are able to legally access, transfer, and share copyrighted contents and corresponding digital rights. From the technical and managerial perspectives, they give a wide survey on state-of-the-art of Digital Rights Management (DRM) system, and start with a generic DRM ecosystem that effectively supports two typical application scenarios, and the ecosystem builds multi-stakeholder trust and maximizes risk management opportunities.

Another defect of DRM is lack a role which acts as a trusted third party (TTP). Song Yang etc. (2012) provided an improved certification contract protocol with a convergence manage method. A TTP based on national government creditability was proposed to conquer the illegally abuse the private rights [5]. Only with the application convergence management of TTP can the powerful solution for meeting the diverse needs of copyrights owners and clients.

The rest of the paper is organized as follows. The second section we will discuss the value which the improved DRM of convergence management mode system is oriented. The third section is description the architecture and the certification algorithm. Finally, we end this paper with a conclusion and the future work.

2 A Fundamental Analysis towards the System of DRM

Although the DRM is a technical convergence manage measure, but its object is the rights set by law. So we have to discuss the digital copyrights in the cyberspace in the context of law and society before studying the DRM. Only by this means can we lay the ground criteria and basic concepts for the technology.

2.1 The Diverse Values of the Digital Copyrights in the Cyberspace

The copyright is a kind of private right, in other words, the copyrights belong to the individual persons. But from another angle, copyrights have the public character as well. Essentially speaking, the object of copyrights is knowledge. And the value of knowledge is for communication. In this sense, the privacy and public transmission conflicts with each other. And this contradiction become more severely after come into being of internet. The transnational and intangible character of the internet crippled the copyrights owners' control capability.

We should also attach the importance to the public right for access the knowledge and fair use the works. In Germany and France, the government view the knowledge accessible for public is same to the private copyrights which can be classified as one of innate human rights. With the technology of Web 2.0, the internet has exceeded the means of information media in a certain extent; it emphasizes the interactivity among the users. The users can not only explore the internet but also construct the web content. For example, the well known game soft company Blizzard developed a game which is named after War craft III. This game is released the source code by the company, soon a lot new mod games based on the original version were programmed by the amateur game team. Among them, there are listed the outstanding representatives such as DOTA. These MOD versions enhanced the game's life unquestionably.

In addition, if we protect the internet digital copyrights according to the traditional standard strictly, the great inconvenience will set back the development of internet and life experience online. The reason is obviously, according to the traditional concept of copy, which should be authorized. In this sense, the "illegal copy" is happening in the computer's cache all the time. Moreover, ordinary Internet users use the works just for personal purpose. This kind of use occurs too frequently and universally, so as to be impossible to get the permission prior to the use. Even the provide as this, the compliance cannot be reached as well. Obviously the "market ineffectiveness" prevails in the digital copyrights circumstance. " Free ride" and" external effects" results in production number of copyrighted works may be less than the social total demand, the market allocation of resources cannot get effective configuration. And the traditional negotiation -- licensing models also have to face the" Coase Theorem in transaction cost". In real life, the transactions need cost, sometimes the cost is expensive. Whether the interests of all parties through private negotiations can solve the problem of externalities depends on the transaction cost quantity.

Obviously, only after the transactions being made, the new added value exceeds the exchange cost, can we say the trade is meaningful. If the exchange gains equal to or less than the transaction cost transaction cost, transaction activity will be abandoned. Because of the existence of transaction costs, private negotiations often fail to solve external effects problem heroin. When the parties with different interests are too many, the effective protocol is more difficult to achieve, because the coordination of each person's is too difficult. Because of work can not meet the society demands, even if the encryption technologies was made by copyright owner by the self-help means, but these techniques also is facing the inevitable fate of crack by more and more sophisticated hackers. Out of market, can the legal measure such as law suit solve the problem? The answer is no. Reason is that the infringing parties are too many. It is impossible to sue every consumer using internet. So the negotiation-license mode cannot fulfill the need of avoiding free riding by individual copy online.

The jurisprudence which appeared in Germany as early as in the last century introduced the system of copyright compensation. In 2007, German Senate passed the copyright law of the "Second part of standardizing the information society ". The law reform include: private copying in digital form, including the network environment will statutory license to the others by law. And after then a convergence packed compensation was paid to the copyrights owners. Under the new law, which bears the payment obligations typically provide replication services to download resources, and manufacturing storage medium enterprises. The law obviates the principle of voluntary licensing out of networked digital media creatively, instead a system of compensation. This kind of system seems to be setting limitations on authors' rights. However, due to the protection of collective management, copyright by the author of the most important value - the economic benefits, compared to only the exclusive right can not control the efficiency is much favorable.

2.2 The Mechanic of Convergence DRM and Its International Application

Early in the last century, the EU proposed to build DRM technology systems, put the digital copyright into a convergence management model. And the European Union, through its powerful transnational legislative competence, unifies the members' domestic law system, such as Copyright Directive in information Society, European Parliament and of the council on certain legal Aspects of Information Society, in particular Electronic Services etc.

In 1992, Japan established a compensation system for digital copy. It requests the media producer and the equipment manufacturer pay some money to the owner of copyrights. The amount of the compensation is determined through consultation and authorized by official.

In summary, the Western developed countries has initially formed the convergence management mode in the management of copyright, and try to coordinate the contradiction between the public knowledge communication and the copyright protection as a private right through the compensation system. The gap is significant between China and the developed, Chinese copyright system is not perfect, social NGO sector development is not perfect, too. With the rapid development in the digital network environment, and promote the establishment of public administration vigorously. The need of construction of service system in order to promote copyright market transactions and the requirements of the development of copyright transaction in copyrights and intellectual industries is imminent.

3 Architecture Design for the Convergence DRM System

According to the principle which has discussed, the first step is to build a neutral convergence copyrights management service facility as the trust third party (TTP). We call the facility as convergence digital copyrights management center (CDCMC). The law should grant the service center the jurisdiction for managing the digital copyrights. And the service center should be imparted the national public credit. The certification, automatic license and compensation should be carried out by the directing of the service center.

A. The copyrights owner establish connection with sever of CDCMC.

B. Copy rights owner exchange the Certificate ID with CDCMC.

C. CDCMC query protocol online to verify the validity of the certificate of the copyright holder, if authenticated, the next step. Otherwise, the content server returns an error message to the user and terminates this Agreement.

D. Copyright owner submit the request and the ID information to the content server for the registration request.

E. The information submitted by the content server to verify the copyright holder. Verify that the content server to save the user's information to their local copyright holder information in the database to register information, and then return to the copyright holder to terminate the agreement; such as the copyright holder of copyright should not submit their true identity or their the claims of copyright does not belong to the user's content server returns an error message to the user and the next step.

F. The copyright holder may cancel the registration and termination of the agreement; or update the information submitted and re-send the content server, go to Step E.

G. After having taken the Copyright and copyright owner's information, CDCMC generate a certificate to restore the identity information of the works. At the same time, the certificate will be accompanied by a download stream counters to record the number and times for download and use. The certificate and the stream counter meter are forced to be implanted into the download stream suppliers.

H. Meanwhile, download user must also register an account in CDCMC. After verification, the downloader was assigned private key, after the works download they need the private key been shown and automatically obtain the corresponding authorization. And some eternal records were kept in the server, in order to accept supervision and pay for the download later.

I. Download stream suppliers offer the download service and provide the appropriate Certificate of Authorization

J. According to stream counter meters, the authorization number of traffic providers will be notified of payments to copyright owners.

The program steps can be seen in the Fig 1.

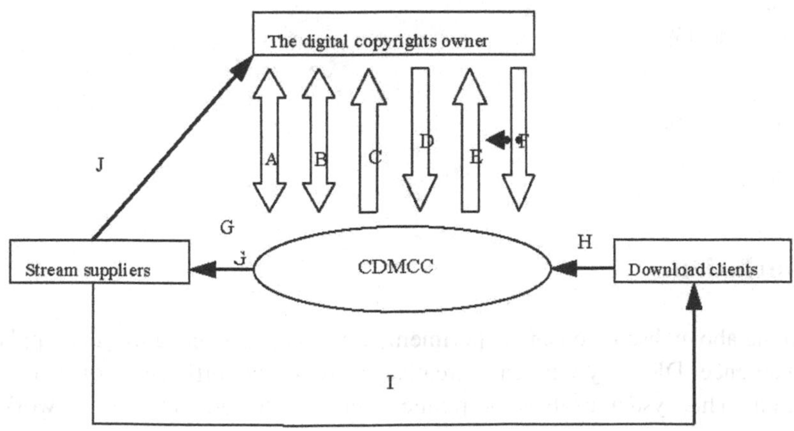

Fig. 1. The steps of the convergence digital copyrights management systems

And the algorithm can be descripting as follows:

Definitions: the two-tuples of (c,r) meets the request of c=H (m || g || g_1 || h_1 || $g^r h^c$ || $g^{r1} h^{c1}$), we call it symmetry information. $m \in (0,1)^*$ is about a knowledge signature of h,h1.Recorded as: SPK $\{\alpha | h=g^{\alpha} \wedge h_1=g_1^{\alpha}\}$ (m). Prove that if you know an integer: $x=\log=\log_g{}^1 h^1$.

And then set parameters, download providers offer CDCMC their own identity information, select $s \in_R Z^*_n$ as the information of registration in CDCMC. Next CDCMC use the Schnorr identification scheme to make sure if the applier has registered in the commercial administration department and the existence of adequate Security deposit in the bank. If the reply is "true", then the respective account should be built in the database of CDCMC. Of course the stream counter meter also will be implanted into the stream supplier's system. After completion of this course, the authorization and deciphering code will be sent to the stream supplier.

The concrete deciphering algorithm is to take decryption key as CK. (Random generate after confirming the identity and having implanted the stream counter meter).

$$D_{kj^{-1}} = (\mathrm{EGPK}(\mathrm{CK})) = (\mathrm{CKemod} \prod_{N_j \in Sr} N_j)^{dj} \mathrm{mod} N_j$$

$$= (\mathrm{CK}^e \mathrm{mod}(N_j \cdot \prod_{t \in s \wedge t \neq jr} N_t) \mathrm{mod} N_j)^{d_j} \mathrm{mod} N_j$$

$$= (\mathrm{CK}^e \mathrm{mod} N)^{d_j} \mathrm{mod} N_j$$

$$= \mathrm{CK}^{ed_j} \mathrm{mod} N_j$$

$$= \mathrm{CK}^{x\Phi(n_j)+1} \mathrm{mod}\ N_j$$

$$= \mathrm{CK}$$

4 Conclusion

Based on the above discussion and experiment, we can clearly recognize that applying the convergence DRM system can greatly improve the efficiency of copyright management. This system enables the public clients download and uses the works in freedom.

References

1. Samuelson, P.: Toward a 'New Deal' for Copyright for an Information Age. Michigan Law Review 121(1), 1–3 (2002)
2. Jessica, L.: Sharing and Stealing. Law of Intellectual Property Journal 22(1), 14–19 (2008)
3. Kalker, T.: On interoperability of DRM. In: Proceedings of the ACM Workshop on Digital Rights Management, DRM 2006, pp. 45–46 (2006)
4. Zhang, Z.: Digital Rights Management Ecosystem and its Usage Controls: A Survey. Jouranl of AICIT, JDCTA 5(3), 255–272 (2011)
5. Song, Y., et al.: Improved E-commerce Certification Security Using Contract Protocol Based on RSA and DSA Algorithm. Journal of AICIT, JCIT 7 (in press, 2012)

Nonlinear Plant Identifier Using the Multiple Adaptive RBF Network Convex Combinations

Xiangping Zeng[1,3], Haiquan Zhao[2], and Weidong Jin[2]

[1] School of Information Science & Technology at Southwest Jiaotong University, Chengdu, 610031, China
[2] School of Electrical Engineering at Southwest Jiaotong University, Chengdu, 610031, China
[3] Center of Electronic Lab, Chengdu University of Information Technology, Chengdu, 610255, China

zxping163@163.com, hqzhao0815@yahoo.com.cn

Abstract. Due to its simplicity and robustness, the stochastic gradient (SG) algorithm has become a very popular learning algorithm for adaptive radial basis function (RBF) network filter. However, a difficulty is its inherent compromise between convergence speed and precision, that is imposed by the selection of a fixed value for the step sizes. In this paper, to solve the problem, we propose to generalize the views of the convex combination, which is combining multiple RBF filters (MCRBF) using the SG algorithm with different step sizes. Simulation works with nonlinear system identification have been carried out to illustrate the effectiveness of this approach.

Keywords: radial basis function, neural network, adaptive filter, stochastic gradient algorithm, convex combination.

1 Introduction

It is well known fact that adaptive RBF network filter is a FIR type nonlinear filter. To train RBF networks, several learning algorithms (Stochastic gradient algorithm (SG) [1-2], Orthogonal least square (OLS) [3], Resource allocation network extended Kalman filter (RANEKF) [4], Growing and pruning algorithm (GAP) [5] and *etc.*) have been proposed in the literature. Among these algorithms, similar to the least mean square (LMS) algorithm of the finite impulse response (FIR) linear filter, the stochastic gradient (SG) algorithm for adaptive RBF filter has become the most popular learning algorithm due to its robustness, good tracking capabilities, and simplicity in terms of computational complexity and easiness of implementation. Therefore, it has been successfully employed in a wide variety of applications such as nonlinear plant identification.

However, like the LMS algorithm, slow convergence speed and high misadjustment are mainly drawbacks of SG algorithm. The selection of fixed step sizes of the SG algorithm imposes the convergence and steady-state error performance. That is, the large step size would lead to rapid convergence but highly steady-state error, while the small step size low steady-state error but slow convergence speed. Therefore, it is very difficult to balance between speed of convergence and final maladjustment. In this paper, we will pay attention to this issue.

D. Jin and S. Lin (Eds.): Advances in CSIE, Vol. 2, AISC 169, pp. 185–191.
springerlink.com

Recently, combining two adaptive linear filters with different step sizes is gaining popularity as a simple solution to mitigate the compromises inherent to adaptive linear filters [6]. A similar combination scheme has been previously used by in the nonlinear signal processing fields of functional link artificial neural network (FLANN) [7-8] and adaptive Volterra filter [9]. The approach of the convex combination is just aimed at improving the speed of convergence vs precision balance of adaptive algorithms [7-9]. As a consequence, generalizing the views of the convex combination approach, a novel nonlinear adaptive filter-thc convex combination of multiple RBF networks (MCRBF) is presented to improve the convergence speed and reduce the steady-state error of the SG algorithm with different step sizes in this paper.

2 Multiple Convex Combination of Radial Basis Function Networks

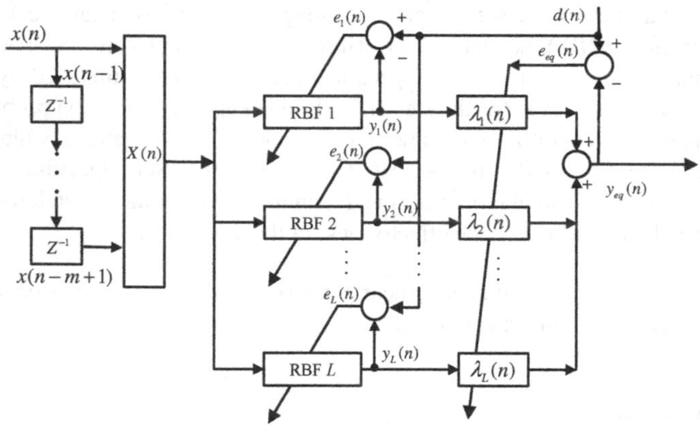

Fig. 1. Schematic of the MCRBF network

Keeping the views of the convex combination, the generalized of the convex combination of multiple RBF network filters is presented as show in Fig. 1. The nonlinear filter allows the combination of an arbitrary number of individual RBF network filters using the SG algorithm with the different step sizes to overcome the disadvantages of slow convergence and low steady-state error. From the figure, it is clearly shown that a novel MCRBF network filter consists of L radial basis function networks, with the input vector $X(n)$ defined by

$$X(n) = [x(n), x(n-1), \cdots, x(n-m+1)]^T \tag{1}$$

where m is the memory length of the nonlinear filter and T denotes the transpose operator.

For the ith RBF network, its hidden layer has k computing units $h_{i,l}(n), l = 1, \cdots, k, i = 1, 2, \cdots, L$, which is the response of the lth hidden neuron to the input $X(n)$. In this paper, we can select a Gaussian function as $h_{i,l}(n)$

$$h_{i,l}(n) = exp(-\frac{\|X(n) - C_{i,l}(n)\|^2}{2\sigma_{i,l}^2(n)}), l = 1, 2, \cdots, k. \tag{2}$$

where $\|\bullet\|$ denotes the Euclidean norm, the centre for the lth hidden neuron is a $m\times 1$ vector $C_{i,l}(n)$ as follows:

$$C_{i,l}(n) = [c_{i,l,1}(n), c_{i,l,2}(n), \cdots, c_{i,l,m}(n)]^T. \tag{3}$$

and $\sigma_{i,l}^2(n)$ represents the width of the Gaussian function.

Thus, the output $y_i(n)$ (i=1,2,…,L) of the ith RBF of the proposed nonlinear filter is expressed by

$$y_i(n) = \sum_{l=1}^{k} w_{i,l}(n) h_{i,l}(n) = W_i^T(n) H_i(n) \tag{4}$$

where $w_{i,l}(n)$ is the weight coefficient of the ith RBF network, $W_i(n)$ and $H_i(n)$ vectors are respectively defined by

$$W_i(n) = [w_{i,1}(n), w_{i,2}(n), \cdots, w_{i,k}(n)]^T \tag{5}$$

$$H_i(n) = [h_{i,1}(n), h_{i,2}(n), \cdots, h_{i,k}(n)]^T \tag{6}$$

Therefore, we can obtain the output of the proposed nonlinear filter as follows:

$$y(n) = \sum_{i=1}^{L} \lambda_i(n) y_i(n) \tag{7}$$

and $\lambda_i(n)$ is a mixing coefficient, and is defined by

$$\lambda_i(n) = \frac{exp(a_i(n))}{\sum_{j=1}^{L} exp(a_j(n))}, i = 1, 2, \cdots, L \tag{8}$$

which guarantees that $0 \le \lambda_i(n) \le 1$, and $\sum_{i=1}^{L} \lambda_i(n) = 1$.

3 Adaptive Learning Algorithm of the MCRBF Filter

To train the MCRBF network, according to the SG algorithm rule, the cost function of the proposed nonlinear filter is defined by

$$E(n) = \frac{1}{2}(d(n) - y(n))^2 = \frac{1}{2} e_{eq}^2(n) \tag{9}$$

where $e_{eq}(n)$ represents the error between the desired response $d(n)$ and the output $y(n)$, and defined by

$$e_{eq}(n) = d(n) - y(n) \tag{10}$$

Furthermore, the error $e_{eq}(n)$ of the overall scheme can be expressed as a convex combination of the errors of the component networks as

$$\begin{aligned} e_{eq}(n) &= d(n) - (\lambda_1(n)y_1(n) + \lambda_2(n)y_2(n) + \cdots + \lambda_2(n)y_2(n)) \\ &= \sum_{i=1}^{L} e_i(n) \end{aligned} \tag{11}$$

where the error $e_i(n)$ of the ith RBF filter is defined by

$$e_i(n) = d(n) - y_i(n) \tag{12}$$

For the MCRBF filter, the weight coefficient $W_i(n)$, the centre $c_{i,l,j}(n)$ and width $\sigma_{i,l}(n)$ of the Gaussian function of the each RBF network are individually adjusted. Thus, by the gradient descent algorithm rule, the update equation of the weights $W_i(n)$ of the RBF i network is given by

$$W_i(n+1) = W_i(n) + u_{wi}\lambda_i(n)e_i(n)H_i(n) \tag{13}$$

where u_{wi} is the step size and controls the convergence performance,

Similarly, the parameters $c_{i,l,j}(n)$ and $\sigma_{i,l}(n)$ $(j = 1,2,\cdots,m; l = 1,2,\cdots,k)$ of the proposed network are adapts by the stochastic gradient algorithm as follows:

$$c_{i,l,j}(n+1) = c_{i,l,j}(n) + u_{ci}e_i(n)W_i(n)h_{i,l}(n)\frac{x(n-j+1) - c_{i,l,j}(n)}{\sigma_{i,l}^2(n)} \tag{14}$$

$$\sigma_{i,l}(n+1) = \sigma_{i,l}(n) + u_{\sigma i}e_i(n)W_i(n)h_{i,l}(n)\frac{\|X(n) - C_{i,l}(n)\|^2}{\sigma_{i,l}^3(n)} \tag{15}$$

where u_{ci} and $u_{\sigma i}$ are the step sizes.

As for the proposed network, to minimize the cost function $E(n)$ of the overall network, the gradient adaptive algorithm is used to calculate the correction $\Delta a_i(n)$ along the negative of the gradient $\nabla_{a_i} E(n)$. So the update equation of the parameter $a_i(n)$ of the mixing coefficient $\lambda_i(n)$ is given by

$$\begin{aligned} a_i(n+1) &= a_i(n) + \Delta a_i(n) \\ &= a_i(n) - u_a \nabla_{a_i} E(n) \\ &= a_i(n) + u_{a_i} e_{eq}(n)\left[y_i(n) - y(n)\right]\lambda_i(n) \end{aligned} \tag{16}$$

where the step size u_a is used to keep $\lambda(n)$ in the interval [0,1]. However, a disadvantage of this scheme is that $a(n)$ stops updating whenever $\lambda(n)$ is close to 0 or 1. To solve this problem, $a(n)$ must be restricted to the interval [-4, 4].

4 Simulation

To illustrate the effectiveness of the proposed adaptive network, nonlinear system identification application is carried out in the simulation.

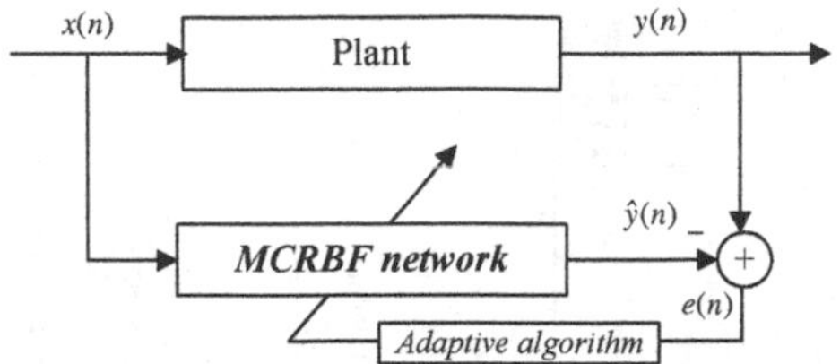

Fig. 2. Identification scheme of a nonlinear system

Fig. 3 depicted the identification scheme of a nonlinear system-based on the MCRBF network filter. The plant is described by the difference equation as follows:

$$y(n+1)=x^{3}(n)+\frac{y(n)}{1+y^{2}(n)} \tag{17}$$

During the test phase, effectiveness of MCRBF networks is analyzed by a sinusoidal signal as follows:

$$x(n)=0.5\sin(\frac{2\pi n}{1000}) \tag{18}$$

In this simulation, the parameter L of the CRBF filter is set to 4. To obtain best results, the learning parameter values of RBF and MCRBF networks are chosen for several trials. All the parameters are summarized in Table 1.

Table 1. The parameters of all the nonlinear filters

Parameters	MCRBF	RBF
Learning rate	$u_{w1}=0.5$, $u_{c1}=u_{\sigma1}=0.1$, $u_{a1}=200$, $u_{w2}=0.3$, $u_{c2}=u_{\sigma2}=0.05$, $u_{a2}=200$, $u_{w3}=0.1$, $u_{c3}=u_{\sigma3}=0.01$, $u_{a3}=200$, $u_{w4}=0.05$, $u_{c4}=u_{\sigma4}=0.005$, $u_{a4}=200$,	u_w=0.03, u_c=0.03, u_σ=0.03.
Number of inputs	2	2
Number of hidden layer neurons	4	4

The results of nonlinear identification with the sinusoidal signal (18) are depicted in Fig. 3. It is clearly shown that the identification of the plant of the MCRBF filter is more satisfactory than that that of RBF network. Meanwhile, we can find that the estimation error of the MCRBF network is less than that of the RBF network. Moreover, the steady-state normalized mean square errors after the convergence of

MCRBF and RBF networks are obtained by -28.1809 dB and -22.1088 dB, respectively. In addition, we can obviously observe that the speed convergence of the MCRBF network is faster than that of the RBF network.

Therefore, the performance of the CRBF network is superior to that of the RBF network.

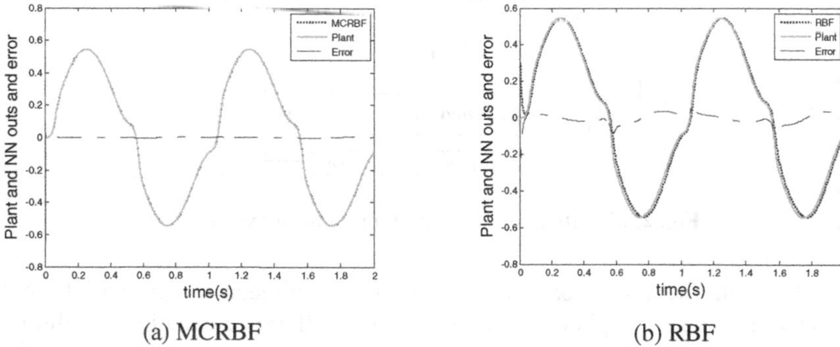

(a) MCRBF (b) RBF

Fig. 3. Identification of the nonlinear plant with the test sinusoidal signal

5 Conclusion

The paper presents a novel adaptive nonlinear filter that uses the convex combination of multiple RBF networks (MCRBF) using the SG learning algorithm with different step sizes. The performance of the proposed network is evaluated by nonlinear system identification. Simulation results demonstrate the effectiveness of the MCRBF filter. The use of the convex combination of adaptive nonlinear filters represents a logical direction for further research in actual applications. Currently, we are working on combining different algorithms for adaptive nonlinear filter in a similar manner.

Acknowledgments. This work was partially Supported by the National Science Foundation of PR China (Grant: 61071183, 60971103 and 61134002) and the Sichuan Provincial Youth Science and Technology Fund (Grant: 2012JQ0046).

References

1. Kassam, S.A., Cha, I.: Radial basis function networks in nonlinear signal processing. In: Proc. IEEE 27th Annu. Asilomar Conf. Signals, Syst., Comput., vol. 2, pp. 1021–1025 (1993)
2. Chng, E.S., Chen, S., Mulgrew, B.: Gradient radial basis function networks for nonlinear and nonstationary time series prediction. IEEE Trans. Neural Networks 7(1), 190–194 (1996)
3. Chen, S., Cowan, C.F.N., Grant, P.M.: Orthogonal least squares learning algorithm for radial basis function networks. IEEE Trans. Neural Netw. 2(2), 302–309 (1991)
4. Kadirkamanathan, V., Niranjan, M.: A function estimation approach to sequential learning with neural networks. Neural Comput. 5, 954–975 (1993)

5. Huang, G.B., Saratchandran, P., Sundararajan, N.: An generalized growing and pruning RBF (GGAP-RBF) neural network for function approximation. IEEE Trans. Neural Networks 16(1), 57–67 (2005)
6. Arenas-García, J., Figueiras-Vidal, A.R.: Adaptive combination of normalized filters for robust system identification. Electron. Lett. 41, 874–875 (2005)
7. Zhao, H.Q., Zhang, J.S.: Adaptively combined FIR and functional link neural network equalizer for nonlinear communication channel. IEEE Trans. Neural Networks 20(4), 665–674 (2009)
8. Zhao, H.Q., Zeng, X.P., Zhang, J.S., Zhang, X.Q., Liu, Y.G., Wei, T.: Adaptive decision feedback equalizer using the combination of the FIR and FLNN. Digital Signal Processing 21, 679–689 (2011)
9. Azpicueta-Ruiz, L.A., Zeller, M., Figueiras-Vidal, A.R., Arenas-Garcia, J., Kellermann, W.: Adaptive combination of Volterra kernels and its application to nonlinear acoustic echo cancellation. IEEE Trans. Audio, Speech, and Language Processing 19(1), 97–110 (2011)

Estimation of Rotor Resistance of Induction Motor Based on Extended Kalman Filter

Guohan Lin and Qin Wan

Department of Electrical and Information Engineering
Hunan Institute of Engineering
Xiangtan, China
lgh2966@126.com

Abstract. Estimation of rotor resistance rapidly and accurately has received a lot of attention due to its significance in improving the performance of induction motor(IM) vector control system. In this paper, algorithm of estimating rotor resistance of IM based on extended Kalman filter(EKF) is presented. The algorithm can estimate the rotor resistance on line by measuring the voltage, current and speed of stator. Using MATLAB/SIMULINK, a simulation model is built and tested. The simulation results show that the presented algorithm is effective and accurate.

Keywords: parameter estimation, extended Kalman filter, rotor resistance, indirect vector control.

1 Introduction

Vector control has been used widely in motor drive systems to improved performance. According to different way by which produce the reference vector in coordinate transformation, vector control can be divided into the indirect and direct vector control. Indirect vector control is one of the most effective method due to its simplicity in designing and construction. Under the rotor flux synchronous rotating coordinate, the excitation of the stator current component and torque component control the torque and rotor flux independently, the flux position angle using in coordinate rotation transformation is calculated from rotor speed and slip rotate speed. The slip gain relate to the rotor time constant which is define by rotor resistance. However, rotor resistance is changing with temperature and excitation conditions and thus affect the control system performance largely. To achieve satisfied performance, estimate the rotor resistance rapidly and accurately is necessary [1],[2].

This paper present a method of estimation of rotor resistance based on EKF. The paper is organized as follows. Section 2 describes the model of induction motor model. In Section 3 an estimator using EKF for the rotor resistance is designed. Simulations of the scheme are carried out in Section 4 and some conclusions are given in Section 5.

D. Jin and S. Lin (Eds.): Advances in CSIE, Vol. 2, AISC 169, pp. 193–198.
springerlink.com

2 Induction Motor Model

To using EKF for estimation of the rotor resistance, various machine models is possible to be used. For example, it is possible to use the equations expressed in the rotor flux oriented reference frame, or in stator flux oriented reference frame. In order to avoid extra calculations and non-linear transformation, stationary reference frame is preferred. A fourth order dynamic model for induction motor developed in stationary reference frame is as following[3]:

$$\frac{di_{s\alpha}}{dt}=-\left(\frac{Rs}{\sigma L_s}+\frac{L_m^2 Rr}{\sigma L_s L_r}\right)i_{s\alpha}+\frac{L_m R_r}{\sigma L_s L_r^2}\psi_{r\alpha}+\frac{L_m}{\sigma L_s L_r}pw_r\psi_{r\beta}+\frac{1}{\sigma L_s}u_{s\alpha} \; . \tag{1}$$

$$\frac{di_{s\beta}}{dt}=-\left(\frac{Rs}{\sigma L_s}+\frac{L_m^2 Rr}{\sigma L_s L_r}\right)i_{s\beta}-\frac{L_m R_r}{\sigma L_s L_r}pw_r\psi_{r\alpha}+\frac{L_m R_r}{\sigma L_s L_r^2}\psi_{r\beta}+\frac{1}{\sigma L_s}u_{s\beta} \; . \tag{2}$$

$$\frac{d\psi_{r\alpha}}{dt}=\frac{R_r}{L_r}L_m i_{s\alpha}-\frac{R_r}{L_r}\psi_{r\alpha}-Pw_r\psi_{r\beta} \; . \tag{3}$$

$$\frac{d\psi_{r\beta}}{dt}=\frac{R_r}{L_r}L_m i_{s\beta}-\frac{R_r}{L_r}\psi_{r\beta}+Pw_r\psi_{r\alpha} \; . \tag{4}$$

$$T_e=3PL_m\left(\psi_{r\alpha}i_{s\beta}-\psi_{r\beta}i_{s\alpha}\right)/(2L_r) \; . \tag{5}$$

3 Rotor Resistance Estimation Algorithm Based on EKF

3.1 EKF Algorithm

Extended Kalman filter is a full-order random observer. Based on system noise and the statistical characteristics of measurement noise, EKF can be used to obtain the multi-input multi-output system such as induction motor model, real-time recursive optimal. The development of the Kalman filter is closely linked to the stochastic systems. The nonlinear stochastic systems are described by relations.[4]

$$\begin{aligned}\dot{x}(k)&=f\left(x(k-1),u(k),w(k)\right)\\ y(k)&=h\left(x(k),v(k)\right)\end{aligned} \; . \tag{6}$$

where x and u are state variable and control variable respectively. $\boldsymbol{w}(k)$ represents the disturbances apply at the system input, while $\boldsymbol{y}(k)$ is the measurable output affected by the random noise $\boldsymbol{v}(k)$ .It will be assumed that $\boldsymbol{w}(k)$ and $\boldsymbol{v}(k)$ are not correlated and zero-mean stochastic processes. From the statistic point of view, the stochastic processes $\boldsymbol{w}(k)$ and $\boldsymbol{v}(k)$ are characterized by the covariance matrices Q and R respectively. It is further assumed that the initial state $\boldsymbol{x}0$(k) is a vector of random variables of mean $\boldsymbol{x}0$ and covariance P0 , not correlated with the stochastic processes $\boldsymbol{w}(k)$ and $\boldsymbol{v}(k)$ over the entire interval of estimation. For nonlinear time invariant systems, the following relations of recurrent computation describe the general step of the implementation algorithm of Kalman filters [5],[6].

step 1: state prediction

$$\dot{x}(k)=f\left(\dot{x}(k\text{-}1),u(k),x0\right)\ . \tag{7}$$

step 2: computation of convariance

$$P(k)=G(k)P(k-1)G(k)^{T}+Q\ . \tag{8}$$

where $G(k)=\frac{\partial f}{\partial x}(x(k-1),u(k-1),w(k-1))$

step 3: computation of Kalman gain

$$\mathrm{K(k)=P(k)M(k)^{T}\left(M(k)P(k)M(k)^{T}+R\right)^{-1}}\ . \tag{9}$$

step 4: state update

$$\mathrm{\dot{x}(k+1)=\dot{x}(k)+K(k)\left(y(k)\text{-}h\left(\dot{x}(k)\right)\right)}\ . \tag{10}$$

step 5: estimation covariance computation

$$\mathrm{P(k)=(I\text{-}K(k))M(k)P^{-1}(k)}\ . \tag{11}$$

3.2 Rotor Resistance Estimation Based on EKF

EKF can be used for combined state and parameters identification by treating selected parameters as extra states and forming an augmented state vector. The rotor resistance R_r is the parameter to be estimated and augmented into state vector, model of IM developed in stationary reference frame is as following:

$$\begin{matrix}\dot{\mathrm{X}}=\mathrm{AX+Bu}\\ \mathrm{y=CX}\end{matrix}\ . \tag{12}$$

as
$$A=\begin{bmatrix} -\left(\frac{R_s}{\sigma L_s}+\frac{1-\sigma}{\sigma\tau_r}\right) & 0 & \frac{L_m}{\sigma L_s L_r \tau_r} & \frac{w_r L_m}{\sigma L_s L_r} & 0 \\ 0 & -\left(\frac{R_s}{\sigma L_s}+\frac{1-\sigma}{\sigma\tau_r}\right) & -\frac{w_r L_m}{\sigma L_s L_r} & \frac{L_m}{\sigma L_s L_r \tau_r} & 0 \\ \frac{L_m}{\tau_r} & 0 & -\frac{1}{\tau_r} & -w_r & 0 \\ 0 & \frac{L_m}{\tau_r} & w_r & -\frac{1}{\tau_r} & 0 \\ 0 & 0 & 0 & 0 & 1 \end{bmatrix},\quad B=\begin{bmatrix} 1/\sigma L_s & 0 & 0 & 0 & 0 \\ 0 & 1/\sigma L_s & 0 & 0 & 0 \end{bmatrix}^{T}$$

$$X=\begin{bmatrix} i_{s\alpha} & i_{s\beta} & \Psi_{r\alpha} & \Psi_{r\beta} & R_r \end{bmatrix}^{T},\ C=\begin{bmatrix} 1 & 0 & 0 & 0 & 0 \\ 0 & 1 & 0 & 0 & 0 \end{bmatrix},\ u=\begin{bmatrix} u_{s\alpha} & u_{s\beta} \end{bmatrix}^{T},\ y=\begin{bmatrix} i_{s\alpha} & i_{s\beta} \end{bmatrix}^{T}$$

Considering the process noise and measure noise, assume the sampling time is T_s ,Equ.12 is discrete and linearized and following formula is obtained:

$$\begin{matrix} X(k+1)=A_d X(k)+B_d U(k)+W(k) \\ Y(k+1)=C_d(k)+V(k) \end{matrix}\ . \tag{13}$$

$$A_d=\begin{bmatrix} 1-T_S\left(\frac{R_S}{\sigma L_S}+\frac{1-\sigma}{\sigma\tau_r}\right) & 0 & \frac{T_S L_m}{\sigma L_S L_r\tau_r} & \frac{T_S w_r L_m}{\sigma L_S L_r} & 0 \\ 0 & 1-T_S\left(\frac{R_S}{\sigma L_S}+\frac{1-\sigma}{\sigma\tau_r}\right) & -\frac{T_S w_r L_m}{\sigma L_S L_r} & \frac{T_S L_m}{\sigma L_S L_r\tau_r} & 0 \\ \frac{L_m}{\tau_r}T_S & 0 & 1-\frac{1}{\tau_r}T_S & -w_r T_S & 0 \\ 0 & \frac{L_m}{\tau_r}T_S & w_r T_S & 1-\frac{1}{\tau_r}T_S & 0 \\ 0 & 0 & 0 & 0 & 1 \end{bmatrix},\ B_d=\begin{bmatrix} \frac{T_s}{\sigma L_s} & 0 \\ 0 & \frac{T_s}{\sigma L_s} \\ 0 & 0 \\ 0 & 0 \\ 0 & 0 \end{bmatrix}$$

Based on equ.13,using EKF algorithm, the state can be estimated at different times and thus estimate rotor resistance of IM.

4 Simulation Result

To verify the effectiveness of the proposed scheme, simulations were conducted in Simulink .Induction machine driven by a VSI with rotor estimation was show in fig.1.

In fig. 1,a squirrel cage induction motor was used to test the algorithm. Induction motor parameter are: Nominal output power PN=1.2kW, Nominal voltage UN=380V, Nominal voltage current IN=2.67A, Nominal frequency fN=50Hz, Rs=5.27Ω,Rr=4.35Ω, Ls=479mH, Lr=479mH, Lm=421mH, σ=0.228,P=2, Nominal speed SN=1440rpm.

Assumed the reference flux is constant with the value of 0.856 Wb, and the reference speed and load is 157rad/s and 10N.m respectively. Fig 2 a) show the real and estimation of rotor resistance when rotor resistance is constant,it can be seen from the figure that it takes less than 0.5 seconds to reach the true value despite small fluctuations around the true value.

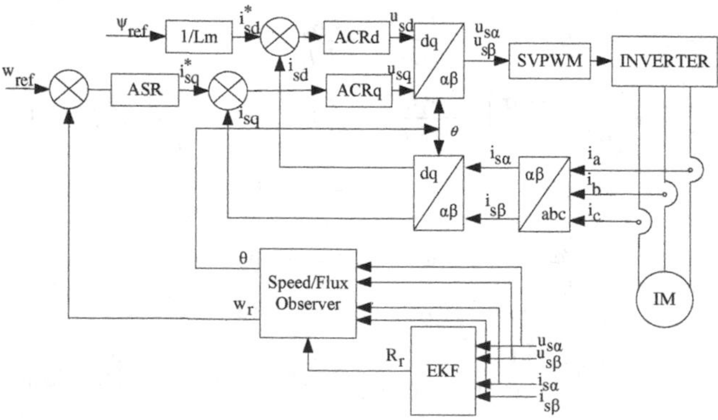

Fig. 1. Structure diagram of rotor resistance estimation based on EKF

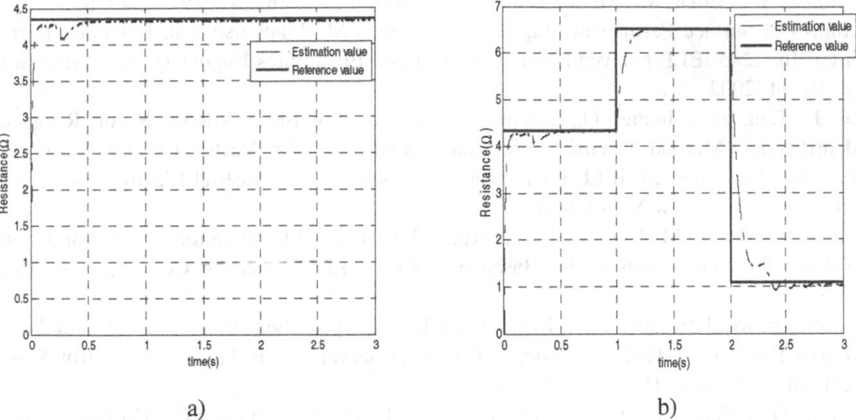

Fig. 2. Rotor resistance estimation with resistance constant and changing

To simulate the change of resistance due to different running condition such as temperature change, assume that the rotor resistance was stepped from 100% to 150% of its rated value in one second and stepped from 150% to 25% of its rated value in two second, reference flux, reference load and reference speed is the same as above discuss. Simulation results of rotor resistance estimation including the convergence process was show in fig.3.b). As seen in the figure, after about 0.5s the estimated value of rotor resistance converges to the true value as expected. When the true value steps up and down, the EKF estimator can estimate the rotor resistance changes quite well.

5 Conclusion

Estimation of the rotor resistance rapidly and accurately can improve greatly the static and the dynamic performance of the indirect vector control systems of IM. In this paper ,a simple estimation algorithm based on EKF in indirect vector control has been presented. Simulation results show that the estimation algorithm is robust to the variation of the rotor resistance which is changing when the IM is running. the effectiveness and the validity of the proposed EKF estimator is verified.

Acknowledgment. This work was support by Hunan Provincial Natural Science Foundation of China (11JJ4049) and Supported by Innovation Platform Open Funds for Universities in Hunan. Province(11K019).

References

1. Yahia, K., Zouzou, S.E., Benchabane, F., Taibi, D.: Comparative Study of an Adaptive Luenberger Observer and Extended Kalman Filter for a Sensorless Direct Vector Control of Induction Motor. Actra. Electrotehnica 50, 99–107 (2009)

2. Comnac, V., Cernat, M., Moldoveanu, F., Draghici, I.: Sensorless Speed and Direct Torque Control of Surface Permanent Magnet Synchronous Machines using an Extended Kalman Filter. In: 22nd IEEE Convention of Electrical and Electronics Engineers, Tel-Aviv, Israel, pp. 39–44 (2002)
3. Li, J., Ren, H., Huang, Q., Zhong, Y.: A Novel On-line MRAS Rotor Resistance Identification Method Insensitive to Stator Resistance for Vector Control Systems of Induction Machines. In: IEEE International Symposium on Industrial Electronics, pp. 541–546. IEEE Press, New York (2010)
4. Zhao, X., Wang, M.-Y., Liu, S.-X.: Rotor Resistance Identification in Vector Control System of Asynchronous Motor Based on EKF. J. Elec. Mach. & Con. App. 36, 18–23 (2009)
5. Montanari, M., Peresada, S.M., Rossi, C., Tilli, A.: Speed Sensorless Control of Induction Motors Based on a Reduced-Order Adaptive Observer. IEEE Trans. on Control System Technology 5, 1049–1064 (2007)
6. Huerta González, P.F., Rivas Rodriguez, J.J., Torres Rodriguez, I.C.: Estimación de la resistencia del rotor usando una red neuronal artificial en el control vectorial indirecto del motor de inducción. IEEE Latin America Trans. 6, 76–183 (2008)

Data Exchange Solution for Seismic Precursory Observation Data*

WeiFeng Shan, Jun Li, QingJie Liu, and Bing Zhang

Dept. of Disaster Information Engineering,
Institute of Disaster Prevention
065201 Sanhe, Hebei, China
{shanweifeng,lijun,liuqingjie,zhangbing}@cidp.edu.cn

Abstract. China seismic precursory network consists of more than 800 precursory monitoring stations and over 30 Oracle database nodes, and it is responsible for collecting, transferring and monitoring the seismic precursory data. There are several approaches to exchange data among database nodes, but they are mostly adapt to two Oracle database nodes and difficult to manage for the usage of database link (DBlink) technology. In this paper, we introduced a novel data exchange solution for seismic precursory observation data using user-defined update operation logs. While an update operation (insert, delete or update) triggers, the relative information of it will be recorded into the update log table. We detail the update log table schemer, exchange task structure and three phase of an exchange task: export, transfer and import phases. This solution is an effective, manageable, flexible solution for exchange of seismic precursory observation data.

Keywords: Seismic Precursory Observation Data, Earthquake, Data Exchange, Data Replication, Oracle Database.

1 Introduction

China [1] is one of the countries which have very high seismicity in their continents and have suffered severe damage from past strong earthquakes. More than half of China's land locates in an area with intensity VII or above. China has built a national network on earthquake observation which includes 1160 stations for strong motion observation, and over 800 precursory monitoring stations that can observe gravity, earth's magnetic field, crustal deformation, earth's electric field and underground fluid. The completion of this advanced modern earthquake observation network provides a solid basis to carry out exploration and application of earthquake prediction research and for accumulation of rich and large volumes of observation datasets for future earthquake research.

* This work is supported by the Teacher Research Fund of China Earthquake Administration (No. 20110110) and Youth Science Fund for Disaster Prevention and Reduction (No. 200905).

D. Jin and S. Lin (Eds.): Advances in CSIE, Vol. 2, AISC 169, pp. 199–205.
springerlink.com

In order to collect, manage and share the earthquake precursory observation data, a management information system based on Oracle database system for earthquake precursory observation network has been developed, which includes data collection module, data service module, data exchange module, configuration module and system monitoring module. Data exchange module is an import one that is responsible for transmitting the collection precursory observation data from area node to national earthquake precursory network center and backup the production data produced by five subject centers.

There are several approaches to realize data exchange and replication between two or more Oracle databases nodes. For example, we may complete data exchange based on Oracle Stream or snapshot log technologies. But these solutions face to database specialist, not for end users, and they are difficult to modify configuration and lack of convenience and manageability. In this paper, we introduce a novel data exchange solution for seismic precursory observation data using our own update log record, which logs all update recorders' relative information. An exchange task will select update recorders according to operation logs and task information, and package them into one or more .zip files, then transfer them to the target database node, all .zip file are extracted and imported into database while transmission completed.

The paper is organized as follows. Section 2, we briefly describes the China seismic precursory network and its data management system. Section 3 discusses several approaches for data exchange between two or more Oracle database nodes. A data exchange solution for seismic precursory observation data is proposed and discussed in section 4. Section 5 concludes the paper.

2 China Seismic Precursory Network Data Management System

As shown in Fig. 1, China seismic precursory network includes over 800 precursory monitoring stations and more than 2000 precursory observation instruments that can observe gravity, earth's magnetic field, crustal deformation, earth's electric field and underground fluid data.

All observation data of instruments are collected by over 30 area seismic precursory network centers (AC node) every day, and area nodes forward them to the national seismic precursory network center (NC node), then NC node distribute the observation data among 5 subject centers (SC node), i.e., national gravity network center, national geomagnetic network center, national crustal deformation network center, national earth's electric field network center and national underground fluid network center, according to the classification of them.

In seismic precursory network, each node, such as AC, SC and NC, is installed Oracle database and other applications, so it is a distributed application.

The exchange data can be classified into several groups, such as observation data, metadata, monitoring data and preprocessing observation data. In this paper, we only consider observation data. All observation data collected by ACs must be transferred to the NC node and then are distributed to SCs.

3 Related Works

Recently, distributed applications have become popular due to the rapid development of internet technology. The distributed applications may have huge amounts of data which are stored in several relational databases, so data exchange and replication is a big issue for these applications for consistence, availability, correctness of data.

China earthquake precursory network uses Oracle 10g database system to store the observation data. Oracle database provides several generic replication models for data replication; now let's talk about them respectively.

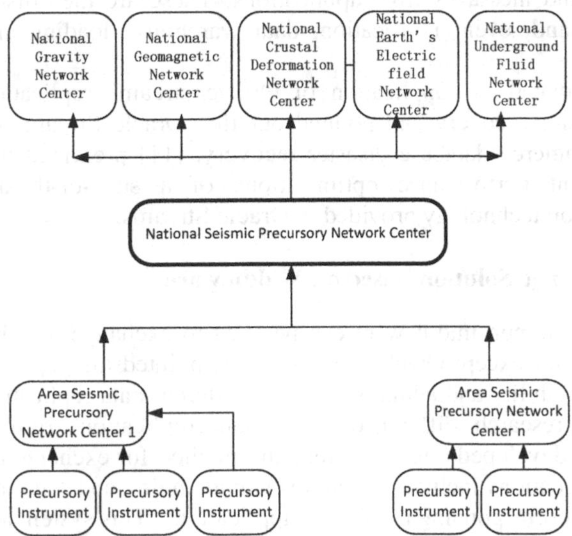

Fig. 1. China Earthquake Precursory Observation Network

3.1 Data Exchange Solution Based on Snapshot Replication

Snapshot replication [6], also known as materialized views, is the basic replication of Oracle and implemented using standard CREATE SNAPSHOT or CREATE MATERIALIZED VIEW statements. It can only replicate data (not procedures, indexes, etc), replication is always one-way, and snapshot copies are read only.

Ref. [8] introduced a data exchange model of earthquake precursor metadata with the help of Oracle snapshot replication technology. Ref. [9] described and tested a data exchange method between two Oracle nodes using materialized view technology and summarized that sequences and tables without primary key are not suitable for replication. Ref. [10] analysis the data structure of snapshot log and present a data exchange solution based on snapshot log.

3.2 Data Exchange Solution Based on Advanced Replication

Oracle Advanced Replication is one of several Oracle database replication technologies and supports bidirectional replication, multiple masters, conflict resolution, etc. Advanced replication [6] can not only be used to replicate tables, but also indexes, views, packages, procedures, functions, triggers, synonyms, etc. Unlike snapshot based data replication, advanced replication allows update operations to each mast and slave nodes.

Ref. [5] gave a data replication example for distributed enterprises application using Oracle advanced replication.

3.3 Data Exchange Solution Based on Oracle Streams

Oracle streams [7] replication is another replication solution. Oracle Streams captures events, updates and messages from applications. These are then distributed for use with replication and, event notification, data warehouse loading and many other applications.

Ref. [2] introduced the application of Oracle streams replication in database backup and disaster recovery and pointed out that Oracle Streams replication is a choice for E-commerce database disaster recovery. [11] presented the architectural design and recent performance optimizations of a state-of-the-art commercial database replication technology provided in Oracle Streams.

3.4 Data Exchange Solution Based on Middleware

There are some common middleware can be used to exchange data between two or more database nodes except Oracle tools. Ref. [4] pointed the gap along three axes: performance, availability and administration of different data replication solution and summarized the research difference about data replication from academia and industry. Ref. [3] developed a navel system and method for exchanging data between two or more applications includes a data exchange engine and a number of adapters associated with a corresponding number of applications. This system allowed users to define their own adapters if need to realize data exchange.

4 Data Exchange Solution for Seismic Precursory Observation Data

From the discussion of above section, there are many data exchange solutions. From the general principle, these solutions may be adapted to seismic precursory observation network, but they have several shortcomings. First of all, most data replication methods except middleware solution strongly rely on database link technology (DBlink), which is a definition of how to establish a connection from one Oracle database to another, this means that we must configure the peer's login user and password in one database which lead to difficult to manage database users, inconvenience of modifying users' password, and insecurity without modifying password. Another shortage of these replication methods is they are used by database administrators (DBAs), not by end users. If a new observation device is deployed, its observation data must be exchange to the related subject center. It is difficult for end users to modify the replication configuration.

It is necessary to build a flexible, configurable, high performance and safe data exchange solution to ensure the consistence, correctness, completeness and security of seismic precursory observation data.

4.1 Update Operation Log Table

Because all database schemes of database nodes in seismic precursory network are same, we only consider the data replication (table replication) and ignore other database objects, such as procedures, functions and triggers, etc. In order to identify each observation data record, we use a universally unique identifier (UUID), which is a unique reference number used as an identifier in computer software and the value of a UUID is a 16-byte (128-bit) number and represented as a 32-character hexadecimal string as the primary key for each observation data tables.

While an update operation to the observation data table coming, a new transaction that contains this update operation and a new insert operation to Operation Log table is composed and executed. As shown in Table 1, an operation log record includes table name, UUID, Instrument's identifier and operation type (insert, delete or update operation).

Table 1. Scheme of Update Operation Log Table

Filed Name	Field Type	Primary Key	Description
RowIndex	Number(38)	Yes	Sequence, step by 1
TableName	Varchar2(50)	No	Table's name that is updated
Operation	Varchar2(10)	No	Insert, Delete or update operation
UUID	Varchar2(36)	No	The updated recorder's UUID
InstrumentId	Varchar2(50)	No	Instrument's identifier

4.2 Data Exchange Task and Data Exchange Task Log

Seismic precursory observation data is replicated from AC nodes to the NC node, and then are distributed to SC nodes according to their category. So the data exchange can be viewed as an incremental data exchange. A data exchange requirement is defined by a data exchange task, which describes source node, target node, exchange moment, exchange cycle, instruments, and etc. User can modify freely exchange moment or relative instruments via web pages.

Data exchange log table record the detail execution info of each task, including task's identifier, starting moment and ending moment of task, try times, first row index and last row index of update operation log table, successful export flag, successfully transfer flag and successfully import flag, and etc. While a task starts, it locates the next record of the last exchange recorder of previous exchange task.

4.3 Data Exchange Flow

As shown in Fig. 2, a data exchange is divided into three phases: export phase, transfer phase and import phase.

In export phase, the data for exchange are selected from Oracle database using Data Access Module which is realized by EJB, and are serialized into byte array and are compressed into a .zip file. While the exported recorder number reaching the configured maximum number, the .zip file is closed and a new .zip file is created to store the next exported records. Each entry's name of .zip file is composed by the update operation name (insert, delete or update), UUID, table name, and order sequence number.

During transfer phase, all .zip files of a task created in export phase will be sent to the target node using socket technology. Firstly, socket client sends the authorization data to the socket server, namely target node, if verification is successful client and server consult whether starting transfer from last break point of transmission. Then client send all .zip files of task, and send an "OK" flag message while all files are transferred successfully. In order to ensure the correctness, consistence and integrity of transmission, we make a digest for all .zip file using MD5 Algorithm that is a widely used cryptographic hash function that produces a 128-bit (16-byte) hash value and is employed in a wide variety of security applications, and is also commonly used to check data integrity, and send it to target server, target server also make a digest for all received .zip files and compared to the received MD5 message, if they are same, server closes the channel to client, otherwise deletes all .zip files and starts to transfer again.

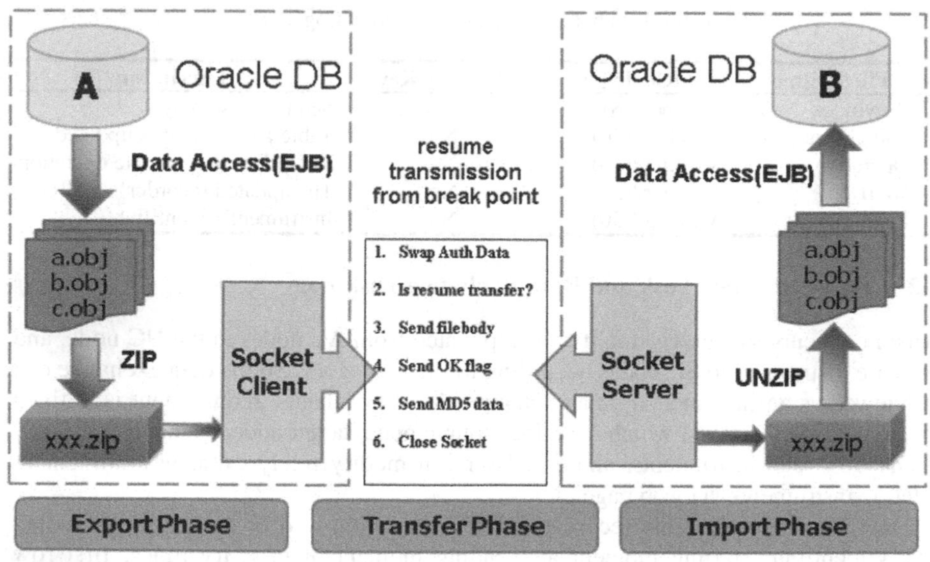

Fig. 2. Data Exchange Solution Architecture

During import phase, a transaction is started firstly, and then all .zip files of a task are unzipped according the time order, each entry of zip file is deserialized to an entity object by the table name and UUID information and is updated into Oracle database. When all update recorders are executed successfully, the transaction is committed.

Only if three phases are completed successfully, a data exchange is successful. Because the network bandwidth between nodes is narrow and instability, while the servers' performance is strong enough, we employ compress technology to enhance the system's performance and stability and resume from break point technology to increase the performance of system.

5 Concluding Remarks

In this paper we firstly introduced China seismic precursory network, and listed the data exchange requirement of precursory observation data. Secondly, we compared several data replication approaches for Oracle database each other and discussed the shortage of them. Finally, we present a novel data exchange solution for seismic precursory observation data, it is easy to configure and flexible to use.

References

1. Chen, J.: Earthquake Disaster Reduction in Developing China. In: The 14th World Conference on Earthquake Engineering, Beijing (2008)
2. Guan, D.: Research and Implementation of Database Backup and Disaster Recovery Based on Oracle Streams. M.S. thesis. Chengdu University of Technology, Chengdu, China (2008)
3. Sheard, N.C., et al.: Data Exchange System and Method. U.S. Patent 6,453,356 B1 (September 17, 2002)
4. Emmanuel, C., George, C., Anastasia, A.: Middleware-based database replication: The gaps between theory and practice. In: Proceedings of the ACM SIGMOD International Conference on Management of Data, pp. 739–752. Association for Computing Machinery, Vancouver (2008)
5. Wang, Z.: Research on the application of oracle advanced replication of distributed database. Journal of East China Jiaotong University 22(5), 115–117 (2005)
6. Orafaq 2011. Advanced Replication FAQ, http://www.orafaq.com/wiki/Advanced_Replication_FAQ (accessed: February 10, 2012)
7. Orafaq 2010. Oracle Streams, http://www.orafaq.com/wiki/Oracle_Streams (accessed: February 10, 2012)
8. Liu, G., Teng, Y., Wang, C., Tang, W.: Oracle Replication and Its Application in Earthquake Precursor Metadata Exhcange. Earthquake Research in China 24(2), 142–149 (2008)
9. Zhou, X.: Research and Implement of Data Synchronous Replication Using Oracle Materialized Views. China Hi-Tech Enterprises (15), 69–70 (2010)
10. Huang, X., Chen, L., Wei, W., Xu, S.: Data Synchronous Method Based on Snapshot Log Analysis. Journal of Logistical Engineering University 22(2), 59–62, 67 (2006)
11. Wong, L., Arora Nimar, S., Gao, L., Hoang, T., Wu, J.: Oracle streams: A high performance implementation for near real time asynchronous replication. In: Proceeding of 25th IEEE International Conference on Data Engineering, pp. 1363–1374 (2009)

Design and Realization of the Online Bookstore System Based on Struts Framework

Yufen Wang[1], Changjiang Li[2], and Chaohua Lu[1]

[1] School of Information Technology, Shangqiu Institute of Technology, Shangqiu, China
[2] School of Information Technology, Henan Institute of Science and Technology, Xinxiang, China

Abstract. Researched the MVC pattern and Struts framework, designed and realized an online bookstore system. Introduced system requirements analysis and design process in detail, realized functions of each module in the system. Basically meet the design requirements, function test results show that the system can run normally. Research results to some extent, make beneficial exploration to the application of design patterns in Web system development and making use of the mature framework technology to make system development.

Keywords: MVC pattern, Struts framework, Online bookstore system, Web application.

1 Introduction

In recent years, with the continuous development of Internet technology and the wide application of J2EE platform, multilayer Web system structure based on the B/S gradually has developed and matured, violently attacked the C/S system structure. B/S structure becomes the current choice architecture of application software. But with the increase of scale and complexity of software systems, in the development of multilayer Web application, have a lot of problems, such as low efficiency, low reusable program, maintenance difficulties, not easy to expand and so on. Therefore, how to write simple and efficient, easy to extend and maintain applications, is an important research issue.

The biggest characteristic of MVC pattern is the separation of display logic and business logic, which is very advantageous for the development of large Web application system, it is adapted to the development tendency which Web application systems become more and more complicated. According to the thought of MVC pattern, Struts framework provide a framework to create Web application, which greatly improves the development efficiency, so Web application system development pattern based on the MVC pattern will be more widely applied.

2 The Design of the Online Bookstore System

2.1 The Design of the System Function

According to the function requirements of the system, the online bookstore system can be divided into two modules: the foreground user module and the background management module. Functional structure as shown in Fig.1 and Fig.2.

D. Jin and S. Lin (Eds.): Advances in CSIE, Vol. 2, AISC 169, pp. 207–210.
springerlink.com

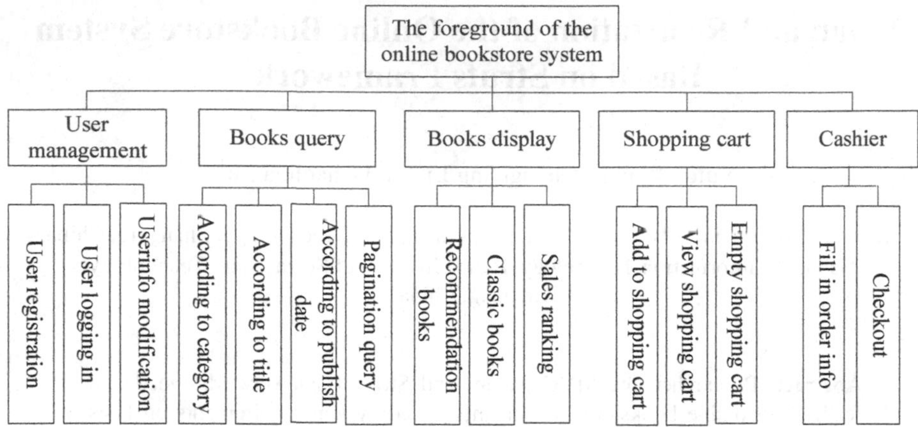

Fig. 1. The function structure of the foreground user module

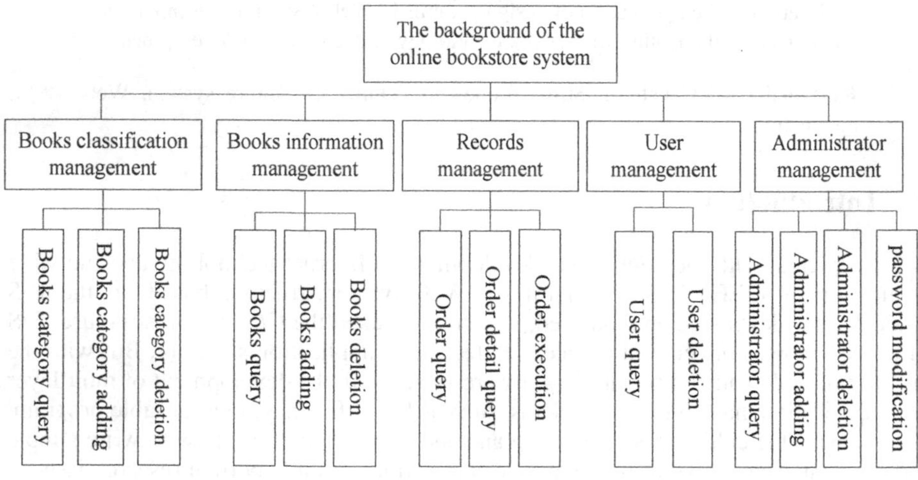

Fig. 2. The function structure of the background management module

2.2 The Database Design

From the requirement analysis can find that database entities used in the system include user information entity, administrator information entity, books entity, member order entity, member order details entity and books category entity. The corresponding database tables are UserInfo, Admin, Books, messageRecord, record and Type. The relationship between the tables as shown in Fig.3.

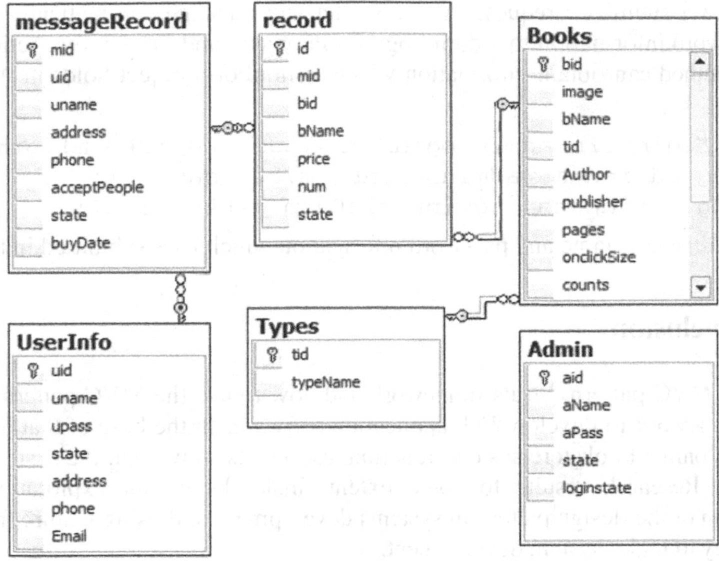

Fig. 3. The relationship between the tables

3 The Realization of the Online Bookstore System

In the development process of online bookstore system, in strict accordance with the MVC design pattern idea, use Struts framework to develop Web application. Due to limited space, here introduce the realization of model module.

Model module includes ActionForm class and Action class, many functions of the online bookstore are realized by these classes.

In the ActionForm class, use custom member variables to describe corresponding attribute in form, and at the same time generate get or set method of the member variable. For example, corresponding ActionForm class name of the administrator login page is AdminLoginForm.java, the configuration and realization code in the Struts-config.xml file is as follows.

The configuration in the Struts-config.xml file:

```
<form-bean name="adminLoginForm" type="com.web.form.AdminLoginForm" />
```

Realization code:

```
public class AdminLoginForm extends ActionForm{
private String apass;
private String aname;
public String getApass(){
return apass;}
Public void setApass(String apass){
This.apass=apass;}
Public String getAname(){
Return aname;}
Public void setAname(String aname){
This.aname=aname;}
}
```

When a user submits a request, ActionServlet will package automatically username and password information to AdminLoginForm object, and pass to the Action. Using the get method can obtain information which ActionForm object holds in Action, for example:

```
AdminLoginForm adminLoginForm=(AdminLoginForm)form;
String adminname=adminLoginForm.getAname( );
String adminpass= adminLoginForm.getApass( );
```

So can get the username and password information which user submitted in the page.

4 Conclusion

Focus on MVC pattern, Struts framework and how to use the MVC pattern based on Struts framework to develop Web application software. In the base of that design and realize an online bookstore system, function test results show that, the system can run normally. Research results to some extent, make beneficial exploration to the application of the design pattern in systems development and using mature framework technology to make system development.

References

1. Xu, X., Fei, X.: Research on the Application Software Development Framework Based on MVC Pattern. Computer Engineering and Application 41(30), 102–104 (2005)
2. Deng, Y.: Research and Application of MVC Design Pattern in Electronic Commerce System. Journal of Nanjing University of Posts and Telecommunications (2), 80–83 (2006)
3. Feng, X., Cui, Y.: Research on Development of E—Commerce Site bv MVC Patterns. Journal of Petrochemical Universities (3), 64–67 (2007)

The Evolution Strategy Implementation of the Least Trimmed Square Algorithm

Yuhua Dong and Jifeng Ding

College of Information & Communication Engineering, Dalian Nationalities University, Dalian, China
dongyuhua@dlnu.edu.cn

Abstract. The least trimmed square algorithm is a robust statistical method. Comparing with the classical least square method, it can eliminate the inference of the abnormal observed data in the process of data analysis. But the least trimmed square algorithm has disadvantages, such as on the condition of nonlinear function fitting. On the other hand, the evolution strategy method has the excellent global searching ability. The paper utilizes the evolution strategy algorithm to realize the solution to least square algorithm, which can overcome the disadvantages of the least square algorithm. Some examples show that the method is feasible.

Keywords: the least trimmed square algorithm, evolution strategy, parameter estimation.

1 Introduction

The least square algorithm is widely used in variously, which is classical statistical method to extract data and interpret data. The criterion of the least square algorithm is to minimum the square sum of estimation residual, but it is sensitive to the abnormal data. In order to overcome the characteristics of sensitive to outliers, many robust regression methods were proposed, such as least median square(LMS). It sorts the residual from high to low, and only make the residual square of median minimum. So when the ratio of median least square in outliers is up to 50%, it can still work rightly. The least trimmed square (LTS) is the improved method of LMS, it sorts the residual from high to low, and it makes the sum of the first h terms minimum, h can be adjustable according to the requirements. So LTS has better statistical characteristics than LMS, and parameter estimation is more robust. But they are helpless for nonlinear fitting. The genetic algorithm [5] was proposed by Karr to solve the least median square, and achieves better results. The method of covariance matrix adaptation evolution strategy (CMA-ES) is proposed in the paper to solve the least trimmed square algorithm, the random searching ability is more stronger. Some examples are given to prove the method feasible.

2 The Least Trimmed Square

Consider the linear regression model

$$y_i = x_{i1}\theta_1 + x_{i2}\theta_2 + \ldots + x_{ip}\theta_p + \sigma e_i = \mathbf{x}_i^t\theta + \sigma e_i \quad i = 1, \ldots, n \tag{1}$$

D. Jin and S. Lin (Eds.): Advances in CSIE, Vol. 2, AISC 169, pp. 211–215.
springerlink.com

Where $\mathbf{x}_i$ is independent variables of p dimensional vector, y_i is dependent variable, σe_i is error item. The n observations are $Z=(X,\mathbf{y})$, for a given estimation parameter $\hat{\theta}$, the residual is $r_i(\hat{\theta})=y_i-\mathbf{x}_i^t\hat{\theta}$.

The classical least square algorithm makes the residual square sum minimum, that is, make the objective function $s_{LS}=\sum_i^n r_i^2$ minimum. From the objective function, it can be seen that even an abnormal data can make the parameter estimation of least square failure. A simple improved method is to turn the objective function into the sum of residual absolute values $s_{LA}=\sum_i^n |r_i|$, then for objective function the outliers residual is amplified squarely no longer, and it can improve the outliers influence on parameter estimation to some extent. On this basis, Huber[1] developed the M estimation, the objective function is the function of residual, but it greatly increased the difficulty of solving. In 1984, Rousseeuw [2] proposed the least median square, the residual square sum are sorted from small to large, $(r^2)_{1:n}\le(r^2)_{2:n}\le\ldots\le(r^2)_{n:n}$, the middle term of residual square sum of objective function is $s_{LTS}=(r^2)_{h:n}$ $(h=[(n+p+1)/2])$. Then it was improved as the least trimmed square, and the objective function is the sum of residual squares and the first h residual squares after sorting, that is, $s_{LTS}=\sum_{i=1}^h (r^2)_{i:n}$, and h can be adjustable. When the outliers are less, set h larger, which can make the LTS performance more close to the classical least square. When $h=[(n+p+1)/2]$, the anti-outliers ability is equal to that of the least median square. Compared with the least median square, LTS objective function has smoother and better statistical ability, and it is more difficult to solve.

3 Evolution Strategy

The evolution strategy (ES) was proposed by Recheberg [3]. Evolution strategies use natural problem-dependent representations, and primarily mutation and selection, as search operators. In common with evolutionary algorithms, the operators are applied in a loop. An iteration of the loop is called a generation. The sequence of generations is continued until a termination criterion is met. As far as real-valued search spaces are concerned, mutation is normally performed by adding a normally distributed random value to each vector component. The step size or mutation strength (i.e. the standard deviation of the normal distribution) is often governed by self-adaptation. Individual step sizes for each coordinate or correlations between coordinates are either governed by self-adaptation or by covariance matrix adaptation (CMA-ES) [4].

Two main principles for the adaptation of parameters of the search distribution are exploited in the CMA-ES algorithm [6]. First, a maximum-likelihood principle is based on the idea to increase the probability of successful candidate solutions and search steps. The mean of the distribution is updated such that the likelihood of previously successful candidate solutions is maximized. The covariance matrix of the distribution is updated (incrementally) such that the likelihood of previously successful search steps is increased. Both updates can be interpreted as a natural gradient descent. Also, in consequence, the CMA conducts an iterated principal components analysis of successful search steps while retaining all principal axes. Estimation of distribution algorithms and the Cross-Entropy Method are based on very similar ideas, but estimate (non-incrementally) the covariance matrix by maximizing the likelihood of successful solution points instead of successful search steps.

Second, two paths of the time evolution of the distribution mean of the strategy are recorded, called search or evolution paths. These paths contain significant information about the correlation between consecutive steps. Specifically, if consecutive steps are taken in a similar direction, the evolution paths become long. The evolution paths are exploited in two ways. One path is used for the covariance matrix adaptation procedure in place of single successful search steps and facilitates a possibly much faster variance increase of favorable directions. The other path is used to conduct an additional step-size control. This step-size control aims to make consecutive movements of the distribution mean orthogonal in expectation. The step-size control effectively prevents premature convergence yet allowing fast convergence to an optimum.

4 The Evolution Strategy Implementation of LTS

LMS and LTS only resolve the linear problem, and they can not solve the nonlinear parameter estimation, or resolve the problem after linearizing the nonlinear problem. The initial solution is given into the objective function, then the function value is get by evolution strategy. The optimal process is determined according to the value, thus the evolution strategy is irrelevant with the linear objective function or not. So the evolution strategy can solve the difficulty of LMS and LTS very well.

5 Examples and Analysis

5.1 Linear Fitting

The simplest linear equation $y = a + bt$ is considered here. The Gauss white noise of $10db$ original signal and outliers are added to the generated simulation data, the generated original simulation data is shown in Fig.1. The fitting results based on evolution strategy LTS is also shown in Fig.1. It can be seen that the fitting exclude completely the influence of abnormal data, and recovery the original linear signal perfectly.

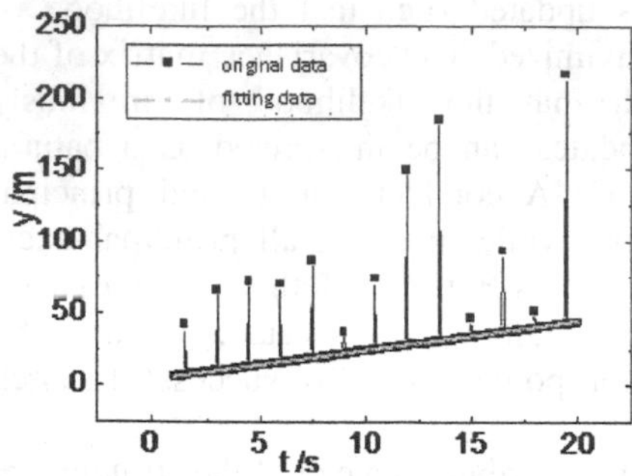

Fig. 1. the original data and fitting data by linear fitting

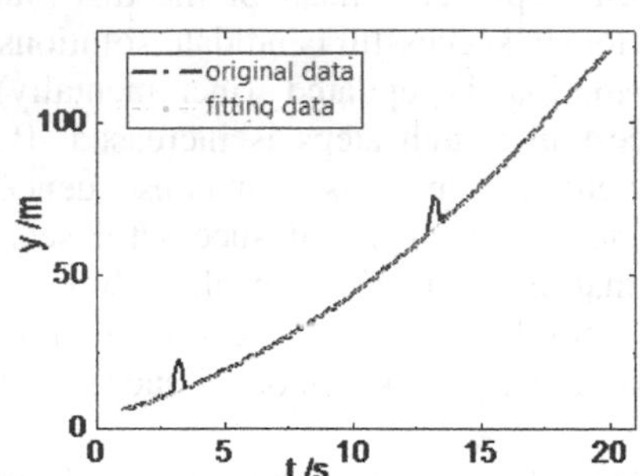

Fig. 2. original data with spotted outliers and fitting data by quadratic function

5.2 Quadratic Function Fitting

The aircraft data generally can be expressed by quadratic function, and the aircraft data may contain data bulge, that is spotted outliers. The quadratic function $y = a + bt + ct^2$ is used in this section, and the Gauss white noise and spotted outlier are added, as shown in Fig.2. and the fitting results by LTS is also shown in Fig.2. It can be seen that the fitting effect is good, and the spotted outliers do not influence the fitting precision. In the telemetry data joint process, there usually contains many data bulge. In actual data LTS fitting processing, h can be determined by the ratio of bulge, which can avoid the bulge influencing on the processing precision.

5.3 Nonlinear Fitting

There are some dynamic systems in control system, such as spring-heavy-damper system with typical two order system, the transient response under the operation of the initial displacement is

$$y(t) = \frac{y(0)}{\sqrt{1-\zeta^2}} e^{-\zeta\omega t} \sin(\omega\sqrt{1-\zeta^2}t + \theta) \tag{2}$$

Where $\theta = \cos^{-1}\zeta$, the initial displacement is $y(0)$, ζ and ω are constant value related to the mass of heavy, the elastic coefficient of the spring, and damping coefficient. Set theses parameters, then the random Gauss white noise and the abnormal data are added to the transient response, the produced data and the obtained fitting data by LTS is shown in Fig.3. From the figure, it can be seen that the LTS algorithm based on the evolution strategy for the nonlinear fitting obtained better fitting precision. As shown in Fig.3, the value of h should be restricted strictly, the data contributing to fitting should be much more. And for data with oscillator attenuation, the ability the LTS algorithm to resist the outliers is limited.

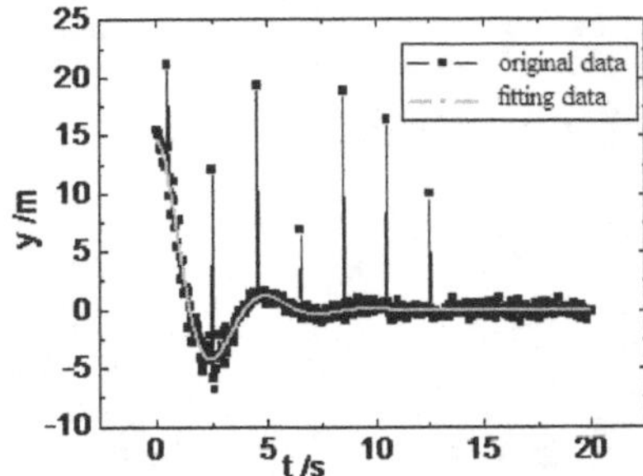

Fig. 3. the original data and fitting data by nonlinear fitting

6 Conclusions

The intelligent algorithm of evolution strategy is used to solve the least trimmed square in the paper, and examples verify that the LTS algorithm has good anti-inference performance to isolated outliers and spotted outliers. And the most important is the method can be well applied for nonlinear parameter estimation and data fitting, which has benefit to error separation of guidance instruments.

References

1. Huber, P.J.: Robust regression: asymptotics, conjectures and Monte Carlo. Ann. Stat. 1, 799–821 (1973)
2. Rousseeuw, P.J.: Least median of squares regression. Am. Statist. Ass. 79, 871–880 (1984)
3. Mitchell, M.: An Introduction to Genetic Algorithms. The MIT Press, Cambridge (1996)
4. Hansen, N., Ostermeier, A.: Completely Derandomized Self-adaptation in Evolution Strategies. Evol. Comput. 45, 159–195 (2001)
5. Karr, C.L., Weck, B., et al.: Least median squares curve fitting using a genetic algorithm. Engng. Applic. Artif. Intell. 44, 177–189 (1995)
6. `http://en.wikipedia.org/wiki/CMA-ES`

Research on the Information Security Technology of University Campus Network

Changyi Wang[1], Zhuo Zhang[1], and Xianhua Song[2]

[1] Power Supply Company of Zhengzhou, 450000, China
[2] Zhengzhou Railway Vocational and Technical College, Zhengzhou, 450000, China
{jack1975,songxianyi}@126.com, zhang-zhuo@msn.com

Abstract. With the rapid development of network technology, the network security has gained more and more attention. The modern university campus network is an open network environment, and many users are using the internal network to achieve office automation and resource sharing, so that the security of the intranet information becomes increasingly important. This paper analyzes and studies the security issues faced by current campus network as well as the firewall technology, data encryption technology, intrusion detection technology, anti-virus technology, VLAN technology and other security technologies that can solve the issues.

Keywords: University Campus, The network security, Information security, Information technology.

1 Introduction

With the development of information technology, especially the network technology, most university campuses have set up the internal local area network (LAN) for internal office and realizing resource sharing, and this has played an invaluable role in speeding up information processing, improving working efficiency, reducing labor intensity and realizing resource sharing [1]. In LAN, there are some confidential data files and information passed over the network every day. For example, files and documents with certain security classification and the financial data within the unit are all confidential data, and to protect the security of these confidential data is very important. Therefore, at the meantime of achieving the rapid information exchange within the network interoperability, how to strengthen the security within the network and prevent the university's critical data from leakage is a major research issue in the network security field.

2 Analysis of Internal Network Security Technology

At present, the network security technologies widely used mainly include the physical security analysis technology, the network structure security analysis technology, the system safety analysis technology, the management security analysis technology as

D. Jin and S. Lin (Eds.): Advances in CSIE, Vol. 2, AISC 169, pp. 217–221.
springerlink.com

well as other security services and security strategies. The following is the analysis of some important technologies.

2.1 The Intrusion Detection Technology

The intrusion detection technology is the system to identity and appropriately handles the malicious uses of the computer and network resources, and it includes the external invasion and the non-authorized behaviors of internal users. Designed and configured for ensuring the security of the computer system, it can promptly detect and report unauthorized or abnormal phenomena in the system, and it is used to detect the behaviors which violate the security strategy in the computer network.

The intrusion detection capability is an important factor to measure whether a defense system is complete and effective. Powerful and complete intrusion detection systems can compensate for the insufficiency of the firewall's relatively static defense.

Given the characteristics of the university campus network, rising's intrusion detection system (RIDS) not only has the traditional intrusion detection capability, but can also interoperate with the Rising firewall through installing the interactive plug-ins [2]. In order to more effectively detect the external and internal attacks of the campus network, corresponding RIDS is installed in each network that needs protection, thus to interoperate with the firewall through a dedicated response mode.

2.2 The Anti-virus Technology

The computer virus is the program which causes computer failure and damages computer data, and it can infect other programs and do self-replication. Especially in the network environment, the computer virus has the immeasurable threat and destructive power. Therefore, to prevent the computer viruses is an important part of the construction of campus network security, and the specific method is to frequently scan and monitor the files in the server with the anti-virus software, or to use anti-virus chips and set the access permission to the network directory and files on the workstation. For example, our university has installed the trend anti-virus software configured by the distance education center for real-time monitoring, and the results are good.

2.3 The Auditing and Monitoring Technology

Audit is the process to record the user's all activities of using the computer network, and it is the important tool to improve security. It can not only identify who has access to the system, but can also point out how the system is used. As to determine whether there is a network attack, the audit information is very important for identifying the problem and the attack source. Meanwhile, the system event log can more quickly and systematically identify the problem, and this is the important basis for dealing with accidents in the subsequent stage. Besides, by continuous collection, accumulation and analysis of the security incidents, to do selective audit trail to some of the sites or users can early detect the possible disruptive behaviors.

Therefore, in addition to using general network management software to systematically monitor and manage the system, the relatively sophisticated network monitoring equipment at present should also be used to do real-time inspection, monitoring, alarm and blocking to the common operations at all levels of LAN, thereby preventing the attacks and crimes against the network.

2.4 The Firewall Technology

The firewall is the particular network interconnection equipment to strengthen the access control between networks, prevent the external network users entering the internal network and having access to internal network resources through external network with illegal means, and protect the internal network operating environment [3]. It inspects the data packets like the lined mode transmitted between two or more networks according to certain security strategy, thus to determine whether the communication between networks is allowed and monitor the network operation.

According to the overall security of university network, two firewalls are considered to be arranged in the interconnection area, that is, the interface between the education network and the campus network as well as the interface between the INTERNET network and the campus network. The configuration scheme is as follows: Firewall 1 is to isolate the education network and the campus network, and Firewall 2 is to isolate the INTERNET network and the campus network, thereinto, WWW, e-mail and other external servers are put in the DMZ area of the firewall to be isolated from the internal networks.

2.5 The Data Encryption Technology

The data encryption technology is mainly to protect the reliability of the network security through network data encryption, and this can effectively prevent the leakage of confidential information. The emergence of encryption technology has provided a guarantee for the global e-commerce, so that the Internet-based electronic trading system becomes possible, therefore, the perfect symmetric encryption and asymmetric encryption technology is still the mainstream of 21st century [4]. The symmetric encryption is the conventional password-based technology, and the encryption and decryption operations use the same key. The asymmetric encryption means that the encryption key is different from the decryption key, and the encryption key is disclosed to the public and can be used by anyone, but the decryption key is only known by the decryption person.

2.6 The Network Host's Operating System Security and Physical Security Technologies

As the first defense line, the firewall cannot fully protect the internal network, and it must be combined with other measures to improve the system's security level. Following the firewall, it is the operating system security and physical security measures based on the network host. According to the level from low to high, it is respectively the host system's physical security, the operating system's kernel security, the system service security, the application service security and the file system security; meanwhile, the host security check, the bug fixes and the system backup security are the supplementary security measures. These constitute the second defense line of the entire network system, mainly to prevent some attacks that breakthrough the firewall and attacks launched within the system. The system backup is the last defense line of the network system, and it is used to restore the system after attacks. After the firewall and the host security measures, it is the overall safety inspection and response measure constituted by the system's safety auditing, intrusion detection and emergency response mechanism. It extracts the network status information from the firewall and network host in the network system or directly from the network link layer as the input to provide to the intrusion detection subsystem. The intrusion detection subsystem determines whether there is an invasion occurring according to certain rules, and if there is, it will start the emergency treatment measures and generate a warning message. Moreover, the system's security auditing can also serve as the information source to handle the attacks and consequences as well as improve the system safety strategy in the future. Firewall structure figure of the system can be seen in Fig 1.

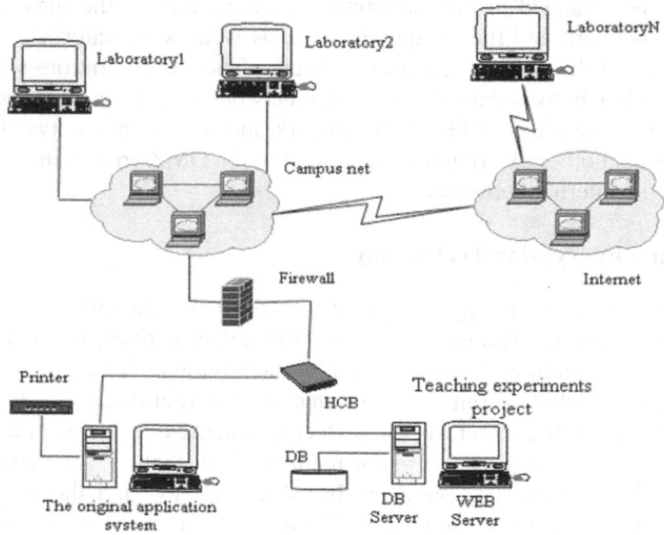

Fig. 1. Firewall structure figure

2.7 The Access Control

The access control includes authentication and access control. The authentication technology is the security technology which is the first to be applied in the computer, and it is still widely used currently, as it is the first barrier for the network information security. The access control is an important aspect of the network security theory, mainly including staffing constraint, data identification, access control, and type control and risk analysis. To use the access control and the authentication technology together can give users with different identities different operating authority, thus to achieve the graded management of information with different security classification.

2.8 The VLAN Technology

In addition to using the above technologies in the campus network security, to strengthen the campus network's security management will play an effective role in protecting the network information security [5]. As for the campus network, VLAN technology can be used to strengthen the management, and to segment and isolate the network according to different security classifications can achieve the inter-user access control, and this can reach the purpose of limiting the users' unauthorized access. The campus network can also be divided into multiple IP subnets, and the subnets are connected to each other via routers, routing switches, gateways or firewalls and other devices, so that the security mechanisms of the middleware can be used to control the access between the subnets.

3 Conclusions

The university network security is a comprehensive issue which involves technology, management, use and many other aspects, and this not only includes the security issues of the information system, but also includes the physical and logical technical measures, while one technology can only solve one kind of problem and it is not a panacea [6]. Therefore, to establish a sound network management system is an important measure for the campus network security, and the healthy and normal campus network needs to be maintained by all the teachers and students together. With the further development of computer network technology, the protection technology for university network security will also develop continuously following the development of network application.

References

1. Hu, D., Min, J.: The network security, p. 32. Tsinghua University Press, Beijing (2008)
2. Luo, B., Zhangjun: The computer virus harm and prevention. Economic and Technological Cooperation Information 1039 (October 2010)
3. Zhang, M.: Network security of actual combat explanation, p. 125. Electronics industry publishing company, Beijing (2008)
4. Case study, use a firewall, IDS, IPS structuring industrial security system [EB/OL], http://articles.e-works.net.cn/Security/Article78132.html
5. Zhang, H., Zhang, X.: Digital campus in the typical safety analysis and prevention measures. China Electrization Education, 56 (June 2009)
6. Wang, D., Gan, J.: Research and practice of constructing campus network security. China Electrization Education, 22 (January 2008)

Robot Reinforcement Learning Methods Based on XCSG

Jie Shao[*], Shuai Chen, and ChengDong Zhao

School of Computer Science, Shangqiu Institute of Technology, Shangqiu 476000, P.R. China
sj012328@163.com

Abstract. This paper proposed a robot reinforcement learning method based on learning classifier system. A Learning Classifier System is a accuracy-based machine learning system with gradient descent that combines reinforcement learning and rule discovery system. The genetic algorithm and the covering operator act as innovation discovery components which are responsible for discovering new better reinforcement learning rules. The reinforcement learning component is responsible for adjusting the fitness of rules in the system according to some reward obtained from the environment. The advantage of this approach is its accuracy-based representation, which can easily reduce learning space, improve online learning ability and robot robustness.

Keywords: Accuracy-based machine learning system with gradient descent (XCSG), Genetic algorithm, Reinforcement learning.

1 Introduction

Reinforcement learning is defined as the problem of a robot that learns to perform a task learns by trial-and-error interaction with its unknown environment which provides feedback in terms of reward and makes it more efficient and more effective in choosing its correct action[1]. However, big learning space, slow learning speed, learning outcomes and other issues of uncertainty still exists.

This paper proposed a novel approach to solve the problem of a robot reinforcement learning. XCSG is a accuracy-based machine learning system with gradient descent that combines reinforcement learning and rule discovery system. The reinforcement learning component is responsible for adjusting the fitness of rules in the system according to numerical reward obtained from the environment. The rule discovery system which includes covering operator and genetic algorithm acts as an innovation discovery component which is responsible for discovering new better learning rules. The advantage of this approach is its accuracy-based representation, which can easily reduce learning space, improve robot online learning ability and robustness.

The rest of this paper is structured as follows. Section 2, we provide the necessary background knowledge on reinforcement learning include RL, RL-GDM, XCS, XCSG and Reinforcement Learning is designed. The experimental design and results are described in Section3. Finally, conclusion and future works are given in Section 4.

[*] Corresponding author.

D. Jin and S. Lin (Eds.): Advances in CSIE, Vol. 2, AISC 169, pp. 223–228.
springerlink.com

2 Background Materials

2.1 Reinforcement Learning (RL)

Reinforcement learning [2-3] (RL) algorithms are very suitable for robot learning by trial-and-error interaction with its dynamic environment. At each time step, the robot perceives the complete state of the environment and takes an action, which causes the environment to transit into a new state. The robot receives a scalar reward signal that evaluates the quality of this transition. This feedback is less informative than in supervised learning, where the robot would be given the correct actions to take. Q-learning is one of the well-known reinforcement learning, the algorithm does not require any model. Q-value and the optimal strategy for the algorithm are updated as follows:

$$Q_{t+1}(s_t, a_t) = Q_t(s_t, a_t) + \beta * (r + \gamma * \max Q_{t+1}(s_{t+1}, a_{t+1}) - Q_t(s_t, a_t)) \tag{1}$$

$$\pi(s_{t+1}, a_{t+1}) = \arg\max Q(s_t, a_t) \tag{2}$$

Where $\beta (0 \le \beta \le 1)$ is learning rate, $\gamma (0 \le \gamma \le 1)$ is discount factor.

2.2 Q-Learning with Gradient Descent Methods (RL-GDM)

While applying gradient descent to cl_j learning, we are trying to minimize estimate error. At each time step t+1, we usually parameterize a function approximation of Q-learning by a weight matrix W using gradient descent.

$$\Delta w = \beta (r + \gamma \max Q(s_{t+1}, a_{t+1}) - Q(s_t, a_t)) \frac{\partial Q(s_t, a_t)}{\partial w} \tag{3}$$

2.3 Accuracy-Based Machine Learning System with Gradient Descent (XCSG)

The XCSG [4-7] classifier system is an LCS that evolves its classifier by an accuracy-based fitness approach. Each XCS classifier contains the usual condition, action, and reward prediction parts. Complementary, XCS contains a prediction error estimate and a fitness estimate, which represents the relative accuracy of a classifier.

2.3.1 Mapping $Q(s_t, a_t)$ to XCS

To improve the learning capabilities of XCS and reduce learning unstable, we add gradient descent methods to XCS. The system prediction is computed as fitness weighted average of classifier predictions. The relationship $Q(s_t, a_t)$ and $P(a_t)$ based on $|A|_{-1}$ is computed as follows:

$$Q(s_{t+1}, a_{t+1}) = P(a_{t+1}) = \frac{\sum cl_j \in [A]_{-1} P_j \times F_j}{\sum cl_j \in [A]_{-1} F_j} \tag{4}$$

Where P_j and F_j are the prediction and the fitness of classifier cl_j respectively.

2.3.2 Mapping Weights to XCS

The weights w of a weight matrix W are used as the function arguments to produce an approximated value in usual methods. XCS consider the value as the classifier's prediction. We estimate the gradient component for a classifier $cl_k \in [A]_{-1}$ by computing the partial derivate of $Q(s_t, a_t)$ about P_k of a classifier cl_k.

$$\frac{\partial Q(s_{t+1}, a_{t+1})}{\partial w} = \frac{\partial}{\partial P_k}\left[\frac{\sum cl_j \in [A]_{-1} P_j \times F_j}{\sum cl_j \in [A]_{-1} F_j}\right] = \frac{1}{\sum cl_j \in [A]_{-1} F_j}\frac{\partial}{\partial P_k}\left[\sum_{cl_j \in [A]_{-1}} P_j \times F_j\right] \tag{5}$$
$$= \frac{F_j}{\sum cl_j \in [A]_{-1} F_j}$$

2.3.3 Reinforcement Component with Gradient Descent

XCS with gradient descent (XCS-GDM) [8-10] works as the traditional XCS except for the update of classifier prediction. When the parameters of classifier in $[A]_{-1}$ are updated, the sum $F_{[A]_{-1}}$ of classifiers' fitness in $[A]_{-1}$ is computed as follows:

$$F_{[A]_{-1}} = \sum_{clj \in [A]_{-1}} F_j \tag{6}$$

In XCS-GDM, we regard classifier prediction as the main role in function approximation approaches. For each classifier $cl_k \in [A]_{-1}$, it's P_k is updated as follows:

$$P_k \leftarrow P_k + \beta(R + \gamma \max P(a) - P_k)\frac{F_k}{F_{[A]_{-1}}} \tag{7}$$

3 Experiments and Simulation

Robot learning environment is composed by a number of grids. If a grid has been occupied by obstacles, the grid marked by "O"; Free grid is marked by" "; The target position is marked by "G". Robot can stop at any free grid and can also move to any free connected grid. The robot has eight sonar sensors, Each is associated with detecting the status of the grid. Sensor information is coded by two binary code, 10 denote obstacles Code, 11 target location coding, 00 free code grid position, so each robot has 16-bit binary sensor input. Eight each robot moves correspond to the eight adjacent grid, assuming that the robot target position is located in the center of Fig.1.

Experimental environment is formed by a number of Grids. Simulation objective is that robot can move from the free grid to the goal G by reinforcement learning. When the robot reach the target position, the entire simulation process is completed. This paper is to test XCSG learning results in robot domain through different experimental environments.

S_0	S_1	S_2	S_3	S_4
S_5	S_6	S_7	S_8	S_9
O	O	G	S_{10}	S_{11}
O	O	O	S_{12}	S_{13}
O	O	O	S_{14}	S_{15}

Fig. 1. Robot positions

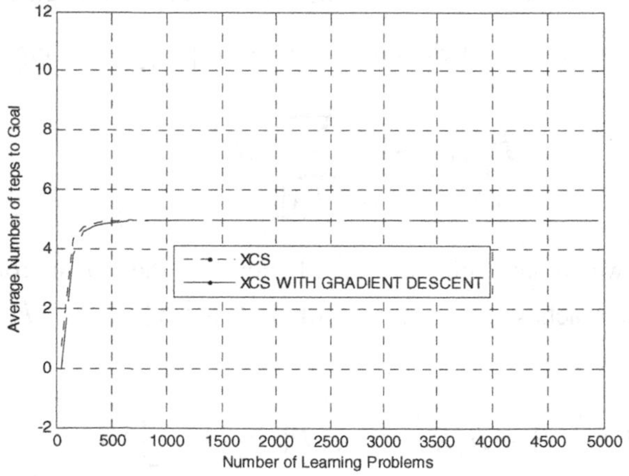

Fig. 2. RL in maze 5 environment （N=3000）

Experiment 1: RL in maze 5 environment

The experimental environment is maze 5(Fig.3), from the curve point of view, when the classifier number of individuals N = 3000, XCS and XCSG can achieve optimal operation in fig.2. However, when the classification number of individuals N = 2500, XCS learning strategies were not all be optimized, but XCSG learn optimization strategies are always stable in fig.5.

Experiment2: RL in maze 6 environment

The experimental environment is maze 6(Fig.4), the target location is more subtle compared to maze 5, which causes the robots walk more random. We can find that XCS can not optimize from the Fig.6, and XCSG can achieve stable and rapid learning optimization. All the individual classifiers can achieve global optimization compared to **Experiment 1 in maze 5 environments**

O	O	O	O	O	O	O	O	O
O							G	O
O			O		O	O		O
O		O						O
O				O	O			O
O		O		O			O	O
O		O			O			O
O						O		O
O	O	O	O	O	O	O	O	O

Fig. 3. Maze 5 environment

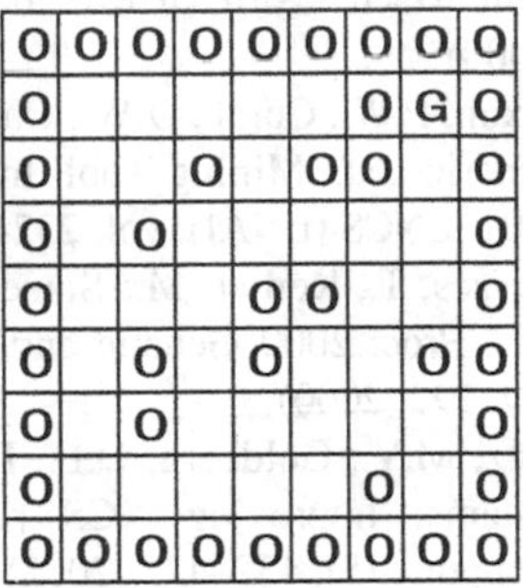

Fig. 4. Maze 6 environment

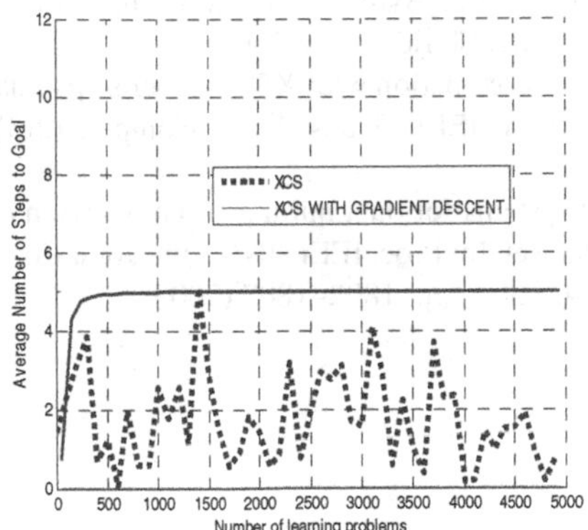

Fig. 5. RL in maze 5 （N=2500）

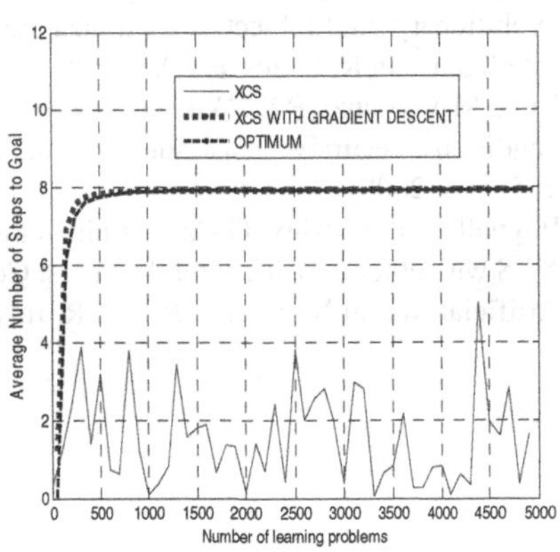

Fig. 6. RL in maze 6 （N=2500）

4 Conclusions and Future Work

We propose a XCSG method to solve multi-robot reinforcement learning problems. XCS can establish low-dimensional approximation functions mapping to the environment return value. Gradient descent technique can use on-line knowledge to establish a stable approximation of functions, so that return values of environmental mapping table has been maintained at a low-dimensional state. Algorithm analysis and simulation show that, XCS with the gradient descent method in multi-robot reinforcement is efficient and stable convergence.

References

1. Barid, L.C.: Residual algorithms: Reinforcement Learning with function approximation. In: Proc. 12th Int. Conf. Mach. Learn., pp. 30–37 (July 1995)
2. Glorennec, P.Y.: Reinforcement learning: An overview. In: Eur. Symp. Intell. Tech., Aachen, Germany, pp. 17–35 (2000)
3. Wiering, M.: Multi-agent reinforcement learning for traffic light control. In: Proc. 17th Int. Conf. Mach. Learn. (ICML 2000), June 29-July 2, pp. 1151–1158. Stanford Univ., Stanford (2000)
4. Dixon, P.W., Corne, D.W., Oates, M.J.: A Preliminary Investigation of Modified XCS as a Generic Data Mining Tool. In: Lanzi, P.L., Stolzmann, W., Wilson, S.W. (eds.) IWLCS 2001. LNCS (LNAI), vol. 2321, pp. 133–150. Springer, Heidelberg (2002)
5. Kovacs, T., Kerber, M.: Some dimensions of problem complexity for XCS. In: Wu, A.S. (ed.) Proc. 2000 Genetic and Evolutionary Computation Conf. Workshop Program, pp. 289–292 (2000)
6. Butz, M.V., Goldberg, D.E., Lanzi, P.L.: Gradient descent methods in learning classifier systems: Improving XCS performance in multistep problems. IEEE Trans. Evol. Comput. 9(5), 452–473 (2005)
7. Bernadó-Mansilla, E., Garrell, J.: Accuracy-based Learning Classifier Systems: Models, analysis and applications to classification tasks. Evolutionary Computation 11(3), 209–238 (2003)
8. Hung, K.-T., Liu, J.-S., Chang, Y.-Z.: Smooth path planning for a mobile robot by evolutionary multiobjective optimization. In: IEEE Int. Symposium on Computational Intelligence in Robotics and Automation, Jacksonville, Florida (June 2007)
9. Butz, M.V., Lanzi, P.L., Wilson, S.W.: Function approximation with XCS: Hyperellipsoidal conditions, recursive least squares, and compaction. IEEE Trans. Evol. Comput. 12(3), 355–376 (2008)
10. Bagnall, A.J., Cawley, G.C.: Learning classifier systems for data mining: A comparison of XCS with other classifiers for the Forest Cover dataset. In: Proc. IEEE/INNS Int. Joint Conf. Artificial Neural Netw., Portland, OR, July 20-24, vol. 3, pp. 1802–1807 (2003)

Research on Multi-robot Path Planning Methods Based on Learning Classifier System with Gradient Descent Methods

Jie Shao*, JunPeng Zhang, and ChengDong Zhao

School of Computer Sciencc, Shangqiu Institute of Technology, Shangqiu 476000, P.R. China
sj012328@163.com

Abstract. This paper deals with the problem of multi-robot path planning based on learning classifier system in a dynamic narrow environment, where the workspace is cluttered with unpredictably moving objects. A Learning Classifier System is an accuracy-based machine learning system with gradient descent that combines reinforcement learning and rule discovery system. The genetic algorithm and the covering operator act as innovation discovery components which are responsible for discovering new better path planning rules. The reinforcement learning component is responsible for adjusting the fitness of rules in the system according to some reward obtained from the environment. The advantage of this approach is its accuracy-based representation, which can easily reduce learning space, improve online learning ability and robot robustness.

Keywords: Learning classifier system, Genetic algorithm, path planning, Multi-robot, Gradient descent, covering operator.

1 Introduction

Multi-robot path planning [1-3] is one of the most important topics in robotics research. The goal of robot path planning is to find a trajectory for all robots from a starting position to a goal position, while avoiding collision with obstacles. Recently, a number of path planning algorithms have been developed, But the main difficulties in finding an optimal path in narrow environment arise from the fact that the analytical methods are too complex to be used in tangible and enumerative search methods are overwhelmed by the size of the search space. On the other hand, many evolutionary algorithms have been shown to be ineffective in path planning when the search space is large.

This paper proposed a novel approach to solve Multi-robot path planning in narrow dynamic environment. XCSG is an accuracy-based machine learning system with gradient descent that combines reinforcement learning and rule discovery system. The reinforcement learning component is responsible for adjusting the fitness of rules in the system according to some reward obtained from the environment. The rule

* Corresponding author.

D. Jin and S. Lin (Eds.): Advances in CSIE, Vol. 2, AISC 169, pp. 229–234.
springerlink.com

discovery system which includes covering operator and genetic algorithm acts as an innovation discovery component which is responsible for discovering new better path planning rules.

2 Accuracy-Based Learning Classifier System (XCS)

The XCS [4-7] classifier system is an LCS that evolves its classifier by an accuracy-based fitness approach. Each XCS classifier contains the usual condition, action, and reward prediction parts. Complementary, XCS contains a prediction error estimate and a fitness estimate, which represents the relative accuracy of a classifier.

2.1 Implementation Component

The initial classifier set is randomly generated. Each classifier is represented as the state, action pair in $|P|$. According to the environmental input, the match set $|M|$ is formed from the population $|P|$, and then action sets $|A|$ composed of classifier's action is generated. The final, a action based on the classifier probability is selected. The prediction array $P(a_i)$ of each action a_i is calculated by the following equation:

$$P(a_i) = \frac{\sum cl_k \in |M| a_i P_k \times F_k}{\sum cl_k \in |M| a_i F_k} \tag{1}$$

2.2 Reinforcement Component

Each classifier in the process of implementation component will obtain a reward from the environment and the reward prediction. Classifier fitness strength will be updated as follows:

$$P \leftarrow R + \gamma \max P(a) \tag{2}$$

Where $\gamma (0 \le \gamma \le 1)$ learning rate, R serves as the reward from the environment.

$$P_j \leftarrow P_j + \beta (P - P_j) \tag{3}$$

$$\varepsilon_j \leftarrow \varepsilon_j + \beta (|P - \varepsilon_j| - \varepsilon_j) \tag{4}$$

$$k_j = \begin{cases} 1 & \text{if } \varepsilon_j \le \varepsilon_0 \\ \alpha(\varepsilon_j / \varepsilon_0)^{-v} & \text{otherwise} \end{cases} \tag{5}$$

$$k_j' = \frac{(kj \times num_j)}{\sum cl_k \in |A|_{-1} (k_k \times num_k)} \tag{6}$$

Where ε_0 ($\varepsilon_0 > 0$) Control prediction errors redundancy, $\alpha(0 < \alpha < 1)$ and $\nu(\nu > 0)$ denote constant, which control decline of accuracy k rate When ε_0 is exceeded. In action set, the absolute accuracy value k is converted to a relative accuracy value k'. XCS's fitness value is to be updated based on the relative precision of the value:

$$F_j \leftarrow F_j + \gamma * (k_j' - F_j) \tag{7}$$

2.3 Rule Discovery System

The task of rule discovery system is to generate new classifiers by using covering operator and genetic algorithm. If the match set $|M|$ is empty, covering operator will generate new classifier whose condition part will be matched with the environmental message / input message and this new classifier will be added to the classifier store by replacing the worst classifier in order to keep the fixed size of the classifier store.

After reaching timed interval, XCS will select two classifiers from action set $|A|_{-1}$ by using roulette wheel for running genetic operators. To sum up, XCS can evaluate the exiting rules and discover high performance rules by using covering operator and genetic algorithm. So its capabilities of implicit parallelism, robust and potential self-adaptive learning will benefit robot reinforcement learning.

3 XCS with Gradient Descent Methods (XCS-GDM)

3.1 Mapping $Q(s_t, a_t)$ to XCS

To improve the learning capabilities of XCS and reduce learning unstable, we add gradient descent methods to XCS. The system prediction is computed as fitness weighted average of classifier predictions. The relationship $Q(s_t, a_t)$ and $P(a_t)$ based on $|A|_{-1}$ is computed as follows:

$$Q(s_{t+1}, a_{t+1}) = P(a_{t+1}) = \frac{\sum cl_j \in [A]_{-1} P_j \times F_j}{\sum cl_j \in [A]_{-1} F_j} \tag{8}$$

Where P_j and F_j are the prediction and the fitness of classifier cl_j respectively.

3.2 Reinforcement Component with Gradient Descent

XCS with gradient descent (XCS-GDM) [8-9] works as the traditional XCS except for the update of classifier prediction. When the parameters of classifier in $[A]_{-1}$ are updated, the sum $F_{[A]_{-1}}$ of classifiers' fitness in $[A]_{-1}$ is computed as follows:

$$F_{[A]_{-1}} = \sum_{clj \in [A]_{-1}} F_j \tag{9}$$

In XCS-GDM, we regard classifier prediction as the main role in function approximation approaches. For each classifier $cl_k \in [A]_{-1}$, it's P_k is updated as follows:

$$P_k \leftarrow P_k + \beta(R + \gamma \max P(a) - P_k)\frac{F_k}{F_{[A]_{-1}}} \tag{10}$$

4 Experiments and Simulation

Experimental scenario 1 Dynamic narrow environment with U-shaped obstacles
Figure 1 shows Multi-robots in the narrow context of the U-shaped trajectory. Dynamic obstacles (No. 1-3) were in their narrow U-shaped environment for up and down movement, four were successfully reach the ultimate goal of the robot point G point.

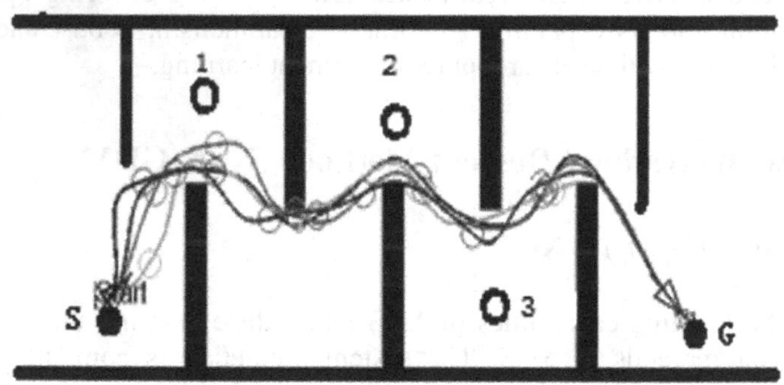

Fig. 1. Multi-robot trajectory in the U-shaped environment

Experimental scenario 2 Multi-barrier in narrow dynamic channels environments
Figure2 is the in the multi-robot trajectory in narrow channel environment. Dynamic obstacles (No. 1-3) were down as a straight line in their environment, all involved in the planning of the robots have achieved satisfactory simulation curve. Presented in this paper XCSG based integration algorithm, because all XCSG have strong ability to forecast returns, all the stability of the robot can achieve a satisfactory convergence effect for multi-robot trajectory in the narrow dynamic environment.

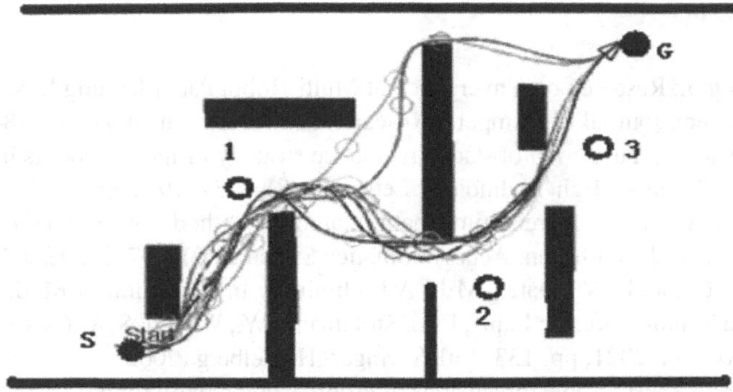

Fig. 2. Multi-channel robot trajectories in narrow channel environment

Experimental scenario 3 Multi- punch obstacles in narrow dynamic environments
Figure 3 is the trajectory of the multi-robot systems with multi-punch obstacles in narrow environment. Although there are several easy to fall into local minimum risk, the robots (R1, R2, R3) can still be very stable to reach their goals.

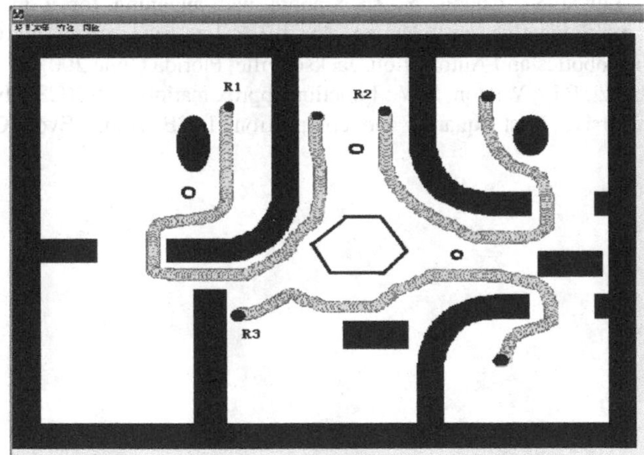

Fig. 3. Trajectories of multi-robot in multi-convex and concave narrow environment

5 Conclusions

This paper presented a novel approach to solving the problem of multi-robot path planning. XCSG is a accuracy-based machine learning system that combines covering operator and genetic algorithm. The covering operator is responsible for adjusting precision and large search space according to some reward obtained from the environment. The genetic algorithm acts as an innovation discovery component which is responsible for discovering new better path planning rules.

References

1. Shao, J., Yang, J.: Research on Cnvergence of Multi-Robot Path Planning Based on Learning Classifier System. Journal of Computer Research and Development 47(5), 948–955 (2010)
2. Zheng, M., Cai, Z., Yu, J.: An obstacle avoidance strategy of mobile robots under dynamic environments. Chinese High Technology Letters 16(8), 813–819 (2006)
3. Zhang, C.G., Xi, Y.G.: A real time path planning method for mobile robot avoiding oscillation and dead circulation. Acta Automatica Sinica 29(2), 197–205 (2003)
4. Dixon, P.W., Corne, D.W., Oates, M.J.: A Preliminary Investigation of Modified XCS as a Generic Data Mining Tool. In: Lanzi, P.L., Stolzmann, W., Wilson, S.W. (eds.) IWLCS 2001. LNCS (LNAI), vol. 2321, pp. 133–150. Springer, Heidelberg (2002)
5. Kovacs, T., Kerber, M.: Some dimensions of problem complexity for XCS. In: Wu, A.S. (ed.) Proc. 2000 Genetic and Evolutionary Computation Conf. Workshop Program, pp. 289–292 (2000)
6. Butz, M.V., Goldberg, D.E., Lanzi, P.L.: Gradient descent methods in learning classifier systems: Improving XCS performance in multistep problems. IEEE Trans. Evol. Comput. 9(5), 452–473 (2005)
7. Bernadó-Mansilla, E., Garrell, J.: Accuracy-based Learning Classifier Systems: Models, analysis and applications to classification tasks. Evolutionary Computation 11(3), 209–238 (2003)
8. Hung, K.-T., Liu, J.-S., Chang, Y.-Z.: Smooth path planning for a mobile robot by evolutionary multiobjective optimization. In: IEEE Int. Symposium on Computational Intelligence in Robotics and Automation, Jacksonville, Florida (June 2007)
9. Butz, M.V., Lanzi, P.L., Wilson, S.W.: Function approximation with XCS: Hyperellipsoidal conditions, recursive least squares, and compaction. IEEE Trans. Evol. Comput. 12(3), 355–376 (2008)

Robot Hybrid Architecture Based on Parallel Processing of Multi-DSP

YongBin Ma, JunPeng Zhang, Jie Shao, and ChengDong Zhao

School of Computer Science, Shangqiu Institute of Technology, Shangqiu 476000, P.R. China

Abstract. This paper proposed a new robot hybrid architecture based on that multi-DSP system. In the system, cooperation-planning layer undertakes the parallel processing, which combine different system structure adopted in different periods for development of robot, established robot hybrid architecture based on Parallel processing of multi-DSP, DSP hardware realizes and concrete application, in order to reduce the system structural design time, the common ability, real-time character and dependability of the security system, and offer the theoretical foundation for the fact that robot cooperated with the design of the control system. And easy to realize the robot signal communication and system's expansion, enhanced the entire robot assembly system's Practicality, Real-time and Reliability.

Keywords: robot, hybrid architecture, multi-DSP.

1 Introduction

Robot technology has already been widely used in many fields, urged by investigation and application, research of robot has been a hot field. Reasonable robot architecture is the important basis of finishing the robot architecture. In some task orientation applications, reasonable robot architecture and efficient navigation are key technologies of robot navigation, and also the requirement of the rapid development of multi-robot system navigation technology. Based on the analysis of traditional robot architecture, a novel control architecture which integrated DSP with artificial potential field method techniques is proposed for robot architecture.

The system structure of multi-intelligent robots DSP is defined between every part of intelligent robot system interaction and function are distributed, determine the information flow relation of robots each other or a lot of intelligent robot systems and logical calculation structure. The system structure is an important content to robot systems study, it mainly studies how to organize and control the hardware of the robot and software system to realize the necessary work [1] [7]. Robot systems are composed of individual robots. As the key part of the individual robot, the control system of the robot has determined the cooperation ability of the robot in many robots systems. This architecture realized the flexible combination of basic function modules of robot architecture, which made the function and knowledge have better expansibility.

D. Jin and S. Lin (Eds.): Advances in CSIE, Vol. 2, AISC 169, pp. 235–240.
springerlink.com

2 Traditional Robot System Structures

2.1 Serial System Structure

The serial system structure follows a serial control line controlled from electrical machinery that the environment is perceived, the modeling, planning, Act, and according to the flow direction of information and behavioral function, resolve the control course of the robot into different function module, the module close the ring chain after making up one, information is flowed into via the sensor by the environment, the environment of returning via the executive body after planning decision is dealt with, thus implement the behavior of controlling. See Fig. 1. The intelligence of this system lies in planners or programmer, but not the system of executing the task, the control between every part flows and information flow is one-by-one in the system, but planning and modeling of such systematic structure are very difficult, and the whole serial connection of system structure is relatively long, can't meet the real-time change of the external environment condition.

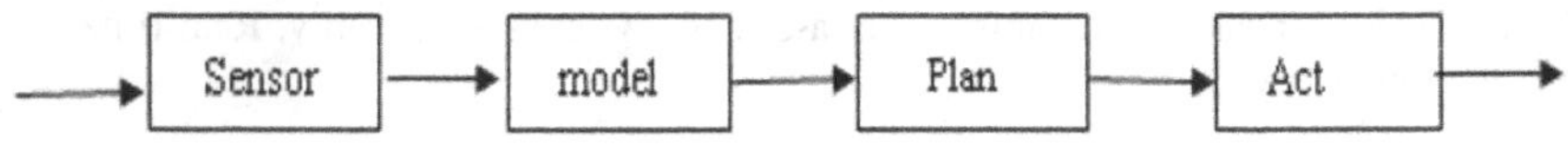

Fig. 1. Serial system structure

2.2 Subsumption System Structure

Subsumption structure proposed by Rodney A Brooks in 1986 [2], while exporting ports in input in the traditional structure, that is to adopt the network level from bottom to top to construct the structure between sensor system and act system, upper layer behavior may behavioral output to lower floor inhibitory activity has produced. Because of setting up avoiding hindering, avoiding hindering roaming, following three control systems, subsumption structure has strong real-time character. Because all control system is designed as independent module, and should interferes the behavioral internal function of lower floor in upper layer behavior each other, it is unable to carry on the independent design to the system, it causes the redesign of the whole control system.

2.3 Pass the Stepwise Structure Hierarchically

G. Saridis proposes passing the stepwise system structure [3][5] hierarchically. The system structure well arranged the structure of this system, and easy to realize, interrelate by way of passing the steps hierarchically between the module, but each layer can only exchange information instead of a link from head to foot storey, the lower floor should wait for the planning of the upper layer, the upper layer should wait for the task of the lower floor to finish, the complex reaction time of the foreign department incident is relatively long.

2.4 Response Type System Structure

The response type system structure based on Motor Schema that Arkin put forward[4], see Fig. 2. The system structure allowed each Motor Schema to correspondent to a basic behavior, strengthened systematic flexibility, and designed the module which can meet different tasks and environmental demand, easy to reconstruct, suitable for the trends, open environment.

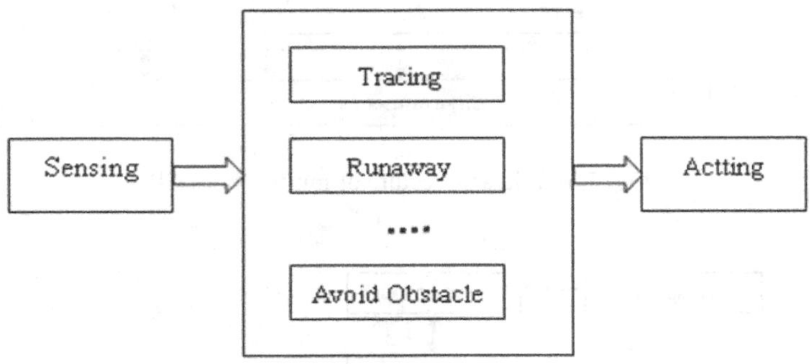

Fig. 2. Response type system structure

3 Robots Hybrid Architecture (Rha)

Robot DSP Mixed control system divide into 4, successively from top to bottom functionally: Task Coordinates Layer (TCdL), Behavioral Planning Layer(BPL), Behavior Control Layer(BCL) and sense and act Controls layer. See Fig.3. Task coordinate layer and behavioral planning layer form upper strata control, behavior control layer and sport control layer form ground floor control. The upper strata control through the analysis in information such as the operation state of environmental information and robot that is perceived to the transducer and deduce, deal with task scheduling and trouble of every function module while controlling in the ground floor effectively, so as to ensure that moves high-efficient and dependability that the robot operates wholly; The ground floor is made up by a lot of DSP function module independently each, finishes each of photo electricity correctly separately.

4 Robot Architecture Based on Multi-DSP

The hardware is controlled and divided into two parts :upper layer controlling and ground floor control, which adopt CAN bus and link some network of the controller between the control system of the upper layer and control system of the ground floor and between DSP process systems in the ground floor, can realize reliable data communication and real-time, high-efficient task scheduling well. The block diagram of the hardware is seen Fig4

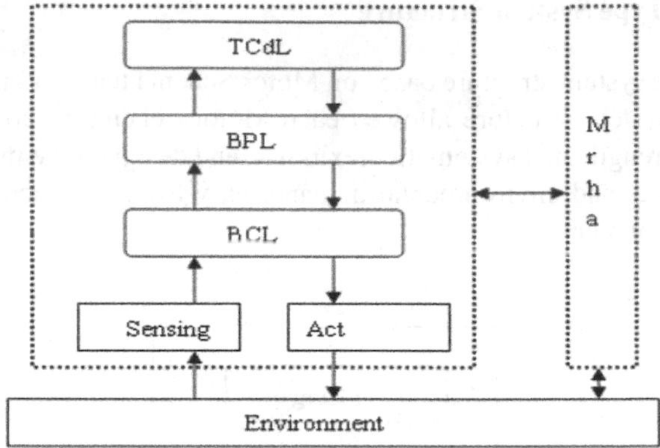

Fig. 3. Hybrid architecture of multi-robot DSP

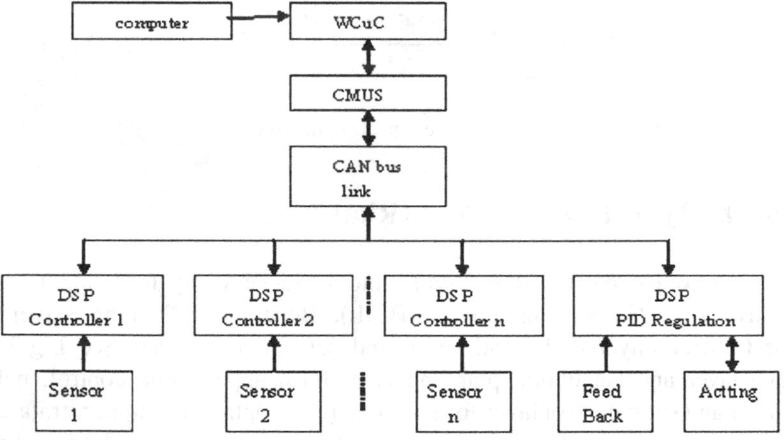

Fig. 4. The block diagram of the hardware

Note: WcuC: Wireless communication card, CMUS: Controlling machine of upper strata,PID: proportion Integration Differentiation

4.1 Upper Layer Control Systems

In order to realize the real-time communication between every layer within the robot, adopt CAN bus card of the industrial standard to connect each job module in the ground floor DSP control system in the control system of the upper strata, the ones that are used for realizing the robot and controling personnel are real-time and mutual, transmit and realize the pronunciation control of the robot through the wireless, can convey other working orders. Meanwhile, the robot can be transmitted the real-time visual scene and vision result by the wireless, infrared, supersound, close to the transducer to survey robot states such as the result, etc. to send the master station, so that controling personnel further assign the task order, realize the good one man-machine harmoniously and mutually.

4.2 Control System of Ground Floor

In the control system of ground floor, every sensor signal sends into corresponding DSP system and deals with respectively. See Fig. 5. These DSP process system adopt TMS320LF2407A that TI Company produced, DSP chip is specially employed to the controlled field, system structure that high-speed signal deals with and the digital control function need, its order carries out the pace and is up to 80MIPS, and most orders can finish during the single cycle of one 25ns. In addition, it also has I / O port and other ancillary equipment in very strong slice, can simplify the peripheral circuit to design, reduce the systematic cost.

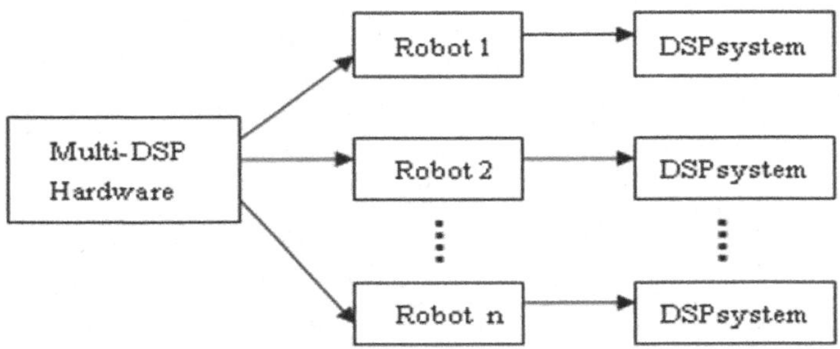

Fig. 5. Robots hardware block diagram

5 Conclusions

Reasonable robot architecture is the important basis of finishing the robot task. Based on the analysis of traditional robot architecture, a new control architecture which integrated DSP with artificial potential field method techniques is proposed for robot hybrid architecture. This architecture realized the flexible combination of basic function modules of robot architecture, which made the function and knowledge have better expansibility.

References

1. Tan, M., Wang, S., Cao, Z.Q.: Muti-robots systems. The publishing house of Tsing-Hua University, Beijing (2005)
2. Brooks, R.A.: A robust layered contol system for a mobile robot. IEEE Journal of Robotics and Automation (1986)
3. Saridis, G.: Toward the realization of intelligent controls. Proceeding of the IEEE (1979)

4. Arkin, R.C.: Motor schema-based mobile robot navigation. The International Journal of Robotics Research (1999)
5. Zhang, P.R., Zhang, Z.J., Zheng, X.D., Zhang, H.B.: The control system of the robot is designed and realized. The publishing house of Tsing-Hua University, Beijing (2006)
6. Chen, X.D., Sun, Y., Jia, W.C.: The robot sports of walking of many feet are planned and controlled. Publishing house of Central China University of Science and Technology, Wuhan (2006)
7. Fang, J.J., He, G.P.: Intelligent robot. The industry of publishing house of chemical industry equips information publishing centre, Beijing

Method of Image Fusion Based on Improved Probability Theory

Chengdong Zhao[1,2], Xuhui Wang[2], and Jie Shao[1]

[1] Shang Qiu Institute of Technology, He Nan Shangqiu 476000
[2] Nan Jing University of Science and Technology, Jiang Su Nan jing 210094

Abstract. Image Fusion Technique is an new research issue. This essay brings forward to a skill of compromising use ratio and image which can efficiently decrease interference made in compromising pictures and provides more abundant, accurate and reliable information according to physical characteristics generated by sensors, so further analysis and management can be done.

Keywords: probability theory, sensor, Image Fusion.

1 Introduction

Image Fusion, an important branch of information fusion, refers to maximatily extract useful information of photo data processed by image-processing and computer technology, and finally synthesize images of high quality to raise the utilization rate of image information. Until now great achievements have been made due to researchers' efforts. Image fusion based on pyramid [1], firstly propounded by Buri and Adelson, and is a method of image fusion of multi-scale and multi-resolution. Its fusion process is carried on the groundwork of different degree, spatial resolution and analysis. Though data in the telescoping-decompounding are redundancy, and the gross value should have been increased by more than one third comparing with original image [2]. While image fusion based on wavelet transform [3], multiple-scales and multires are much better, whereas it may extend output-time when decomposition progression is large [4]. Yet affine algorithm can automatically carries out medical image fusion with accordance with the gray level medical images [5], using pixel fusion algorithm.

Owing to different physical characters produced by sensors, problems maybe come up in image fusion procedure. For instance, probable complementation of different sensors' informational features, polarity contrary of local contrast of visible lights and infrared rays and interference that sensor may suffer. To overcome the above flaws, this essay propounds a method of image fusion based on improved probability theory [6]. Its procedure can be showed as below:

Step1: Sensors images are needed to be modeled,

Step2: Image should be multires—dissected in image fusion based on pyramid Laplacian.

D. Jin and S. Lin (Eds.): Advances in CSIE, Vol. 2, AISC 169, pp. 241–246.
springerlink.com

Step3: Fuse them using BM DOA Estimator
Step4: Images can be gotten by reversing pyramid.

The new point of the research is that all sensor images are changed into standard image by additive noise affine transformation. And image fusion is turned into likelihood estimation of all kinds of sensors. Yet BM DOA framework affords ways of approximate evaluation, maximum likelihood estimation and maximum posteriori probability estimation

2 Analysis and Design of Image Fusion Based on Probability

2.1 Design and Realization of Fusion Process

Under the condition of same illumination, unlimited visibility and noiseless sensors, supposedly take imaging real scene where image made by real sensor may be transformed as s, while in

$$a_i(l,t) = \alpha_i(l,t) + \beta_i(l,t)s(l,t) + \varepsilon_i(l,t) \tag{1}$$

a_i as image in sensor number I, $l \equiv (x,y,k)$, (x,y) as pixel position, k as the k layer pyramid, t as time, α as sensor offset, β as sensor gain(including local adverse effect and complement effect), ε as zero-mean noise.

Imaging parameter and noise location are supposed to be very slow, so some parameter in small space region is real and available. Parameter estimation can be simplified taking advantage of this hypothesis.

Next, the Bayesian Estimation fusion [7] will be carried on. Supposedly, s as Gaussian normal distribution of mean value $s_0(l,t)$ which can be gotten from image in fine day, variance $\sigma_s^2(l,t)$. Yet conditional probability density function $P(a|s)$ is supposed to normal location, mean value to $\beta s + \alpha$, variance to $\Sigma_\varepsilon = diag[\sigma^2_{\varepsilon_1}, \sigma^2_{\varepsilon_2}, \cdots, \sigma^2_{\varepsilon_q}]$, therefore marginal probability density function $P(a)$ should be normal location, mean value should be $\mu_m = \beta s_0 + \alpha$, variance

$$C = \sum\varepsilon + \sigma_s^2 \beta\beta^T \tag{2}$$

Posteriori probability density function $P(a|s)$ should be normal location as well, mean value should be $M^{-1}[\beta^T \Sigma \varepsilon^{-1}(a-\alpha) + s_0 / \sigma^2_s]$, variance $M^{-1} = (\beta^T \sum \varepsilon^{-1} \beta + 1/\sigma^2_s)^{-1}$. With the above probability density definition, obviously there are two fusion methods that can be chosen—maximum likelihood estimate $ML\,\hat{s} = \max_s P(a \mid s)$ and maximum posteriori probability estimation

MAP posteriori fusion estimation shall be its mean value

$$\hat{s} = (\beta^T \sum_{\varepsilon}^{-1} \beta + 1/\sigma_s^2)^{-1} [\beta^T \sum_{\varepsilon}^{-1} (a - \alpha) + s_0 / \sigma_s^2] \tag{3}$$

If $\lim \sigma_s^2 \to \infty$, the ML fusion estimation can be received.

2.2 Model Parameter Estimation

Parameters are needed to estimated including $\alpha(l,t)$, $\beta(l,t)$ and $\sum\varepsilon(l,t)$. In order to estimate β、α, β、α、$\sum_\varepsilon$、s_0 and σ_s^2 are supposedly close to constant in a small space region. If α is seen as the extreme value of its logarithm likelihood function $L = \sum_{n=1}^{N} \lg[P(a_n)]$, the maximum likelihood estimation can be gotten.

$$\alpha_{ML} = \mu_a - \beta_{s_0} \tag{4}$$

(Σ_a Is variance of this region 5×5), the solution is

$$\beta_{ML} = \Sigma_\varepsilon^{\frac{1}{2}} \tilde{U} \frac{(\tilde{\lambda} - 1)^{\frac{1}{2}}}{\sigma_s} r \tag{5}$$

$\tilde{U}$、$\tilde{\lambda}$ are the feature vectors of matrix $\Sigma_a = \Sigma_\varepsilon^{\frac{1}{2}} \Sigma_a \Sigma_\varepsilon^{-\frac{1}{2}}$ and eigenvalue r=±1

A way except maximum likelihood estimation [8] is the method of least square. Least squares estimate α_{LS} can be realized by making of α minimal. β_{LS} By making Minimal the solution is

$$\beta_{LS} = \frac{\lambda^{\frac{1}{2}}}{\sigma_s^2} U r \tag{6}$$

in which U, λ are the feature vectors of co variation matrix $(\sum_a - \sum_\varepsilon)$ and eigenvalue r=±1.

2.3 Defect Analysis and Assumption about Solution

The parameter estimation mentioned above do not afford the value of s_0 and σ_s^2. Supposedly, we can set up a single well-work solid pattern, however it has to strengthen constraint to combine all part models. Making $\|\beta\| = 1$ tenable all the time or $\sigma_s^2 = \lambda$ in the formula (8). λ Is the main eigenvector of $\sum_a - \sum_\varepsilon$, moreover change rate a depends on the change of s. Therefore we hope $\lambda \propto \sigma_s^2$ and each model is restrict and ensured to have the same rate. Then notice that changing s_0 also causes $\hat{s}$'s change. s_0 Have to be always zero in order to get the consistency of areas, so choose limit parameter estimation of s_0 and σ_s^2 is $\boldsymbol{\beta}_{LS} = r\boldsymbol{U}$, further

$$\boldsymbol{\alpha}_{LS} = \boldsymbol{\mu}_a \tag{7}$$

space average is used to calculate sampling mean value and variance to estimate $\boldsymbol{\beta}$ 、 $\boldsymbol{\alpha}$, which is not consistent because s_0 is space change attributing to σ_s^2 and leading to its over-estimation (for instance ,there are regions in scene) can not record clearly model's mean value and variance. Although a more integrated model should imitate s_0's space change, though it can produce bigger change.

Finally, parameter r does not show clearly in each super-pixel. Each super-pixel is chosen carefully to combine part models and super-pixel's show shift to another $\boldsymbol{\beta}$. Arbitrary direction signs leaded to by large location change are not permitted, yet a simple heuristic rule can do.3 Comparisons between MAP Algorithms and PCA Algorithms.

MAP and ML fusion law is quite close to PCA. Noise is supposed to be variance $\sum_\varepsilon = \sigma_s^2 I$.

$$\hat{s} = \frac{1}{1+\sigma_\varepsilon^2/\sigma_s^2} U_a^T (a-\mu_a) + \frac{1}{1+\sigma_s^2/\sigma_\varepsilon^2} s_0 \tag{8}$$

3 Experiment Result and Analysis

Applying the fusion method to runway with ALG system of visible light and infrared rays, experiment result will be show as below

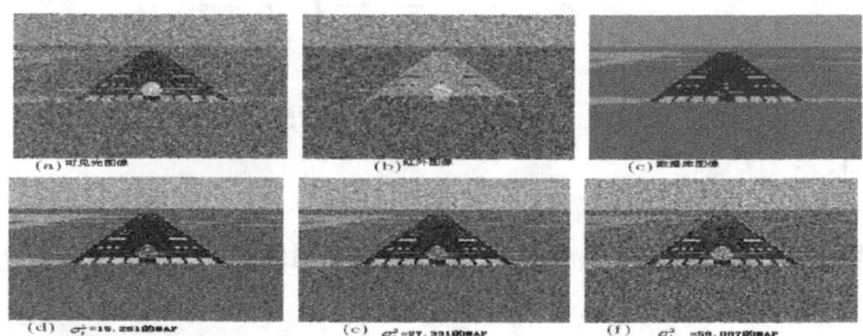

Fig. 1. Image fusions of Visible Light and Infrared Rays Imitated by Data Image

Experiment 1 Fig 1 is result of experiment fused by data image. Fig2 (a) and Fig 2 (b) imitate original images of noise of visible light and infrared ray, the scene is runway in airport. Fig2 (c) is the image of same scene from topographic database. Although the Fig can be seen clearly, It does not show the real shape of runway. We can take image luminance in database as mean value s_0 of MAP fusion law in formula (5), what measure in image as variance σ_s^2 (parameter β 、 α and scene noise variance have to be accurately estimated before). Fig 2 (d) (e) (f) are MAP fusion image of σ_s^2 12.251, 27.331, 58.087. The higher the value of σ_s^2 is, the more contribution the original image emphasizes, while low σ_s^2 shows the greater contribution.

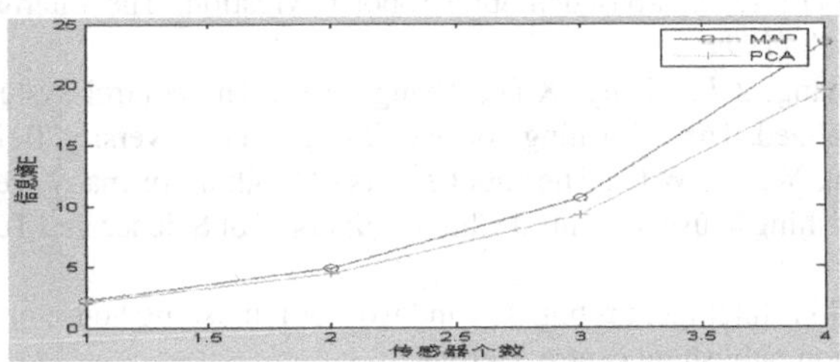

Fig. 2. Comparisons between Amount of Information of MAP and PCA Algorithm

Experiment 2 Fig 2 is the comparison of the amount of information E in the final image fusion of MAP and PCA algorithm

$$E = -\sum_{q=1}^{n} p_q \log 2 p_q \tag{9}$$

Pq As conditional probability [9] that sensors put on expected image, q=1.....n as the number of sensor.

From the result of analogue experiment, the amount of information of fusion image of MAP algorithm is more than PCA algorithm, the more the amount of information is, the better the fusion quality is (8),(9). As the number of sensors increases, the differences of this two amount of information are great, which can indicate that the MAP algorithm indeed ensure efficiently the amount of information of image fusion in imaging system of many sensors.

4 Conclusions

Image probability fusion methods put forward in this essay applies to imaging system of many kinds of sensors with visible light, infrared ray and radar. It not only combines firm framework like PCA algorithm, but also affords a method of fusing information of imaging form database. The result of simulation experiment shows that this method can fuse image efficiently, enhance obviously image definition of final image and information capacity.

In recent years, unceasing emergence of efficient image fusion methods make full cognition and being recognized of application of image fusion in medical science, remote sensing, computer vision, weather prediction and military target recognition, which is believed to overcome research direction of some technical difficulty

References

1. Tan, M., Wang, S., Cao, Z.Q.: Muti-robots systems. The publishing house of Tsing-Hua University, Beijing (2005)
2. Brooks, R.A.: A robust layered contol system for a mobile robot. IEEE Journal of Robotics and Automation (1986)
3. Saridis, G.: Toward the realization of intelligent controls. Proceeding of the IEEE (1979)
4. Arkin, R.C.: Motor schema-based mobile robot navigation. The International Journal of Robotics Research (1999)
5. Zhang, P.R., Zhang, Z.J., Zheng, X.D., Zhang, H.B.: The control system of the robot is designed and realized. The publishing house of Tsing-Hua University, Beijing (2006)
6. Chen, X.D., Sun, Y., Jia, W.C.: The robot sports of walking of many feet are planned and controlled. Publishing house of Central China University of Science and Technology, Wuhan (2006)
7. Fang, J.J., He, G.P.: Intelligent robot. The industry of publishing house of chemical industry equips information publishing centre, Beijing

Advanced Incremental Mining Algorithm for Risk Analysis Based on the Association Rules

Ying Mei

Computer and Information Engineering Department
Guangzhou Maritime High College
Guangzhou, 510725, China
yzsmcg@sina.com

Abstract. This paper introduces improving rate and proposes the incremental mining algorithm with the weighted model for optimizing association rules based on CBA mining algorithm. The risk analysis of the strong association rules is proposed for trend forecasting. And the risk degree of the lost rules based on the incremental mining is also analyzed. Comparing with the traditional algorithm, the improved algorithm is fast, efficient in incremental data mining and can find trends in association rules. The decision making reliability is enhanced by the association rules obtained from the improved algorithm. The algorithm was used in bank cost analysis with test results showing that the prediction precision of the algorithm is better than that of the traditional algorithm.

Keywords: incremental mining, risk analysis, bank cost analysis.

1 Introduction

Apriori algorithm is the way that is commonly used for mining association rules' frequent item set. But it mostly finds association rules from the static information database. In factual application, this algorithm has some shortcoming. The paper analyzes the change of the strong association rules with data updating in database by incremental mining algorithm according to factual need of decision-making. We analyze the strong association rules by the weighted model and consider the risk degree of the lost association rules. Accordingly, we improve the significance of incremental updating association rules mining and efficiently forecast the trend of rules.

2 Traditional CBA Algorithm

CBA is one traditional mining algorithm using in bank cost analysis. CBA algorithm produces classifier through two processes. The first process, find classification association rules(CAR). The second process, choose high prior degree rules for cover training set from the discoverable CAR, that is, if the left of the associations rules are the same and the right of them are different, we choose high-cof. rules as possible

D. Jin and S. Lin (Eds.): Advances in CSIE, Vol. 2, AISC 169, pp. 247–252.
springerlink.com

rules. After obtaining association rules that satisfy the min-sup and min-cof. between the conditional attributes and the decision attributes, the importance of the rules is defined.

Traditionally, CBA was used in bank cost analysis. In part 2 and part 3, we'll discuss a new mend of the original algorithm.

3 Introduction of the Improving-Rate Factor

The higher the cof. is, the stronger the association rules is, namely, the strong association rule. But sometimes it is wrong. In circumstance of precondition and conclusion both with high-sup, although the precondition and conclusion are irrelative, their association cof. is high. One better way to judge the intensity of association rules is to compare the cof. with the standard value of the rule, here we suppose that relative item set's generate probability created with every rule and generate probability of former item set is independent. We can use frequent item set's frequency to calculate the standard value. Standard cof. is the concomitantly generate association's sup divided by the number of transaction in database. And this can help us calculate "the improving-rate" of the rule. "The improving-rate" is the cof. of the rule that is divided by the cof. of generate association supposed to be independent. If the rate is bigger than 1, it means that the rule is useful. And the bigger "the improving rate'' is, the higher the association rule's intensity is.

4 Incremental Mining Algorithm

Apriori algorithm is the way that mines and analyzes the static information data. When the data in database keeps invariability, it is useful. But when the data is changed, it needs to scan the new database over again for adapting the increase of data in database. With the updating of the database, the original rules are not likely to keep identical. In the modern data management circumstance that information increases rapidly, it results in much repeated work and over-load of database's operation. And it wastes the original result obtained by old database mining.

We use the incremental mining algorithm to mend the original algorithm. The incremental mining algorithm takes full advantage of the old mining results and mines updating association rules with the incremental portion of the database. And it can efficiently decrease the database scanning times and sufficiently improved the efficiency of data mining. Suppose D for the original database, and d for changing data set. The incremental updating of the association rules mostly means how to get association rules of $D \cup d$ when the new data set d is added to the original database D or d is deleted from D with the changeless min-sup and the changeless min-cof (sometimes they can changed). For the incremental updating database, incremental mining concerns the following four instances of item set: (1) frequent in D , frequent in d , then frequent in $D \cup d$; (2) frequent in D, not frequent in d , then uncertain in $D \cup d$; (3) not frequent in D, frequent in d , then uncertain in $D \cup d$; (4) not frequent in D , not frequent in d , then not frequent in $D \cup d$.

For satisfying high efficiency updating of association rules, it creates many kinds of data mining algorithm as FUP, IUA, AIUA, NEW FUP and FUFIA.

FUP algorithm solves (1), (2) and (4), but it can not solve the third. IUA and NEW FUP algorithm solve (3), but the efficiency isn't high.

5 Improvement of Original Algorithm by Incremental Mining Algorithm

In Apriori algorithm of original association rules mining, if the database is updated, the rules created by original database mining will be not appropriate. The ecumenical way is to mine the new database again using the algorithm. In the modern data management circumstance that information increases rapidly, it results in much repeated work and over-load of database's operation. And it wastes original result obtained by old database mining. The ameliorative algorithm pays attention to the trends of rules development when mines database by association rules algorithm. The new algorithm reduces the number of database scan, and improves efficiency of data mining. It has better precision in decision-making support.

We use incremental mining algorithm to mend the original algorithm. For protecting rules, we present a rule protection model with weight. This rule protection model with weight will exert forecast function in the latter work of incremental mining. Suppose W_1 and W_2 are the weight of D and d respectively. For association rule $X \Rightarrow Y$, we define the sup and cof. as the following[5]:

$$Supp\ w(X \cup Y) = W_1 \bullet Supp_1(X \cup Y) \\ + W_2 \bullet Supp_2(X \cup Y), confw(X \Rightarrow Y) = \\ Supp\ w(X \cup Y) / Supp\ w(X)$$

Here, $Supp_1(X \cup Y)$ and $Supp_2(X \cup Y)$ are the rule's $X \Rightarrow Y$ sup. in D and d respectively. $Suppw(X \cup Y)$ and $confw(X \Rightarrow Y)$ are the rule`s $X \Rightarrow Y$ sup. and con. in $D \cup d$ respectively.

For association rule $X \Rightarrow Y$, its sup. and cof. are defined as the following[5]:

$$Supp\ w(X \cup Y) = W_1 Supp_1(X \cup Y) + \cdots + \\ W_n Supp_n(X \cup Y), \\ Confw(X \Rightarrow Y) = W_1 Conf_1(X \Rightarrow Y) + \cdots + \\ W_n Conf_n(X \Rightarrow Y),$$

Here, $Supp_i(X \cup Y)$ and $Conf_n(X \cup Y)$ are rule's ($X \Rightarrow Y$) sup. and cof. in incremental set D_i.

First, we use AIUA(advanced incremental updating algorithm) to mend incremental association rules when min-sup isn't changed. The algorithm need respectively scan only once the original database and the new incremental database. The algorithm is as the following [6]:

Input: DB, the original transaction database

L, the set of DB's frequent item set db;

S_0, the min-sup

Output: L'', the set of $DB \bigcup db$'s frequent item set

(1) $L'' = \varphi$

for each $X \in L$ do begin

if X .count $L \geq S_0 \times (|DB| + |db|)$, then $L'' = L'' + \{X\}$;

$L_F = L - \{X\}$;Else $S_{D-\max} = \max\{x_i \sup D \mid x_i \in L\}$

end

(2) $S_{d-\min} = S_0 - t \times (S_{D-\max} - S_0)$;for each $x \in db$ do begin

if X . count $db \geq S_0 \times (|DB| + |db|)$ then $L'' = L'' + \{X\}$; $db = db - \{x\}$;

else

$L' = \{x_i \in db \mid x.count\ db \geq S_{d-\min} \times$

$(|DB| + |db|)\}$

end

(3) for each $X \in L_F$ do begin

for each $x \in L'$ do begin

if X .count $L_F + X$ .count $L' \geq S_0 \times (|DB| + |db|)$

then $L'' = L'' + \{X\}$; $L' = L' - \{X\}$

else $L_F = L_F - \{X\}$

end

end

(4) $L'_d = L' - \{x \mid S_{d-\min} \leq x, Supd \leq S_0\}$

(5) if $L'_d = \varphi$ then output L''

else for each $x \in L'_d$,do begin

$S_{d-\max} = \max\{x_i \sup d \mid x_i \in L\}$; $S_{D-\min} = S_0 - 1/t \times (S_{d-\max} - S_0)$

end

end

(6) $L_D = aproiri - gen(L_k, S_{D-\min})$

(7) for each $x \in L'_d$ do begin ;for each $X \in L_D$ do begin

if X .count $L_D + X$ .count $L_d^{'} \geq S_0 \times (|DB| + |db|)$ then $L^{''} = L^{''} + \{X\}$

else delete X

end

end

(8) output $L^{''}$

Here: X .count L , X .count db , X .count L_F , X .count L' , X .count L_D and X .count L_d' are item set X sup in L , db , L_F , L' , L_D and L_d' respectively.

Then, we calculate significative association rules' sup and cof in new database using the rule protection model with the weight. Calculate sup(x) in $D \cup d / \sup(x)$ in $D-1$ and cof(x) in $D \cup d / cof(x)$ in $D-1$, then we get T_1 and T_2 . Here we call them T. T reflects change trend of the strong association rules on the basis of incremental mining. We analyze and forecast the useful association rules by the change trends with database updating every time. Thereby, it supports the decision-making better.

On the other side, we study the lost rules that were the strong association rules in the original database after database updating. Calculate sup(x) in $D \cup d$ and cof(x) in $D \cup d$, and get T . The absolute value of T shows the eliminated risk when the strong association rules in the original database are changed with updating data. The absolute value of T for the lost association rules, it is defined as risk degree of the lost. In practical application, people often concern the risk of the strong association rules in the original database, which is the risk of being weak with periodic increase of database. We can also regard the risk degree as the risk that the strong rules turn into the weak rules or the weak rules turn into the strong rules. Higher the risk degree is, lower the universality of the rule is. The introduction of risk degree is for the need of solving the practical problem. It forecasts the trends of association rules based on incremental mining from the different aspect and increases the reliability of forecasting.

6 The Application of Improved Algorithm in Bank Cost Analysis

In process of bank cost analysis, we analyze the basic product attributes and product cost. Based on it, we get the rule that product's profit is relative to one or some of the attributes and obtain idiographic association rules. The data mining process of bank cost analysis includes: gathering of data, data pretreatment, model training and model evaluating.

We analyze the six months statistic information of bank product. The former five months' data is the original database. And the sixth months' data is the incremental data. We use the original algorithm and the improved algorithm to mine the bank cost analysis database.

The result shows that the improved algorithm is better than the original CBA algorithm. The precision of cost analysis is improved 30 percent. So, with the improved algorithm, we can efficiently mine association rules and mend incremental data mining in database, then improve precision of decision-making.

7 Conclusion

We introduce the improving-rate factor to increase the precision for optimizing the original algorithm, and present the improved algorithm for mining incremental data in database. With the data of one bank's six months statistical information, we test the new algorithm and compare it with the original algorithm. The result shows that the improved algorithm improves the reliability of data mining, and provides more information for decision-maker.

References

1. Han, J., Kamber, M.: Data Mining: Concepts and Techniques. Morgan Kauffman Publishers (2001)
2. Liu, Y.-A., Yang, B.: Research of an Improved Apriori Algorithm in Mining Association Rules. Computer Application, 418–420 (February 2007)
3. Ding, W.-P., Shi, Q., Guan, Z.-J.: Algorithm of Effective Association Rules Mining Based on Transaction Rule-tree. Application Research of Computers, 83–86 (May 2007)
4. Ren, X.-L., Shi, Z.-Z.: Bank Cost Analysis Based on Data Mining. Application Research of Computers, 53–57 (September 2007)
5. Wang, J., Wang, X., Pang, G.: Association Rules Mining Algorithm for Cold-rolling Processes. Tsinghua Univ. (Sci. & Tech.) 47(S2), 1761–1765 (2007)
6. Meng, R., Su, Y.-J., Zhu, X.-F., Zhang, J.-L.: An Efficient Incremental Updating Algorithm in Data Mining for Maintaining Association Rules. Journal of Guangxi Academy of Sciences, 125–128 (May 2006)
7. Johnson, R., Wichern, D.: Applied Multivariate Statistical Analysis, 5th edn. Prentice-Hall (2002)
8. Otey, M.E., Parthasarathy, S., Wang, C., Veloso, A., Meira Jr., W.: Parallel and Distributed Methods for Incremental Frequent Itemset Mining. IEEE Transactions on Systems, Man, and Cybernetics—part B: Cybernetics, 2439–2450 (December 2004)
9. He, Y.: Dynamic Growing Data Mining of More Frequent Itemset in Large Database. Computer Engineering, 76–78 (January 2006)
10. Ma, Z., Lu, Y.: Exploding Number of Frequent Itemsets in the Mining of Negative Association Rules. J. Tsinghua Univ. (Sci. & Tech.) 47(7), 1212–1215 (2007)
11. Zhang, H., Wang, J.: Research of Multiple Minimum Supports Frequent Item Sets Minging. Computer Applications, 2290–2293 (September 2007)
12. Zou, L., Zhang, Q.: Algorithm of Weighted Association Rules Mining with Multiple Minimum Supports. Journal of Beijing University of Aeronautics and Astronautics, 590–593 (May 2007)

LEACH-Based Security Routing Protocol for WSNs

Jianli Wang, Laibo Zheng, Li Zhao, and Dan Tian

School of Information Science and Engineering, Shandong University,
27 Shanda Nan Road, Jinan, P.R.China, 250100
glhwjl@163.com.cn, zhenglaibo@sdu.edu.cn,
zhaoli0525@sina.com, td200412327@126.com

Abstract. LEACH is the first protocol of wireless sensor networks based on clustering and layered structure technology, but the security problem related to the wireless sensor network is not considered. Therefore, it is necessary to provide an efficient security routing algorithm for practical wireless sensor networks. In this paper, we proposed a LEACH-based key management scheme for wireless sensor networks based on Exclusion Basis Systems and μ TESLA. We use EBS for key generation and distribution, and use μ TESLA to guarantee the cluster head can update security key after the first round. The proposed algorithm decreases the storage requirements of keys, and the network communications load for updating cluster keys. The key management scheme can enhance the survivability and ensure the security of WSNs.

Keywords: WSNs, Exclusion Basis Systems, LEACH, Security protocol.

1 Introduction

Due to the fact that in certain applications of sensor networks, like military applications, diplomatic communications, e-learning and air traffic control etc., security of WSNs becomes more and more important. These systems process data gathered from multiple sensors to monitor events in an area of interest. Sensors in such systems are typically disposable and expected to last until their energy drains. Some recent researches have focused on managing secure wireless communications in such networks. When a large group of sensors are constrained in energy, computation and communication resource, an efficient key management procedure becomes critical. LEACH (Low Energy Adaptive Clustering Hierarchy) has very important significance for wireless sensor network routing protocol, so design appropriate key management scheme to strengthen its security is an important issue of current research.

1.1 System Model

The system architecture of the WSN is depicted in Fig. 1. In this model, a sensor network consists of a large number of sensors distributed over an area of interest. LEACH is a self-organizing, adaptive clustering protocol that uses randomization to distribute the energy load evenly among the sensors in the network. The nodes organize themselves into local clusters, with one node in each cluster acting as the cluster-head.

D. Jin and S. Lin (Eds.): Advances in CSIE, Vol. 2, AISC 169, pp. 253–258.
springerlink.com

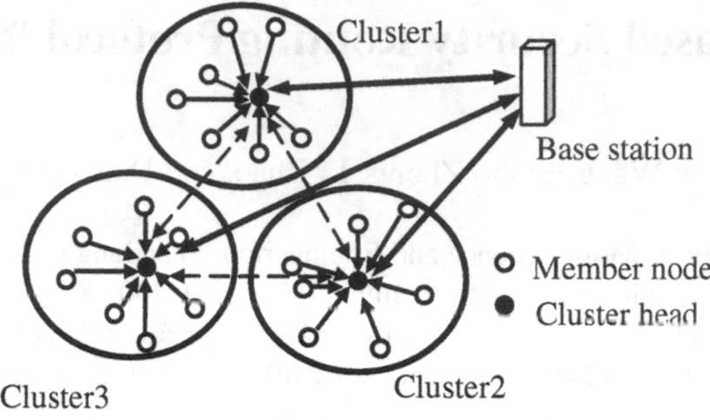

Fig. 1. Hierarchical architecture of Wireless Sensor Network

The cluster-heads fuse the data from the different sensors, perform mission-related data processing, and send it to the base station via long-haul transmission. The cluster-head nodes are not fixed but self-elected at different time intervals.

1.2 Related Work

Wireless sensor networks have a wide use on civil and military applications that call for security, e.g., target surveillance in hostile environments. The resource constrained environment has motivated extensive research that addresses energy-aware hardware and software design issues [1], [2]. Much effort has been done on the energy-efficient communication protocols [3], [4]. The energy-constrained nature of the sensor networks makes the problem of incorporating security very challenging. The design of the security protocols for sensor networks should be developed towards conservation of the sensor resources. The level of security versus the consumption of energy, computation and memory resources constitute a major design trade-off [10].

Recently, a number of solutions for securing WSNs have been proposed [5-9]. Ferreira A. C etc. have proposed SLEACH, which use μ TESLA to authorize the cluster head through the base station [6]. Leonardo B Oliveira presented a random key distribution solution for securing clustered sensor networks which implemented the authentication of cluster head to the members of the cluster, so as to ensure the legitimacy of the cluster members [7]. Ding Han-cheng proposed a key management for dynamically clustering WSN (KMDC) which adopts the EBS optimal combination group key management algorithm, and can reduce the storage burden of the management of the key and the key communication load of networks [8].

2 LEACH-Based Security Routing Protocol

A key management procedure is an essential constituent of network security. It is necessary to maintain a balanced security level with respect to those constraints. In this section we propose a LEACH-based security routing protocol for sensor networks, whose objective is to minimize the sensor's computation, communications and storage overhead due to the key management operations.

2.1 Exclusion Basis Systems

EBS was proposed by Eltoweissy in 2004, which based on a combination of group communication principles [5]. Exclusion Basis Systems provide a general framework for the investigation of key management systems. An EBS is defined as a collection of Γ that denotes subsets of the set of members. Each subset corresponds to a key, and the elements of a subset $A \in \Gamma$ are the nodes have that key. An EBS Γ of dimension (n, k, m) represents a situation in a secure group, n means the number of sensor nodes; k means the number of keys assigned to each node and k+m means the total number of keys. So it can be proved to be that:

1) When $C_{k+m}^{k} \geq n$, any n combinations from C_{k+m}^{k} can constitute an EBS (n, k, m), and then form a key distribution scheme.
2) By broadcast at most m packets, we can dynamically cancel and update the whole key of any node, then drive out this node.

The base station is the only authority for key generation. It will be the direct service for the administration keys for the cluster head and communication keys for the inter-cluster head interaction and for message traffic. Although the head of a cluster would be responsible for the key management for the sensors in the cluster, the keys still have to be generated by the base station. Each cluster will be assigned a set of distinct communication keys for data encryption. Sharing the same key among sensors in a cluster will enable selective decryption of data messages for the purpose of aggregation if instrumented in the network operation. The base station will periodically renew keys for the inter-cluster head and cluster head to sensor nodes communication to counter potential on-going spoofing.

It is proved in [5] that the overhead of an optimum EBS is half of that a binary key tree. However EBS may suffer from collusion attacks. In [9], a special kind of polynomial, the common trivariate polynomial, is presented, which can guarantee that all the nodes having the same polynomial can get the same key. The common trivariate polynomial keys are used in stead of the normal keys in EBS system and a new dynamic key management scheme is designed for clustered wireless sensor networks, which can solve the collusion attacks problem effectively. This problem is out the scope of our research.

2.2 System Initialization and Operation

The basic terminology in Table 1 is used for describing the key management protocol.

Table 1. Notation used in the Key management Protocols

Notation	Description	Notation	Description
B	Base station	K_{int}	Initialization key
C_j	Cluster j	$K_a(C_j)$	The authentication key of cluster j
N_i	Sensor node i	$Mem(C_j)_i$	Member node i of cluster j
H_j	Cluster-head of cluster j	K_{ebs}	EBS key management set
$ID(N_i)$	Sensor node i identifier	$\mathrm{E}(K, Data)$	Encryption function of data with key K
$K_{master}(N_i)$	Key shared between node i and BS	$\Vert$	Concatenation operator

We assume that the base station is secure. In the cluster establishment phase, sensor nodes customize cluster head according to LEACH's selection method and, then, the cluster head broadcasts a hello packet. The Hello packet must be authenticated, and encrypted through the initial key K_{int} in the first cluster formation process,

$$H_j \rightarrow broadcast : \mathrm{E}(K_{int}, Hello) \ . \tag{1}$$

Member nodes receive the packet and then return the response to the cluster. Response content includes their ID and Ack,

$$N_i \rightarrow H_j : \mathrm{E}\left(K_{int}, ID(N_i) \ \| Ack\right) \ . \tag{2}$$

And then cluster head sent the identification of all the members who want to join this cluster to the base station,

$$H_j \rightarrow \mathrm{B}: \mathrm{E}\left(K_{master}\left(H_j\right), ID(N_i) \| ID(N_m) \cdots\right) \ . \tag{3}$$

The base station construct EBS, each cluster is assigned a set of distinct communication keys for data encryption,

$$\mathrm{B} \rightarrow H_j : \mathrm{E}\left(K_{master}\left(H_j\right), ID\left(C_j\right) \| K_a\left(C_j\right) \| K_{ebs} \| \mathrm{E}\left(K_{master}\left(N_i\right) \| K_a\left(C_j\right) \| K_{ebs}\right) \| \cdots\right) \tag{4}$$

At the network stable working stages, the member node i of cluster j sent the authentication key of cluster j, identifier of the member node i of cluster j, EBS key management set and data to the cluster-head of cluster j. Cluster-head of cluster j sent $K_a\left(C_j\right), ID\left(C_j\right), K_{ebs}, ID\left(H_j\right), K_{master}\left(H_j\right)$ and the fused information received from cluster member nodes to base station,

$$Mem\left(C_j\right)_i \rightarrow H_j : \mathrm{E}\left(K_a\left(C_j\right), ID\left(Mem\left(C_j\right)_i\right) \ \| K_{ebs}, Data\right) \ . \tag{5}$$

$$H_j \rightarrow \mathrm{B}: \mathrm{E}\left(K_a\left(C_j\right), ID\left(C_j\right) \| K_{ebs} \| K_{master}\left(H_j\right), \| ID\left(H_j\right), Data\right) \ . \tag{6}$$

Once the clusters are created and the TDMA schedule is fixed, data transmission can begin. Assuming nodes always have data to send, they send it during their allocated transmission time to the cluster head. The first round security LEACH cluster protocol working process is shown in Fig. 2.

After a round of operation, the system reselects cluster head. Different from the initial condition, the key K_{int} will be erased for security reasons. Before the broadcast, new cluster head request a radio key from base station, the μ TESLA (a combinatorial optimization of the group key management problem) is to broadcast K_{mac} packet first, and then announced the keys by base station [11], to conform the

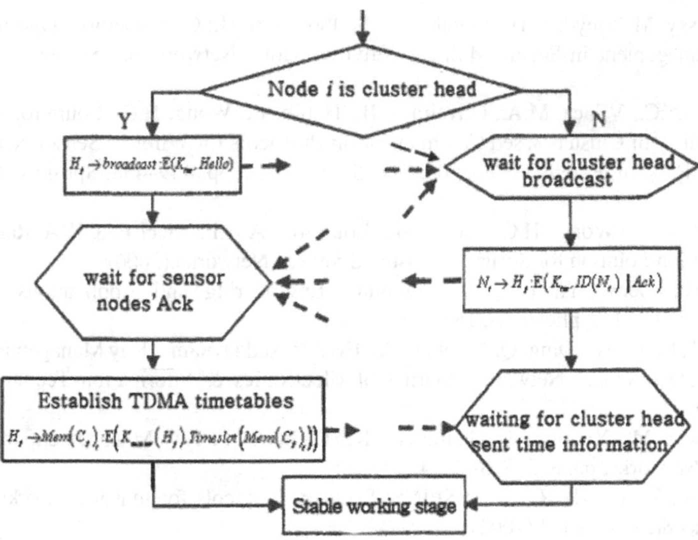

Fig. 2. Flow map of the first round security LEACH cluster protocol constructing

attacker can't forge the correct broadcast packets before the key is announced. The K_{int} is replaced by K_{mac}, and a new round of security LEACH become work.

3 Conclusion

In this paper, we present a LEACH-based security routing protocol for WSNs. We use exclusion basis system for key generation and distribution. Exclusion Basis Systems provide a general framework for the investigation of key management systems. By using EBS, the storage requirements of storing keys is decreased, the network communications load for updating cluster keys is reduced. The use of μ TESLA guaranteed the cluster head update security after the first round. Such flexibility is invaluable for the resource constrained sensor nodes.

References

1. Akyildiz, I.F., et al.: Wireless sensor networks: a survey. Computer Networks 38, 393–422 (2002)
2. Pottie, G.J., Kaiser, W.I.: Wireless integrated network sensors. Communications of the ACM 43(5), 51–58 (2000)
3. Heinzelman, W.R., Wendi, B., et al.: Energy-Efficient Communication Protocol for Wireless Microsensor Networks. Paper Presented at the Proceedings of the 33rd Annual Hawaii International Conference on System Sciences, Maui, HI, pp. 1–10 (2000)
4. Younis, M., Youssef, M., Arisha, K.: Energy-Aware Routing in Cluster-Based Sensor Networks. Energy-Awarerouting, 129–136 (2002)

5. Eltoweissy, M., Heydari, H., Morales, L., Sudborough, H.: Combinatorial Optimization for Key Management in Secure Multicast Environments. Network and System Management (2004)
6. Ferreira, A.C., Vilaça, M.A., Oliveira, L.B., Habib, E., Wong, H.C., Loureiro, A.A.F.: On the Security of Cluster-Based Communication Protocols for Wireless Sensor Networks. In: Lorenz, P., Dini, P. (eds.) ICN 2005. LNCS, vol. 3420, pp. 449–458. Springer, Heidelberg (2005)
7. Oliveira, L.B., Wong, H.C., Bern, M., Loureiro, A.A.F.: SecLEACH-A Random Key Distribution Solution for Securing Clustered Sensor Networks (2006)
8. Ding, H., Geng, Y., Bin, L.: Computer Engineering and Applications. Computer Engineering and Applications, 157–160 (2008)
9. Kong, F., Li, C.-W., Ding, Q.-Q., et al.: An EBS-Based Dynamic Key Management Scheme for Wireless Sensor Networks. Journal of Electronics & Information Technology 31(5) (2009)
10. Eltoweissy, M., Youois, M., Ghumman, K.: Lightweight Key Management for Wireless Sensor Networks, pp. 813–818 (2004)
11. Perrig, A., Szewczyk, R., et al.: SPINS: Security protocols for sensor networks. Wireless Networks 8(5), 521–534 (2002)
12. Chorzempa, M., Park, J.-M., et al.: Key management for long-lived sensor networks in hostile environments. Computer Communications 30(3), 1964–1979 (2007)
13. Cheng, H., Wang, J., Yang, G.: Research of Secure Frame Protocol Based on SPINS. Computer Science 33(8) (2006)
14. Wang, J., Yang, G., Chen, S.: Secure LEACH Routing Protocol Based on Optimal Number of Head Nodes for W ireless Sensor Network. Journal of Nanjing University of Posts and Telecommunications (Natural Science) 28(3) (2008)

A System of Robot Detecting and Tracking Moving Objects Based on Modified Optical Flow

GuangZhao Cui, KeKe Liang, and JinChao Guo

Department of Electric & Information Engineering, Zhengzhou University of Light Industry, Zhengzhou, 450002, Henan Province, P.R. China
cgzh@zzuli.edu.cn

Abstract. Robot's adaptability to surrounding as an essential subject of robot research is based on detecting environment precisely and acting timely. Moving object detection and tracking is an important approach of robot detecting external environment. Moving object detection and tracking based on computer vision means building a system with vision part and motor drive of the robot. In this paper, the robot estimates the motion of object and tracks it by driving motors according to the changes of optical flow fields got from modified optical flow calculation of collected images. In the experiment, this system responded quickly, stably and accurately. And it could make the robot move exactly and timely to respond to changes of surrounding.

Keywords: Modified optical flow, Moving object detection and tracking, Computer vision.

1 Introduction

As an essential research of computer vision, there are many analysis methods, such as frame, optical flow and the method of reducing background [1]. And in these analysis methods, optical flow algorithms have advantages of follows: first, the optical flow is better for comparatively complicated movement, such as scaling and rotation. Second, optical flow can correctly detect the sub-pixel mobile distance with higher precision. Optical flow algorithms have three mainly kinds: differential optical flow algorithm, the frequency domain optical flow algorithm and correlation optical flow algorithm [2]. LK (Lucas-Kanade) and HS (Horn-schunk) [3] are two typical differential optical flow algorithms. Compare to LK, HS has advantages of visual geometric meaning and smaller amount of calculation.

This paper discusses one system which associated the robot vision part and motor drive part. Through the optical flow algorithm, the robot gets optical flow fields of collected images from vision part, and will drive motors to track the moving target detected by analyzing changes of optical flow fields. In the experiment, we use Pioneer3-DX robot from Activ Media Company. The paper is organized as follows: Section2 presents a brief description of moving target detection and tracking system and modified optical flow. Section3 talks about the detail presentation of moving target detection and tracking system. Section 4 provides experiments and results analysis. Section 5 gives conclusions and future work.

D. Jin and S. Lin (Eds.): Advances in CSIE, Vol. 2, AISC 169, pp. 259–265.
springerlink.com

2 Description of the System

The visual system is playing a more and more important role in the robot obstacle avoidance, target detection and tracking and so on. Because of demand of the complex data and accuracy, obstacle avoidance mainly uses binocular vision system. And monocular vision system is mainly used in target detection and tracking, navigation, robot path planning, etc.

If call the part through which the robot get external information as visual part, The optical flow calculation and analysis will be called calculative part; And the robot motor drive system will be called motor part. The robot target tracking system is the combination of visual part, calculative part and motor part. After analysis and calculation of the data collected by visual system, the robot will get direction and speed of movement of the target need to be tracked. And according to the direction and speed the robot can drive motor to track the target. As shown in Fig. 1.

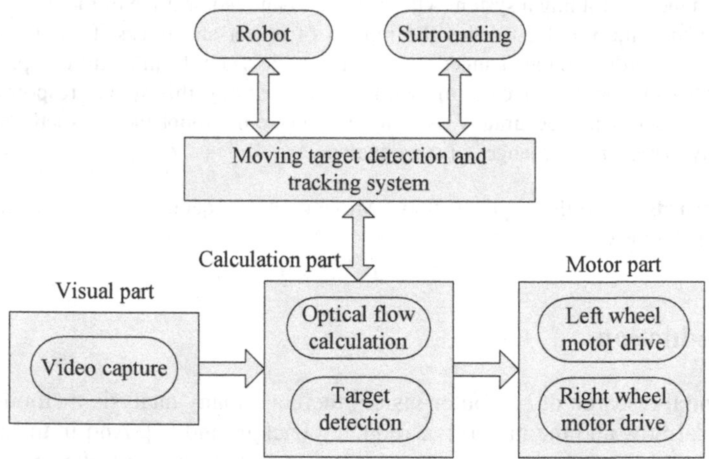

Fig. 1. Flowchart of the moving target detection and tracking system.

In the system, the visual part is video capture. Calculative part consists of optical flow calculation and target detection. Motor part is comprised of left wheel motor drive and right wheel motor drive. Among these three parts, the calculative part is the foundation of this system. HS optical flow algorithm can detect the complete contour of the moving object, but the amount of the computation is large. A combination of pyramid LK optical flow and HS optical flow is proposed for moving object detection and tracking. The result of experiment shows that this algorithm can effectively improve the speed of computation of the moving object detection and tracking system.

3 Realization of the System

The three-dimensional object in the real-world environment to be observed projects the velocity vector to the image. And this projection is the two-dimensional motion vector. These two-dimensional instantaneous velocity vectors together constitute a velocity field called optical the flow field. Easy to know that, the optical flow field not only embodies the information of moving object, but also embodies the surrounding environment three-dimensional data. Moving targets detection is the premise of the robot's target tracking, also is a hot issuc of computer vision. There are two main methods to detect moving object: motion compensation method [4] and optical flow method [5]. Motion compensation method is mainly dependent on the accuracy of the compensation. And this method cannot describe the complete contour of target. HS optical flow algorithm [6] can detect the complete contour of the moving target. And HS optical flow method [7] has been widely used in the static background moving target detection.

3.1 Video Capture

The video capture system is monocular system adopting CanonVC-C50i camera in the experiment. Image acquisition card is the VCE-B5A01 image acquisition card produced by IMPERX Company. and sampling frequency 60 frames per second. The images are collected in size 320×240. This size affects the computation amount of the optical flow algorithm mentioned below.

3.2 Modified Optical Flow Algorithm

The optical flow algorithm is a calculation method, which has to solve the problem of optical flow equation as the first priority, looking for two-dimensional instantaneous velocity vector field to reconstruct the three-dimensional object information. Optical flow has been used widely in the target segmentation, identification and tracking, robot navigation, the information of shape recovery, and other important fields [8].

The HS optical flow algorithm is a kind of typical differential optical flow calculation methods. Differential method [9] is also named Spatio - temporal gradient method. It is a method to calculation each image point velocity vector on the basis of Spatio - temporal differential in the time-varying image grey space.

Let I (x, y, t) is the image brightness of image point (x, y) at t moment. if u (x, y) and v (x, y) are this point optical flow vector level and vertical components. if this point moves to (x + δx, y+ δy) at the time t +δt. And the brightness is unchanged, so that

$$I(x+\delta x, y+\delta y,t+\delta t) =I(x, y,t). \qquad (1)$$

In the equation above, $\delta x = u\times\delta t$ and $\delta y = v\times\delta y$. After series development of the equation with Tarlor and a series of operations, u and v can be given by the following iterative formula:

$$u^{k+1} = \bar{u}^{k} - \frac{I_x(I_x\bar{u}^{k} + I_y\bar{v}^{k} + I_t)}{\alpha + I_x^2 + I_y^2}. \tag{2}$$

$$v^{k+1} = \bar{v}^{k} - \frac{I_y(I_x\bar{u}^{k} + I_y\bar{v}^{k} + I_t)}{\alpha + I_x^2 + I_y^2}. \tag{3}$$

In the equations, $\bar{u}$ and $\bar{v}$ are the local average. k is iteration times. α is weighted value. By iterative formulas above, it is known that the calculation amount of HS optical flow depends on the size of the image and iteration times. When the size of images initial value is more adjacent the real value, iteration times should be reduced.

For this reason, we calculate the sparse optical flow field of the images using the pyramid LK optical flow algorithm which has a high speed of operation. After that, we remove the moving targets from the sparse optical flow field and extract the motion vector of the background as the initial value of HS optical flow. After a few iteration times, HS algorithm can detect the complete contour of the moving target. Algorithm flow chart is shown in Fig. 2

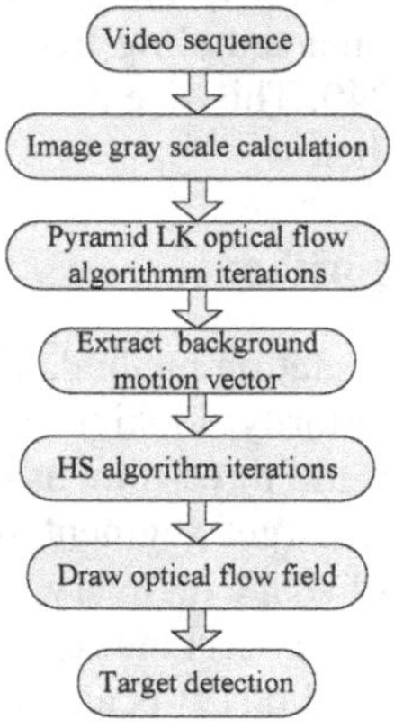

Fig. 2. Flow chart of algorithm in this paper.

3.3 Motor Drive

Most of the robots move by driving motor. Especially, step motor and servo motor are used to robots widely. Pioneer3-DX robot uses two step motors. And there are two sets of step motor control system that can control two wheels independently. If v denotes the robot translational speed; ω denotes the angular velocity of robot; ω_l and ω_r denote left and right step motor angular velocity respectively. Their relationship is as follow:

$$\begin{bmatrix} v \\ \omega \end{bmatrix} = \begin{bmatrix} \frac{R_r}{2} & \frac{R_l}{2} \\ \frac{R_r}{T} & -\frac{R_r}{T} \end{bmatrix} \begin{bmatrix} \omega_r \\ \omega_l \end{bmatrix}. \tag{4}$$

In the equation 4, R_r and R_l are the radius of two wheels respectively. Optical flow field got from the HS optical flow algorithm can't be the parameters directly to drive motors. It still needs to be singular value decomposition. After the singular value decomposition, the robot will get combination of principal component linear. In the optical flow field each column could expressed as a_j. And a_j can be corresponding to column vector of the matrix A. After the singular value decomposition, it has the following formula:

$$A_{m\times n} = P_{m\times n} Q_{n\times n} E^T_{n\times n}. \tag{5}$$

In the equation 5, P is columns related orthogonal matrixes. Q matrix is diagonal matrix. And the diagonal element of matrixes Q is the singular value of matrixes A. E^T is the characteristic vector matrixes namely the orthogonal vectors matrixes. Every movement of the robot can be seen as the combines of two basic movements: translation and rotation. So that, the optical flow field a_j can be seen as

$$a_j = p_{j1} q_1 e_1^T + p_{j2} q_2 e_2^T. \tag{6}$$

The two components in the equation 6 correspond to the translation and rotation vector respectively. This is used to drive stepping motor as the parameters for robot to make a movement to realize the moving target tracking.

4 Results Analysis

Video sequences used in the experiments are collected by the CanonVC-C50i camera on the mobile robot Pioneer3-DX. Video sequence has its background of white plastic plate, and has a target moving independently. The size of the image sequence is 320×240 as mentioned above. Image sequences given in Fig. 3 are collected by the camera in the situation that the robot moves only, camera doesn't. Moving object is the paper box. And the paper box moves in a straight line.

Figure 3 (a) and (b) are two consecutive frames of the image sequences. Figure 3 (c) is the optical flow result of traditional HS optical flow algorithm. And it has the iterative times 180. Figure 3 (d) is the algorithm result of modified optical flow algorithm used in this paper. And it has the iterative times 10. Figure 3 (e) and (f) are the figures enlarged of traditional HS optical flow and modified optical flow respectively for illustrating the detail of optical flow field.

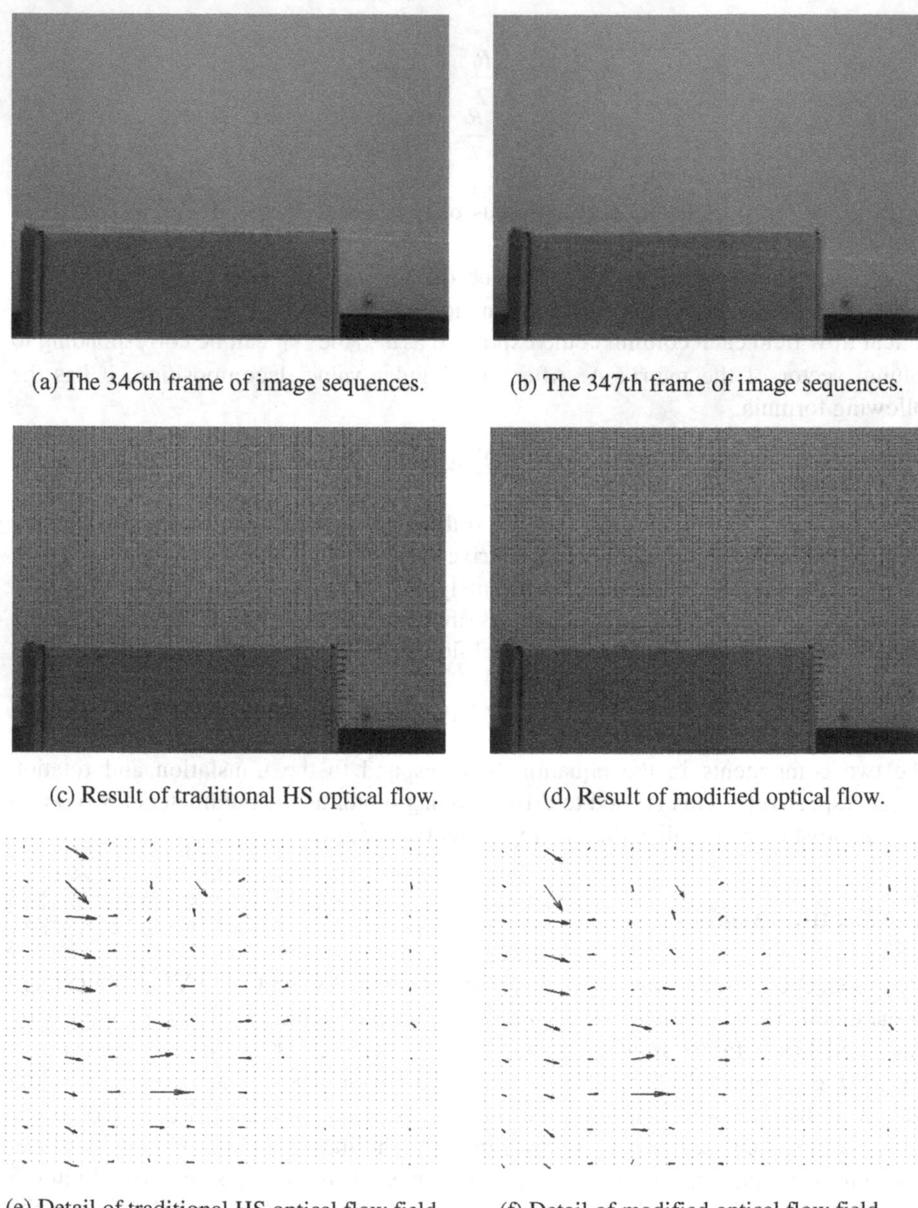

(a) The 346th frame of image sequences.

(b) The 347th frame of image sequences.

(c) Result of traditional HS optical flow.

(d) Result of modified optical flow.

(e) Detail of traditional HS optical flow field.

(f) Detail of modified optical flow field.

Fig. 3. Comparison of traditional HS optical flow algorithm result and modified optical flow algorithm result.

We can see that modified optical flow has a similar result with traditional HS optical flow. But the iterative times of traditional HS optical flow is at least 15 times than modified optical flow. From the detail, we know that the motion vector of modified optical flow is not accurately as traditional HS optical flow. And the texture smooth area of the background has no optical flow vector been detected in these two methods. It makes troubles to moving object detection and tracking. The algorithm should be improved again in the future.

5 Conclusion

Modified HS optical flow algorithm is simple and can describe complete contour of moving object. This algorithm used to moving target detection and tracking system of the robot can estimate the contour of moving targets accurately and track the contour. In the laboratory, after many times experiment with the system, the robot moved more accurately, stably, and smoothly in the target tracking process. But, in dynamic background moving target detection, Modified HS optical flow algorithm is sensitive to noise and can not detect the optical flow vector in the texture smooth area of the background. At present, we are seeking a method to make a further improvement for these problems.

References

1. Liu, M.X., Meng, F.: Research Progress of Moving Object Detection and Tracking Algorithm. Software 12, 85–88 (2010)
2. Beauchemin, S.S., Barron, J.L.: The Computation of Optical Flow. ACM Computing Surveys (CSUR) 27, 433–466 (1995)
3. Horn, B.K.P., Schunck, B.G.: Determining Optical Flow. Artificial Intelligence (1981)
4. Lin, C.C., Wolf, M.: Detecting Moving Objects Using a Camera on a Moving Platform. In: IEEE International Conference on Pattern Recognition, pp. 460–463 (2010)
5. Shi, G.D., Wang, J.Z.: Dynamic Background Moving Targets Detection and Tracking. Journal of Beijing University of Technology: Natural Science Edition 29, 858–860 (2009)
6. Wang, L.: Application of Optical Flow Technology in the Moving Target Detection and Tracking. National University of Defense Technology (2007)
7. Shi, G.D., Wang, J.Z., Wang, G.R.: Real-time Detection of Human Movement Based on the Optical Flow. Journal of Beijing University of Technology: Natural Science Edition 28, 794–79 (2008)
8. Tao, L.: Technology Image Information Acquisition Based on Optical Flow. Huazhong university of science and technology (2005)
9. Liu, G.F.: Optical Flow Calculation Technology. Journal of Southwest Jiaotong University 12, 856–863 (1997)

The Presentation of the Quarternary Super-Wavelet Wraps and Applications in Computer Science

Ping-An Wang*

Department of Fundamentals, Xijing University, Xi'an 710123, P.R. China
zxs123hjk@126.com

Abstract. Wavelet analysis has become a developing branch of mathematics for over twenty years. In this paper, the notion of orthogonal nonseparable quarternary variate wavelet packs, which is the generalization of orthogonal univariate wavelet packs, is proposed by virtue of analogy method and iteration method. Their biorthogonality traits are researched by using time-frequency analysis approach and variable separation approach. Three orthogonality formulas regarding these wavelet wraps are obtained. Moreover, it is shown how to draw new orthonormal bases of space $L^2(R^4)$ from these wavelet wraps. A procedure for designing a class of orthogonal vector-valued finitely supported wavelet functions is proposed by virtue of filter bank theory and matrix theory.

Keywords: Nonseparable, binary wavelet packs, wavelet frame, Bessel sequence, orthonormal bases, time-frequency analysis approach.

1 Introduction and Notations

Computer science or computing science designates the scientific and mathematical approach in information technology and computing. A computer scientist is a person who does work at a professional level in computer science and/or has attained a degree in computer science or a related field. Wavelet analysis is nowadays a widely used tool in applied mathematics. The advantages of wavelet wraps and their promising features in various application have attracted a lot of interest and effort in recent years. Aside from the straightforward construction of Daubechies' wavelets, only a few, specific construcion of multivariate orthonormal wavelet systems exist presently in the literature. The main advantage of wavelets is their time-frequency localization property. Already they have led to exciting applications in signal processing [1], fractals, image processing [2] and so on. Sampling theorems play a basic role in digital signal processing. They ensure that continuous signals can be processed by their discrete samples. Vector-valued wavelets are a sort of generalized multiwavelets [3]. Vector-valued wavelets and multiwavelets are different in the following sense. For example, prefiltering is usually required for discrete multiwavelet transforms but not necessary for discrete vector-valued wavelet

* Corresponding author.

D. Jin and S. Lin (Eds.): Advances in CSIE, Vol. 2, AISC 169, pp. 267–273.
springerlink.com

trans-forms [4]. In real life, video images are vec engineering [2,3]. Coifman R. R. and Meyer Y. firstly introduced the notion for orthogonal wavelet packets which were used to decompose wavelet components. Chui C K. and Li Chun L. [4] generalized the concept of orthogonal wavelet packets to the case of non-orthogonal wavelet packets so that wavelet packets can be employed in the case of the spline wavelets and so on. Tensor product multivariate wavelet packs has been constructed by Coifman and Meyer. The introduction for the notion on nontensor product wavelet packs attributes to Shen Z [5]. Since the majority of information is multidimensional information, many researchers interest themselves in the investigation into multivariate wavelet theory. But, there exist a lot of obvious defects in this method, such as, scarcity of designing freedom. Therefore, it is significant to investigate nonseparable multivariate wavelet theory. Nowadays, since there is little literature on biorthogonal wavelet wraps, it is neces-sary to investigate biorthogonal wavelet wraps.

In the following, we introduce some notations. Z and Z_+ denote all integers and all nonnegative integers, respectively. R denotes all real numbers. R^4 denotes the 4-dimentional *Euclidean* space. $L^2(R^4)$ denotes the square integrable function space. Let $x=(x_1,x_2,x_3,x_4)\in R^4$, $\omega=(\omega_1,\omega_2,\omega_3,\omega_4)\in R^4$, $k=(k_1,k_2,k_3,k_4)\in Z^4$, The inner product for any functions $\hbar(x)$ and $g(x)$ ($\hbar(x),\, g(x)\in L^2(R^4)$) and the Fourier transform of $g(x)$ are defined, respectively, by

$$\langle \hbar, g\rangle = \int_{R^4} \hbar(x)\overline{g(x)}\,dx, \quad \hat{g}(\omega) = \int_{R^4} g(x)e^{-i\omega\cdot x}dx,$$

where $\omega\cdot x=\omega_1x_1+\omega_2x_2$ and $\overline{g(x)}$ denotes the complex conjugate of $g(x)$. Let R and C be all real and all complex numbers, respectively. Z and N denote, respectively, all integers and all positive integers. Set $Z_+=\{0\}\cup N, a,s\in N$ as well as $a\geq 2$ By algebra theory, it is obviously follows that there are a^4 elements $d_0,d_1\ \cdots,d_{a^4-1}$ in $(d_1+mZ^4)\cap(d_2+mZ^4)=\phi$, where $\Omega_0=\{d_0,d_1,\cdots,d_{a^2-1}\}$ denotes the aggregate of all the different representative elements in the quotient group $Z^4/(mZ^4)$ and order $d_0=\{\underline{0}\}$ where $\{\underline{0}\}$ is the null element of Z_+^4 and d_1,d_2 denote two arbitrary distinct elements in Ω_0 .Let $\Omega=\Omega_0-\{\underline{0}\}$ and Ω,Ω_0 to be two index sets. Define, By $L^2(R^4,C^s)$, we denote the set of all vector-valued functions $L^2(R^2,C^s):=\{\hbar(x)\ =(h_1(x)),h_2(x),\cdots,h_u(x))^T : h_l(x)\in L^2(R^4), l=1,2,\cdots,s\}$,where T means the transpose of a vector. For any $\hbar\in L^2(R^2,C^s)$ its integration is defined as follows $\int_{R^4}\hbar(x)dx=(\int_{R^4}h_1(x)dx, \int_{R^4}h_2(x)dx,\cdots, \int_{R^4}h_s(x)dx)^T$.

Definition 1. A sequence $\{\lambda_n(x)_{n\in Z^4}\subset L^2(R^4,C^s)\}$ is called an orthogonal set, if

$$\langle \lambda_n, \lambda_v\rangle = \delta_{n,v}I_s, \quad n,v\in Z^4, \tag{1}$$

where I_s stands for the $s \times s$ identity matrix and $\delta_{n,v}$, is generalized Kronecker symbol, i.e., $\delta_{n,v} = 1$ as $n = v$ and $\delta_{n,v} = 0$, otherwise.

2 The Quarternary Multiresolution Analysis

Firstly, we introduce multiresolution analysis of space $L^2(R^4)$. Wavelets can be constructed by means of multiresolution analysis. In particular, the existence theorem[8] for higher-dimentional wavelets with arbitrary dilation matrice has been given. Let $h(x) \in L^2(R^4)$ satisfy the following refinement equation:

$$f(x) = m^4 \cdot \sum_{k \in Z^4} b_k f(mx - k) \tag{2}$$

where $\{b(n)\}_{n \in Z^2}$ is real number sequence which has only finite terms and $f(x)$ is called scaling function. Formula (1) is said to be two-scale refinement equation. The frequency form of formula (1) can be written as

$$\widehat{f}(\omega) = B(z_1, z_2, z_3, z_4)\, \widehat{f}(\omega/m), \tag{3}$$

where

$$B(z_1, z_2, z_3, z_4) = \sum_{(n_1, n_2, n_2, n_2) \in Z^4} b(n_1, n_2, n_3, n_4) \cdot z_1^{n_1} \cdot z_2^{n_2} \cdot z_3^{n_3} \cdot z_4^{n_4}. \tag{4}$$

Define a subspace $X_j \subset L^2(R^4)$ $(j \in Z)$ by

$$V_j = clos_{L^2(R^4)} \left\langle m^j f(m^j x - k):\ k \in Z^4 \right\rangle. \tag{5}$$

Definition 2. We say that $f(x)$ in (2) generate a multiresolution analysis $\{V_j\}_{j \in Z}$ of $L^2(R^4)$, if the sequence $\{V_j\}_{j \in Z}$ defined in (4) satisfy the following properties: (i) $V_j \subset V_{j+1}$, $\forall\, j \in Z$; (ii) $\bigcap_{j \in Z} V_j = \{0\}$; $\bigcup_{j \in Z} V_j$ is dense in $L^2(R^4)$; (iii) $\psi(x) \in V \Leftrightarrow \psi(mx) \in V_{k+1}, \forall\, k \in Z$ (iv) the family $\{f(m^j x - n): n \in Z^2\}$ forms a *Riesz* basis for the spaces V_j.

Let $Y_k (k \in Z)$ denote the complementary subspace of V_j in V_{j+1}, and assume that there exist a vector-valued function $G(x) = \{g_1(x), g_2(x), \cdots, g_{m^2-1}(x)\}$ constitutes a *Riesz* basis for Y_k, i.e.,

$$Y_j = clos_{L^2(R^4)} \left\langle g_{\lambda:j,n} : \lambda = 1, 2, \cdots, m^4 - 1;\ n \in Z^4 \right\rangle, \tag{6}$$

where $j \in Z$, and $g_{\lambda:j,k}(x) = m^{j/2} g_\lambda(m^j x - k)$, $\lambda = 1, 2, \cdots, m^4 - 1$; $k \in Z^4$. Form condition (5), it is obvious that $g_1(x), g_2(x), \cdots, g_{m^4-1}(x)$ are in $Y_0 \subset X_1$. Hence there exist three real number sequences $\{q_n^{(\lambda)}\}(\lambda \in \Delta = \{1, 2, \cdots, m^4 - 1\}$, $n \in Z^4)$ such that

$$g_\lambda(x) = m^4 \cdot \sum_{k \in Z^4} q_k^{(\lambda)} f(mx - k), \tag{7}$$

Formula (7) in frequency domain can be written as

$$g_\lambda(\omega) = Q^{(\lambda)}(z_1, z_2) \hat{f}(\omega/m), \quad \lambda = 1, 2, \cdots, m^2 - 1. \tag{8}$$

where the signal of sequence $\{q_k^{(\lambda)}\}(\lambda = 1, 2, \cdots, m^4 - 1,\ k \in Z^4)$ is

$$Q^{(\lambda)}(z_1, z_2) = \sum_{(n_1, n_2) \in Z^2} q_{(n_1, n_2)}^{(\lambda)} \cdot z_1^{n_1} \cdot z_2^{n_2}. \tag{9}$$

A bivariate function $f(x) \in L^4(R^4)$ is called a semiorthogonal one, if

$$\langle f(\cdot), f(\cdot - k) \rangle = \delta_{0,k}, \quad n \in Z^4. \tag{10}$$

We say $G(x) = \{g_1(x), g_2(x), \cdots, g_{m^4-1}(x)\}$ is anorthogonal bivariate vector-valued wavelets associated with the scaling function $f(x)$, if they satisfy:

$$\langle f(\cdot), g_\nu(\cdot - k) \rangle = 0, \quad \nu \in \Delta, \ k \in Z^4, \tag{11}$$

$$\langle g_\lambda(\cdot), g_\nu(\cdot - n) \rangle = \delta_{\lambda,\nu} \delta_{0,n}, \ \lambda, \nu \in \Delta, \ n \in Z^4 \tag{12}$$

3 The Traits of Nonseparable Bivariate Wavelet Packs

To construct wavelet packs, we introduce the following notation: $a = 2, h_0(x) = f(x)$, $h_\nu(x) = g_\nu(x), b^{(0)}(n) = b(n),\ b^{(\nu)}(n) = q^{(\nu)}(n)$, where $\nu \in \Delta$ We are now in a position of introducing orthogonal bivariate nonseparable wavelet wraps.

Definition 3. A family of functions $\{h_{mk+\nu}(x) : n = 0, 1, 2, 3, \cdots,\ \nu \in \Delta\}$ is called a nonseparable bivariate wavelet packs with respect to an orthogonal scaling function $\Lambda_0(x)$, where

$$\Lambda_{mk+\nu}(x) = \sum\nolimits_{n \in Z^4} b^{(\nu)}(n) \Lambda_k(mx - n), \tag{13}$$

where $\nu = 0, 1, 2, \cdots 15$. By taking the Fourier transform for t (12), we have

$$\hat{h}_{nk+\nu}(\omega) = B^{(\nu)}(z_1, z_2, z_3, z_4) \cdot \hat{h}_k(\omega/2). \tag{14}$$

where

$$B^{(\nu)}(z_1, z_2, z_3, z_4) = B^{(\nu)}(\omega/2) = \sum_{k\in Z^4} b^{(\nu)}(k)\, z_1^{k_1} z_2^{k_2} z_3^{k_3} z_4^{k_4} \tag{15}$$

Lemma 1 [6]. Let $\phi(x) \in L^2(R^2)$. Then $\phi(x)$ is an orthogonal one if and only if

$$\sum_{k\in Z^2} |\hat{\phi}(\omega + 2k\pi)|^2 = 1. \tag{16}$$

Lemma 2. Assuming that $f(x)$ is an semiorthogonal scaling function. $B(z_1, z_2)$ is the symbol of the sequence $\{b(k)\}$ defined in (3). Then we have

$$\Pi = |B(z_1, z_2)|^2 + |B(-z_1, z_2)|^2 + |B(z_1, -z_2)|^2 + |B(-z_1, -z_2)|^2 \tag{17}$$

Proof. If $f(x)$ is an orthogonal bivariate function, then $\sum_{k\in Z^2} \left|\hat{f}(\omega + 2k\pi)\right|^2 = 1$. Therefore, by Lemma 1 and formula (2), we obtain that

$$1 = \sum_{k\in Z^2} |B(e^{-i(\omega_1/2 + k_1\pi)}, e^{-i(\omega_2/2 + k_2\pi)}) \cdot \hat{f}((\omega_1, \omega_2)/2 + (k_1, k_2)\pi)|^2$$

$$= |B(z_1, z_2) \sum_{k\in Z^2} \hat{f}(\omega + 2k\pi)|^2 + |B(-z_1, z_2) \cdot \sum_{k\in Z^2} \hat{f}(\omega + 2k\pi + (1,0)\pi)|^2$$

$$+ |B(z_1, -z_2) \cdot \sum_{k\in Z^2} \hat{f}(\omega + 2k\pi + (0,1)\pi)|^2 + |B(-z_1, -z_2) \cdot \sum_{k\in Z^2} \hat{f}(\omega + 2k\pi + (1,1)\pi)|^2$$

$$= |B(z_1, z_2)|^2 + |B(-z_1, z_2)|^2 + |B(z_1, -z_2)|^2 + |B(-z_1, -z_2)|^2$$

This complete the proof of Lemma 2. Similarly, we can obtain Lemma 3 from (3), (8), (13).

Lemma 3. If $\psi_\nu(x)$ ($\nu = 0,1,2,3$) are orthogonal wavelet functions associated with $h(x)$. Then we have

$$\sum_{j=0}^{1} \{B^{(\lambda)}((-1)^j z_1, (-1)^j z_2) \overline{B^{(\nu)}((-1)^j z_1, (-1)^j z_2)} + B^{(\lambda)}((-1)^{j+1} z_1, (-1)^j z_2)$$

$$\cdot \overline{B^{(\nu)}((-1)^{j+1} z_1, (-1)^j z_2)}\} := \Xi_{\lambda,\mu} = \delta_{\lambda,\nu}, \quad \lambda, \nu \in \{0,1,2,3\}. \tag{18}$$

For an arbitrary positive integer $n \in Z_+$, expand it by

$$n = \sum_{j=1}^{\infty} \nu_j 4^{j-1}, \quad \nu_j \in \Delta = \{0,1,2,3\}. \tag{19}$$

Lemma 4. Let $n \in Z_+$ and n be expanded as (17). Then we have

$$\hat{h}_n(\omega) = \prod_{j=1}^{\infty} B^{(\nu_j)}(e^{-i\omega_1/2^j}, e^{-i\omega_2/2^j}) \hat{h}_0(0).$$

Lemma 4 can be inductively proved from formulas (14) and (18).

Theorem 1. For $n \in Z_+$, $k \in Z^3$, we have

$$\langle h_n(\cdot), h_n(\cdot - k)\rangle = \delta_{0,k}. \tag{20}$$

Proof. Formula (20) follows from (10) as n=0. Assume formula (20) holds for the case of $0 \le n < 4^{r_0}$ (r_0 is a positive integer). Consider the case of $4^{r_0} \le n < 4^{r_0+1}$. For $\nu \in \Delta$, by induction assumption and Lemma 1, Lemma 3 and Lemma 4, we have

$$(2\pi)^2 \langle h_n(\cdot), h_n(\cdot - k)\rangle = \int_{R^2} \left|\hat{h}_n(\omega)\right|^2 \cdot \exp\{ik\omega\} d\omega$$

$$= \sum_{j \in Z^2} \int_{4\pi j_1}^{4\pi(j_1+1)} \int_{4\pi j_2}^{4\pi(j_2+1)} \left|B^{(\nu)}(z_1, z_2) \cdot \hat{h}_{[n/8]}(\omega/2)\right|^2 \cdot \mathrm{e}^{ik\omega} d\omega$$

$$= \int_0^{4\pi} \int_0^{4\pi} \left|B^{(\nu)}(z_1, z_2)\right|^2 \sum_{j \in Z^2} |\hat{h}_{[n/8]}(\frac{\omega}{2} + 2\pi j)|^2 \cdot e^{ik\omega} d\omega$$

$$= \int_0^{4\pi} \int_0^{4\pi} \left|B^{(\nu)}(z_1, z_2, z_3)\right|^2 \cdot e^{ik\omega} d\omega = \int_0^{2\pi} \int_0^{2\pi} \Pi \cdot e^{ik\omega} d\omega = \delta_{o,k}$$

Thus, we complete the proof of theorem 1.

Theorem 2. For every $k \in Z^2$ and $m, n \in Z_+$, we have

$$\langle h_m(\cdot), h_n(\cdot - k)\rangle = \delta_{m,n}\delta_{0,k}. \tag{21}$$

Proof. For the case of $m = n$, (20) follows from Theorem 1.As $m \ne n$ and $m, n \in \Omega_0$, the result (20) can be established from Theorem 2, where $\Omega_0 = \{0,1,2,3\}$. In what follows, assuming that m is not equal to n and at least one of $\{m, n\}$ doesn't belong to Ω_0, rewrite m , n as $m = 4m_1 + \lambda_1$, $n = 4n_1 + \mu_1$, where $m_1, n_1 \in Z_+$, and $\lambda_1, \mu_1 \in \Omega_0$.**Case 1** If $m_1 = n_1$, then $\lambda_1 \ne \mu_1$. By (17), formulas (21) follows, since

$$(2\pi)^2 \langle h_m(\cdot), h_n(\cdot - k)\rangle = \int_{R^2} \hat{h}_{4m_1+\lambda_1}(\omega)\overline{\hat{h}_{4n_1+\mu_1}(\omega)} \cdot \exp\{ik\omega\} d\omega$$

$$= \int_{[0,4\pi]^2} B^{(\lambda_1)}(z_1, z_2) \sum_{s \in Z^2} \hat{h}_{m_1}(\omega/2 + 2s\pi) \cdot \overline{\hat{h}_{m_1}(\omega/2 + 2s\pi)}\, \overline{B^{(\mu_1)}(z_1, z_2)} \cdot e^{ik\omega} d\omega$$

$$= \frac{1}{(2\pi)^2} \int_{[0,2\pi]^2} \Xi_{\lambda_1,\mu_1} \cdot \exp\{ik\omega\} d\omega = O.$$

Case 2 If $m_1 \ne n_1$ we order $m_1 = 4m_2 + \lambda_2$, $n_1 = 4n_2 + \mu_2$, where $m_2, n_2 \in Z_+$, and $\lambda_2, \mu_2 \in \Omega_0$. If $m_2 = n_2$, then $\lambda_2 \ne \mu_2$. Similar to Case 1, we have (21) follows. That is to say, the proposition follows in such case.

As $m_2 \neq n_2$, we order $m_2 = 2m_3 + \lambda_3$, $n_2 = 2n_3 + \mu_3$, once more, where $m_3, n_3 \in Z_+$, and $\lambda_3, \mu_3 \in \Omega_0$. Thus, after taking finite steps (denoted by r), we obtain $m_r, n_r \in \Omega_0$, and $\lambda_r, \mu_r \in \Omega_0$. If $\alpha_r = \beta_r$, then $\lambda_r \neq \mu_r$. Similar to Case 1, (21) holds. If $\alpha_r \neq \beta_r$, Similar to Lemma 1, we conclude that

$$\langle h_m(\cdot), h_n(\cdot - k)\rangle = \frac{1}{(2\pi)^2}\int_{R^2} \hat{h}_{4m_1+\lambda_1}(\omega)\overline{\hat{h}_{4n_1+\mu_1}(\omega)} \cdot e^{ik\omega} d\omega$$

$$= \frac{1}{(2\pi)^2}\int_{[0,2^{r+1}\pi]^2} \{\prod_{t=1}^{r} B^{(\lambda_t)}(\omega/2^t)\} \cdot O \cdot \{\prod_{t=1}^{r} B^{(\mu_t)}(\omega/2^t)\} \cdot e^{ik\omega} d\omega = O.$$

Theorem 3. If $\{G_\beta(x), \beta \in Z_+^2\}$ and $\{G_\beta(x), \beta \in Z_+^2\}$ are vector-valued wavelet packs with respect to a pair of biorthogonal vector-valued scaling functions $G_0(x)$ and $G_0(x)$, then for any $\alpha, \sigma \in Z_+^2$, we have

$$\langle G_\alpha(\cdot), \tilde{G}_\sigma(\cdot - k)\rangle = \delta_{\alpha,\sigma}\delta_{\underline{0},k} I_s, \quad k \in Z^2. \tag{22}$$

References

1. Telesca, L., et al.: Multiresolution wavelet analysis of earthquakes. Chaos, Solitons & Fractals 22(3), 741–748 (2004)
2. Iovane, G., Giordano, P.: Wavelet and multiresolution analysis: Nature of ε^∞ Cantorian space-time. Chaos, Solitons & Fractals 32(4), 896–910 (2007)
3. Zhang, N., Wu, X.: Lossless Compression of Color Mosaic Images. IEEE Trans. Image Processing 15(16), 1379–1388 (2006)
4. Chen, Q., et al.: A study on compactly supported orthogonal vector-valued wavelets and wavelet packets. Chaos, Solitons & Fractals 31(4), 1024–1034 (2007)
5. Shen, Z.: Nontensor product wavelet packets in $L_2(R^s)$. SIAM Math. Anal. 26(4), 1061–1074 (1995)
6. Chen, Q., Qu, X.: Characteristics of a class of vector-valued nonseparable higher-dimensional wavelet packet bases. Chaos, Solitons & Fractals 41(4), 1676–1683 (2009)
7. Chen, Q., Wei, Z.: The characteristics of orthogonal trivariate wavelet packets. Information Technology Journal 8(8), 1275–1280 (2009)
8. Chen, Q., Huo, A.: The research of a class of biorthogonal compactly supported vector-valued wavelets. Chaos, Solitons & Fractals 41(2), 951–961 (2009)

Predicting Gas Emission Based on Combination of Grey Relational Analysis and Improved Fuzzy Neural Network

Xiaoyan Tang[1] and Zhengguo Wang[2]

[1] College of Geology and Environment, Xi'an University of Science and Technology, Xi'an, China, 710054
[2] Trans-Asia Gas Pipeline Company Limited, Beijing, China, 100007
lzdtxy2004@sina.com

Abstract. Gas emission is controlled by various factors; it is a very complex problem how to predict gas emission according to these factors. When by using geological information and winning technical data to predict it, there are lots of uncertainty, ambiguity and highly nonlinear characteristics. This feature is difficult using traditional mathematical expressions to describe. This paper combines grey relational analysis with improved fuzzy neural network (IFNN) to predict gas emission. Based on grey relational degree these gas emission-sensitive factors are optimality collected. Take these factors as input of IFNN model, the prediction model of gas emission is established. Results show that it is reliable. Recognition precision is high, and practicability is better.

Keywords: Grey Relation, Improved Fuzzy Neural Network, Gas Emission, Prediction.

1 Introduction

Gas occurs in the coal seam, almost everywhere there is in the mine. It can be disturbed in all mining activities every time. Gas emission, gas outburst and gas disaster involve complex geological information and winning technical factors. Gas emission has fuzziness and randomness and is controlled by geological information and winning technical factors, these factors tend to have non-linear characteristics. Therefore, the prediction of mine gas emission is an important and indispensable step in new mine, mine reconstruction and productive mine. The prediction accuracy of mine gas emission is directly related to mine safety and economic benefit. So the prediction method research and the improvement of its prediction accuracy has been one of important subject all over the world's major coal-producing countries.

In the actual study, when more factors, it will increase the network complexity, reduce the network performance and affect the calculation accuracy. Grey relational analysis provides a better solution. It can conduct a comprehensive quantitative analysis of all factors and determine the main factors. Fuzzy neural network has powerful knowledge expression and powerful learning ability [1-2]. Not only it expands the application scope of fuzzy technology, but also it is better than the conventional neural network in learning time, training step and precision [3]. Because

D. Jin and S. Lin (Eds.): Advances in CSIE, Vol. 2, AISC 169, pp. 275–280.
springerlink.com

of its advantages, fuzzy neural network technologies are widely applied in petroleum exploration and development. During predicting gas emission by using conventional data, there are some characteristics of uncertainty, ambiguity and highly nonlinear problem between these factors and gas emission. Thus, the method of combining grey relational analysis and improved fuzzy neural network is applied to predict gas emission in this paper.

2 Predicting Model of Gas Emission Based on Grey Relational Analysis and Improved Fuzzy Neural Network

2.1 Grey Relational Analysis Method

Grey relational analysis is a multi-factor analysis method with good and broad adaptability [4]. It is a new method developed on the basis of grey system theory. By comparing the grey relational degree, the primary and secondary factors can be found out in the system and the primary factors of affecting on one variable can be obtained.

Based on the qualitative analysis of the researched matter, a dependent variable and multi-independent variable are determined.

Assume that the dependent variable is the sequence of mine gas emission.

$$x_0 = [x_0(1), x_0(2), \cdots, x_0(n)]^T \tag{1}$$

The corresponding independent variable is the sequence of influencing factor:

$$x_i = [x_i(1), x_i(2), \cdots, x_i(n)]^T \text{ (i=1,2,...,m)} \tag{2}$$

The relational coefficient of the sequence of mine gas emission affected on the "i"th factor sequence at the "k"th moment is:

$$\xi_i(k) = \frac{\min\limits_i \min\limits_k |x_0(k) - x_i(k)| + \rho \max\limits_{1\le i\le m} \max\limits_{1\le k\le n} |x_0(k) - x_i(k)|}{|x_0(k) - x_i(k)| + \rho \max\limits_{1\le i\le m} \max\limits_{1\le k\le n} |x_0(k) - x_i(k)|} \tag{3}$$

Where: $0 < \rho < 1$, if the ρ value is smaller, the difference between the relational coefficients is bigger. In this paper, $\rho = 0.5$.

The average value of relational degree at each moments is:

$$r(x_0, x_i) = \frac{1}{n} \sum_{k=1}^{n} \xi_i(k) \tag{4}$$

$r(x_0, x_i)$ is the relational degree of mine gas emission affected on the "i"th factor sequence. Based on the value, the relational degree of mine gas emission affected on each factor sequence can be determined.

2.2 Improved Fuzzy Neural Network

The improved fuzzy neural network shows in Fig.1. There are four layers, namely, the input layer, the fuzzy layer, the fuzzy inference layer and the anti-fuzzy layer.

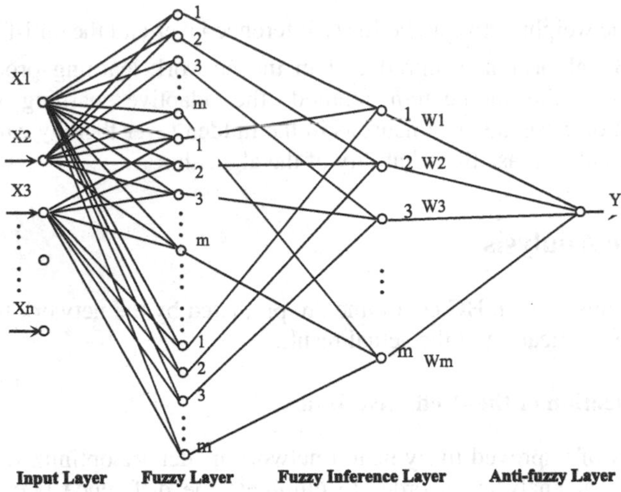

Fig. 1. The improved four layers fuzzy neural network structure

In the first layer, each neuron node is connected directly to each input component. Input variable x is an n-dimensional feature vector, n is determined by the specific problems.

In the second layer, each neuron node represents a fuzzy linguistic variable, which is to calculate the memberships of each component of the input vector belongs to the language variable corresponding to the fuzzy sets. The number of nodes of the fuzzy layer is $m \times n$ (m is the number of clusters). The input variables are x_i , the output is the degree of membership that each variable belongs to cluster.

The membership function is the following formula, as follows,

$$u_{ij} = \exp(-(x_i - m_{ij})^2 / \sigma_{ij}^2) \qquad 1 \le i \le n \, ; \, 1 \le j \le m \tag{5}$$

Where, x_i is input variable, m_{ij} , σ_{ij}^2 corresponds to each node of the fuzzy layer.

In the third layer, each neuron node represents a fuzzy rule, which is to match the fuzzy rules and calculate the practical degree of each rule. The node output π_j is the multiply of membership. The number of nodes m is obtained by sample cluster analysis based on K-means method, which can also adjust the value of this function according to actual needs.

$$\pi_j = u_{1j} \bullet u_{2j} \bullet \cdots u_{nj} = \prod_{i=1}^{n} u_{ij} \qquad 1 \le j \le m \tag{6}$$

In the fourth layer, each neuron node is carried on anti-fuzzy calculation, and the exact value of network output is obtained. The node output is respectively all input algebraic multiply. y is the ultimate anti-fuzzy output as follows,

$$y = \omega_1\pi_1 + \omega_2\pi_2 + \cdots + \omega_m\pi_m \tag{7}$$

Where, ω_i is the weight between the fuzzy inference layer and the anti-fuzzy layer.

Improved BP algorithm is introduced in the network learning process [5]. The strategy of using the momentum method, the adaptive learning rate and the self-adjustment of the number of neurons in the hidden layer thereby can enhance the learning speed and increase the reliability of the algorithm.

3 Instance Analysis

The mine gas emissions in HC coal mine are predicted by the network training of the method and the verification of the actual result.

3.1 Quantification of the Predictive Index

When the index of improved fuzzy neural network model are optimized by the use of the grey relational analysis, in order to eliminate the difference between unit and dimension of all factors, firstly the unitary method must be carried out to the initial data. The mine gas emission rates and the influencing factors in HC coal-mine are shown in Table 1. Roof lithology, geological structure and mining method are treated as qualitative variable. The initial data will be converted to the numerical values [0,1] range with equation (8) in input layer, the date in exchange with equation (9) in output layer.

Table 1. The mine gas emission rate and the influencing factors in HC coal-mine

Sample	GZ	MCHD/m	MCMS/m	TBYX	BBYX	Romax	MTJG	XPXS	SYZH	$Q/m^3 \cdot t^{-1}$
1	2	7	582.85	4	3	1.58	3	1.62	69.79	0.54
2	5	14	439.76	3	1	2.05	5	0.05	27.03	0.95
3	5	12	525.62	3	2	1.88	4	0.02	15.78	0.74
4	3	8	658.13	3	2	1.73	4	0.76	8.44	0.65
5	5	17	397.20	3	1	2.05	5	0.70	32.04	0.93
6	4	12	414.52	3	2	1.78	4	0.43	25.36	0.72
7	4	10	544.38	3	2	1.78	4	0.70	27.72	0.70
8	5	6	553.55	4	2	1.65	3	0.16	30.19	0.60
9	5	13	536.85	3	1	1.92	5	0.15	12.55	0.80
⋮	⋮	⋮	⋮	⋮	⋮	⋮	⋮	⋮	⋮	⋮
30	2	7	343.80	4	2	1.64	3	0.95	41.75	0.50

$$x_i^- = \frac{x_i - x_{min}}{x_{max} - x_{min}} \tag{8}$$

$$x_i = x_i^-(x_{max} - x_{min}) + x_{min} \tag{9}$$

Where, x_i^- is the normalized data, x_i is the un-normalized data, x_{min} is the minimum value of data change, x_{max} is the maximum value of data change.

3.2 The Optimization of Predictive Index

Treat mine gas emission as a reference sequence, the influencing factor as a comparison sequence, in accordance with (3) the relational coefficient between mine gas emission and each factor at the "i"th factor sequence at the "k"th moment are calculated, and then the relational degree of influencing factor are computed and played in sequence. The main factors of influencing on mine gas emission are determined by the sequence of relational degree. The results are shown in table 2.

Table 2. The relational degree and the factors sequence of influencing on mine gas emission

Influencing factor	GZ	MCHD	MCMS	TBYX	BBYX	Romax	MTJG	XPXS	SYZH
Relational coefficient	0.754	0.761	0.631	0.537	0.458	0.427	0.385	0.615	0.672
The sequence	2	1	5	6	7	8	9	4	3

As can be seen from Table 2, in many of these factors, the largest factor of influencing on the mine gas emission is the coal seam thickness (MCHD); the results show that the relational degree is 0.761, followed by geological structure (GZ). Combined with the law of gas geology, the characteristic of gas emission and winning technology, the five parameters (MCMD, GZ, SYZH, MCMS, XP) are determined as the prediction model index.

3.3 The Improved Neural Network's Computation Based on the Optimized Index

Based on the grey relational analysis of the factors of influencing on mine gas emission, mine gas emission is predicted by using improved fuzzy neural network. According to the calculation and the sequence of grey relational degree, the above-mentioned five factors are treated as the input factors in improved fuzzy neural network model. The network input layer consists of five input variable, and the output layer is the mine gas emission. At the same time the errors are computed between the actual output and the expected output of the network. The results are shown as followed in table 3.

As can be seen from Table 3, the maximum prediction error is 10.81% and the average error is 8.01%. With a view to the actual production of coal mine, the improved fuzzy neural network prediction method can be used as a reference method for predicting mine gas emission.

Table 3. The network testing results

Training sample	1	2	3	4	5	6	7	8	9	10
Measured gas emission / $m^3 \cdot t^{-1}$(Q)	0.54	0.95	0.74	0.65	0.93	0.72	0.7	0.6	0.8	0.5
Networks output /$m^3 \cdot t^{-1}$(Q)	0.59	1.02	0.69	0.71	1	0.8	0.66	0.65	0.72	0.54
Relative error /%	8.34	7.25	6.48	8.79	7.65	10.81	5.38	7.62	9.38	8.35
Relative average error /%	8.01									

4 Conclusion

Mine gas emission is affected by a variety of factors. If never take these factors to analysis, it increases the difficulty of analyzing problem. The model stability will be affected; the results have the larger prediction error and low accuracy. In this paper the main factors of influencing on mine gas emission are determined more precisely by grey relational analysis; thereby the prediction of mine gas emission has much more pertinence and forecast accuracy.

The five main factors of influencing on mine gas emission are determined through the sequence of grey relational degree. These factors are treated as the input of improved fuzzy neural network model, and the corresponding mine gas emission as the model output. Predicted results show that the model is reliable. Recognition precision is relatively high, and practicability is better. It provides a new way for the mine gas emission projections.

Acknowledgement. This research is supported by Scientific Research Program Funded by Shaanxi Provincial Education Department (Program No. 11JK0784).

References

1. Shitong, W.: Fuzzy Neural System, Fuzzy Neural Network and Application Design. Shanghai Science and Technology Literature Publishing House, Shanghai (1998)
2. Shaohua, X., Jiuzhen, L.: Study on Pattern Selection and Generalization Ability for Neural Networks. Computer Science 28(6), 94–96 (2001)
3. Lixin, W.: Fuzzy Systems and Fuzzy Control Tutorial. Tsinghua University Press, Beijing (2003)
4. Songmei, Z., Ji, W., Feng, W., et al.: Grain Yield Prediction Model of Liaoning Province Based on Grey Relational Analysis. Water Saving Irrigation (5), 64–66 (2011)
5. Tang, X., Liu, Z.: Igneous Rocks Recognition Based on Improved Fuzzy Neural Network. In: Shen, G., Huang, X. (eds.) CSIE 2011, Part II. CCIS, vol. 153, pp. 338–342. Springer, Heidelberg (2011)

Calibration for Zooming Image Shrink-Amplify Center

Hongwei Gao, Changyi Luan, Fuguo Chen, Guang Yang, and Kun Hong

School of Information Science & Engineering, Shenyang Ligong University,
Shenyang, 110159
ghw1978@sohu.com

Abstract. According to the depth estimation, the calibration technology of zooming cameras, which combines the general method of camera calibration with characteristics of zooming cameras, is researched in this paper. By the least squares method, two coordinates of shrink-amplify center which have been generated under two different focal length are calibrated and analyzed. Thus, the foundation of depth estimation for zooming image would be laid. The experimental results show that the depth estimation can be done by shrink-amplify center which replaces the principal point and the coordinate of shrink-amplify center has good stability after calibrated many times.

Keywords: Zooming image, Depth estimation, Shrink-amplify center, Calibration.

1 Introduction

The depth estimation of the two images is a fundamental problem in computer vision. It is the key step towards the goal of image understanding and has important application in robotics, scene understanding and 3-D reconstruction. Zooming image, as a kind of monocular visual depth cues [1, 2] has a wide range of application in visual surveillance, visual tracking, robot context awareness and map building and so on [3]. Ma and Olsen [4] first proposed the method of depth estimation by using zoom lens. In theory, zooming image can provide the information about depth. Lavest etc. [5, 6, 7, 8] did accurate research on the optical properties of zoom lens and put forward that the zooming lens must be described by thick lens model in depth estimation. Asada and Baba etc. [9, 10, 11] put forward three-parameter model of zoom lens which includes zoom, focus and aperture that based on the actual structure of lens. Fayman, etc.[12] applied depth estimation of zooming image to the visual tracking and came up with a active vision technology of zooming tracking which broaden the application of depth estimation for zooming image. All above of studies are based on the calibration for zooming lens.

2 Calibration for Zooming Cameras

Camera calibration is a very important question in computer vision. The so-called camera calibration, getting interior parameters and external parameters according to the

D. Jin and S. Lin (Eds.): Advances in CSIE, Vol. 2, AISC 169, pp. 281–286.
springerlink.com

given camera model, is the process that establishes the relationship between camera image point position and 3D space point position. In a fixed parameter of zoom lens, there is no much different between zoom camera calibration and ordinary camera calibration. In order to establish the accurate model of the zoom camera, it is need to set up a table and record a series of calibration results which are obtained under different settings of zoom lens. Calibration of zoom camera has some characteristics. For example, main point drift, relation between internal and external parameters which under different focal length. And in the specific application, only use the right camera internal and external parameter model can we obtain the accurate results.

Optical properties of zoom lens are determined by three parameters, they are: zoom, focus and aperture. Through the zoom, we can get images of target in different resolution. Focus can make focus to different distance of targets. When using aperture, we can adjust brightness of image according to the light condition. The calibration of zoom lens is indispensable for depth estimation technology. Limited to experimental conditions, normal manual zoom camera is used for getting image in this paper. The hardware of calibration system includes manual zoom camera, plane calibration board and camera support.

2.1 Collection for Zooming Camera

Acquisition of zooming image need at least a focusing process and interior parameters of camera are often changed. So how to ensure the repeatability of manual zoom must be concerned. For manual zoom lens, repeatability of focusing can be guaranteed by scale of zoom ring. But more secure method is to use the minimum and maximum limit focal length.

On the basis of availability of calibration data, the clarity of the image must be considered. Because it directly affect the precision of image matching and depth reconstruction. For this reason, we do a simple test about the clarity of zooming image, as shown in figure 4. Firstly, we do focusing in different focal length and then get image. There is a set of consecutive zoom images in figure 1(a) and two images under maximum and minimum focal length in figure 1(b). Although having done single focus, the obtained images are very clear.

(a) (b)

Fig. 1. Collection for zooming images

From what has been discussed above, we can infer that:

1) Zooming cameras, based on manual focusing, uses the method of off-line calibration in this paper.
2) The stability of the calibration data can be ensured though the fixed aperture, single focusing and limit focusing.

3) A distinct image can be got though not every image at the best focus location and making only single focusing.

2.2 Least Squares Estimation of Shrink-Amplify Center

Shrink-amplify center is a very important parameter for matching and depth estimation. Firstly, it can optimize matching and eliminate mismatch; secondly, it is reference point for calculating radial parallax. Every image has a shrink-amplify center. This is determined by the position of the optical axis when do imaging. For high-grade zoom lens, the change of main point can be neglected and main point or its mean can be used as shrink-amplify center. But for most of ordinary zoom lens, the change or stability of main point must be concerned. The zooming images used for depth estimation reflect the same scene in different scale. And there is always a relative shrink-amplify center between different zooming images. When calculating the radial parallax for depth estimation, we need the shrink-amplify center. So the calibration and calculation method about shrink-amplify center are crucial.

Firstly, the calibration of a pair of zooming images should be investigated. Assume that n pairs of match points are obtained from a zooming image. A pair of match point can determine a line. So there are n linear equations:

$$\begin{cases} a_1 x + b_1 y = c_1 \\ a_2 x + b_2 y = c_2 \\ \cdots \\ a_n x + b_n y = c_n \end{cases} \tag{1}$$

Set the coordinate of shrink-amplify center for (Z_x, Z_y). In ideal state, shrink-amplify center is in every line. So there is a matrix equation:

$$\begin{bmatrix} a_1 & b_1 \\ a_2 & b_2 \\ \vdots & \vdots \\ a_n & b_n \end{bmatrix} \begin{bmatrix} Z_x \\ Z_y \end{bmatrix} = \begin{bmatrix} c_1 \\ c_2 \\ \vdots \\ c_n \end{bmatrix} \tag{2}$$

The foregoing formula can be simplified for:

$$A\theta = b \tag{3}$$

Obviously, shrink-amplify center $\theta = (Z_x, Z_y)$ can be converted into the least squares parameter estimation. That is when the N>2, shrink-amplify center θ is only determined by the following formula:

$$\hat{\theta}_{LS} = \left(A^T A\right)^{-1} A^T b \tag{4}$$

3 Calibration Results and Analysis

Firstly, a pair of shrink-amplify center of zooming image should be estimated by least squares. Taking calibration board as structured scene, we can extract scene feature point easily. As shown in figure2 (a), we should collect a pair of zooming image of at a fixed position in static scene. With the calibration kit, it is accurate to extract the feature corner of each image. The extraction results are shown in Figure 2 (b), where a pair of feature points of zooming image are put under the same image coordinate system.

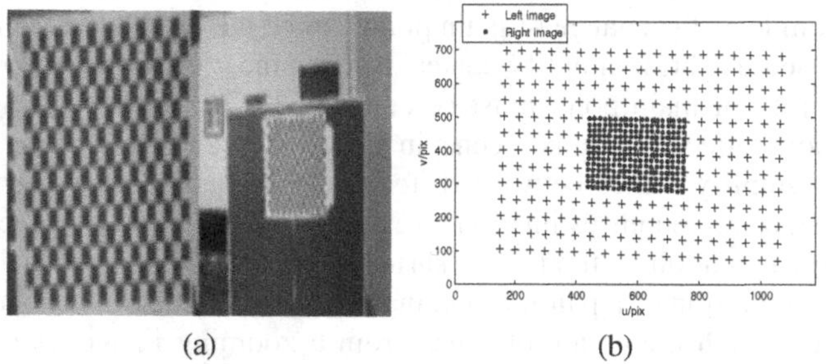

Fig. 2. Image corner extraction

The experiment can extract 247 pairs of feature matching points from the scene, which can determine 247 linear equations. And then use the least squares estimation method proposed in the last sections to calculate the shrink-amplify center of two images. As shown in figure 3(a), all straight lines, determined by feature matching points, are drew in the figure by black lines. In figure 3(a), least squares estimation results of shrink-amplify center are marked with a red cross lines. And the calculation results of shrink-amplify center, shown in figure 3(b), are marked in the original image of this zooming image. In order to observe the shrink-amplify center position clearly, the partial enlarged display are carried out, as shown in figure 3(c). Shrink-amplify center is almost the same point in real scene. In fact, the zoom camera is centered on this point for the zooming image of the scene.

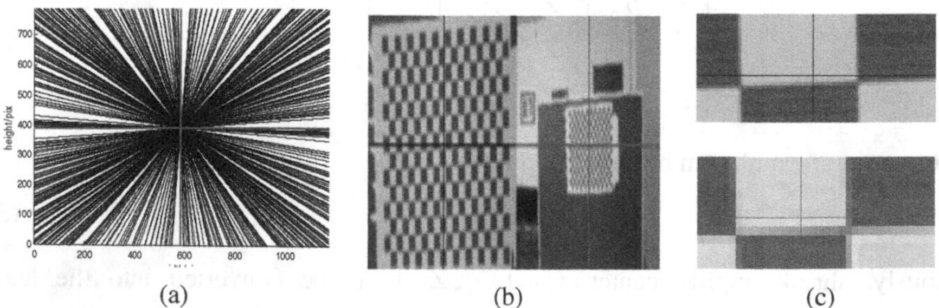

Fig. 3. Least squares estimation of shrink-amplify center

For the zoom camera, it is necessary to consider the repeatability and stability of the parameters. The following would describe repeatability and stability of shrink-amplify center from two aspects. (1) The changes of shrink-amplify center under two fixed focal lengths. (2) The drift of shrink-amplify center in continuous zoom.

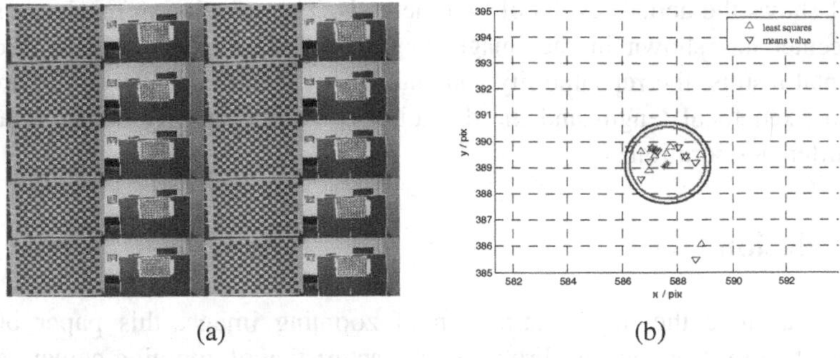

(a) (b)

Fig. 4. Calibration for shrink-amplify center under two fixed focal length

For the changes of shrink-amplify center in two fixed focal length, the experiment collects 10 pairs of zooming image under F18 and F55, as shown in Figure 4(a). From left to right, top to bottom, zooming images are composed of a pair of images. In addition to using the least square method to estimate the shrink-amplify center of each pair of zooming image, it also uses means value algorithm. The means value algorithm is that calculates all the points of intersection directly and then calculates the mean. By comparing the two kinds of calculation results, it shows that the least square method obtains better results. As shown in figure 4(b), red marks show the calculation results of least square method and the blue marks show the calculation of means value method. The circle's center in the chart is the means value of shrink-amplify center calculation results. The standard deviation is radius.

Table 1. Results of Repeated Calibration for Shrink-amplify Center

serial number	least square method		means value method	
01	586.7117	389.6217	586.2630	389.7364
02	587.6138	389.5541	588.6589	389.1921
03	587.7742	389.8048	586.9955	389.2346
04	587.0007	388.9011	586.7202	388.5516
05	588.8428	389.4934	588.0648	389.7696
06	587.2123	389.4562	587.2798	389.6083
07	588.8829	386.0627	588.6782	385.4847
08	587.0944	389.7135	587.1269	389.7077
09	587.1181	389.7392	587.2309	389.6722
10	588.3253	389.4170	588.2958	389.4255
mean value	587.6576	389.1764	587.5314	389.0383
variance	0.7817	1.1229	0.8384	1.3027

Table 1 shows the actual results of two methods. The calculation results of shrink-amplify center has shown in the image coordinates. The unit is pixels. From the experimental results, the repeatability and stability of shrink-amplify center are good under two fixed focal length and the deviation of shrink-amplify center is about 1 pixel in different operation.

4 Conclusion

In order to achieve the depth estimation of zooming image, this paper builds a monocular stereo vision system. From the characteristics of zooming camera and the method of camera calibration, the stable shrink-amplify center of zooming image can be got by least square method. Taking into account the characteristics of manual zooming lens, the stability of the calibration parameters are investigated, so as to complete the depth estimation of zooming image.

Acknowledgement. This work is supported by China Liaoning Provice Educational Office Fund (No.L2011038).

References

1. Wang, J., Wang, Y.Q.: A Monocular Stereo Vision Algorithm Based on Bifocal Imaging. Robot 33(6), 935–937 (2007)
2. Liu, X.X., Wang, Y.Q.: Analysis of the Monocular Stereo System Based on Bifocal Imaging. Computer Measurement & Control 16(9), 1316–1318 (2008)
3. Xia, T.K., Yang, M., Yang, R.Q.: Progress in monocular vision based mobile robot navigation. Control and Decision 25(1), 1–6 (2010)
4. Ma, J., Olsen, S.I.: Depth from zooming. Journal of the Optical Society of America 7(10), 1883–1890 (1990)
5. Lavest, J.M., Rives, G., Dhome, M.: Three-dimensional reconstruction by zooming. IEEE Transactions on Robotics and Automation 9(2), 196–207 (1993)
6. Lavest, J.M., Rives, G., Dhome, M.: Modeling an object of revolution by zooming. IEEE Transactions on Robotics and Automation 11(2), 267–271 (1995)
7. Delherm, C., Lavest, J.M., Dhome, M., Lapresté, J.T.: Dense Reconstruction by Zooming. In: Buxton, B.F., Cipolla, R. (eds.) ECCV 1996. LNCS, vol. 1065, pp. 427–438. Springer, Heidelberg (1996)
8. Lavest, J.M., Delherm, C., Peuchot, B., Daucher, N.: Implicit Reconstruction by Zooming. Computer Vision and Image Understanding 66(3), 301–315 (1997)
9. Asada, N., Baba, M., Oda, A.: Depth from blur by zooming. In: Proc. Vision Interface, pp. 165–172 (2001)
10. Baba, M., Asada, N., Oda, A., Migita, T.: A thin lens based camera model for depth estimation from defocus and translation by zooming. In: Proceedings of ICVI, pp. 274–281 (2002)
11. Baba, M., Asada, N., Oda, A., Yamashita, H.: Depth from Defocus by Zooming Using Thin Lens-Based Zoom Model. Electronics and Communications in Japan, 53–62 (2006)
12. Fayman, J.F., Sudarsky, O., Rivlin, E., Rudzsky, M.: Zoom tracking and its applications. Machine Vision and Applications 13(1), 25–37 (2001)

Research of Circuit Board Fast-Scanning Detecting Device

ZhiLiang Chen[1], Yue Liu[2], LiGuo Tian[1], Meng Li[1], and JiePing Zhang[1]

[1] Tianjin Key Laboratory of Information Sensing &Intelligent Control,
Tianjin University of Technology and Education, Tianjin, China, 300222
[2] Tianjin Modern Vocational Technology College, Tianjin, China
czl_tj@sina.com, {tlg1234,limeng-3260711}@163.com

Abstract. Along with the development of electronic technology, it is more and more important to design a printed circuit board (PCB) and test it for electronic product. At first it analyzed the PCB auto-test system, and then designed the contact-type PCB detection system. It is mainly composed of a needle plate to fix the tested circuit board, power supply module, A/D conversion module, single-chip microcomputer (SCM), display module and the keyboard module. With the core of AVR SCM ATmega16, it transmitted the measured point voltage gathered by A/D conversion module to the SCM, and then displayed it by the displaying module. After tested, the device works well and meets the requirements.

Keywords: Printed circuit board (PCB), Single-chip microcomputer (SCM), ATmega16, A/D.

1 Introduction

The PCB has been widely used in the modern electronic industry, various methods for detecting the quality of circuit board are also emerged as the times require because of the urgent need for industry [1]. The paper designed a printed circuit board testing system. With the core of the single-chip microcomputer, the paper designed the circuit board fast-scanning detecting device, which can simultaneously detect several points voltages on the circuit board, also has the properties of low cost and simple operation, and saves the time for workers during the course of detection of a large number of circuit boards, greatly improving the work efficiency and reducing the manpower and material resources.

2 Function of Fast-Scanning Detecting Device for Circuit Board

Function of fast-scanning detecting device for circuit board as shown in Fig. 1, it is made up of the needle board to fix the tested circuit board, input signal and variables acquisition module (A/D conversion), centralization and control processing module (SCM), parameter setting and manual control module (keyboard operation), detection result display module (nixie tube display) and result and memory communication

D. Jin and S. Lin (Eds.): Advances in CSIE, Vol. 2, AISC 169, pp. 287–291.
springerlink.com

module. With the core of ATmega16 SCM, the device transmits the measured point voltage gathered by A/D conversion module to the SCM, and then displays it by the nixie tube displaying module or data storage or transmission by result and memory communication module.

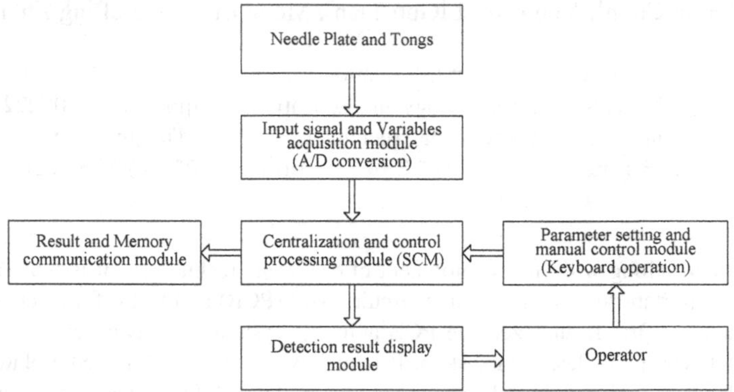

Fig. 1. The functional diagram of fast-scanning device

3 Hardware Design

3.1 Power Supply Module

After voltage transformation, the AC voltage connected to the POWER interface is rectified by the bridge rectifier and filtered by the capacitor, and then connected to the chip 7805, so as to get the 5V DC voltage from the output terminal of 7805 [2]. The 5V DC signal filtered by the capacitor provides the ATmega16 single-chip microcomputer with the power supply, so as to ensure ATmega16 SCM work normally. With parallel connection of a resistor and LED light-emitting diode branch on the output terminal of the power circuit is used to display whether power supply circuit could work normally (bright: normal ;contrary: abnormal), circuit is shown in Fig. 2.

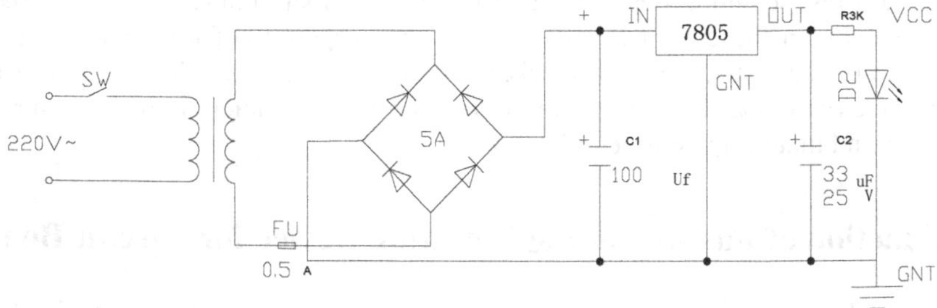

Fig. 2. The power supply circuit

3.2 A/D Module

A/D module uses the ADC of Atmega16, the ADC is a kind of 10-bit successive approximation, and connected to a 8-channel analog multiplexer which allows 8 single-ended voltage inputs constructed from the pins of Port A, the single-ended voltage inputs refer to 0V (GND); The ADC contains a sample and hold circuit which ensures that the input voltage connected to the ADC is held at a constant level during the course of conversion; The device also supports 16 differential voltage input combinations [3]. At first, Atmega16, from 8-way data channel of Port A, takes sample from the needle plate, and then, after the single-chip processing, finally the result is sent to LED digital tube to display it.

3.3 Storage and Communication Design

In the AVR microcontroller, there are 512 bytes of data memory E^2PROM and serial peripheral interface SPI, USART and Two-wire serial interface (TWI) and other communication interfaces. The user can set a full duplex asynchronous communication mode to communicate with some equipments, such as CRT terminal, personal computer and so on.

4 Software Design

4.1 Main Program

After putting the measured circuit board on the needle plate, AVR SCM starts to work, on the LED digital display the initial state “00" can be seen. At first, keyboard will be detected, at this time the system can be turned into the setting state or the stop state. After the keyboard test is finished, SCM begins to check whether the circuit board to be tested is connected to the needle plate correctly. If the connection isn’t correct LED digital tube will display an error code, contrarily, SCM will detect the voltages of the circuit. If the detection voltages are different from the given parameters, SCM will trigger the buzzer; on the contrary, LED digital tube will display the code "88". The flow chart of the main program is shown in Fig. 3.

4.2 Keyboard Control Program

When entering the work state, keyboard detection will be started, pressing the key K1 could turn into the setting state, also pressing the key K2 could run the program of voltage detection. After entering the settings state, the key K2 is used to plus 1 for detecting period and the key K3 is used to reduce the detecting cycle, each pressing could change the cycle, also can quit and stop the detection program. During the course of running the program, the key K2 is used to switch to the stop state and then regain to detection state. Besides, during the course of detecting, the key K3 is used to terminate the test program or enter the ready state. The flow chart of keyboard control program is shown in Fig. 4.

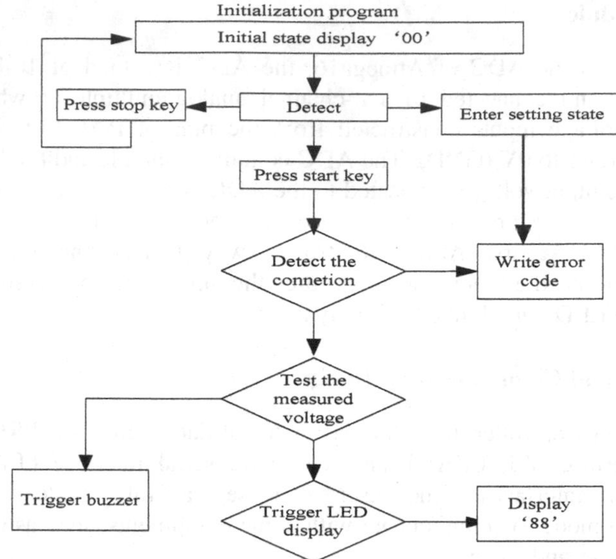

Fig. 3. The flow chart of the main program

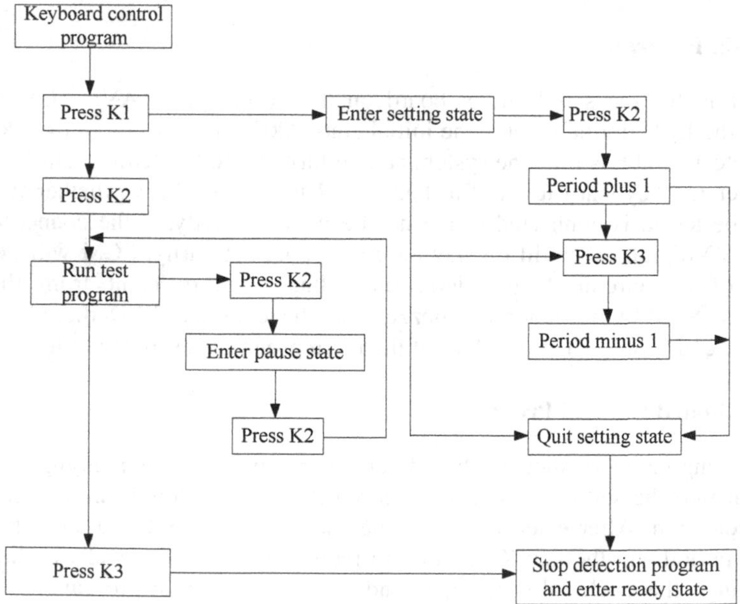

Fig. 4. The flow chart of the keyboard control program

5 Conclusion

With the development of the electronics and printed circuit board manufacturing technology, modern electronic products become increasingly complex, the density of printed circuit board is increasing, followed by detection and repair of printed circuit board has become even more difficult [4]. The fast-scanning detecting device which uses single-chip microcomputer as the core has simple operation, low cost, high efficiency and so on. After tested, the device works well.

Acknowledgments. The authors thank all the people who had given helps to them in the work. This work is supported by Tianjin University Science & Technology Fund (Grant No.20100721).

References

1. Yang, S.: Realization of Contact Printed Circuit Board Detection System. Electronic Science and Technology Journal 7, 12–15 (2006)
2. Sun, Y.: Power supply design and skills training tutorials. Publishing House of Electronics Industry Press, Beijing (2007)
3. Liu, H.: AVR Microcontroller Theory and measurement and control engineering applications. Beihang University Press, Beijing (2008)
4. PCB Resources Network, http://www.pcbres.com

Design of Farmland Information Intelligent Collection Node Based on GPS

Yue Liu[1], LiGuo Tian[2], Meng Li[2], and ZhiLiang Chen[2]

[1] Tianjin Modern Vocational Technology College, Tianjin, China, 300222

[2] Tianjin Keylaboratory of Information Sensing&Intelligent Control, Tianjin University of Technology and Education, Tianjin, China

{liuyue201202,tlg1234}@163.com

Abstract. Timely and accurate acquisition of the farmland information is the foundation of the implementation of precision agriculture. Farmland information acquisition technology based on the GPS is an important means of precision agriculture information acquisition. According to the characteristics of field information collection, such as high density, high accuracy, low cost and so on, it designed intelligent farmland information acquisition node based on GPS. The node with the portable GPS positioning module, used the embedded processor STM32F103VC as the core, integrated the farmland information acquisition sensor, could acquire, process and transmit temperature and humidity in the field information, so as to realize to monitor the farmland information parameter in the long distance. After tested, the node works well and has achieved the design requirements.

Keywords: Precision agriculture, Information acquisition, GPS positioningm, Intelligent node, Humiture sensor.

1 Introduction

Precision agriculture is a rising and interdisciplinary synthesis technology developed in the international agriculture field at the early 1980s [1], it is an important modern agriculture form developed on the basis of the modern information technology, biotechnology, engineering technology and a series of high and new technology, provided with some characteristics, such as the reasonable use of agricultural resources, increase of crop yield, lower production costs, improvement of the ecological environment [2]. Farmland information collection technology plays an important role in the implementation of precision agriculture. Application of the low cost, high efficient intelligent equipment in farmland information collection to get temperature, humidity, light and CO_2 concentration which could affect the crop growth and soil moisture and other real-time dynamic microenvironment information based on precise space position can better advance the implementation of precision agriculture. On the basis of characteristics of farmland information data acquisition, the paper designed a farmland information intelligence collection node based on GPS,

D. Jin and S. Lin (Eds.): Advances in CSIE, Vol. 2, AISC 169, pp. 293–298.
springerlink.com

which can better acquire temperature and humidity parameters in the farmland microenvironment information and implement the corresponding data display, storage and transmission, thus, provide the farmland information data support for implementation of precision agriculture.

2 Application of GPS Technology in the Farmland Information Acquisition

GPS (Global Positioning System) is the English abbreviation of navigation satellite timing and Ranging/global positioning system, referred to as the" global positioning system". The system which is a satellite-based radio navigation and positioning system, has totipotency(land, sea and aerospace), global, all-weather, continuous and real-time navigation, positioning and timing function, provides all types of users with precise three-dimensional coordinates, speed and time [3].

In order to increase the soil utilization rate to the hilt, Precision agriculture technology based on the geographical coordinates is to manage various measures effectively in crop production (fertilizer, seeds and pesticides) according to the soil characteristics and crop physiological state, optimize the crop production, so as to achieve the minimal resource consumption and environmental pollution, maximize economic benefits, and make agriculture develop sustainably. GPS equipment can provide the accurate information of spatial position. On the basis of the combination of GPS technology and farmland information acquisition technology, in the acquisition of farmland information, spatial location information could be acquired synchronously at the same time, thus dynamic microenvironment information for each spatial position could be obtained in real time, which could provide data support for the precision agriculture [4].

3 Overall Design of System

Farmland information collection of precision agriculture mainly obtains crop growth microenvironment information, including humiture, illumination, CO_2 concentration and soil moisture, which are important for crop growth. According to those above information, the paper designed the farmland information intelligent acquisition node based on GPS, the node used the embedded processor STM32F103VC as the core, collocated with the portable GPS positioning module, Nokia5510 LCD module, humiture sensor module, ZigBee wireless transmission module and SD card data storage module, can be used alone for data acquisition, processing and display as well as constitutes wireless sensor network together with the ZigBee module and the other node for field information data collection. The general diagram of the node is shown in Fig.1.

4 Hardware System

The hardware system of the farmland information intelligent acquisition node mainly comprises a microprocessor module, GPS positioning module, humiture sensor module, LCD display module, wireless transmission module and power supply module.

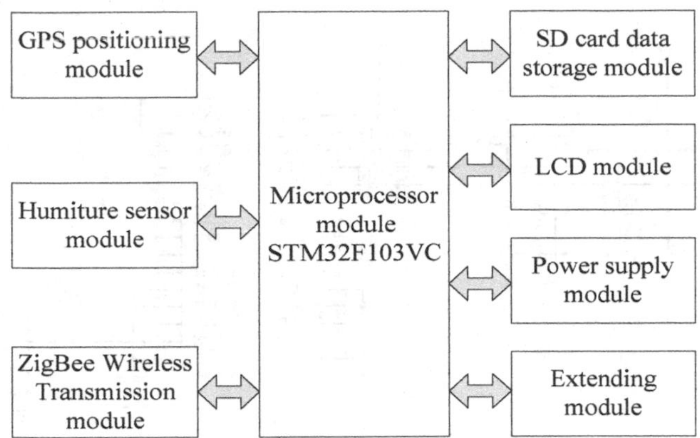

Fig. 1. The system overall block diagram

4.1 Microprocessor Module

The microcontroller adopts 32-bit processor STM32F103VC used ARM Cortex-M3 as core , its working frequency is 72MHz, built-in high-speed memory (high 128K byte flash and 20K byte SRAM); it has a wealth of enhanced I/O port and peripheral connected to the two APB bus; contains two 12-bit ADC, three general 16-bit timer and a PWM timer, also contains a standard and advanced communication interfaces: up to 2 I^2C and SPI, 3 USART, one USB and one CAN [5].

4.2 GPS Positioning Module

GPS positioning module use the global satellite positioning receiving chip GS-89m-J which belongs to the Gstar series. GS-89m-J is a high performance, low-power consumption intelligent satellite receiver module, which is the third generation satellite positioning receiving module made by the United States Rifi company as well as a complete satellite positioning receiver. The module takes a digital /analog separate-design method, which effectively improves the anti-interference ability; besides, it has 32-channel capture engine, maximum update rate is 1s, accuracy range is less than 10 meters. The module connect with the microprocessor through the serial interface , the GPS data receiving circuit as shown in Fig. 2.

4.3 Humiture Sensor Module

SHT71 is a high-integrated chip which integrates many functions such as temperature sensing, humidity sensing and signal conversion, A/D conversion and heater, and output the digital signal. The SHT71 has high antijamming capability, higher cost performance and lower power dissipation; additional, all the technical index of the SHT71 is identical to the design requirement.

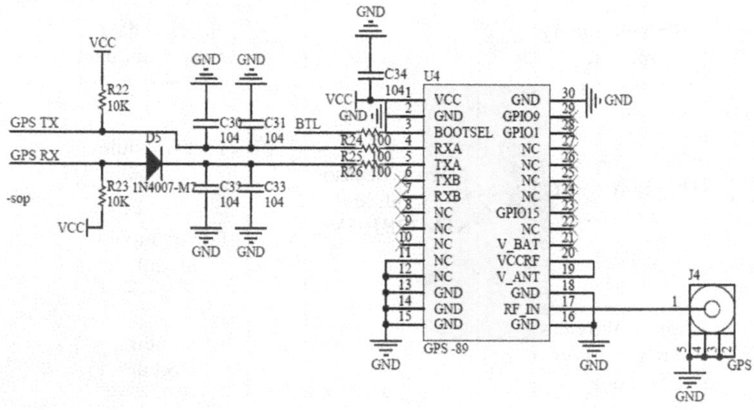

Fig. 2. GPS data receiving circuit

4.4 LCD Module

LCD module using Nokia3310 LCD, can display English characters and Chinese and graphics. The control driver of LCD is PCD8544. PCD8544 is a low-power consumption CMOS LCD control driver, designed to drive the 48 rows and 84 columns of the graphical display, all the necessary display function are integrated on a chip, including LCD voltage and the bias voltage generator, only need a few external components and small power consumption.

4.5 ZigBee Wireless Transmission Module

Wireless transmission module is mainly used for data transmission when the node constitutes wireless sensor network with other nodes, here ZigBee technology is adopted. The chip used the second generation of ZigBee®/IEEE 802.15.4 RF SoC of TI Company CC2530, which is mainly used for 2.4 GHz free-license ISM band. CC2530 integrates enhanced 8051 microcontroller core, 2.4GHz RF radio transceiver in accordance with the IEEE802.15.4 standard and the 256k Flash program memory, supports the newest ZigBee2007pro protocol stack, equips with two powerful USART with support for multi-group serial protocol, one MAC timer in accordance with IEEE 802.15.4, one 16-bit timer, and two 8-bit timer.

5 Software System

The software system of the farmland information intelligent collection node includes a board-level initialization procedure, driver, embedded operating system and applications, these parts combined together organically form specific integration software system. The bottom-up software architecture of the node includes: driver layer, operating system layer and application layer, software architecture as shown in Fig. 3.

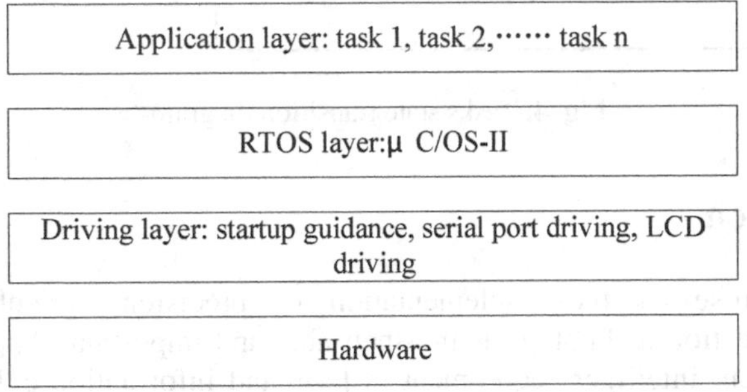

Fig. 3. The structure chart of the software architecture

5.1 The Imbedded Operating System

μ COS-II is a real-time and be-deprived kernel of operating system. The operating system supports up to 64 tasks, but every task's priority should be different each other; the task provided with small priority number has higher priority than the task provided with large priority number, also, the system always dispatches the ready task provided with the highest priority to run.

5.2 Task Management and Communication of μCOS-II

A task is in four different conditions, namely, dormancy state, ready state, running state, wait or suspend state. The conversion of the different task status is shown in Fig.4. Once the task is established, it will be in ready state and waiting for running, and when the concurrent events emerge, the task provided with the highest priority will be immediately given the right to CPU, and after the task is finished and suspended or the interrupt service routine is finished, next task provided with the highest priority than others will get the right to use CPU. If the running of system results in that the priority of the task which in the ready state is higher than the priority of the task which in running state, the system will call the scheduling function, so the task which in the running state will lose the occupation of the CPU and return to the ready state, however, the task in ready state which has the highest priority will be turned to the running state.

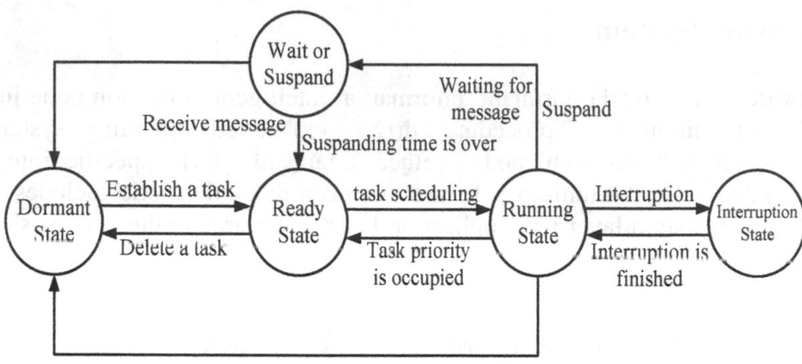

Fig. 4. Tasks state transition diagram

6 Conclusion

During the course of the implementation of precision agriculture, farmland information collection technology is indispensable and important. Application of low cost, high-efficient intelligent equipment in farmland information collection and the use of information technology to promote and transform traditional agriculture could push forward Chinese traditional agriculture to develop towards precision agriculture steady and rapidly.

Acknowledgments. The authors thank all the people who had given helps to them in the work. This work is supported by Tianjin University Science & Technology Fund (Grant No.20100721).

References

1. Wang, M.: Development of Precision Agriculture and Innovation of Engineering Technology. Agricultural Engineering Journal 15, 1–8 (1999)
2. Liu, A., Feng, Z., Xu, L.: Modern Precision Agriculture and Development of Precision Agriculture in China. Journal of China Agricultural University 5, 20–25 (2000)
3. Liu, J.: GPS Satellite Navigation Positioning Principle and Method. Science Press, Beijing (2008)
4. He, Y., Fang, H.: Research on Precision Agriculture Information Processing System Based on GPS and GIS. Journal of Agricultural Engineering 18, 145–149 (2002)
5. Li, N.: Application Development of STM32 Processor Based on MDK. Beijing University of Aeronautics and Astronautics Press, Beijing (2008)
6. Shao, B.: Embedded Real-time Operating System μC/OS-II, 2nd edn. Beijing University of Aeronautics and Astronautics Press, Beijing (2003)

An Improved Shuffled Frog-Leaping Algorithm for Knapsack Problem

Zhoufang Li, Yan Zhou, and Peng Cheng

School of Information Science and Engineering
Henan University of Technology
ZhengZhou, China
lzhf1978@126.com

Abstract. Shuffled frog-leaping algorithm (SFLA) has long been considered as new evolutionary algorithm of group evolution, and has a high computing performance and excellent ability for global search. Knapsack problem is a typical NP-complete problem. For the discrete search space, this paper presents the improved SFLA (ISFLA), and solves the knapsack problem by using the algorithm. Experimental results show the feasibility and effectiveness of this method.

Keywords: shuffled frog-leaping algorithm, knapsack problem, optimization problem.

1 Introduction

Knapsack problem (KP) is a very typical NP-hard problem in computer science, which was first proposed and studied by Dantzing in the 1950s. There are many algorithms for solving the knapsack problem. Classical algorithms for KP are the branch and bound method (BABM), dynamic programming method, etc. However, most of such algorithms are over-reliance on the features of problem itself, the computational volume of the algorithm increasing by exponentially, and the algorithm needs more searching time with the expansion of the problem. Intelligent optimization problem for solving NP are the ant colony algorithm, greedy algorithm, etc. Such algorithms do not depend on the characteristics of the problem itself, and have strong global search ability. Related studies have shown that it can effectively improve the ability to search for the optimal solution by combining the intelligent optimization algorithm with the local heuristic searching algorithm.

Shuffled frog-leaping algorithm is a new intelligent optimization algorithm, it combines the advantages of meme algorithm based on genetic evolution and particle swarm algorithm based on group behavior. It has the following characteristics: simple in concept, few parameters, the calculation speed, global optimization ability, easy to implement, etc. and has been effectively used in practical engineering problems, such as resource allocation, job shop process arrangements, traveling salesman problem, 0/1 knapsack problem, etc. However, the basic leapfrog algorithm is easy to blend into local optimum, and thus, this paper improved the shuffled frog-leaping algorithm to

D. Jin and S. Lin (Eds.): Advances in CSIE, Vol. 2, AISC 169, pp. 299–303.
springerlink.com

solve combinatorial optimization problems such as knapsack problem. Experimental results show that the algorithm is effective in solving such problems.

2 The Mathematical Model of Knapsack Problem

Knapsack problem is a NP-complete problem about combinatorial optimization, which is usually divided into 0/1 knapsack problem, complete knapsack problem, multiple knapsack problem, mixed knapsack problem, the latter three kinds can be transformed into the first, therefore, the paper only discussed the 0/1 knapsack problem. The mathematical model of 0/1 knapsack problem can be described as:

$$\begin{cases} \max \sum_{i=0}^{n} x_i v_i \\ \sum_{i=0}^{n} x_i w_i \le C \quad (x_i = 1 \ \ or \ \ 0, \ \ i = 1, 2, ..., n) \end{cases} \tag{1}$$

where: n is the number of objects;

w_i is the weight of the *i-th* object(i=1,2…n);

v_i is the value of the *i-th* object;

x_i is the choice status of the *i-th* object; when the *i-th* object is selected into knapsack, defining variable x_i=1,otherwise x_i=0; C is the maximum capacity of knapsack.

3 The Basic Shuffled Frog-Leaping Algorithm

It generates P frogs randomly, each frog represents a solution of the problem, denoted by U_i, which is seen as the initial population. Then it calculates the fitness of all the frogs in the population, and arranges the frog according to the descending of fitness. Then dividing the frogs of the entire population into m sub-group of, each sub-group contains n frogs, so $P = m*n$. Allocation method: in accordance with the principle of equal remainder. That is, by order of the scheduled, the *1, 2, ..., n* frogs were assigned to the *1,2, ..., N* sub-groups separately, the *n+1* frog was assigned to the first sub-group, and so on, until all the frogs were allocated.

For each sub-group, setting U_B is the solution having the best fitness, U_W is the solution having the worst fitness, U_g is the solution having the best fitness in the global groups. Then, searching according to the local depth within each sub-group, and updating the local optimal solution, updating strategy is:

$$S = \begin{cases} \min\{\text{int}(rand(U_B - U_W)), S_{\max}\}, U_B - U_W \ge 0 \\ \max\{\text{int}(rand(U_B - U_W)), -S_{\max}\}, U_B - U_W \le 0 \end{cases} \tag{2}$$

$$U_q = U_w + S \tag{3}$$

where, S is the adjustment vector of individual frog, S_{max} is the largest step size that is allowed to change by the frog individual. *rand* is a random number between 0 and 1.

4 The Improved Shuffled Frog-Leaping Algorithm for KP

A frog is on behalf of a solution, which is expressed by the choice status vector of object, then frog $U=(x_1, x_2, \ldots, x_n)$, where, x_i is the choice status of the *i-th* object; when the *i-th* object is selected into knapsack, defining variable x_i=1,otherwise x_i=0; f (i), the fitness function of individual frog can be defined as:

$$f(i) = \sum_{i=1}^{n} x_i v_i \tag{4}$$

4.1 The Local Update Strategy of Frog

The purpose of implementing the local search in the frog sub-group is to search the local optimal solution in different search directions, after searching and iterating a certain number of iterations, making the local optimum in sub-group gradually tend to the global optimum individual.

Definition 1. Giving a frog's status vector *U*, the switching sequence *C(i,j)* is defined:

$$C(i, j) = \begin{cases} 1, \text{ when i = j and } U_i \text{ changed} \\ 0, \text{ when i} \neq \text{j and } U_i = U_j \\ 2, \text{ when i} \neq \text{j and } U_i \neq U_j \end{cases} \tag{5}$$

where, U_i said the state of object *i* becomes from the selected to the cancel state, or in turn; U_i= U_j, object *i* and object *j* exchange places, that object *i* and object *j* are selected or deselected at the same time. $U_i \neq U_j$, object *i* is selected or canceled, or in turn. Then the new vector of switching operation is: $U' = U + C(i, j)$

Definition 2. Selecting any two vectors U_i and U_j of frog from the group, *D*, the distance from U_i to U_j is all exchange sequences that U_i is adjusted to U_j.

$$D(U_i, U_j) = \{C(i_1, j_1) \to C(i_2, j_2) \to \ldots \to C(i_m, j_m)\} \tag{6}$$

where, m is the number of adjusting.

Based on the above definition, the update strategy of the individual frog is defined as follows:

$$l = \min\left\{\text{int}[rand \times |D_{B,W}|], l_{\max}\right\} \tag{7}$$

$$s = \{D(U_B, U_W)\} \tag{8}$$

$$U_q = U_w + S \tag{9}$$

where, l is the number of switching sequence $D(U_B, U_W)$ for updating U_W; l_{max} is the maximum number of switching sequence allowed to be selected; S is the switching sequence required for updating U_W.

4.2 The Global Information Exchange Strategy

During the execution of the basic shuffled frog-leaping algorithm, the operation of updating the feasible solution was is executed repeatedly, it is usually to meet the situation that updating fail, the basic shuffled frog-leaping algorithm updates the feasible solution randomly, but the random method often falls into local optimum or reduces the rate of convergence of the algorithm.

Obviously, the key that overcoming the shortcomings of basic SFLA in evolution is: it is necessary to keep the impact of local and global best information on the frog jump, but also pay attention to the exchange of information between individual frogs. In this paper, first two jumping methods in basic SFLA are improved as follows:

$$Pn = PX + r1*(Pg - Xp1(t)) + r2*(PW - Xp2(t)) \tag{10}$$

$$\mathbf{P_n = P_b + r3*(P_g - Xp3(t))} \tag{11}$$

where, $Xp1(t)$, $Xp2(t)$, $Xp3(t)$ are any three different individuals which are different from X. Meanwhile, we remove the sorting operation according to the fitness value of frog individual from basic SFLA, and appropriately limit the third frog jump. Thus, we get an efficient modified SFLA based on the improvements of above. In the modified algorithm, the frog individual in the subgroup generates a new individual (the first jump)by using formula(10), if the new individual is better than its parent entity then replacing the parent individual. Otherwise re-generating a new individual (the frog jump again) by using (11). If better than the parent , then replacing it. Or when $r_4 \leq FS$ (the pre-vector, its components are $0.2 \leq FSi \leq 0.4$), generating a new individual (the third frog jump) randomly and replacing parent entity.

The new update strategy will enhance the diversity of population and the search through of the worst individual in the iterative process, which can ensure communities' evolving continually, help improving the convergence speed and avoid falling into local optimum, and then expect algorithm both can converge to the nearby of optimal solution quickly and can approximate accuracy, improved the performance of the shuffled frog-leaping algorithm.

5 Simulation Experiment

Two classical 0/1 knapsack problem instances were used in the paper, example 1 was taken from the literature_[4], example 2 was taken from the literature_[5]. The comparison algorithm used in the paper was branch and bound method for 0/1 knapsack problem. Under the same experimental conditions, two instances of simulation experiments were conducted 20 times, the average statistical results were shown in Table 1.

Table 1. Comparison of two algorithms

instance	The set of solutions	Capacity of value		Average running time (ISFLA/BABM)
		ISFLA	BABM	
1	101110101111101	1042/878	1037/878	1. 83/2. 35
2	1101110101101001101110010001101000 1011	3025/989	3025/989	6. 73/8. 21

6 Conclusion

The shuffled frog-leaping algorithm is a kind of search algorithm with random intelligence and global search capability, this paper improved shuffled frog-leaping algorithm and solved the 0/1 knapsack problem by using the algorithm. Experiments show that the improved algorithm has better feasibility and effectiveness in solving 0/1 knapsack problem.

References

1. Eusuff, M.M., Lansey, K.E.: Optimization of water distribution network design using the shuffled frog leaping algorithm. Water Resource Planning and Management 129(3), 210–225 (2003)
2. Li, Y.-H., Zhou, J.-Z., Yang, J.-J.: An improved shuffled frog-leaping algorithm based on the selection strategy of threshold. Computer Engineering and Applications 43(35), 19–21 (2007)
3. Luo, X.-H., Yang, Y., Li, X.: Improved shuffled frog-leaping algorithm for TSP. Journal of Communication 30(7), 130–135 (2009)
4. He, Y.-C.: Greedy genetic algorithm and its application for KP. Computer Engineering and Design 28(11), 19–22 (2007)
5. Wu, Z.-H.: Algorithm Design and Analysis, pp. 251–252. Higher Education Press, Beijing (1993)

Bounding Box Modeling in Virtual Orthodontics Treatment System

Zhanli Li and Youlan Chen

College of Computer Science and Technology
Xi'an University of Science and Technology
Xi'an, China
Chenyoulancy1@163.com

Abstract. The objective of this paper was establishing the bounding box applied on 3D dental models which reflected the characteristics of single-tooth shapes accurately. First, according to the Gaussian curvature of each tooth's surface and the definition of feature points on single-tooth in oral medicine, the feature points are extracted. Second, based on the feature points, established a local coordinate system which reflect the direction of the inertia axis, mesi-distal and buccolingual. Finally, according to the maximum value of the local coordinate system, constructed a bounding box for each tooth adopted the Axis-Aligned Bounding Box (AABB) method. Based on the model and method, a software is developed by using Visual C++ and OpenGL for the bounding box establishment of single tooth. Experiments show, that method meet the requirements of oral medicine, more reliable and effective compared with traditional methods.

Keywords: Bounding Box, Gaussian Curvature, feature point, Orthodontics.

1 Introduction

In traditional dental treatment, treatment facilities need to adjust to achieve the correct path for tooth movement repeatedly. The treatment process is usually time consuming and expensive. Therefore, developing a computer-aided simulation system will be an important goal of orthodontics [1].

A computer-aided virtual surgery system employs the image data to help doctors to decide operation program reasonably, to evaluate all surgical techniques before the actual operation, and improve the success rate, security and accuracy. Therefore, virtual surgery [2] had been an important role in the medical treatment.

The virtual orthodontics treatment system is the basis of the virtual surgery. In the virtual orthodontic system, from the arrangement of the teeth to the tooth movement path planning, we need to use a single-tooth which has the original feature of tooth. Therefore, the establishment of a bounding box for a single-tooth is necessary. This paper was analyzed the process of the construct a bounding box with the combination of the characteristics of single tooth and oral medicine. A method for construct a bounding box based on the feature points was presented.

D. Jin and S. Lin (Eds.): Advances in CSIE, Vol. 2, AISC 169, pp. 305–311.
springerlink.com

2 Feature Point Extraction

The dental model is composed of a number of vertices which is the most basic primitive geometric element of the surface. Feature points are an important factor in describing the characteristics of tooth. Therefore the extraction of feature points plays a significant role in the surface segmentation and integration of measurement data in Reverse Engineering (RE)[3].

The feature point will not variable with the variation of the coordinate system, it's a core for establishment of coordinate system for single tooth, so the extraction of feature points referred as one crucial indicator for dental description as well as the key to represent the direction of the teeth. Three-dimensional dental cast as an irregular shape and it is impossible to use precise mathematical function to describe the corresponding three-dimensional mesh surface, also there is no continuous curvature. The discrete Gaussian curvature of the curve, which can describe the feature points of tooth surface, has adopted to reflect the degree of bend. To calculate the curvature values of vertices of the triangular patch using the following formula:

$$K = \frac{2\pi - \sum_{j} \theta_j}{A} \tag{1}$$

The geometric meaning of this formula is simple and clear. The Gaussian curvature value of the vertex in triangular mesh surface is regards as the ratio of the numerator and the denominator, representing the curvature of the point in a field. The numerator is the difference of the 2π and union of the angles which is adjacent neighborhood of the point corresponding to the vertex and triangles. The denominator is the corresponding area of the adjacent neighborhood of the point.

Today, there are several methods to estimate the discrete curvature of triangular surface mesh, but this paper adopted the following metho.

In the triangular grid, define the edge $e_j = p_j - p_i$ of two end points p_i and p_j, and angle $\alpha_j = \angle(e_j, e_{j+1})$ of both edges e_j, e_{j+1}, the intersection angle λ_i of both sides $p_i p_j$, $p_j p_{j+1}$, the intersection angle γ_j of both sides $p_i p_{j+1}, p_j p_{j+1}$, defined the triangle p_i, p_j, p_{j+1} which surrounded by e_j, e_{j+1} as T_i ,the unit normal vector as follows:

$$n_j = \frac{e_j \times e_{j+1}}{\| e_j \times e_{j+1} \|} \tag{2}$$

Define the angle between the normal vector of two adjacent triangles T_{j-1} and T_j to be $\beta_j = R(n_{j-1}, n_j)$.

Every two triangles T_{j-1} and T_j with mutual edge e_j make a small internally tangent cylinder, so that the triangular mesh surface will be substituted by a series of small cylinders which make up the surface more smooth than before. Therefore, direct from the theorem of differential geometry can be Gaussian curvature and the discrete mean curvature formula. The A is in the formula (3) is showed in the Fig1.

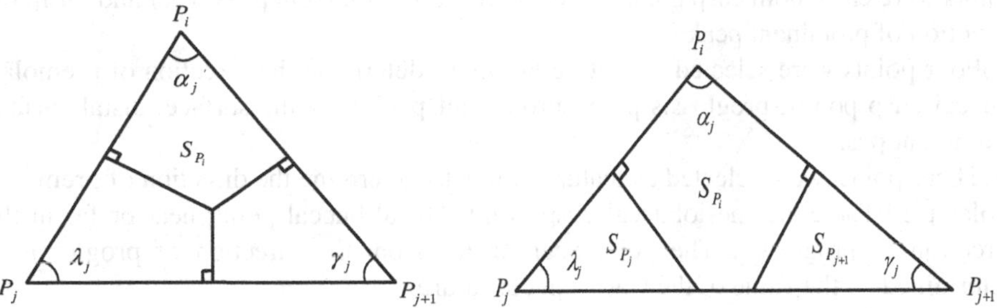

Fig. 1. Area calculation (the left is the graph of the acute triangle and the right is the graph of the obtuse triangle)

$$K = \frac{2\pi - \sum_j \alpha_j}{A} \qquad |H| = \frac{1}{4A}\sum_{j=1}^{n} \| e_j \| |\beta_j| \tag{3}$$

When the p_i is an acute triangle, the area was calculated by the formula(4) :

$$S_{pi} = \frac{1}{8}\{\| p_i p_j \|^2 \cot \lambda_j + \| p_i p_{j+1} \|^2 \cot \gamma_j\} \tag{4}$$

When the p_i is an obtuse angle, the area was calculated by the formula (5) to (8):

$$S_{p_i} = S\Delta_{p_i p_j p_{j+1}} - S_{p_j} - S_{p_{j+1}} \tag{5}$$

$$S\Delta_{p_i p_j p_{j+1}} = \frac{1}{2} \cdot P_i P_j \cdot P_i P_{j+1} \cdot \sin(\pi - \alpha) \tag{6}$$

$$S_{p_j} = \frac{1}{8} \| p_j p_{j+1} \|^2 \tan \lambda_j \tag{7}$$

$$S_{p_{j+1}} = \frac{1}{8} \| p_i p_j \|^2 \tan \gamma_j \tag{8}$$

Solving Gaussian curvature of the triangular mesh surface we need to use discrete differential geometry, using the following formula:

$$K(v) = \tfrac{1}{A}(2\pi - \sum \theta_i) \tag{9}$$

Based on the method mentioned above, the feature points are extracted by the following steps of different teeth.

(1) Three points were selected as feature points in the central incisors and lateral incisors, namely: mesio-incisal- edge point, distal- incisal- edge point and a point in line of intersection with crowns and gums.

(2) There points are selected to determine the direction of the canine teeth. These three points were each tooth cusp point in the direction of prominent peak, near and far in the direction of prominent peak.
(3)Four points were selected as feature points to determine the direction of premolar. Buccal cusp points, proglossis point, prominent peak near the surface, distal surface prominent peak.
(4)There points were selected as feature points to determine the direction of premolar. molar teethThere are mesiobuccal cusp point, Distal buccal point, near or far in the direction of proglossis. The point near or far from the direction of proglossis is determined by the value of the Gaussian curvature.

The extraction result of the feature points as shown in Fig 2.

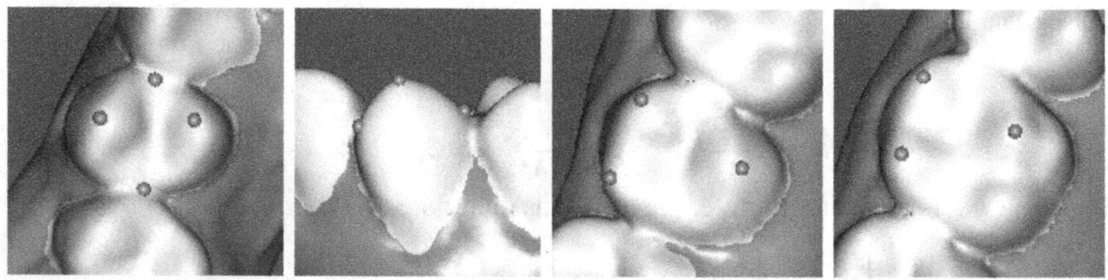

Fig. 2. Extraction result(where from left to right are feature point extraction result of left canine, left first premolar, left first molar and the left second molar)

3 Local Coordinate System Establishment

In this paper, the following method was adopted to establish the tooth local coordinate system under the restriction of the feature points, with three axes to reflect the direction of upper and lower jaws direction, mesio-distal direction and buccolingual direction.
(1) Define the mesial feature point of the cutting edge known as A, distal feature point named B, and the point in the intersection line of crown and gums as a feature point C .The line AB and the cutting edge of incisor are parallel lines, so defined the direction of AB as the Z axis, namely the direction of mesio-destial of incisors; use the direction of cross product which determined by BC and AB as the X-axis, it also the direction of labiolingual of incisors. Finally, defined the direction of Y-axis use the cross product of X-axis and Z-axis, namely the direction of upper and lower jaws.
(2) Define the cusp point of canine teeth as A, recorded the most prominent peak near and away the direction of mesio-distal as B and C. The direction of the line BC is defined as the Z axis coordinate system, namely the direction of mesio-destial of canine teeth; used the direction of the cross product of AB and BC defined as the coordinates of the X-axis, it also the direction of ligula of canine teeth; define the direction of cross product of X-axis and Z-axis as Y-axis, namely the direction of the upper and lower jaws.
(3) Premolar teeth has four feature points, defined the cusp of the buccal as A, recorded the spire point of longue as B, close to the most outstanding point of the name C, far the

most outstanding point of the name D. The line AB reflects the direction of the buccal and lingual tooth, defines the direction of the X-axis coordinate system; define the direction of Y axis used the direction of the cross product of AB and CD; Finally, define the direction of the cross product of X and Y as the Z-axis.

(4) The mesial-buccal cusp point of the name recorded as A, distal-buccal cusp point of the name recorded as B, near the distal tip of the tongue tip or apex vertex names recorded as C. Point where the line AB and the tooth parallel to the direction of nearly far in, so define where the line AB direction of Z axis coordinate system; definition of straight line AB and BC lines where the direction of the cross product for the Y-axis coordinate system, and the upper molars mandibular direction; Finally, the definition of Y axis and Z axis direction of the cross product line where the X-axis coordinate system, and the molar buccolingual.

Fig.3 shows the feature points under the surface of the teeth were established incisor, premolar, canine and molar process of local coordinate system.

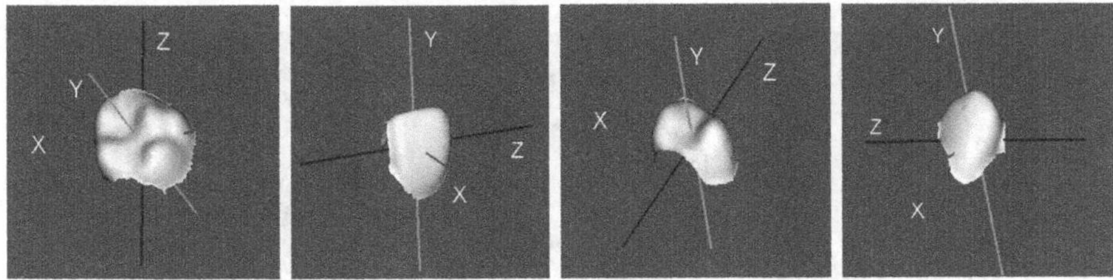

Fig. 3. The establishment of the teeth Local coordinate(where from left to the right are local coordinate system of the Incisor teeth, premolars, canine teeth, molar teeth)

4 Bounding Box Construction

By the previous description, we had known that according to morphological characteristics, local coordinate system of every tooth was defined. The pose description of the tooth space can be converted to the description of the corresponding teeth coordinate system. Traditional bounding box was divided int the following categories: bounding sphere, axis-aligned bounding box (AABB), fixed oriented bounding box, oriented bounding box (OBB). Different bounding box have different characteristics. Therefore, choosing the different bounding box according application is needed.

This paper drawn the bounding box of incisors, molars and canine teeth with the AABB method, close degree of tolerance is better, and simple calculation, the calculation of the intersection test not much, to better meet the collision detection process to achieve real-time requirements and procedures easier, but also to meet the arrangement of teeth in the tooth mesial to solve the calculation of the width requirements. Therefore this paper was adopted AABB as the method to establish the bounding box which is the basis for collision detection and teeth arrangement.

Part of the implementation code are as follows

```
BA = face.NodeList[indexs[0]].Node -
face.NodeList[indexs[1]].Node;
BC = face.NodeList[indexs[2]].Node -
face.NodeList[indexs[1]].Node;
Direction_Z = BA;
Direction_X - CrossProduct(&BA,&BC);
Direction_Y=CrossProduct(&Direction_X,&Direction_Z)
Vertex_Calculation( i );
// Calculate the maxmum value of the vertex i in the coordinate system
Draw_Axes_Box(i,indexs,offset,Line_Direction_X,Line_Direc
tion_Y,Line_Direction_Z);
//draw the coordinate system and bounding box of teeth i.
```

The result of the bounding box as Fig.4.

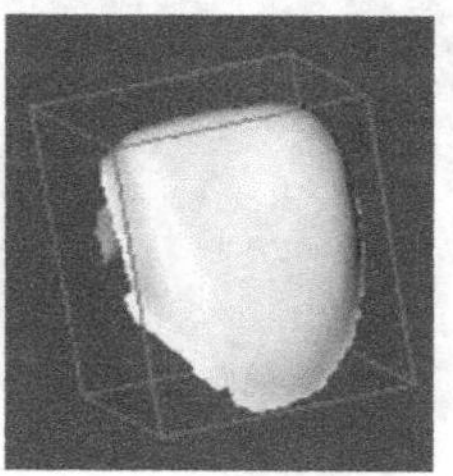
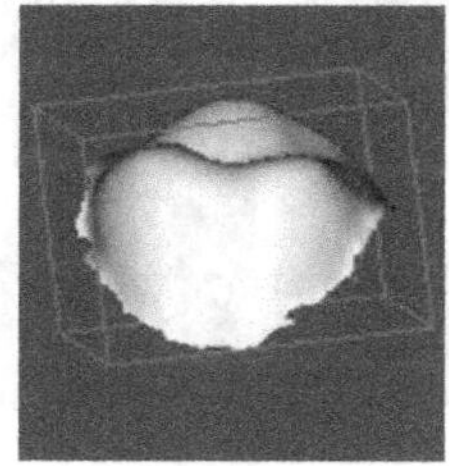
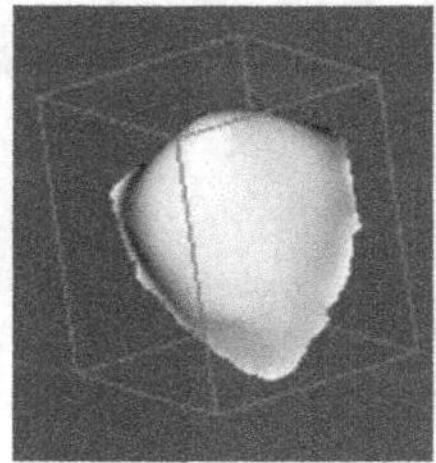

Fig. 4. Bounding box(where from left to the right are the bounding box of the premolars, Incisor teeth and molar teeth)

The establishment of the bounding box of teeth provides the basis for the teeth arrangement. First, according to the bounding box can determine the width of direction of the mesial, it is the basis of teeth arrangement. Second, it is also the key of the collision detection in the process of Path planning for teeth movement.

5 Conclusions

Single-teeth is the basis for teeth movement, therefore establish the single-tooth model accurately is very important. In this paper, Gaussian curvature was used to described the feature points of the teeth and establish the coordinate system of teeth based on the feature points, and finally adopted the AABB to determine the bounding box of the single-tooth, achieved excellent results. The coordinate system of the teeth is a new exploration to provide a theoretical basis for path planning and teeth arrangement, and it boosts the development of virtual orthodontics treatment system.

References

1. Xing, S.-Y., Zhang, S.-H., Liu, X.-B.: The Present Situation of Virtual Surgery System Technology. Computer Engineering and Applications 7, 88–90 (2004)
2. Ji, F.: Research and Development of Virtual Orthodontic System (2006)
3. Wang, Y., Lu, P.: Establishment of artificial tooth coordinate for full denture. J. Pract. Stomatol. 21(5), 9 (2005)
4. Fang, H., Wang, G.: Comparison and Analysis of Discrete Curvatures Estimation Methods for Triangular Meshes. Journal of Computer Aided Design & Computer Graphics 11(17) (2005)
5. Wang, L.-W., Liu, B.-Y., Han, J.-W.: Survey of Box-Based Algorithms for Collision Detection. Journal of Civil Aviation University of China 25(4) (August 2007)
6. Jeongc, D.: Testing neural network crash avoiding system in mobile robot, pp. 21–30. Case Western Reserve University (2001)

Learning to Observation Matrices of Compressive Sensing

Guosheng Gu and Jie Ling

Faculty of Computer, Guangdong University of Technology, Guangzhou, China
gugs@163.com, jling@gdut.edu.cn

Abstract. In this paper, a binary sparse observation matrix for compressive sensing is deterministically constructed via a pseudo-random sequence generated by the sub-shift mapping of finite type on the chaotic symbolic space. Analysis and experimental results demonstrate the proposed matrix's simplification can be regarded as a reliable method and is usable in compressive sensing applications.

Keywords: compressive sensing, deterministic observation matrix, pseudo-random sequence.

1 Introduction

In the recent years, Compressive Sensing [1, 2], which can represent sparse signals at a rate significantly below the Nyquist rate, has drawn much of attention in the field of signal processing. Such an optimum compression technique seeks to acquire as less as possible linear measurements by a non-adaptive observation matrix while maintaining the ability to reconstruct the original signal. Briefly Speaking, Compressive Sensing has an outstanding capability in reducing the number of project samples which is far below the original signal dimension.

Suppose $x \in R^n$ denote an original signal with at most $k << n$ nonzero components, $y \in R^m (m < n)$ the compressed measurements and $\Phi \in R^{m\times n}$ the observation matrix. The main process of Compressive Sensing can be described as to solve the following inverse problem:

$$\min\|x\|_0 \qquad s.t. \qquad y = \Phi x \tag{1}$$

Where $\|\cdot\|_0$ means ℓ_0 norm that counts the number of non-zero components of x. Note that most natural signal α is not sparse in temporal or space domain but in other transform domain (e.g. Fourier, discrete cosine transform (DCT), discrete wavelet transform (DWT)). In such cases the notation changes to

$$y = \Phi x = \Phi \Psi \alpha \tag{2}$$

Where Ψ is called sparsifying matrix.

D. Jin and S. Lin (Eds.): Advances in CSIE, Vol. 2, AISC 169, pp. 313–318.
springerlink.com

Generally, the minimization problem (1) is NP-hard and needs a combinatorial search through any possible sparse signal x. Donoho and Candes [1, 2] firstly provided the basis pursuit algorithm to find the accurate approximation by relaxing the non-convex ℓ_0 condition loss to a convex optimization task such as ℓ_1 minimization. Another kind of proposed reconstructing method is the greedy iteration algorithms such as orthogonal matching pursuit [3], Gradient pursuits [4], iterative thresholding algorithm [5], and Bayesian method [6], etc.

One sufficient condition to specify for which kinds of observation matrices the perfect recovery is feasible is the restricted isometry property (RIP) [7] given by Baraniuk. RIP implies that any subset of columns in $\Phi\Psi$ with cardinality less than the sparsity level of x is orthogonal. Some observation matrices have been explored and investigated. This paper tries to make a study on the observation matrices of Compressive Sensing. In section 2, we give fundamental theory and existing observation matrix ensembles. In section 3, we give the chaotic based observation matrix and comparing it with others. Section 4 is the Numerical results and section 5 is the conclusion.

2 Observation Matrices

For Compressive Sensing, the coherence between rows of the observation matrix Φ and columns of sparsifying matrix Ψ ought to be as small as possible, which can be measured by the inner product.

2.1 Random Observation Matrices

According to the smallest coherence principle, random observation matrices are firstly investigated, such as Gaussian random matrix, Bernoulli matrix, Hadamard matrix and partial Fourier matrix, et al. The entries of Gaussian random matrix [8] or Bernoulli matrix [9] are with independent and identically distribution therefore almost universally incoherent with any fixed sparsifying basis, leading to most least number of samples needed for perfect construction. However, it results in high cost to realize such random matrices in practical applications as they require very high computational complexity and large storage capacity due to their inherent random structure. Hadamard matrix and partial Fourier matrix [10, 11] are procured by randomly selecting arbitrary m rows from an $n \times n$ orthogonal matrix and then renormalizing the columns to be unit-normed. Although partial FFT has fast and efficient implementation, it only performs well in the case that the sparsifying basis is the identity matrix. For Hadamard matrix cases, it only works when the dimension of m is equal to a power of 2.

2.2 Deterministic Observation Matrices

For the sake of making Compressive Sensing applications more practicable, a mount of variable techniques also have been presented for constructing deterministic observation matrices, including complex-Valued chirp sequences [12], second order Reed-Muller codes [13], dual BCH codes [14], Kerdock and Delsarte-Goethals codes [15] and Additive character matrices with small alphabets [16]. Yu [17] construct a Toeplitz-structured observation matrix with digital logistic chaotic sequence and indicate the RIP of this kind of matrix is guaranteed.

Meanwhile, some researchers [18, 19] put a loft of effort in sparse observation matrices, with a view to designing good encoding and recovery algorithms with short sketch lengths and low computational complexities.

Although the theoretical recovery bounds may worse than those of random matrices or even difficult to gain in some cases, the deterministic observation matrices may guarantee the recovery performance that is empirically reliable, allowing fast processing and low complexity.

3 Deterministic Sparse Chaotic Matrix

To facilitate the benefits of deterministic construction, this section presents how to construct a deterministic sparse observation matrix for Compressive Sensing via a pseudo-random chaotic sequence. Precisely, we construct the matrix by employing a sequence generated by sub-shift mapping of a finite type of symbolic chaotic space as its elements.

Chaotic maps are one of the most approaches for pseudo-random sequences generation. Three main advantages are easily addressed to chaotic signals. One lies in its noise-like observation and can be quantified to independent-identical-distribution sequences. Secondly, the generated sequences are sensitive to the given initial-values and parameters. Thirdly, it owns the perfect non-periodic and ergodic properties. What's more, generating of chaotic signal is often of low computational cost and storage.

Most widely used chaotic pseudo-random sequences generator is the digital non-linear maps such as the well known Logistic Map, which is represented by the following formula:

$$x_{n+1} = rx_x(1-x_n) \tag{3}$$

Where r is a positive constant. For the cases $3.99465 < r \le 4$, the system is in chaotic state and produces very complicated pseudo-random sequences.

However, in practical applications, all the digital chaotic sequences will degenerate to periodic ones because they are only allowed to be accomplished in subsistent mechanisms with finite precision. As it happens, excellent properties of digital chaotic sequences are weakened.

To overcome the period degeneration issue, in [20] we have proposed a pseudo-random sequence generation method through the sub-shift mapping of finite type on the chaotic symbolic space. Comparing with the digital chaotic maps, the sub-shift of finite type is determined by a 16-order square matrix and complete nonlinear. The produced sequences are with elements within the integer subset {0, 1, 2, ···, 15} and have proved overwhelming non-periodic, non-convergent advantages over digital chaotic sequences.

In this paper, we consider a binary observation matrix Φ that is associated with the pseudo-random sequence which is mentioned above.

Suppose $\{z_l\}$ denotes a pseudo-random sequence generated by the sub-shift mapping of finite on the chaotic symbolic space, with elements within the set {0, 1, 2, ···, 15}. Construct a observation matrix $\Phi = (\phi_{ij})_{m \times n}$ column by column with this sequence, written as

$$\phi_{ij} = \begin{cases} 1 & if \quad i = 16l + z_l \bmod m \qquad and \qquad j = 16l + z_l \operatorname{div} m; \\ 0 & otherwise. \end{cases} \tag{4}$$

Where $\operatorname{mod}(\cdot)$ denotes the remainder function and $\operatorname{div}(\cdot)$ the quotient function after a dividend is divided by a divisor. From the definition, the location of the nonzero component of the matrix is chosen uniformly by the value and the subscript of z_l. According to the ergodic property of chaotic sequences, one can see the binary sparse matrix Φ containing almost same amount of nonzero values per column and per row, where the nonzero components take the value 1 with probability 1/16. Note that only non-zero elements of Φ will lead to computational operations in encoding process. Therefore, the encoding complexity of Compressive Sensing with the proposed structural matrix has been decreased. Using such a structure, the matrix can be efficiently implemented in practice.

4 Numerical Simulations

In the experiments, we compare the properties of the proposed measurement matrix, the Bernoulli matrix and the Toeplitz-structured matrix. A one dimension discrete sparse signal of length 1024 with only 32 non-zero elements was chosen as test data. The reconstruction algorithm is orthogonal matching pursuit (OMP) algorithm.

The performances of these measurement matrices are compared by the accurate reconstruction rate-the proportion of correct data numbers of the reconstruction signals to all data numbers of original signal. The results based on experiments are shown in Fig. 1.

When increasing the number of samples in the tests, accurate reconstruction rate will ascend correspondingly. When the number of samples is larger than a certain threshold, the signal finally can be perfectly reconstructed. As can be seen in Fig. 1, the proposed observation matrix outperforms the two other candidates. It provides attractive feasibility for implementation in practice.

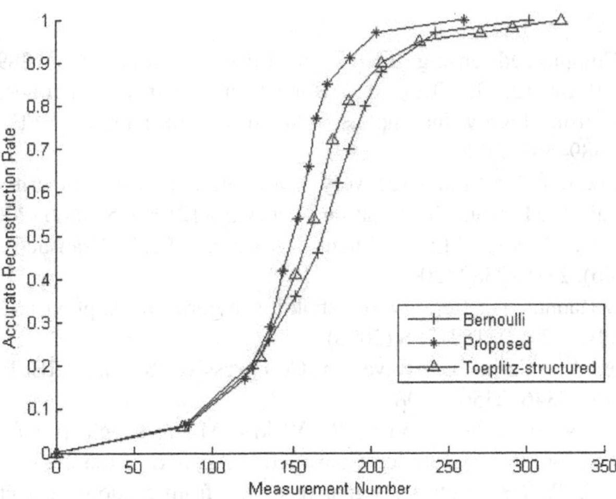

Fig. 1. Comparison of accurate reconstruction rate of the tested observation matrices

Table 1 illustrates the MSE (Mean Square Error) performance between the original signal and the reconstructed signals of different observation matrices under different measurement number. It demonstrates the proposed observation matrix outperforms the others once again.

Table 1. Results of the reconstructed errors

Number of measurements	Bernoulli matrix	Toeplitz-structured matrix	The proposed Matrix
128	0.2661	0.2493	0.2078
256	0.0496	0.0480	0.0466

5 Conclusion

In this work, we explored a non-periodic pseudo-random sequence to form an observation matrix for Compressive Sensing. Simulation results show that matrix perform quite well to two other matrices. Since such matrix is sparse reducing the computation and storage complexity, it is desirable for practice of applications.

Acknowledgement. This work is supported by the Industry-Education-Research Cooperation Project of Guangdong Province and the Ministry of Education (No.2011A090200068) and the Science and Technology Planning Project of Guangdong Province (No.2011B010200003).

References

1. Donoho, D.: Compressed sensing. IEEE Trans. Inform. Theory 52(4), 1289–1306 (2006)
2. Candes, E., Romberg, J., Tao, T.: Robust uncertainty principles: Exact signal reconstruction from highly incomplete frequency information. IEEE Trans. Inform. Theory 52(2), 489–509 (2006)
3. Tropp, J., Gilbert, A.C.: Signal recovery from random measure-merits via orthogonal matching pursuit. IEEE Trans. Information Theory 53(12), 4655–4666 (2007)
4. Blumensath, T., Davis, M.: Gradient pursuits. IEEE Transactions on Signal Processing 56(6), 2370–2382 (2008)
5. Fornasier, M., Rauhut, H.: Iterative thresholding algorithm. Applied and Computational Harmonic Analysis 25(2), 187–208 (2008)
6. Ji, S.H., Xue, Y., Carin, L.: Bayesian Compressive Sensing. IEEE Trans. Signal Processing 56(6), 2346–2356 (2008)
7. Baraniuk, R., Davenport, M., DeVore, R., Wakin, M.: A simple proof of the restricted isometry property for random matrices. Constructive Approximation 28(3), 253–263 (2008)
8. Candès, E., Tao, T.: Near optimal signal recovery from random projections: Universal encoding strategies? IEEE Trans. Inf. Theory 52(12), 5406–5425 (2006)
9. Mendelson, S., Pajor, A., Tomczak-Jaegermann, N.: Uniform uncertainty principle for Bernoulli and sub-Gaussian ensembles. Construct. Alg. 28, 269–283 (2008)
10. Donoho, D., Tsaig, Y.: Extensions of compressed sensing. Signal Processing 86(3), 533–548 (2006)
11. Candès, E.: Compressive sampling. In: Int. Congress of Mathematic, Madrid, Spain, vol. 3, pp. 1433–1452 (2006)
12. Applebaum, L., Howard, S.D., Searle, S., Calderbank, R.: Chirp sensing codes: deterministic compressed sensing measurements for fast recovery. Appl. and Comput. Harmon. Anal. 26, 283–290 (2009)
13. Howard, S., Calderbank, R., Searle, S.: A fast reconstruction algorithm for deterministic compressive sensing using second order Reed-Muller codes. In: Conference on Information Systems and Sciences (CISS), Princeton, NJ (March 2008)
14. Ailon, N., Liberty, E.: Fast dimension reduction using Rademacher series on dual BCH codes. In: Annual ACM-SIAM Symposium on Discrete Algorithms (SODA), pp. 215–224 (January 2008)
15. Calderbank, R., Howard, S., Jafarpour, S.: Construction of a large class of deterministic sensing matrices that satisfy a statistical isometry property. IEEE J. Sel. Topics in Signal Process. 4(2), 358–374 (2010)
16. Yu, N.Y.: Additive character sequences with small alphabets for compressed sensing matrices. In: IEEE Conference on Acoustics, Speech and Signal Processing (ICASSP), Czech Republic, Prague (May 2011)
17. Yu, L., Barbot, J.P., Zheng, G., Sun, H.: Toeplitz-structured Chaotic Sensing Matrix for Compressive Sensing. In: 2010 7th International Symposium on Communication Systems Networks and Digital Signal Processing (CSNDSP), Newcastle, pp. 229–233 (2010)
18. Berinde, R., Indyk, P.: Sparse recovery using sparse random matrices. MIT-CSAIL Technical Report (2008)
19. Berinde, R., Gilbert, A., Indyk, P., Karloff, H., Strauss, M.: Combining geometry and combinatorics: A unified approach to sparse signal recovery. In: Proc. 46th Annu. Allerton Conf. Commun. Control Comput., pp. 798–805 (2008)
20. Gu, G.S., Han, G.Q.: Design of Stream Cipher Based on Sub-shift of Finite Type on Symbolic Space. Journal of Chinese Computer Systems 6, 1003–1007 (2007)

Steady Data Judgment Algorithm and Neutral Network Robust Training in Boiler Combustion Optimization

XuGuang Wang and Jie Su

School of Control & Computer Engineering, North China Electric Power University, 619 YongHua North Street, Baoding, China
wang_xuguang@163.com

Abstract. This paper proposes a steady data judgment algorithm, which includes three major parts, i.e. gross error detection, steady data judgment and steady data re-sampling. In order to minimise the impact caused by noise when training neutral network, the paper introduces the RANSAC algorithm and put forward a RANSAC-BP algorithm. In this algorithm a t-distribution based confidence interval is constructed to replace the distance threshold, which is selected by experience; the algorithm culls noisy data during neutral network training and then re-train the neutral network with noise free data. The algorithm has been validated by a simulation experiment.

Keywords: Combustion optimization, Steady data, RANSAC-BP, Confidence interval.

1 Introduction

In China at the current stage, the thermal power generation is the major approach for energy acquisition. In the overall thermal power generation cost, the fuel generally accounts for over 70%, thus to improve the efficiency of the boiler combustion system is of great significance to decreasing the operation cost of the electric power system. The coal-dominant energy structure results in serious air pollution in China, which costs high economic cost as well as environment cost. Therefore, the high efficiency and low emission combustion of coal-fired boiler has become the objective of the boiler combustion optimization research, which could effectively increase the operation efficiency, decrease the power generation cost, reduce the emission of pollutants, and monitor the safe operation of the boiler [1], thus attracting more and more attention.

The boiler combustion optimization techniques include three categories [2]. The first category detects on line the important parameters of the boiler combustion, and then directs the operators to regulate the combustion, which takes the domestic predominance at present [3,4]. The second category is a kind of boiler operation monitoring and control system based on DCS, which adopts advanced control logic, control algorithm or artificial intelligence technology [5] to realize the boiler combustion optimization. This category has been developing rapidly with the gradual maturity of the advanced control and artificial intelligence technology and its

D. Jin and S. Lin (Eds.): Advances in CSIE, Vol. 2, AISC 169, pp. 319–329.
springerlink.com

successful application in industry [6]. The third category optimizes the boiler combustion through improving the combustion equipment, such as combustor, heating surface, etc. The three categories mentioned above have their respective advantages in practice, however, the second one doesn't need to transform the boiler equipment, instead it takes full advantage of the boiler operating data, takes DCS control as the basis, applies the advanced modeling, optimization and control techniques, and directly increases the operating efficiency of the boiler and decreases the NOx emission. For its advantages of small investment, low risk, and obvious effect [7,8], it has become the most preferred combustion optimization technique.

To apply the second category of combustion optimization technique should be based on the acquisition of reliable data which are able to reflect the working condition of the boiler combustion system, i.e. steady data. The working condition data acquired from direct measurement or DCS include not only steady data but unsteady data as well. That's because it is a dynamically regulating process for the boiler combustion system shifting from a steady state to another steady state. The working condition data acquired in the unsteady state are of no significance for boiler combustion optimization. The traditional criterion for steady data judgment is relatively simple, which is generally based on the distance between neighboring working conditions. But the problem with this criterion lies in that the judgment threshold's acquisition relies on experience, which brings a lot of inconvenience in application. Besides, neutral network is usually used for boiler combustion modeling. However, in the working conditions used for neutral network training there normally exist noises which will seriously affect the training results. Thus it is necessary to find an approach to cull noises before training neutral network.

Assuming the measuring parameters obey the Gaussion distribution, this paper proposes a steady data judgment algorithm, which includes three major steps, i.e. gross error detection, steady data judgment and re-sampling. Inspired by the RANSAC algorithm and combining the BP network training algorithm, the paper puts forward a RANSAC-BP neutral network training algorithm, which reduces the impact caused by noises during neutral network training. In addition, a simulation experiment is conducted to validate this algorithm.

2 Steady Data Judgment Algorithm

The power generation in a thermal power plant is a process where firstly the pulverized coal is burnt to turn chemical energy to thermal energy, and then through steam doing work, the thermal energy is turned to electric energy finally. At different stages of the boiler combustion, the parameters (working conditions) are always transiting from one steady state to another. The steady state refers to a state before and after which in short intervals the parameters (working conditions) vary little. And the corresponding working conditions are called steady data.

Let $D \subset R^{m+n}$ denotes the set of all working conditions, and working condition $d = [X, Y] = [x_1, x_2, \ldots, x_m, y_1, y_2, \ldots, y_n] \in D$, which could be acquired from DCS or relevant measuring equipment. Hereinto, Y refers to the output parameters of boiler combustion, e.g. combustion efficiency, emission load of NOx, exhaust smoke temperature, etc. X denotes the other combustion parameters besides the output parameters, which could be grouped into the improvable parameters, i.e. variable parameters (e.g. the primary air quantity, the secondary air quantity, combustion angle, etc.) and the un-improvable parameters, i.e. constant parameters (e.g. pulverized coal composition, etc.). Without loss of generality, let

$$X = [X_{\mathrm{var}}, X_{con}] = [x_1, \ldots, x_r, x_{r+1}, \ldots, x_m]. \tag{1}$$

Hereinto, $X_{\mathrm{var}} = [x_1, \ldots, x_r]$ refers to variable parameter and $X_{con} = [x_{r+1}, \ldots, x_m]$ refers to constant parameters.

Among all the working conditions, only the steady data have significance to the combustion optimization, the reason is that one working condition is usually the data acquired at the same time, but in the unsteady data what X corresponds to is not Y, but the one later than Y. Thus, the steady data must be selected for the acquisition of the relation between X and Y, i.e. the acquisition of steady data is the premise of boiler combustion optimization.

For the acquisition of the boiler combustion data, the common practice is to obtain working conditions at a regular interval Δt. Thus the working conditions comprise of both steady data and unsteady data, which makes it necessary to have a criterion to judge the steady data. To make it easy to state, all the working conditions is noted in chronological order as $D = \{d_i | i = 1, \ldots, N\}$, $d_i = [x_1^i, x_2^i, \ldots, x_m^i, y_1^i, y_2^i, \ldots, y_n^i]$.

2.1 Gross Error Detection

Gross errors refer to the errors beyond the expectation under stated conditions. Due to subjective or objective reasons, in the original working condition set D, there inevitably exist measuring errors. If the errors are too big, then they will be gross errors, i.e. noises. If the working conditions with gross errors are not be removed, it will influence not only the judgment of steady data but also the modeling of boiler combustion, even the combustion optimization effect eventually.

This paper assumes the variable components of the working conditions obey the Gaussion distribution. Take x_j as an example, and let $x_j \sim N(u_j, \sigma_j)$, $j \le r$. Hereinto, u_j and σ_j are respectively the mathematical expectation and the standard deviation of x_j. According to the 3σ criterion for gross error, if the x_j satisfy

$$u_j - 3\sigma_j \le x_j \le u_j + 3\sigma_j. \tag{2}$$

the working condition component x_j can be considered to be free of gross errors. Thus, if every working condition component satisfies the formula above, then it is can be considered that the working condition has no gross errors.

The exact values of x_j's mathematical expectation u_j and standard deviation σ_j are unknown to us, but they can be estimated through the unbiased statistics below.

$$\hat{u}_j = \frac{1}{N} \sum_{i=1}^{N} x_j^i. \tag{3}$$

$$\hat{\sigma}_j = \sqrt{\sum_{i=1}^{N}(x_j^i - \hat{u}_j)^2 \Big/ (N-1)} \,. \tag{4}$$

2.2 Steady Data Judgment Criterion

$2p-1$ of neighboring working conditions $d_1, d_2, \ldots, d_{2p-1}$ are taken to form the working condition matrix below

$$S = \begin{bmatrix} d_1 - \mathbf{u} \\ d_2 - \mathbf{u} \\ \vdots \\ d_{2p-1} - \mathbf{u} \end{bmatrix} \times \begin{bmatrix} 1/\sigma_1 & 0 & \cdots & 0 \\ 0 & 1/\sigma_2 & \cdots & 0 \\ & & \ddots & \\ 0 & 0 & \cdots & 1/\sigma_{m+n} \end{bmatrix} \in R^{(2p-1)\times(m+n)} \,. \tag{5}$$

Hereinto vector $\mathbf{u} = [\mu_1, \mu_2, \ldots, \mu_{m+n}]$ is the mathematical expectation of working condition, and $\sigma_1, \sigma_2, \ldots, \sigma_{m+n}$ are the standard deviations of working condition components. Element of the working condition matrix S

$$s_j^i = \frac{x_j^i - \mu_j}{\sigma_j} \sim N(0,1) \,. \tag{6}$$

Thus $\sum_i (s_j^i)^2 \sim \chi^2(2p-1)$. According to the working condition matrix, a judgment criterion can be formed. Let

$$\|S\| = \max_j \sum_i (s_j^i)^2 \,. \tag{7}$$

If

$$\|S\|^2 \le \chi^2(2p-1, \alpha) \,. \tag{8}$$

then d_p is considered as steady data. This is the Chi-distribution based steady data judgment criterion. Here α is the remarkable level of Chi-square distribution, and the threshold $\chi^2(r+n, \alpha)$ means

$$P(\xi \le \chi^2(2p-1, \alpha)) = 1 - \alpha \,. \tag{9}$$

where random variable $\xi \sim \chi^2(2p-1)$.

Likewise, according to the working condition matrix (5), the statistic below can be constructed

$$\eta_j = \left(\frac{x_j^p - \mu_j}{\sigma_j}\right) \Big/ \sum_{i \ne p} (s_j^i)^2 \,. \tag{10}$$

Since $\frac{x_j^p - \mu_j}{\sigma_j} \sim N(0,1)$ and $\sum_{i \neq p}(s_j^i)^2 \sim \chi^2(2p-2)$, then $\eta_j \sim t(2p-2)$. If

$$\max_j |\eta_j| \le t(2p-2,\alpha). \tag{11}$$

then d_p is considered as steady data. This is the t-distribution based steady data judgment criterion. Here α is the remarkable level of t-distribution, and the threshold $t(2p-2,\alpha)$ indicates

$$P(|\xi| \le t(2p-2,\alpha)) = 1-\alpha. \tag{12}$$

and random variable $\xi \sim t(2p-2)$.

The value of the remarkable level in the two judgment criteria can be changed in order to get sets of steady data satisfying different requirements, i.e. the bigger the remarkable level value is, the more reliable the obtained steady data will be.

2.3 Steady Data Re-sampling

Filtered by the steady data judgment criterion, the steady data still need to be re-sampled according to the time at which they were obtained. The reason is that when the boiler combustion comes into a relatively stable condition, it is possible that many neighboring working conditions are all steady state data. But the information reflected by these working conditions are almost the same. As a result if all the working conditions are stored, the data will be excessively redundant, causing low efficiency in data processing and waste of storage space.

In practical operation, we can cluster the steady state data according to the time when they were obtained, and pick up a subset of each cluster randomly to store. In this paper, we do it along a simple way: Firstly, sequence all the steady data according to their obtainment time, and then select a positive integer k based on experience. Secondly, randomly generate a positive integer I. If I is an odd (even) number, then observe all the steady data at odd (even) parity. If the minimum sampling interval between a steady data and its two neighboring steady data is less than $k\Delta t$, then the working condition should be deleted from the steady data set. The process won't be stopped until no steady data need to be deleted.

One point to mention: if the steady data to be observed has only one neighbor, e.g. the first or the last steady data, then we only need to observe the interval between the working condition and its only neighboring steady data. If the interval is less than $k\Delta t$, delete it from the steady data set.

2.4 Judgment Algorithm

The steady data judgment algorithm is summarized as follows:

The inputs of the algorithm are the set of all the working conditions sequenced by sampling time $D = \{d_i | i = 1,\ldots,N\}$ and the positive integer k (based on experience).

① For all the $d_i \in D$, use the 3σ criterion to judge whether d_i contains gross error,

- If d_i contains gross error, then delete it from D;

② For all the $d_i \in D, p \le i \le N-p$, construct the statistic $\max_j \sum_i (s_j^i)^2$ according to formula (7), and construct the statistic $\max_j |\eta_j|$ according to formula (10);

- If $\max_j \sum_i (s_j^i)^2$ does not satisfy the judgment criterion of formula (8), label d_i;
- If $\max_j |\eta_j|$ does not satisfy the judgment criterion of formula (11), label I;

  Delete the labeled d_i from D

③ Generate a positive integer I randomly.

- If I is an odd number, then observe all the steady data d_i at odd parity from the first steady data;
- If I is an even number, then observe all the steady state data d_i at even parity from the first steady data;

Record the minimal sampling interval $\min(T(d_{i-1}, d_i), T(d_i, d_{i+1}))$ between d_i and its neighboring steady data.

④ Label d_i, which satisfies $\min(T(d_{i-1}, d_i), T(d_i, d_{i+1})) \le k\Delta t$ and record the number of labeled d_i as num;

- If $num > 0$, then delete the labeled d_i from D, and go to ③;
- If $num = 0$, then stop.

Output D.

3 BP Network Robust Training

The problem this paper discussed is a complicated multivariable coupling nonlinear one. Based on experience, we should take a BP network with at least two hidden layers as the learning machine to obtain the relation between X and Y. This paper takes the network model with two hidden layers. Generally, the output of the neutral network is the boiler combustion efficiency and the NOx discharge. Experiments showed that the relation between X and Y is better expressed by two neutral networks with multi inputs and single output than one neutral network with multi inputs and binary outputs. In this paper we separately train two neutral networks with multi inputs and single output as Figure 1 demonstrates.

Although the steady data can be obtained through the steady data judgment criterion, there inevitably exist noises in the steady data due to the measuring equipments precision, human operation and other factors. If the neutral network is trained with data containing noises, the training result will not be satisfying. Thus it is necessary to find a method to cull the data noises before training.

RANSAC[9] algorithm is a robust estimation algorithm. Its key idea is selecting appropriate data subset to train the model, and use the obtained model to judge whether the data in the data set satisfy the given conditions. All the data satisfying the given conditions form a consensus set. The ultimate model is fitted with the largest consensus set.

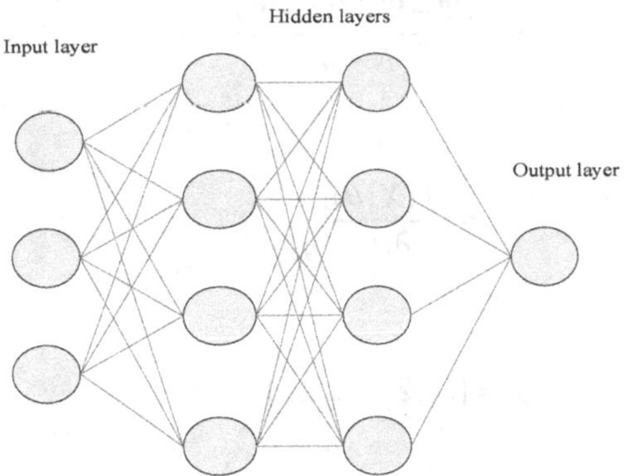

Fig. 1. A neutral network with two hidden layers

Based on the RANSAC algorithm, we put forward the RANSAC-BP algorithm as follows:

① ite=0. Randomly sample a subset of the steady data set $D_{sel}^{ite} \subset D$ and train BP network $f(\bullet)$, $ite = ite+1$; $D_{con}^{ite} = \phi$.

② For all the $d \in D$, calculate $f(\bullet)$'s confidence interval $CI_{\alpha}(X, D_{sel}^{ite})$ determined by d and D_{sel}^{ite}. If $Y \in CI_{\alpha}(X, D_{sel}^{ite})$, then $d \in D_{con}^{ite}$.

- If $|D_{con}^{ite}| \geq T_{num}$, then go to ③; otherwise, if $ite \geq T_{ite}$, then go to ④, otherwise go back to ①;

③ Use all the steady data in D_{con}^{ite} for training and get the BP network $f(\bullet)$.

④ Let $D_{\max} = \arg\max\{|D_{con}^{ite}|\}$, Train $f(\bullet)$ with $D_{\max}$.

Parameter Specification

α is the remarkable level, $CI_{\alpha}(X, D_{sel}^{ite})$ is the confidence interval of $100(1-\alpha)\%$ of $f(\bullet)$. According to [10], $CI_{\alpha}(X, D_{sel}^{ite})$ can be represented in the form below

$$f(X, \theta_0) \pm t_{N-q}(1-\frac{\alpha}{2}) s\sqrt{Z^T(\mathbf{Z}^T\mathbf{Z})^{-1}Z}\,. \tag{13}$$

Hereinto, θ_0 is the parameter vector to determine the neutral network $f(\bullet)$; N is the number of data in D_{sel}^{ite}, i.e. $N = |D_{sel}^{ite}|$; q is the dimension of θ_0, i.e. $\theta_0 \in R^q$; t_{N-q} refers to the t-distribution with degree of freedom $N-q$.

If $\boldsymbol{\Psi}$ denotes actual output vector of neutral network, then the expected output vector $\boldsymbol{\Psi}$ can be expressed as

$$\widehat{\boldsymbol{\Psi}} = [f(X_1,\theta_0), f(X_1,\theta_0), \ldots, f(X_N,\theta_0)]^T \quad ,k = 1\ldots N\ . \tag{14}$$

Thus in formula (13)

$$s - \frac{(\boldsymbol{\Psi} - \widehat{\boldsymbol{\Psi}})^T(\boldsymbol{\Psi} - \widehat{\boldsymbol{\Psi}})}{N-q}\ . \tag{15}$$

Column vector

$$Z = \left.\frac{\partial f(X,\theta)}{\partial \theta}\right|_{\theta=\theta_0}\ . \tag{16}$$

Matrix

$$\mathbf{Z} = [Z_1, Z_2, \ldots, Z_N]^T\ . \tag{17}$$

In formula (17)

$$Z_k = \left.\frac{\partial f(X_k,\theta)}{\partial \theta}\right|_{\theta=\theta_0}\ ,\quad k = 1\ldots N\ . \tag{18}$$

The reason we use the form of confidence interval described in formula (13) instead of fixed threshold to evaluate the training result is that when remarkable level is given, the confidence interval is adaptive, this is significant.

Meanwhile, one point worthy notice is that in the step ② of the above algorithm, $d \in D_{con}^{ite}$ when remarkable level α is given, the actual outputs of the neutral networks should fall within the respective confidence interval to which X responds.

4 Simulation Experiment

In order to test the validity of the RANSAC-BP algorithm, this section designs a simulation experiment.

Take the simple neutral network with single input and single output as example, then the functional relation to be trained is a simple function $y = x^x/3$. To takes samples at a interval of 0.05 between 1-3, then 41 input points can be obtained, as demonstrated in Table 1 (Lines 2 and 6). If there is no noise, according to the function $y = x^x/3$, the ideal outputs are showed in Lines 3 and 7. Now we randomly select 15 ideal output data from Line 3 and 7, add Gaussian noise with standard deviation 2.5, the stained data is shown in Lines 4 and 8.

If there are noises, we use the original BP algorithm to train the neutral network, i.e. using the data x (Lines 2 and 6 in Table 1) and y added with noises (Lines 4 and 8 in Table 1) to training the network, then we get the results shown in Figure 2.

In Figure 2, the blue curve represents function $y = x^x/3$, the black triangles refer to locations of the data with noises, and the red diamonds indicate the expected outputs of the neutral network. Obviously, when noises exist, there is big deviation between the expected outputs and the actual data, which illustrates that the noise has great influence on the result of the neutral network training. Thus, it is necessary to find a way to cull the noisy data first and then train the neutral network.

Table 1. Training data

Index of sampling point	x	y without noises	y added with noises	Index of sampling point	x	y without noises	y added with noises
1	1	0.3333	0.3333	*22*	2.05	1.4520	1.452
2	1.05	0.3509	0.6357	*23*	2.1	1.5832	3.369
3	1.1	0.3702	0.3702	*24*	2.15	1.7283	1.7283
4	1.15	0.3915	0.3915	*25*	2.2	1.8889	1.8889
5	1.2	0.4149	3.0818	*26*	2.25	2.0668	6.1257
6	1.25	0.4406	0.5888	*27*	2.3	2.2639	2.2639
7	1.3	0.4688	0.4688	*28*	2.35	2.4825	0.7531
8	1.35	0.4998	0.4998	*29*	2.4	2.7251	4.8701
9	1.4	0.5339	0.2948	*30*	2.45	2.9946	6.1296
10	1.45	0.5713	0.5713	*31*	2.5	3.2940	3.2940
11	1.5	0.6124	0.6124	*32*	2.55	3.6271	-0.357
12	1.55	0.6575	0.6575	*33*	2.6	3.9977	3.9977
13	1.6	0.7071	-1.373	*34*	2.65	4.4104	0.808
14	1.65	0.7616	0.7616	*35*	2.7	4.8704	4.8704
15	1.7	0.8216	1.5576	*36*	2.75	5.3832	5.3832
16	1.75	0.8876	0.8876	*37*	2.8	5.9555	7.3834
17	1.8	0.9602	0.9602	*38*	2.85	6.5946	6.5946
18	1.85	1.0403	1.0403	*39*	2.9	7.3086	7.3086
19	1.9	1.1285	-2.212	*40*	2.95	8.1069	8.1069
20	1.95	1.2259	1.2259	*41*	3	9	9
21	2	1.3333	1.3333				

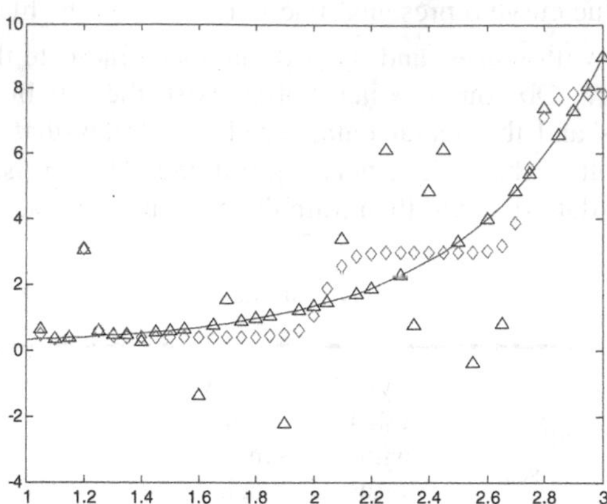

Fig. 2. Neutral network trained by BP

With the same noisy data, we use the RANSAC-BP algorithm to train the neutral network and show the training result in Figure 3. Here the blue curve represents the function $y = x^x/3$, the black triangles (overlapped with red diamond) refer to locations of the data without noise, the red diamonds indicate the expected outputs of the neutral network, and the green "*" mark the locations of the identified noisy data. As the RANSAC-BP algorithm culled the noises first and then trained the network with data without noises, the noise-free data agree well with the expected data of the neutral network.

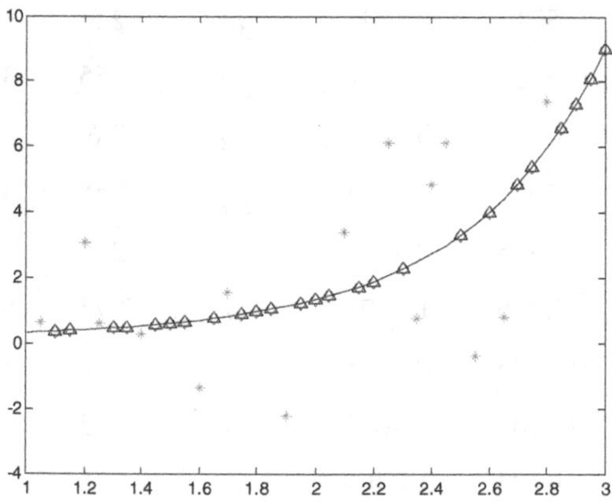

Fig. 3. Neutral network trained by RANSAC-BP

5 Conclusion

Assuming the measuring parameters obey the Gaussion distribution, this paper proposes a steady data judgment algorithm, which includes three major parts, i.e. gross error detection, steady data judgment criterion and steady data re-sampling. In order to minimise the impact caused by noise when training neutral network, the paper introduces the RANSAC algorithm into the neutral network training and put forward the RANSAC-BP algorithm. This algorithm is robust to data noises, and has been validated by a simulation experiment.

To find the optimal working condition is the ultimate objective of the boiler combustion optimization. The traditional way is to realize the optimization by genetic algorithm and other optimization algorithms. However, there always exist complicated coupling relations between different components of a working condition, but these coupling relations are not taken into consideration Thus, further study will be needed on how to measure the coupling degree between working condition components and how to apply this coupling relation to benefit the combustion optimization.

Acknowlegements. This work was supported by Youth Foundation of North China Electric Power University (No.916031105).

References

1. Wang, P., Li, L., Chen, Q., et al.: Optimal algorithm research on high efficiency and low emission combustion of coal-fired boiler. Power Engineering 24(4)_4 (2004)
2. Zhou, J., Fan, Z., Si, F., et al.: A review about the development of boiler combustion optimization technology. Boiler Technology 39(5)_5 (2008)
3. Li, Z.: Visual method of flame's development and application. Institute of Engineering Thermophysics, Cas (2006)
4. Hill, S.C., Douglas Smoot, L.: Modeling of nitrogen oxides formation and destruction in combustion systems. Progress in Energy and Combustion Science 26 (2000)
5. Daqi, G., Yi, W.: An optimization method for the topological structures of feed-forward multi-layer neural networks. Pattern Recognition (31), 1337–1342 (1998)
6. Wang, P., Li, L., Chen, Q., et al.: Research on applications of artificial intelligence to combustion optimization in a coal-fired boiler. Proceedings of Chinese Society for Electrical Engineering 24(4)_5 (2004)
7. Schnell, U., Kaess, M., Brodbek, H.: Experimental and numerical investigation of NOx formation and its basic interdependencies on pulverized coal flame characteristics. Combustion Science and Technology (1993)
8. Lwendt, J.O.: Mechanisms governing the formation and destruction of NOx and other Nitrogenous species in Low NOx coal combustion systems. Combustion Science and Technology (1995)
9. Fisher, M.A., Bolles, R.C.: Random sample consensus: a paradigm for model fitting with application to image analysis and automated cartography. Communications of the ACM 24(6), 381–395 (1981)
10. Rivals, I., Personnaz, L.: Construction of confidence intervals for neural networks based on least squares estimation. Neural Networks 13, 463–484 (2000)

Triangulation Inequality Violation in Internet Delay Space

Yifei Zhang and Hongli Zhang

Research Centre of Computer Network and Information Security Technology,
Harbin Institute of Technology, Harbin 150001, China
yifeizhangever@163.com, zhl@pact518.hit.edu.cn

Abstract. Some large distributed systems benefit from the implementation of network coordinates. However, triangulation inequality violation (TIV) will degrade the performance of network coordinate systems. In fact, TIV is an attribute of the Internet delay space. In this paper, we explore TIV in Internet delay space, and its impact on Vivaldi firstly. Then, we propose a detection mechanism to identify nodes with serious violations of triangulation inequality. Furthermore, we apply our method to filter reference nodes with severe violations on Vivaldi and evaluate its performance. The experimental results show its effectiveness.

Keywords: Triangle Inequality Violation, TIV, Network Coordinate, Network Measurement.

1 Introduction

Currently, a number of large distribution applications, such as content service, VoIP, video on demand, are deployed widely on the Internet [1], [2], [3]. Their performances depend on the underlying network topology. Therefore, some methods based on active measurements are proposed to obtain distance information to improve performance of these applications. However, the utility of active measurement is generally limited since it is time-consuming and may result in network congestion in the Internet. Also, methods based on network infrastructure have drawbacks that a large number of infrastructure nodes will increase maintenance cost and network overhead [4], [5]. Therefore, network coordinates systems (NCSs) are applied to distributed network applications to estimate network proximity [6], [7], [8]. In NCSs, network delays are mapped into Euclidean space and can be estimated by Euclidean distances.

However, network delays do not necessarily satisfy triangle inequalities, and this will decrease the accuracy of estimating network distances [9]. In the Internet the triangle inequality states that for any three hosts A, B, and C, the distance AC is greater than the sum of distances AB and BC. This means that the direct route to a host is less than that through an intermediate node, and triangulation inequality violation (TIV) appear. In fact, TIV, which are mainly caused by routing policies, e.g. commercial interests, are common in Internet delay space [10], [11]. Since it is

D. Jin and S. Lin (Eds.): Advances in CSIE, Vol. 2, AISC 169, pp. 331–337.
springerlink.com

impossible to change the existing architecture or routing policies of the Internet, TIV will inevitably exist in Internet delay space.

In [12], researchers have attempted to identify the severity of violating triangulate inequality of an edge and improve the performance of NCSs. However, it is inefficient as a large number of edge delays are needed to denote the severity which result in the complicated calculations.

In this paper, we aim at exploring the TIV of Internet delays and reducing its impact on network coordinate systems. The rest of the paper is organized as follows. In section 2, we analyze characteristics of TIV of Internet delays and the impact on Vivaldi. In section 3, we propose a method to detect TIV, and evaluate it. In section 4, the related work is presented. In section 5, we summarize the paper and propose recommendations for future work.

2 Related Work

2.1 Triangulation Inequality

In [13], the triangulated heuristic rules are applied to estimate network distances. The key idea is to assume the triangulation inequality holds and select n base nodes B_i ($1 < i < n$), and a normal node H being assigned a coordinate according to distances between H and the base nodes. Then, coordinate of H is denoted as (d_{HB1}, d_{HB2}, ... , d_{HBn}). Given two nodes H_a and H_b, the distance between them is greater than $L = \max_{i\in\{1,2,\ldots,n\}}\left(\left|d_{H_aB_i} - d_{H_bB_i}\right|\right)$ and less than $U = \min_{i\in\{1,2,\ldots,n\}}\left(\left|d_{H_aB_i} + d_{H_bB_i}\right|\right)$. Various weighted averages of L and U is used to estimate the distance between H_a and H_b.

In [14], the authors believe that most of the delays are consistent with the triangulation inequality rules. And the triangulation inequality rules are used to filter malicious nodes.

2.2 Vivaldi

Vivaldi is a decentralized scheme that estimates network proximity. The key idea is that each node in Vivaldi is corresponding to a few reference nodes and assigned a Euclidean coordinate. Any node whose coordinate is known can be selected as reference node. Their coordinates are calculated by simulating a spring system, in which the measured latencies are modeled as the extensions of a spring between bodies with unit mass. Since a spring always tries to reach a low-energy state, the coordinates can be calculated by minimizing the error function created according to errors between actual distances and predicted distances.

3 TIV in the Internet

The online version of the volume will be available in LNCS Online. Members of institutes subscribing to the Lecture Notes in Computer Science series have access to all the pdfs of all the online publications. Non-subscribers can only read as far as the

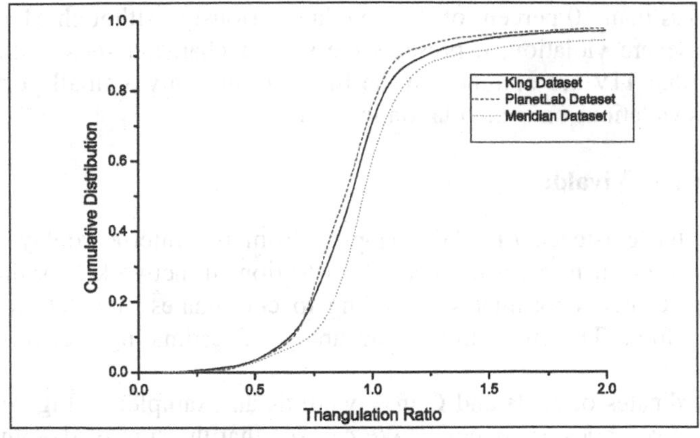

Fig. 1. The cumulative distribution of TIV in the Internet delay datasets

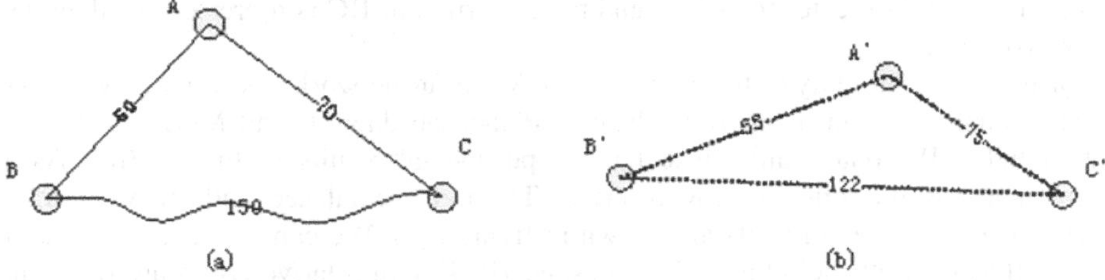

Fig. 2. Distortion of the map from delay space to Euclidean space

abstracts. If they try to go beyond this point, they are automatically asked, whether they would like to order the pdf, and are given instructions as to how to do so.

Please note that, if your email address is given in your paper, it will also be included in the meta data of the online version.

3.1 Characteristics of TIV in the Internet

As Internet route also reflect the social relationships, a selected route does not be selected with least hop count or smallest delay. In the process of route selection, economic interests and political relations, as well as network performance (e.g., the shortest path routing), have to be considered [10]. Therefore, it is inevitable for Internet delays that violate triangulation inequality.

We use King dataset, PlanetLab dataset, and Meridian dataset to analyze the characteristics of TIV on Internet routes. Given any three nodes A, B and C, edge AC causes a violation if $d_{AC} > d_{AB} + d_{BC}$. We use triangulation ration to characterize edges with TIV on Internet delays. Triangulation ration is the ration of one side length in a triangle to the sum of the other two. We compute the triangulation ratio of any three nodes in these datasets. The cumulative distribution is shown in Figure 1. We can see that nearly 30 percent of king data and PlanetLab data violate triangulation inequality.

Moreover, less than 10 percent of them violate seriously. Although Meridian dataset has a more severe violation, it presents the similar characteristics as the other two. This shows that TIV are common in the Internet, and only a small portion of paths have serious violations on triangulation inequality.

3.2 Impact on Vivaldi

Because of the existence of TIV, mapping from the Internet delay space to the Euclidean space will necessarily leads to distortion. In network coordinate systems, nodes compute their coordinates according to coordinates and delays of reference nodes. Therefore, TIV will impact accuracy of estimating network coordinate systems.

Take coordinates of A, B and C in Vivaldi as an example. In Figure 2(a), delays between any two nodes are denoted. We can see that the sum of d_{AB} and d_{AC} is 120 which is less than d_{BC} (150). Therefore, it means that the route from B to C is longer than that through A, and they violate the triangulation inequality. When they are mapped into Euclidean space using Vivaldi (in Figure 2(b)), we can see that the sum of d'_{AB} and d'AC is greater than d'_{BC}, and the distortion of BC is more serious than the other two edges.

Further, we will analyze the impact of TIV on an network coordinate system—Vivaldi. Firstly, we run it on King dataset, PlanetLab dataset, and Meridian dataset that include TIV routes, and present the experimental results in Figure 3(a). As a comparison, we filter out routes with serious TIV in these datasets and run Vivaldi on them. The experimental results are shown in figure 3(b). We can see that for Vivaldi in PlanetLab datasets including TIV routes, nearly 90% of relative errors are less than 0.5. However, for that without TIV routes, nearly 95% of relative errors are less than 0.5. And in the other two datasets the similar results can be seen. This shows that routes with TIV reduce the performance of Vivaldi.

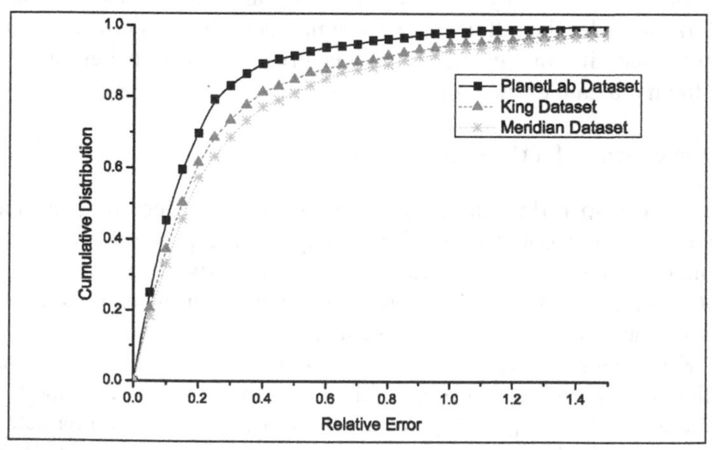

(a) Performance of Vivaldi with TIV edges

Fig. 3. The cumulative distribution of TIV in the Internet delay datasets

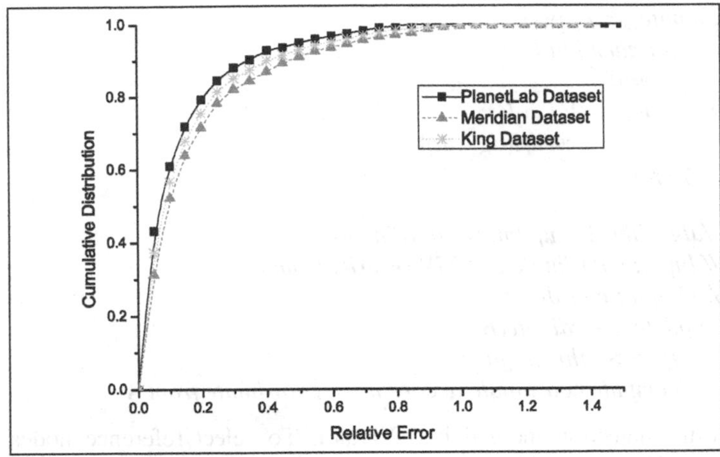

(b) Performance of Vivaldi without TIV edges

Fig. 3. *(continued)*

4 Detecting TIV

It is difficult to ascertain whether a path violates triangulation inequality. In NCSs, selecting a reference node with smaller severity of TIV would improve the accuracy of estimating network distances. To illustrate severity of TIV on a node we define a new metric γ as follows:

$$\gamma = \frac{\sum\sum d_{MN} / (d_{AM} + d_{AN})}{n^2} \tag{1}$$

where n is the number of nodes in a given collection R that includes some selected nodes. And M, N denotes any two nodes in R. The ratio of dMN to the sum of dAM and dAN which is greater than 1 indicates that TIV occurs on MN. The greater the value of γ is, the more likely the path violates triangulation inequality.

As γ indicate severity of TIV of a node, we can use it to identify TIV. In NCS each nodes select n nodes and calculate the corresponding γ. The parameter would be sent to its peers with its coordinates and RTTs. Then, its peers would determine whether it can act as the reference nodes by the parameter γ. The one that has a serious violation will not be selected as reference nodes. In non-centric NCS like Vivaldi, nodes can select the k nodes that often communicate with them as their reference nodes.

In the following, the algorithm for detecting TIV in NCS is presented. For each node A in the system, compute_severity computes its parameter γ. And node B judge whether update its coordinate according to TIV severity of node A in update_coordinate.

//algorithm1: detecting TIV in Node A
// Input: latencies from reference nodes
// Output: Severity of TIV of Node A

```
compute_severity (L)
    for each i in L
        γ=0
        for each j in L
                γ=d_ij/(d_ik +d_kj)
    γ=γ/n^2

//algorithm2:  updating coordinates
// Input: coordinate and TIV severity from A
// Output: coordinate
  update_coordinate(B)
     If  γ >  threshold exit
     compute coordinate according to coordinate from A
```

We apply the detection method into Vivaldi. To select reference nodes with less severity of TIV, nodes will eliminate those with severely violation of triangulation inequality. The results of running Vivaldi on the datasets are shown in Figure 4. The accuracy of predicting network distance is improved, and is as good as datasets without TIV routes.

5 Discussions

In this paper, we explored the characteristics of triangulation inequality violations (TIV) in Internet delay space, and its impact on Vivaldi. Then, we presented a detection mechanism to identify nodes that violate triangulation inequalities severely. Moreover, we applied our scheme in Vivaldi. The experiments show the effectiveness of our scheme.

Our detecting method does not take into account exceptions. For instance, a reference node with low severity on TIV may send unexpected delay. In the future, we will make further study on TIV in the Internet, and attempt to exclude TIV from network coordinate systems and improve their performance in more complex situations.

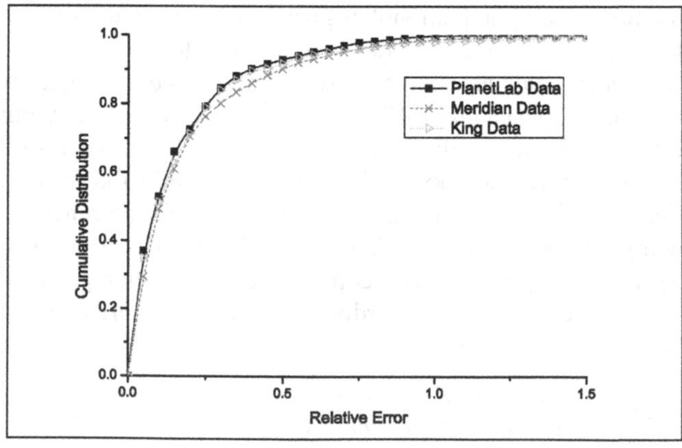

Fig. 4. Severity of TIV with detection mechanism

Acknowledgment. This work is partially supported by the National Grand Fundamental Research 973 Program of China (Grant No. 2011CB302605); the National Natural Science Foundation of China (Grant No. 61173145); High-Tech Research and Development Plan of China (Grant No. 2010AA012504, 2011AA010705).. The authors also gratefully acknowledge the helpful comments and suggestions of the reviewers, which have improved the presentation.

References

1. Liebeherr, J., Nahas, M.: Application-layer Multicast with Delaunay Triangulations. In: Proceedings of IEEE Global Telecommunications Conference, San Antonio, TX, United States (2001)
2. Lumezanu, C., Baden, R., Spring, N., Bhattacharjee, B.: Triangle Inequality and Routing Policy Violations in the Internet. In: Moon, S.B., Teixeira, R., Uhlig, S. (eds.) PAM 2009. LNCS, vol. 5448, pp. 45–54. Springer, Heidelberg (2009)
3. Maymounkov, P., Mazières, D.: Kademlia: A Peer-to-Peer Information System Based on the XOR Metric. In: Druschel, P., Kaashoek, M.F., Rowstron, A. (eds.) IPTPS 2002. LNCS, vol. 2429, pp. 53–65. Springer, Heidelberg (2002)
4. Francis, P., Jamin, S., Jin, C.: IDMaps: A Global Internet Host Distance Estimation Service. IEEE/ACM Transactions on Networking 9(5), 525–540 (2001)
5. Guyton, J.G., Schwartz, M.F.: Locating nearby copies of replicated Internet servers. Computer Communications Review 25(4) (1995)
6. Frank, D., Russ, C., Frans, K.: Vivaldi: A Decentralized Network Coordinates System. In: Proc. of ACM SIGCOMM Conference on Computer Communications, Portland, vol. 34(4), pp. 15–26 (2004)
7. Ledlie, J., Gardner, P., Seltzer, M.: Network Coordinates in the Wild. In: Proceedings of the 4th USENIX Conference on Networked Systems, pp. 299–311 (2007)
8. Tang, L., Crovella, M.: Virtual Landmarks for the Internet. In: Proceedings of the ACM SIGCOMM Internet Measurement Conference, Miami Beach, FL, United States (2003)
9. Zheng, H., Lua, E.K., Pias, M., Griffin, T.G.: Internet Routing Policies and Round-Trip-Times. In: Dovrolis, C. (ed.) PAM 2005. LNCS, vol. 3431, pp. 236–250. Springer, Heidelberg (2005)
10. Lehman, L.W., Lerman, S.: PCoord: network position estimation using peer-to-peer measurements. In: Proc. of the 3rd IEEE International Symposium on Network Computing and Applications, pp. 15–24 (2004)
11. Ng, T.S.E., Zhang, H.: Predicting internet network distance with coordinates-based approaches. In: Proc. of IEEE INFOCOM, New York, vol. 1, pp. 170–179 (2002)
12. Wang, G., Zhang, B., Ng, T.S.E.: Towards Network Triangle Inequality Violation Aware Distributed systems. In: Proceedings of the 7th ACM SIGCOMM Internet Measurement Conference, San Diego, CA, United States (2007)
13. Hotz, S.M.: Routing information organization to support scalable interdomain routing with heterogeneous path requirements. Philosophiae Doctor Thesis (1994)
14. Costa, M., Castro, M., Rowstron, A.: PIC: Practical Internet Coordinates for Distance Estimation. In: Proceedings of the 24th International Conference on Distributed Computing Systems, Hachioji, Tokyo, Japan (2004)

Towards the Design of Network Structure of Data Transmission System Based on P2P

Jian Yu and Hua Wang

School of Information and Electronic Engineering, Zhejiang University of Science and Technology, 310023 Hangzhou, China
yujian9770@sohu.com, wanghua96@126.com

Abstract. The emerging technology of P2P realizes the peer computing by utilizing nodes in Internet. The applications of P2P lie in many fields such as file sharing, distributed computing, collaborative work, Internet storage and so on. The overload of servers and the occupation of bandwidth are mitigated. This presents a wide and deep promise of applications. However, the ever-increasing nodes present some research and technical challenges, such as the stability of the target system, the latency of forward of nodes among multiple layers, the dependability of communications among nodes and so on. The paper presents a novel hierarchy of data transmission system based on P2P by the management of nodes. The super node is coined to optimize the structure of network of P2P.

Keywords: Data Transmission, P2P, Super Node, End Agent.

1 Introduction

The emerging technology of P2P realizes the peer computing by utilizing nodes in Internet. P2P fully taps the idle resources of Internet and has the potential advantages in terms of utilization of resources, extensibility and fault tolerance. The applications of P2P lie in many aspects such as file sharing, distributed computing, collaborative work, Internet storage and so on. If P2P was applied into the data transmission system under the environment of game playing, the role of client machines would be found by buffering a part of information acting as a role of server and thus scattering the services. By this way, the overload of servers and the occupation of bandwidth are mitigated. This presents a wide and deep promise of applications.

However, the ever-increasing nodes present some research and technical challenges as follows.

- The system should adjust and monitor the action of nodes.
- To ensure the stability of the target system.
- To decrease the latency of forward of nodes among multiple layers.
- To guarantee the dependability of communications of nodes.
- To provide the good extensibility of the target system.

The key approach to resolve these challenges is to design better networking structure of data transmission system based P2P technology. Some literatures contribute to the

D. Jin and S. Lin (Eds.): Advances in CSIE, Vol. 2, AISC 169, pp. 339–344.
springerlink.com

field for these challenges. Chi-Feng proposes a DLNA-based Multimedia Sharing System [1]. A method for classifying broadband users into a P2P- and a non-P2P group based on the amount of communication partners ("peers") is presented in [2]. The self-structured organization combines the benefits of both unstructured and structured P2P information systems [3]. Relevant research is presented in [4-6]. Gorodetsky proposes virtual peer-to-peer (P2P) emulation environment intended for testing and verification of implemented mobile P2P agent applications in [7]. An algorithm which creates a random overlay of connected neighborhoods providing topology awareness to P2P systems is put forward in [8].

The paper presents a novel hierarchy of data transmission system based on P2P by the management of nodes. The super node is coined to optimize the structure of network of P2P.

2 The Hierarchy of Data Transmission System

The hierarchy of the data transmission system based on P2P is divided into four aspects: application layer, notification layer, networking transmission layer and physical layer as illustrated in Figure 1.

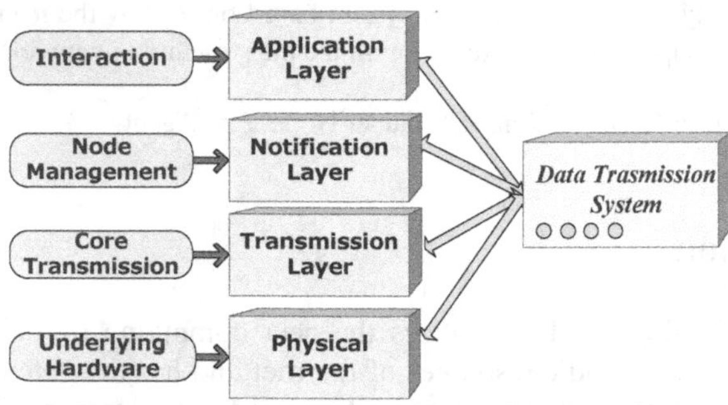

Fig. 1. The Hierarchy of Design of Data Transmission System based on P2P

(1) Application Layer
The application layer provides the interface between users and the system. Users share a virtual file space, which provides some basic functions associated with files. Application Layer shields the technical detail of routing and transmission.

(2) Notification Layer
The layer implements the node and directory management. The structure of the transmission system is designed and realized. The primary function is the choice of super node. When a normal node logs on, basic information is written into the super node. The inner networking is constructed by utilizing the hashed IP address. The indexing data is stored into the relevant node after computing key words by Hash functions.

(3) Transmission Layer
The layer is the core one in the whole system. The functions include that nodes search resources and upload and download resources. The overload balance, QoS

(Quality of Service) and security model is added into this layer to improve the efficiency and security of the system.

(4) Physical Layer

Physical layer is composed of computers with storing space and computing ability, nodes and underlying physical elements. The user node constitutes the elementary unit of P2P storage system by contributing the storage space and computing resources. The layer is the lower entity of file storage and the physical basis of the entire system.

2.1 The Management of Nodes

In fact, the management of nodes is the process of communications among nodes, which includes several steps.

(5) The logon of nodes

When a new node wants to join in the target P2P system, it is necessary for it to logon the super node, i.e., to establish the relationship between the super node and the new node. The IP address and username of the new node will be registered into the super node to get service and relevant management provided by the super node. On occasions, the new node will evolve into a super node and waits for the request of another nodes. The super node maintains a local table that records all the information of another nodes having connected into the super node. The necessary information contains IP address, username, connection status, the reputation value and the factor of resource consuming.

(6) The construction of the structure of shared files in the same cluster

Once added into the target system, the node uses the hash function to compute the key words of files existing in the shared directory. The index information is placed into the matched node. The information is dynamic with different size and the shared content can be modified at any time. Accordingly, the index information is not stored in the database, but the new memory space is employed with table-driven method recording the information of key words. The method decreases the time consuming of read data and improves the searching efficiency.

(7) Establishing the connection while nodes are downloading

After registration of nodes, they can use the downloading function. Users enter the filename to be searched. Then, the system computes the keywords and queries the routing table. At the same time, the keywords are submitted to super nodes who forward the query to other super nodes. The super node firstly searches the local database. If not matched, the super node sends the request package to another super node by corresponding routing algorithm. The address of source node having the resource requested is returned to the query node. If the source node does not exist, the timeout information is informed to the query node.

(8) The exit of nodes

The exit of nodes is classified into two categories of normal nodes and super nodes. When super nodes exit, the factor of resource consuming is queried to judge the node that can replace the exit node. The information of memory database and the routing information of other super nodes are sent to the replacing node. The system broadcasts the exit of the node and the change of the super node.

When normal nodes exit, the information of key words is sent to the neighbor nodes. In the meanwhile, the relevant super node is informed to disconnect with the exit node. The super node deletes the relevant entries of the exit node and sends the information to other nodes to update the routing tables.

3 The Architectural Design of Data Transmission System

The data transmission system based on P2P consists of *Web service*, *Controller*, *Media Server* and *End Agent*. Also, *Stream Reporting Bus* and *Aggregate Bus* are designed based on bus technology, as illustrated in Figure 2.

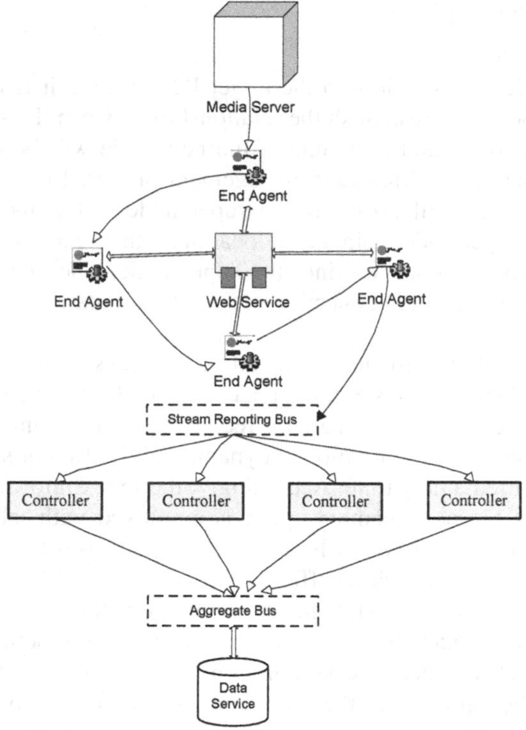

Fig. 2. The Architectural Design of Data Transmission System based on P2P

Controller server is employed to monitor user behavior and is charge of the networking structure of nodes. Controller server is composed of Controller program and Date Service. Media server intercepts the media stream and furthermore coding them. The real stream is provided for the root user node by the Media Server. The user node has two categories: End Agent and Stream Consumer. The End Agent is a basic unit, which forms the entire system networking. An End Agent is responsible for its father node to get the real stream and provides for child nodes. Furthermore, the Agent has the transmission function of control information between user and controller. In order to balance the overload of the underlying system, controllers are composed of Controller Front and Data Service center. The Streaming Reporting Bus reports the streaming information from End Agent to controllers and the Aggregate Bus collects stream resulting from controller interacting with Date Service.

In the context of game playing, the media data will be handled. The decoding of media data is integrated into the Player. The End Agent is distributed into the networking structure acting as a communicating module. Agents are the main unit composing the structure of networking tree. Tasks of agents are the receipt and forward of media stream and the instruction interaction with controller servers. The End Agent will buffer the stream of father nodes and retain the service status. Controlling instructions should be added between agents and control servers to strengthen the management of users.

When a user wants to join the network, some tasks must be done as follows. Firstly, the potential user send a request named Get to Web Service to get the server address of the responding controller. Secondly, the TCP connection is established according to the address. If the user is a new one, registration activity is invoked. If the user is an old one, a validation process is invoked. This information is stored in the public Data Service center. After the success of validation, the data provided by the Media Server can be requested through the Controller Server. The Controller can choose 10 neighbor node addresses, which are nearer to the node chosen, according to the IP address. The user node can get a data stream required by a TCP connection.

In order to find the optimized father node, the ping sensor method is employed to select the ideal father node. The ping sensor algorithm works as in the Figure 3.

```
Algorithm: Ping Sensor
TRACE_ADDRESS=NULL; //Initialize the address list to be null
N0=Current user node;
Ns=Neighbor node set; //the number of neighbor nodes is 10
i=2;
mintc=ComputeTimeConsuming(Ns1,N0);
//compute the ping time of the current to N0
while(true){
  if(i=9){
     break;
   }else{
      tci=ComputeTimeConsuming(Nsi,N0);
      //compute the ping time of the Nsi node to N0
      if (tci<mintc){
         mintc=tci;
         minnode=i;} //maintain the minimal ping time
     i++;
 }
return minnode; //return the node with the minimal ping time
```

Fig. 3. The Algorithm of Ping Sensor

The algorithm returns the node information that has the minimal ping time between the initial node and the target node. The computing process is done by the iterating all the 10 father nodes. The algorithm detect the time of returning ping package. The node with the minimal time of returning time is the optimized node. The exception is that nodes protected by firewalls cannot be detected. This can be resolved by detecting nodes when they apply to join the system. The detection is executed during the user online.

Acknowledgments. This work has been supported by Science and Technology Planning Project of Zhejiang Province under grant number 2006C30031 and Research Project of Zhejiang Province Educational Department under grant number Y201119886. Thank colleagues for carefully reviewing my work to strengthen and elevate its quality.

References

1. Chin-Feng, L., Yueh-Min, H., Han-Chieh, C.: DLNA-Based Multimedia Sharing System for OSGI Framework With Extension to P2P Network. IEEE Systems Journal, 262–270 (2010)
2. Kettig, O., Kolbe, H.: Monitoring the Impact of P2P Users on a Broadband Operator's Network over Time. IEEE Transactions on Network and Service Management, 116–127 (2011)
3. Forestiero, A., Mastroianni, C.: A Swarm Algorithm for a Self-Structured P2P Information System. IEEE Transactions on Evolutionary Computation, 681–694 (2009)
4. Xiaosong, L., Kai, H.: Collusive Piracy Prevention in P2P Content Delivery Networks. IEEE Transactions on Computers, 970–983 (2009)
5. Kantere, V., Skiadopoulos, S., Sellis, T.: Storing and Indexing Spatial Data in P2P Systems. IEEE Transactions on Knowledge and Data Engineering, 287–300 (2009)
6. Chuan, W., Baochun, L., Shuqiao, Z.: On Dynamic Server Provisioning in Multichannel P2P Live Streaming. IEEE/ACM Transactions on Networking, 1317–1330 (2011)
7. Gorodetsky, V., Karsaev, O., Samoylov, V., Serebryakov, S., Balandin, S., Leppanen, S., Turunen, M.: Virtual P2P Environment for Testing and Evaluation of Mobile P2P Agents Networks. In: The Second International Conference on Mobile Ubiquitous Computing, Systems, Services and Technologies, UBICOMM 2008, pp. 422–429 (2008)
8. Papadakis, H., Roussopoulos, M., Fragopoulou, P., Markatos, E.P.: Imbuing unstructured P2P systems with non-intrusive topology awareness. In: IEEE Ninth International Conference on Peer-to-Peer Computing, P2P 2009, pp. 51–60 (2009)

FPGA Implementation and Verification System of H.264/AVC Encoder for HDTV Applications

Teng Wang, Chih-Kuang Chen, Qi-Hua Yang, and Xin-An Wang

Key Lab of Integrated Microsystems Science and Engineering Applications, Peking University Shenzhen Graduate School, Shenzhen 518055, China
wangteng@sz.pku.edu.cn, anxinwang@pku.edu.cn

Abstract. For huge systems like video processing, FPGA prototyping plays an important role before taping out. In this paper, a verification system for H.264/AVC encoders with FPGA prototyping is proposed and implemented. An H.264 encoder with baseline profile of Level 3.2 was carried out with a clock frequency of 200MHz on a Xilinx Virtex-6 FPGA connected with DDR3 memory, which could satisfy real-time encoding for HDTV applications (720P@60fps) with a PSNR around 34 db. The encoder was finally implemented with SMIC 65nm CMOS technology for silicon verification.

Keywords: FPGA Prototyping, Verification, H.264/AVC, HDTV.

1 Introduction

Moore's low shows that the processing capability of IC manufacturing has been improved increasingly, and that the average annual growth rate can approach 58% [1]. At the same time, FPGA leading companies Xilinx and Altera announced FPGA products with 28nm process technique on 2011 [2][3]. It makes whole verification system integrated on FPGA with real-time running becomes possible. Fig. 1 shows the trade-off between performance and flexibility of a variety of verification methods [4]. It shows that FPGA prototype takes good balance between performance and flexibility. Therefore, for complex IC such as ASIC or SoC (System on Chip), FPGA verification is still an effective method before IC tape-out.

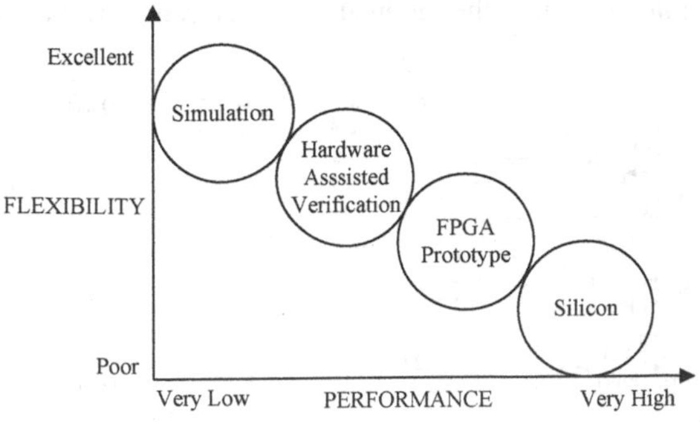

Fig. 1. Trade-off between performance and flexibility of a variety of verification methods

D. Jin and S. Lin (Eds.): Advances in CSIE, Vol. 2, AISC 169, pp. 345–352.
springerlink.com

Nowadays FPGA is of higher and higher integration and many modern applications use FPGAs to implement complex systems. In order to design quickly and with more flexibility, FPGA-based development board becomes more and more popular. A development board is comprised of one or more piece of latest high-capacity and fast-speed FPGA, varieties of proven peripherals, industry standard interfaces, power supply circuits, status indicators, control switches and debug interface to make it easy to create a prototype system for most complex applications.

A facility and effective verification approach is proposed for an H.264/AVC encoder design and then implemented on a highly integrated FPGA platform. The system is first simulated for function debugging and then implemented on the FPGA platform for real-time verification with a clock frequency of 200MHz. The paper is organized as follows: Section 1 is a general introduction, section 2 presents an overview of H.264 encoder and the FPGA platform used in this paper, section 3 introduces the proposed schemes and the architecture and the verification system, section 4 shows the implementation results and section 5 draws the conclusion.

2 Overview of H.264 Encoder and FPGA Prototype

For providing better compression of video images, H.264/AVC standard is jointly developed by ISO/IEC Moving Picture Experts Group (MPEG) and ITU-T Video Coding Experts Group (VCEG) and has been published in 2003 [5]. Fig .2 shows the H.264 encoder system block [6].

In Fig. 2, current frame (named as Fn) is processed in units of MBs (macro block), which are encoded in intra or inter mode. A predicted MB (marked as 'P') is formed based on reconstructed frame which is unfiltered (named as uF`n). The reference picture use the previously encoded frames named as F`n-1. In Intra mode 'P' is formed from samples in the current frame that have previously reconstructed, while in Inter mode 'P' is formed by MC (motion compensation) from F`n-1. The 'P' is subtracted with Fn to produce a residual block named as Dn, which is then processed by DCT and Quantization to obtain a set of quantized transform coefficients marked as 'X', which is then encoded with Entropy and NAL encode for transmission or storage. The coefficients 'X' sent to Inverse Quantization and IDCT to produce a differential block named as D`n. 'P' is added to D`n to create a reconstructed block named as uF`n, which is then filtered by DB (de-blocking) filter to reduce the effects of blocking distortion and obtain the reconstructed reference frame named as F`n.

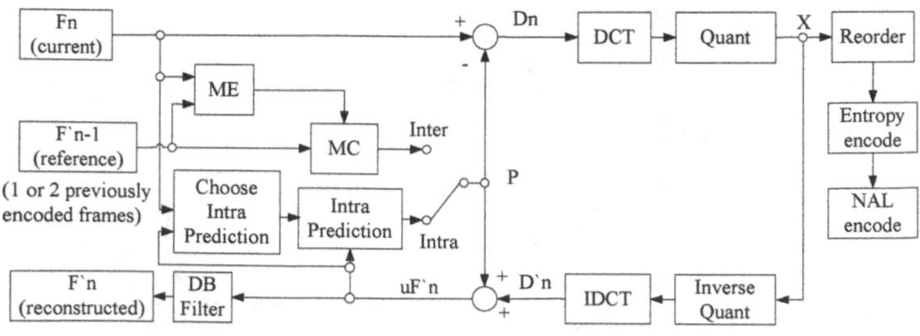

Fig. 2. The architecture of H.264 Encoder

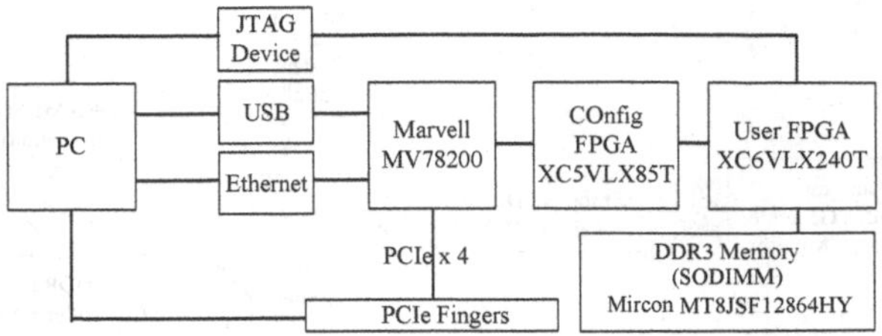

Fig. 3. The architecture of purposed FPGA prototype

Fig. 3 shows the architecture of FPGA platform [7] used in this paper. The Marvell MV78200 CPU connected with the Configuration FPGA (Xilinx Virtex-5 XC5VLX85T) is used to configure the two user FPGAs (Xilinx Virtex-6 XC6VLX240T). The User FPGA XC6VLX240T uses 40 nm copper CMOS process technology and the platform can emulate up to 5 million gates of logic as measured by a reasonable ASIC gate counting standard. The Marvell MV78200 can interact with PC through either the USB, Ethernet or PCIe interface. Xilinx ISE Project Navigator12.3 suite tool with Chipscope provides simulation and debug environment. A Linux kernel (Linux 2.6.22.18) provides the basic services and device drivers used on the Marvell CPU. External memory of the FPGA platform is DDR3 SODIMM, which can be used to store the RAW data and encoded data of the H.264 encoders. The JTAG device is connected between PC with FPGA, which provides a configuration and debug interface via JTAG chain.

3 Architecture of Purposed Scheme

An H264/AVC encoder for HDTV (720P@60fps) application with target spec of baseline profile at Level 3.2 is designed. The proposed design is conducted with a pipelining architecture of 5 stages, which is controlled by a control unit. The first stage is ME (Motion Estimation); the second stage is FME (Fraction Motion Estimation); the third stage consists intra prediction, DCT/IDCT, quantization and inverse quantization; the fourth stage is composed of VLC (variable length coding) and DB (de-blocking filer) and the fifth stage is NAL (network abstract level) coding.

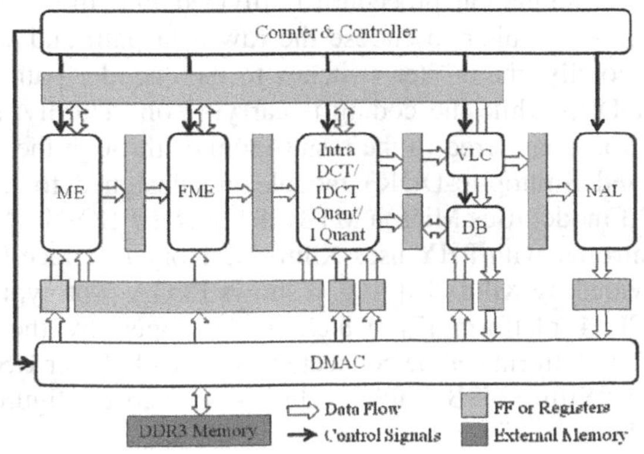

Fig. 4. Architecture of purposed H264 Encoder

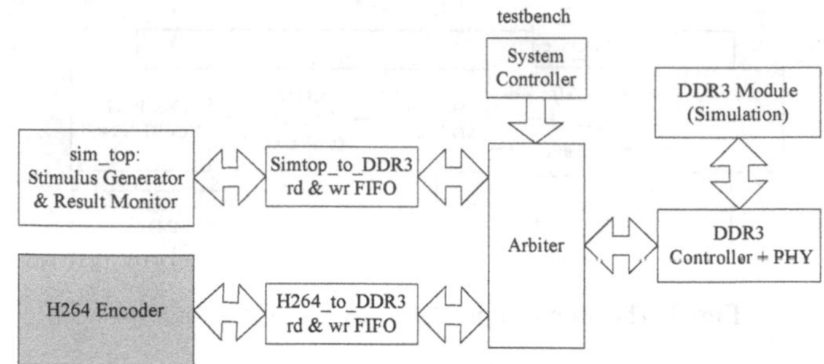

Fig. 5. The proposed simulation and debugging system for H.264 encoder

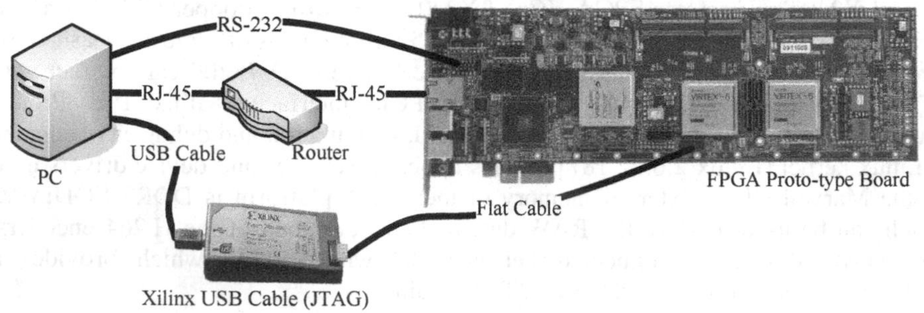

Fig. 6. FPGA prototyping system based on DN-DualV6-PCIe-4 platform

Shared memories and distributed memories are used among the pipeline stages for data transfer, which means the flowing of the pipeline is data-driven. This can increase the utilization of memories and obtain better timing performance. Controller modules provide control signals which dominate the coding behaviors in the H.264 encoder. Direct Memory Access controller (DMAC) module plays as interface with external memory of DDR3.

Fig. 5 shows the architecture of the proposed simulation and debugging systems for the encoder. The simulation procedure is divided into three steps. Firstly, the system controller control arbiter to choose the raw data path and write raw data to DDR3 memory. Secondly, the arbiter switches to the encoder path and the encoder interacts with the DDR3 while the coding is carrying on. Thirdly, the encoded bit-stream in the DDR3 is transferred to the result monitor through the arbiter. A H264-to-DDR3 module and Simtop-to-DDR3 module are designed to match the DDR3 protocol. The DDR3 model uses Micron MT8JSF12864HY SODIMM for simulation. And the DDR3 controller with PHY uses Xilinx Memory Interface Generator (MIG) tools, which is provided by Xilinx [8]. Fig. 6 shows FPGA prototyping system based on DN-DualV6-PCIe-4 platform [7], which is developed by the DINI group of California, USA. The platform can be connected to PC with either USB or Ethernet or PCIe interface. The "Xilinx USB Cable" cable is used to configure and debug the FPGAs via JTAG chain.

Table1 shows the hardware and software used in proposed system. ModelSim and Xilinx ISE suit are used for simulation, debugging and implementation to the FPGA. A Linux kernel is installed on FPGA prototype board, which provides the FPGA platform a configuration and communication environment with PC.

Table 1. (a) Hardware and (b) Software environment in the proposed scheme

Software	Description	Note
Window 7	Win7 Professional 64	OS of PC
Linux	Linux 2.6.22.18	OS of FPGA Prototype Board
Xilinx ISE	Xilinx ISE Navigator v12.3	Xilinx FPGA design suit v12.3
	Xilinx ChipScope	Xilinx Debug tools
ModelSim	ModelSim SE PLUS 6.5	Simulation tool

(a)

Hardware	Description
PC	MB: ASUS M4A88TD-M
	CPU: AMD Athlon™ II X2 250 processor 3.0 GHz
	RAM: DDR3-1333, 4GB x 4
FPGA Prototype	Marvell MV78200 CPU (Dual)
Board	Configuration FPGA: Xilinx Virtex-5 LX85
	User FPGA: Xilinx Virtex-6 XC6VLX240T x 2
Programming Cable	Xilinx USB Cable
Router	801.11 b/g router, for Ethernet interface use

(b)

4 Results and Discussion

Before FPGA implementation, function simulation is first conducted with Xilinx ISE based on the simulation system proposed in Fig.5 and the simulation waveform is demonstrated in Fig.7 (a). The design is then implemented on the platform presented above and Fig.7 (b) shows the debug waveform with Xilinx ChipScope. Both the simulation and implementation results provide good evidence for the correctness of the encoder.

The overall system is implemented on a Xilinx Virtex-6 xc6lx240t-1156 FPGA with 200MHz working frequency, which can support real-time HDTV applications. 92,109 slices of the FPGA are occupied and Table 2 shows the FPGA utilization summary. Table 3 presents the results comparison of the proposed implementation and that in Ref. [9] and [10]. It has to be noted that the 92K slices occupied includes the DDR3 MIG and the H264-to-DDR3 module. Furthermore, the proposed design is implemented with SMIC 65nm CMOS technology and the core size is about 3.24 mm^2 with a clock constraint of 350MHz. Fig.10 shows the layout of the ASIC design.

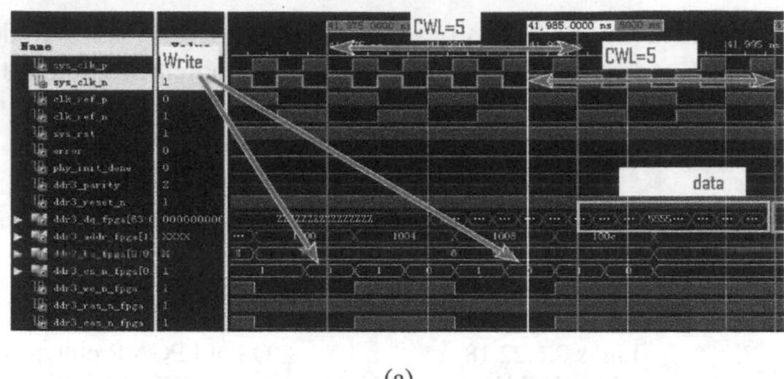

(a)

(b)

Fig. 7. (a) Simulation waveform with Xilinx ISE; (b) Debug waveform with Xilinx ChipScope

Table 2. FPGA utilization summary of H.264 encoder implementation

Slice Logic Utilization	xc6vlx240t (Used/Available)	Utilization
Slice Registers	77,646/ 301,440	25%
Slice LUTs	92,109/ 150,720	61%
Occupied Slices	33,718/37680	89%
RAMB36E1/FIFO36E1s	92/416	22%
DSP48E1s	28/768	3%
Bonded IOBs	183/600	30%

Table 3. Specification and performance comparison with other work

Items	Ref.[9]	Ref.[10]	Proposed
Spec	Baseline profile, Level 3.0; 1024 x 768@30fps	Baseline profile, Level 3.1; 1280 x 720@30fps	Baseline profile, Level 3.2; 1280 x 720@60fps
Platform & performance	Xilinx Virtex II with SDRAM; 12K slice size; 50MHz.		Xilinx Virtex6 with DDR3 SDRAM; 92K slice size; 200MHz.
ASIC Design	UMC0.18um, 3.88mm^2, 100MHz	UMC0.18um, 31.7mm^2, 108MHz	SMIC65nm, 3.24mm^2, 350MHz

The encoded bit-stream from the real-time coding is decoded with a standard H.264 decoder and the comparison of original frame and the decoded frame is demonstrated in Fig. 8. Fig. 9 presents the PSNR of the Y, U and V under 3.7Mbps bitrates with a QP of 32. The average PSNR of the first 30 frames is about 34 db.

Fig. 8. Comparison of original frame with decoded frame after the H.264 encoded

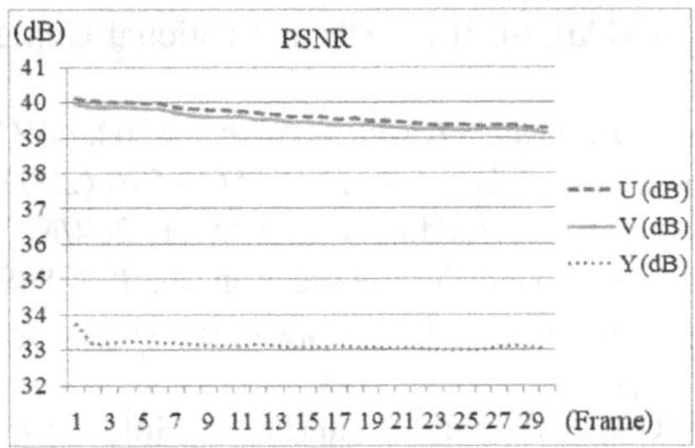

Fig. 9. PSNR of the first 30 frames within the testing case

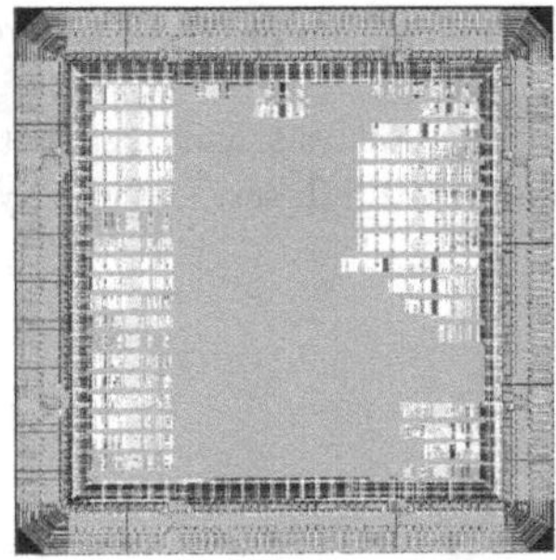

Fig. 10. The layout of ASIC implementation of the proposed encoder (with pads)

5 Conclusion

In this paper, a verification system for H.264/AVC encoders with FPGA prototyping is proposed and implemented based on the Dini DN-DualV6-PCIe-4 platform. An H.264 encoder of baseline profile at Level 3.2 with a DMAC interface that can interact with DDR3 SDRAM was carried out with a clock frequency of 200MHz on a Xilinx Virtex-6 FPGA, which could satisfy real-time encoding for HDTV applications

(720P@60fps) with a PSNR around 34 db. The encoder was finally implemented with SMIC 65nm CMOS technology for silicon verification with a core size of 3.24 mm^2 and working frequency of 350 MHz.

Acknowledgment. This work is supported by National High Technology Research and Development Program ("863" Program, Grant No.2009AA01Z127) of China and by Shenzhen Science & Technology Program (Grant No.JC201005270278A) .

References

1. Wang, Y., Wang, Y.: China's IC Industry Development - from the country of consumption to the power industry, p. 241. Science Press, Beijing (2008)
2. Xilinx website, `http://www.xilinx.com`
3. Altera website, `http://www.altera.com`
4. Huang, W., Wang, X., et al.: Implementation of high-speed verification platform based on emulator for reDSP & reMAP. In: IEEE 8th International Conference on AISC, pp. 682–685 (2009)
5. Wiegand, T., Sullivan, G.J., et al.: Overview of the H264/AVC Video Coding Standard. IEEE Trans. Circuits Syst. Video Technol. 13(7), 560–576 (2003)
6. Puri, A., Chen, X., et al.: Video Coding Using the H.264/MPEG-4 AVC Compression Standard. Signal Processing: Image Communication 19, 793–849 (2004)
7. DN-DualV6-PCIe-4 User Manual, `http://www.dinigroup.com/new/DN-DualV6-PCIe-4.php`
8. Xilinx Virtex-6 FPGA Memory Interface Solutions User Guide V3.91, `http://www.xilinx.com/support/documentation/ip_documentation/mig/v3_91/ug406.pdf`
9. Babionitakis, K., Lentaris, G., et al.: An Efficient H.264 VLSI Advanced Video Encoder. In: 13th IEEE International Conference on Electronics, Circuits and Systems, pp. 545–548 (2006)
10. Chen, T.C., Chien, S.Y., et al.: Analysis and architecture design of an HDTV720p 30frames/s H. 264/AVC encoder. IEEE Trans. Circuits Syst. Video Technol. 16(6), 673–688 (2006)

Image Quality Assessment Method Based on Support Vector Machine and Particle Swarm Optimization

Xiang Li and Yuan-yuan Wang

College of Computer Engineering, Huaiyin Institute of Technology, Huai'an 223003, China
Largepearstudio@gmail.com, 461044170@qq.com

Abstract. In order to improve the assessment accuracy of white noise, Gauss blur, JPEG2000 compression and other distorted images, this paper puts forward an image quality assessment method based on support vector machine and particle swarm optimization. Firstly, it extracts the sample image data and determines the assessment indexes. Secondly, it pre-treats the sample data, including normalized and PCA (Principal Component Analysis) dimensionality reduction process. Thirdly, it uses particle swarm optimization to select the optimal parameters. Fourthly, it uses the best parameters to train the training set data. Finally, it predicts and analyzes the predictive set data and establishes the image quality assessment model. The experimental results show that the image quality assessment method has a higher accuracy than traditional method and it can accurately reflect the image visual perception of the human eye.

Keywords: support vector machine, particle swarm optimization, image quality assessment.

1 Introduction

Image quality assessment is an important study topic in the image processing area, which has a higher practical application value. Image quality assessment method is usually divided into two classes: subjective quality assessment and objective quality assessment. Subjective quality assessment depends on the subjective feelings of perceivers for the image quality, and as the method is subject to the perceivers' background, psychological and physiological factors and other factors, it is more difficult to be applied directly for actual image processing.

Image quality objective assessment uses the mathematical model to quantitative the assessment index and simulates human visual perception system to assess the image quality. Common image quality objective assessment indexes include PSNR (Peak Signal to Noise Ratio), SSIM (Structural Similarity), and FSIM (Feature-Similarity). Related study finds that the objective assessment deviates from the subjective assessment, hard to be consistent because the understanding and analysis process of human brain for image is nonlinear and image information itself is complicated and diversified. To solve this problem, it puts forward an objective assessment method based on the Human Visual System (Human Visual System, HVS). Because of the complexity of HVS and imperfection of visual error theory, the method is more difficult to be applied practically.

D. Jin and S. Lin (Eds.): Advances in CSIE, Vol. 2, AISC 169, pp. 353–359.
springerlink.com

Support vector machine has an obvious advantage in the treatment of the highly nonlinear classification, regression and other practical problems, which is an important method to solve the "Dimension Disaster" and "Learning" [1]. LibSVM is the SVM pattern recognition and regression packages developed by the professor Jen Lin of Taiwan University, the software can solve C-SVC, nu-SVC, epsilon-SVR, nu-SVR and other problems [2].

This paper puts forward an image quality assessment model, which uses LibSVM image to link the subjective and objective assessment of image quality, and uses Particle Swarm Optimization looking for the best parameters, to improve the accuracy of image quality assessment.

2 Realization of SVM in the Subjective and Objective Assessment of Image

2.1 Realization of SVM in the Subjective and Objective Assessment of Image

The principle of Support vector machine is to establish a classification hyper plane as a decision surface, making the separation edge between positive cases and counter cases be maximized [3]. According to the input data, SVM prediction finds the nonlinear mapping function: $\varphi(x)$:$R^d \rightarrow F$ mapping to the output space, and its principle is as shown in Fig.1, higher dimensional space effect as shown in Fig.2.

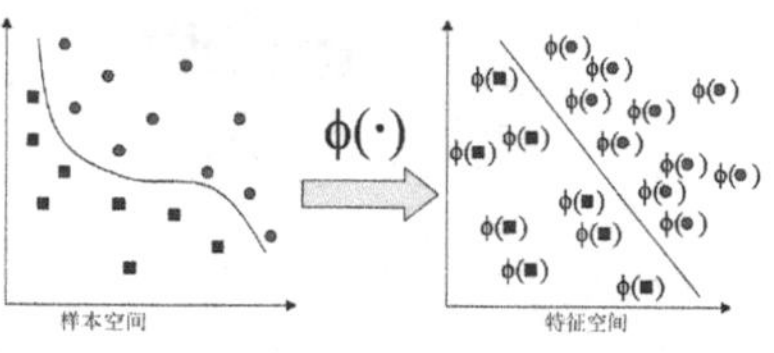

Fig. 1. Mapping Schematic Diagram from Sample Space to Feature Space

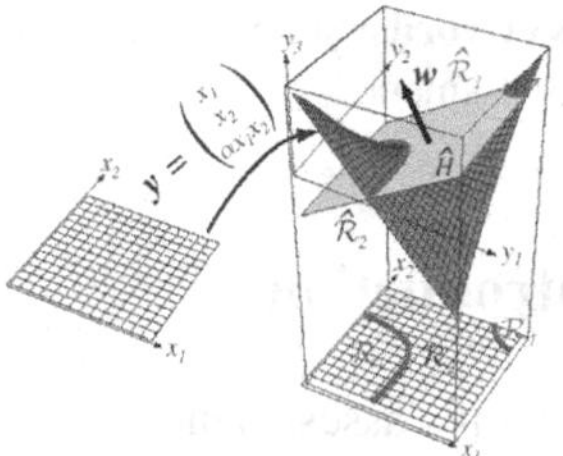

Fig. 2. Higher Dimensional Feature Space Schematic Diagram of Support Vector

And then it makes the linear estimates in the higher dimensional feature space, and its estimating function is:

$$f(x) = (w \cdot \varphi(x)) + b \qquad w \in F \tag{1}$$

Where, b is the bias term, $(w\cdot)$ is the inner product operation, w and b are minimized through the generalization function and are estimated to get the Eq. (2).

$$R(w) = \frac{1}{2}\|w\|^2 + C\sum_{i=1}^{n} J^{\varepsilon}(y_i, d_i) \tag{2}$$

Where, C is the penalty factor, representing the penalty for misclassification, the bigger C is the more powerful the penalty for misclassification is; ε is the insensitive loss function, and d^i represents the true output of SVM [4].

Because the dimensionality of feature space is higher, and the target function is no differentiable, it will be more difficult to directly solve the Eq. (2). SVM solves this problem through the introduction of dot product function $K(x_i, x_j)$ and the duality of Wolfe, and the solution of dual problem for Eq. (2) is:

$$\begin{aligned} &\min\left(\frac{1}{2}\|w\|^2 + C\sum_{i=1}^{n}\xi_i^2 + \hat{\xi}_i^2\right) \\ &s.t.\, W^T\varphi(x_i) + b - y_i \le \varepsilon + \hat{\xi}_i \quad i = 1,2...n \\ &y_i - (W^T\varphi(x_i) + b - y_i \le \varepsilon + \hat{\xi}_i \; i = 1,2...n \\ &\hat{\xi}_i, \xi_i, i = 1,2...n \end{aligned} \tag{3}$$

Where ξ_i and $\hat{\xi}_i$ are slack variables, considering some samples cannot be correctly classified by hyper planes, the introduction of Lagrange multiplier method can get the weight vector.

$$w = \sum_{i=1}^{n}(a_i - a_i^*)\varphi(x_i) \tag{4}$$

Get $f(x)$ from Eq. (1) and Eq. (4):

$$f(x) = \sum_{i=1}^{n}(a_i - a_i^*)K(x_i, x) + b \quad (0<a_i<C,\; 0<a_i^*<C) \tag{5}$$

$K(x_i, x) = (\varphi(x_i)\cdot\phi(x))$ represents the kernel function of SVM.

2.2 Introduction of SVM Image Quality Assessment Method

Main factors influencing the application effects of SVM:

1) The selection of samples. The sample set affects most the final trained model, if the selection range of samples is smaller, the application generalization ability of the trained model will be lower, and if the selection range of samples is larger, the time of constructing model will be longer.
2) The selection of kernel function. The models trained from different kernel functions have different effects, so need to find the kernel function with the best effect. Because this paper uses samples as natural or character image, the kernel function uses general radial basic function, as Eq. (6) shows:

$$K(x_i, x) = \exp\left(-\frac{\|x - x_1\|^2}{\delta^2}\right) \tag{6}$$

3) The selection of the optimal parameters for SVM. The parameters of SVM affect most the model precision, so selecting the suitable SVM parameters is an important step of the algorithm process.

3 The Optimal Parameter Selection of SVM Based on PSO

3.1 Search Algorithm Selection

From Eq. (5) and Eq. (6), the optimal parameter selection in this method is mainly to determine the optimal penalty factor C and the width of radial basic function δ^2. The traditional grid search and cross validation method are simple, but to achieve high precision of prediction needs a long time, so this paper will adopt heuristic optimization algorithm. Commonly used heuristic optimization algorithms include ACA (Ant Colony Algorithm), GA (Genetic Algorithm), PSO algorithm, and SA (Simulated Annealing). ACA easily falls into local optimum. In GA, chromosomes share information with each other, and the whole population moves uniformly to the optimal area. In PSO, only gBest gives information to other particles, this is one-way flow of information, the whole search update process follows the current optimum process, compared with GA, in most cases all the particles of PSO may quickly converge at optimum, so this paper choose PSO algorithm.

3.2 The Optimal Parameter Selection of PSO Based on SVM

The optimal parameters selection algorithm process in this paper is as shown in Fig.4:

1) LibSVM initialization setting includes the population quantity, evolution generations, cross validation broken number, the search range of the optimal parameters.
2) Construct the initial population to generate a group of population randomly, in which each individual represents a group of LibSVM parameters.
3) Use the accuracy of CV (Cross Validation) in LibSVM toolbox as the standard of assessing each model error, determining the fitness of each individual to reflect the generalization and predictive ability of the model, the higher the accuracy is, the higher the fitness is.
4) For each particle, compare its history optimal fitness with the fitness value of the best position experienced in groups, and if better, take it as the current global best position.
5) Judge whether it can meet the termination condition, if it meets determine LibSVM model, or else update the speed and position of particles and then turn to step 3, until it gets the satisfactory LibSVM model parameters.

4 Image Quality Assessment Method Test Based on LibSVM and PSO

4.1 Sample Selection and Parameter Configuration

The trial samples use LIVE Image Quality Assessment Database 2 provided by the image and video engineering lab from American TEXAS University, which has 29 high-accuracy and high-quality color images as the original images, and simulates the Gaussian blur, Rayleigh channel distortion, Gaussian white noise, JPEG compression, and JPEG2000 compression, and these distortion types form five chart galleries, with a total of 779 images. These chart galleries provide every image with DMOS (Differential Mean Opinion Score), as the comparative standard of objective quality assessment. The bigger DMOS value is, the worse the image quality is, the smaller DMOS value is, the better the image quality is, and the value range of DMOS is [0,100]. Randomly select a portion of images from the database as a sample set, and then take the rest images as a test set.

In this experiment, select these eight characteristic values PSNR (X1), (X2), FSIM SSIM (X3), VIF (X4), VSNR (X5), NQM (X6), IFC (X7), UQI (X8) of sample images as the input of LibSVM [5], use DMOS (Y) corresponding with database as the output of LibSVM, to construct the sample data set, part of data as shown in table 1. Select 400 pieces of images with the serial number of 001-400 as the sample of training set, and 379 pieces of images with the serial number of 401-779 as the sample of test set.

Table 1. Index Value of LIVE Image Quality Assessment Database 2

Image	Y	X_1	X_2	X_3	X_4	X_5	X_6	X_7	X_8
001	15.513	31.983	0.982	0.991	0.655	37.397	35.009	7.717	0.931
002	24.090	31.330	0.940	0.975	0.484	29.081	31.309	4.148	0.766
003	68.991	22.857	0.836	0.902	0.226	18.698	22.885	2.088	0.540
004	74.398	23.620	0.826	0.892	0.183	17.309	20.634	1.503	0.471
005	69.229	22.280	0.814	0.891	0.200	17.913	22.193	1.847	0.510
…	…	…	…	…	…	…	…	…	…
777	35.233	30.728	0.965	0.986	0.651	33.542	38.483	6.116	0.733
778	56.341	28.776	0.926	0.963	0.464	28.056	25.732	3.821	0.633
779	34.050	31.271	0.968	0.986	0.673	29.761	38.119	6.352	0.747

This paper uses Matlab (2011 b) combining with Libsvm-3.1 toolbox to make programs. The selection of initial parameters is as follows: speed update parameters c1 = 1.5, c2 = 1.7; maximum number of evolution maxgen = 200; the greatest number of population sizepop = 20; the maximum penalty factor popcmax = 100, the minimum popcmin = 0.1; the maximum nuclear function parameter g popgmax = 100, the minimum popgmin = 0.01; cross validation brokennumber v = 5.

4.2 Results and Analysis

Use Libsvm-3.1 and the above initial parameters for optimizing parameters of samples, and the experiment results show that the optimal penalty factor C = 0.90654, the optimal kernel function parameter g = 2.4493, mean square error MSE = 0.0036734. After analysis, C 、g and MSE value all meet the standard of optimal parameters, parameters optimization effect is very good. The final fitness curve is as shown in Fig.3, the fitting effect of training set is as shown in Fig.4, and the fitting effect of test set is as shown in Fig.5 below.

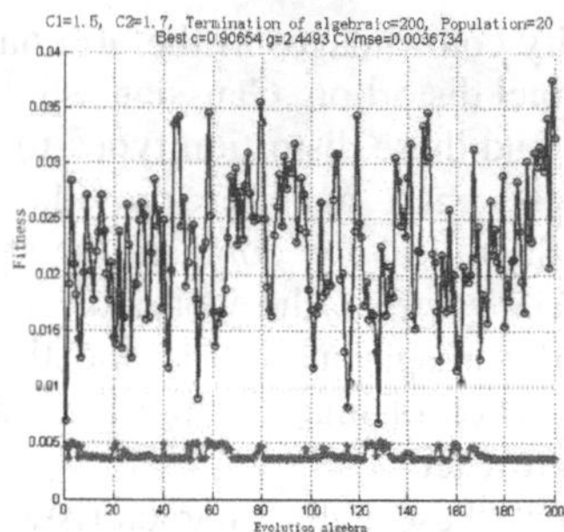

Fig. 3. PSO Fitness (Accuracy Rate) Curve of Searching Optimal Parameters

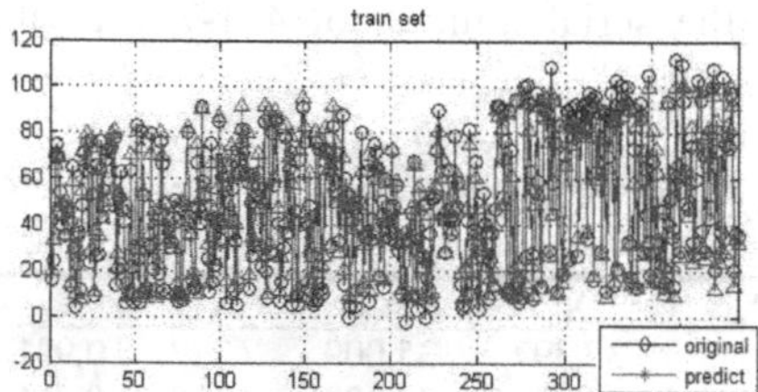

Fig. 4. Training Set Fitting Effects

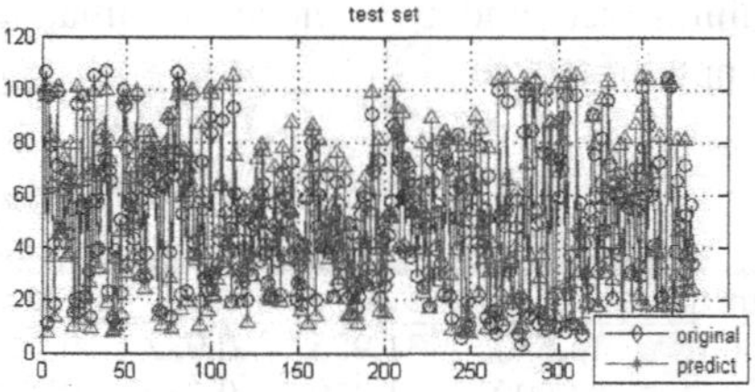

Fig. 5. Test Set Fitting Effects

In order to validate the relative merits of the model this paper puts forward, carry out the contrast test, and the reference models are: linear regression model, BP artificial neural network model, LibSVM model based on the grid search, LibSVM model based on GA, and predictive analysis results of each model are shown in Table 2.

Table 2. Each Model Prediction Result Square Error Sum

Reference Models	Error sum of square	Correlation coefficient of the model
linear regression model	100.6922	0.7863
BP neural network model	793.9206	0.5172
LibSVM model based on grid search	151.4513	0.8017
LibSVM model based on GA	62.412	0.8315
LibSVM model based on PSO	45.6232	0.8816

Table 2 results show that the square error sum of LibSVM model based on PSO is the smallest, and the correlation coefficient of model is the highest. It also proves that SVM for the image quality assessment based on PSO algorithm this paper puts forward is not only feasible, but also can get a better assessment result, which is an effective image quality assessment method.

5 Conclusions

This paper puts forward an image quality assessment method based on SVM and PSO on the basis of studying support vector machine and particle swarm optimization, and overcomes the shortcoming of traditional image quality subjective and objective assessment method when dealing with higher dimension, nonlinear problems. This method first uses SVM to establish image quality subjective and objective correlation function, and then uses PSO to optimize the parameters for SVM, making the assessment method more accurate and reliable. The experimental results show that this method has a higher advantage in the prediction accuracy and correlation compared with the traditional method, the correlation between subject and object is accurate, so it is an effective image quality assessment method.

Acknowledgments. This work was supported by the Technology Support Programs of Huai'an under grant HAG2011041 and the Education Teaching and Research of Huaiyin Institute of Technology under grant JYC201108.

References

1. Shi, F., Wang, X.-C., Yu, L., Li, Y.: 30 Cases Analysis of MATLAB Neural Network, pp. 112–113. Beijing University Press (2010)
2. Chang, C.-C., Lin, C.-J.: LibSVM: A Library for Support Vector Machines[EB/OL], http://www.csie.ntu.edu.tw/~cjlin/libsvm/
3. Cai, K.-S.: Study on Cultivated Land Prediction Base on Support Vector Machines. Computer Simulation 28, 199–202 (2011)
4. Chen, Y., Xiong, Q.: Support vector machine application tutorial, pp. 231–232. Meteorological Press (2011)
5. Okarma, K.: Colour Image Quality Assessment Using Structural Similarity Index and Singular Value Decomposition. In: Bolc, L., Kulikowski, J.L., Wojciechowski, K. (eds.) ICCVG 2008. LNCS, vol. 5337, pp. 55–65. Springer, Heidelberg (2009), http://www.springerlink.com/content/x416178822t70x52/

Research and Design of Heterogeneous Data Exchange System in E-Government Based on XML

Huaiwen He, Yi Zheng, and Yihong Yang

School of Computer, University of Electronic Science and Technology of China, Zhongshan Institute, 528402 Zhongshan, Guangdong

Abstract. To solve the "information island" problem in e-government system, this paper presents a heterogeneous data exchange model based on XML standard and Web Service transport technology. Using SOAP encapsulation mechanism and XML encryption to provide data transport security from data exchange center system to each bureau branch OA system. A dynamic configurable adaptor was implemented to translate message between data center system and each bureau branch OA system. This model elevated data exchange level from simple data sharing to directly government business exchange and it has been applied and tested in practice.

Keywords: e-government, heterogeneous data exchange, XML, Information Island.

1 Introduction

In the process of e-government construction over past ten years, some achievements have been obtained. Accompany with the rapid development of e-government, there have been many problems, in which the most serious problem is that it is extreme difficult in data sharing and exchanging among government agencies because of the differences of applications, construction, system structure and data resources in original departments. For the reason of scattered data resources, many "information islands" were formed and information can not effectively share on online e-government system, especially in the process of business needed parallel approval among various departments, which business data flow between various departments. Therefore, the data exchange among departments has become the first problem to be solved for parallel approval.

For these reasons, a centralized heterogeneous exchange model base on XML, Web Service technology was proposed in this paper. By using XML data exchange standards and data conversion adapter, data exchange between client and central system can be implemented which has great practical significance and is easy applied in China.

2 The Design of Heterogeneous Data Exchange System

Now many cities have built one-stop parallel approval system to improve government office efficiency. The one-stop parallel approval system uses centralize exchange way

D. Jin and S. Lin (Eds.): Advances in CSIE, Vol. 2, AISC 169, pp. 361–366.
springerlink.com

which includes a data exchange center as middleware of data exchange. The main function of data exchange center is managing data which flow between different departments, and ensure data uniqueness, data availability and data submit in workflow process, which application model is shown as figure 1:

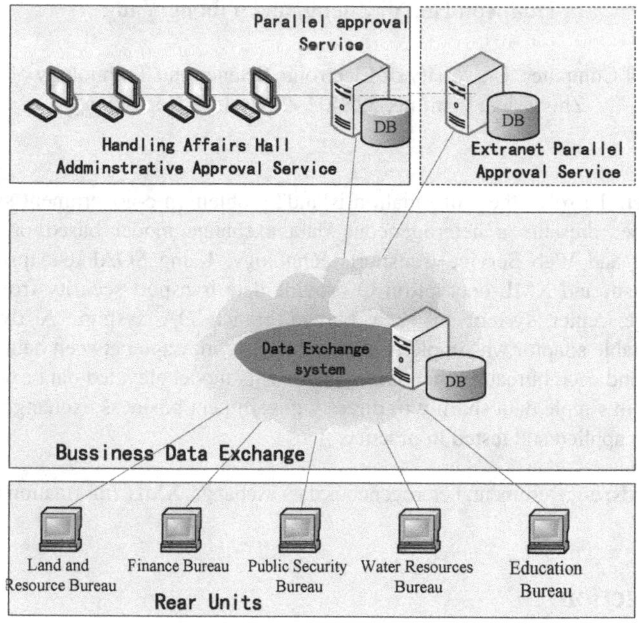

Fig. 1. Application model of data exchange system

By data exchange center in one-stop parallel approval system, each department can simply share data and exchange file. But because of fundamental difference of content of exchange data, and without uniform standard in each department OA system, therefore, in order to ensure the universal and independence of data exchange center in one-stop parallel approval system, it need to define a data exchange standard based on XML[1]. Therefore, when each department exchanges data with data exchange center, it need to convert its own business data format to data exchange center predefined data format, then uses data exchange service provided by data exchange center.

Data exchange system as a data center system matching system, provides service to each department which need exchange data, and integrate it into workflow system in one-stop system, can be seem as relatively independent system, which forms a loose coupling relation with data exchanging center. To implementing business data exchange, we design a centralized exchange model and its structure is shown as figure 2:

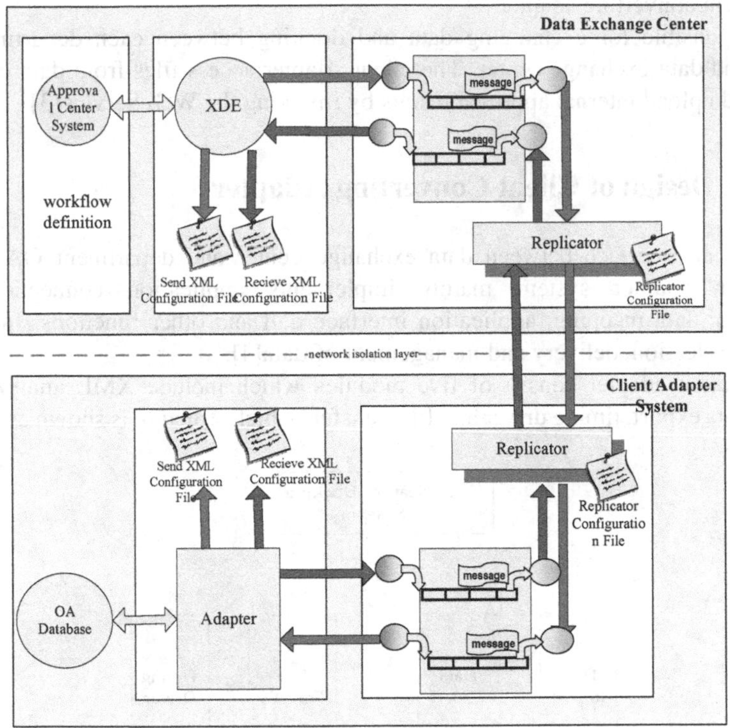

Fig. 2. Architecture of data exchange system

The data exchange system consists of the following four modules:

(1) Data exchange center system.

Be responsible for transforming data flow and converting data format between different departments. Through the center adapter module, data exchange center system obtains business data according to the configuration file from extranet. Then a new business approval will generate which is analyzed and managed by process tracking module and process handling module in data exchange center and the new approval workflow will be corresponds to the proper department.

(2) Network transmission module.

Be Implemented by Web Service technology. This module uses soap messages to encapsulate and transport XML business data and binary file upload and download, which is independent with concrete physical network condition and firewall[2].

(3) Security module.

Use the XML SOAP message encryption and XML electronic signature technology based on PKI technology to replace traditional SSL encryption.

(4) Client converting adapter.

Be responsible for exchanging data and docking between each department OA system and data exchange center. The client adapter access files from data exchange center and upload internal approval results by invoking the Web Service[3].

3 The Design of Client Converting Adapter

Acting as an interface between data exchange center and department OA system, application adapter system mainly implements application connection, data conversion, data mapping, application interface call and other functions. In another word it is selection, delivery and management of data[4].

Application adapter consist of five modules which include XML analyzer, data import, data export, timing driver and file transfer, which structure is shown as figure 3:

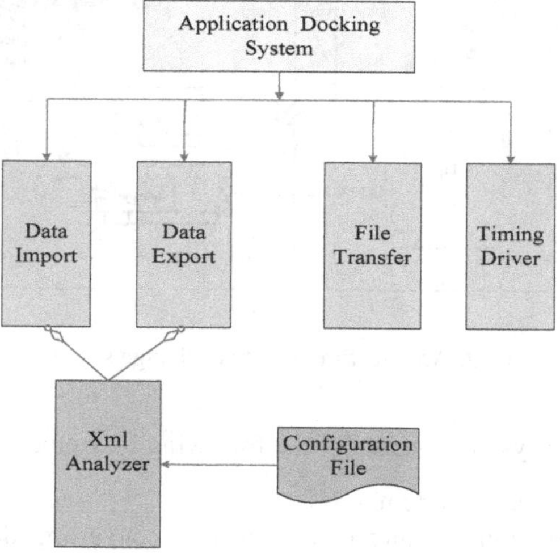

Fig. 3. Functional structure of application adapter

Details of each module are as following:

(1) XML analyzer: mainly be used to analyze the structure of configuration file, so that the data exchange center system can obtain the business system interface of specific department.

(2) Data import: be use to obtain interfaces of its own department business systems by analyzing XML configuration file, and mapping data from other departments, making data effective , then write data into its own background database systems in specific sectors, and store attachment in specified location, at last generate a new business process in its own business systems.

(3) Data Export: be use to obtain interfaces of its own department business systems by analyzing XML configuration file and periodically check its own background database system to obtain the state of pending processes. After gaining approval

state, the approval data finished will convert into the format of data exchange system and be written it into a specific directory on the server.

(4) File transfer: by using specific network protocol to transfer files with other departmental and periodically obtain data from other departments, Meanwhile, this module specific directory use for uploading files to servers of other department. This module mainly provides an interface, the actual model file transfer services will be provided by Web services transport layer[5][6].

(5) Timing driver: work like an agent, which periodically drive data export module check the status of pending process and periodically drive file transfer module go to the business system which had sent agent requests, and query the approval results[7].

In the subsystems of different department, XML is used when data interfaces were offered to external, and a unified set of XML data specification was followed, which resolved the problem of inconsistent data structure and data dictionary in multi-level networking[8].Figure 4 was the standard of XML packet received by city council, including process ID, approving opinions and so on[9].

Example: municipality, virescence cutting, pruning,migration seedlings application, receiving data

```
<?xml version="1.0" encoding="gb2312"?>
<XDE xmlns="http://www.kingsoft.com/egov/xde">
        <flowID>1672</flowID>
        <instanceID>5457</instanceID>
        <stepID>3</stepID>
        <practicer>szj:0</practicer>
        <docFile></docFile>
        <exFile></exFile>
        <field><name>JLDYJ</name><value>1</value></field>
        <field><name>YLKPS</name><value>approve</value></field>
        <field><name>JLDPS</name><value>approve</value></field>
        <field><name>YLKYJ</name><value>1</value></field>
</XDE>
```

Fig. 4. Receiving XML data standard by municipality

4 Application of Data Exchange Model

Heterogeneous Data Exchange System in E-Government Based On XML was well applied and tested in the one-stop parallel approval E-government system of Zhaoqing City, Guangdong province. Multiple administration department data exchange has been completed, for example data exchange of urban and rural planning bureau(C/S architecture, SQL Server Database), City council (J2EE architecture, Oracle database), Health bureau (Lotus Domino). This system realized the seamless heterogeneous data exchange among government agencies and parallel approval data center.

5 Conclusions

This paper is focus on current "Information Island" background and difficulty of heterogeneous data exchange among government agencies in e-government system. A centralized heterogeneous data exchange model based on XML and Web Service technology was designed and realization. XML was used as standard of data exchange and Web Service technology was used in transmission, which resolved the problems of data mapping and data transfer among heterogeneous database. An adapter subsystem was designed to achieve the government data exchange by the conversion of XML documents. Finally, the joint work of scattered government agencies was realized.

References

1. Liu, Y., Yao, S.: Relational storage of XML and data exchange between XML and RDBMS. Computer Engineering and Design 25, 2039–2043 (2004)
2. Widener, P., Eisenhauer, G., Schwan, K.: Open metadata formats: efficient XML-based communication for high performance computing. In: Proceedings of 10th IEEE International Symposium on High Performance Distributed Computing, pp. 371–380 (August 2001)
3. Sakamuri, B.C., Madria, S.K., Passi, K., Chaudhry, E., Mohania, M., Bhowmick, S.S.: AXIS: A XML Schema Integration System. In: Song, I.-Y., Liddle, S.W., Ling, T.-W., Scheuermann, P. (eds.) ER 2003. LNCS, vol. 2813, pp. 576–578. Springer, Heidelberg (2003)
4. Seligman, L., Rosenthal, A.: XML's Impact on DataBases and Data Sharing. IEEE Computer (37), 59–67 (2002)
5. Greunz, M., Schopp, B., Haes, J.: Integrating e-Government Infrastrcutures through Secure XML Document Containers. In: Proceeding of the 34th Hawaii International Conference on System Sciences 2001, pp. 202–210 (2001)
6. Palavra, D., Lisnjic, D.: Web services as standard of connecting heterogeneous information systems. In: 26th International Conference on Information Technology Interfaces, June 7-10, vol. 1, pp. 201–206 (2004)
7. Naedele, M.: Standards for XML and Web services security. Computer 36(4), 96–98 (2003)
8. Woodman, S., Morgan, G., Parkin, S.: Portal replication for Web application availability via SOAP. In: Proceedings of the Eighth International Workshop on Object-Oriented Real-Time Dependable Systems (WORDS 2003), January 15-17, pp. 122–130 (2003)
9. Yang, C.S., Liu, C.Y., Chen, J.H., Sung, C.Y.: Design and Implementation of Secure Web-based LDAP Management System. In: Proceedings, 15th International Conference on Information Networking, January 31-February 2, pp. 259–264 (2001)

Study on Soft Measurement of Moisture Eliminating Ratio in Tobacco Leaf Re-drying Line

Jingjie Dong[1], Lihui Feng[1,*], Wen Chen[2,*], Wei Song[1], and Wenyong Yang[3]

[1] Information Engineering and Automation College, Kunming University of Science and Technology, KunMing, China
[2] Metallurgy and Energy College, KunMing University of Science and Technology, Kunming, China
[3] Haitian Information Technology Limited Company, KunMing, China
{dongjingjie1026,flh6206,chenwen63}@126.com, songwei515@163.com, kmhitech@gmail.com

Abstract. Moisture eliminating ratio of heating zone in tobacco leaf re-drying line is one of key parameters to measure how much water contained in the tobacco leaf, which hardly is directly measured online. Therefore, this paper introduces how soft-sensing technology is applied to the process of tobacco leaf's re-drying, in order to achieve online soft-sensing measurement of moisture eliminating ratio. The paper adopts multiple linear regression analysis method and multiple stepwise regression analysis method, establishing the soft-sensing model of moisture eliminating ratio. It shows that the multiple stepwise regress analysis method have significant prediction effect in the modeling of moisture eliminating ratio, by comparison the analysis results that multiple linear regression and multiple stepwise regress analysis method get.

Keywords: Tobacco Leaf re-drying Line, Moisture Eliminating Ratio, Multiple Linear Regression, Multiple Stepwise Regression, Soft-Sensing.

1 Introduction

Tobacco leaf re-drying is one of important links in the production of tobacco and moisture eliminating ratio of heating zone of tobacco leaf re-drying line is one of key parameters to measure how much water contained in the tobacco leaf. Because of many affected factors, moisture eliminating ratio is hardly directly measured online. Soft-sensing technique as a new technology rises up in recent years, may be used for measuring indirect measurement physical quantity. This paper uses multiple linear regression analysis method and multiple stepwise regression analysis method to establish the soft-sensing model of moisture eliminating ratio. In order to get the best method, two models are used for analysis and prediction.

---
* Corresponding authors.

D. Jin and S. Lin (Eds.): Advances in CSIE, Vol. 2, AISC 169, pp. 367–373.
springerlink.com

2 The Craft General Situation of Tobacco Leaf Re-drying Machine

Tobacco leaf re-drying machine can regulate moisture of the tobacco leaf and remove impurity gas. This machine is a key device that can provide fine quality material. Tobacco leaf re-drying production line is composed of stockroom–style feeder, electronic belt conveyor scale, five drying regions, one cooling region, two reversion regions and belt conveyor [1]. The production line is shown as Fig.1.

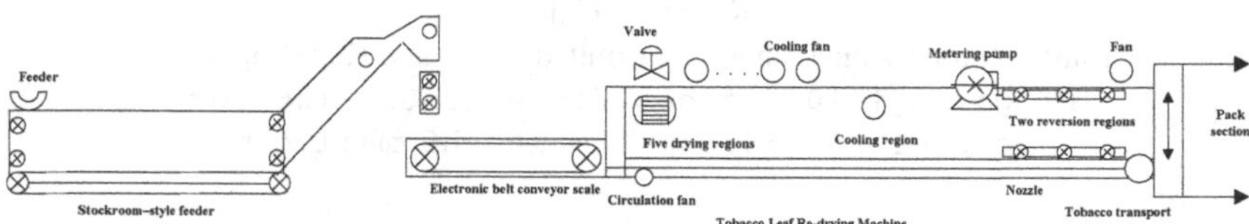

Fig. 1. Construction diagram of tobacco leaf re-drying line

The process is that tobacco leaf is taken up by feeder, then leaf falls down to belt conveyor. First, tobacco leaf is dried and circulating air is heated by saturated steam in the drying region. Saturated steam enters into the box by heat exchanger. At the same time, partial humidity of tobacco leaf is exhausted .Second, when tobacco leaf enters into cooling region, its temperature reduces to 40°C. Last, in the reversion region, dry and cool tobacco leaf reach finally moisture content and temperature by high pressure water spray system, steam spray system and forced convection used with gas circulator. Therefore, the drying region is an important link in tobacco leaf re-drying production line.

3 Data Processing

The data used by this paper has 150 groups that were measured by Real-time production line derived from real tobacco leaf re-drying company. In order to obtain reasonable, typical data, this paper establishes the ideal soft-sensing model of moisture eliminating ratio and the data is dealt with correlation analysis and pretreatment.

3.1 Correlation Analysis

Correlation analysis is one kind of numerical value measurement, which is the relationship between two random variables. Correlation analysis can portray the liner relationship precisely by calculating correlation coefficient. Considering the actual production, there are many influenced factors of moisture eliminating ratio rate y(kg/h) This paper selects six variables, which are x_1 environment temperature(^0C), x_2 environment absolute humidity(g/m^3), x_3 temperature of the ventilation of region(^0C), x_4 absolute humidity of the ventilation of region(g/m^3), x_5 exhaust air rate (m^3/h) and x_6 atmospheric pressure(Pa).

Based on 150 groups from field data, this paper selects 120 groups data which have 6 influenced factors. Then, the liner relationship between different influenced factors and moisture eliminating ratio can be got by calculating correlation coefficient .There are 5 regions that constitute heating zone of tobacco leaf re-drying line. The operating principle of these regions is similar. Here is as the third region for example to illustrate specific analytic process. The result of correlation analysis is shown as Table 1.

Table 1. Correlation coefficient matrix of influencing factors for row wave ratio

	x_1	x_2	x_3	x_4	x_5	x_6	y
x_1	1.00						
x_2	0.94	1.00					
x_3	-0.30	-0.37	1.00				
x_4	0.25	0.27	-0.51	1.00			
x_5	-0.21	0.20	-0.43	0.13	1.00		
x_6	-0.63	-0.58	0.35	-0.18	0.05	1.00	
y	0.23	0.25	-0.54	0.99	0.21	-0.17	1.00

The table.1 shows that the 3th region correlation coefficient between atmospheric pressure and moisture eliminating ratio is minimum, that is -0.17, so the atmospheric pressure is removed. The rest of five factors are used to establish the regression model.

3.2 Data Pretreatment

Before building the model, data should be pretreated and process mainly includes detecting, eliminating abnormal data, data normalization or normalization processing.

This paper has 103 groups samples after processing. Among this data, 65 groups samples are used to build model and 38 groups samples are used to verification.

4 Soft-Sensing Model of Moisture Eliminating Ratio

This paper analyses influenced factors about moisture eliminating ratio of tobacco leaf re-drying line by multiple linear regression and multiple stepwise regression analysis and establishes the soft-sensing model of moisture eliminating ratio. By comparison and analysis research results, the significant prediction effect will be got.

In five regions, this paper select x_1 environment temperature, x_2 environment absolute humidity, x_3 temperature of the ventilation of region, x_4 absolute humidity of the ventilation of region and x_5 measured value of exhaust air rate as input and moisture eliminating ratio y as output . The structure of this model is shown as Fig.2.

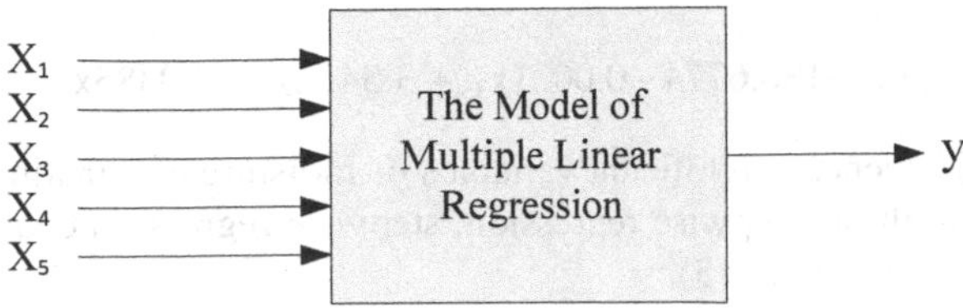

Fig. 2. Soft-sensing model structure of moisture eliminating ratio

4.1 Multiple Linear Regression

Regression analysis is the most common experience modeling method, which provides an effect method to search function relation and correlation ship. The coefficient is obtain from least squares method.

Multiple linear regression model of moisture eliminating ratio is shown as formulae (1) and $i=(1\sim5)$ is drying region number.

$$Y_i = b_{0i} + b_{1i}x_{1i} + b_{2i}x_{2i} + b_{3i}x_{3i} + b_{4i}x_{4i} + b_{5i}x_{5i}. \quad (1)$$

Least squares method calculates coefficient in the formula (1). Multiple linear regression model of the third region is displayed as formula (2), and other regions have the same structure but coefficient is different. Because space is limited, other coefficients don't list out anymore.

$$Y_3 = -136.9805 + 0.0033\, x_{13} - 0.0011\, x_{23} - +0.0023x_{33} + 3.5460x_{43} + 0.0386x_{53}. \quad (2)$$

4.2 Multiple Stepwise Regression

Stepwise regression algorithm is on the basis of each independent variable which influences dependent variable, using Sum of squares of partial regression to calculate and check F value. Then, we can get less terms, better correlation and higher accuracy optimal regression equation.

From the point of view of the correlation coefficient, influenced effect of moisture eliminating ratio from big to small in proper order is x_4 absolute humidity of the ventilation of region, x_3 temperature of the ventilation of region, x_2 environment absolute humidity, x_5 exhaust air rate and x_1 environment temperature. Utilizing the idea of stepwise regression, considering marked degree of factors to moisture eliminating ratio, impact factors are introduced into the equation. The comparison between evaluation index and appropriate statistical variable F are showed as Table 2.

Table 2. Summary statistics updated with each step based on multiple linear regression model

	x_4	x_4, x_5	x_3, x_4, x_5
R^2	0.991829	0.999996	0.999997
Adj-R^2	0.99157	0.999996	0.999997
RMSE	0.233185	0.004869	0.00439
F	7647.54	8.8424E+6	7.2303E+6

$$Y_3 = -136.6774 - 0.0021x_{33} + 3.546x_{43} + 0.0385x_{53}. \quad (3)$$

Table 2 shows influenced significant variables of moisture eliminating ratio are x_3, x_4, x_5.Through the multiple stepwise regression, stepwise regression equation of the third region is shown as formulae (3).

5 Inspection of the Soft-Sensing Model of Moisture Eliminating Ratio

5.1 Statistical Analysis and Inspection

There are four evaluation indices to inspect and analysis of two models, which are MSE, R2, F variable and p. The result of text and analysis about multiple linear regression and multiple stepwise regression are shown as table 3.

Table 3. Error analysis of moisture eliminating ratio for multiple linear regression model and multiple stepwise regress model

	multiple linear regression model				
	1#	2#	3#	4#	5#
R^2	0.937	0.963	0.974	0.980	0.947
MSE	0.875	2.287	1.852	6.389	11.196
F	3.50E+6	3.21E+6	4.32E+6	4.33E+6	2.83E+6
p	0	0	0	0	0
	multiple stepwise regress model				
	1#	2#	3#	4#	5#
	0.9996	0.9992	0.9997	0.9997	0.9996
	0.875	2.287	1.852	6.389	12.074
	5.93E+6	4.34 E+6	7.23E+6	1.08E+6	6.41E+6
	0	0	0	0	0

The table.3 shows that R^2=0.974 at the model of multiple linear regression, the correlation is very little. So, the result is not good. But R^2=0.9997, the value approach 1 at the model of multiple stepwise regression. This illustrate liner relation is closed and the fitting of model is better. At the same F=4.7724E+6, P<0.01, the model of multiple stepwise regression is more meaningful.

5.2 Model Fitting Effects

According to simulate the 38 groups moisture eliminating ratio. It is observed that not only at the model of multiple linear regression but also at the model of multiple stepwise regression two curves have basically the same trend. But at the model of multiple stepwise regression, two curves are essentially coincident, which have better simulated effect.

5.3 Model Predicting Effects

The significance of designing model is to predict moisture eliminating ratio of heating zone of tobacco leaf re-drying line. The measured value about environment temperature and absolute humidity of the ventilation of region are generated into multiple linear regression equation (2) and multiple stepwise regression equation (3)

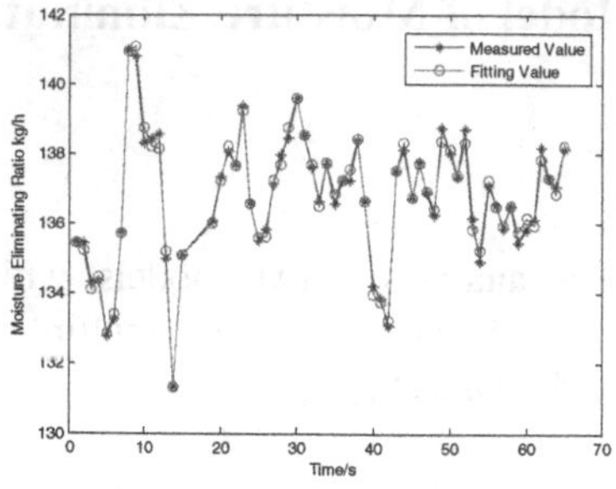

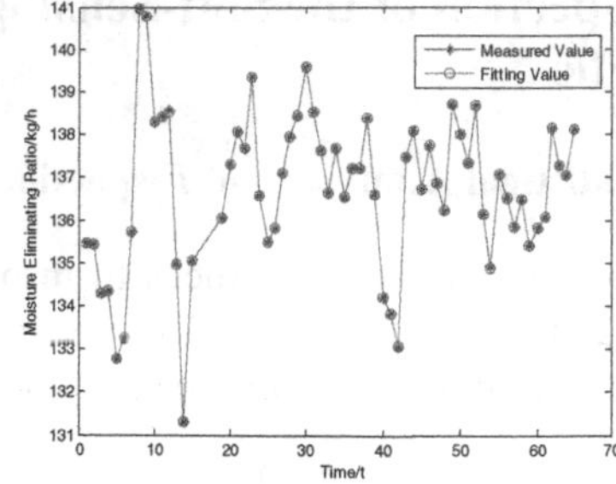

(a) Fitting curve of multiple linear regression (b) Fitting curve of multiple stepwise regression

Fig. 3. Fitting curves of two models

can predict moisture eliminating ratio. Figure.4 shows prediction results of regression model of moisture eliminating ratio and Table.4 shows parts of prediction results and relative error.

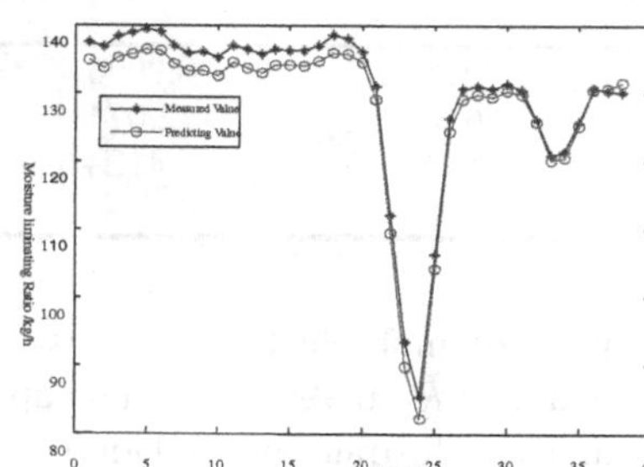

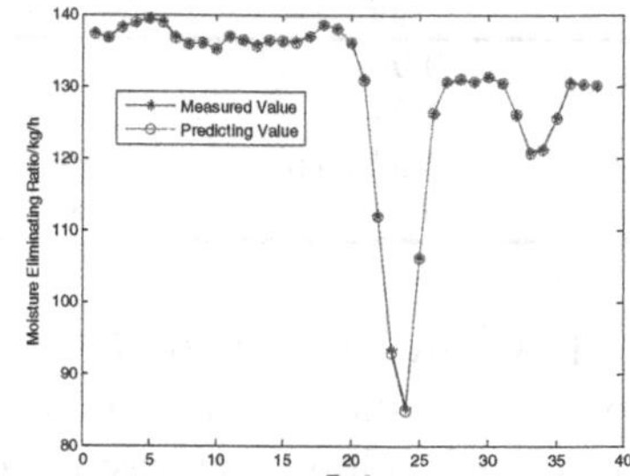

(a) Forecasting curve of multiple linear regression (b) Forecasting curve of multiple stepwise regression

Fig. 4. Forecasting curve of two models

Table 4. Model prediction analysis

Measured value kg/h	Predicted value kg/h		relative error %	
	multiple linear	multiple stepwise	multiple linear	multiple stepwise
135.4455	137.4431	137.3735	1.47	1.42
135.4383	136.3009	136.7716	0.64	0.98
134.2859	137.9576	138.2755	2.73	2.97
134.3378	138.4963	138.8642	3.09	3.36
132.7683	139.2057	139.3630	4.84	4.96
133.2480	138.9814	138.9515	4.30	4.28
135.7316	137.0075	136.8452	0.94	0.82
140.9610	135.8832	135.9562	3.60	3.55
140.7746	135.7336	136.0626	3.58	3.34
138.3019	135.0132	135.1663	2.37	2.26

Figure.4 and Table.4 show that forecasting curve and actual curve have basically the same trend. Two curves of multiple stepwise regression model are essentially coincident but relative error is less than multiple linear regression model. So multiple stepwise regression model has higher prediction accuracy.

According to contrast data statistics test analysis, fitting effect and prediction effect about multiple linear regression model and multiple stepwise regression model, this paper can obtain that data statistics test analysis, fitted value and prediction effect at the model of multiple stepwise regression are better than the model of multiple linear regression.

6 Conclusion

This paper focuses on establishing the model about moisture eliminating ratio of heating zone of tobacco leaf re-drying line. Multiple linear regression method and multiple stepwise regression method are used to establish the model. From the result of the model, conclusions are as follows:

1) Heating zone of tobacco leaf re-drying line is a complex process and simple multiple linear regression could not satisfied with demand.
2) According to lead impact factor to regression equation gradually, the multiple stepwise regression model is established. Better fitting and predicting effects can achieved by analysis and simulation.
3) There are three factors influence moisture eliminating ratio of heating zone of tobacco leaf re-drying line. They are air absolute humidity, temperature and flow rate of the ventilation gate at the region. According to analysis influencing factors and establishing soft-sensing model, online measurement can be realized.

References

1. Jin, J., Li, K.: Fieldbus-based Control System for Tobacco Leaf Re-drying. Instrument and Meter for Automation 10(27), 49–53 (2006)
2. Guo, S.M.: The Soft Measurement Technology in the Application of Wine Fermentation Process. Kunming University of Science and Technology (2004)
3. Ying, X., Xiao, C., Wang, L.: Tepwise Regression Arithmetic for the Gauging Equation of Vibrational Canister Barograph. Journal of PI.A University of Science and Technology 3(1), 86–89 (2002)
4. Lu, Y.M., Yue, T.X., Chen, C.F.: Radiation Modeling Based on Stepwise Regression Analysis in China. Journal of Remote Sensing 14(5), 858–863 (2010)

Registration Algorithm of Multi-sensor System Errors Based on High-Precision Navigation Equipments

Pixu Zhang, Hangyu Wang, and Zhangsong Shi

Electronic Engineering College, Naval University of Engineering, Wuhan 430033, China
pxzhang_531@126.com

Abstract. In order to eliminate the influences of time wrap and sensor system errors on precision of multi-sensor data fusion in maritime cooperation of multi-platform, the transfer relation between Earth-Center Earth-Fixed coordinate system and Cartesian coordinate frame described by navigation information is modeled, in which Earth-Center Earth-Fixed coordinates are transitional characters and to be eliminated. Combined with innovation correlation function sequence, a self-adaptive filter is presented to register time and sensor system errors for multi-platform, which identifies model characters and error characters on line. Contrastive simulations show that the registration algorithm based on high-precision navigation equipments can converge quickly with high precision.

Keywords: Multi-sensor, Navigation information, Self-adaptive filter, System error registration.

1 Introduction

For maritime cooperation of multi-platform, the precision of multi-sensor data fusion is obviously influenced by sensor system errors and time wrap coming from communication delay, information desertion and clock inconsistency [1]. At the same time, navigation equipments as GPS or COMPASS are common on many platforms, and in the process of error registration position errors of the sensor are piddling if navigation data is used to transform sensor observation coordinates into Earth-Center Earth-Fixed coordinates [2-3]. The paper presents a transfer model of Earth-Center Earth-Fixed coordinate system and Cartesian coordinate frame described by navigation information, in which Earth-Center Earth-Fixed coordinates are transitional characters and to be eliminated, and a self-adaptive filter, which identifies model characters and error characters on line by innovation correlation function sequence, to register time and sensor system errors.

2 The Sensor System Error Model for Multi-platform

Supposed that the precision of satellite navigation data is high, the position errors of sensors are piddling. The time t is a local time for every platform. The maritime

D. Jin and S. Lin (Eds.): Advances in CSIE, Vol. 2, AISC 169, pp. 375–380.
springerlink.com

platform T_1 observes the aerial platform T_3 and obtains the observation $(r_{31}(t), q_{31}(t), \varepsilon_{31}(t))$ as distance, azimuth and elevation. The maritime platform T_2 observes the aerial platform T_3 and obtains the observation $(r_{32}(t), q_{32}(t), \varepsilon_{32}(t))$. The maritime platform T_1 observes the maritime platform T_2 and obtains the observation $(r_{21}(t), q_{21}(t), \varepsilon_{21}(t))$. The aerial platform T_2 moves horizontally along a straightaway with the velocity as V_3 and the course as C_3. The position of T_1 is $(B_1(t), L_1(t), H_1(t))$ as latitude, longitude and altitude, and the position of T_2 is $(B_2(t), L_2(t), H_2(t))$. Set $(\Delta r_1, \Delta q_1, \Delta\varepsilon_1)$ and $(\Delta r_2, \Delta q_2, \Delta\varepsilon_2)$ as distance, azimuth and elevation system error of sensors in maritime platforms and Δt as time wrap clock in T_2 quicker than T_1.

Taking $x_{21g}(t)$ in the time t of its clock as an example, the Cartesian coordinates for T_1 to observe T_2 are obtained [4-5].

$$\begin{aligned}
x_{21g}(t) &= (r_{21}(t)+\Delta r_1)\cdot\sin(q_{21}(t)+\Delta q_1)\cdot\cos(\varepsilon_{21}(t)+\Delta\varepsilon_1) \\
&= r_{21}(t)\cdot\sin q_{21}(t)\cdot\cos\varepsilon_{21}(t)\cdot\cos\Delta q_1\cdot\cos\Delta\varepsilon_1 - r_{21}(t)\cdot\sin q_{21}(t)\cdot\sin\varepsilon_{21}(t)\cdot\cos\Delta q_1\cdot\sin\Delta\varepsilon_1 \\
&\quad + r_{21}(t)\cdot\cos q_{21}(t)\cdot\cos\varepsilon_{21}(t)\cdot\sin\Delta q_1\cdot\cos\Delta\varepsilon_1 - r_{21}(t)\cdot\cos q_{21}(t)\cdot\sin\varepsilon_{21}(t)\cdot\sin\Delta q_1\cdot\sin\Delta\varepsilon_1 \\
&\quad + \sin q_{21}(t)\cdot\cos\varepsilon_{21}(t)\cdot\cos\Delta q_1\cdot\cos\Delta\varepsilon_1\cdot\Delta r_1 - \sin q_{21}(t)\cdot\sin\varepsilon_{21}(t)\cdot\cos\Delta q_1\cdot\sin\Delta\varepsilon_1\cdot\Delta r_1 \\
&\quad + \cos q_{21}(t)\cdot\cos\varepsilon_{21}(t)\cdot\sin\Delta q_1\cdot\cos\Delta\varepsilon_1\cdot\Delta r_1 - \cos q_{21}(t)\cdot\sin\varepsilon_{21}(t)\cdot\sin\Delta q_1\cdot\sin\Delta\varepsilon_1\cdot\Delta r_1 \\
&= r_{21}(t)\cdot\sin q_{21}(t)\cdot\cos\varepsilon_{21}(t) - r_{21}(t)\cdot\sin q_{21}(t)\cdot\sin\varepsilon_{21}(t)\cdot\Delta\varepsilon_1 \\
&\quad + r_{21}(t)\cdot\cos q_{21}(t)\cdot\cos\varepsilon_{21}(t)\cdot\Delta q_1 + \sin q_{21}(t)\cdot\cos\varepsilon_{21}(t)\cdot\Delta r_1 + e_{1x}(t)
\end{aligned} \tag{1}$$

$y_{21g}(t)$ and $z_{21g}(t)$ are transformed homoplastically.

$$\begin{aligned}
y_{21g}(t) &= r_{21}(t)\cdot\cos q_{21}(t)\cdot\cos\varepsilon_{21}(t) - r_{21}(t)\cdot\cos q_{21}(t)\cdot\sin\varepsilon_{21}(t)\cdot\Delta\varepsilon_1 \\
&\quad - r_{21}(t)\cdot\sin q_{21}(t)\cdot\cos\varepsilon_{21}(t)\cdot\Delta q_1 + \cos q_{21}(t)\cdot\cos\varepsilon_{21}(t)\cdot\Delta r_1 + e_{1y}(t)
\end{aligned} \tag{2}$$

$$z_{21g}(t) = r_{21}(t)\cdot\sin\varepsilon_{21}(t) + r_{21}(t)\cdot\cos\varepsilon_{21}(t)\cdot\Delta\varepsilon_1 + \sin\varepsilon_{21}(t)\cdot\Delta r_1 + e_{1z}(t) \tag{3}$$

In the time t of its own clock, the Cartesian coordinates for T_1 and T_2 to observe T_3 are obtained homoplastically.

3 The System Error Registration Model and Self-adaptive Filter

After modeling transfer relation of time and sensor system errors, a transfer matrix of Earth-Center Earth-Fixed coordinate system and Cartesian coordinate frame is described and calculated by navigation information, in which Earth-Center Earth-Fixed coordinates are transitional characters and to be eliminated, and the innovation correlation function sequence of observations is obtained to establish self-adaptive filter.

Set $(x_{1e}(t), y_{1e}(t), z_{1e}(t))$, $(x_{2e}(t), y_{2e}(t), z_{2e}(t))$ and $(x_{3e}(t), y_{3e}(t), z_{3e}(t))$ as Earth-Center Earth-Fixed coordinates of T_1, T_2 and T_3 which are obtained based on World Geodetic System.

$$\begin{cases} x = (N+H)\cos B\cos L \\ y = (N+H)\cos B\sin L \\ z = [N(1-e^2)+H]\sin B \end{cases} \tag{4}$$

Thereinto, $N = a/\sqrt{1-e^2\sin^2 B}$. a is long radius of the elliptic earth, and e is the first eccentricity.

Referring to the literature [6], the transfer relation between Earth-Center Earth-Fixed coordinate system and Cartesian coordinate frame is obtained as showed in formulas (5)-(7), with $\Phi_{gei}(t)$ as the transfer matrix from Cartesian coordinate frame to Earth-Center Earth-Fixed coordinate system and $\Phi_{egi}(t)$ as the transfer matrix from Earth-Center Earth-Fixed coordinate system to Cartesian coordinate frame.

$$\begin{bmatrix} x_{2e}(t) \\ y_{2e}(t) \\ z_{2e}(t) \end{bmatrix} = \Phi_{ge1}(t) \cdot \begin{bmatrix} x_{21g}(t) \\ y_{21g}(t) \\ z_{21g}(t) \end{bmatrix} + \begin{bmatrix} x_{1e}(t) \\ y_{1e}(t) \\ z_{1e}(t) \end{bmatrix} \tag{5}$$

$$\begin{bmatrix} x_{3e}(t) \\ y_{3e}(t) \\ z_{3e}(t) \end{bmatrix} = \Phi_{ge2}(t) \cdot \begin{bmatrix} x_{32g}(t) \\ y_{32g}(t) \\ z_{32g}(t) \end{bmatrix} + \begin{bmatrix} x_{2e}(t) \\ y_{2e}(t) \\ z_{2e}(t) \end{bmatrix} \tag{6}$$

$$\begin{bmatrix} x_{31g}(t) \\ y_{31g}(t) \\ z_{31g}(t) \end{bmatrix} = \Phi_{eg1}(t) \cdot \begin{bmatrix} x_{3e}(t) - x_{1e}(t) \\ y_{3e}(t) - y_{1e}(t) \\ z_{3e}(t) - z_{1e}(t) \end{bmatrix} \tag{7}$$

Thereinto,

$$\Phi_{gei}(t) = \begin{bmatrix} -\sin L_i(t) & -\sin B_i(t)\cdot\cos L_i(t) & \cos B_i(t)\cdot\cos L_i(t) \\ \cos L_i(t) & -\sin B_i(t)\cdot\sin L_i(t) & \cos B_i(t)\cdot\sin L_i(t) \\ 0 & \cos B_i(t) & \sin B_i(t) \end{bmatrix} \tag{8}$$

$$\Phi_{egi}(t) = \Phi_{gei}(t)^T \tag{9}$$

The formula (10) is obtained by taking formulas (5)-(6) into (7).

$$\begin{bmatrix} x_{31g}(t) \\ y_{31g}(t) \\ z_{31g}(t) \end{bmatrix} = \Phi_{eg1}(t)\cdot\Phi_{ge2}(t) \cdot \begin{bmatrix} x_{32g}(t) \\ y_{32g}(t) \\ z_{32g}(t) \end{bmatrix} + \begin{bmatrix} x_{21g}(t) \\ y_{21g}(t) \\ z_{21g}(t) \end{bmatrix} \tag{10}$$

The discrete observation equation is obtained by taking sensor system error model showed as formulas (1)-(3) into formula (10), which is described by matrix equation with time and sensor system error vector, $X = [\Delta r_1 \quad \Delta q_1 \quad \Delta\varepsilon_1 \quad \Delta r_2 \quad \Delta q_2 \quad \Delta\varepsilon_2 \quad \Delta t]^T$.

$$Z(k) = H(k)\cdot X(k) + V(k) \tag{11}$$

Thereinto, $V(k)$ is the observation error.

$$Z(k) = \begin{bmatrix} r_{31}(k)\cdot\sin q_{31}(k)\cdot\cos\varepsilon_{31}(k) \\ r_{31}(k)\cdot\cos q_{31}(k)\cdot\cos\varepsilon_{31}(k) \\ r_{31}(k)\cdot\sin\varepsilon_{31}(k) \end{bmatrix} - \Phi_{eg1}(k)\cdot\Phi_{ge2}(k)\cdot \begin{bmatrix} r_{32}(k)\cdot\sin q_{32}(k)\cdot\cos\varepsilon_{32}(k) \\ r_{32}(k)\cdot\cos q_{32}(k)\cdot\cos\varepsilon_{32}(k) \\ r_{32}(k)\cdot\sin\varepsilon_{32}(k) \end{bmatrix} - \begin{bmatrix} r_{21}(k)\cdot\sin q_{21}(k)\cdot\cos\varepsilon_{21}(k) \\ r_{21}(k)\cdot\cos q_{21}(k)\cdot\cos\varepsilon_{21}(k) \\ r_{21}(k)\cdot\sin\varepsilon_{21}(k) \end{bmatrix} \tag{12}$$

$$H(k) = [H_1(k) \quad \Phi_{eg1}(k)\cdot\Phi_{ge2}(k)\cdot H_2(k)] \tag{13}$$

$$\Phi_{eg1}(k) = \begin{bmatrix} -\sin L_1(k) & -\sin B_1(k)\cdot\cos L_1(k) & \cos B_1(k)\cdot\cos L_1(k) \\ \cos L_1(k) & -\sin B_1(k)\cdot\sin L_1(k) & \cos B_1(k)\cdot\sin L_1(k) \\ 0 & \cos B_1(k) & \sin B_1(k) \end{bmatrix} \tag{14}$$

$$\Phi_{ge2}(k)=\begin{bmatrix} -\sin L_2(k) & \cos L_2(k) & 0 \\ -\sin B_2(k)\cdot\cos L_2(k) & -\sin B_2(k)\cdot\sin L_2(k) & \cos B_2(k) \\ \cos B_2(k)\cdot\cos L_2(k) & \cos B_2(k)\cdot\sin L_2(k) & \sin B_2(k) \end{bmatrix} \tag{15}$$

$$H_1(k)=\begin{bmatrix} \sin q_{21}(k)\cdot\cos\varepsilon_{21}(k) & r_{21}(k)\cdot\cos q_{21}(k)\cdot\cos\varepsilon_{21}(k) & -r_{21}(k)\cdot\sin q_{21}(k)\cdot\sin\varepsilon_{21}(k) \\ \cos q_{21}(k)\cdot\cos\varepsilon_{21}(k) & -r_{21}(k)\cdot\sin q_{21}(k)\cdot\cos\varepsilon_{21}(k) & -r_{21}(k)\cdot\cos q_{21}(k)\cdot\sin\varepsilon_{21}(k) \\ \sin\varepsilon_{21}(k) & 0 & r_{21}(k)\cdot\cos\varepsilon_{21}(k) \end{bmatrix} - \begin{bmatrix} \sin q_{31}(k)\cdot\cos\varepsilon_{31}(k) & r_{31}(k)\cdot\cos q_{31}(k)\cdot\cos\varepsilon_{31}(k) & -r_{31}(k)\cdot\sin q_{31}(k)\cdot\sin\varepsilon_{31}(k) \\ \cos q_{31}(k)\cdot\cos\varepsilon_{31}(k) & -r_{31}(k)\cdot\sin q_{31}(k)\cdot\cos\varepsilon_{31}(k) & -r_{31}(k)\cdot\cos q_{31}(k)\cdot\sin\varepsilon_{31}(k) \\ \sin\varepsilon_{31}(k) & 0 & r_{31}(k)\cdot\cos\varepsilon_{31}(k) \end{bmatrix} \tag{16}$$

$$H_2(k)=\begin{bmatrix} \sin q_{32}(t)\cdot\cos\varepsilon_{32}(t) & r_{32}(t)\cdot\cos q_{32}(t)\cdot\cos\varepsilon_{32}(t) & -r_{32}(t)\cdot\sin q_{32}(t)\cdot\sin\varepsilon_{32}(t) & V_3\cdot\sin C_3 \\ \cos q_{32}(t)\cdot\cos\varepsilon_{32}(t) & -r_{32}(t)\cdot\sin q_{32}(t)\cdot\cos\varepsilon_{32}(t) & -r_{32}(t)\cdot\cos q_{32}(t)\cdot\sin\varepsilon_{32}(t) & V_3\cdot\cos C_3 \\ \sin\varepsilon_{32}(t) & 0 & r_{32}(t)\cdot\cos\varepsilon_{32}(t) & 0 \end{bmatrix} \tag{17}$$

The system equation is described as formula (18).

$$X(k+1)=\Phi(k+1|k)\cdot X(k)+W(k) \tag{18}$$

Thereinto, $W(k)$ is the system disturbance.

$$\Phi(k+1|k)=\begin{bmatrix} 1&0&0&0&0&0&0\\ 0&1&0&0&0&0&0\\ 0&0&1&0&0&0&0\\ 0&0&0&1&0&0&0\\ 0&0&0&0&1&0&0\\ 0&0&0&0&0&1&0\\ 0&0&0&0&0&0&1 \end{bmatrix} \tag{19}$$

Because of uncertainty of system disturbance $W(k)$ and observation error $V(k)$ in this estimation model of time and sensor system errors, the self-adaptive filter estimating and adjusting the gain matrix K is necessary [7], which identifies model characters and error characters on line by innovation correlation function sequence of observations.

(1) calculate the innovation correlation function sequence.

$$\overline{C}_i^k=\frac{1}{k}\cdot\sum_{l=i+1}^{k}\tilde{\tilde{Z}}(l|l-1)\cdot\tilde{\tilde{Z}}^T(l-i|l-i-1)=\overline{C}_i^{k-1}+\frac{1}{k}\cdot[\tilde{\tilde{Z}}(k|k-1)\cdot\tilde{\tilde{Z}}^T(k-i|k-i-1)-\overline{C}_i^{k-1}] \tag{20}$$

(2) calculate transitional variables A, $\overline{P}\cdot H^T$, $\hat{R}$, $H\cdot\overline{P}$ and δP, and estimate the gain matrix K.

$$A=\begin{bmatrix}H\cdot\Phi & H\cdot\Phi^2 & \cdots & H\cdot\Phi^n\end{bmatrix}^T \tag{21}$$

$$\overline{P}\cdot H^T=(A^T\cdot A)^{-1}\cdot A^T\cdot\begin{bmatrix} \overline{C}(1)+H\cdot\Phi\cdot\overline{K}\cdot\overline{C}(0) \\ \overline{C}(2)+H\cdot\Phi\cdot\overline{K}\cdot\overline{C}(1)+H\cdot\Phi^2\cdot\overline{K}\cdot\overline{C}(0) \\ \cdots\cdots \\ \overline{C}(n)+H\cdot\Phi\cdot\overline{K}\cdot\overline{C}(n-1)+\cdots+H\cdot\Phi^n\cdot\overline{K}\cdot\overline{C}(0) \end{bmatrix} \tag{22}$$

$$\hat{R}=\overline{C}(0)-H\cdot\overline{P}\cdot H^T \tag{23}$$

$$\delta P = \Phi \cdot \{\delta P - [\overline{P} \cdot H^T + \delta P \cdot H^T] \cdot [\overline{C}(0) + H \cdot \delta P \cdot H^T]^{-1} \cdot [H \cdot \overline{P} + H \cdot \delta P] + \overline{K} \cdot H \cdot \overline{P} + \overline{P} \cdot H^T \cdot \overline{K}^T - \overline{K}\overline{C}(0)\overline{K}^T\} \cdot \Phi^T \quad (24)$$

$$\hat{K} = (\overline{P} \cdot H^T + \delta P \cdot H^T) \cdot (\overline{C}(0) + H \cdot \delta P \cdot H^T)^{-1} \quad (25)$$

(3) calculate the filter value of state variable.

$$\hat{X}(k \mid k) = \Phi \cdot \hat{X}(k-1 \mid k-1) + \hat{K} \cdot [Z(k) - H \cdot \Phi \cdot \hat{X}(k-1 \mid k-1)] \quad (26)$$

4 Simulations

In order to verify feasibility and validity of the registration algorithm of sensor system errors for multi-platform based on high-precision navigation equipments, a scenario with two maritime platforms and an aerial platform is set as follows.

Initial positions of the maritime platforms T_1, T_2 and the aerial platform T_3 are $(22.7640°, 118.2694°, 0\,m)$, $(23.0335°, 118.1172°, 0\,m)$ and $(23.0966°, 118.0408°, 5000\,m)$. T_1 and T_2 are immobile and T_3 moves horizontally along a straightaway with a constant velocity of 400 m/s and a course of 100°. Set simulative system errors, distance, azimuth, elevation and time, as $(500\,m, 5°, 5°, 50\,s)$ for maritime platforms T_1 and T_2.

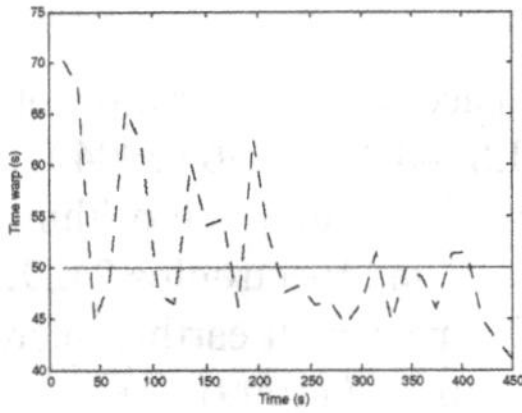

Fig. 1. Estimation of time warp (*dashed curve*) in contrast to the true value (*real curve*)

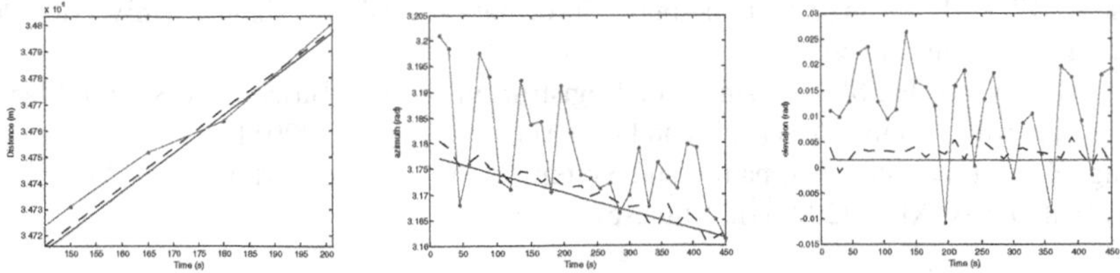

Fig. 2. Registration of distance, azimuth and elevation (*dashed curve*) in contrast to the true value (*real curve*) and the observation (*dotted real curve*)

Due to the influence of time and sensor system errors, disturb scatter of distance, azimuth and elevation obviously excurse from true values. As contrastive simulations shown in Fig. 1-2, the effect of filter is improved by identifying model characters and error characters and adjusting gain matrix based on innovation correlation function sequence on line, and the registration algorithm based on high-precision navigation equipments estimates system errors with higher precision.

5 Conclusions

For maritime cooperation of multi-platform, sensor system errors and time wrap because of the clock frequency error and communication delay between two platforms on the sea are obviously influential on precision of multi-sensor data fusion. Thanks to navigation equipments as GPS or COMPASS, position errors of the sensor are piddling in the process of error registration if navigation data with high precision is used to transform sensor observation coordinates into Earth-Center Earth-Fixed coordinates. Based on navigation data and real-time innovation correlation function sequence, a transfer model between Earth-Center Earth-Fixed coordinate system and Cartesian coordinate frame, in which Earth-Center Earth-Fixed coordinates are transitional characters and to be eliminated, and a self-adaptive filter, which identifies model characters and error characters and adjusts gain matrix on line, are presented to register sensor system errors. The feasibility and validity of this registration algorithm are verified by simulations.

References

1. Li, W., Leung, H., Zhou, Y.: Space-time registration of radar and ESM using unscented kalman filter. IEEE Trans. on AES. 40, 824–836 (2004)
2. Li, H., Feng, X.: Least square registration algorithm based on the ECEF coordinate transformation. System Engineering and Electronics 24, 92–95 (2002) (in Chinese)
3. Zhou, Y., Leung, H.: Sensor alignment with earth-centered earth-fixed (ECEF) coordinate system IEEE Trans. on AES 35, 410–417 (1999)
4. Pan, P., Feng, X., Sun, P., Liu, J., Liu, Y., Li, F.: Registration algorithm of radar and IR sensor based on two stage filtering of UKF and KF. Opto-Electronic Engineering 35, 28–34 (2008) (in Chinese)
5. Han, C., Zhu, H., Duan, Z.: Multi-source Information Fusion. Tsinghua University Press, Beijing (2006) (in Chinese)
6. Wang, C., Wang, H., Shi, Z.: An Error Registration Method Based on UKF for Maritime Multi-platforms. Applied Mechanics and Materials 48, 192–197 (2011)
7. Wang, Z.: Optimal State Estimation and System Identification. Northwestern Polytechnical University Press, Xi'an (2004) (in Chinese)

Research and Design of Universal and Open Software Development Platform for Digital Home

CaiFeng Cao

School of Computer Wuyi University, Jiangmen 529020, China
cfcao@126.com

Abstract. With the development of digital home system, building a universal, independent, and open design platform can help to promote the efficiency of the system development and is beneficial to interconnection and inter-communication among devices in it. In this paper, the core technologies of Android system are investigated; referencing Android, it puts forward the architecture and key technologies of the platform. The platform can configure flexibly different hardware structures and operating systems, fully integrate the third party softwares, then provides the development support of system and application software for different electronic equipments.

Keywords: digital home, general independent open platform, Android.

1 Introduction

There are many kinds of electronic equipments in digital home system, which widely use embedded software development techniques. Building unified development platform becomes the focus of research and development of the industry.

Android is the open source platform of intelligence mobile phone launched by Google. It includes operating system, user interface, middleware and application software[1]. Its openness, portability, rich application software, convenient development environment and many other advantages make it as the one of the world's most popular smartphone development and operation platforms just in two or three years. Its embedded software development technology has achieved great success. The research and application on Android is never stop. This paper studies Android design technologies, makes full use of its existing resources and system mechanism, then expands the development area, finally makes a new strong open software design platform for digital home. The platform supports different hardware environment and integrates the third party software. Through this platform, users can develop various kinds of systems on independent hardware, such as differentiation middleware, individual application. The platform is called UOSDP.

2 Analysis of Android System

Android system integrates Java and C/C++ advantages, forms a easy to use and efficient smartphone development platform. This platform applies to both system developers and application developers and can be conveniently referred.

D. Jin and S. Lin (Eds.): Advances in CSIE, Vol. 2, AISC 169, pp. 381–386.
springerlink.com

Firstly, Android system has a good architecture, which can be divided into 5 levels from bottom to top. The first layer is Linux kernel, completes operation system function. This layer cuts Linux OS and adds related driver program, such as YAFFS2 Flash, Binder IPC driver, WiFi, bluetooth driver etc. The second is hardware abstraction layer, which provides hardware call interface for upper native framework. It realizes the calls of kernel hardware drivers by the methods of hardware module, hardware_legacy, or C++ inheritance realization .The third layer includes native libraries with C/C++ interface, JNI (Java Native Interface) and Java runtime environment. The libraries such as SQLite, WebKit, OpenGL etc[2]. can communicate with the upper layer through JNI. The base Java library and virtual machine Dalvik form Java runtime environment. The fourth is application framework layer, includes SDK and some class libraries[3]. It is the embodiment of the core mechanism of Android platform. The fifth is application layer, including applications provided by the system and the third party.

Secondly, Dalvik virtual machine provides an running environment for Android java program, which executes .dex file. Google's Dalvik has characteristics of efficiency, conciseness and saving resources, is very suitable for embedded system with limited resources. Dex format is a compressed format, suitable for the system of limited memory and processor with limited speed [4]. Dalvik can run multiple processes, each application thread is corresponding to a Linux thread, each application is running in its own sandbox. Dalvik virtual machine is a highly modular system, which can be replaced by a separate different realization. Dalvik bases Linux, but has a certain commonality. At present it can support systems based on UNIX, Linux, BSD, Mac OSX.

3 UOSDP and Its Architecture

With the popularity of Android, people hope Android can adapt to different hardware environment, combine with the third party software, and support different customers and other network environments. So far more than 170 kinds of Android devices are on the global market, which has proved the Android's high openness and expansibility. There is no doubt that Android developers will face the expansion demand of more bottom modules and the upper frames in the future. Therefore, to build general, open and independent design platform is necessary and urgent.

General independent design platform has flexibility and modular structure, the user can combine system modules autonomously, then form development platform and real machine running environment, which can be suitable for different electronic products and different type developers. We can use and expand Android original ecological system, set up new Android software stack, then provide the industry general framework and application development platform irrelevant to equipment. Using its dynamic open source development community around it, developers can easily and quickly create new applications for their devices; OEM (Original Equipment Manufacturer) will be able to use more and more applications. Using the new ecological system, OEM will be able to fast rebuild and optimize particular development platform.

For large, complex software system, a clear system structure is very important. This structure will have features of open platform oriented service, open kernel and framework, open hardware platform. Using new platform, the third party software developers can develop applications based on their own core functions, can customize middleware system and operating system.

UOSDP system architecture makes full use of the Android system success mode, still is divided into five layers.

The first layer includes the OS based on Linux kernel, other OS and various device drivers. It realizes process scheduling, memory management, network communication, process communication and the function of virtual file system. The Binder process communication, YAFFS2 file system, Low Memory management (Low Memory Killer), anonymous Memory sharing (Ashmem) are increased in it. Device drivers include general equipment drivers such as flash driver, USB driver, and personalized device drivers such as display driver, keyboard driver in digital home system. This layer realizes free choosing OS, loading external equipments and device drivers, provides open hardware interface, supports different processor system structure. Especially for OS based on Linux, the processor system code is installed in the folder /arch /, equipment node in the folder /dev/. All equipments register in Platform_device file, device drivers in Platform_driver file.

The second layer is hardware abstraction layer, which realizes driver calls and provides hardware calls interface for upper native frame. Same function realization can have different drivers. Hardware abstraction layer includes dynamically loadable libhardware.so (DLL) and its interface, different hardware abstract module (*. so or *.dll) and its interface. Android native frame gets hardware modules id through libhardware interface, then dynamically opens hardware module and its interface, and calls kernel drivers through the method open( ) of hardware module.

This layer realizes different hardware support in the way of adding hardware abstraction module, which embodies the openness and expansibility.

The third layer is C++/C class library and Java runtime environment. In addition to the original SQLite, WebKit, OpenGL, it will expand the native library and service by using the existing mechanism to meet the development needs of different electronic equipment in digital home. The running environment still adopts Dalvik virtual machine. Expanding Dalvik is nessary to meet the needs of different operating systems and processors.

This layer openness is reflected in the native software integration and the support to different processors and operating systems.

The fourth is multiapplication framework layer. This layer contains different SDK to support the application development of different electronic equipment. The existing application frameworks will be expanded through new subclasses, overloading methods or new packages. This layer is developed by Java language, which realizes the aggregation and richness of application frameworks in the open-source way.

The fifth is application layer, which contains various kinds of applications about different electronic equipments.

After reconstruction, the UOSDP system architecture is as shown in figure1.

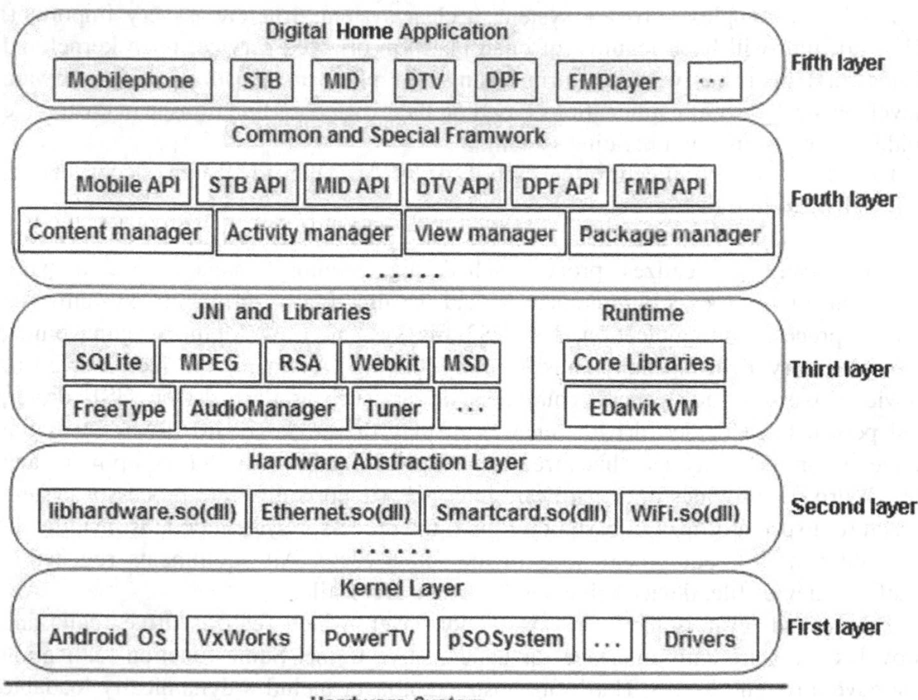

Fig. 1. Platform system architecture

4 Key Technology of UOSDP

4.1 JNI Expansion

As adding a new native library, its DLL with JNI mechanism must be constructed. JNI needs to be expanded. Android simplifies JNI development. Its main support library is nativehelper. The head file JNI.h defines the structure type JNINativeMethod which fulfills functions mapping. The structure is as follow[4]:

```
Typedef struct {Const char * name; Const char *
signature; Void * fnPtr;} JNINativeMethod;
```

The first variable stores a Java function name. The second variable stores function parameters and return value in string type. The third variable fnPtr points to C function.

Extension JNI algorithm for native library is as follows.

1) to build up native library API file, such as api1.CPP, in which the JNINativeMethod array is defined, each element realizes a pair of functions mapping.

2) to establish registration file, such as, using function registerNativeMethods owned by AndroidRuntime registers the array.
3) by using C++ development tools, the API file and related file will be compiled into DLL, then storing it in the module's JNI directory.
4) to write Java class with native methods , so that C++ native codes can be called in Java program .

4.2 Combination of Android System and Different Hardware Environment

At present Dalvik supports platform Linux, BSD, Mac OSX and the system based on UNIX. For different processors under these OS, the first is to construct the interpreter for specific architecture CPU. Assumed that the name of the CPU architecture is called Myarch, the main procedure is as follows.

1) building a new compiling configuration file config-Myarch for Myarch. To the file content we can copy existing file config-allstubs, and make the necessary changes.
2) using the tool gen-mterp. py provided by the system , through the configuration file config-Myarch, then generating interpreters InterpAsm-Myarch.S of assembly language edition and InterpC-Myarch.C of C language version.
3) expanding the file /dalvik/vm/Dvm. mk, adding processor architecture code files and interpreter files. The statements are described as follows:

```
Ifeq ($(dvm_arch), Myarch)
  # dvm_arch_variant: = Myarch
   MTERP_ARCH_KNOWN: = true
   LOCAL_SRC_FILES + = \
   # processor architecture interface code files
   Arch / $(dvm_arch_variant) /CallMyarchABI. S
   Arch / $(dvm_arch_variant) /HintMyarchABI. C
   # interpreter file
   mterp/out/InterpAsm-Myarch. S
   mterp/out/InterpC-Myarch. C
  endif
```

4.3 Combination of Different Operating Systems and Android Middleware

For different operating systems, such as operating system supporting set-top box, Dalvik virtual machine needs transplantation and modification. Dalvik virtual machine is based on Apache Harmony technical architecture with stratified and modular structure[4]. In the structure, the top is Java library, the middle is virtual machine, the lower is operating system. In virtual machine, the portability layer encapsulates the differences of different operating system, which provides unified interface accessing low-level system. Specifically, Dalvik virtual machine has process mechanism. On the one hand, it retains traditional Java process control API, on the other hand it combines the characteristics of Linux operating system and adds special process control API [4]. Dalvik provides the class dalvik.system.Zygote, Zygote has the zygote interface that is used to access Dalvik virtual machine. Zygote packages the function fork( ) owned by Linux system, which is used to create a new instance of virtual machine process and run Android application. So, for different operating system, we only need to modify Zygote, rewrite fork( ) function and interface class. The expanded Portability layer can adapt to different operating systems.

5 Conclusion

After achieving initial success in the mobile phone market, Android now start entering into other digital equipment field. In digital home system, the traditional DTV, set-top box and blue light have finite application, Android can easily extend the applications taking network as the center, make equipments have Internet connectivity. They not only have local functions, but also have new network services. Basing on Android, the open design platform UOSDP will provide a device-independent, flexible structure, modular customization and commonality support system. The platform structure in this paper proposes opening strategy in every level for different kinds developers to participate in and build it.

Acknowledgement. This work is supported by National Engineering Research Center of Digital Life, Guangzhou 510006, China, under Grant No. 2011NERCDL001.

References

1. Chenjing, Chen, P., Li, W.: Analysis of Android Kernel. Modern Computer, 112–114 (November 2009)
2. Hanchao, Lianquan: Android Deep Development in System Level, pp. 12–13. Publishing House of Electronics Industry, Beijing (2011)
3. Wang, Y.: Development Travel on Android Platform, pp. 3–4. Publishing House of Electronics Industry, Beijing (2011)
4. Yang, F.: Android Internals: System, pp. 468–524. China Machine Press, Beijing (2011)
5. Yi, P.: Research on Structure and Performance of Dalvik Virtual Machine[EB/OL]. Jiling University (April 2011), http://dlib.cnki.net/kns50/detail.aspx?dbname=CMFD2011&filename=1011100372.nh
6. Zou, G.: Porting and Optimization of Dalvik Virtual Machine of Android on Loongson-based platform[EB/OL]. China University of Petroleum (May 2011), http://d.g.wanfangdata.com.cn/Thesis_Y1875804.aspx
7. Android Open Source Project[EB/OL], http://source.android.com/
8. McFadden, A.: Dalvik Porting Guide[EB/OL], http://android.git.kernel.org

A 3D Clipping Simulation Based on Virtual Surgery

Peng Wu[1], Kai Xie[1,2], Houquan Yu[1,2], Yunping Zheng[3], and Chao Wu[1]

[1] School of Electronics and Information, Yangtze University, Jingzhou, China
[2] Key Laboratory of Oil and Gas Resources and Exploration Technology of Ministry of Education, Yangtze University, Jingzhou, China
[3] School of Computer Science and Engineering, South China University of Technology, Guangzhou, China
peng_wu@189.cn, pami2009@163.com, hq_yu@163.net

Abstract. On the basis of classic surface reconstruction algorithms, a surface reconstruction based 3D clipping simulation is proposed. The algorithm employs implicit function to represent the clipping object and utilizes clipping object construction tree to construct complex clipping object. The results show that the method realizes 3D medical clipping simulation accurately. Then, we designed and implemented the 3D Orthopedic Operation Simulation Platform, which involves 3D clipping simulation, 3D modeling, 3D marking and measuring. Clinical case demonstrated the great value of the platform in medical applications.

Keywords: Surface Reconstruction, 3D Cutting, Surgical Simulation.

1 Introduction

Plastic surgery is difficult and dangerous, and the appearance and function must be taken into account. Orthopedic surgery put forward higher requirements for imaging science such as the preoperative accurate diagnosis, determined extent of disease, 3D stereoscopic vision, the complex spatial structure, accurate measurement of 3D morphological parameters, surgical path and plan, simulation of surgery, the predicted impact of surgery on the face, etc. In order to meet the clinical demand for orthopedic surgery, the system of CT 3D reconstruction assisted orthopedic began to appear, It is an emerging cross-disciplinary, the rise in the last 10 years, has been the hot spot of research and application at home and abroad.

The research of this paper is focused on in-depth study reconstructive surgery virtual surgery technique, and put forward a three-dimensional cutting simulation algorithm for reconstructive surgery virtual surgery technique in its basis.

2 Surface Reconstruction Scheme Based 3D Clipping Simulation

This paper presents a simulation algorithm based on 3D medical cutting of surface reconstruction[1,2,3]. Flow chart of surface reconstruction based clipping simulation as shown below:

D. Jin and S. Lin (Eds.): Advances in CSIE, Vol. 2, AISC 169, pp. 387–392.
springerlink.com

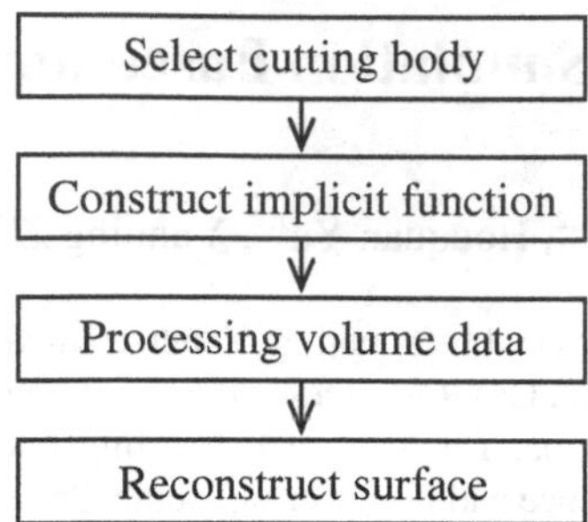

Fig. 1. Flow Chart of Surface Reconstruction based Clipping Simulation

Construct a suitable cutting body, such as cube, ball, NURBS surfaces, etc. Place the surface on the appropriate cutting position; get the location of the surface relative to the model. According to the selected cutting body and the relative position of the cutting body with the model, use the implicit function of volume data space to express the cutting body.

In the model space, we use implicit function to express the cutting body. Assume the implicit function of that cutting body is $f(X)$, and $X = (x\ y\ z)^T$ is the model space coordinates, then $P = \{X \mid f(X) = 0\}$ means the set of all the points on the surface of cutting body; and $P = \{X \mid f(X) > 0\}$ means the set of all the points in the internal of cutting body; $P = \{X \mid f(X) < 0\}$ means the set of all the points in the external of cutting body.

After getting the implicit function of volume data space, implicit expressions must be converted from model space to volume data space. There are two transformations form volume data space to model space. First is scale transformation (the interval in X axis, Y axis and Z axis are different), used to recover the original proportion of the data, assuming the transformation matrix is

$$M_{spacing} = \begin{pmatrix} X_{spacing} & 0 & 0 \\ 0 & Y_{spacing} & 0 \\ 0 & 0 & Z_{spacing} \end{pmatrix}$$

In the matrix, $X_{spacing}$, $Y_{spacing}$, $Z_{spacing}$ are the grid intervals in X, Y, Z axis of volume data respectively. Second is geometry transformation (rotate, translate and transform) in model space, transformation matrix is M_g. Assuming the coordinate of model space corresponding to the coordinate $X' = (x'\ y'\ z')^T$ within volume data is $X = (x\ y\ z)^T$, then

$$X = M_g M_{spacing} X' \tag{1}$$

So, the implicit function corresponding to cutting body in volume data space is

$$F(X') = f(M_g M_{spacing} X') \tag{2}$$

If the shape of cutting body is more complex that cannot be represented while using a single function, it can be expressed through a combination of functions. For example, if the cutting body is a hemisphere, it is a combination of a complete ball and a plane. The following are part of general rules of implicit function combination.

Suppose there are two cutting body in volume data space can be expressed in implicit function $F_1(X)$ and $F_2(X)$.

The intersection of the cutting body is

$$P = \{X \mid F_1(X) > 0\} \cap \{X \mid F_2(X) > 0\} = \{X \mid \min(F_1(X), F_2(X)) > 0\} \tag{3}$$

Then the corresponding implicit function is $F(X) = \min(F_1(X), F_2(X))$.

Cutting body's sum aggregate is

$$P = \{X \mid F_1(X) > 0\} \cup \{X \mid F_2(X) > 0\} = \{X \mid \max(F_1(X), F_2(X)) > 0\} \tag{4}$$

The corresponding implicit function is $F(X) = \max(F_1(X), F_2(X))$.

Cutting body's complementary set is

$$P = \{X \mid F'(X) < 0\} = \{X \mid -F'(X) > 0\} \tag{5}$$

The corresponding implicit function is $F(X) = -F'(X)$.

3 The Simulation Platform of 3D Orthopedic Surgery

Craniofacial structure where many vital organs are concentrated is complex, simultaneously relates the appearance, orthopedic surgery involving skull and face is difficult, dangerous, and the appearance and function must be taken into account. Craniofacial surgery put forward higher requirements for imaging science. It requires that the preoperative diagnosis must be accurate, determine the extent of disease, provide a three-dimensional stereoscopic vision, display the complex spatial structure, accurately measure three-dimensional morphological parameters, design surgical path and plan, simulate surgery, predict the impact of surgery on the face, etc. 3D surgical simulation based on CT images is a big hot topic today, because it overcomes the limitations of previous methods. Using human-computer interaction, that is using the mouse to move the osteotomy block, osteotomy block can move and rotate in 3D coordinate direction at the same time. The system can automatically output the changed data corresponding osteotomy block in the direction of three dimensions.

This system uses 3D surgery simulation based on CT images. The system platform is using the Visual C++ 6.0 and the Pixel Shader 2.0 [4,5]. It was tested in PC with a Pentium 4 (2.4 GHz) CPU and the NVIDIA Geforce 6600 graphic card. The surgery simulation is executed on the three-dimensional model close to realistic anatomical morphology. The osteotomy simulation and the actual surgical procedure is roughly the same. According to the measurement of diagnostic results choose the type of orthopedic surgery, and pre-install parameters in three-dimensional axial distance and the angle of rotation. The system automatically simulates the classic craniofacial orthopedic osteotomy by default parameter. The following clinical case of orbital hypertelorism illustrates the application of three-dimensional orthopedic surgery simulation platform.

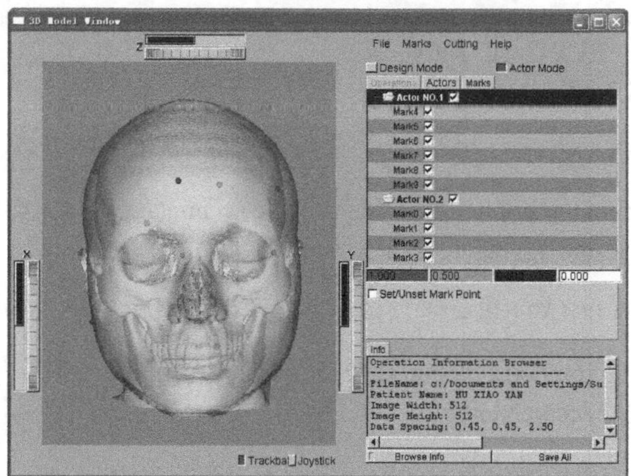

Fig. 2. 3D Orthopedic Surgery Simulation Platform

3.1 Orbital Hypertelorism

Orbital hypertelorism is a severe craniofacial deformity. Orthopedic surgery is the only treatment method. Osteotomy methods as shown in Figure 3(a) and Figure 3(b), after the correction, the patient's inside medial orbital distance is 20mm. In 1967 in France, Tessier[6] first reported the implementation of the intracranial path orbital distance widened orthotics and achieved satisfactory effect. At present, there are a few report to carry out such surgery at home and abroad, Shanghai Ninth Hospital successfully carried out the first cases of intracranial and extra cranial joint path orbital hypertelorism corrective surgery in 1977. Till late 1990s, the hospital has

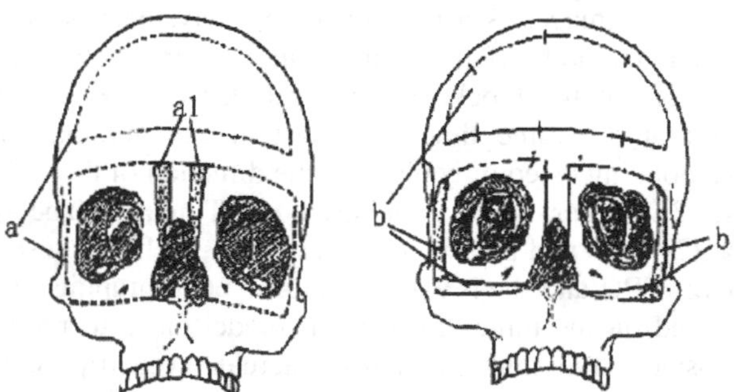

Fig. 3. Skull of pre-Osteotomy and Skull of post-operative correction. a. indicates the line for Osteotomy. a1. Indicates the bone removed. b. indicates the bone graft in the gap.

accumulated dozens of surgical experience[7]. It is the first domestic hospital which has the most reports of such surgeries.

3.2 The Experimental Results of Cases

Through cooperation with the Shanghai Ninth Hospital, we simulated 3D osteotomy of orbital hypertelorism. The schematic diagram of 3D osteotomy simulation is shown in figure 4. From the deformity of preoperative bone tissue, we can see that the patient's orbital, nasal, the central of pear-shaped bone have widespread bone tissue proliferation. Orbital spacing is 49.06mm (normal is no more than 28mm). In addition, this patient's both sides of the orbit outward, backward, downward reverse, the orbital diameter and the standard plane inclined angle respectively are -0.8° and -1.5° (normal is 4°-8°). Therefore, our surgery's goal is to eliminate the central excess bone and reverse back both sides of the orbital to the normal form. In. Figure 4(b) the per-orbital is divided into three parts, the red part of the middle takes down to another use. Figure 4(c) split the middle piece of bones longitudinally. Figure 4(d) move both sides of the yellow bone to the middle, the split bones of the middle pad in both sides of the eyes and cheekbones, the remnants of the lateral orbital edge, thus the design is basically completed. Figure 5 is the patient's front view and side view of the pre-operative and post-operative rectification of orbital hypertelorism. We can see that the appearance of the patient has been improved significantly.

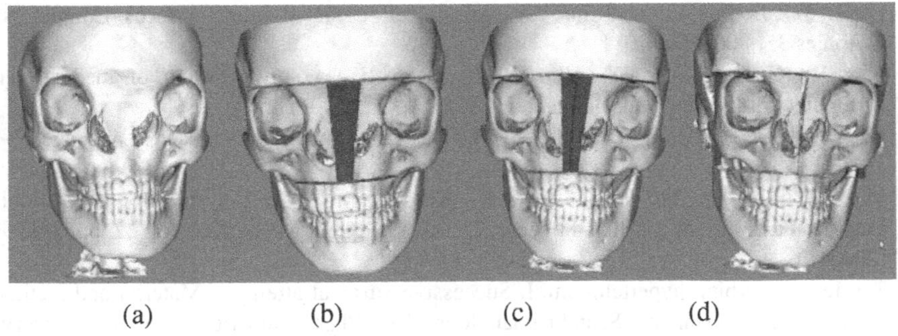

(a) (b) (c) (d)

Fig. 4. Three-dimensional Clipping Simulation of Orbital Hypertelorism.

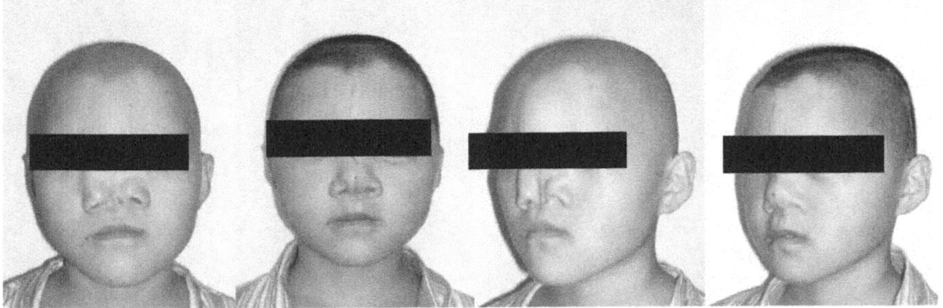

Fig. 5. Rectification of orbital hypertelorism.

4 Conclusions

This paper presents a simulation algorithm based on surface reconstruction of 3D medical clipping. The 3D Orthopedic Operation Simulation Platform is designed and implemented based on that algorithm. The platform contains many functions based on surface reconstruction such as 3D clipping simulation, 3D modeling, 3D marking and measuring. Clinical cases demonstrated the great value of the platform which can effectively complete planning, simulation and other functions of plastic surgery in medical applications.

Acknowledgements. This work has been partially supported by CNPC Innovation Foundation (2010D-5006-0304), Specialized Research Fund for the Doctoral Program of Higher Education of China (20070532077), Natural Science Foundation of Hubei Province of China (2009CDB308), Educational Fund of Hubei Province of China (Q20091211, B20111307).

References

[1] Jun, W., Zhouwang, Y., Liangbing, J., Jiansong, D., Falai, C.: Parallel and adaptive surface reconstruction based on implicit PHT-splines. Computer Aided Geometric Design 28, 463–474 (2011)

[2] Feifei, W., Jieqing, F.: Real-time ray casting of algebraic B-spline surfaces. Computers & Graphics 35, 800–809 (2011)

[3] Xie, K., Gang, S., Jie, Y., Zhu, Y.M.: Interactive volume cutting of medical data. Computers in Biology and Medicine 37, 1155–1159 (2007)

[4] Kessenich, J., Baldwin, D., Rost, R.: The OpenGL Shading Language. Language, 1–29 (2010)

[5] Mark, W.R., Glanville, R.S., Akeley, K., Kilgard, M.J.: Cg: a system for programming graphics hardware in a C-like language. In: ACM SIGGRAPH 2003, vol. 22, pp. 896–907 (2003)

[6] Tessier, P.: Orbital hypertelorism. I. Successive surgical attempts. Material and methods. Causes and mechanisms. Scandinavian Journal of Plastic and Reconstructive Surgery 6, 135–155 (1972)

[7] Disheng, Z.: Surgical treatment of orbital hypertelorism. Chinese Journal of Surgery 21, 32 (1983)

Ranking Software Risks Based on Historical Data

Yu Wang[1], Shun Fu[2,*], and Teng Zhang[3]

[1] College of Information Science and Technology, Chengdu University of Technology, China
[2] College of Management Science and Technology, Chengdu University of Technology, China
[3] College of Tourism and Urban-Rural Planning, Chengdu University of Technology, China

Abstract. Risk ranking is a key step for risk management. This study divides ranking methods of software risks into two steps. The first step initializes a risk set from historical data. The second step achieves metric values for the risks in the set as ranking references. Although historical data has gained prominence when getting initial risks for ranking software risks, the existing ranking methods only use results of the second step. The ranked values of the initial risks from the first step are not exploited. Actually, ranked values of historical data are sometimes beneficial. Another kind of ranking method by this study utilizes the metric values or positions from both steps (RHD-2). The method exploits historical data properly and gets suitable ranked results for decision-makers by utilizing proper empirical coefficients according to the practical situations.

Keywords: historical data, risk analysis, software risk.

1 Introduction

Risk ranking [1] [2] results are based on metric values, which include generally occurrence probability and impact. Risk exposure [3] is the main method of risk ranking utilizing the two metric values. Risk exposure is the product of probability and loss values. Most risk management approaches rely on risk estimation approaches that are either impractical or theoretically questionable [4]. The expected value of risk exposure is often impractical because the accurate estimate for probability and loss is seldom available. How to solve the limitations? One solution is to provide more ranking methods for decision-makers to select according to practical needs.

This study analyzes and presents new ideas based on historical data. Historical data has got attention in risk relative studies [5], such as estimating the probability [6], flood risk estimation [7], risk factors in medical science [8], geo-risk assessment [9], and risks of grassland fire [10]. However, historical data is used only for initialization purpose in existing software risk ranking processes, so does not exploit historical data. This study focuses on ranking methods exploiting known information from historic data. We name these methods as RHD methods. Ranking processes of the RHD methods have two steps. There are two kinds of RHD methods according to the utilization of values from the two steps. The first kind is RHD-1; the other is RHD-2. The RHD-1 depends on metric values from the second step. Most existing ranking

* Corresponding author.

D. Jin and S. Lin (Eds.): Advances in CSIE, Vol. 2, AISC 169, pp. 393–398.
springerlink.com

practices first get initial risks from historical data (HD), and then ranking metric values are gained for the risks. Ranked results are then gained based on the metric values. The RHD-2 is a new idea introduced in this study. It uses ranked results from both steps. The results include either metric values or ranked positions.

2 RHD-1

2.1 Practices

The RHD-1 is the currently used ranking idea utilizing historical data. Risks are generally listed out of historical data. Article [11] presents a list of risk factors based on the literature at home and abroad. The preliminary list is further refined and many items are reworded through the pilot survey and interviews with academic experts and practicing professionals. Answers to the questionnaire of the refined list are given on the scale of 1 to 5. Ranked results are gained based on the scale values.

We can see that risk-ranking practices begin with risk factors from historical data, such as the literature or former studies, questionnaires or lists by experts [12] [13] [14]; and then, ranked results are gained from them.

2.2 Two-Steps-Mathematical Model of the RHD-1

The ranking process of software risks can be divided into two steps. We introduce the following notations for analyzing the ranking process:

notation	meaning
s	*step 1*
$f(s)$	*a metric value from step 1*
$p(s)$	*a ranked position from step 1*
t	*step 2*
$f(t)$	*a metric value from step 2*
$p(t)$	*a ranked position from step 2*

1) Step 1: initialization from historical data.

This step initializes a set of risks from historical data. Let s stand for step 1, $f(s)$ for the initial metric value of a risk in step 1, and $p(s)$ for the ranked position of the risk based on the $f(s)$ value. The $f(s)$s and $p(s)$s can be set zero when the risk factors from historical data have no initial value.

2) Step 2: ranking risks.

In step 2, a new metric value is bestowed to each risk according to the present condition, and then a position value is decided based on the metric value. Let t stand for step 2, $f(t)$ for the new metric value and $p(t)$ for the ranked position from step 2.

Table 1 presents the domestic risk factors rating from [15]. The final ranks of the risks, same with their ranked positions, are in the first column.

Table 1. Domestic risk factors rating in [15]

Final rank	Risk factor	metric values
1	Original set of requirements is miscommunicated	9.4
2	Lack of communication	8.5
3	Poor change controls	8.5
4	Lack of top management support	8.4
5	Lack of required technical know-how by vendor	8.3
6	Lack of vendor commitment	8.3
7	Failure to manage end-user expectations	7.7
8	Lack of project management know-how by client	7.7
9	Inadequate user involvement	7.6
10	Inadequate staffing by vendor	7.6
11	Vendor viability	7.3
12	High turnover of vendor employees	7.3
13	Failure to consider all costs	7.1
14	Differences in ongoing support and maintenance	7.1
15	Differences in development methodology/processes	7.0
16	Difficulties with integration	6.6
17	Lack of business know-how by vendor	6.4
18	Lack of knowledge of new technology	6.3
19	Conflicts between user groups	6.1
20	Negative impact on employee morale	5.9

3 RHD-2

3.1 Ranking with Metric Values of Both Steps

The RHD-2 method is also divided into two steps. However, the ranked results of the RHD-2 method are based on values from both steps, not just from a single step.

Let $q1$ be the relation value of the metric values from both steps:

$$q1 = f(s) \cdot f(t) \tag{1}$$

where "·" stands for a kind of compositional operation.

A concrete formation to (1) is

$$q1 = \alpha \times f(s) + \beta \times f(t) \tag{2}$$

where α and β are coefficients, $\alpha \in [0, 1]$, $\beta \in [0, 1]$, and

$$\alpha + \beta = 1 \, .$$

Given $\alpha = 1$ and $\beta = 0$, we get

$$q1 = f(s) \, .$$

This equation means that, we can sometimes get ranking result directly from historical data. This is acceptable in agile development, little development teams, or teams who do not allot enough specialists for risk management.

Given $\alpha = 0$ and $\beta = 1$, we get

$$q1 = f(t) \, .$$

This equation can be applied where metric values from the second step are trustable. The results are trustable when specialists presents them aiming at a practical and familiar project in the second step. The RHD-1 conforms to the model of $q1 = f(t)$.

Existing methods attach importance to the results from step 2. Actually, there are drives for us to attach importance to the results from step 1. Most of the time, results from step 1 are from authoritative historical data. It is a pity that usable values of the $f(s)$ and $p(s)$ from historical data are ignored in risk ranking.

In step 2, the $f(t)$ is always decided by specialists. There is a question: are the results by these specialists more reliable than the results from historical data? We introduce an empirical coefficient μ to reflect the relation of current data with historical data. The coefficient reflects the reliability relation of the data from both steps. The historical data from step 1 is more reliable when $\mu > 1$. The values from step 2 are more reliable when $\mu < 1$. The coefficients in (2) are then:

$$\alpha = \mu / (\mu+1). \tag{3}$$

$$\beta = 1 / (\mu+1). \tag{4}$$

The results of step 2 are equally reliable to those of step 1 when $\mu = 1$, or $\alpha = \beta = 0.5$. At this time, we get

$$q1 = (f(s) + f(t)) / 2. \tag{5}$$

3.2 Ranking with Positions of Both Steps

The second ranking method resorts to the $p(s)$ and $p(t)$. The equation is

$$q2 = p(s) \cdot p(t)$$

where $q2$ is a positional metric to rank risks, and "·" stands for a kind of compositional operation.

A concrete form similar to (2) is

$$q2 = \alpha \times p(s) + \beta \times p(t). \tag{6}$$

Given $\alpha = 0.5$ and $\beta = 0.5$, then

$$q2 = (p(s) + p(t))/2. \tag{7}$$

3.3 A Property

Suppose ranking positions start from 1, or the $p(s)$ and $p(t)$ are greater than 0. The $q1$ and $q2$ are also greater than 0.

The empirical coefficient μ embodies which of the $p(s)$ and $p(t)$ has greater effect upon the $q2$. However, the $p(s)$ and $p(t)$ do not necessarily have the same effect upon the $q2$ when μ equals to 1.

Theorem 1. *Define deviation as the absolute value difference of the $p(s)$ or $p(t)$. Given $\mu=1$, the $p(s)$ or $p(t)$ that has greater deviation has greater effect upon the $q2$.*

Let i and j stand for two different risks, deviation of $p(s)$ is

$$|p(s_i) - p(s_j)|$$

and deviation of $p(t)$ is

$$|p(t_i) - p(t_j)|$$

4 An Example

Software risks in this section are initiated with the historical data shown in table 1. The values of *f(s)* and *p(s)* are in the third column and the first column of table 1, and also in the second column and first column of table 2 respectively.

The *f(t)* values in the third column of table 2 are according to the scale: 10 = very important, 7 = important, 4 = slightly important, 1 = unimportant. It is time-consuming for a project to conduct a complex Delphi process like [15] when assessing software risk. We think that it will be rewarding sometimes for a company to get the *f(s)* or *p(s)* results from former practices and get a quick answer to the *f(t)* values by authoritative means.

The *q1* results by (5) are shown in the fourth column of table 2.

Table 2. Risk ranking

risk ID or p(s)	*f(s)*	*f(t)*	*q1=(f(s)+f(t))/2*	*f(t)*	*p(t)*	*q2= (p(s)+p(t))/2*	*ranked by q1*	*ranked by q2*
1	9.4	10	9.7	10	*1*	*1*	*1*	*1*
2	8.5	10	9.25	10	2	2	2	2
3	8.5	4	6.25	4	*7*	*5*	*7*	*3*
4	8.4	7	7.7	7	*13*	*8.5*	*13*	*7*
5	8.3	4	6.15	4	*15*	*10*	*15*	*8*
6	8.3	7	7.65	7	*20*	*18*	*20*	*4*
7	7.7	10	8.85	10	*4*	*5.5*	*4*	*9*
8	7.7	1	4.35	1	*6*	*7*	*6*	*15*
9	7.6	7	7.3	7	*9*	*9*	*9*	*5*
10	7.3	4	5.65	4	*11*	*10.5*	*11*	*10*
11	7.3	7	7.15	7	*16*	*13.5*	*16*	*16*
12	7.3	1	4.15	1	*17*	*14.5*	*17*	*11*
13	7.1	10	8.55	10	*18*	*15.5*	*18*	*17*
14	7.1	4	5.55	4	*19*	*16.5*	*19*	*19*
15	7.0	10	8.5	10	*3*	*9*	*3*	*12*
16	6.6	7	6.8	7	*5*	*10.5*	*5*	*13*
17	6.4	7	6.7	7	*10*	*13.5*	*10*	*18*
18	6.3	7	6.65	7	*14*	*16*	*14*	*20*
19	6.1	7	6.55	7	*8*	*13.5*	*8*	*14*
20	5.9	10	7.95	10	*12*	*16*	*12*	*6*

The ranked orders of the *p(t)* based on the *f(t)* values are in the third column of table 2. They differ from the orders of the *p(s)* values in the first column, which are ranked by the *f(s)* values. The ranked positions by *q1* in the eighth column are the same as those *f(t)* values in the sixth column. This is because the *f(s)* values from [15] differ less but the *f(t)* values given in the project differ more. The ranked results in the ninth column by *q2* differ from the results ranked by the *p(t)* and the *p(s)*. The values in the seventh column show that different risks may have a same *q2* value sometimes.

5 Conclusion

This study presents the RHD-2 ranking method integrating values from both steps with adjustable empirical coefficients. The integrated values may be metric values or ranked positions from both steps. The coefficients present weight to metric values from both steps. The idea of the RHD-2 uses historical data with weight, thus gives decision-makers choices and acts as a reference for ranking activities.

Acknowledgments. This research is supported by the study of the National Natural Science Foundation of China under Grant No. 61001069, the multi-online web game project of the Chengdu University of Technology under Grant No. HD0067, the fund of the CDUT Online of Chengdu University of Technology under Grant No. 2009-640 and the fund of Education Ministry of Sichuan Province under Grant No. 10ZA111.

References

1. Kontio, J.: The riskit method for software risk management, version 1.00. Computer Science Technical Reports. University of Maryland, College Park, MD, USA (1997)
2. Haimes, Y.Y.: Risk modeling, assessment, and management, 3rd edn. Wiley publishing (2009)
3. Boehm, B.W.: Software risk management: principles and practices. IEEE Software 8(1), 32–41 (1991)
4. Kontio, J., Getto, G., Landes, D.: Experiences in improving risk management processes using the concepts of the Riskit method. In: Sixth International Symposium on the Foundations of Software Engineering (FSE-6), vol. 23(6), pp. 163–174 (1998)
5. Meyer, P.H., Visser, J.K.: Improving project schedule estimates using historical data and simulation. South African Journal of Industrial Engineering 17(1), 27–37 (2006)
6. French, S.: Decision Theory: an Introduction to the Mathematics of Rationality. Ellis Horwood, Chichester (1986)
7. Benito, G., Lang, M., et al.: Use of systematic, palaeoflood and historical data for the improvement of flood risk estimation. Natural Hazards 31(3), 623–643 (2004)
8. Wanhainen, A., Bergqvist, D., et al.: Risk factors associated with abdominal aortic aneurysm: a population-based study with historic and current data. Journal of Vascular Surgery 41(3), 390–396 (2005)
9. Friedrichv, B., Andreas, P.: How historical data can improve current geo-risk assessment. Zeitschrift fur Geomorphologie 51(1), 31–43 (2007)
10. Liu, X.-P., Zhang, J.-Q., Fan, J.-B.: Historical data-based risk assessment of fire in grassland of northern China. Journal of Natural Disasters 16(1), 61–65 (2007)
11. Song, J., Li, J., Wu, D.: Modeling the Key Risk Factors to Project Success: A SEM Correlation Analysis. In: Shi, Y., Wang, S., Peng, Y., Li, J., Zeng, Y. (eds.) MCDM 2009. CCIS, vol. 35, pp. 544–551. Springer, Heidelberg (2009)
12. Nah, F.F.-H., Lau, J.L.-S.: Critical factors for successful implementation of enterprise system. Implementation of Enterprise Systems 7(3), 285–296 (2001)
13. Ghosh, S., Jintanapakanont, J.: Identifying assessing the critical risk factors in an underground rail project in Thailand: a factor analysis approach. International Journal of Project Management 22, 633–643 (2004)
14. Reed, A.H., Knight, L.V.: Effect of a virtual project team environment on communication-related project risk. International Journal of Project Management 28, 422–427 (2010)
15. Nakatsu, R.T., Iacovou, C.L.: A comparative study of important risk factors involved in offshore and domestic outsourcing of software development projects: A two-panel Delphi study. Information & Management 46, 57–68 (2009)

The Application of the Computer Simulation in the Stability Analysis of Roadway

Fengzhen Liu[1] and Weiguo Qiao[1,2]

[1] Shandong University of Science and Technology, Qingdao 266590, Shandong, China
[2] Laboratory of Civil Engineering and Disaster Prevention and Mitigation, China
lfz828@126.com

Abstract. The higher tectonic stress in deep mine brought more difficult for the roadway layout, and the selection supporting form and parameter. In order to make sure about the influencing of the tectonic stress on the roadway stability, According to the intersection angle change between the tectonic stress and the roadway axis, We established different numerical models by using FLAC3D being based on the actual geological conditions on DongE Coal mining, The results indicate that the deformation is the largest when the roadway axial is vertical to the tectonic stress direction, and the tectonic stress also determines the failure pattern of roadway.

Keywords: computer simulation, tectonic stress, stability, numerical model.

1 Introduction

FLAC is based on explicit difference method to solve motion equation and dynamic equation. After ascertained the geometric shape of the research area, this area will be discretized at first, and divided into several grid units, each grid unit through nodes between connected, after applied load on a node, the stress of the node and the external force change and timestep enables obtain the unbalance force of node using the virtual work principle, then the unbalanced force will be applied on a node again for the next iterative process, until the unbalance force is small enough or node displacement

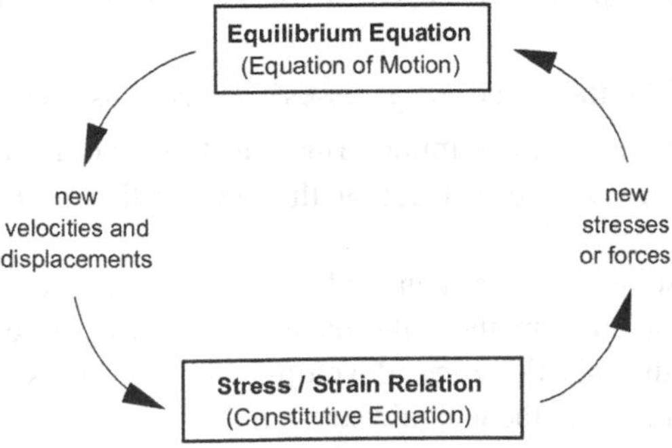

Fig. 1. Basic explicit calculation cycle [1], This shows the general calculation sequence in FLAC

D. Jin and S. Lin (Eds.): Advances in CSIE, Vol. 2, AISC 169, pp. 399–404.
springerlink.com

tend to the equilibrium. The general calculation sequence is illustrated in Figure 1. This procedure first invokes the equations of motion to derive new velocities and displacements from stresses and forces. Then, strain rates are derived from velocities, and new stresses from strain rates. Every cycle around the loop is taken one timestep; each box in Figure 1 updates all of its grid variables from known values that remain fixed while control is within the box.

2 Theoretical Analysis of Tectonic Stress

Under the influence of all kinds of the earth's crust tectonic movement force, the stress produced in crust is called tectonic stress, the measured data show that, due to the existence of tectonic stress, the level stress is more than gravity stress. Level of stress is an important factor causing roof caving, floor heave, the two ribs infolding. In recent years, with the increase of mining depth, tectonic stress also is very complex, after excavating roadway , due to the influence of the tectonic stress it becomes quite difficult to support and maintain roadway, it seriously affected the deep mine mining. Because the tectonic stress has obvious directional, therefore, it is important that for the laying direction of roadway to study the effect of the tectonic stress on the stability of surrounding rock in deep.

Calculation the elastic stress field and displacement field of roadway surrounding, it can be regarded as three dimensional space problem to using the theory of elastic mechanics, it has certain limitations simplified as plane strain problem. in the course of calculation, The plane strain the problem was in the plane strain based on the analysis of the problems, and superposition the shear stress and an unidirection compression stress . Considering the stress components in the axis of roadway, the stress periphery roadway is summarized in the equation below [2]:

$$\sigma_\theta = (\sigma_3 \cos^2 \alpha \cdot + \sigma_1 \sin^2 \alpha)(1 - 2\cos 2\theta) + \sigma_2 (1 + 2\cos 2\theta) \tag{1}$$

$$\tau_{\theta z} = \sin 2\alpha(\sigma_1 - \sigma_2)\sin\theta \tag{2}$$

$$\sigma_z = \sin^2 \alpha \cdot (\sigma_3 - 2\mu\sigma_1 \sin^2 \alpha) + \cos^2 \alpha(\sigma_1 - 2\mu\sigma_3 \cos 2\theta) - 2\sigma_2 \cos 2\theta \tag{3}$$

Where σ_1, σ_2, σ_3 is the principal stresses, μ is Poisson's ratio, u , α is the intersection angle between the maximum principal stress and the axial of roadway, θ is the intersection angle between a dot of the surrounding rock and the horizontal plane

In order to make sense of the influence of the tectonic stress on the laying roadway, by the numerical simulation method, the paper analyzed the rule of the surrounding rock deformation, the plastic zone development characteristics, and the stress distribution characteristics in the tectonic stress field.

3 Analysis of Roadway Stability

3.1 Establishment of Model

The simulation roadway is layout along the coal floor, coal floor is sandstone, integrity is better, the uniaxial compressive strength is 60 ~ 85*MPa*. The buried depth of roadway is 820 m; the roadway by vertical stress $\sigma_z = \gamma H = 21.3MPa$, the largest horizontal principal stress is $\sigma_h = 31.9MPa$ by the actual measurement.

Basic assumptions

(1) In the course of the simulation, the deep roadway is considered as the problem of space, physical model using elastic-plastic model, the failure criterion using Mohr-Coulomb model. Assuming the rock is homogeneous, isotropic, and don't consider the influence of fractures, weak layers on the strength of surrounding rock

(2) Considering the boundary effect, Model should have enough size, the stress changes can be negligible far from the excavation site by St.Venant principle [3], according to the simulation and the theory of the practical mining, and take 10 to 12 times to the excavation scope, the roadway is laying in the central part of model.

(3) Boundary conditions: the horizontal displacement is limited on the left and right boundary ($u_x = 0$), the horizontal and the vertical displacement are limited on bottom boundary ($u_x = 0$, $u_y = 0$), applying the stress equivalent to the weight stress of overlying rock on the upper boundary [4].

Simulation model

Considering the calculation speed and the accuracy of the simulation results, the meshing is properly dense in the roadway of roof and floor; the rest is sparse, on the basis of the generation of finite difference grid. Model was divided 5800 units, a total of 7224 nodes.

In the process of simulation, the problem of model initialization is difficult because the changing angle between the applying direction of the principal stress and the roadway axial, during the establishing model, the roadway axial can be rotated a certain angle, aiming at this situation, four representative angle are chosen, respectively 30°、45°、60°、90° , taking model size 40 m × 40 m × 20 m (long × high × width).

3.2 Simulation Results

(1) The deformation of surrounding rock

Fig 3 and Fig 4 shows the displacement of the roof and floor of roadway, it seems that the displacement is increasing with the steps at first, and the growth is bigger, then it tends to a limit, the roadway comes to the stability, Compared to the displacement, When the angle between the roadway axial and tectonic stress direction is changing from 30°、45°、60° to 90°, the ribs displacement is increasing with the increasing angle.

From fig 4, it shows that, excavated roadway in the same rock, due to the influence of tectonic stress, the deformation of floor is relatively large; the floor heave has a large proportion in the roadway deformation. When the angle increases to 90°, the deformation of roof and floor come up to the most. It shown that the angle between the roadway axial and tectonic stress direction has a larger influence on deformation of the surrounding rock, when the roadway axial and the tectonic stress direction is approximate parallel, the deformation is minimum amount; When the roadway axial and the tectonic stress direction is vertical ,the deformation is the largest.

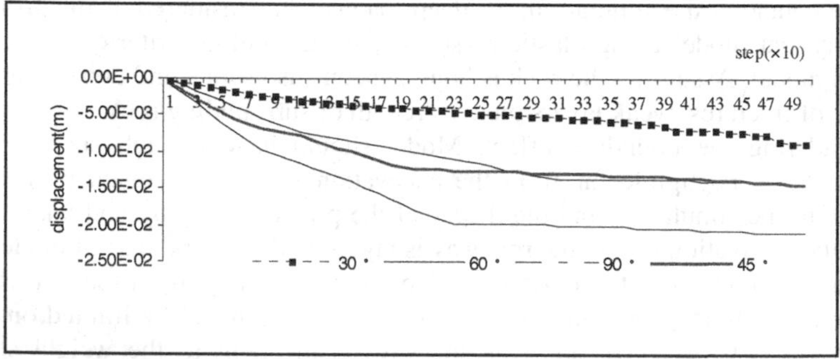

Fig. 3. Displacement of in the roof of roadway. The curves correspond to respectively the angle between the roadway axial and tectonic stress direction 30° ,45 °,60 °,90 °

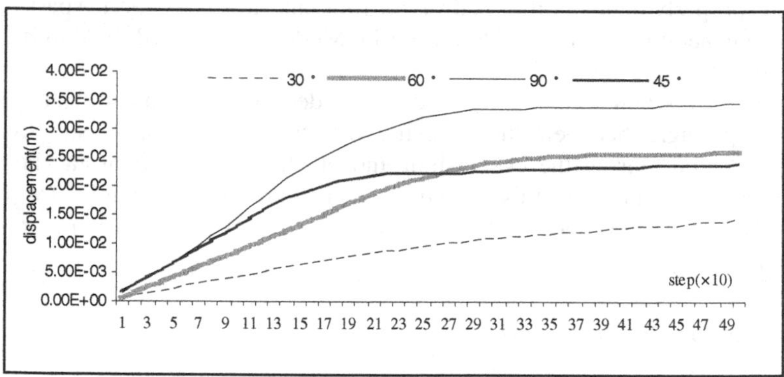

Fig. 4. Displacement of in the floor of roadway. The curves correspond to respectively the displacement of the angle between the roadway axial and tectonic stress direction is *30 ° ,45 ° ,60 ° ,90 °*

(2) The plastic zone of surrounding rock

Fig 5 is the range of plastic zone, when the angle is 0 °and 90°, it shows that the range of deformation of surrounding rock is increasing with the growth of the angle between the roadway axis and the direction of maximum horizontal principal stress, especially in the ribs. When the angle is 0°, the position of the plastic damage area of roadway appears in the two ribs of roadway; when the angle increases, the position transfer to the central of the floor and roof.

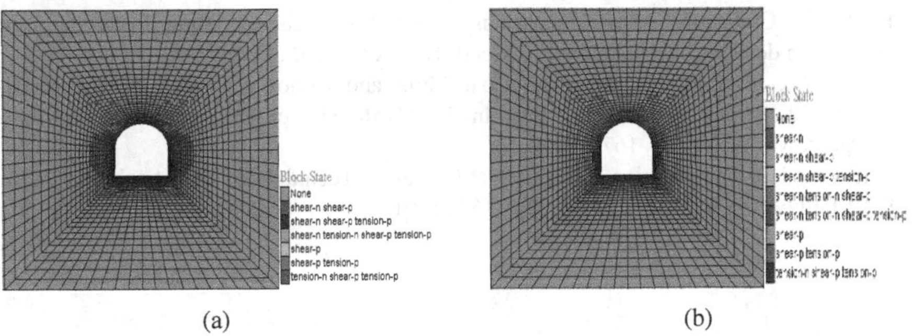

Fig. 5. The plastic zone distribution state, (a) the angle is *0°*; (b) the angle is *90°*

4 Conclusions

(1) It can forecast the roadway deformation, and provide a reference for the choice of supporting scheme and parameters to use computer simulation in the stability analysis of roadway; At the same time also save the planning cost, it showed that the computer simulation used in coal mining has the broad prospect;

(2) The tectonic stress is one of the main factors of influencing the stability of roadway, It shown that the angle between the roadway axial and the tectonic stress direction has a larger influence on deformation of the surrounding rock, when the roadway axial and the tectonic stress direction is approximate parallel, the deformation is minimum amount; When the roadway axial and the tectonic stress direction is vertical, the deformation is the largest;

(3) With the angle between the axial and the maximum horizontal principal stress direction increasing, the deformation range of the roof and the floor increase, the deformation transfers to the deep of surrounding rock;

(4) Research shows that it has an important signification for the reasonable layout of roadways to study the effect of the tectonic stress on the stability of roadway.

Acknowledgments. This study has been funded by the National Natural Science Foundation of China (No. 51009086), the National Program on Key Basic Research Project (No. 010CB226805), the Research Award Fund for Outstanding Middle-aged and Young Scientist of Shandong Province China (No.BS2010HZ015) and the Program for Changjiang Scholars and Innovative Research Team in University (NO. IRT0843).

References

[1] Itasca consulting group inc.: FLAC3D (fast lagrangian analysis of continua in dimensions) version 2.0 user's manual, USA (1997)
[2] Zhou, X., Li, X.: The theoretical analysis of the influence of the Tectonic stress on layout roadway. J. Coal Engineering 10, 17–19 (1998)
[3] Hoek, E., Grabinsky, M.W., Fiederichs, M.S.: Numerical modeling for underground excavation design. J. Trans. Inst Min. Metal. 100, A22–A30 (1991)
[4] Hanna, K., Haramy, K.: Stress on Coal Mine and Conover, (1986); Effect of High Horizontal Entry Intersection Stability. In: 5th Conference on Ground Control in Mining, Morgantown. W. Va, pp. 167–182
[5] Kripakov, N.P.: Application of Numerical Modeling Techniques to the Analysis of Cutter Roof Failure. US Bureau of Mines 28, 4–52 (1991)

Research on Electric Differential for Steering Electric Vehicles

Zitong Wang, Wei Yao, and Wei Zhang

College of Electrical Engineering, Zhejiang University, Hangzhou, China
21010181@zju.edu.cn

Abstract. Without traditional mechanical differentials, electric vehicles need electric differentials to avoid slipping. This paper proposed a novel electric differential control strategy to solve the problem. According to the given speed and road conditions, target slip ratios can be calculated. Based on the closed-loop control, this strategy regulates torques of the driving wheels to follow the required slip ratios strictly. The simulation results indicate that it could realize differential for the two driving wheels. Meanwhile, compared with traditional mechanical differentials, the slip ratios of driving wheels and the risk of slipping are minimized.

Keywords: Electric vehicles, electric differential, slip ratio.

1 Introduction

In traditional cars, driving wheels are fixed on the same shaft. When the car steers, the longitudinal speeds of the driving wheels are in proportion to their steering radiuses. There is a differential in the driving shaft to provide different rotation speeds to the driving wheels, in order to avoid the danger of slipping. In electric vehicles, in-wheel motors are used to drive the wheels, and there is no mechanical differential or driving shaft. In other words, electric vehicles need electric differentials to provide different rotation speeds to the driving wheels as the mechanical differentials do. Furthermore, traditional mechanical differentials do not change the torque distribution to the driving wheels; it may cause steering trouble on some circumstances. This issue proposes a control strategy to accomplish the function of electric differential and overcome these shortcomings.

2 The Steering Model of Electric Vehicles

Figure1 shows the model of an electric vehicle with front wheels steering and rear wheels driving. The overall width is W, and wheel space is L(1)+L(2)(L(1) represents the distance between the center of gravity and the front wheel shaft, L(2) represents the distance between the center of gravity and the rear wheel shaft). To simplify the analysis, we do not take the motions of rolling and pitching into consideration, since they have little influence on the process of steering.

D. Jin and S. Lin (Eds.): Advances in CSIE, Vol. 2, AISC 169, pp. 405–411.
springerlink.com

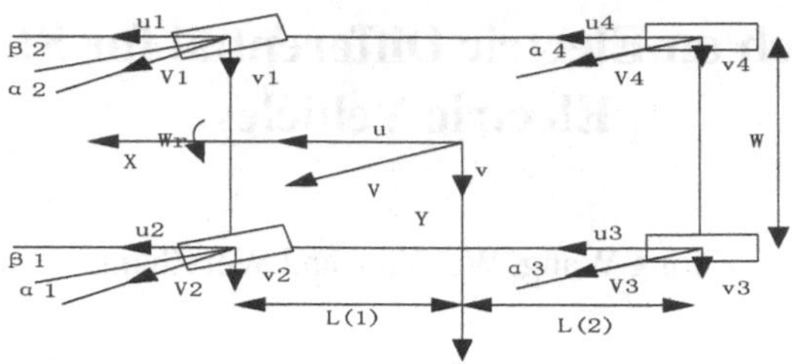

Fig. 1. The speed model of the steering of EV

Define the X-axis as the direction of the longitudinal motion of the EV, and the Y-axis of the lateral motion. The motions of a steering EV include: motion on longitudinal direction with a speed of u; motion on lateral direction with a speed of v; yaw angle velocity: Wr. The resultant speed of EV is $V=\sqrt{u^2+v^2}$, and the speeds of four tyres are: $\begin{cases} u1 = u - Wr * W/2 \\ u2 = u + Wr * W/2 \\ u3 = u - Wr * W/2 \\ u4 = u + Wr * W/2 \end{cases}$, $\begin{cases} v1 = v + Wr * L(1) \\ v2 = v + Wr * L(1) \\ v3 = v - Wr * L(2) \\ v4 = v - Wr * L(2) \end{cases}$;ui is the speed on the X-axis, and vi is the speed on the Y-axis of tyre i.[1]

The steering angles of front wheels are β1=β2=β, and for the rear wheels β3=β4=0 There is a wheel slip angle α for each tyre, because of the lateral force between ground and the tyre. [2] Angle α can be calculated by the following equation:

$$\alpha=\arctan\frac{u}{v}-\beta\,. \tag{1}$$

The driving wheels receive circumferential forces Fi (i=3, 4) caused by the driving torque, while the steering wheels only receive forces of rolling friction fi (i=1, 2) on the plane of tyre. The longitudinal slip ratio is defined as:

$$S=1-\frac{u_w}{\omega R}\,. \tag{2}$$

u_w is the speed represented on the plane of tyre, ω is the rotation angular speed of the tyre, and R is the radius of the tyre. For the steering wheels, we can conclude $u_w=u*\cos\beta+v*\sin\beta$, and for the driving wheels, $u_w=u$.

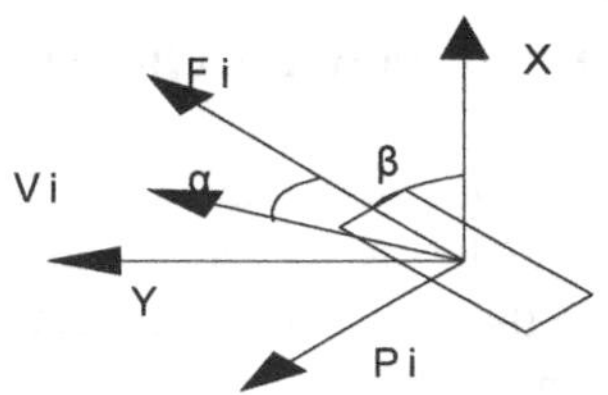

Fig. 2. The relationship between forces, angles and speeds of a tyre

The circumferential force of a driving wheel is F=μ*N. μ is the coefficient of adhesion, that is a function of slip ratio S. The value of the coefficient of adhesion depends on the condition of the grounds. N is the vertical load of the tyre. Define lateral slip ratio as tanα. The lateral force can be calculated by $P=\mu_v*N$. μ_v is lateral adhesion coefficient, a function of lateral slip ratio. For a driving wheel, the torque equation is:

$$Te\text{-}Mr\text{-}F*R=\frac{d\omega}{dt} . \tag{3}$$

Te is the electromagnetic torque from motor; Mr is the rolling friction torque, F is circumferential force, J is inertia moment of the tyre, and ω is rotational angular speed. For a steering wheel, the force of rolling friction is f=Kr*N. Kr is the coefficient of rolling friction, and it ranges from 0.15 to 0.005, according to the road condition. [3]

The angle between f and X-axis is the steering angle β, as well as the angle between P and Y-axis. So we can deduce the following equations:

$$Fx1=-f1*\cos\beta1-P1*\sin\beta1 . \tag{4}$$

$$Fx2=-f2*\cos\beta2-P1*\sin\beta2 . \tag{5}$$

$$Fy1=P1*\cos\beta1-f1*\sin\beta1 . \tag{6}$$

$$Fy2=P2*\cos\beta2-f1*\sin\beta2 . \tag{7}$$

Fx and Fy of a steering wheel are the resultant force represented on X-axis and on Y-axis. Circumferential forces F3 and F4 of driving wheels are on X-axis and their lateral forces P3 and P4 are on Y-axis.

When the EV is keeping rectilinear motion, it receives circumferential forces from driving wheels, forces of rolling friction from steering wheels, and longitudinal wind resistance. Longitudinal wind resistance Fa is calculated by:

$$Fa=Ka*Sa*V^2 . \tag{8}$$

Ka represents wind resistance coefficient, its value are from 0.3 to 0.6, and Sa is front face area. When the EV begins to steer, all four tyres get lateral forces, and EV receives centrifugal force $Fr=M*V^2/R$. Centrifugal force represents M*V*Wr on X-axis, and M*U*Wr on Y-axis. [4]

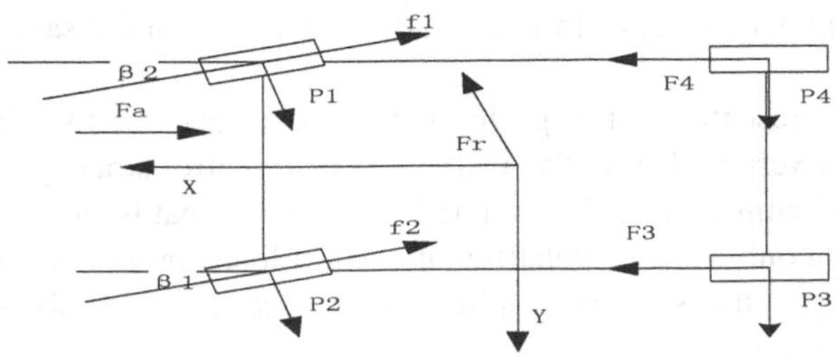

Fig. 3. The force model of the steering EV

When EV is steering, the forces of tyres trigger a torque of X-axis, which changes vertical loads of four tyres. The value of the torque is Fr*h, h is height of the center of gravity. The static vertical load of four tyres are $N_1^0= N_2^0=\frac{Mg * L(2)}{2 * (L(1) + L(2))}$, $N_3^0=$ $N_4^0=\frac{Mg * L(1)}{2 * (L(1) + L(2))}$. In the process of steering, N1= N_1^0+ΔN1, N2= N_2^0-ΔN1; N3= N_3^0-ΔN2, N4= N_4^0+ΔN2. ΔN1+ΔN2=Fr*h/W. Usually, we can use the following equation for simplicity: $\Delta N1=\Delta N2=\frac{Fr * h}{2 * W}$. [5]

To summarize the modeling of steering EV, we could deduce the dynamics equations:

$$Fx1+Fx2+F3+F4+M*v*Wr=M*\dot{u} . \quad (9)$$

$$Fy1+Fy2+P3+P4-M*u*Wr=M*\dot{v} . \quad (10)$$

$$(Fy1+Fy2)*L(1)+\frac{(Fx2 + F4 - Fx1 - F2) * W}{2} -(P3+P4)*L(2)=Iz*\frac{dWr}{dt} . \quad (11)$$

Iz is the inertia moment of the whole EV. With three equations above, we are able to establish the dynamics model of steering EV in Matlab Simulink to simulate and analyze.

3 Electric Differential Strategy Based on Closed Loop Control of Slip Ratios

If we do not take the condition of steering braking into consideration, we should keep the longitudinal coefficient of adhesion as low as possible to lower the risk of slipping and remain a high value of lateral adhesion coefficient in order to provide enough lateral forces, since the lower longitudinal adhesion coefficient the higher lateral adhesion coefficient with the same value of lateral slip ratio. [6] In the book [5], there is a conclusion that cars could get the highest circumferential resultant force if the longitudinal slip ratios of the four tyres get the same value. It is also applicative for the lateral resultant force and lateral slip ratios. We can also deduce that we could minimize the maximum value of four slip ratios if they are in the same value with a given resultant force.

The thesis [7] presented a strategy to control the torques of two driving wheels based on different vertical loads. The method of that control strategy is also used to minimize the maximum value of four slip ratios. However that is an open loop control of torques, and can only take the volatility of vertical loads into consideration. As the author acknowledged, that strategy simplifies many factors and is only applied for the condition of low speed.

This issue presents a new method of electric differential control strategy. It is a double closed loop control strategy. Compared with former method, it sets the longitudinal slip ratio as the controlled variable, rather than the torques, so it is more straightforward.

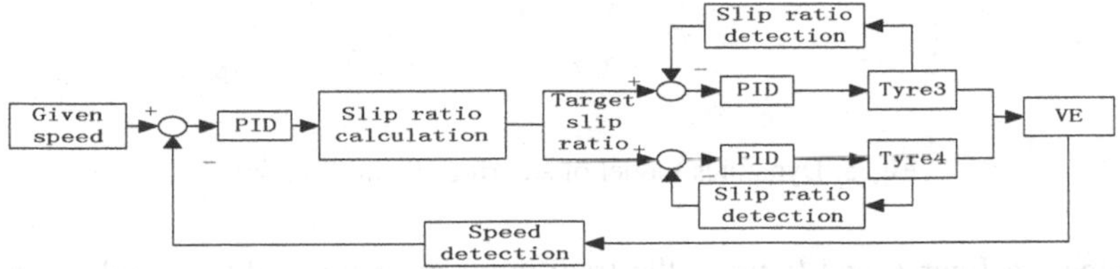

Fig. 4. Block diagram of the closed loop electric differential control strategy

As shown in figure4, in the process of steering, we should first insure the longitudinal speed constant u by controlling the total sum of the torques, and considering the volatility of resultant force on longitudinal X-axis. It is a classical feedback control loop with a PID regulator. The given speed is calculated by the accelerator pedal, and the actual speed is detected by a car speed sensor. The regulator outputs the total sum of torques which determines the longitudinal speed. Then the external loop provides total sum of the torques for calculation module to calculate two target slip ratios. The given total sum of torques divided by the sum of rear wheels vertical loads is the target slip ratio for both driving wheels.

$$S3^*=S4^*=\frac{Ttotal}{Ku*Rk*(N3+N4)} \quad (12)$$

N3 and N4 have been calculated before, $N3+N4=\frac{Mg*L(1)}{L(1)+L(2)}$. Ku is the scale factor of the adhesion coefficient.

Then with the help of the internal control loop, the actual slip ratios are managed to follow the target slip ratios. It is also a classical feedback control loop, receiving the target slip ratio S* from the calculation module and detected slip ratio S from the system. A PID regulator is put into use, for better control performance. Since the EV is a large inertia element, Kp in the PID regulator should be large enough to meet the demand of rapidity. In the control applications, we need vehicle speed sensor to calculate slip ratios.

4 Simulations and Analyses

Establish the dynamics model in Matlab Simulink as figure5, the system variables are defined as follow: vehicle weight M=800Kg, L (1) =L (2) =0.8m, vehicle width W=1.2m, radius of tyre Rk=0.32m, inertia moment of tyre $J=13.7kgm^2$, and the vehicle inertia moment $Iz=600kgm^2$. Simulation begins with start longitudinal speed u=8m/s, and the steering angle is 2.8 degrees. Simulation results are shown below.

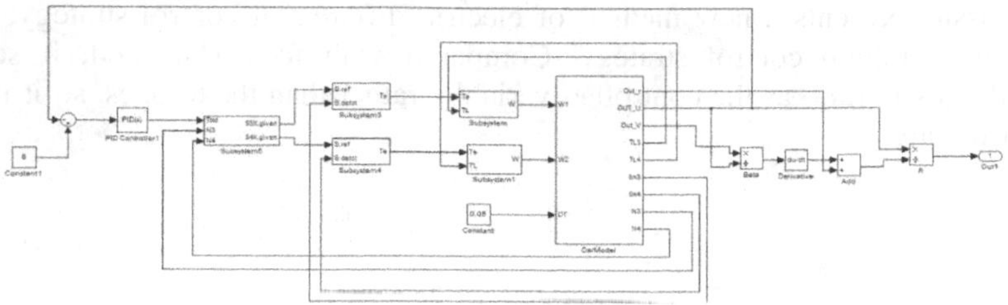

Fig. 5. Dynamics model of steering EV in Simulink

Comparing figure6 and figure7, the traditional mechanical differential distributes torque equally, it brings out different slip ratios between two driving wheels, and the inner wheel has a larger one, which means higher risk of slipping. With the new control method the two slip ratios get very close, much smaller than the slip ratio of inner wheel with traditional differential. It reduces the possibility of slipping and can bring larger lateral force for the inner driving wheel.

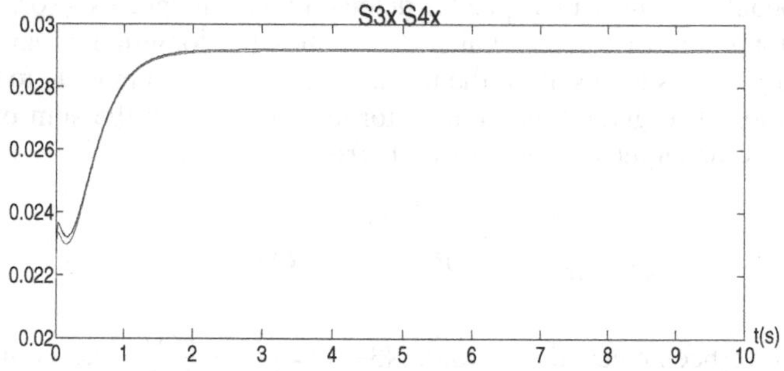

Fig. 6. Slip ratios of driving wheels with closed-loop control of slip ratios

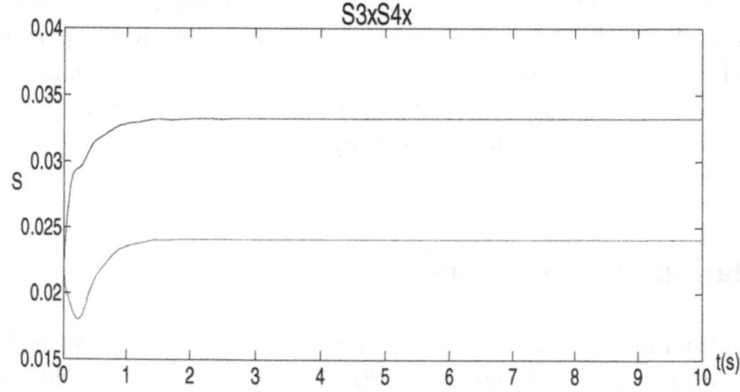

Fig. 7. Slip ratios of driving wheels with traditional distribution of torques

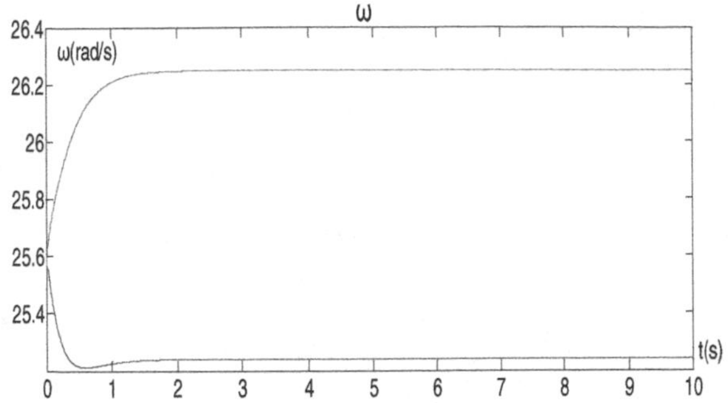

Fig. 8. Different rotation angular speeds of driving wheels

From figure 8, we can see that the electric differential achieves differential of the rotation angular speeds for two driving wheels.

5 Conclusion

This issue presents a new electric differential control strategy for steering EV. It contains two closed control loop, one for the longitudinal speed, and the other for slip ratios. Using two PID regulators, it manages to keep the slip ratios of driving wheels on a target value when the electric vehicle is steering. From the result of simulation, we can see that it realizes differential for two driving wheels, and reduces the risk of slipping for the inner driving wheel as well.

References

1. Ge, Y., Ni, G.: Novel Electric Differential Control Scheme for Electric Vehicles. Journal of Zhejiang University 39(12), 1973–1979 (2005)
2. Jin, L., Wang, Q., Song, C.: Dynamic Simulation Model and Experimental Validation for Vehicle with Motorized Wheels. Journal of Jilin University 37(4), 745–750 (2007)
3. Li, C.: Study of The Driving System for Independent Drive EV. Zhejiang University Master Thesis, Hangzhou (2004)
4. Li, G.: Vehicle Antilock Brake, Theory and Application of Brake Control. Defense industry press, Beijing (2009)
5. Mitschke, M.: Dynamics of automotive. Mechnaical industry press, Beijing (1980)
6. Liu, G.: The Modularization Modeling of Vehicle and the Simulation Research on the Control Method of Anti-lock Braking Sysytem during the Steering Braking. Suzhou University Master Thesis, Suzhou (2007)
7. Ge, Y.: The Control System for In-wheel Driven Electric Vehicle. Hangzhou Zhejiang University Ph.D. Thesis (2004)

Research on Learning Resources Grid Information Retrieval Based on Metadata

Li Yang, Luo Zhong, and Bo Zhu

School of Computer Science and Technology
Wuhan University of Technology
Wuhan 430070, China
yangli_lc@yahoo.com.cn

Abstract. Huge, dynamic and heterogeneous resources are difficult to achieve full sharing because of lacking comprehensive and uniform description format and interaction norms. In this paper, the grid and metadata technology is adopted to solve this problem. First, the layered metadata model is proposed on the basis of hierarchical structure of LRG. All irregular resources are divided into five layers using metadata technology: resource metadata layer, information transmission metadata layer, virtual organization metadata layer, service metadata layer and user metadata layer. Second, referring to domestic and foreign experiences of metadata norms, the description schemes about resource metadata, service metadata and user metadata are presented. Finally, the middleware tool-Alchemi is adopted to simulate and build the LRG system. Retrieval of resource, service and user information in LRG is realized and the proposed metadata schemes are testified to be feasible, valid, practical and significant in theory and application.

Keywords: Learning Resource Grid (LRG), Metadata, Information Retrieval, Alchemi.

1 Introduction

With the high development of computer and network technology recent years, great, dynamic and heterogeneous learning resources on the internet are difficult to share fully and efficiently because of lacking comprehensive and uniform resource description format and interaction norms.

Grid [1,2] is the extent form of network, its goal is to achieve all resources connecting, sharing and collaborating among virtual organization and provide transparent and uniform access interfaces for resource users. Metadata [3,4] is basic information unit of data. It describes data with abstract structure and shows what users want to know, such as data connotation, data quality, data state and how to get them.

In this paper, LRG system is built based on grid technology, learning resources are described uniformly, metadata standard is established and heterogeneity of resources is shielded. Eventually, users can enjoy more convenient, safe, efficient and all-directional service.

D. Jin and S. Lin (Eds.): Advances in CSIE, Vol. 2, AISC 169, pp. 413–418.
springerlink.com

2 Layered Structure of LRG

The registered user in grid includes three categories: teacher, student and administrator. Teacher can be divided into ordinary and inspection two kinds further. When registered user login LRG system, system should carry on identity authentication and provide corresponding service [5].

As a result, the layered structure of LRG is designed as shown in Fig. 1.

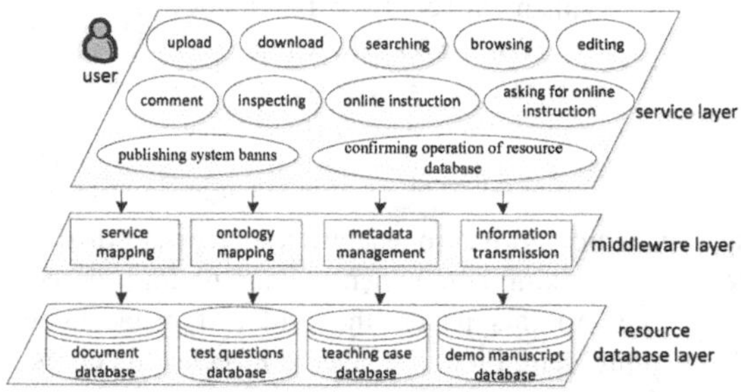

Fig. 1. Layered Structure of LRG

The bottom layer of LRG is resource database which stores a mass of learning resources and their description information. The physical resources can be mainly placed in document database, test questions database, teaching case database and demo manuscript database and so on. The middle layer of LRG is grid middleware which is the hub of whole grid and provides some services, such as ontology mapping and information transmission and so on. Service layer which is oriented to grid registered users directly and provides kinds of services is at the top layer of LRG.

3 Metadata Description Schemes of LRG

Learning resources needed to be standardized by using metadata norms are divided into five layers based on structure of LRG, as shown in Fig. 2.

user metadata	service layer
service metadata of LRG	
virtual organization metadata	organization layer
information transmission metadata	basis layer
resource metadata	

Fig. 2. Layered Model of LRG's Metadata

In this paper, we only describe the metadata on resource layer, service layer and user layer. The metadata description on virtual organization layer and information transmission layer is not discussed.

On the basis of domestic and foreign learning resource metadata norms [6] and referring to the experiences in "education resource construction technology"[7], metadata element can be classified into three kinds: key metadata set, supplemental metadata set and characteristic metadata set.

3.1 Description Scheme of Resource Metadata Layer

The key metadata set of resource information in LRG consists of identifier, title, language, description, creator, contributor, publisher, data, format, type. The specifications are shown in Table 1.

Table 1. Key Metadata Set of Resource Information

ID	Name	Specification
1	Identifier	unique ids of resources in LRG
2	Title	name of resources
3	Language	language which the resources use
4	Description	content which the resources contain
5	Creator	the id of teacher who uploaded the sharing resources
6	Contributor	inspection teacher's id or administrator's id
7	Publisher	name of units or individuals who published the resources
8	Date	the time to share the resources in grid
9	Format	the type of resources data storage in grid
10	Type	the type which the resources belong to in grid

The supplemental metadata set of resource information in LRG consists of lifecycle, version, size, application targets, primary user, cost, annotation and typical learning time.

The characteristic metadata set of resource information in LRG consists of text, picture, audio, video and courseware according to different type of storage.

3.2 Description Scheme of Service Metadata Layer

The metadata set of service information consists of service tag, service type, service name, service object, service target, service time, service duration. The specifications are shown in Table 2.

Table 2. Metadata Set of Service Information

ID	Name	Specification
1	ServiceTag	unique ids of services
2	ServiceType	the type of services
3	ServiceName	name of services
4	ServiceObject	ids of users who used the services
5	ServiceTarget	ids of resources which were operated when using services
6	ServiceTime	the time when services beginning
7	ServiceDuration	the length of time from the start of services until end

3.3 Description Scheme of User Metadata Layer

The key metadata set of user information consists of user tag, username, user type and email. The supplemental metadata set of user information consists of name, gender, nationality, hometown and age to describe the private information of user. Users can fill out optionally when registering and are not required to possess all this five attributes. The characteristic metadata set of student user is composed of grade and highest degree and this two attributes are necessary for student user. The characteristic metadata set of teacher user is composed of teaching age, teaching course, teaching grade, highest degree and professional title and this five attributes are necessary for teacher user.

4 Information Retrieval of LRG

4.1 Experiment Environment and Method

The LRG system is simulated by five hp workstations which are installed respectively with Microsoft.NET Framework and Microsoft SQL Server 2000. One is installed with Alchemi Owner (AO) and the others are installed Alchemi Executor (AE). Alchemi [8], free and open source software, is a grid computing framework based on WINDOWS. The information of resources, services and users is stored according to the metadata scheme of LRG in every Database Server. Every node in distributed database is regard as executor node in Process Design.

Firstly, user submits a search request to AO which runs information search client and submits search thread to Alchemi Manager (AM). AM schedules search thread from AE. AE executes the search thread and search the required information in local database. Secondly, the search results are fed back to AO from AE passing AM. At last, AO shows the results to user.

4.2 Information Retrieval about Resource, Service and User

Users can choose appropriate resource type, user type and retrieval method to search by using combo boxes. As shown in Fig. 4 (a), resource type which can be chosen includes text, picture, audio, video and courseware and retrieval method includes by title, identifier and publisher. As shown in Fig. 4 (b), retrieval method includes by service name and service tag. As shown in Fig. 4 (c), user type which can be chosen includes teacher, student and administrator and retrieval method includes by user tag and user name. After making appropriate choice, users input key words and click “Search” to find required information.

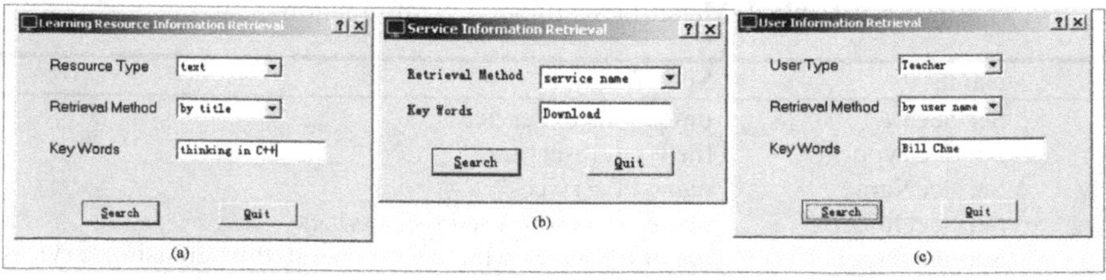

Fig. 4. Information Retrieval about Resource, Service and User

The results suitable for users would be shown in retrieval results window as shown in Fig. 5. Four key attributes like identifier, title, publisher and format are list in Fig. 5 (a). Four key attributes like service tag, service type, service name and service time are list in Fig. 5 (b). Four key attributes like user tag, user name, user type and email are list in Fig. 5 (c).

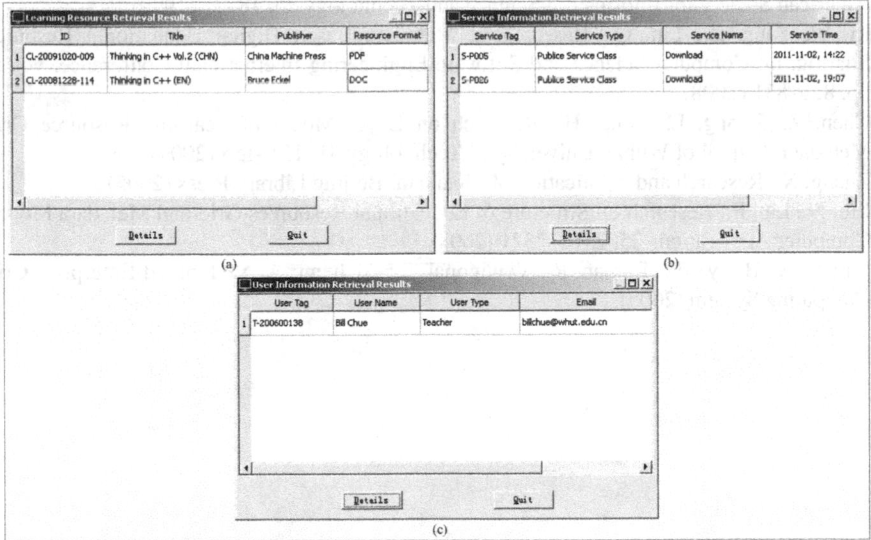

Fig. 5. Information Retrieval Results about Resource, Service and User

User can click “Details” for detailed instructions.

5 Conclusion

Aiming at condition of huge learning resources were built redundantly and not shared fully and efficiently, a layered LRG system was designed. Referring to domestic and foreign data description norms, all resource need to be regulated in LRG was divided hierarchically. Metadata description scheme was proposed by describing resource metadata, service metadata and user metadata in layered model. At last, the LRG system was simulated and built by using Alchemi. Distributed database was deployed according to the metadata description scheme. User can search the information of resource, service and user correctly and efficiently. It was confirmed that the proposed metadata description scheme is feasible and practical under the grid environment. Excepting for information retrieval service, every service in LRG would be improved one by one In future.

References

1. Foster, I., Kesselman, C.: The Grid, Blueprint for a New Computing Infrastructure. Morgan Kaufmann Publishers Inc., San Francisco (2002)
2. Foster, I.: What is the Grid? A three point checklist. GRID Today (2002)
3. Weibel, S.: The State of the Dublin Core Metadata Initiative: April 1999. Bulletin of the American Society for Information Science and Technology 25, 18–22 (1999)
4. Li, S., Yang, Z., Liu, Q.: Research of Metadata Based Digital Educational Resource Sharing. In: Computer Science and Software Engineering International Conference, vol. 5, pp. 828–831 (2008)
5. Chen, Z., Zhong, L., Song, H.: Research on Layer Model of Learning Resource Grid Metadata. Journal of Wuhan University of Technology 31, 151–155 (2009)
6. Zhang, X.: Research and Application of Metadata. Beijing Library Press (2005)
7. Liu, N., Liu, F.: Research on Structure of Educational Resources Grid and Metadata Model. Computer Applications 25, 2518–2521 (2008)
8. Luther, A., Buyya, R., Ranjan, R., Venugopal, S.: Alchemi: A. NET-based Enterprise Grid Computing System (2005)

Fault Feature Extraction of Cylinder-Piston Wear in Diesel Engine with EMD

Fengli Wang, Shulin Duan, and Hongliang Yu

College of Marine Engineering, Dalian Maritime University, Dalian 116026, P. R. China
wangfl@dlmu.edu.cn, duanshulin66@sina.com, yhl1202@dl.cn

Abstract. Aiming at the characteristics of the vibration signals measured from the diesel engine, a novel method combining empirical mode decomposition (EMD) and lifting wavelet denoising is proposed, and is used for feature extraction and condition evaluation of diesel engine vibration signals. Firstly, the original data was preprocessed using the lifting wavelet transformation to suppress abnormal interference of noise, and avoid the pseudo mode functions from EMD. Obtaining intrinsic mode functions(IMFs) by using EMD, the instantaneous frequency and amplitude can be calculated by Hilbert transform. Hilbert marginal spectrum can exactly provide the energy distribution of the signal with the change of instantaneous frequency. The vibration signals of diesel engine piston-liner wear were analyzed. The analysis results show that the method is feasible and effective in fault feature extraction and condition evaluation of diesel engine.

Keywords: empirical mode decomposition, lifting wavelet, Hilbert transform, feature extraction, wear, diesel engine.

1 Introduction

The cylinder-piston wear has long been recognized as an important influence on the performance of internal combustion engines in terms of power loss, fuel consumption, oil consumption, blow-by and harmful exhaust emissions. In addition to causing the change of the surface vibration response of diesel engine [1]. In consideration of the cylinder-piston wear will lead to change of the surface vibration response of diesel engine, it is possible to monitor the changes in wear of the piston ring that take place with running time in the engine by analyzing the change of the surface vibration response of diesel engine. When the cylinder-piston wear occurs, vibration signals of diesel engine is non-stationary. The spectrum based on Fourier transform represents the global rather than any local properties of the signals. For measured signals in practical application with finite data length, a basic period of the data length is also implied, which determines the frequency resolution. Although non-stationary transient signals can have a spectrum by using Fourier analysis, it resulted spectrum for such signals is broad band. For example, the spectrum of a single pulse has a similar spectrum to that of white noise. Consequently, the information provided by Fourier analysis for transient signals were limited.

D. Jin and S. Lin (Eds.): Advances in CSIE, Vol. 2, AISC 169, pp. 419–424.
springerlink.com

In this paper, Hilbert marginal spectrum is introduced. Instead of relying on convolution methods, this method use empirical mode decomposition (EMD) and the Hilbert transform [2]. For a non-stationary signal, the Hilbert marginal spectrum offers clearer frequency energy decomposition than the traditional Fourier spectrum. However, the cylinder-piston wear characteristics are always submerged in the background and noise signals, which will cause the mode mixture and generate undesirable intrinsic mode functions (IMFs). In order to decrease unnecessary noise influence on EMD, it is important to denoise first before decomposing. In the denoising of traditional wavelet transform, the result of wavelet decomposing is related with wavelet basis function. Moreover, an inappropriate wavelet will overwhelm the local characteristic of vibrating signal, and lost some useful detail information of original signal. To circumvent these difficulties, we present a lifting scheme to construct adaptive wavelets by the design of prediction operator and update operator. The simulation and application analysis results show that the method is feasible and effective.

2 Hilbert Marginal Spectrum

Hilbert marginal spectrum analysis is performed into two steps. First, the EMD decomposes the time-series into a set of functions designated as IMFs, and secondly applying Hilbert transform to those IMFs for generation of the Hilbert marginal spectrum [2]. For any signal, to get a meaningful instantaneous frequency using Hilbert transform, the signal has to decompose a time-series into IMFs $c_1(t)$, $c_2(t)$, . . . , $c_n(t)$, and a residue $r_n(t)$,

$$x(t) = \sum_{i=1}^{n} c_i(t) + r_n(t) \,. \tag{1}$$

Applying the Hilbert transform to each IMFs, the original data can be expressed as,

$$x(t) = \mathrm{Re} \sum_{j=1}^{n} a_j(t) e^{i\varphi_j(t)} \tag{2}$$

This frequency–time distribution of the amplitude is designated as Hilbert time–frequency spectrum

$$H(\omega,t) = \mathrm{Re} \sum_{j=1}^{n} b_j a_j(t) e^{i\int \omega_j(t dt} \quad . \tag{3}$$

We can also define Hilbert marginal spectrum

$$h(\omega) = \int_0^T H(\omega,t) dt \,. \tag{4}$$

where T is the total data length. The Hilbert marginal spectrum offers a measure of the total amplitude distribution from instantaneous frequency.

3 Lifting Wavelet Denosing

3.1 The Lifting Scheme

The lifting scheme can be used to construct adaptive wavelets by the design of prediction operator and update operator [3-5]. It does not rely on the Fourier transform. The principle of lifting scheme wavelet transform is described as,

(1) Split: Split the original signal $\{X(k), k \in Z\}$ into even sets $X_e(k) = \{x(2k), k \in Z\}$ and odd sets $X_o(k) = \{x(2k+1), k \in Z\}$.

(2) Update: Using a one-point update filter, the approximation signal is computed, $c(k) = (X_e(k) + X_o(k))/2$

(3)Select prediction operator: Design three different prediction operators

$$N = 1: d(k) = X_o(k) - c(k) \tag{5}$$

$$N = 1: d(k) = X_o(k) - c(k) \tag{6}$$

$$N = 5: d(k) = X_o(k) - \{3[c(k-2) - c(k+2)]/128 + c(k) - 11[c(k-1) - c(k+1)]/64 - c(k+2)\}. \tag{7}$$

Where, N is the number of neighboring $c(k)$ while applying the prediction operator , $k = 1 \sim L/2$. An optimal prediction operator is selected for a transforming sample according to minimizing the $[d(k)]^2$.

(4) Predict: Compute the detail signal $d(k)$ by using the optimal prediction operator.

Because we update first and the transform is only iterated on the low pass coefficients $c(k)$, all $c(k)$ depend on the data and are not affected by the nonlinear predictor. Then reuse these low-pass coefficients to predict the odd samples, which gives the high-pass coefficients $d(k)$. We use a linear update filter and let only the choice of predictor depend on the data. The selection criterion of minimizing the squared error, an optimal prediction operator is selected for a transforming sample so that the used wavelet function can fit the transient features of the original signal.

In the signal denoising, apply various thresholds to modify the wavelet coefficients at each level. The wavelet coefficients are modified via soft-thresholding with universal threshold at each level [6].

3.2 The Wavelet Denosing

The wavelet denoising follows three operations:

(1) Decompose a signal into several levels via the lifting wavelet transform.

(2) Apply various thresholds to modify the wavelet coefficients at each level.

(3) Inversely synthesize the approximation signal and the modified detail signal iteratively from a lower level to an upper level via the wavelet transform, and finally get the denoised signal.

In the signal denoising, the universal threshold is adopted to modify the wavelet coefficients at each level. The wavelet coefficients are modified via soft-thresholding with universal threshold at each level [6].

$$\text{thr}=\sqrt{2\ln M}\cdot\frac{Med(|d|)}{0.6745} \tag{8}$$

where, d={d(k)},M is the length of d, and Med(·) is median function. The wavelet cocfficients are modified via soft-thresholding with universal threshold at each level.

4 Feature Extraction Based on Lifting Wavelet Denoising and EMD

The feature extraction of vibration signals follows three operations:

(1) The lifting scheme is employed to construct wavelet for denoising the given signal. The corresponding predictor and update operator are designed, and the predefined soft-thresholding is used to modify the wavelet coefficients.

(2) After the wavelet denoising, any complicated signal can be decomposed into a finite and often small number of IMFs by using EMD, and the modulated signal with fault information can be separated from the vibration signal.

(3)Hilbert envelop spectrum analysis has been used in the IMFs and extract the fault characteristics of the vibration signal.

5 Cylinder-Piston Wear Monitoring

The proposed method is applied to diagnosing the diesel engine cylinder-piston wear faults. According to the fundamentals of diesel engines, vibrations have a close relationship with the impact of the cylinder-piston. The characteristics of vibrations generated by a diesel engine were measured by accelerometer mounted on the cylinder body of cylinder 3 correspond to the top dead center, we collected three kind vibration signals from the same cylinder, which represent the engine in normal state, slightly wear, and serious wear state condition. All data were sampled at 25.6 kHz, and the analyzing frequency is 10 kHz. The rotating speed of the diesel engine is 1100 r /min around. Fig.1 a) ~ c) show the vibration signals of the engine cylinder-piston in normal, slightly wear, and serious wear state conditions. From the comparison in the time domain, we can see that the amplitude peaks in normal state and slightly wear signals are about the same in the time domain, no distinctness features. But the serious wear signal's is the highest.

From the Hilbert marginal spectrum shown in Fig.2 a) ~ c). we can see that the marginal spectrum offers a measure of the amplitude distribution from each instantaneous frequency. For cylinder-piston in normal state condition, the energy of the signal obvious distributes in a lower frequency area which is limited to a range of 2kHz. For cylinder-piston in slightly wear state condition, the lower frequency energy content is low due to leakage of combustion, and much energy distributes in a higher frequency area (5kHz~7kHz) generated by the occurrence of piston slap.

For cylinder-piston in serious wear state condition, the peaks of energy of the signal are concentrated on the higher frequency area due to increasing the strength of piston slap generated by the cylinder-piston wear, whereas the lower frequency energy content decrease due to increasing the leakage of combustion.

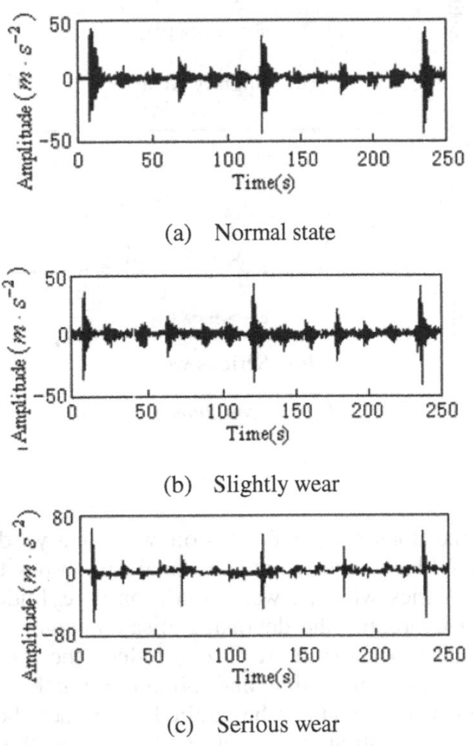

(a) Normal state

(b) Slightly wear

(c) Serious wear

Fig. 1. Vibration signals of diesel engine

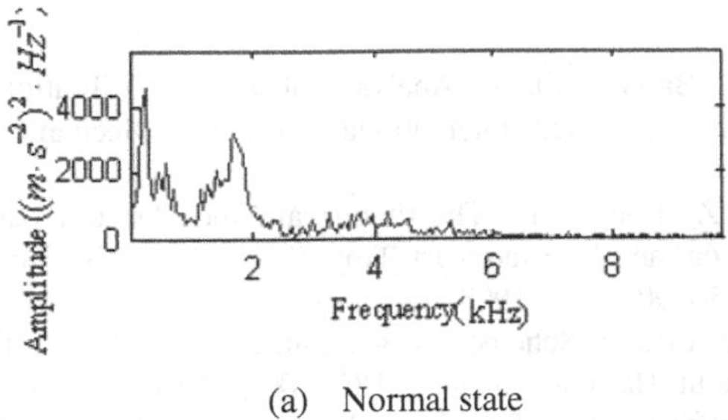

(a) Normal state

Fig. 2. Hilbert marginal spectrum of vibration signals of diesel engine

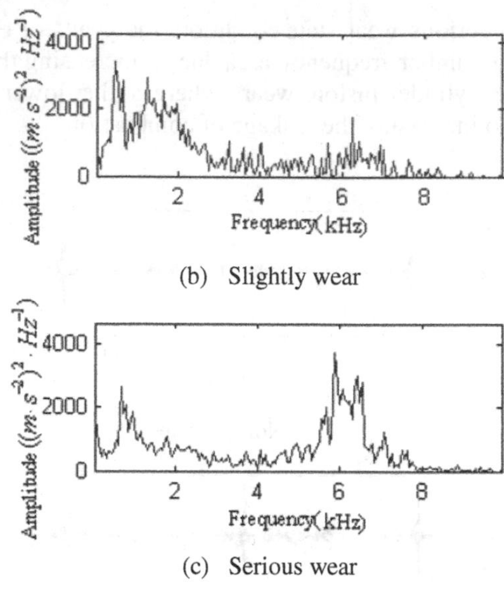

(b) Slightly wear

(c) Serious wear

Fig. 2. *(continued)*

6 Summary

The different wear conditions of cylinder-piston were analyzed from the vibration information of cylinder block surface, and the analysis shows that the vibration of cylinder block surface varies with the wear condition of cylinder-piston. The lifting wavelet transform can overcome the denoising disadvantage of traditional wavelet transform and is adopted to remove noise. It can reduce the mode mixture in EMD, improve the quality of decomposition and obtain a much better decomposition performance. The proposed method can be applied to extract the fault characteristic information of the surface vibration signal, and monitor the wear condition of cylinder-piston effectively.

References

1. Geng, Z., Chen, J., Barry Hull, J.: Analysis of Engine Vibration and Design of An Applicable Diagnosing Approach. International Journal of Mechanical Sciences 45, 1391–1410 (2003)
2. Huang, N.E., Shen, Z., Long, S.R.: The Empirical Mode Decomposition and The Hilbert Spectrum for Nonlinear and Nonstationary Time Series Analysis. Proceedings of the Royal Society of London 454, 903–995 (1998)
3. Sweldens, W.: The Lifting Scheme: A Custom-design Construction of Biorthogonal Wavelet. Appl. Comput. Harmon. Analo. 2, 186–200 (1996)
4. Claypoole, R.L., Geoffrey, M.D., Sweldens, W.: Nonlinear Wavelet Transforms for Image Coding via Lifting. IEEE Transactions on Image Processing 12, 1449–1459 (2003)
5. Kong, G.J., Zhang, P.L., Cao, J.J.: Signal Denoising Based on Lifting Wavelet Transform and Its Application. Computer Engineering and Applications 44, 234–237 (2008)
6. Donoho, D.L.: De-noising by Soft-thresholding. IEEE Transactions on Information Theory 41, 613–627 (1995)

The Personalized Recommendation Method Based on Improve-Collaborative Filtering Algorithms

TingZhong Wang[1,2] and XiHu Zhi[1]

[1] College of Information Technology, Luoyang Normal University, Luoyang, 471022, China
wangtingzhong2@sina.cn

[2] Master of Henan University of Science and Technology, Research Area: Web Mining, Data Mining, Collaborative Filtering

Abstract. Collaborative filtering recommender technology is also known as user-oriented recommendation techniques, is currently the most successful personalized recommendation. This paper describes the collaborative filtering recommendation techniques, collaborative filtering recommendation of three steps, and generation of a neighbor recommended. The traditional collaborative filtering algorithm has been improved. This paper presents personalized recommendation technology in web log mining based on improve-collaborative filtering algorithms in order to get novel information of interest changes. The experimental results show that the method can improve the recommendation of the timeliness and accuracy.

Keywords: collaborative filtering, personalized recommendation, web log mining.

1 Introduction

Web log mining, use of the data set consisting of data mining technology on site data and other relevant data to analyze the mining, in order to obtain valuable website access mode of knowledge. Web log mining can understand the browsing patterns of users of the site, browsing habits and browsing behavior found behavior similar to user groups, access to the Web page will have the same features of the page grouping. Therefore, the use of the Web log mining technology design a personalized recommendation feature intelligent site is increasingly becoming the issues of concern of the researchers.

The basic assumption of collaborative filtering (collaborative filtering) users interested in similar regular access to similar resources and similar interested users will access similar resources [1]. In the analysis and comparison of existing collaborative filtering algorithms based on existing personalized recommendation method less time-sensitive problem, the introduction of a weighted function of time in order to improve the priority of the latest access information in the existing collaborative filtering algorithms to solve the user interest transfer. This paper presents personalized recommendation technology in Intelligent Web Site based on Improve-collaborative filtering arithmetic in order to get novel information of interest changes.

D. Jin and S. Lin (Eds.): Advances in CSIE, Vol. 2, AISC 169, pp. 425–430.
springerlink.com

2 Collaborative Filtering Technology

Collaborative filtering recommender technology is also known as user-oriented recommendation techniques, is currently the most successful personalized recommendation. Higher accuracy of user-based collaborative filtering recommendation, and can make a strange discovery. This chapter first introduces the definition of collaborative filtering; Second, the common method of analysis of collaborative filtering, given the collaborative filtering algorithm based on user clustering process; improvement collaborative filtering algorithm for the novelty of the recommended information is given the implementation process and experimental analysis.

Most collaborative filtering algorithms for these two areas to make recommendations, it is of which the second recommendation is fully available to the first prediction [2]. Global numerical algorithms in collaborative filtering, users of similar user-focused evaluation of a particular value, as well as a series of the right weight to predict the degree of the target user's interest, assuming that the target user a, a value of Pa,, I, that is equation 1:

$$P_{a,i} = \overline{v_a} + k \sum_{b \in neighborhood(a,T_i)} r(a,b)(v_{b,i} - \overline{v_b}) \tag{1}$$

Ti in which the user evaluation of item i set, neighborhood (a, Ti) Ti in said target user a neighbor sets of users, including the neighbor sets of users can be for all users in the Ti , or k-thnearest neighbor user query results, or the scope of the query results (range query) . Indicates that the user a average evaluation value of the evaluation, the evaluation value of the user b of item i, and k is a specification factor, often set to r (a, b) the sum of the reciprocal. r (a, b) the similarity between two users a and b, more applications is the use of the Pearson correlation coefficient (Pearson correlation coefficient) export.

$$r(a,b) = \frac{\sum_{i \in I_{ab}} \left(v_{ai} - \overline{v_a}\right)\left(v_{bi} - \overline{v_b}\right)}{\sqrt{\sum_{i \in I_{ab}} \left(v_{ai} - \overline{v_a}\right)^2} \sqrt{\sum_{i \in I_{ab}} \left(v_{bi} - \overline{v_b}\right)^2}} \tag{2}$$

Which, on behalf of users a and b are evaluated items. For the characterization of user similarity: the constrained Pearson correlation coefficient (constrained Pearson correlation coefficient), the vector similarity. Studies have shown that the similarity between users using the Pearson correlation coefficient or its variants characterized relatively well: equation 3.

$$P_{a,j} = E(r_{a,j}) = \sum_{i=0}^{m} P(r_{a,j} = i \mid A)i \tag{3}$$

This paper uses a collaborative filtering algorithm based on user clustering. Collaborative filtering recommendation system to achieve the core target user a referral service to find the most similar to the "nearest neighbor sets, namely: to generate a user of the

Tu, one arranged by the similarity of the size of" neighbors "collection= {N1, N2,, ... Nn}, and sim (Tu, N1)> the sim (Tu, N2)> ...> sim (Tu, Nn), from N1 to Nn, indicating the similarity of the user Tu and neighbor usersin decreasing order of size of the value of sim (Tu, Ni), as is follows equation 4.

$$sim(i,j) = \frac{\sum_{c\in I_{ij}} \left|(R_{i,c} - \overline{R_i})(R_{j,c} - \overline{R_j})\right|}{\sqrt{\sum_{c\in I_{ij}} (R_{i,c} - \overline{R_i})^2} \sqrt{\sum_{c\in I_{ij}} (R_{j,c} - \overline{R_j})^2}} \quad (4)$$

3 Improvement and Implementation of Collaborative Filtering Algorithms

Traditional collaborative filtering methods can not be interested in the changes shown. For example, the user's interests are constantly changing: a parent may be the introduction of various universities, enrollment, and enrollment information interested students admitted to the University, the interest in this regard may be waning; a woman may be interested in parenting knowledge, are more interested in the past but then Korean [3]. In order to improve the sensitivity of prediction for new projects, real-time the desired results, we discuss the current page of the target user browse to reflect their preferences in the future rather than using the browser. To this end, this paper proposes time-weighted collaborative filtering algorithms to novel information of interest changes.

In collaborative filtering algorithm in the time-weighted, taking into account the "time effect", that is, the sooner browsing interest, the importance of the smaller to reduce the impact of its recommended, this paper introduces a time-weighted function f (t) (t istime variable) to the interest in prediction, and target users Tu weighted prediction score of the project to improve to, as is shown by equation5.

$$P'_{Tu,i} = \overline{R_{Tu}} + \frac{\sum_{n\in Neighbor_{Tu}} sim(Tu,n) * (R_{n,i} - \overline{R_n}) f(t_{ni})}{\sum_{n\in Neighbor_{Tu}} (|sim(Tu,n)|) f(t_{ni})} \quad (5)$$

The sim (Tu n) said that the target user Tu and the similarity of user n of the nearest neighbor, said user n interest in the resource i (n is the "nearest neighbor" concentration of users). And denote the target user Tu and user n degrees of the average interest on resources. That user n on item i the interest of time, we assume that the time function f (t) is a monotonically decreasing function, formulas, such as 6, it has been in reducing the time t, and the value of the time the right to remain in (0,1) range, that is, all data are beneficial to recommend the project, contribute more to the latest data, old data to reflect the user's previous preferences, it is recommended forecast to account for the smaller weights. Time selection index to obtain the target, the exponential time function is widely used in practice; it is more hope to get the progressive past behavioral trends. Function of time, as is shown by equation6.

$$f(t_{ni}) = e^{-t_{ni}} \tag{6}$$

Recommended to use the data pre-processing and clustering collaborative filtering personalization, specific to this article is to table user table the table below shows the user's IP address and user belongs to that class and other related information. The table url_table control of url_id (that uniquely identifies the URL) and URL addresses, to provide support to recommend a specific URL address. Table UrlVisitTimes show that the number of users to access Web pages, time and size of the page to access the page for the user's first visit to the recommended content, its recommended frequency of visits of the page when a user first visits.

Use of improved collaborative filtering TWCF (Time Weight Collaborative Filtering) recommended method to predict the target user evaluation of new projects based on the user's time-weighted points, and accordingly recommended.

Algorithm TWCF (Tu, N, i, ξ)

Input: The target user Tu, the recommended number of items N, the project i, the score threshold value of ξ.

Output: The target users of N recommended items.

(1) For any user Tu and project i predict user Tu projects evaluation points, note the predictive value $P'_{Tu,i}$;

(2) By the algorithm UserClustering come to the target user Tu neighbor sets of users $Neighbor_{Tu}$;

(3) Similarity $sim(Tu,n)$ calculation of the target user Tu and user n ($n \in Neighbor_{Tu}$);

(4) Ask a neighbor users on item i points, the target user Tu evaluation of sub-item i by the formula 5 forecast $P'_{Tu,i}$;

(5) Arranged according to the size of $P'_{Tu,i}$ the order of the score, the use of one of the following two methods to determine the most interesting projects of the user Tu.

First get the IP of the target user, and then determine that the user is the first visit or have visited the site. If the user has previously visited the site, find the closest class with the user and all users within a class based on the similarity with the cluster center, then through Simi Coefficient find out the similarity coefficient between the target users and other users within a class, from table User Ratings elected class nearest neighbor user interest in the page, and finally by the formula predict the ratings of the target user is not accessing a relatively high score of the first N items recommended to the user as a recommendation results.

4 Experimental Results and Analysis

In order to test the efficiency of this method, Set the target user for the Tu entire user space with the U said, first of all on the whole user space for nearest neighbor queries, this article select the most recent neighbor number is 10, the query results in mind as a collection, In order to achieve statistically significant with sex in this article in the entire user spatial statistics indicators.

In the clustering process, we see that the number of clusters specified is critical, the number of clusters to specify, and then calculate the similarity of the target user and the cluster centers need to spend a lot of time, can not effectively improve the recommendation system real-time response speed; the number of clusters specified is too small, each cluster contains more users, even in the most similar to the target user clustering, nearest neighbor queries also need to scan a large number of candidate sets of users, so that the recommendation system the real-time play is very much affected. As the number of clusters to 5, 10 422 users, clustering, and then find the nearest neighbors of the target user on the basis of the clustering, the experimental results are shown in Figure 1, as shown in Figure 2.

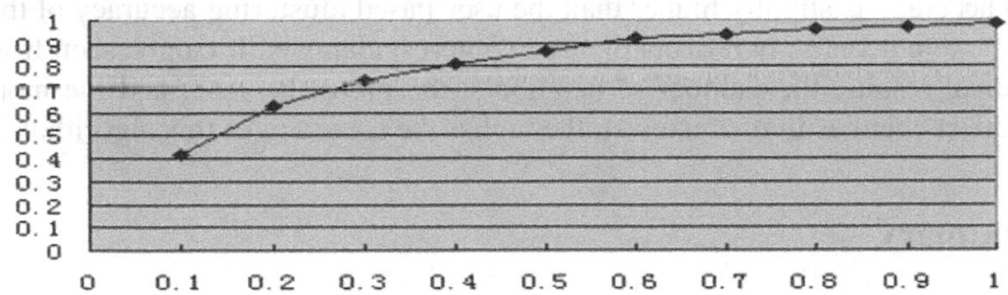

Fig. 1. The Searching efficiency(k=5)

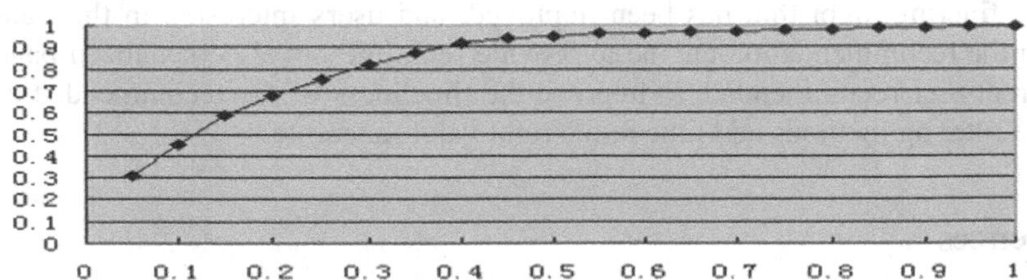

Fig. 2. The Searching efficiency(k=10)

In addition to real-time analysis algorithm, this also added time the weight of this paper, collaborative filtering algorithm with the traditional collaborative filtering algorithm based on user recommendations accuracy. In this study, we hope that the final algorithm is able to accurately predict the evaluation project evaluation points for the user to make a more accurate recommendation algorithm accuracy (Precision) is to evaluate a major indicator of the recommendation algorithm.

Predicting user interest scores of resources, the number of the nearest neighbor involved in the calculation affects the MAE algorithm. Experiment, we take the nearest neighbor number of the target users from 5 to 40 intervals of 5 to view the nearest neighbor sets of different sizes, weights of user clustering collaborative filtering algorithm and time collaborative filtering algorithm to forecast accuracy changes in the experimental results shown in Figure 3.

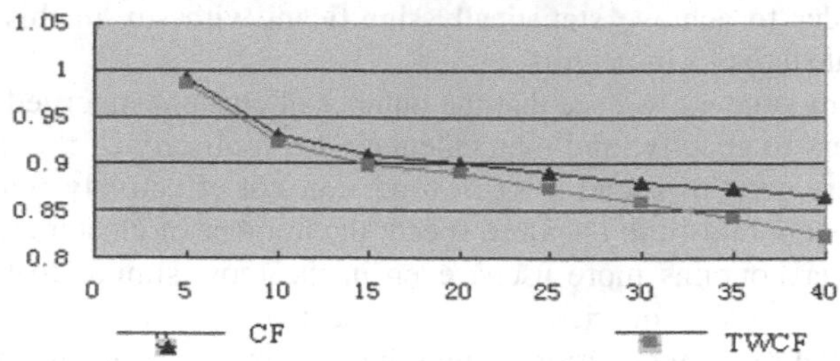

Fig. 3. The result of MAE comparisons of recommendatory arithmetic

From the experimental results, in most cases, the added time the weight of the CF (TWCF) accuracy is slightly higher than the user-based clustering accuracy of the CF, because it more accurately reflects the user's interest changes. It can be seen from the experimental results, the number of neighbors, the more the case, and the more the user to reflect changes in user interest, the higher the accuracy of this algorithm.

5 Summary

This paper describes the collaborative filtering recommendation techniques, collaborative filtering recommendation of three steps, data representation, the formation and generation of a neighbor recommended a detailed description. And traditional collaborative filtering algorithm has been improved, and users interested in the transfer, through the recommendation engine to provide real-time novel personalized page for the current user recommended, to improve the timeliness of the recommended collaborative filtering methods add time to the right value, accurate.

References

1. Breese, J.S., Heckerman, D., Kadie, C.: Empirical analysis of predictive algorithms for collaborative filtering. In: Proceedings of the 14th Conf. on Uncertainty in Artificial Intelligence (UAI 1998), San Francisco, July 24-26, pp. 43–52 (1998)
2. Pennock, D.M., et al.: Collaborative Filtering by Personality Diagnosis: A Hybrid Memory and Model-based Approach. In: Proceedings of the 16th Conference on Uncertainty in Artificial Intelligence, pp. 473–480 (2000)
3. Schafer, J.B., Konstan, J.A., Riedl, J.: E-Commerce recommendation applications. Data Mining and Knowledge Discovery 5(1-2), 115–153 (2001)

Multi-step Predictions Based on TD-DBP ELMAN Neural Network for Wave Compensating Platform

Zhigang Zeng

Automation College of Guangdong Polytechnic Normal University, Guangzhou, 510665, China
zengzhigang1975@sina.com

Abstract. The gradient descent momentum and adaptive learning rate TD-DBP algorithm can improve the training speed and stability of Elman network effectively. BP algorithm is the typical supervised learning algorithm, so neural network cannot be trained on-line by it. For this reason, a new algorithm (TD-DBP), which was composed of temporal difference (TD) method and dynamic BP algorithm (DBP), was proposed to overcome the restriction. TD-DBP algorithm can make Elman network train on-line incrementally. Using the collected real time data, the modified TD-DBP algorithm was able to realize direct multi-step predictions for vertical displacement of wave compensating platform.

Keywords: wave compensating platform, TD-DBP algorithm, multi-step predictions.

1 Introduction

The purposes of shipping movement prediction are optimizing navigating state of the ship. For vertical displacement of wave compensating system, it belongs to passive control because the traditional feedback control would start compensation machine after wave motion, and it is not ideal for the big delay system. But we can obtain ship motion trend in advance by motion prediction. Means used in short-term prediction of shipping movement includes ways based on hydrodynamics and ways based on non-hydrodynamics [1-4]. Convolution algorithm and linear Kalman filter algorithm are the means based on hydrodynamics. Convolution algorithm uses the ocean wave's time-history and the convolution of ship impulse response function to get the time-history of ship navigating state. The computational load of this means is heavy, and it needs to get time-history in advance, so it's not easy to implement. Linear Kalman filter algorithm needs the equations of shipping movement state, but accurate equations of the state is difficult to find [5]. The means based on non-hydrodynamics includes time sequence analysis, periodogram and neural network algorithm, etc. Many scholars use neural network to predict shipping movement, and the study uses neural network based on supervised learning [6].

The artificial neural network algorithm uses movement data during the time of the past for network input and the network will study it, and then modulate the threshold of neuron and connected weight of the neurons by some rules. This means can get anticipant network output with the given network output, and the output is the correct

D. Jin and S. Lin (Eds.): Advances in CSIE, Vol. 2, AISC 169, pp. 431–436.
springerlink.com

prediction value. This means can leave out data analysis and modeling process, and so it's convenient to solve the prediction problem of shipping movement stance. Vertical displacement of wave compensating platform can be only discovered after many steps. With BP algorithm of supervised learning we can't directly train on-line using the neural network. For solving the problem, the paper uses TD-DBP algorithm that combines temporal difference (TD) method[7-10] driven by the deviation between actual outputs with dynamic back-propagation learning algorithm (DBP) to directly train thc Elman network on-line.

2 Elman Neural Network Structure and Its Mathematical Modeling Description

Elman network is composed of four layers: input layer, hidden layer, connection layer and output layer. The number of neurons in the connection layer is the same as the number of neurons in the hidden layer. The hidden layer outputs ... accordingly the process of hidden layer output makes the network sensitive to the data of historical state and beneficial to modeling of dynamic process. Furthermore, the dynamic characteristic of the network is only offered by inner connection, and it needn't using state to be the input or training signal. This is the preponderance of Elman network comparing with static feed-forward network.

The mathematical modeling of Elman neural network is as follows:

$$O_k(t)=\sum_{j=1}^{q}\omega_{kj}H_j(t)$$

$$H_j(t)=g\left(V_j(t)\right)$$

$$V_j(t)=\sum_{a=1}^{q}\omega_{ja}H_{ja}(t-1)+\sum_{i=1}^{p}\omega_{ji}I_i(t)$$

Here, p , q and x are the number of neurons respectively in the input layer, the hidden layer and output layer, and the number of neurons in connection layer is the same as the number in the hidden layer. $I_i(t)$ is the $i-th$ input of Elman neural network; $V_j(t)$ is the sum of the $j-th$ input of hidden neurons; $H_j(t)$ is the $j-th$ output of hidden neurons; $O_k(t)$ is the $k-th$ output of neurons in output layer; ω_{kj} is the connected weights of hidden neurons and output layer neurons; ω_{ja} is feedback connected weights of connection layer neurons to hidden neurons; ω_{ji} is connected weights between input neurons and hidden neurons; $H_{ja}(t-1)$ is one-step delay output of hidden neurons. $g(\cdot)$ is a non-linear variation function, and generally taken as the Sigmoid function $g(x)=1/[1+\exp(-x)]$.

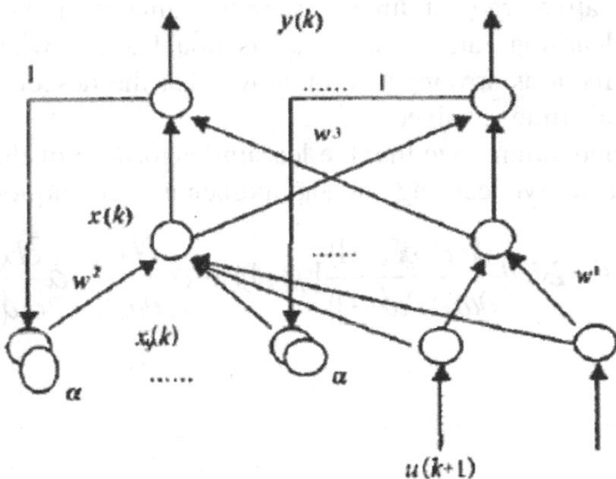

Fig. 1. The structure of Elman neural network

3 Learning Algorithm of Elman Neural Network

3.1 Improved TD-DBP Learning Algorithm

The traditional Elman neural network uses the classical back-propagation algorithm, and the algorithm has two main problems in actual application: slow convergence speed and local minimum point in the target function.

The gradient descent algorithm with self-adaptive learning rate and momentum techniques combines momentum gradient descent algorithm with self-adaptive learning rate gradient descent algorithm, and it can improve network's training speed and stability, and meanwhile it effectively avoid the appearance of local minimum point. Therefore, the paper uses the gradient descent algorithm with self-adaptive learning rate and momentum techniques to improve the TD-DBP algorithm [8].

3.2 Gradient Descent with Self-adaptive Learning Rate Algorithm

The Learning rate has great effect on the whole training process. The smaller learning rate is, the slower convergence speed is. And if the learning rate is too big, it can be modified too much to be oscillatory and divergent. We can avoid this by reasonably changing the learning rate in the training process. Self-adaptive learning rate is good for shortening learning time, and its learning algorithm is:

$$\omega(t+1) = \omega(t) - \eta(t)\frac{\partial E(t)}{\partial \omega(t)}$$

$$\eta(t) = 2\lambda\eta(t-1)$$

$$\lambda = sign[\frac{\partial E(t)}{\partial \omega(t)}\frac{\partial E(t-1)}{\partial \omega(t-1)}]$$

If two successive iterative gradient directions are the same, it shows that the descent is too slow, and the learning rate at this time is double. But when two successive iterative gradient directions are opposite, it shows that the descent is too much, and the learning rate at this time is halved.

We can analyze and summarize that the learning algorithm of the gradient descent algorithm with self-adaptive learning rate and momentum techniques is as follows:

$$\omega(t+1)=\omega(t)-2sign[\frac{\partial E(t)}{\partial\omega(t)}\frac{\partial E(t-1)}{\partial\omega(t-1)}]\eta(t-1)[(1-\alpha)\frac{\partial E(t)}{\partial\omega(t)}+\alpha\frac{\partial E(t-1)}{\partial\omega(t-1)}]$$

4 Testing

4.1 Collecting and Processing of Testing Data

There is a wave maker on the one side of testing pool, and a wave dissipating slope (1:7) on the other side, it can dissipating 90% reflected wave. The testing use regular wave to simulate, wave period is 1.4s, wave height is 0.06m,interval of data acquisition is 50ms. By measuring vertical displacement of 4 sensors who arranged in the model's different places, A/D conversion and computer collection and further conversion calculation, we get the motion data (including rolling, pitching and heaving) of the stabilized platform model.

Because the testing environment and precision of measuring equipments and transmission equipments, the data we have gotten would have missing data and false data, even phase shift and period distortion. Thus, after data processing of rejecting wild dots and phasing, we make the data curves smoothing and keeping continuous.

4.2 Multi-step Predictions for Vertical Displacement of Wave Compensating Platform

Using Elman network to make multi-step predictions for vertical displacement of wave compensating platform, the number of input layer, hidden layer, connection layer and output layer is 5-15-15-4.The hidden layer neurons activation function is sigmoid function. The activation function of input layer neurons, connection layer neurons and output layer neurons is the linear function [1,2].

The connected weight value of neurons is the random number in (-1, 1). Network training's main parameters are as follows: training times, 500; learning rate, 0.01; increase and decrease ratio factor of the learning rate, 1.05 and 0.7; momentum constant, 0.9; network capability target, 1e-8; the minimum gradient of network capability function, 1e-15; and network training stops when the minimum gradient of the training times, capability target or capability function reaches setting requirements.

According to the experiment data after pre-treatment, the direct multi-step predictions results for vertical displacement of wave compensating platform based on Elman neural network are as the following graphs:(Fig.2realline means actual measuring value and dashed line means predictive value):

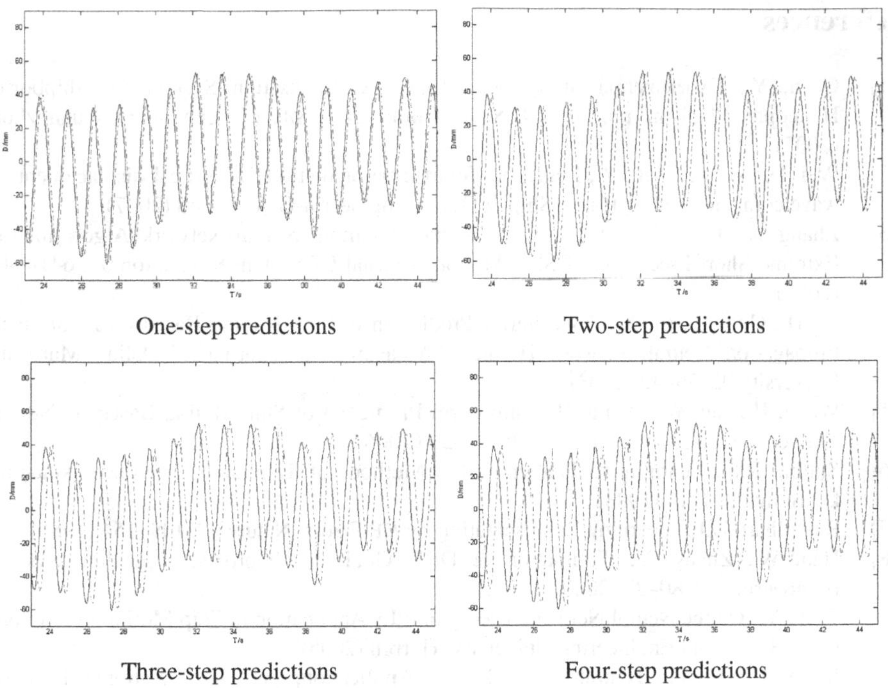

One-step predictions

Two-step predictions

Three-step predictions

Four-step predictions

Fig. 2. Multi-step predictions curves

These curves show that: Elman network can make 4-step(step size is 0.05s) predictions for vertical displacement of wave compensating platform effectively after online-training by the improved TD-DBP algorithm, satisfying requirement of predictions for vertical displacement of wave compensating platform. So, it is feasibility to use Elman network training by the improved TD-DBP algorithm make multi-step predictions for vertical displacement of wave compensating platform online.

5 Conclusion

The TD-DBP algorithm that is TD algorithm combined with DBP algorithm improves the traditional BP algorithm's limitation, and it can make on-line training based on Elman network, and consequently Elman network can implement real-time direct multi-step prediction. The improved TD-DBP algorithm improves study velocity of the Elman network and increases the algorithm's reliability. We implement simulation calculation using data obtained in the experiment. The results indicate that the Elman network based on improved TD-DBP algorithm can effectively implement direct real-time multi-step predictions for vertical displacement of wave compensating platform. Furthermore, this prediction means also can be applied to the other prediction systems such as the forecast of traffic flow and forecast in stock market etc. to play a vital role in wider areas.

References

[1] Chen, Y.: Experimental Study on a Heave Compensation System for Shipborne Helicopter Platform, pp. 90–104. South China University of Technology, Guangzhou (2007)

[2] Wei, D., Ye, J., Chen, Y., Liang, Y.: Short time prediction for vertical displacement of wave compensating platform. Ship & Ocean Engineerin-G 06, 45–48 (2007)

[3] Zhang, X., Peng, X., Zhaoxiren: Diagonal Recurrent Neural Network Algorithm for Extreme Short Prediction of Ship Motion. Journal Of System Simulation 05, 641–642 (2005)

[4] Li, H., Guo, C., Li, X.: Time Series Prediction of Ship Rolling Based on Error Back Propagation Neural Network Humidity Measurement. Journal of Dalian Maritime University 02, 39–42 (2003)

[5] Wang, H., Liu, W., Wang, H.: Short-term Prediction of Ship Motion Based on Neural Networks. Computer Simulation 05, 18–20 (2006)

[6] Sutton, R.S.: Learning to Predict by the Methods of Temporal Difference. Machine Learning 03, 9–14 (1983)

[7] Rumelhart, D.E.: Learning Representation by BP Errors. Nature 07, 149–154 (1986)

[8] Han, W., Zhang, Y.: Estimation of DRE Grade by Improved alogrithm. System Engineering 09, 80–82 (2000)

[9] Shen, Y.: On the Neural Network Theory and Its Application in Ship Motion Prediction, pp. 5–83. Harbin Engineering University, Harbin (2005)

[10] Dong, C.: Matlab Neural Ntework and Application, pp. 1–76. National Defense Engineering Press, Beijing (2005)

Soft-Sensing Based on Genetic Algorithm for Auto-Product-Line

Zhigang Zeng

Automation college of Guangdong Polytechnic Normal University, Guangzhou, 510665, China
zengzhigang1975@sina.com

Abstract. This paper deduces the optimum work point and input, and realizes static control and optimum of energy using inner variable and feed-back control. Consequently, constructs a soft-sensing model by genetic algorithm for Al-product auto-product-line. At last, the author illuminates that genetic algorithm has a great advantage comparing with the other traditional ways by application of genetic algorithm in the example.

Keywords: soft-sensing, genetic algorithm, optimum of energy.

1 Introduction

Al-product auto-product-line includes the following process: releasing, aligning, cleaning, drying, pasting film and punching. After cleaning of material for paste film, the surface of Al-belt must be kept dried. This system designs optimum work point and optimum control algorithm by adopting the checking of outside surrounding variable and inner status variable feed-back to realize optimum of energy and drying degree. We find that the humidity sensor can not always check the humidity of drying model accurately for compelling convection surrounding in drying box, then, it can not realize humidity feed-back. The paper uses genetic algorithm to build model and uses the other variables that are easier for checking to realize optimizing design for estimated variable, thereby, we achieve the goal of humidity feed-back control by soft-sensing.

Genetic algorithm is a random search algorithm based on life nature option and algorithm mechanization. Genetic algorithm breaks a new way for optimizing design since it had been raised by Pro. John Holland in mid-1960s. The algorithm forms a self-adapt overall optimizing probability search algorithm by simulating the heredity and evolution of life in natural among the strong search ways and weak search ways. Genetic algorithm is better than the strong search ways in not tending to part optimizing and better than weak search ways in inspiring self-adapt search lease area, and it has overall optimum by genetic operator. Now the algorithm is extensively used in every field for its self-adapt in optimizing, implied parallel, not depending on problem model and robust in complex non-liner problem. The author illuminates that genetic algorithm has a great advantage comparing with the other traditional ways by application of genetic algorithm in example.

D. Jin and S. Lin (Eds.): Advances in CSIE, Vol. 2, AISC 169, pp. 437–442.
springerlink.com

2 Control of Drying System

Drying system is a procedure for the surface drying of AL-belt before pasting film. After the procedure of cleaning and drying the AL-belt must reach these standards as below:

The surface must have even metal luster, and there is no metal oxide and other things;

The surface must have not oil dirt and spot, droplet should not be hanging on the belt after sprinkled some water on it;

The surface must be dried, facial tissue should not be moist after wipe on it.

Below figure is the flow chart of control system of drying:

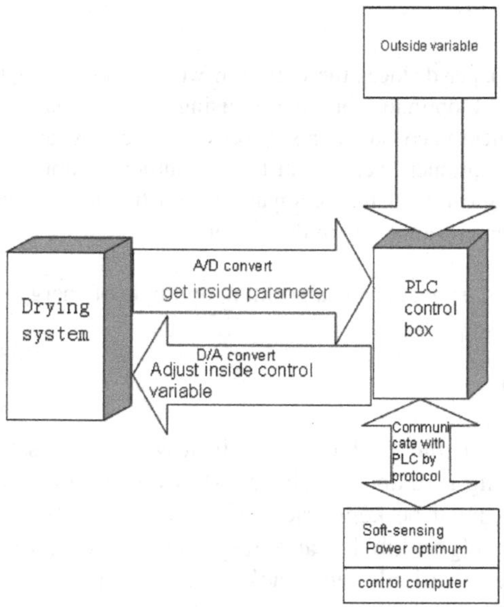

Fig. 1. Block diagram of control system

Control computer computes the optimum work point and input value by outside surrounding variables (humidity and temperature), and realizes static control and optimum of power by inner status variables feed-back control.

3 Math Model

In order to research connection among the inner factors during the course of evaporation, this paper adopt 3 math models as below.

3.1 Evaporation Model

There are 3 main factors work on vaporizing speed: temperature, humidity, wind speed. There are not current water formula of evaporation to describe connection of function between the vaporizing speed and the 3 parameters. It is reported that there have 3 current formulas in using, and we decide to select MinQian`s vaporizing formula[1].

$$E = \Delta P[0.09 + 0.2567(1-\varphi^2)^{1/2}] \cdot (0.348 + 0.5V^{1.8-1.137V^{0.05}}) \tag{1}$$

E means water vaporizing capacity, mm/d;

$\Delta P = P_w - P$, ΔP means saturation vapor press difference, hPa; P means part press of water vaporizing; P_w means saturation vapor press on the same temperature; φ means relative humidity; V means average of wind speed, m/s; P_w is a T-function:

$$P_w = E_0 \times 10^{\frac{7.35T}{235+T}} \tag{2}$$

T means temperature; $E_0 = 6.11hPa$; P_w`s dimension is hPa; The error of the formula meets the required standard in this drying system.

Define of relative humidity:

$$\varphi = \frac{p}{p_w} \times 100\% \tag{3}$$

Then,

$$\Delta P = P_w - P = P_w(1-\varphi) \tag{4}$$

Put formula(2)、(4) into (1), we can get the equation of water vaporizing capacity, the wind speed and the temperature as below:

$$E = 6.11 \times 10^{\frac{7.35T}{235+T}} (1-\varphi)[0.09 + 0.2567(1-\varphi^2)^{\frac{1}{2}}] [0.348 + 0.5V^{1.8-1.137V^{0.05}}]$$

To predigest humidity function in project application

If equation:

$$f(\varphi) = (1-\varphi)[0.09 + 0.2567(1-\varphi^2)^{1/2}]$$

During [0, 1], above formula can be replaced by the equation below.

$$f(\varphi) = \begin{cases} -0.37\varphi + 0.346\left[\varphi \in (0, 0.86)\right] \\ -0.2\varphi + 0.2\left[\varphi \in (0.86, 1)\right] \end{cases}$$

Thereby,

$$E = 6.11 \times 10^{\frac{7.35T}{235+T}} \times f(\varphi) (0.348 + 0.5V^{1.8-1.137V^{0.05}}) \tag{5}$$

3.2 Humidity Model

This system controls wind flow speed by the power of fan and open degree of valve, furthermore, realizes control of humidity. This paper adopts the way of mechanism modeling to build humidity model.

We adopt absolute humidity to build humidity model because absolute humidity is more convenience than relative humidity on system analyzing. Absolute humidity can be donated by mass ratio r and general mass ratio g . The physics meaning of r is the ratio between the pure vapor quality and clean air, and g is the ratio between the pure vapor quality and air quality.

Equation of relative humidity and absolute humidity as below[2]:

$$\varphi = \frac{\frac{r}{0.622 + r} \cdot p}{611 \times 10^{\frac{7.35T}{235+T}}} \times 100\%$$

p is a standard atmosphere; r is quality ratio; T is temperature
Static humidity model:

$$r_{in} = \rho SE(1 + r_{out}) \cdot \frac{1}{m} + r_{out} \tag{6}$$

Dynamic humidity model:

$$\begin{aligned} &M(t)\dot{g}(t) + [m(t) + \rho SE]g(t) \\ &= m(t)g_{out} + \rho SE \end{aligned} \tag{7}$$

ρ is the density of water; S is vaporizing area; E is vaporizing speed; m is wind flow speed; M is quality of mix air in box; g is general mass ratio.

3.3 Temperature Model

This paper adopts the method of mechanism modeling to build temperature model.
Static temperature model:

$$(T_{in} - T_{out})C_{out} \cdot m + K\rho SE = P_{hot} \tag{8}$$

Dynamic temperature model:

$$(T_{in}(t) - T_{out})Cm(t) + K\rho SE + M(t)C\dot{T}_{in}(t) = P_{hot}(t) \tag{9}$$

P_{hot} means the power of heater at the time t; T is temperature; C means specific heat of mix air; K is hidden heat coefficient of water vaporizing.
If $T = 60^{\circ}C$, $K = 2.36 \times 10^{6}$, unit: J/kg [3].

4 Adopt Genetic Algorithm to Realize Optimum of Power

We can not get the static value of g because the state of heavy wind flow in drying box. So we need to use the other variables that can replace humidity to realize feedback control. We can get equation about temperature as flow:

$$MCT + CmT - T_{out}Cm + K\rho SE = P_{hot}$$

In the system, the heater consume maximal energy, so we must find the optimum of the power to heater, i.e, seeking minimum of temperature in the static temperature equation as follow.

So at static state:

$$P_{hot} = \frac{(T - T_{out})\rho SE(1 + r_{out})}{0.0043T - 0.21 - r_{out}} C + K\rho SE \tag{10}$$

We can return the problem to solute optimizing problem of genetic algorithm. This paper uses MATLAB toolbox to solute the problem, and steps are as below [4-5]:

At first, build math model after analyse the target, secondly, define target function, and also define bind conditions if exist. At last, compile a M-file to return function value by file editor, i.e, write function in MATLAB, and transfer optimize program in order window, so we can get optimum. Equation (10) is target function.

4.1 Bind Conditions

Put constant into both vaporizing speed and wind speed function, we get condition:

$$6.363 = 10^{\frac{7.35T}{235+T}}(-0.37\varphi + 0.346)$$

Humidity equation at static state: $r \approx 0.0043T - 0.21$

Temperature scope: $49 \le T \le 60$

4.2 Construct Fitness Function

Construct fitness function by punish function must consider bind conditions above. Compile file of fitness.m by M-editor as follow(part program).

function [sol,eval]=fmy(sol,options)

4.3 Genetic Optimizing Setting

Use floating point coding to realize genetic optimizing, seed scale: m=25; across probability, Pc=0.65; aberrance probability, Pm=0.06; evolution ages, T=100. Compile M-file to transfer genetic algorithm main function: ga.m(part program):

[x,endPop,bestPop,trace]=ga(bounds,'fmy',[],iniPop,[1e611],'maxGenTerm',150,' normGeomSelect',[0.08],['arithXover'],[2,0],'nonUnifMutation',[2, 100 ,3])

4.4 Results

Results as follow table:

Table 1. Result of simulation

Method	T	r	Target function
Tradition method	60	0.43	$2.548 \mathrm{x} 10^6$
Genetic algorithm	58.3342	0.3842	$1.788 \mathrm{x} 10^6$

Results make clear that genetic algorithm is better than traditional ways in overall seek optimum

5 Conclusion

Usually, soft-sensing is based on full-blown sensor of hardware, and computer technology is the core of technical, and be realized by model operation at last. Comparing with traditional method, we have several advantages to construct model by genetic algorithm [6]. At first, genetic algorithm is not easy to put in part optimizing, even the fitness function is not series、abnormity and having noise conditions. Secondly, genetic algorithm has inherent parallel, it is very fit to distributing dispose. Furthermore, genetic algorithm is easy to combine to other technic to form the best solution. There is no fixed solution of genetic algorithm, we can confirm fitness function by the requirement, but only select right operate method of genetic algorithm can we light advantages of genetic algorithm, including selection of coding rule、seed scale and so on. By the way, appearance of MATLAB supply a good exploiting condition. It is very significant for genetic algorithm to apply in control technology domain.

References

[1] Minqian: Research of Calculation Model For Water Surface. In: Water and Electricity Technology Evolve, pp. 41–44. Jiangxi water conservancy bureau, Nanchang (2003)
[2] Li, Y., Fan, J.: Humidity Measurement. Weather book concern, Beijing (1990)
[3] Wu, H., Liu, B., Zhang, G.: Physic Concept Formula and Computer. Sanxi technology book concern, Xian (1982)
[4] Zhou, Z., Ding, D., Tian, Y., Wang, X.: Research and Application of MATLAB GAOT. In: Mechanical Research & Application, pp. 1049–1051. Ganshu technology bureau, Lanzhou (2006)
[5] Feng, X., Chenfang: Realization and Programming of Genetic Algorithm. In: PC Application, pp. 48–50. Shanghai PC application acad., Shanghai (1997)
[6] Liangke, Xia, D.: Solution For Programming and Optimizing of Genetic Algorithm in MATLAB. In: Computer Knowledge and Technology, pp. 1049–1051. China computer correspondence school, Hefei (2006)

An Improved Particle Swarm Optimization Algorithm Based on Two Sub-swarms

Zhihui Yu[1], Wenhuan Wu[2], and Lieyang Wu[3]

[1] School of Mathematics and Computer Science, ShangRao Normal University, ShangRao 334001, China
[2] School of Computer Science and Technology, Zhoukou Normal University, ZhouKou 466001, China
[3] JiangXi Highway Network Management Center, NanChang 330003, China
{yuwisdom,wuwenhuan15,lieyang_wu}@163.com

Abstract. In order to improve performance of particle swarm optimization algorithm (PSO) in global optimization, the reason of premature convergence of the PSO is analyzed, and a new particle swarm optimization based on two sub-swarms (TSS-PSO) is proposed in this paper. The particle swarm is divided into two identical sub-swarms, that is, the first sub-swarm adopts basic PSO model to evolve, whereas the second sub-swarm iterates adopts the cognition only model. In order to enhance the diversity and improve the convergence of the PSO, the worst fitness of the first sub-swarm is exchanged with the best fitness of the second sub-swarm in each iterate for increasing the information exchange between the particles. Compared with other two sub-swarms algorithms, the idea of this algorithm is readily comprehended, and its program is easy to be realized. The experimental results display that the convergence of TSS-PSO evidently gets the advantage of basic particle swarm optimization, as well as its competence of finding the global optimal solution is better than the basic PSO.

Keywords: Particle Swarm Optimization, Precocity Converges, Global Optimization, Sub-Swarm.

1 Introduction

The particle swarm optimization algorithm (PSO), originally introduced by Dr.Eberhart and Dr.Kennedy, is simulate the social behavior of bird flocking [1] [2]. Because the particle swarm optimization is easy to carry into practice and be able to resolve many real problems. Therefore, a lot of scholars have paid close attention to immediately just after the algorithms being arisen, and large amount of research results appeared within almost annual short time [3-6].

The same as other rand optimization algorithm, particle swarm optimization algorithm is prone to be limited into local optimal solution especially in complicated optimization problems [7]. In a PSO system, each particle is "flown" to the current best particle, so the swarm diversity may reduce promptly. Within complicated optimal solution space, present discovered optimal point may be a local optimal point, such swarm extremely leads to converge in local optimal point.

D. Jin and S. Lin (Eds.): Advances in CSIE, Vol. 2, AISC 169, pp. 443–448.
springerlink.com

According to particle swarm optimization algorithm's precocity in every iteration, an improved particle swarm optimization algorithm based new two sub-swarms (TSS-PSO) is arisen. To improve swarm diversity, lessen the possibility of being limited into local optimal point and make the population swarm evolve imperceptibly, the particle swarm is divided into two identical sub-swarms, each swarm search the overall space and exchanges the information. The ability of algorithm has got rise by the fact, and avoided the particles converging prematurely.

2 Particle Swarm Optimization Algorithms Based Two Sub-swarms

2.1 Basic Particle Swarm Optimization Algorithm

Particle Swarm Optimization is a population based search process where individuals, referred to as particles, are grouped into a swarm. Each particle in the swarm represents a candidate solution to an optimization problem. The performance of each particle is measured using a predefined fitness function which encapsulates the characteristics of the optimization problem. We can write the best position found as gBest. In each iteration, the current position is evaluated as a problem solution. The velocity vector decides whose direction and distance when the particle moves in solution space. Each particle updates the velocity and position according to gBest and its own pBest. The optimal solution can be found after iteration.

In a D-dimensional search space, there are N particles in the swarm. As in t iteration, the position vector of individual i can be indicated as:

$$x_i^t = (x_{i1}, x_{i2}, \cdots, x_{id}, \cdots, x_{iD})^T$$

the velocity vector as :

$$v_i^t = (v_{i1}^t, v_{i2}^t, \cdots, v_{id}^t, \cdots, v_{iD}^t)^T$$

the personal best position as

$$p_i^t = (p_{i1}^t, p_{i2}^t, \cdots, p_{id}^t, \cdots, p_{iD}^t)^T$$

the global optimal fitness as:

$$p_g^t = (p_{g1}^t, p_{g2}^t, \cdots, p_{gd}^t, \cdots, p_{gD}^t)^T$$

for each iteration of PSO, the dth-dimension of particle i's velocity vector, and its position vector is updated as follows:

$$v_{id}^{t+1} = \omega v_{id}^t + c_1 r_1 (p_{id}^t - x_{id}^t) + c_2 r_2 (p_{gd}^t - x_{id}^t) \tag{1}$$

$$x_{id}^{t+1} = x_{id}^t + v_{id}^{t+1} \tag{2}$$

The inertia weight w controls the influence of a particle's previous velocity, resulting in a memory effect. The use of the inertia weight improved performance in a number of applications. Originally, it was linearly decreased during a run, providing a balance between exploration and exploitation.

Study factor c_1, c_2 play an role in adjusting a particle's social swarm experience and its own experience. When c_2 is equal to zero, the particle i dosen't have the cognition experience and only has the social experience. In this case, the convergence speed may be very quickly. But dealing with the complicated problems, the optimal solution could be just the local optimal solution. The particle only has the cognition experience when c_2 is equal to zero. At this time, the particle may not exchange the information with others, and its any information is only influenced by itself. But it can't converge quickly. Among others, Kennedy called the cognition only model [8]. This model uses only a particle's personal best position in the velocity update.

v_i of the t iteration is then updated as:

$$v_{id}^{t+1} = \omega v_{id}^{t} + c_1 r_1 (p_{id}^{t} - x_{id}^{t}) \tag{3}$$

r_1, r_2 ∈ [0, 1] represents a vector of random variables following the uniform distribution between 0 and 1.The significance of v_{max} is initialized at the algorithm, If the $\left|v_{id}\right| > V_{\max}$,then $\left|v_{id}\right| = V_{\max}$.

2.2 Two Sub-swarms PSO Algorithm

The Basic Particle Swarm Optimization algorithms can be achieved simply and easily, however, when optimizing the complex multi-function, it will be easily limited into local optimum. Because in fundamental particle swarm algorithm, all particles are influenced by gBest, and flying to gBest particles, and gBest particles may be a local optimum value.

In this paper we introduce a two sub-swarms PSO algorithm in which particles will be divided into two clusters with the same size. The first cluster adopts standard PSO model, namely, evolution on the basis of the formulas (1) and (2), the second cluster adopts the Model of Cognition Only, namely, evolution on the basis of the formula (3) and (2). After iteration when searching, the worst adaptation of particles in the first cluster will be exchanged with the optimal adaptation of the particles in the second cluster. In the entire search process, the inferior particles in the first cluster will be continuously exchanged with the superior particles in the second cluster in this way. The first cluster obtains the superior particles continuously, and the inferior particles exchanged from the first cluster will be evolved as the Model of Cognition Only in the second cluster, because of without influences by the gBest in this model, the particles may be given a new vitality through self-study. For the entire group, which maintain a good diversity in this way, and don't become premature convergence due to meeting local extreme point. To better balance the relationship of exploration and development, the inertia weight w of the first cluster decreases linearly with the iterative process while the inertia weight of the second cluster increases with which.

The processes of two sub-swarms particle swarm optimization algorithm are as follows:

Step1: Initialization the number of each sub-swam particles, the largest number of iterations, inertia weight, acceleration constant, the initial position of particles and initial velocity.

Step2: In the respective groups, calculating the adaptation value of each particle, preserving the best initial location and the optimal adaptation of the initial value.
Step3: For the first sub-swam particles, calculating the new speed of each particles according to equation (1), calculating the new location of each particles according to equation (2), while limiting the new position of particles to handle; For the second sub-swam particles, calculating the new speed of each particles according to equation (3), calculating the new location of each particles according to equation (2), while limiting the new position of particles to handle.
Step4: For each particle, updating the adaptation value and the position of the best; for each sub-swam particles, updating the adaptation value and the position of the best.
Step5: Exchanging the worst particles of the first sub-swam particles and the optimal particles of the second sub-swam particles. Updating the adaptation value and the position of the best of the first sub-swam particles.
Step6: If the cease meet the condition that the number of iterations over the maximum allowable number of iterations or search for the optimal location to meet a predetermined threshold.

3 Experimental Results

3.1 Test Functions

$$F_1 = 21.5 + x_1 \sin(4\pi x_1) + x_2 \sin(20\pi x_2)$$

as $-3.0 \le x_1 \le 12.1, 4.1 \le x_2 \le 5.8$

$$F_2 = \frac{1}{2}\sum_{i=1}^{2}(x_i^4 - 16x_i^2 + 5x_i)$$

as $-5.0 \le x_1 \le 5.0, -5.0 \le x_2 \le 5.0$

$$F_3 = \sum_{i=1}^{5} i\cos[(i+1)x+i]\sum_{i=1}^{5} i\cos[(i+1)y+i]$$

$$+0.5[(x+1.42513)^2 + (y+0.80032)^2]$$

as $-10.0 \le x, y \le 10.0$

$$F_4 = -\sum_{i=1}^{2} x_i \sin(\sqrt{|x_i|})$$

as $-500.0 \le x_i \le 500.0$

F_1 has a global maximum with a function value of 38.8503. Maximum is located at (11.6255, 5.725). The global maximum of F_2 is located at (5.0, 5.0) with a function value of 250. F_3 is Shubert function, and its global minimum located at (-1.42513, -0.80032) is -186.7309. F_4 is Schwefel function. Maxima is located at (420.9687, 420.9687), with a function value of -837.9658.When we use the basic particle swarm optimization to test these functions; the global maxima may not be found.

3.2 Empirical Approach

The particle number of each sub-swarm is 30.In the first cluster, the inertia weight w was scaled linearly from 0.95 to 0.4 with c_1=c_2=2.However, the inertia weight w of the second sub-swarm scaled linearly from 0.4 to 0.95 with c_1=2, c_2=0. For each of the 7 test functions, 500 simulations were done over a maximum of 2000 iterations of the TSSEPSO algorithm. The V_{max} is 0.5 and accuracy class is 0.0001.

3.3 Results and Analysis

The conclusion from this table is that TSS-PSO algorithm of overall situation optimizing ability is obvious better than that the standard particle swarm optimization algorithm. The TSS-PSO is better to find the global optimal solution. The results show that the TSS-PSO is a valid optimization algorithm especially in the complicated multimode functions.

Table 1. Comparison of optimization performances between two optimization algorithms

Test Functions	TSS-PSO		PSO		Best fitness
	Optimal rate (%)	Mean of fitness	Optimal rate (%)	Mean of fitness	
F_1	99.6%	38.8443	1.2%[6]	38.7323[6]	38.8503
F_2	44%	219.2269	28.4%	195.9189	250.0
F_3	60.2%	-182.4708	48%	-184.1234	186.7309
F_4	41.8%	-762.0110	16.6%	-750.8282	837.9658

4 Conclusions

Basic particle swarm optimization will be easily limited into local optimum, optimizing to complicated problem. So a new particle swarm optimization (TSS-PSO) algorithm is proposed in this paper. TSS-PSO is an improved particle swarm optimization algorithm. Each sub-swarm iterates using the different model, and that the particles between two sub-swarms can exchange the information according to the rule. By exchanging the information, the diversity of population can be improved and global optimal solution may be found easily. The overall situation optimizing ability has obviously risen than the fundamental particle group optimization. That method making a clear concept, simple procedure, the strong overall situation optimizing ability, has very good practical value.

References

1. Kennedy, J., Eberhart, R.: Particle Swarm Optimization. In: Proceedings of IEEE Int'l Conf. on Neural Networks, pp. 1942–1948. IEEE Service Center, Piscataway (1995)
2. Eberhart, R., Kennedy, J.: A New Optimizer Using Particle Swarm Theory. In: Proceedings of the 6th International Symposium on Micro Machine and Human Science, pp. 39–43. IEEE Service Center, Piscataway (1995)
3. Shi, Y., Eberhart, R.: A modified particle swarm optimizer. In: Proceedings of IEEE World Congress on Computational Intelligence, pp. 69–73. IEEE Press, New York (1998)
4. Shi, Y., Eberhart, R.: Fuzzy Adaptive Particle Swarm Optimization. In: Proceedings of the Congress on Evolutionary Computation, pp. 101–106. IEEE Press, Seoul (2001)
5. Van den Bergh, F.: An Analysis of Particle Swarm Optimizers. PhD thesis. Department of Computer Science. University of Pretoria, South Africa (2002)
6. Chen, G.-C., Yu, J.-S.: Two sub-swarms particle swarm optimization algorithm and its application. Control Theory & Applications 24(2), 294–298 (2007) (in Chinese)
7. Eberhart, R., Shi, Y.: Particle Swarm Optimization: Developments, applications and resources. In: Proceedings of the IEEE Congress on Evolutionary Computation (CEC 2001), pp. 81–84. IEEE Service Center, Piscataway (2001)
8. Kennedy, J.: Small Worlds and Mega-Minds: Effects of Neighborhood Topology on Particle Swarm Performance. In: Proceedings of the IEEE Congress on Evolutionary Computation, pp. 1931–1938. IEEE Press, New York (1999)

Research and Design of Personalized Information Retrieval Based on Users' Clustering

Yan Hu and Baohong Yu

School of Computer Science and Technology of Wuhan University of Technology,
Wuhan, Hubei, China
huyan168@sina.com, adhduy0221@163.com

Abstract. With the description of the shortcomings of the existing popular search engines and the personalized information retrievals based on individuals, the personalized information retrieval based on users' clustering was proposed, of which the key technologies of users' clustering, user group dictionary, user search and sorting and so on were described in detail and it was proved by experiments that the personalized information retrieval based on users' clustering improved the efficiency of the information retrieval.

Keywords: Search engines Personalized Users' clustering User search Sorting Information retrieval.

1 Introduction

With the increasing popularity and the rapid development of Internet, the amount of information on the network is growing bigger and bigger, whether can we get access to the useful information on the Internet accurately and efficiently plays a great part on the efficiency of information retrieval. Undoubtedly, the search engine is the most beneficial help, but large numbers of the major search engines are for public services rather than for individuals' personalized services and the personalized information retrieval based on individuals with low rate of recall also need a large number of machines learning, in addition, the quality of the services of both are unsatisfied.

Taking into account that there are different degrees of similarity among the individual users with similar or even identical query intentions on the same keyword, choose users with great similarity between each other together as a single user group according to multiple observations. The way of the personalized Information based on users' clustering putting the individual users in the context of a user group in information retrieval is in favor of mutual cooperation between individual users and can obtain the maximum degree of the true intentions of the users' queries to the largest extent, improved the retrieval efficiency.

2 Structure of the System

The system used the distributed three-tier architecture as shown in Fig. 1:

The first tier is the client , mainly to provide interfaces for users' search.

D. Jin and S. Lin (Eds.): Advances in CSIE, Vol. 2, AISC 169, pp. 449–453.
springerlink.com

The second layer is the proxy server. Primarily, it's responsible for handling user-related information, such as user's clustering, the establishment of user group dictionary.

The third layer is the server which is mainly responsible for dealing with interaction with a variety of commercial information retrieval systems on the www, integrating and merging information from different search engines to create the index library.

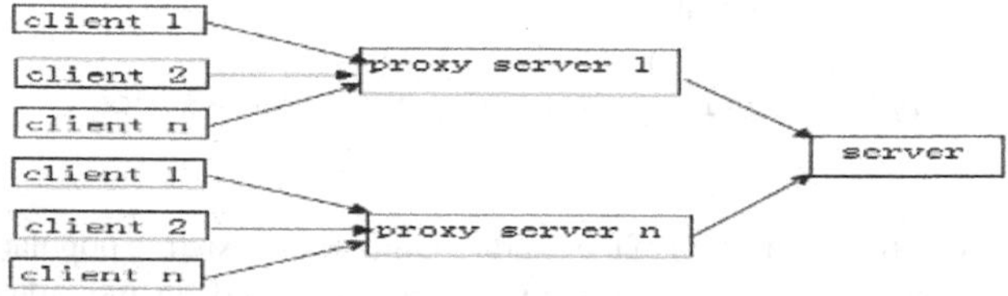

Fig. 1. Structure of the system

3 Key Technologies

3.1 Users' Clustering

Step 1: The establishment of the user eigenvectors

The system got basic information of users by their registration which was made up of username, password, age, education, gender, occupation, interests, and group number and so on, and then put them into vector form with the vector space model. Specifically, they can be expressed as follows:

$$U=(w_1,w_2,\ldots,w_j,\ldots w_n) \tag{1}$$

U Represents the user, w_j represents the weight of item j of the user.

Step 2: Similarity calculation

The similarity is calculated based on the way of distance measurement as follows:

$$sim = \left[\frac{|w_i - w_j|^2}{|w_i|^2 + |w_j|^2} \right] \tag{2}$$

In the above formula, sim can take into account the two factors of the angle between the vectors and the length of them. when the web document eigenvector and the query words in the query vector are not related (that is, there is no intersection of their collection), W_i and W_j is vertical, accordingly the value of sim is 1; when the two pages be the same, accordingly the value of sim is 0; when they are similar but not the same, the value of sim is between 0 and 1.

Step 3: User clustering

We used the single clustering method which is easy to achieve, do not need to compare all the similarity between the sets, performs faster, and is for collections of large numbers of documents and is with high practicability. Meanwhile, it does not require pre-determined values of k in the clustering process, reducing dependence of the domain knowledge and improving the flexibility. The algorithm is described as follows:

Suppose the user set is P = $\{u_1, u_2, \ldots u_i, \ldots, u_n\}$:

(1) See each user of P u_i as a member of the set with a single member $c_i = \{u_i\}$;
(2) Choose one of a set with only one member c_i as a the starting point for clustering
(3) Find the u_j between which and c_i the distance meeting the conditions (maybe the nearest point to c_i, that is the u_j with the largest similarity of sim (c_i, u_j), it can also be a point between which and c_i the distance above the threshold u, that is any u_j with the similarity sim $(c_i, u_j) \geq d$), in the remaining sets which did not cluster. Make u_j be included in ci to create a new set of ck = sim $c_i \cup u_j$;
(4) Repeat process(3) until the distance between c_i and u_k which is nearest to c_i beyond the threshold u, At this point it is considered that a class has been finished clustering
(5) Select a set of only one member to cluster, repeat process (3) and process (4) to start a new round of clustering until all sets with only one member c_i are involved in the clustering.

3.2 User Search

Step 1: Extract the Web document Characteristic information

Server analysis the structure and content of the sets of web documents which were returned by search engine first, and then focused on the TFIDF analysis of them, such as formula (3) shows, and extracted n critical Words with the highest TFIDF value of each document and their frequency to represent as a document vector with vector space model, such as (4) below.

$$W(t,d) = \frac{tf(t,d) \times \log(N / n_t + 0.01)}{\sqrt{\sum_{t \in d} [tf(t,d) \times \log(N / n_t + 0.01]^2}} \tag{3}$$

In formula (3), W (t, d) stands for the weight of the word t in the text d, and tf (t, d) is the frequency of the word t in the text d, N is the total number of the training text, n_t is the total number of the training text in which the word t appears concentrated, The denominator is the normalization factor.

$$d = (t_1, w_1, t_2, w_2, \ldots, t_i, w_i, t_n, w_n) \tag{4}$$

In formula (4), each dimension of the vector was made up of the key and its weight. For each term t_i, $t_i \in T$, T was the set of characteristic words, w_i means the weight of the keyword t_i in the document d .

Step 2 : Establish eigenvectors of the users' query terms

Considering that the clustering processes of the users belonging to the same User group are also very similar in information retrieval, establish a simple dictionary for each user group.

Record the query terms searched of all users and their corresponding web documents to constitute a set of query keywords (Query) and the web document set (Result), and then put this information into the database to constitute the user group dictionaries.

Step 3 : Establish the table of Characteristic words of the user group

Extract some representative words in the user group dictionary to build a table of characteristic words. Consider the two factors of frequency and concentration on the choice of characteristic words. Only when both weight and concentration of a word are larger than the corresponding threshold, the word will be chosen as the characteristic word to add it into the table of characteristics words of the user group .If in a certain period of time, its weight decreases, does not meet the requirements of the system, it will be removed from the table of characteristics words.

Extract each query word in the query set Q of the user groups into a query vector .Deal with each user group in the system in the same way to create the database of user group information in the form shown in Fig. 2 below:

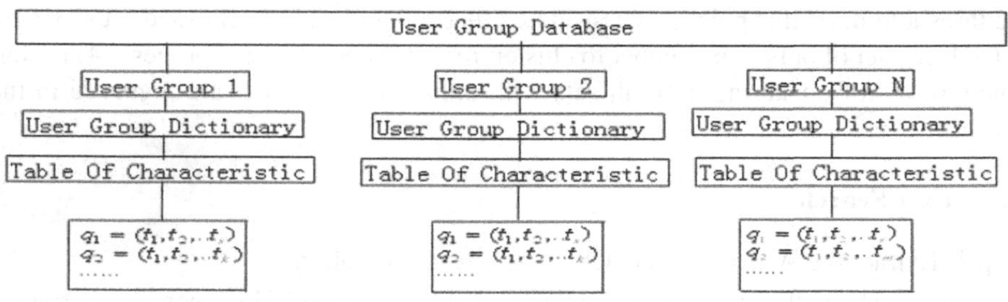

Fig. 2. User Group Database Schema

3.3 Sorting

When the user submits a query, the client passes it to the proxy server, then the proxy server query the table of characteristic words of the user group first, if there is the corresponding characteristic word, it get the corresponding query vector q, and calculate the similarity of the query vector q and the web document vector d, then select all the documents of which the similarity is greater than the threshold in descending order according to the similarity value and returned the results to the user. If there is not, it would be submitted to the server, the server started to search, download, analysis, and then do in the above situation.

4 Testing and Analysis

Experiment was made on a LAN which were made up of three P4s (clocked at 3.0 GHz, memory 1G) PCs. Mysql was used as the database and Java was used as the programming language.

4.1 Evaluation Criterion of Experimental Results

The evaluation of the Information retrieval system's performance is mainly reflected in three areas of the precision, recall and response time. The precision rate and recall which are used as the criterion of the test are defined as follows:

Precision (%) = number of relevant documents / number of all documents retrieved

Recall (%) = number of relevant documents/ number of all documents of the System

4.2 Experimental Results and Analysis

In the experiment, I compared the retrieval performance of the two ways (as shown in Tab. 1) in which all conditions are the same except that one is based on the user clustering and the other is not. To simplify the search algorithm without affecting the accuracy, 5 groups of selected query words were tested in both cases to get the rate of the precision and recall of them. In the following experimental results, performance of personalized information retrieval based on the users' clustering which provided search results that users were interested in accurately and fast was greatly improved compared to personalized information retrieval based on individuals.

Table 1. Experiments Result

	based on users' clustering			based on individuals		
User Search	Precision	Recall	Response	Precision	Recall	Response
Personalized	92.85%	95.25%	1.25(s)	92.35%	34.82%	2.32(s)
Arithmetic	89.92%	98.32%	1.37(s)	89.10%	38.45%	2.51(s)
Cluster	91.26%	93.47%	1.17(s)	91.15%	42.39%	1.92(s)
Reverse Index	98.75%	97.52%	1.56(s)	98.45%	47.92%	2.89(s)
Search Engine	94.23%	92.63%	1.65(s)	93.89%	32.06%	3.1 (s)

References

[1] Manning, C.D., Raghavan, P., Schutze, H.: Introduction to Information Retrieval. People's Telecon Publishing House, Beijing (2010)

[2] Jain, A.K., Dubes, R.C.: Algorithms for Clustering Data. Prentice-Hall, Inc. (1998)

[3] Henzinger, M.R.: Hyperlink analysis for the Web. IEEE Internet Computing 5(1), 45–50 (2001)

[4] Zhu, M., Wang, J., Wang, J.: The Study of Feature Selection in the Web Page Classification. Computer Engineering 26(8), 35–37 (2000)

[5] Wang, G., Xu, J.: TCBLSA: A New Method of Chinese Text Clustering. Computer Engineering 30(5), 21–22, 37 (2004)

[6] Song, J., Wang, Y.: Improvement of the Robot Search Algorithm. Journal of The China Society For Scientific and Technical Information 21(2), 130–133 (2002)

[7] Xing, C., Gao, F., Zhan, S.: A Collaborative Filtering Recommendation Algorithm Incorporated with User Interest Change. Journal of Computer Research and Development 44(2) (2007)

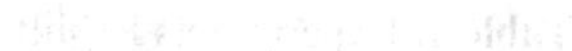

The Application Analysis of Clustering and Partitioning Algorithm in Web Data Mining

GuoFang Kuang* and MingLi Song

College of Information Technology, Luoyang Normal University, Luoyang, 471022, China
xhs_ls@sina.com

Abstract. Web Data Mining is the primary solution source model of semi-structured data and semi-structured data model, query and integration issues. The basic idea of clustering is to define the similarity between the distance, the distance that represents the data between the data to measure the similarity of the size of the data are classified, until all the data gathering is completed. This paper proposes the web data mining based on clustering and partitioning algorithm. Finally, the paper verifies the proposed algorithm, and the results show the new method to compensate for the previous clustering algorithms in the analysis of the data type shortcomings.

Keywords: web data mining, clustering, partitioning algorithm.

1 Introduction

Data mining is a new kind of data processing technology from the vast amounts of data automatically and efficiently extracts useful knowledge. In the initial stages of the development of data mining, researchers more attention focused on the mining of the data stored in the database of KDD (Knowledge Discovery in the Database, access to knowledge from the database) concept is proposed in this case the. In recent years, rapid development and widespread use of the Internet, the amount of information on the Web at an alarming rate growth, the current information resources have become an astronomical figure [1]. How people from this mass of data to get useful data and information, all this urgent need to an automatically found from the Web resources, access to information, and will not get lost in a sea of data the direction of new technology, so web mining technology came into being.

The basic idea of clustering is to define the similarity between the distance, the distance that represents the data between the data to measure the similarity of the size of the data are classified, until all the data gathering is completed. The World Wide Web is a huge, widely distributed, global information service center, covering news, advertising, consumer information, financial management, education, government, e-commerce and many other information services, which includes a rich and dynamic hyperlinks and page views message. Web is a centralized control, absence of a unified structure, integrity constraints, transaction management, non-standard query language

* Author Introduce: Kuang GuoFang (1974-), Male, lecturer, Master, College of Information Technology, Luoyang Normal University, Research area: web data mining, clustering.

D. Jin and S. Lin (Eds.): Advances in CSIE, Vol. 2, AISC 169, pp. 455–460.
springerlink.com

and data model, loosely distributed information systems, unlimited expansion of their mining very difficult to obtain knowledge reliable. Internet era, the problem is not difficult to obtain information, but to grasp the hidden behind the truly valuable information from huge amounts of data or users to access information useful knowledge is a breakthrough in the human limit. Web Data Mining and pointed out a way to resolve this problem.

Web Data Mining is the primary solution source model of semi-structured data and semi-structured data model, query and integration issues. Solve the problem of heterogeneous data integration and query on the Web; you must have a model to clearly describe the data on the Web. On the Web Data semi-structured features, the key to solving the problem is to find a semi-structured data model. Therefore, we must first establish a semi-structured data model to describe the data on the Web; also need a semi-structured model extraction techniques and technology of semi-structured model is automatically extracted from existing data. Web-oriented data mining semi-structured data model extraction techniques to semi-structured model. This paper proposes the web data mining based on clustering and partitioning algorithm. Finally, verify that the proposed algorithm, the results show the new method to compensate for the previous clustering algorithms in the analysis of the data type shortcomings.

2 The Research of Clustering and Partitioning Algorithm

The basic idea of clustering is to define the similarity between the distance, the distance that represents the data between the data to measure the similarity of the size of the data are classified, until all the data gathering is completed.

Partition clustering is the first data classification, and according to some principle of the amendment until the classification is more reasonable. The basic idea is: given a data set of data, based on experience to set the class number, generated according to some provisions of a cluster center, followed by calculation of the distance of each data center, select the distance the center of the minimum, the data classified in this class to get a cluster.

There are two representative algorithms: Partition-based clustering K-means algorithm and K-medoid algorithm [2]. K-means algorithm, it is the mean of each class of data in each class to represent. In the K-medoid algorithm, one of the data near the cluster center to each class is shown by equation 1.

$$x_i = \left(x_{i1}, x_{i2}, \ldots, x_{ip}\right)^{\mathrm{T}}, \quad i = 1, 2, \ldots, n. \tag{1}$$

Partitioning algorithm, run faster, able to effectively handle large data sets; but must determine in advance, and the center of the election to obtain a good or bad to have a significant impact on the clustering structure, is shown by equation 2.

$$D_{rk}^2 = \frac{n_p}{n_r} D_{pk}^2 + \frac{n_q}{n_r} D_{qk}^2 - \frac{n_p}{n_r}\frac{n_q}{n_r} D_{pq}^2 \tag{2}$$

At this time, it is each data as a meta-space of a data element space. Set the distance between the data generally should meet the following requirements:

(1) The similarity measure should be non-negative, that is $d\left(x_i, x_j\right) \geq 0$.

(2) Data between the similarity measures should be the maximum.

(3) Similarity measure should satisfy the symmetry $d\left(x_i, x_j\right) = d\left(x_j, x_i\right)$.

Hierarchical clustering The basic idea is: first data into classes, then the distance between the provisions of class and class, choose the smallest distance to merge into a new class, this time for a class, calculate the new class other types of distance, and then merge the nearest two categories, so that each reduction of one class until all subclasses are clustered into one class, as is shown by equation 3.

$$D_{rk}^2 = \alpha_p D_{kp}^2 + \alpha_q D_{kq}^2 + \beta D_{pq}^2 + \gamma \left| D_{kp}^2 - D_{kq}^2 \right| \tag{3}$$

CURE (Cluster Using REprentatives) is a bottom-up hierarchical clustering algorithm. In CURE, the center-based algorithm and distance calculation algorithm does not apply to non-spherical or clusters of arbitrary shape. The CURE Therefore, a fixed number of points represents a cluster, thereby improving the algorithm for mining the ability of the clustering of arbitrary shape. For the case of low-dimensional data, the complexity of the CURE algorithm, the number of data. When dealing with large amounts of data, the algorithm must be based on sampling by technology.

Two classes respectively, and set them contain data. If it is a merger, contains data from the class with the class. The problem to be solved, it is $n_r = n_p + n_q$ and other types of distance, that is $G_k (k \neq p, q)$, calculate $G_k (k \neq p, q)$ the distance between classes and the recurrence formula. The similarity is calculated as follows equation 4.

$$\hat{f}_n = f(u_n) = f[\sum_{k=1}^{K} w_k \sum_{m=1}^{M} x_m \psi(\frac{x_m - b_k}{a_k})] \tag{4}$$

The classification rules can dig out some common features, this feature can be used on the new added to the database data to classify. Web data mining classification techniques can get access to these users on personal or shared access mode characteristics of the user to access server documentation. The clustering technique is the regular features of a visit to the user characteristics excavation [3]. Finally, the model analysis, dig out the people's understandable knowledge model to explain.

Web mining is related to the integrated technology of Web technologies, databases, machine learning, data mining, statistics, computer languages and other subjects of interest is extracted from Web documents and Web activity, potentially useful patterns and hidden information (or knowledge) process. Web mining in general is defined as: interest, useful patterns and implicit information extracted from Web-related resources and behavior. Different researchers from different angles, Web mining has a different understanding, as is shown by equation 5.

$$d(x_i, x_j) = [\sum_{k=1}^{p} (x_{ik} - x_{jk})^m]^{\frac{1}{m}} \tag{5}$$

Clustering process to construct a spanning tree, which contains the class hierarchy information, and the similarity between all classes and class, it generates a hierarchical nested class, and can be used for any feature type, high accuracy. But because each merger, the global compare all the similarity between the class and choose from the smallest two classes, so run slower, not suitable for large-scale data sets.

Algorithm 1. Mean by class-based clustering algorithm

Input: Data sets $X = \{x_i\}_{i=1}^{n}, x_i \in R^d$, the number of clusters.

Output: Meet the square error criterion is the minimum disjoint cluster $\{X_h\}_{h=1}^{k}$.

(1) Choose k data as the central point of the initial;

(2) Web page collection P = {p1, p2, ..., pn}, of xij (i = 1,2, ..., n; j = 1,2, ..., m), said the Web pages within a given time pi and pj number of common access to the same user session;

(3) IF C=0 OR ||D||≤1 THEN

(4) Assign each of the remaining data from it's recently the center of represented class;

(5) Take λ2 for the matrix element values directly from the R 'to find where the degree of similarity (xi, xj) (rij = λ2,), λ2, elements corresponding corresponds to λ1 = 1 is equivalent to classification xiclass, and xj where the merger, all of these cases the merger after the equivalent classification corresponds to λ2;

(6) Randomly select a center point data O_{random};

(7) if $S < 0$, then O_{random} replaced to O_j form a new center point of the set;

Many data mining algorithms trying to minimize the impact of isolated points, it is exclude them. But one person's noise is another man's signal, which may lead to the hidden information is lost. In other words, the isolated point is very important. For example, intrusion detection, outlier may indicate intrusion. In this way, the outlier detection and analysis is an interesting data mining tasks, known as outlier mining.

Web Data Mining is the primary solution source model of semi-structured data and semi-structured data model, query and integration issues. Solve the problem of heterogeneous data integration and query on the Web; you must have a model to clearly describe the data on the Web. On the Web Data semi-structured features, the key to solving the problem is to find a semi-structured data model. Therefore, we must first establish a semi-structured data model to describe the data on the Web; also need a semi-structured model extraction techniques and technology of semi-structured model is automatically extracted from existing data.

3 The Web Data Mining Based on Clustering and Partitioning Algorithm

Web mining is to find hidden patterns from a large number of Web documents in the collection process. Will be seen as the output as input, the Web mining process is a mapping from input to output: $\xi : C \to p$.

Current information on the Web is mainly divided into three categories: 1) Web page content, including text and multimedia information; 2) Web page hyperlinks between reference data; 3) Web server on the user login Web site access log data. The resulting Web mining is divided into three categories: Web content mining, Web structure mining, and Web access log mining [4].

Web content mining is a web-based content, Web mining, Web data, extract useful knowledge of the process. Web content mining targeted at Web document information and multimedia information content in terms of its mining, can be divided into a text document on the Web (including Text, HTML and other formats) and multimedia documents (including Image, Audio and Video media mining type).The basic Web content mining is a text mining.

Algorithm 2. the web data mining based on clustering and partitioning algorithm

Input: Data sets $X = \{x_i\}_{i=1}^{n}, x_i \in \mathrm{R}^d$, the number of clusters.

Output: k Classes so that all data and recently the center of the dissimilarity minimize the sum of it.

(1) Web Data Matrix data compression to the interval [0, 1].Let, U = {x1, x2, ..., xn} for Web object collections, each Web object by the m indicators that its properties: xi = {xi1, xi2, ..., xim from} (i = 1,2, ..., n);

(2) Disjoint partition data sets, Constrained K-means objective function optimal;

(3) To calculate the cluster data for the mean time, if the data is assigned a class (set $X_h^{(t+1)}$); when the data will be assigned to h^* class (collection $X_{h^*}^{(t+1)}$)

$$h^* = \arg\min_{h \in \{1,\dots,k\}} \left\| x - \mu_h^{(t)} \right\|^2$$;

(4) Take λ = λ1 = 1, the matrix elements of the maximum similarity class [xi] R = {xi | rij =, λ1}, for each xi is about to meet rij = 1 xi and xj is divided into a class, constitute similar class;

(5) Take the elements of the degree of similarity λ3 matrix element value from the R 'directly to the (xi, xj) (rij = λ3), similar to the equivalent category corresponds to λ2, xi where class and xj where the class merge;

(6) if $S < 0$, then O_{random} replaced to O_j form a new center point of the set;

Web structure mining is found from the web page hyperlink structure and their mutual relations. You can use to find hidden links on a page after the structural model, this model, the authoritative site on the web page classification or similarity measure for web page creation, and help users find related topics. Web structure mining and content mining are closely linked; both are mining raw data on the Web, usually in an application to combine these two data mining tasks. The paper is using WindowsXP operating system, and using Visual C + +6.0 to achieve the above clustering sets and computing algorithms, comparing the results shown in Figure 1.

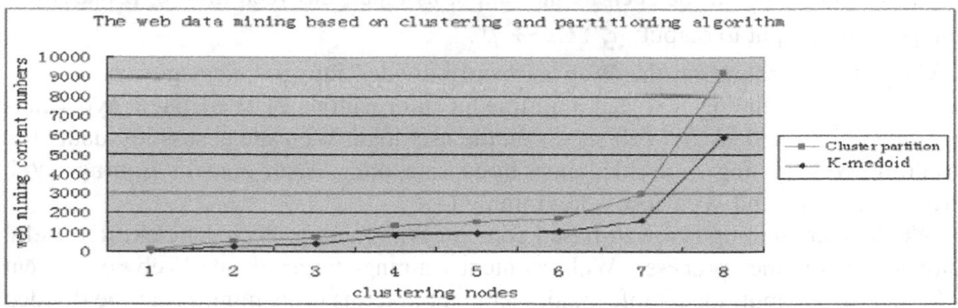

Fig. 1. The compare of web data mining based on clustering and partitioning algorithm with K-medoid

4 Summary

This paper first introduces the clustering used in mathematical knowledge, then the clustering algorithm to classify and explain the characteristics of each algorithm, followed by the introduction of a partition clustering algorithms for web mining, combined with outlier analysis algorithms on the traditional clustering the problems of the algorithm in web mining. This paper proposes the web data mining based on clustering and partitioning algorithm. Finally, verify that the proposed algorithm, the results show the new method to compensate for the previous clustering algorithms in the analysis of the data type shortcomings.

References

1. Han, J., Kamber, M., Tung, A.K.H.: Spatial clustering methods in data mining: A survey. In: Geographic Data Mining and Knowledge Discovery, pp. 188–217. Taylor and Francis, London (2001)
2. Xing, E.P., Ng, A.Y., Jordan, M.I., Russell, S.: Distance metric learning, with application to clustering with side-information. In: Advances in Neural Information Processing Systems, vol. 15, pp. 505–512. MIT Press, Cambridge (2003)
3. Kosala, R., Blockeel, H.: Web mining research: A survey. SIDKDD Explorations (July 2000)
4. Han, J., Kamber, M., Fan, M., Meng, X.-F., et al. (Transl.): Data Mining: Concepts and Techniques. China Mechine Press, Beijing (2001)

Digital Temperature Measurement System Based on Oscillator*

Ling Zhu

Electrical Engineering Automation Department of Zhejiang University
Lilac9152@163.com

Abstract. To solve the conventional temperature measurement system's problems, such as lots of hardware, high cost, poor Anti-interference ability, a new temperature measurement system based on oscillator is designed. This system, which takes negative temperature coefficient thermistors as temperature-sensing and goes through ring oscillator RC, will reflect the change of the temperature to the change of the pulse cycle. Error compensation and then LED display is done through the single-chip microcomputer timing/counter measuring pulse cycle and then processed by the software. The system, using the minimum hardware circuit design, is characterized by its simple circuits, low power consumption, low cost and reliable function. It is especially suitable for all kinds of intelligent household appliances.

Keywords: negative temperature coefficient thermistors (NTC), RC ring oscillator, microcontroller, temperature measurement.

1 Introduction

Different temperature sensors, such as sensor thermocouple, thermal resistance, thermal resistance of semiconductor are commonly used to measure temperature while negative temperature coefficient thermistor (NTC) is used on the room temperature zone. Negative temperature coefficient thermistor is characterized by high sensitivity, small size, fast response and low cost. The resistance will change obviously as the temperature changes. The paper presents a new method of temperature measurement, using NTC as the foundation and oscillator RC as the key.

The conventional measurement, according to different methods of analog to digital conversion, can be divided into two categories; one is analog-to-digital conversion with AD chip, which is composed of testing circuit, amplifying circuit, A / D conversion circuit and single-chip microcomputer. Through the NTC and the detection circuit, output the simulation of weak voltage corresponding to the temperature, which is converted to digital signal by A/D after being amplified. The temperature is measured through the single chip computer data processing of the digital signal. The other is to

* 2011 Zhejiang University Students' Scientific Research Training Plan(7672).

D. Jin and S. Lin (Eds.): Advances in CSIE, Vol. 2, AISC 169, pp. 461–466.
springerlink.com

use V / F chip to replace the A / D chip. Their shared defects are complex structure, high cost and lots of hardware.

2 Principle of Temperature Measurement

The principle of RC ring oscillator [1] is shown as in Figure 1. In the figure, Rt is negative temperature coefficient thermistor, Rs current-limiting resistor. The output voltage waveform of the V_O is shown as in Figure 2. It is a rectangular pulse, period T = 2.2RtC [2]. C is a constant. So the resistance of the Rt will change as the temperature changes, making period T of the oscillator output pulse change. By detecting the pulse period T, the environment temperature of the thermal resistance can be measured, which is called the realization of temperature. Compared with the conventional measurement circuit, the new method has the advantages of less hardware, reduced cost and greatly improved reliability of resisting disturbance because of the pulse signal.

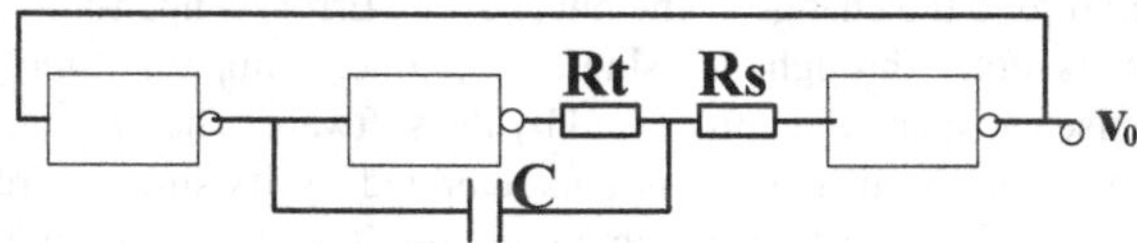

Fig. 1. RC ring oscillator schematic diagram

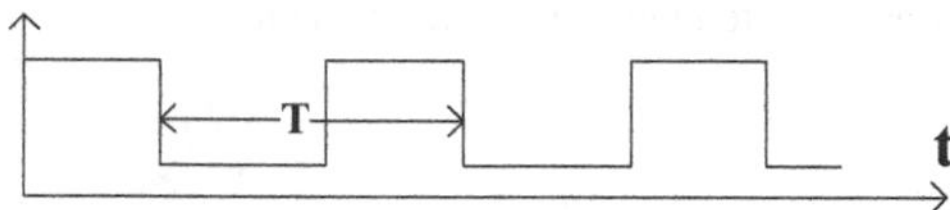

Fig. 2. V_O waveform voltage

3 Hardware Circuit Design

Shown as in Figure 3, hardware circuit is composed of an SCM, RC ring oscillation circuit, display, keypad and watchdog. For a minimal optimization system, chip P87LPC764BN, a 20-foot encapsulated chip, made in Philips Company is chosen, which fit in with the requirement of high integration and low cost.

Accelerated kernel 80C51 is used to speed up by two times as fast as the standard MCU 80C51. In the chip, there is a 32Byte user code and RAM of 28 bytes which can meet the needs of the system. The 32Byte user code can be used for storing sequence code and set parameters; With the watchdog circuit, external watchdog is not needed. No external components are needed with the power-on reset in the chip. There are 2 16 digital timers / counters; All architraves have 20mA drive ability. The voltage VDD = 4.5 ~ 6.0V.

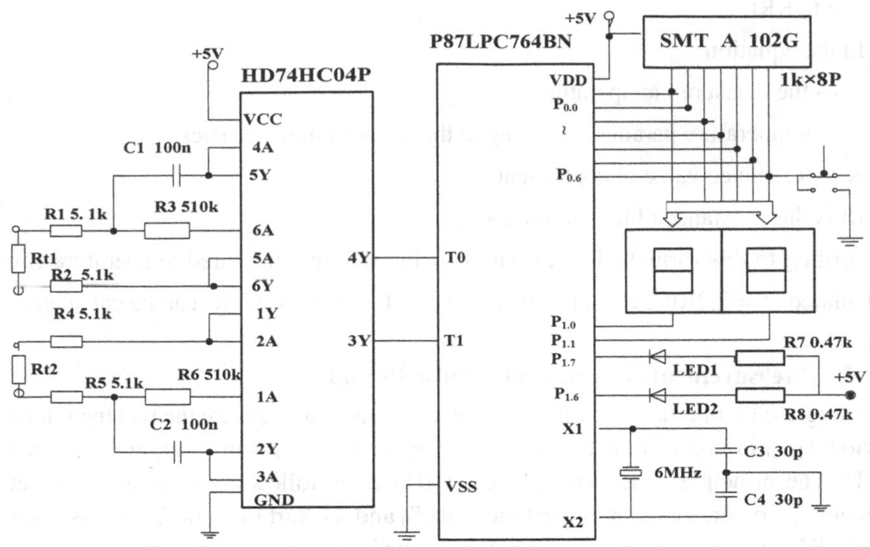

Fig. 3. System hardware circuit diagram

The system has two temperature measuring points. They are the negative temperature coefficient thermistor Rt1 and Rt2, and the temperature measurement range is 0 ~ 99 C. The chip HD74HC04P has six negaters, forming two RC oscillators. A rectangular pulse which is generated through the RC oscillator were transmitted to input pins of P87LPC764BN's counter T0 and T1 and then temperature can be calculated. As the line P0 has driving power, which directly drives LED device. When LED1 is on, the temperature (t1) of the measuring point1 is displayed; when LED2 lights, it shows the temperature (t2) of measuring point 2. Human-computer dialogue is achieved and parameter is set by the keyboard interrupt function of the chip P87LPC764BN's mouth P0. To improve the reliability of the system, the watchdog timer used. The figure is cleared to zero in the subprogram. The base pin is to be developed.

4 Software Design

4.1 Interrupt Service Program of Temperature Measurement

4.1.1 Mathematical Model of Temperature Measurement

Resistance value of NTC is an exponential non-linear curve, which in general can fully meet the requirements of precise measurement after linear processing according to the temperature range. In order to improve precision, a compensation link is added in software. The temperature calculation formula is as follows:

$T = t_0\text{-}KRt$

In the equation,

T --- the measured temperature

t_0 --- temperature parameter relating to thermistor characteristics

K ---resistance value of coefficient

Rt is the resistance of the thermistor

According to this formula, if Rt, t_0 and K is known, the measured temperature can be calculated. T = 2.2RtC; C is known, as long as T is measured, Rt can be calculated.

4.1.2 Measurement of Rectangular Pulse Period T

The single-chip internal two timers / counters are used to measure the rectangular pulse period T[3] in order to track in real-time temperature measurement, and occupy less CPU. The principle of measuring point 1 (t1) is as follows: supposing T_0 is set as counter, T1 timer, T0, at the same time both T_0 and T_1 start to work. T_0 begins to count on the RC oscillating circuit output pulse 1 while T_1 begins to count frequency signal in the internal, i.e. using twelve internal clock frequency signal for timing pulse.

When timing is up, overflow trap T_1 makes T_0 stop counting, so that a period of measurement is completed. Rectangular wave period $T = N_1T_1 / {}_{12} / N_0$, in which N_0, N_1 are respectively count value of T_0 and T_1, $T_1 / {}_{12}$ is the frequency signal period of the internal clock. Apparently, N_0 changes real time, N1 (preset) and $T_1 / {}_{12}$ are known. If T_0 and T_1 exchanges, we can know the measuring principle.

4.1.3 Interrupt Service Program of Temperature Measurement

In the initial program, T_0 is set as counter T_1 timer, and T_0 and T_1 start to work at the same time. T_1's interrupt service program flow diagram is shown as in figure 4. In the T_1 timer interrupt service procedures, first make T_0 stop counting and read count value N_0, which is compared with the previous measured value N_0. If there is a change, subroutine is started. Calculate T, Rt_1 and t_1, convert t_1 into BCD and store it in digital display buffer. Then set T_1 as counter, T_0 timer, both T_0 and T_1 start to measure the temperature on point 2 (T_2). Return CPU to the main program. T_0's program is the same as T1's procedure, thus completing the measurement of T_2, and start the next period's measurement.

4.2 Main Program

The main program is mainly composed of initial program and display subroutine.

In initial procedure, automatic reset in the internal chip is used. T_0 is set as counter T_1 timer, and T_0, T_1, keyboard interrupt and watchdog circuit begin to work. After initialization of CPU, the value of T1 and T2 is dynamically displayed by starting subroutine. Whenever a dynamic scanning is completed, the watchdog is cleared to

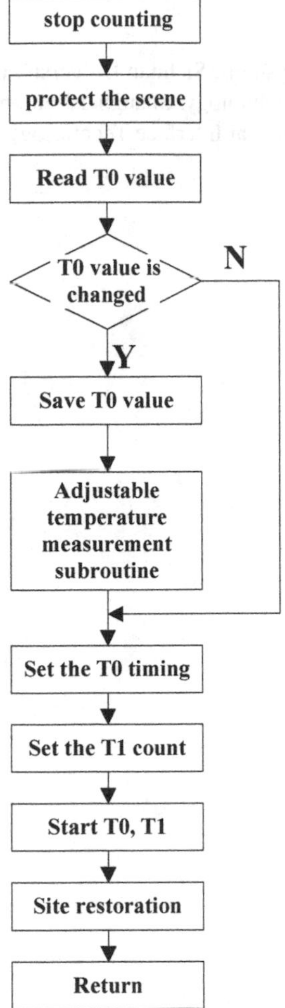

Fig. 4. T1 interrupt service program

zero. If executive agencies are needed, the only thing to do is to control subroutine; real-time temperature measurement can be realized.

5 Conclusion

Digital temperature measurement system based on oscillator, using the minimum hardware circuit design, is characterized by its. simple circuits, low power consumption, low cost and strong anti interference capability. It is especially suitable for all kinds of intelligent household appliances. Besides it can also be used to measure intelligent instrument in far distance.

References

1. Fu, Y.: Digital Circuits And Systems. Sichuan University press (2003)
2. Tang, Y.: Digital Electronic Technology. Shanghai Jiaotong University Press (2001)
3. Li, H.: CS-51 Series MCU Practical Interface Technology. Beihang University press (1993)

Numerical Simulation Research of Weld Stress Field after Welding Trailing

Qinghua Bai*, Yuejin Ma, and Weilian Sun

College of Mechanical and Electrical Engineering,
Agriculture University of HeBei, Baoding, 071000, China
baiqinghua@eyou.com

Abstract. Welding is a complex process including of physics, conducting heat, metallurgy, mechanics. There are some cracks, air holes and residual stress in the parts after welding, which will reduce welding strength and welding quality. Though there are many methods to relif residual stress, it is a difficult problem about how to improve it. In the article, it uses the software ANSYS to simulate muti-welding seam process of flat, knows temperature field and residual stress field in the parts after welding, so we can know some parameters such as area of hammerhead, peening temperature, peening force, and the peening frequency and control them to redcue residual stress, then optimize welding parameters, improve welding quality of the parts.

Keywords: residual stress, temperature field, simulation, ANSYS software, strength.

1 Introduction

Welding is used in all kinds of regions as a basic method and technology of manufacture. Being a process of a quick and partly course which be warmed and cooled [1,2], there are to cause non-uniform temperature field and stress field and have residual stress to remain in the part.

There are many methods to eliminate residual stress such as heat treatment, vibration method, explosion method, mechanical strength method, peening method [2-6], but have some limitations. Trailing peening is a usually method to reduce, to eliminate welding stress, to avoid welding crack. Welding with trailing peening is a new method which has a development in several years, it has some advantages such as easy to operate, simple equipment, high efficiency [5-10]. There are some important parameters to eliminate welding stress in the process of peening such as peening force, peening temperature, peening frequency, the area of peening head [11].

In the article, we use ANSYS software to simulate multi-welding seams of a flat of 60CrMnMo material with limited elements method. With the time transmission in welding process, we can find temperature field variation, relif residual stress field, and

* The author: Qing-hua Bai (1972-), Male, born in Heibei Province Cang County, Ph.D. Candidate, study in the theory, design of Automobile.

D. Jin and S. Lin (Eds.): Advances in CSIE, Vol. 2, AISC 169, pp. 467–472.
springerlink.com

can decide when to begin to peening welding, so we can eliminate welding stress in maximum degree.

2 Principle to Relif Welding Stress Using Welding With Trailing Peening

2.1 Principle to Welding With Trailing Peening

When metal of welding seam have a tensile deformation stronger than theirs CST in range of the brittleness temperature (BTR), there will produce welding cracks.

$$\frac{d\varepsilon}{dT} > \text{CST}$$

ε-the inner deformation, real deformation when welding seam was cooled

T- Temperature

Through trailing peening, those crystal which have high temperature extrude each other, so it make welding seam become intensive. The other hand, under the action of electric arc heat, peening can make a metal in BTR have a transversal extrusion strain. So it can neutralize tensile strain which happen when welding metal was non-uniform to warm up, and prevent to produce welding hot crack. The third, the direction of welding crystalline grain will be disarranged, form isometric crystal, it also can prevent to produce welding hot crack.

3 Mathematic Model of the Flat with Multi-welding Seams

3.1 Non-linear Transient Heat Conducting Control Equation

Welding temperature field is a typical non-linear transient problem, the equation [12]:

$$\frac{\partial}{\partial x}\left(k_{xx}\frac{\partial T}{\partial x}\right)+\frac{\partial}{\partial y}\left(k_{yy}\frac{\partial T}{\partial y}\right)+\frac{\partial}{\partial z}\left(k_{zz}\frac{\partial T}{\partial z}\right)+q=\rho c\cdot\frac{\partial T}{\partial t}$$

ρ— material density

c — specific heat

k — material conductor

q — welding resource intensity

3.2 The Boundary Condition

We look part's bottom surface, up surface, surrounding boundary as exchange heat boundary condition.

$$k\cdot\frac{\partial T}{\partial n}+\alpha(T-T_0)=0$$

n — normal direct of boundary surface

α— surface exchange coefficient

T0 — surrounding medium temperature (20°C)

3.3 The Type of Heat Source

The typical heat sources are point source, surface source, solid surface. We use Guass surface source here, the mathematic model is:

$$q = q_m \exp\left(-3\frac{r^2}{R^2}\right)$$

q_m— maximum heat flux
R — heat radius of electric arc
r — the distance to center of heat source

In the process of simulation, we divide welding seam area into elements alone, use the method of dead and live element, kill the second welding seam firstly, then put Guass heat source which writes with APDL language on first welding seam, live the second welding seam and put Guass heat source on it. So, we can get temperature distribution and have a good base for stress analysis, easily to decide when to begin peening.

3.4 The Problem of Enthalpy

We need think enthalpy because there is change from liquid to solid, which it absorb heat in this process. It may define different enthalpy with variable temperature to solve in ANSYS software.

$$H = \int \rho c(T) dT$$

C (T) — specific heat

4 Temperature Field Distribution of Physics Model

4.1 The Physics Model of Multi-welding Seams Flat

The material of part is 60CrMnMo, its physics parameters see Table 1. Dimension is 120mmx120mmx8mm, welding seam is in the middle of it. Being a axial symmetry, we just need analysis half of part.

Table 1. Thermo physical performance parameter

Temp °C	200	400	600	800	1000	1200	1400	1600
Conductivity coefficient (W/mm. °C)	0.0419	0.0394	0.0323	0.0291	0.0265	0.0294	0.0294	0.0294
Density (g/mm3)	7.85	7.85	7.85	7.85	7.85	7.85	7.85	7.85
Specific heat c(J/kg. °C)	502	536	586	695	674	670	670	670
Exchange coefficient (W/mm. °C)	18.85	33.44	55.95	88.92	134.6	195.4	273.4	370.9

4.2 Temperature Field Distribution of Physics Model

In the analysis, the welding current is 250A, welding voltage is 25V, welding time is 6S, welding efficiency is 0.9. When the time is 50S, We chose six points which are in the middle of part including A (0, 0, -0.004) 、B (0.02, 0, -0.004) 、C (0.04, 0, -0.004) 、 D (0.06, 0, -0.004) 、E (0.08, 0, -0.004) 、F (0.1, 0 ,-0.004)、G (0.12, 0, -0.004). See Fig. 1:

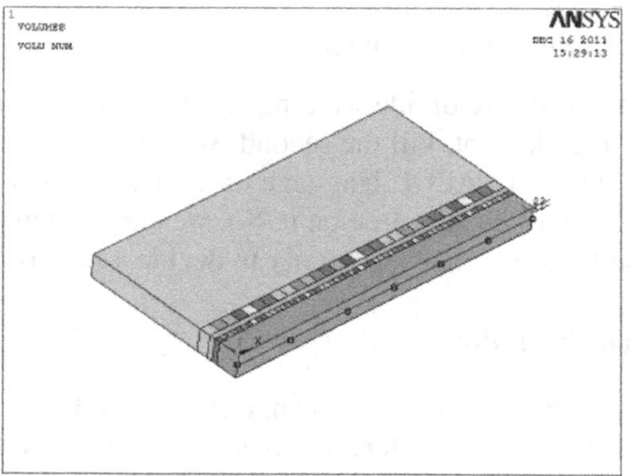

Fig. 1. Points A-G in the model

From Figure2 we can see, when the time is 80S, the temperature of second welding seam is evidently higher than the first one, the most high temperature is 2102°C. In this figure we can know, the temperature in the welding seam is most highest than other points.More far to welding seam, the lower temperature is. We also can know, the temperature of second welding seam is 2102°C which is high almost 100°C than the first one.

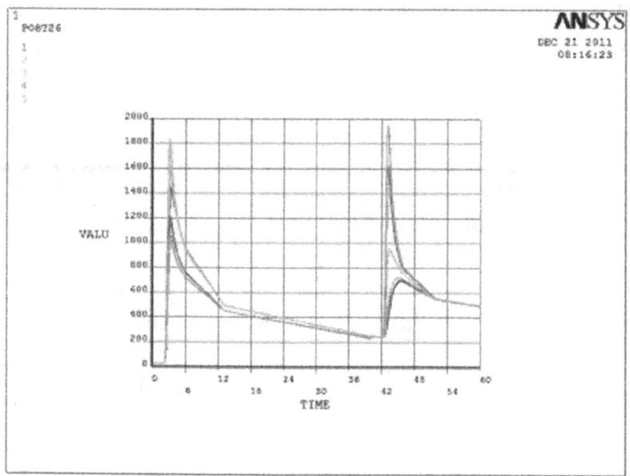

Fig. 2. Points A-G distribution of temperature

4.3 Stress Field Distribution of Physics Model

After analyzing temperature field distribution of the model, we use the software to analyze the stress field distribution. We can have several simulating results of stress in Fig. 3 and Fig.4.

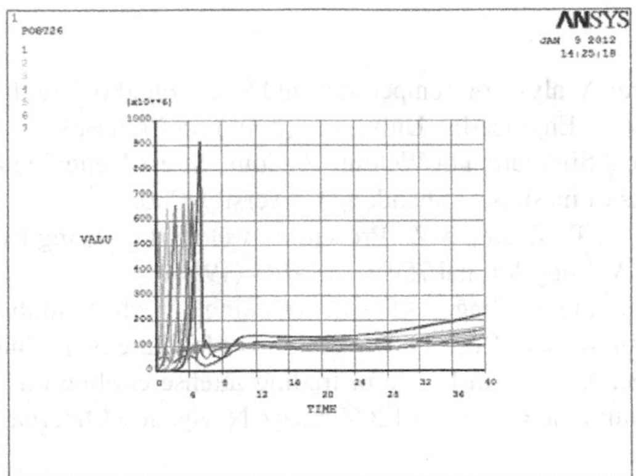

Fig. 3. Points A-G distribution of Stress in 40s

In fig.3, we can see stress distribution of those points which are in the middle of part(A,B,C,D,E,F,G). when welding the first time, the point F have a biggest stress about 915.6Mpa, the stress will increase when welding it second time after cooled 40 sec.

From Fig. 4, in welding process of first 6 sec(40sec~46sec), we can see the stress of points which first increasing suddenly to 300Mpa, and after a little second descending, then increasing to 320Mpa. We also can know the second seam stress less than the first one, and the stress will increase with the process of cooling.

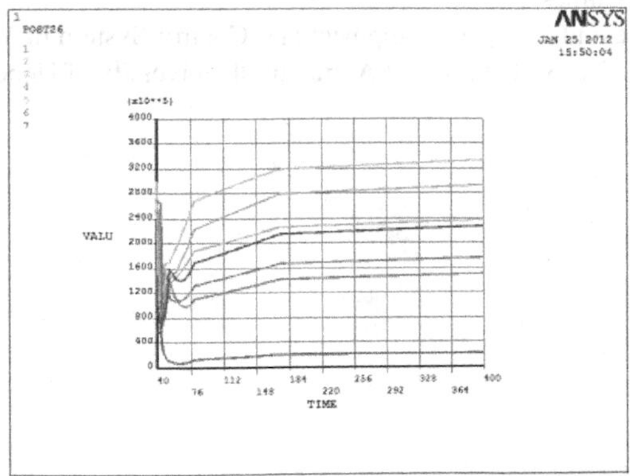

Fig. 4. Points A-G distribution of Stress in 400s

5 Conclusions

We can have a conclusion from the result which analysis with ANSYS software, the distribution of welding temperature field we had is coincide to measured result.

References

1. Xia, P.: Numerical Analysis of Temperature and Stress Field of Welding, pp. 22–27. Thesis of masters of Harbin Engineering University (2005) (in Chinese)
2. Liu, X.: Numerical Simulation of Welding Residual Stress Relieving by Hammer Peening, pp. 11–13. Thesis of masters of Shandong University (2005)
3. Liu, W.P., Tian, X. T., Zhang, X.Z.: Preventing weld hot cracking by synchronous rolling during welding. Welding Journal 75(9), 297–304 (1996)
4. Yang, Y.: A study of preventing welding hot cracking of high strength aluminum alloy sheet with inverse strain method. Doctoral Dissertation of Institute of Technology, Harbin (1995)
5. Tian, X., Guo, S., Xu, W.: Influence of trailing intense cooling on welding hot cracking susceptibility of aluminum alloy LY12CZ. Space Navigation Material & Technology 28(5), 48–52 (1998)
6. Guo, S.: A study of controlling welding hot cracking and distortion of thin aluminum alloy plate with high susceptibility to hot cracking. Doctoral Dissertation of Harbin Institute of Technology, Harbin (1999)
7. Lobanov, L.M.: Heat abstracting paste reduces distortions. Welding and Metal Fabrication (3), 65–70 (1982)
8. Kopsov, L.E.: The influence of hammer peening on fatigue in high-strength Steel. Welding Journal 13(6), 479–482 (1991)
9. Bell, R., Militaru, D.V.: Effect of peening and grinding on the fatigue strength of fillet welded joints. British Welding Journal 3(3), 13–21 (1969)
10. Knight, J.W.: Improving the fatigue strength of fillet welded joints by grinding and peening. Welding Research Intemational 8(6), 519–539 (1978)
11. Yan, H.: Influence of Welding with Trailing Peening by Electromagnetic Hammer on Welding Stress and Microstructure, pp. 4-5. Thesis of masters of Agricultural University of Hebei (2005) (in Chinese)
12. Wang, X.: Devise and Study on Equipment and Control System of Welding with Trailing Peening, pp. 4-9. Thesis of masters of Agricultural University of Hebei (2005) (in Chinese)

Technology Roadmap of Electric Vehicle Industrialization

Qinghua Bai*, Shupeng Zhao, and Pengyun Xu

College of Mechanical and Electrical Engineering,
Agricultural Universityof Hebei, Baoding 071001, China
baiqinghua@eyou.com

Abstact. Through the understanding of the development of the domestic and foreign electric vehicle dynamic and trend, we can know the state new energy vehicles encouraging policies and development strategies, combine with the auto industry distribution in the city, to make electric vehicle industry as a breakthrough, and Look for the city of the electric car development road; To grasp new energy development and Strategic opportunities for the development of new industries, we can make of the electric car industry and the solar energy industry together, to be Perfect the electric car infrastructure construction, strengthen the construction of the electric car local standards, and Make the encouragement and support policies, Eventually develope baoding characteristics of the electric car industry layout, the technology roadmap, industrial supporting model and policy measures, etc.

Keywords: electric Vehicle, new energy, solar energy, development strategy, Technology Roadmap.

1 Introduction

The electric car is in power since battery for, rely on the powerful motor powered new traffic tools. Electric cars have the pollution, the power supply diversification, clean energy conversion efficiency high, simple structure and convenient maintaining, called by "the 21 st century green car". The wide application of the electric car, not only reduce the pollution of the environment, and relieve the energy pressure, but also gives China the auto industry to catch up behind the world advanced level of opportunities [1].

2. The Electric Car the Classification and Development Situation

At present, the new energy automobile generally include a battery-powered electric vehicles (BEV), hybrid electric vehicle (HEV) and fuel cell vehicles (FCV) three mainstream technology [2].

* The author: Qing-hua Bai (1972-), Male, born in Heibei Province Cang County, Ph.D. Candidate, study in the theory, design of Automobile.

D. Jin and S. Lin (Eds.): Advances in CSIE, Vol. 2, AISC 169, pp. 473–478.
springerlink.com

2.1 Battery Electric Vehicles(BEV)

Battery electric vehicles in vehicle battery powered can really realize zero exhaust, completely solve the problem of energy and pollution, the trip range mileage can meet the urban traffic has basic requirements, technology has gradually mature and began to commercialization, but battery performance restrict battery electric car industrialization process. Electric cars can't reach the traditional power, power requirements; Limited storage can't meet the needs of the long-distance driving, battery life is short,charging time is so long,Lack of social supporting charging infrastructure, etc.

2.2 Hybrid Electric Vehicle (HEV)

Perfect integration of driving system is a hybrid car to the practical application, to need to have high aspect ratio of energy and power energy storage device, low cost and high efficiency of the power electronic equipment and fuel economy high emissions low of the engine. Need to solve the key technology includes three aspects: One is engine and electric motor optimal coupling power allocation of control and realize, and power allocation and transmission device and its integration of design [3].

2.3 Fuel Cell Vehicles (FCV)

Fuel cell vehicles to hydrogen and oxygen from the burning of energy as a motor power, is recognized as the car is energy structure adjustment of the final solution, now the automobile companies invest in power, committed to the development of the fuel cell vehicles and test run, and has made many technical breakthrough [4].

3 The Development of Domestic and Foreign Electric Vehicles and Popularization of Strategy and Policy Measures

Many countries and regions have formulated incentive policy, promote the industrialization of the electric vehicles. In the early 1990 s, Japan MITI through the establishment of "the third electric cars popular plan" and so on the electric vehicles development strategy, and encourage the development of the industrialization of the electric vehicles [6]. Barack Obama came to power in 2009 after made it clear that, by 2015, the United States will have 1 million cars charging type hybrid cars on the road. Europe in 1990, established the "city of electric vehicles" association, there were 60 in the European Union organization in cities, the feasibility of the electric vehicle research and guide the operation of the electric vehicle.

Beginning in the 1990 s in China to study the electric vehicle, in April 2001, ministry of science &technology convened the electric vehicle development strategy first national experts seminar, discuss and make through the 863 electric vehicle major projects of the project research content and the project's guidelines, give the electric vehicle industry policy and financial support.

The electric car industrialization has gradually formed. By 2000, Europe already has 16255 electric vehicles [6]. Japan's Toyota prius, lexus EX cars, etc by the end of January 2009 has total sales of 1.7 million vehicles. China also accelerate the electric car industrialization, in three to five years China or become the world's largest electric vehicle quantities country, China will most likely be electric vehicle kingdom.

4 The Feasibility of the Industrialization of the Electric Vehicle on BaoDing City Research

In 2011, the national development and reform commission announced the drafts fo car and new energy auto industry development planning, sure to plug-in hybrid vehicles, pure electric cars and fuel cell cars for the direction of development, set up to 2020 new energy vehicles industrialization and the size of the market to the global the first, new energy auto possession to 5 million cars goal.

At present, the country had been Beijing, Shanghai, more than 40 local electric vehicle for city development planning, and submit to the relevant departments of the state build electric vehicle production base of the report.

There are nine automobile mabufactors in Baoding city such as The Great Wall, HeBei changan, zhongxing automobile etc, have production capacity of 900000 cars annually. It has auto parts enterprise many 300 including Fengfan Stock Limited Company, Lingyun Stock Limited Company etc. So it has unique advantage and the industrialization of the electric vehicle hardware conditions in development the electric vehicle of Baoding.

5 Electric Vehicle Industrialization Technology Roadmap of Baoding City

Through making the city the electric car technology roadmap for the industrialization of Baoding, guide the electric car industrialization process, promote the unoin of the electric car industry and photovoltaic industry, walk a way of with local characteristics of new energy vehicle development road. We can do it like Fig.1.

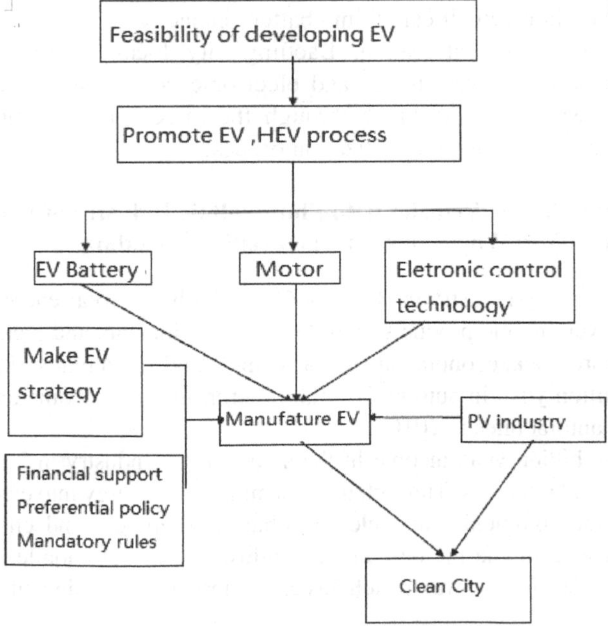

Fig. 1. Electric vehicle industrialization technology roadmap

5.1 Promote the Fuel Cell Vehicle Research and Development

Through making electric vehicle development strategy for local conditions, it is a breakthrough with hybrid cars for the industrialization of the breach and speed up the pure electric vehicle industrialization.

From the existing features selective and orderly development of the electric vehicle industry, considering the advantage and practical factors of hybrid vehicles, it can be in hybrid electric vehicles and pure electric vehicle field to take the lead to achieve the breakthrough, and then to a higher fuel cell vehicles field.

5.2 Make a Base for the Electric Vehicle of the Industrialization Process

Through the government guidance, integrated software and hardware resources, establish the electric vehicle technology research committee to enterprise to rely on, to focus their efforts on the electric vehicle the technical challenges [7].

The Great Wall motor company cooperates with many famous international professional parts company, succeeds development of seven electric vehicles, including Oula BEV, Spirit EV,Hafe HEV, Xuanli car, Deer car, Tengyi C20EV,Tengyi V80 Plug-in hybrids. Some new vehicles is arrivalling including Tengyi C20R BEV,Tengyi C30 etc. The Great Wall of pure electric car system platform, hybrid system platform, will be made and speed up the $1 billion into electric car research and development and industrialization in the near future. Fengfan Stock Limited Company is one of the leading enterprises battery industry, through enhancing the ability of independent innovation, by technical innovation optimization product layout, the company has rich scientific research ability in the electric vehicle batteries and sets up a 30 million cylindrical lithium battery capacity.

Through the resources integration, Baoding city focus on scientific research strength, solves power battery, motor and electronic control system of technology electric vehicles which need to break through the three core technology direction, promotes the electric vehicle industrialization process.

5.3 Combinate Electric Vehicles with Photovoltaic Industry of Baoding, and Develop the Way Which Has Characteristic of Baoding

There are two mainstay industry in Baoding City which are Solar energy, automobile manufacturing. Automobile products include small-engine cars and pickup trucks and SUV, subminiature truck, economical special vehicle and so on many kinds of models. There is 60.5 billion yuan in automotive manufacturing sales income and the vehicle sales of 731500 automobiles in 2010.

There is 45.66 billion yuan income in the solar energy industry in 2010, more than 2005 years increased 6.6 times. Through transforming solar energy into electrical energy, then storing in the community into electric charge equipment and charging electric vehicle, so we can combine the new energy industry with automobile manufacturing industry together, develop the way which has characteristic of baoding city.

5.4 To Speed up the Local Electric Vihecle Industrialization Process

Through measures of economic support, compulsory preferential policy and regulations, take demonstration operation, favorable lease, strengthen the construction of infrastructure, and other means to to promote the development of electric vehicle industry [8].

Financial support: in the research and development period of the electric vihecle, the government gives a higher intensity funding to support the research of enterprise, develops economical practical electric vehicle as soon as possible. Offering more funds, make good economic basis for the development of the electric vihecle.

Preferential policy: the users can get more preferential policy and benefit, and are willing to use an electric vehicles. The preferential policy of the electric vehicles vigorously promote, it is power for producers and users provides development and buying.

Mandatory regulations: there are two aspects in mandatory regulations. One is making restrictive regulations for manufactures, sales, uses, and exhaust of fuel vehicles; the other is planning application for electric vehicles. Among them about the strict limit fuel automobile emission regulations, it is a direct role in promoting in the development of electric vehicles.

6 Conclusions

As a production base of pickups, Miniature truck, economical suvs and special function of vehicles, it has a unique prerequisite for Baoding city in the progress of electric car research and industrialization; Through the research and promotion of residential solar charging device, we can realize Baoding city of solar energy and motor industry in the concept of strategic seamless connection; As a way of the government the appropriate policy guide and strongly support, promote China's electric vehicles industrialization process, at the same time, develop low carbon industry as the leading city economy, carry forward to low carbon for the concept of life concept, make with low carbon for guidance of government construction blueprint, advocate low energy consumption and low emission, low pollution economic model, make Baoding city be a clear water blue sky, ecological livable charm city.

References

1. Bai, M., Zhou, J.: Review of Electric Vehicle Industry Development in Our Country. Beijing Automibile 2, 15–17 (2004) (in Chinese)
2. Hu, S-H., Yangwei: Electric Vehicle Industrialization Strategy Analysis of Our Country. Beijing Automibile 3, 20–25 (2004) (in Chinese)
3. Zhang, W.: Research Actuality and Key Technologies of Hybrid Electric Vehicle. Journal of Chongqing Institute of Technology 20(5), 19–22 (2006) (in Chinese)
4. Li, P., Yi, X., Hou, F.: Current Situation of Foreign Automobile Development and Its Enlightment in China. Journal of Beijing University of Technology 30(1), 49–54 (2004) (in Chinese)

5. Fan, Y., Zhang, W., Chen, Y.: Analysis of Foreign Electric Vehicle Development. Central China Power 23, 8–12 (2010) (in Chinese)
6. Wang, Y., Yao, L., Wang, Y.: Foreign electric car development strategy. The Auto Industry Research 9, 35–40 (2005) (in Chinese)
7. Chen, Z., Wang, X.: Research on Electric Vehicles Industrial Development Strategy in China. Shanghai Energy Conservation 8, 1–6 (2010) (in Chinese)
8. Wang, H., Huang, J.: According to the government how to boost the electric car industrialization. Journal of Consumption 2, 233–234 (2010) (in Chinese)

Visualization Method of Trajectory Data Based on GML, KML

Junhuai Li, Jinqin Wang, Lei Yu, Rui Qi, and Jing Zhang

School of Computer Science & Engineering, Xi'an University of Technology,
Xi'an 710048, China
{lijunhuai,wangjq,yulie,zhangjing}@xaut.edu.cn

Abstract. With the advantages of GML and KML, we propose a new method of visualization of trajectory data. To achieve the integration and sharing of geographic data of different sources and different formats, XSLT and XPath technologies are utilized firstly to convert the original trajectory data to GML format for storage and transmission. And then GML is converted to KML to achieve the rapid visualization of trajectory data in browser. Finally, using KML files, the trajectory data can be visualized rapidly in Google Maps and Google Earth.

Keywords: Visualization of trajectory, GML, KML.

1 Introduction

Attribute to the rapid development of geographic information technology and the increasingly sophisticated of mobile communication technologies, the implementations of GIS have become more and more diversified, e.g., GIS, Web GIS, Mobile GIS. Simultaneously Meanwhile, with the development of positioning technology and the update of mobile devices, it is very easy to gather a variety of location information quickly, and store the collected trajectory data into database or file. Actually, the trajectory records the individual's activities in the real world, which to some extent will reflect the individual's intentions, preferences and behavior patterns [1, 2]. People can get their location information by using some mobile devices along with a certain kind of positioning method. Then through the visualization of the location information, they can analyze their own activities and moving trajectories [3]. E.g., to manage and schedule the vehicles effectively, GIS has connected the trajectories data together in the order of time to form the moving trajectories of vehicles; Travel enthusiasts can associate their trajectories with the photographs taken during the travel to memory the journey, and also can share the travel experiences with friends.

2 Related Works

GML provides a kind of "open" standard for the GIS field in the network era, its starting point is the encoding of spatial data (including distributed spatial data

D. Jin and S. Lin (Eds.): Advances in CSIE, Vol. 2, AISC 169, pp. 479–484.
springerlink.com

encoding); As the wide application of XML encoding in various fields, which further involves the integration coding of the spatial data and non-spatial data: The applications of XML in description of style (such as XSL), 2D graphics (such as SVG), voice (such as Voice XML) fields etc. have made GML to provide basis for the reasonable performance of spatial data; The many advantages of GML have laid the foundation for the interoperability of heterogeneous GIS invisibly [4].

As more and more organizations and software developers use XML/GML as the criterion of spatial data representation, transmission, storage, the unification of spatial data encoding and data interoperability and sharing will eventually become realities, and thus truly realize the open access to spatial information. Currently, there are three methods to parse GML files [5]. The first one is based on the visualization strategy of raster image. In this strategy, the client user sends GML map request to the Web server, Web server accepts the request and call the GML map service program to handle user's request, and finally sends the got raster map (GIF or JPG format) to the client display in a static page form. The second one is component-based visualization strategy. In this strategy, Web server does not handle all of the user's requests, but sends the client GML data and GML data processing components(such as Java applet, ActiveX components or Plug-in) through Web server, users can operate the requested GML data in client through the component, these operations including map window, zoom, roam, and information inquiries, etc. The third one is SVG-based visualization strategy. SVG (Scalable Vector Graphics, Scalable Vector Graphics Markup Language) is an open 2-dimensional vector graphics format, as the same as Mathematical Markup Language (MathML), it is an application of Extensible Markup Language XML. SVG features include nested transformations, clipping paths, transparency processing, filter effects, and other extensions. Meanwhile, SVG supports animation and interaction, and also supports the holistic XML-DOM interfaces [6]. SVG can be embedded in other XML documents, and SVG documents can also be embedded by other XML content, the various SVG graphics can be easily combined to form new SVG graphic, thereby stepping forward a big step on the graphical reusability.

The above three typical GML visualization methods have their own advantages and disadvantages. SVG-based visualization strategy has great improvements compared to the former two visualization strategies. First, compared to the component-based strategy, as the same it needs to request data only once, but the map operation can be done on the client. Meanwhile, the strategy does not need a special component design for the client, so that its requirements to client are very low, as the same as the strategy based on raster image. But different from the strategy based on raster image, it does not need to complete the operation of the map through the server. SVG-based visualization strategy has its own disadvantages, i.e., SVG does not support the topology, geographic coordinate systems and 3D data, which greatly increased its limitations.

3 Trajectory Data Visualization Model

GML is a recognized as the intermediary-convert file by most GIS software. Through converting the trajectory data collected from devices in different locations into a

GML file, we can carry on a variety of selective visualizations, such as converting GML format file into KML to visualize.

In order to visualize trajectory data from various sources, formats and data models in GIS rapidly, this paper proposed a trajectory data visualization model (TD-Visual Model)combining the advantages of GML and KML. Firstly, converting trajectory data in different formats into a unified GML file to achieve the integration and sharing of heterogeneous data. And then, utilize the XSLT technology to generate KML file from GML file. Visualization of trajectory data can be achieved based on the advantages of the KML visualization finally.

In the model, when a client sends a request to the Web server, Web server accepts the request and converts the trajectory data that queried from the database into GML file; then convert GML file into KML file through XSLT, and finally it will be presented to the user through the Google Earth or Google Map. The structure of TD-Visual model is shown in Fig. 1.

The achievement of TD-Visual model mainly consists of five steps: GML Schema modeling, trajectory data conversion, the establishment of XSLT template, the conversion from GML to KML and trajectory data visualization.

(1) GML Schema Modeling

GML consists of three basic XML Schemata, namely feature.xsd, geometry.xsd and xlink.xsd. Feature.xsd defines an abstract model of geographical features; geometry.xsd defines the information of specific geometry; and xlink.xsd defines various function links [4].

Geographic data is stored based on the layer; each layer is composed of point, line, and surface. Therefore, a layer model is necessary when storing the trajectory data. According to the criterion of UML diagrams and GML, this paper established a GML Schema mode which conformed to the trajectory data after conversion

(2) Trajectory Data to GML

In order to achieve the conversion from trajectory data to GML, a schema must be obeyed as for the operations of trajectory data DBMS and the middle layer data. There are three ways to convert trajectory data to GML documents in .NET: The first is to use push model SAX (Simple API for XML); the second is to use DOM (Document Object Model); the third is to use flow model XMLTextReader class and XMLTextWriter.

Taking the efficiency into account, this paper adopted the third method which is a flow model to parse GML document. When using C# to generate GML document, it will use XMLTextWriter class, and WriteStartDocument (), WriteStartElement (), WriteAttributeString (), WriteElementString (), WriteEndDocument () etc. methods in the class. It can be easily observed in Fig. 1 that the trajectory data from the database is firstly stored, then an instance of XMLTextWriter class is created, and finally GML documents are generated based on the Schema specification as well as the properties and methods of XMLTextWriter class.

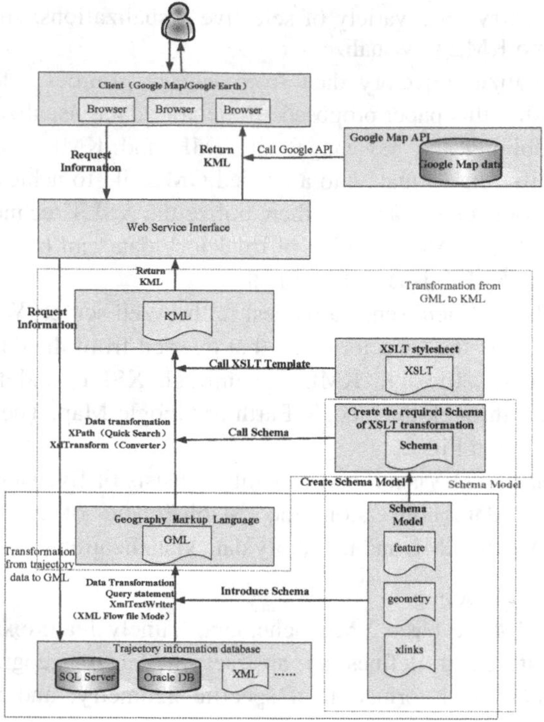

Fig. 1. TD-Visual Model

(3) The Conversion from GML to KML

The conversion from GML to KML is the most important part in trajectory data model. As both GML and KML are based on XML, we use XSLT as the data conversion technology [5]. During the conversion process, the following technologies will be utilized: Xpath Quick Search technology which can quickly retrieve the nodes that need matching; create an XSLT style sheet, which defines the matching style of each node and the target document's style; XSLT Processor means XSLT parser, through which converting GML to KML can be achieved.

(4) Trajectory Data Visualization

KML is mainly used to describe the marks, such as the positioning mark, commercial label, etc. The converted KML file can be loaded directly from Google Earth to achieve KML visualization, but in order to achieve KML visualization by calling JavaScript, the KML must be imbedded in a Web site [6]. To visualize the trajectory data in KML file, points and lines must be visualized.

The KML file that contains a <kml> label, a <Document> label named Paths, which describes the properties of line that after visualization, and a <Placemark> label named Absolute Extruded. When the file is read by Google Earth client, <Placemark> will display the latitude and longitude specified by <coordinates> according to the properties described in <Document>.

4 Trajectory Data Visualization Platform

Based on the abovementioned methods, this paper develops a trajectory data visualization platform, which can be divided into three-tier structure: the data access layer, the business logic layer and the presentation layer.

4.1 Application Model

Figure 2 describes the major class model of business logic layer. A series of point objects are the main elements of a trajectory principal. Each trajectory instance is accompanied with a corresponding Object ID and Trajectory ID.

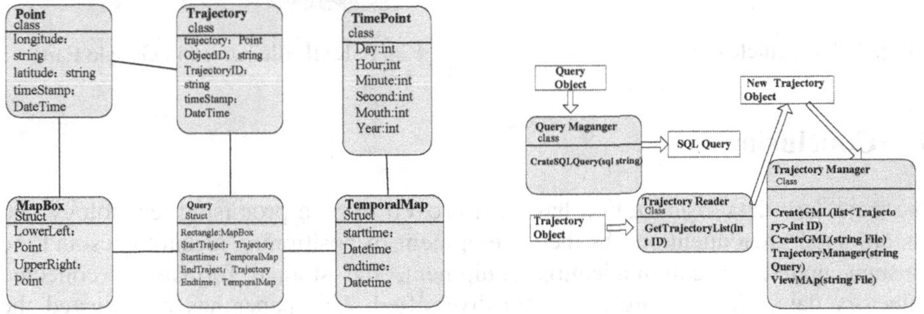

Fig. 2. Business logic object class

Fig. 3. Business logic layer class

A Map Box instance is used as a parameter of Query instance. Two point objects are used to model the corner coordinates of a space form, representing the top right corner and bottom left corner of the form. Time Point principal demonstrates a specific time stamp, which is composed of several elements including year, month, day, hour, etc. Similarly, two Time Point instances are used to model a Temporal Map, representing the start time and end time of an interval. This Temporal Map instance is utilized as the parameter of Query instance. A Query instance represents an actual query, and will be sent to the database after conversion. The corresponding element indicates the actual value of the query, and the query type will be instanced to a Query Type. Business logic layer owns a range of entities, as shown in Figure 3. The figure contains the modification of object state that the administrator can get all of the methods by changing some of the states in this layer. Parser class is used to provide a validation and extraction of the value of GML query file. It converts GML query file into SQL statements, and constitutes a Query instance through Query Manager Entities.

Finally, a trajectory instance can be converted into a GML/KML format file by the Trajectory Manager.

4.2 System Implementation

Based on the trajectory visualization platform, we can record the trajectories of vehicles according to the GPS data and query the trajectories of each vehicle as well as the detailed information in the map. We will get corresponding GML file, KML file and map display by visualizing the queried trajectory data. Figure 4 shows the parsed trajectory in Google Map and Figure 5 presents the driving route parsed by KML in the Google Map.

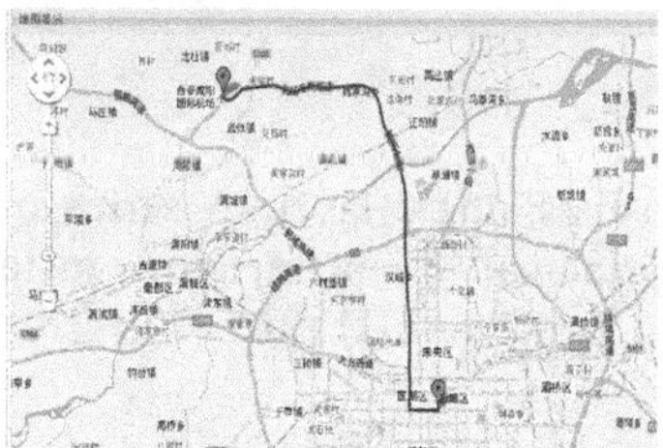

Fig. 4. The vehicle trajectory search function

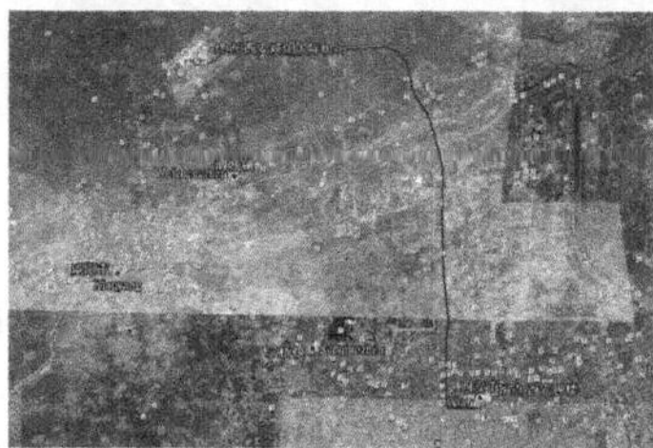

Fig. 5. KML file parsed in Google Earth

5 Conclusion

The trajectory data visualization has been proved to be a promising technology and has attracted great attention. As the development of positioning technologies and the constant update of communication equipment, the storage formats of collected trajectory data will be more and more diversified. This paper has investigated the visualization of trajectory data from various sources and formats, and future work will mainly focus on the trajectory data sharing and integration, visualization of trajectory data in multi-platforms and trajectory data mining.

Acknowledgments. This work was supported by a grant from the Natural Science Foundation of China (No. 61172018), the Science & Research Plan Project of Shaanxi Province (No. 2011NXC01-12) and Science & Research Plan Project of Shaanxi Province Department of Education (No.2010JC15). The authors are grateful for the anonymous reviewers who made constructive comments.

References

1. Yu, Z., Xing, X.: Enable Smart Location-Based Services by Mining User Trajectories. Communication of CCF 6(06), 23–30 (2010)
2. Kritzler, M., Raubal, M., et al.: A GIS Framework for Spatio-temporal Analysis and Visualization of Laboratory Mice Tracking Data. Transactions in GIS 11(5), 765–782 (2007)
3. Buliung, R.N., Kanaroglou, P.S.: A GIS toolkit for exploring geographies of household activity/travel behavior. Journal of Transport Geography 14(1), 35–51 (2006)
4. Open GIS Consortium: Geography Markup Language (GML) 1.0, http://www.opengis.org
5. Gkoutsidis, I.S.: Trajectory Data Visualization: The VisualHERMES Tool. MSc THESIS, ME/0567 (2008)
6. Wang, Z., Zhang, Y., Ren, J.: On the KML Markup Language Specifications and Their Exchange with GML. Standardization of Surveying and Mapping 26(1), 10–13 (2010)

Study on Improved Rijndael Encryption Algorithm Based on Prefix Code

Zhiqiang Xie[1,2], Pengfei Gao[1], Yujing He[1], and Jing Yang[2]

[1] College of Computer Science and Technology, Harbin University of Science and Technology, Harbin 150080, China

[2] College of Computer Science and Technology, Harbin Engineering University, Harbin 150001, China

xzq0111@tom.com, {gpf_84,heduo1987524}@163.com, yangjing@hrbeu.edu.cn

Abstract. Aiming at problem that Rijndael algorithm can be attacked by Square attacks, an improved Rijndael algorithm method which is based on prefix codes is put forward. Due to the good Characteristics in decoding of prefix codes, the method will change the order of sub-keys through the procedure of decrypting different plaintext inputted by prefix codes, and make the order relate with the plaintext, in other words, it makes the system be variable. With the improvement it has no effect on the efficiency, but it can resist the Square attack fundamentally.

Keywords: Rijndael algorithm, prefix codes, encryption, decryption, improvement.

1 Introduction

Rijndael cryptogram is an iterative grouping cryptogram, and its all transformations are based on the state matrices [1].

Since Rijndeal algorithm has been proposed, it has been seeing as a new generation data cryptogram standard and is used to protect various data all the while. Its security performance is far higher than DES algorithm. But with the technological advancement, there are a number of security issues of Rijndael algorithm which has been revealed. They are as follows:

(1) Rijndael algorithm belongs to symmetry cipher, it is easy to analyze its structure and it cannot resist energy attack and error attack effectively [2].

(2) There is equivalent transform relationship between component functions in S-box which is nonlinear component of Rijndael algorithm. Though it is helpful to reduce the hardware realization cost for S-box, it also can cause attack which aims at this algorithm.

(3) The Rijndael algorithm will appear again after 16 times overlaps of linear expanded layer for any plaintext and plaintext differential.

(4) The S-box of Rijndeal algorithm has short cycle track, which will be the key to penetrate the Rijndeal algorithm.

D. Jin and S. Lin (Eds.): Advances in CSIE, Vol. 2, AISC 169, pp. 485–491.
springerlink.com

(5) Rijndael cannot resist the Square attack effectively [3]. The limit-quantity of breaking the Rijndael algorithm which is encrypted 5 times is only $2^{32}+4\times2^{40}$.

Presently there are many improved Rijndael algorithms, for example:

(1) Improve S-box. The improved S-box has better algebra character and stronger algebra attack resistant ability. But the ability for resisting Square attack is weak, because Square attack is a kind of selecting plaintext attack and the attack intension doesn't depend on S-box and column mixed matrix.
(2) Improve key generated algorithm and enhance the performance of confusion. This kind of improved algorithm enhances the encrypted intension of algorithm in a certain extent, but encryption system of improved algorithm is still not changed to encrypt different plaintext, so it cannot resist the Square attack radically, and it reduces the algorithm efficiency to a certain extant [4].

According to these problems, if the Square attack is wanted to be resisted thoroughly, the encryption system must be changed by a certain condition while encrypting, and make Square attack not assault. So this paper presents an improved blueprint based on prefix code. According to the decoding character of prefix code, for plaintext inputted, apply the prefix code selected random to decode it to change the using sequence of round key. And it can make the using of round key relate to the plaintext. Thereby the encryption system can be changeable. So it can radically have the ability to resist Square attack under the condition of not reducing the Rijndael algorithm efficiency.

2 Introduction and Analysis of Related Knowledge

2.1 Prefix Code

Given a sequence set, if there is not one sequence which is the prefix for any other sequences, this sequence set is called as prefix code. For instance, {000, 001, 01, 1} is prefix code, {1, 0001, 000} is not prefix code.

Thereinto, prefix code is one to one correspondence with binary tree, namely, there is one corresponding prefix code for any binary tree, the same, there is one corresponding binary tree for any prefix code. Through this kind of corresponding relationship, it can be obtained that the prefix code can be decoded, if the corresponding binary for given prefix code is a complete binary tree [5]. And this paper takes advantage of the decoding feature to decode the sub-keys and to change the sequential of sub-keys according to the difference of plaintext inputted. So it can resist Square attack better.

2.2 Encryption and Generating Round Keys Process of Rijndael Algorithm

Rijndael algorithm adopts Square structure, and bytes (8bits) and words (32bits) are regarded as process units. The plaintext is separated to N_b words, and each word is composed by 4 bytes. Let $N=\max\{N_b, N_k\}+6$, the algorithm will do one initial transform and N-1 transforms and one end transform. Let T_0 denotes initial transform, the middle transforms are denoted by T_1, T_2, ..., T_{N-1}, so the encryption transform is

Rijndael=$T_N \cdot T_N$-1·...T_1 ·T_0. Each transform result of Rijndael is called as state, and each state is a matrix of 4×N_b order bytes [6].

The initial transform is composed by key adding, the end transform is composed by byte transform (S-box), shift rows and key adding. The other N-1 transforms are compounded with the 4 kinds of transform above.

In the process of encryption using Rijndael algorithm, the key is the critical part of the algorithm. The algorithm security depends on keys to a great extant, but it doesn't use keys directly in practical encryption. It uses the seed keys to generate round keys by key generated algorithm. Input these seed keys and round keys to key schedule table, and use the round keys which are in key schedule table sequentially in the encryption process. The process of generating round keys and generating key schedule table are shown as follows.

Rijndael algorithm and decryption algorithm use a key schedule table which is composed by the number of seed key bytes. For example, value of seed key is 128bits/ 4word:

The key schedule table of 00 01 02 03 04 05 06 07 08 09 0a 0b 0c 0d 0e 0f is shown as Table.1.

Table 1. Key Schedule Table.

00	01	02	03
04	05	06	07
08	09	0a	0b
0c	0d	0e	0f
58	46	f2	19
5c	43	f4	fa
	…	…	
c4	18	c2	71
e3	a4	1d	5d

The dimension of key schedule table is 4 columns and N_b×(N_r+1)=4×(10+1)=44 rows. Thereinto, the value of N_b is the grouping length dividing 32, and the value of N_r is the number of transform rounds from plaintext to ciphertext. First, copy the value of seed keys to key schedule table. As the seed keys are 128bits bytes and the key schedule table is 4 columns, the seed keys are copied to the former 4 rows of key schedule table. The round keys of the other rows are calculated according to the following algorithm [7]. Thereinto, the value of N_k is that the key length divides 32.

RotWord() function is to rotate a row of key schedule table to a position leftward. SubWord() function is to replace a row of given key schedule table one byte by one byte by replacing table S-box. Rcon[] is a constant array, these constant are 4 bytes, and each is matching with one row of key schedule table.

2.3 Square Attack Method and Security Analysis

Square attack is an attack method which is proposed by using block operation characteristic of Square cryptogram and the reversibility of each transform in SPN cryptogram structure. It is effective to attack Square cryptogram which has low number round transforms.

Square attack is an attack which selects plaintext and builds on two important concepts which are Λ set and balance character.

The basic attack process is as follows:

(1) Select one Λ set.
(2) Encrypt this Λ set and observe the diffusion path of balance bytes in the encryption process (this process includes key hypothesis).
(3) Sum all possible values of balance bytes at the end of path, and according to this to decide whether keep or delete the hypothesizing keys. It can select multiple Λ set to repeat the process above to validate the correctness of keys kept.
(4) Output right keys.

It is known through analyzing the basic principle of Square attack that it depends on the invariable of encryption system if you want to use Square attack method. So if make the Rijndael algorithm have ability of resisting Square attack, it must make the encryption system be alterable according to a certain condition and make Square attack be unable to attack. This paper presents an improvement algorithm based on prefix code. The prefix code has good decoding features and it is random to obtain the prefix code, so the change law cannot be found. Decode the plaintext inputted according to prefix code to change the using sequence of round keys, and make these be related to plaintext, therefore make the encryption system change along with different plaintext inputted. So Rijndael algorithm can have the ability to resist Square attack radically under the situation that the efficiency of the algorithm is not influenced.

3 Improvement Blueprint and Example

3.1 Sequential Changing Process of Round Keys

The following example has group length and cryptogram length of 128, and introduces the process of changing round keys by prefix code.

(1) Input 128 bytes plaintext A, change bytes for A through S-box, and obtain data B of 128 bytes.
(2) Make data B of 128 bytes XOR with seed keys, obtain data C of 128.
(3) Select prefix code randomly, and use this prefix code to decode data C. If the last part of decoded information cannot be a sequence of prefix code, add 0 to the end of the decoded information, and obtain a coding set $\{C_0, C_1, \ldots\ldots C_{43}\}$. Thereinto, $C_0, C_1, \ldots\ldots C_{43}$ is the decimal representation for corresponding coding, and the first 44 of sequence is availability sequence because the row number of key schedule table is 44.
(4) Provided that, sequence of the original round keys is $Key_0Key_1 \ldots Key_{42}Key_{43}$, the following algorithm is adopts to exchange the sequence. If value of C_i (i is from 0 to 43) is j, exchange the position of Key_i and Key_j. Repeat this process, and start from C_0 till C_{44}. A new round keys sequence can be obtained.
(5) Start encryption process of Rijndael algorithm according to new round keys sequence.

(6) As it needs to decode, the changed round keys sequence code should be encrypted. The encryption principle is that the sequence code is 1001 0011 1100 0101...... of 64 bytes, if new round keys sequence is $Key_9Key_3Key_{12}Key_5$...... Then these codes are encrypted by original DES encryption, and at this time the encryption keys can be original keys. Results can be saved with cryptogram to make management of resources easy and reduce the effect of algorithm efficiency.
(7) Repossess round keys sequence while decoding, then the encryption data can be obtained by doing the reverse process of Rijndael algorithm.

3.2 Example

Example shows the change process of round keys as follows:

(1) Select 128 bytes plaintext A=00 11 22 33 44 55 66 77 88 99 aa bb cc dd ee ff. Replace it by bytes through S-box replacement table, and obtain 128 bytes data B=63 82 93 c3 1b fc 33 f5 c4 ee ac ea 4b 55 28 16.
(2) Select Key=00 01 02 03 04 05 06 07 08 09 0a 0b 0c 0d 0e 0f. Make data B XOR with Key, and obtain 128 bytes data C= 63 83 91 c0 1f f9 35 f2 cc e7 a6 41 47 58 26 19.
(3) Select prefix code randomly {000, 001, 01, 10, 11}. Make 128 bytes data C be transformed to binary data, and decode, then obtain {01, 10, 001, 11, 000, 001, 11, 001, 000, 11, 10, 000, 000, 001, 11, 11, 11, 11, 10, 01, 001, 10, 10, 11, 11, 10, 01, 01, 10, 01, 10, 01, 11, 001, 11, 10, 10, 01, 10, 01, 000, 001, 01, 000, 11, 10, 10, 11, 000, 001, 001, 10, 000, 11, 001}. The corresponding first 44 decimal is {1, 2, 1, 3, 0, 1, 3, 1, 0, 3, 2, 0, 0, 1, 3, 3, 3, 3, 2, 1, 1, 2, 2, 3, 3, 2, 1, 1, 2, 1, 2, 1, 3, 1, 3, 2, 2, 1, 2, 1, 0, 1, 1, 0}, because only the first 44 sequence is available sequence.
(4) According to above sequence, and C_0=1, so Key_0 is changed position with Key_1. The round keys sequence after first time changing is $Key_1Key_0Key_2$... Key_{43}. Repeat this process and after 44 times changing the round keys sequence is Key_{41} Key_{43} Key_{39} Key_{35} Key_1 Key_0 Key_2 Key_5 Key_4 Key_6 Key_3 Key_8 Key_{11} Key_7 Key_9 $Key_{14}Key_{15}$ Key_{16} Key_{10} Key_{13} Key_{19} Key_{20} Key_{18} Key_{22} Key_{17} Key_{24} Key_{23} Key_{21} Key_{27} Key_{26} Key_{28} Key_{29} Key_{30} Key_{25} Key_{32} Key_{33} Key_{31} Key_{36} Key_{34} Key_{37} Key_{38} Key_{12} Key_{40} Key_{42}.
(5) It is encrypted by Rijndael algorithm according to this round keys sequence, and at the same time round keys sequential code 01000001 01000011 00111001...01000010 which has been changed is encrypted by original Rijndael algorithm, and this can be convenient for decoding.

4 Improvement Analysis

After improvement, the result analysis is as follow:

(1) Round keys sequential transform depends on prefix code, and the selection of prefix code is random. After changing the order of round keys through prefix code, attacker cannot hypothesize the changing state of order according to plaintext.

(2) The round keys sequence transform process of this algorithm adopts simple algorithm, add an algorithm which encrypts round keys sequence which has been changed. At this process of encryption, apply the process quantity of algorithm which encrypts original plaintext. In this way, the algorithm complexity can be reduced. And the running result shows that encrypting 3.77M document under running environment of Windows2000, CPU 2.2G HZ, EMS memory 256M, the encryption speed is 5.027Mb/s before being improved, but after being improved, the speed is 5.021Mb/s. So this method does not influence the running efficiency of Rijndael algorithm basically.

(3) As using sequence of round keys is related with the inputted plaintext in this algorithm, it has strong resistant ability against Square attack. Square attack method means that attacker selects a Λ set which includes 256 plaintexts that different numbers are only on one byte position. After three times encrypting, the difference of this kind of values will diffuse to all 16 bytes of cryptogram and the corresponding bytes are different. Then make 256 cryptograms modular 2 add by bytes and the sequence of summation is the sequence whose elements are all 0. This is one Square character of Rijndael algorithm. Square attack depends on the invariable of encryption system, but this improved algorithm proposed by this paper changes the sequence of round keys, and make it be related to plaintext. While the inputted plaintexts are different, the using sequence of round keys is also different. So the encryption system is changing with the inputting of different plaintext, the Square attack is resisted radically, and the security of Rijndael algorithm is increased to a great extant. The Square attack limited-quantity comparing analysis before and after improved is as Table.2 shown.

Table 2. Table of Comparison and Analysis.

	Original Rijndael encryption algorithm	Rijndael encryption algorithm improved by changing sub-key generated algorithm	Improvement Rijndael encryption proposed by this paper
The limited-quantity of Square attack	$2^{32}+4\times2^{40}$	$2^{32}+4\times2^{104}$, though the limited-quantity is more than original algorithm, there is still potential being attacked as the algorithm cannot make encryption system change constantly.	Make the encryption system change with the changing of plaintext, so it resists Square attack radically and invalidates square attack.

5 Conclusion

For Rijndael algorithm, the most effective attack method is Square attack at present. It needs to improve Rijndael algorithm. This paper proposed an improvement blueprint which imports prefix code, because Square attack method depends on invariable of encryption system to attack. Through decoding character of prefix code, decode the plaintext to change the using sequence of round keys. So it can make encryption system change with different plaintext inputted, and therefore increase security and make the algorithm have resistant ability against Square attack radically. Analysis shows that the proposed blueprint can improve Rijndael algorithm effectively under precondition that it will not influence algorithm efficiency.

Acknowledgments. This project is supported by the National Natural Science Foundation of China (No. 60873019 and No.61073043), the Natural Science Foundation of Heilongjiang Province of China (No.F200901 and No.201101), China Postdoctoral Science Foundation (No.20090460880), Heilongjiang Province Postdoctoral Science Foundation (No.LBH-Z09214) and Harbin Outstanding Academic Leader Foundation (No.2010RFXXG054 and No.2011RFXXG015).

References

1. Liu, J.M., Wei, B.D., Cheng, X.G., Wang, X.M.: Cryptanalysis of Rijndael S-box and improvement. J. Applied Mathematics and Computation 170(2), 958–975 (2005)
2. Zhang, P., Wei, Z.H.: The Research of AES Algorithm and the Structural Analysis. Science & Technology Information 18(67), 91–92 (2008)
3. Duo, L., Li, C.: The Inverse Square Attack of Rijndael Cipher. Electronics & Information Technology 26(1), 65–71 (2004)
4. He, M.W., Liu, R.: An Improved Method Based on AES Rijndael Algorithm. Microcomputer Information 22(6), 94–96 (2006)
5. Millidiu, R.L., Labe, E.S.: Bounding he Inefficiency of Length-Restricted Prefix codes. Algorithmica 31(4), 513–519 (2007)
6. Xiao, P.: The Application of AES and ECC in the Transport of Data Encryption. Network Security Technology & Application 10, 87–89 (2008)
7. Duo, L., Li, C., Zhao, H.W.: The Relation between Cyclic Shift Structure and Rijndael's Safety. Journal of China Institute of Communications 24(9), 153–161 (2003)

An Integrated Flexible Scheduling Algorithm by Determining Machine Dynamically

Zhiqiang Xie[1,2], Yuzheng Teng[1], Yujing He[1], and Jing Yang[2]

[1] College of Computer Science and Technology, Harbin University of Science and Technology, Harbin 150080, China
[2] College of Computer Science and Technology, Harbin Engineering University, Harbin 150001, China
xzq0111@tom.com, tengyzh@ufida.com.cn, heduo1987524@163.com, yangjing@hrbeu.edu.cn

Abstract. Recently the existing integrated flexible scheduling algorithm of complex product processing and assembling which are processed simultaneously cannot make full use of the features that there is multiple machines and multiple utility time, and machines selection is not flexible to obtain the optimum scheduling results. Aiming at this problem, an integrated flexible scheduling algorithm of processing and assembling by determining machines dynamically which can advance the finishing time of procedures is presented. Analysis and examples show that this algorithm can realize the optimum result of the integrated flexible scheduling problem of complex products processing and assembling under the situation that the algorithm complexity is not improved.

Keywords: Complex product, Integrated flexible scheduling of processing and assembling, Flexible processing tree, Adaptive and flexible scheduling strategy, Scheduling algorithm.

1 Introduction

Flexible job Shop scheduling problem (FJSSP) is more complex NP-hard problem [1]. As there are many available machines for processing procedures, machines constraints are reduced and the optimal space is enlarged. However the difficulty of problem is increased and the problem complexity is further raised [2].

In order to solve the flexible scheduling problem of complex products processing and assembling, Reference [3] proposed an optimum scheduling algorithm which had considered this problem. It used the machines selection method which reduces the flexible problem by using the shortest process time. But because this algorithm prematurely restricts the flexible machines selection, it cannot make full use of the features that there are multiple machines and multiple utility time and machines can be selected flexible. So optimum scheduling results cannot be obtained.

This paper makes full use of the feature of flexible machines selection, and presents building a flexible processing tree based on the existing achievements to solve the complex products scheduling problem that there are constraints between

D. Jin and S. Lin (Eds.): Advances in CSIE, Vol. 2, AISC 169, pp. 493–499.
springerlink.com

procedures. Allied critical path method [4-6] is adopted to determine the processing sequence of flexible processing tree. An adaptive and flexible scheduling strategy is presented which can adjust machines dynamically by computing the finishing time of procedures. The reasonable machines of processing procedures can be determined and the shortening of total working hours can be realized. At end, examples show the optimum results of this algorithm.

2 Build Flexible Processing Model

In order to solve the flexible scheduling problem of complex products processing and assembling, the processing machines and assembling machines are defined as processing machines together, and so does the processing and assembling. For each machine can be allocated to each procedure reasonable and constraints relationships between procedures also be considered, build mathematical model according to reference [3].

In order to adapt the feature of flexible scheduling better, this paper presents to build flexible processing tree. The flexible processing tree can reflect the features of flexible scheduling of complex products processing and assembling better based on expand processing tree [3]. The procedure number, machine number and process time which are the attribute of node from flexible processing tree are sequenced by the process time from short to long. This is different from simply listing machines and process time which is the feature of expand processing tree. In the flexible processing tree different process time and different machines are separated by commas.

3 The Designing of Flexible Scheduling Algorithm

Method of dividing the integrated flexible scheduling problem of complex products processing and assembling into two sub-problems is adopted by this paper. One sub-problem is to determine the scheduling order of procedures; the other one is to determine the machines for every procedure. For controlling the algorithm complexity, allied critical path method is adopted to determine the processing order. For finishing procedure earlier, adaptive and flexible scheduling strategy is designed to determine machines on which procedures are processed. So the shortest total working hours can be realized. Specification is as follow:

3.1 Determine Scheduling Order of Procedures

The critical path is a path that the sum of process time is maximum of all procedures from all paths which are obtained by traversing the flexible processing tree from leaf node to root node. Procedures on critical path are important for the total working hours. The allied critical path method is adopted in the paper proposed by reference [4]. As the procedures classification is adopted to schedule at the same time, the complexity of algorithm can be controlled in quadratic.

For flexible scheduling problem, in order to make the total working hours as short as possible, the minimal process time is computed to determine the critical path. At the basis of this, scheduling order can be determined by allied critical path method.

3.2 Determine Scheduling Order of Procedures

As flexible scheduling problem has feature that machines can be selected flexible, machines which can make product finish earliest can be selected from machines set. Aiming at finishing product earlier, procedures must be finished as early as possible. Design adaptive and flexible scheduling strategy which can select machine on which the procedures can be finished earliest. For machines with the same finishing time, the machine balanced strategy is adopted. So the overall optimization scheduling can be realized.

The thought of adaptive and flexible scheduling strategy is to select machines with earliest finishing time while scheduling procedures. If there is machine on machine set with shortest process time which is not used at the earliest process time period of procedure, this machine can be selected to process this procedure; otherwise, the machine is selected which may be of relatively shorter finishing time from other machines from the machine set with longer process time. Repeat this selecting process till the machine with relatively earlier finishing time is found.

If all machines from the machine set cannot process procedure at the earliest process time, the finishing time which obtained by using the optimization strategy of procedures scheduling from each machine set are compared. Machine with the earliest finishing time is selected to be the process machine. If machines which can be selected are not sole, select the machine with the shortest process time.

If available machines are not sole, occupy machine balanced strategy to choose machine.

Adaptive and flexible scheduling strategy makes full use of the features that there are multiple machines and multiple utility time and machines selection is flexible to make the finishing time of part of procedures can be earlier than the finishing time on machines from machine set with shortest process time. So make the total working hours of complex product flexible scheduling than the algorithm proposed by reference [3] can be realized.

Machine balanced strategy which calculates the total working hours of all scheduled procedures on each optional machine separately is adopted by. Then the machine with minimal total working hours is selected [3]. This strategy takes full account of the processing load of each machines, make every machine balanced using. Thereby the parallel working hours of whole processing systems can be enhanced.

The optimization strategy of procedure scheduling is that after the procedures processed order of flexible scheduling problem is determined, the flexible scheduling problem is converted into the scheduling problem of procedures with single machine. At this time the allied critical path method is adopted to optimize the procedures scheduling [3].

4 Scheduling Algorithm

Step 1: build flexible processing Tree A, n=0, select the minimal process time of every node to compose Tree A(n).

Step 2: compute the sum of every path of Tree A(n), then choose the maximal value. If the value is not sole, turn to Step 3. Otherwise go to Step 4.

Step 3: calculate the number of machine on maximal path of Tree A(n), choose the path with minimum of machines, then go to Step 4.

Step 4: sequence procedures by allied critical path method and select the first procedure. Go to Step 5.

Step 5: n=n+1, set the initial value of this node of Tree A(n) the machine number from machines set with minimal value and process time. If there is no machine which need wait, turn to Step 11. Otherwise go to Step 6.

Step 6: record the minimal value of finishing time of all machines from machine set with minimal value by using the optimization strategy of procedure scheduling. Go to Step 7.

Step 7: find out the machine with the second minimal value from Tree A (the initial tree), the corresponding location value of this node of Tree A(n) is changed into the machine with second minimal value and process time. Go to Step 8.

Step 8: there is machine with no waiting from machine set of this procedure from Tree A(n), then record the earliest finishing time of procedure on this machine. Compare the finishing time with the earliest finishing time of procedure on scheduled machine. Select the machine with shorter time, then turn to Step 11; otherwise go to Step 9.

Step 9: all machines need waiting, record the minimal value of completion time by using the optimization strategy of procedure scheduling. Go to Step 10.

Step 10: the process time on machine from machines\ set is the maximal value of the procedure process time. Select machine from machine set with shorter time firstly. Turn to Step 11; otherwise go to Step 7.

Step 11: the machine from machine set which should be selected is sole, turn to Step 13; otherwise go to Step 12.

Step 12: select machine by the machine balanced strategy, then go to Step 13.

Step 13: process this procedure. If there are procedures after this procedure, turn to Step 5; otherwise end.

5 Scheduling Example

The data of product A of example is shows as Table 1. And its flexible process tree is as Fig.1.

Its flexible Tree A(0) which just contain the machines with minimal process time is as Fig.2. In Fig.2, the allied critical path method is adopted to determine the scheduling order of product A(0). Then adaptive and flexible strategy is adopted to select machines for processing procedures and the result is shown like Table 2.

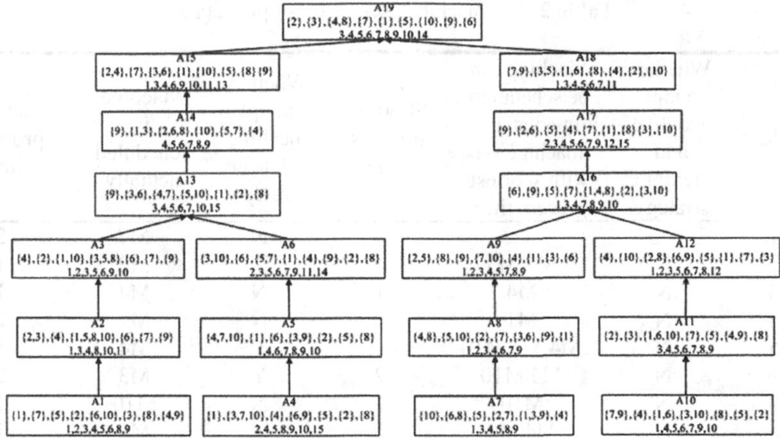

Fig. 1. Flexible processing tree of product A (Tree A)

Table 1. Flexible processing data of product A

	M1	M2	M3	M4	M5	M6	M7	M8	M9	M10
A1	1	4	6	9	3	5	2	8	9	5
A2	4	1	1	3	4	8	10	4	11	4
A3	3	2	5	1	5	6	9	5	10	3
A4	2	10	4	5	9	8	4	15	8	4
A5	4	8	7	1	9	6	1	10	7	1
A6	6	11	2	7	5	3	5	14	9	2
A7	8	5	8	9	4	3	5	3	8	1
A8	9	3	6	1	2	6	4	1	7	2
A9	7	1	8	5	1	9	4	2	3	4
A10	5	10	6	4	9	5	1	7	1	6
A11	5	3	4	8	7	5	6	9	8	5
A12	7	3	12	1	6	5	8	3	5	2
A13	7	10	4	5	6	4	5	15	3	6
A14	5	6	5	9	8	6	8	6	4	7
A15	6	1	4	1	10	4	3	11	13	9
A16	8	9	10	8	4	1	7	8	3	10
A17	7	3	12	5	4	3	6	9	2	15
A18	4	7	3	6	3	4	1	5	1	11
A19	7	3	4	5	8	14	6	5	10	9

Table 2. The machines selection of product A

Procedure	Whether adopt adaptive and flexible strategy	Machine can be scheduled immediately in machine set with shortest process time	Shortest process time	Whether adopt machine balanced strategy	Machine is scheduled actually	Actual process time
A1	Y	--	--	N	M7	2
A2	N	M2,M3	1	Y	M2	1
A3	N	M4	1	N	M4	1
A4	N	M1	2	N	M1	2
A5	N	M4,M7,M10	1	Y	M4	1
A6	N	M3,M10	2	Y	M3	2
A7	N	M10	1	N	M10	1
A8	N	M4,M8	1	Y	M8	1
A9	N	M5	1	N	M5	1
A10	N	M9	1	N	M9	1
A11	Y	--	--	Y	M2	3
A12	N	M4	1	N	M4	1
A13	N	M9	3	N	M9	3
A14	N	M9	4	N	M9	4
A15	N	M2,M4	1	Y	M2	1
A16	N	M6	1	N	M6	1
A17	Y	--	--	Y	M6	3
A18	N	M7,M9	1	N	M7	1
A19	N	M2	3	N	M2	3

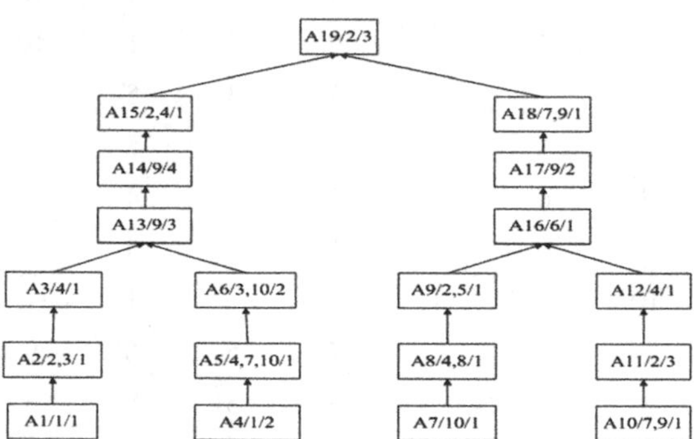

Fig. 2. Flexible processing Tree A(0) of product A

The flexible scheduling Gantt chart of product A is as the left figure in Fig.3 which adopts the algorithm which is proposed by reference [3]. And the flexible scheduling Gantt chart of product A is as the right figure in Fig.3 which scheduling order is determined by allied critical path method and machines are selected by adaptive and flexible scheduling strategy.

Compared the left figure with right figure in Fig.3, the result obtained by this algorithm proposed is better than the result obtained by the algorithm proposed by reference [3].

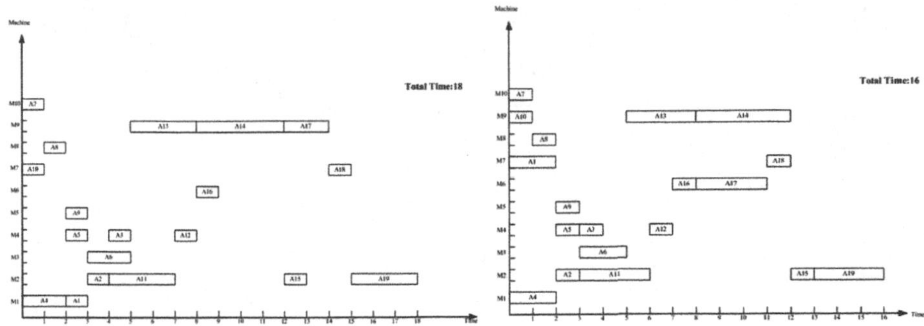

Fig. 3. Scheduling Gantt chart of according to algorithm proposed by reference [3] and proposed by this paper

6 Conclusions

The integrated flexible scheduling algorithm of complex product presented by this paper takes full use of the features of flexible scheduling problem, and makes procedures finish early though selecting more machines for procedures, whereas the algorithm complexity is stable. The total working hours are shorter than existing flexible scheduling algorithm of complex product.

Acknowledgments. This project is supported by the National Natural Science Foundation of China (No. 60873019 and No.61073043), the Natural Science Foundation of Heilongjiang Province of China (No.F200901 and No.201101) and Harbin Outstanding Academic Leader Foundation (No.2010RFXXG054 and No.2011RFXXG015).

References

1. Garey, M.R., Johnson, D.S.: Computers and Intractability: A Guide to the Theory of NP-Completeness. Freeman, San Francisco (1979)
2. Jansen, K., Mastrolilli, M., Solis-Oba, R.: Approximation Algorithms for Flexible Job Shop Problems. International Journal of Foundations of Computer Science 16(2), 361–379 (2005)
3. Xie, Z.Q., Hao, S.Z., Ye, G.J., Tan, G.Y.: A new Algorithm for Flexible Job-shop Scheduling with Constraints between Jobs. Computers & Industrial Engineering 57(3), 766–772 (2009)
4. Xie, Z.Q., Teng, Y.Z., Yang, J.: Integrated Scheduling Algorithm with No-wait Constraint Operation Group. Acta Automatica Sinica 37(3), 371–379 (2011)
5. Xie, Z.Q., Ye, G.J., Zhang, D.L., Tan, G.Y.: Study of a New Nonstandard Job Shop Scheduling Algorithm. Chinese Journal of Mechanical Engineering 21(4), 97–100 (2008)
6. Xie, Z.Q., Yang, J., Yang, G., Tan, G.Y.: Dynamic Job-Shop Scheduling Algorithm With Dynamic Set of Operation Having Priority. Chinese Journal of Computers 31(3), 502–508 (2008)

An Optimum Design to the Simulation Platform of Combustion Process of Diesel Engines

Jing Wang[1,2], Qian Wang[2], and Cisheng Lin[1]

[1] Zhenjiang Watercraft College of the PLA, Zhenjiang 212013, Jiangsu, P.R. China
[2] School of Energy and Power Engineering, Jiangsu University, Zhenjiang 212013, Jiangsu, P.R. China
kingarrow10@sina.com

Abstract. This thesis provides an optimum design to the simulation platform of diesel engine working process. It designed on KIVA-3V source program, the in-cylinder hydrokinetics and the multidimensional model of in-cylinder combustion. A comparison between the simulated results and the experimental results of a diesel engine shows that the errors do exist, because the former is influenced by the accuracy of computation mesh and the latter is influenced by the clearance at the top of the combustion chamber. However, the errors are within the allowable limit. Therefore, this platform can be well adopted to simulate the real working conditions of diesel engines.

Keywords: Diesel Engine, Simulation Platform, KIVA-3V, Combustion, Emission.

1 Introduction

During the whole working process of the diesel engine, gases in the cylinder are in constant and complicated turbulent motion, with strong compressibility, transiency, rotational flow and aeolotropism. The rapid development of numerical simulation theory and technology makes it possible to carry out the multidimensional numerical simulation to the in-cylinder working process of diesel engines. The thesis makes use of the KIVA-3V source program developed by the Los Alamos National Laboratory of the USA [1] to design a simulation platform for the working process of diesel engines, and proves its reliability via experiments. The result shows that this platform can be used to carry out research on the working performance of diesel engines.

2 Design of the Simulation Platform

2.1 Checking the PDF File

The block diagram of the simulation platform is shown in Fig 1. The platform includes a layer of hardware structure and a layer of software structure, and the latter is composed mainly of the layer of application software and the layer of operating system. With the cooperation of the hardware structure and the operating system, the

D. Jin and S. Lin (Eds.): Advances in CSIE, Vol. 2, AISC 169, pp. 501–506.
springerlink.com

KIVA-3V source program can be compiled in Visual Fortran 6.5. When various basic technical parameters and boundary conditions of the diesel engine are inputted and calculated, the changing rule of the in-cylinder pressure, temperature and emission are outputted and displayed via Fieldview 8.0, OriginPro 7.5 etc.

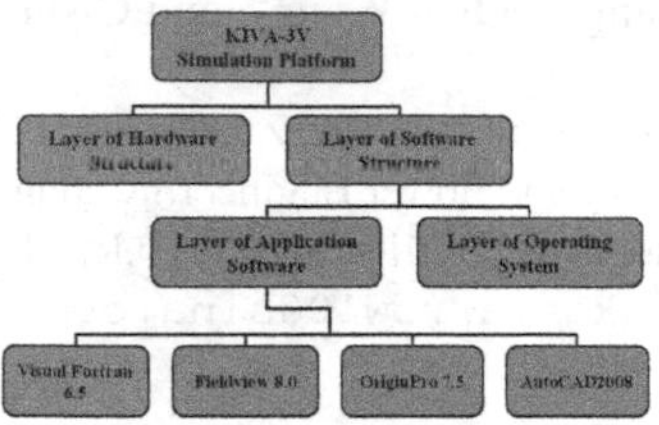

Fig. 1. Block diagram of the simulation platform

2.2 Main Program Structure of Numerical Calculation

This numerical calculation simulation platform was based on the KIVA-3V program and composed of three functional modules, i.e. the Preprocessor, the Hydrocode and the Postprocessor [2]. The structural relationships between them are shown in Fig 2.

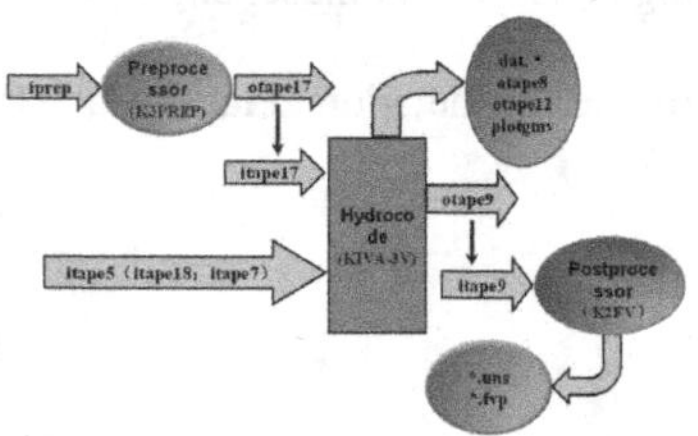

Fig. 2. Block diagram of the main program

Based on the source program of KIVA-3V, this thesis adds some new program codes to calculate the mass emission and emission per cycle of NOx, so as to disclose the variation rule of NOx with the change of parameters.

2.3 Multidimensional Computation Model and Method

The combustion process is a flow process comprising chemical reactions. No matter how complicated it is, the fluid motion, as a form of the motion of matter, has to follow the basic rules of the nature.

The computation models used in this thesis are: the turbulence model – RNG k-ε model [3]; the division and nebulization model of oil spray –KH-RT model [4]; the turbulent combustion model – a time model of the turbulent combustion features [5]; and the NOx emission model – an extended Zeldovich model [6]. The computer simulation of flow and combustion process also includes some basic equations such as the mass conservation equation, momentum conservation equation, energy conservation equation and state equation etc.

ALE (the Arbitrary Lagrangian-Eulerian) Method [7], is used in this thesis as the computation method of the numerical simulation. ALE is suitable for fluid problems in cylinders of the internal combustion engine, where the geometrical shape is irregular and the volume is in constant change.

3 Experimental Study of the Simulation Platform

3.1 Subject of the Experiment

The subject of the experiment in this thesis is CA6DL1-30 supercharged diesel engine, whose structure of combustion chamber is shown in Fig 3 and basic specifications are listed in Tab 1 [8].

Table 1. Specifications of the diesel engine

Type of Engine	CA6DL-30
Cylinder Bore [mm]	110
Stroke [mm]	135
Length of Connecting Rod [mm]	217
Discharge Capacity [L]	7.7
Compression Ratio	17.5
Combustion Chamber	ω
Rated Speed [r/min]	2300
Rated Power [KW]	220
Maximum Torque [N·m]	1100
Number of Cylinders	6
Number of Jet Hole * Diameter [mm]	6×0.167
Convexity of Nozzle [mm]	2.5

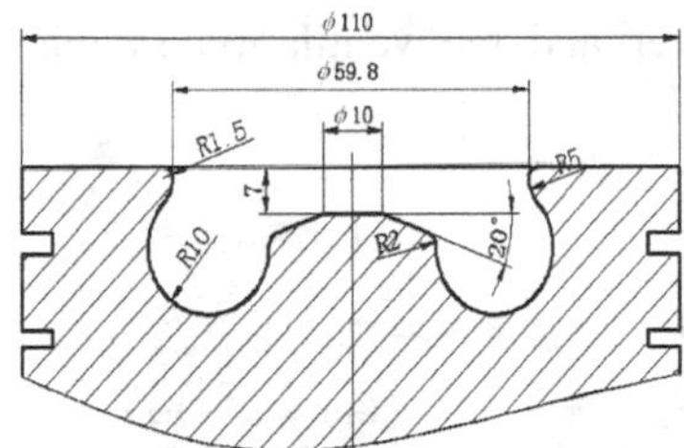

Fig. 3. The structure of combustion chamber

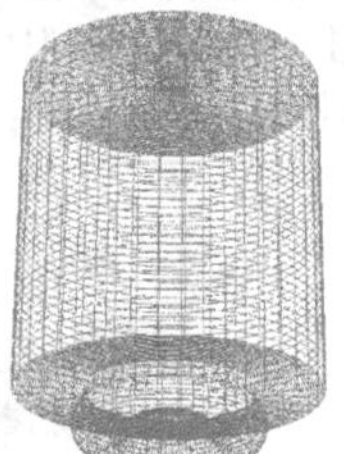

Fig. 4. Computation mesh

The three-dimensional computational mesh generated by KIVA-3V mesh partition rules [1] according to the basic specifications of CA6DL1-30 engine is shown in Fig 4.

The initial and boundary conditions of the experimental subject are shown in Tab 2.

Table 2. Initial and boundary conditions

Intake Air Temperature [K]	325
Supercharged Pressure [Mpa]	0.156
Initial Swirl Ratio	1.6
Temperature of the Cylinder Head [K]	503
Temperature of the Cylinder Wall [K]	403
Temperature of the Piston Crown [K]	553
Initial Temperature of the Fuel [K]	353
Injection Pressure [Mpa]	150
Fuel Injection Duration Period [°CA]	20
Volume of Injection per Cycle [g]	0.01989
Angle of Contact of Injection [°]	110
Taper Angle of Injection [°]	25
Closure of Intake Valve [°CA]	234.0
Opening of Exhaust Valve [°CA]	478.0

3.2 Test Equipment of the Experiment

In order to check the accuracy of the numerical simulation platform to the in-cylinder working process of the diesel, a test was carried out to the changing of the in-cylinder pressure and emission of CA6DL1-30 diesel engine.

The structure of the test equipment is shown in Fig 5. During the test, the temperature is 20°C and the outer atmospheric pressure is 0.1MPa. The PM (Particulate Matter) emission is measured via the CVS-7400S Constant Volume Sample System manufactured by HORIBA, Japan. The Exhaust gas dilution is made by full flow dilution system. The emission of gaseous pollutants is measured by MEXA-7100D exhaust gas analyzer produced by HORIBA, Japan. Among them, the NOx is measured by the CLD with NO_2/NO converter or the HCLD, and the CO is measured by the NDIR. The air flow meter and the ventilation system are also included in the text equipment.

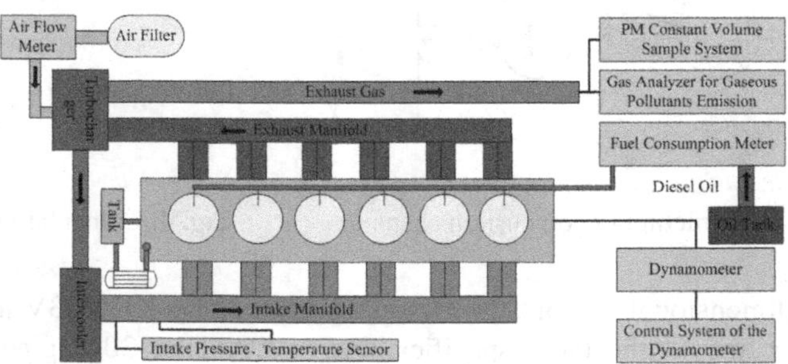

Fig. 5. Structure of the Test Equipment

3.3 Result Analysis of the Experimental and Computed Values

Values of in-cylinder pressure in computation and experiment with the change of crank angles are shown in Fig 6. It can be seen from the figure that the maximum in-cylinder explosion pressure in computation is 14.7501MPa and it takes place at 363.009°CA, while the maximum in-cylinder explosion pressure in experiment is 15.0386MPa and it takes place at 364.01°CA. It shows that the computed value is well in correspondence with the experimental value, either in the maximum values or in the crank angles.

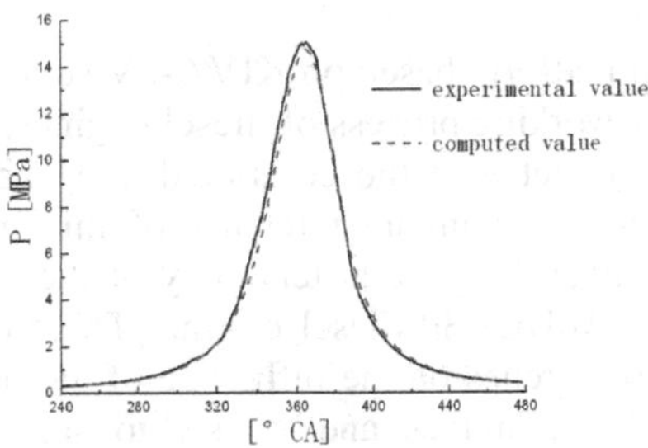

Fig. 6. Computed and Experimental Values of In-cylinder Pressure

Fig 7 shows the experimental values of five groups of NOx emission per cycle. The sampling period of the experiment is 52ms, that is to measure the NOx emission per cycle of the diesel engine. Altogether five groups of experimental values are obtained, and the NOx emissions of 10 operation cycles are taken successively in each group. According to the five groups of data, the average value of the NOx emission per cycle is 4.32025×10-4g/cycle. Compared with the computed value 4.72455×10-4g/cycle, the error is 8.56%.

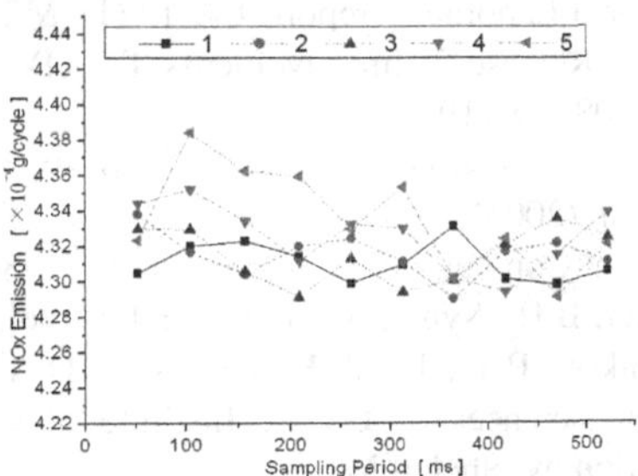

Fig. 7. Experimental Values of NOx Emission

The comparison of the experimental and computed values of the in-cylinder average pressure and the NOx emission shows that some error still exist between them. The main reasons lie in that the initial temperatures of the cylinder head, cylinder wall and piston crown are empirical values, that there are some differences between the computation mesh and the real conditions of the combustion chamber, and that the movement of the inlet and outlet valves are not considered. What's more, the clearance at the top of the combustion chamber hinders the complete emission of the exhaust gases, which may result in the error of experimental values. Of course, there are still many other reasons that negatively affect the accuracy of computation. However, generally speaking, the computed values are still reliable, and the numerical simulation platform established in this thesis can be used in the research of the working performance of diesel engines.

4 Conclusion

In this thesis, a simulation platform based on KIVA-3V source program is established to simulate the in-cylinder working process of diesel engines. According to the results of comparison and analysis between the computed and experimental values of the CA6DL1-30 diesel engine, the computed results of this platform is reliable to a certain degree, and can reflect the general tendency of the real working process and emission features of the CA6DL1-30 diesel engine. Therefore, the platform can be used to carry out further researches on the influence of the parameter changes on the working performance of diesel engines, and be used to find out effective measures to improve the combustion performance and to reduce the emission of diesel engines.

Acknowledgements. This research was supported by the National Natural Science Foundation of China (No.51076060), the Project Funded by the Priority Academic Program Development of Jiangsu Higher Education Institutions and Industrial Support Project of Jiangsu province(No.BE2010198).

References

1. Amsden, A.A.: A Block-Structured KIVA Program for Engines with Vertical or Canted Valves. Los Alamos National Laboratory report, LA-13313-MS (1997)
2. Amsden, A.A.: KIVA-3V, Release 2, Improvements To KIVA-3V. Los Alamos National Laboratory report, LA-13608-MS (1999)
3. Xie, M.Z.: Computational Combustion of Internal Combustion Engines. Dalian University of Technology Press, Dalian (2005)
4. Hiroyasu: J. Foreign Internal Combustion Engine 6, 44–47 (1986)
5. Magnussen, B.F., Hjertager, B.H.: Symp. (Int.) on Combust vol.16 (1977)
6. Zeldovieh, Y.B., Sadovnikov, P.Y., Frank-Kamenetskii, D.A.: Oxidation of Nitrogen in Combustion. Academy of Seienees of USSR. Institute of Chemical Physies, Moseow-Leningrad (1947); translation by, Shelef, M.
7. Hirt, C.W., Amsden, A.A., Cock, J.L.: Journal of Computational Physics 14 (1974)
8. Lu, C.Q.: Multi-dimensional Numerical Simulation for Combustion Process of CA6DL1-30 Diesel Engine. Jiangsu University, Jiangsu (2009)

Optimization of Test Points at Circuit Board Based on Network Topology Structure

Ying Liu[1], HaiLiang Ma[2], and FangJun Zhou[3,4]

[1] College of Electric and Information Engineering, Naval University of Engineering, Wuhan 430033, China
[2] The Equipment Research Institute of PLA's Second Artillery, Beijing 100085, China
[3] College of Naval Architecture & Power, Naval University of Engineering, Wuhan 430033, China
[4] P.O.Box 91278, Lüshun 116041, China

Abstract. Selection of test points is the foundation of fault diagnosis for circuit board. On the basis that test point set includes fault source information as far as possible, principle of optimized selection for test point is expanded in this paper. Optimal test point set of complicated system is generated by use of improved association matrix for circuit board.

Keywords: Circuit board, fault diagnosis, improved association matrix, optimized selection.

1 Introduction

With the development of electronics technique, circuit structure becomes more and more complicated. We rely more on computer and graph theory to analyze and design circuit. The more complicated the circuit is, the more test points are needed for circuit board fault diagnosis. Connector of a circuit board is not enough to diagnose fault, because pins of the connector are limited. Test points in the circuit board are needed to get more information. Test signal and output signal are driven from circuit within. For example, needle bed technique leads out all nodes of a circuit to measure. But this technique won't work for Large-scale integrated circuit or super-large-scale integrated circuit.

If you optimize test point set and select limited node to meet the need of measure, you can promote measure efficiency and reduce work capacity. Reference [1] defines that the optimal fault detection point includes most information of fault source, which considers that test point set covers fault source information as far as possible. Principle of optimized selection for test point is expanded in this paper.

2 Topology Structure Model of Circuit

Topology structure of a network can be described by a graph, or a matrix, which is association matrix, loop matrix, or cut-set matrix usually.

D. Jin and S. Lin (Eds.): Advances in CSIE, Vol. 2, AISC 169, pp. 507–511.
springerlink.com

Let a electrical network composed of l branching-offs and n nodes, as shown in figure 1. All branching-offs and nodes are numbered arbitrarily. Therefore the association properties between branching-offs and nodes can be expressed by a $n \times b$ order matrix M_c. A row of M_c is corresponding for a node, and a column of M_c for a branching-off. Element m_{jk} of matrix M_c is defined as follows:

(1) $m_{jk} = 1$, if k-th branching-off is dependent on j-th node, and k-th branching off departs from j-th node;
(2) $m_{jk} = -1$, if k-th branching-off is dependent on j-th node, and k-th branching-off directs towards j-th node
(3) $m_{jk} = 0$, if k-th branching-off is independent on j-th node.

M_c is called association matrix of a electrical network between nodes and branching-offs. Association matrix of a electrical network figure shown in figure 1 is formula (1).

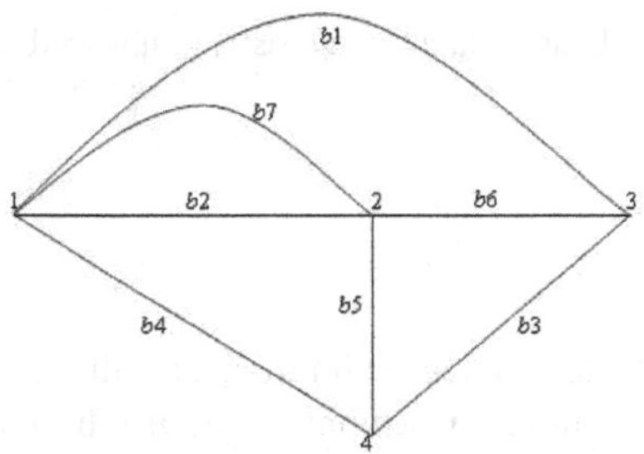

Fig. 1. Structure of simple circuit network

It can be found that every column of a association matrix has only two non-zero elements (+1) and (-1), and $\sum_{j=1}^{n} M_j = [0.0.0...]_{1\times l}$. Therefore association matrix is linear correlation. Giving a $(n-1)\times l$ order association matrix can determine a directed graph which is composed of l branching-offs and n nodes

$$M_c = \begin{matrix} \\ n1 \\ n2 \\ n3 \\ n4 \end{matrix} \begin{matrix} \begin{matrix} b_1 & b_2 & b_3 & b_4 & b_5 & b_6 & b_7 \end{matrix} \\ \begin{pmatrix} 1 & 1 & 0 & 1 & 0 & 0 & 1 \\ 0 & 0 & 0 & 0 & 1 & 1 & -1 \\ -1 & -1 & 1 & 0 & 0 & -1 & 0 \\ 0 & 0 & -1 & -1 & -1 & 0 & 0 \end{pmatrix} \end{matrix} \tag{1}$$

3 Improved Association Matrix Model

On the basis of discussion above-mentioned, definition of association matrix is modified as follow:

(1) $m_{jk} = 1$, if k -th branching-off is dependent on j -th node, and k -th branching-off has component or load;
(2) $m_{jk} = i$, if k -th branching-off is dependent on j -th node, and k -th branching-off has no component or load. That is to say, nodes at the end of a branching-off are direct short.
(3) $m_{jk} = 0$, if k -th branching-off is independent on j -th node.

$M_{c'}$ is called improved association matrix of a electrical network between nodes and branching-offs. Let branching-off $b3$ in figure 1 direct short, we get

$$M_c = \begin{array}{c} \\ n1 \\ n2 \\ n3 \\ n4 \end{array} \begin{array}{c} \begin{matrix} b_1 & b_2 & b_3 & b_4 & b_5 & b_6 & b_7 \end{matrix} \\ \begin{pmatrix} 1 & 1 & 0 & 1 & 0 & 0 & 1 \\ 0 & 1 & 0 & 0 & 1 & 1 & 1 \\ 1 & 0 & 1 & 0 & 0 & 1 & 0 \\ 0 & 0 & 1 & 1 & 1 & 0 & 0 \end{pmatrix} \end{array} \tag{2}$$

Let branching-off $b5$ broken, $b3$ and $b7$ direct short in figure 1, we get

$$M_c = \begin{array}{c} \\ n1 \\ n2 \\ n3 \\ n4 \end{array} \begin{array}{c} \begin{matrix} b_1 & b_2 & b_3 & b_4 & b_5 & b_6 & b_7 \end{matrix} \\ \begin{pmatrix} 1 & 1 & 0 & 1 & 0 & 0 & i \\ 0 & 1 & 0 & 0 & 0 & 1 & i \\ 1 & 0 & i & 0 & 0 & 1 & 0 \\ 0 & 0 & i & 1 & 0 & 0 & 0 \end{pmatrix} \end{array} \tag{3}$$

$M_{c'}$ has following characteristics compared with M_c:

(1) M_c corresponds to directed graph. For timing sequence, direction and strong logicality of working procedure, directional graph is necessary in thermodynamics system and so on. Nevertheless, reference direction is presumed artificially and is of little physical significance or topology significance. $M_{c'}$ corresponds to undirected graph, which reflects prominently definite physic relationship among branching-off, node and component of a circuit.
(2) $M_{c'}$ distinguishes direct short from loaded between two nodes.
(3) $M_{c'}$ reveals broken circuit, short-circuit and common status. If a column is all zero, the corresponding branching-off is open. If there are two elements i in a column, the corresponding branching-off is direct short, and the two nodes equal to one node.
(4) Modules of i equal to one.
(5) $M_{c'}$ can reflect not only normal topology relationship of a circuit, but also fault status. Fault of a circuit is inserted by setting one element zero or i, and by

multiplying a $b \times b$ order matrix on the right side of $M_{c'}$. The dimension of $M_{c'}$ needn't changing. The topology structure of a circuit remains original.

4 Optimized Selection of Test Points at Circuit Board

4.1 Principle for Optimized Selection

The set N is composed of all nodes on the circuit, and the set L is composed of all branching-offs on the circuit. There exists a mapping $T(N) = L$. Let V is a non-empty optimal set of test points, and $num[T(n_i)]$ is all branching-offs associated with node n_i, where $V = \{(n_a, n_b, \ldots, n_t) | \forall n_i \in N\}$. The principle for optimized selection is described in mathematical method as listed below:

(1) $V \subset N$;
(2) $n_a \neq n_b \neq n_c \neq \ldots \neq n_t$;
(3) $L - T(V) \rightarrow 0$;
(4) $\cap(T(n_a), T(n_b), T(n_c) \ldots T(n_t)) \rightarrow 0$;
(5) $\text{num}(T(n_a)) \approx \text{num}(T(n_b)) \approx \text{num}(T(n_c)) \approx \ldots \approx \text{num}(T(n_t))$;
(6) $\text{num}(V) \leq \text{num}_{\max}$;

Explanation of these principle is as follows:

(a) Principle (1) and (2) prevent invalid node from selecting;
(b) The association degree demanded in principle (3) points out that detecting should cover all components as far as possible.
(c) Principle (3) and (4) describe the isolation degree of node sets to improper characteristic of every component. If a node connects with many components, the fault cannot be determined yet, even some fault information can be detected from the node.
(d) Principle (6) weighs up the measure quantity;
(e) Principle (7) restricts selecting too little test points.

4.2 Realization of Optimized Test Point Model Based on Improved Association Matrix

In terms of discussion above-mentioned, optimal test point set of a circuit board can be established by building a improved association matrix. The improved association matrix is realized by this method below.

Adding together correlation rows of improved association matrix corresponding to nodes in set V gives $\sum_{j=a}^{t} M_j = [0 \cdot 1 \cdot 3 \cdot i \cdots]_{1 \times l}$.

The fewer elements '0' in $\sum_{j=a}^{t} M_j$, the more node set V covers branching-off. The more elements '1' in $\sum_{j=a}^{t} M_j$, the less repetition degree of branching-off associated with nodes in set V. The more elements '2' , '3' in $\sum_{j=a}^{t} M_j$, the greater repetition degree of associated branching-off in node set V. The smaller the variance $Var\{num[T(n_a)], num[T(n_b)], \text{L} num[T(n_t)]\}$, the more average the number of branching-off associated with a node.

5 Conclusion

Building a optimal test set is a kind of set covering problem. Truck dispatch, [2] resource distribution [3] and location of equipment [4] [5] are also set covering problem. Principles of building test set condition each other. A set does not exist meeting all condition. Stressing one principle does not accord with engineering practice. Neglecting branching-off coverage won't detect fault on circuit board. Neglecting repetition degree of branching-off associated with nodes will reduce precision of fault location. Too many branching-off associated with a node will damage status resolution of different branching-off. Few nodes will gather many faults on one node. With the electrical network spreading, too many nodes will make processing difficult and consume much resource.

This paper not only takes source information coverage by test set fault into account, but also improves optimization principle. Improved association matrix can build a optimal test set and regenerate topology structure of electrical network. Although improved association matrix is of high order, it is a sparse matrix and takes not too much memory. Improved association matrix can be applied to complicated system

References

1. Yuan, H., Chen, G.: Research on testability problem and fault diagnosis method in analog circuit. Journal of Electronic Measurement and Instrument 20(5), 17–20 (2006)
2. Balinski, M.: Integrate programming: method, uses and computation. Management Science 12(3), 253–313 (1965)
3. Garey, M., Johnson, D.: Computers and Intractability: A Guide to the Theory of NP-Completeness, pp. 50–55. W H Freeman, New York (1979)
4. Valenta, J.: Capital equipment decisions: A model for optimal systems interfacing, pp. 30-33. Master thesis, Masszchusetts Institute of Technology (1969)
5. Walker, W.: Using the set-covering problem to assign fire companies to fire house. European Journal of Operational Research 22, 275–277 (1974)

Selection Method of Minimum Test Set at Circuit Board Based on Improved Node Matrix

Ying Liu[1], HaiLiang Ma[2], and FangJun Zhou[3,4]

[1] College of Electric and Information Engineering, Naval University of Engineering, Wuhan 430033, China
[2] The Equipment Research Institute of PLA's Second Artillery, Beijing 100085, China
[3] College of Naval Architecture & Power, Naval University of Engineering, Wuhan 430033, China
[4] P.O.Box 91278, Lüshun 116041, China

Abstract. Graph theory is applied in this paper to optimization design of test point set in circuit board fault diagnosis. The optimal detection node set of a circuit board can be established by a improved node matrix. The improved node matrix can also regenerate topology network graph of a circuit, which reflects the relationship among every node and the working status of all components in the circuit.

Keywords: Circuit board, fault diagnosis, minimum test set.

1 Introduction

For restriction of limited nodes which can be reached, the fault information for diagnosis is not sufficient and the fault position can not be determined precisely. Selection of test points is directly related to how much fault information can be obtained, diagnosis precision and efficiency. The optimal fault detection point defined in Reference [1] includes most information of fault source. Reference [1] also considers that test point set should cover fault source information as far as possible. Principle of optimized method for test point is expanded in this paper.

2 Topology Structure Model of Circuit

Topology structure of a network can be described by a graph, or a matrix. Let a electrical network composed of l branching-offs and n nodes, as shown in figure 1. All branching-offs and nodes are numbered arbitrarily. Therefore the association properties between branching-offs and nodes can be expressed by a $n \times b$ order matrix M_c. A row of M_c is corresponding for a node, and a column of M_c for a branching-off. Element m_{jk} of matrix M_c is defined as follows:

(1) $m_{jk} = 1$, if k-th branching-off is dependent on j-th node, and k-th branching-off departs from j-th node;

D. Jin and S. Lin (Eds.): Advances in CSIE, Vol. 2, AISC 169, pp. 513–518.
springerlink.com

(2) $m_{jk} = -1$, if k-th branching-off is dependent on j-th node, and k-th branching-off directs towards j-th node;

(3) $m_{jk} = 0$, if k-th branching-off is independent on j-th node.

M_c is called association matrix of a electrical network between nodes and branching-offs. Association matrix of a electrical network figure shown in figure 1 is formula (1).

It can be found that every column of a association matrix has only two non-zero elements (+1) and (-1), and $\sum_{j=1}^{n} M_j = [0 \cdot 0 \cdot 0 \cdots]_{1\times l}$. Therefore association matrix is linear correlation. Giving a $(n-1)\times l$ order association matrix can determine a directed graph which is composed of l branching-offs and n nodes.

$$M_c = \begin{array}{c} \\ n1 \\ n2 \\ n3 \\ n4 \end{array} \begin{array}{c} \begin{matrix} b_1 & b_2 & b_3 & b_4 & b_5 & b_6 & b_7 \end{matrix} \\ \begin{pmatrix} 1 & 1 & 0 & 1 & 0 & 0 & 1 \\ 0 & -1 & 0 & 0 & 1 & 1 & -1 \\ -1 & 0 & 1 & 0 & 0 & -1 & 0 \\ 0 & 0 & -1 & -1 & -1 & 0 & 0 \end{pmatrix} \end{array} \quad (1)$$

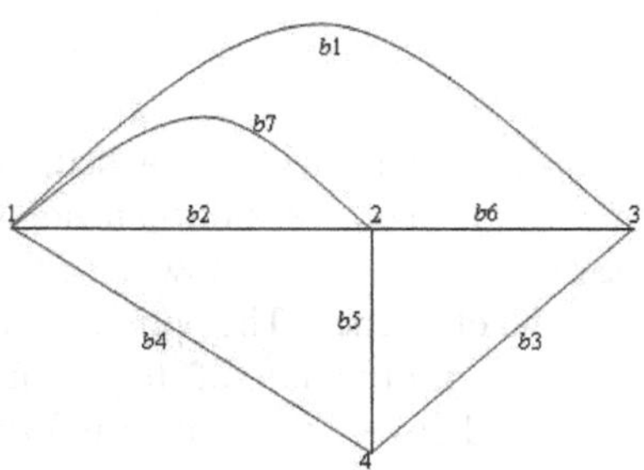

Fig. 1. Structure of a simple circuit network

3 Improved Node Matrix Model

Rows and columns of adjacent node matrix G defined in reference [1] rank identically. If two nodes are connected, the corresponding element of matrix is 1. Otherwise, the corresponding element is 0. Topology structure of an electrical network can be described by this way, but whether branch has element or not can not be determined. The improved node matrix is shown as follow.

$$G = \begin{array}{c} \\ n_1 \\ n_2 \\ n_3 \\ n_4 \end{array} \begin{array}{c} \begin{array}{cccc} n_1 & n_2 & n_3 & n_4 \end{array} \\ \begin{pmatrix} i & 2 & 1 & 1 \\ 2 & i & 1 & 1 \\ 1 & 1 & i & 1 \\ 1 & 1 & 1 & i \end{pmatrix} \end{array} \tag{2}$$

$$G_{\triangle I} = \begin{array}{c} \\ n_1 \\ n_2 \\ n_3 \\ n_4 \end{array} \begin{array}{c} \begin{array}{cccc} n_1 & n_2 & n_3 & n_4 \end{array} \\ \begin{pmatrix} i & 2 & 1 & 1 \\ & i & 1 & 1 \\ & & i & 1 \\ & & & i \end{pmatrix} \end{array} \tag{3}$$

$$G_{\triangle II} = \begin{array}{c} \\ n_1 \\ n_2 \\ n_3 \\ n_4 \end{array} \begin{array}{c} \begin{array}{cccc} n_1 & n_2 & n_3 & n_4 \end{array} \\ \begin{pmatrix} & 2 & 1 & 1 \\ & & 1 & 1 \\ & & & 1 \\ & & & \end{pmatrix} \end{array} \tag{4}$$

A matrix composed of n nodes has n^2 elements. Because the matrix is symmetrical, $\frac{n\cdot(n+1)}{2}$ elements are duplicated. All n diagonal elements are i. Number of the effective matrix elements is $\frac{n\cdot(n-1)}{2}$. The matrix can be simplified to a triangular matrix.

If a branch is broken, 1 should be subtracted from the element $G_{(a,b)}$ corresponding to two nodes (n_a, n_b) associated with the branch.

If a branch is direct shot, 1 should be subtracted from the element $G_{(a,b)}$ corresponding to two nodes (n_a, n_b) associated with the branch and add i.

Let branching-off $b5$ broken, $b3$ and $b7$ direct short in figure 1, we get

$$G = \begin{array}{c} \\ n_1 \\ n_2 \\ n_3 \\ n_4 \end{array} \begin{array}{c} \begin{array}{cccc} n_1 & n_2 & n_3 & n_4 \end{array} \\ \begin{pmatrix} i & 1+i & 1 & 1 \\ 1+i & i & 1 & 0 \\ 1 & 1 & i & i \\ 1 & 0 & i & i \end{pmatrix} \end{array} \tag{5}$$

$$G_{\Delta I}=\begin{array}{c} \\ n_1 \\ n_2 \\ n_3 \\ n_4 \end{array}\begin{array}{c} \begin{array}{cccc} n_1 & n_2 & n_3 & n_4 \end{array} \\ \begin{pmatrix} i & 1+i & 1 & 1 \\ & i & 1 & 0 \\ & & i & i \\ & & & i \end{pmatrix}\end{array} \tag{6}$$

$$G_{\Delta II}=\begin{array}{c} \\ n_1 \\ n_2 \\ n_3 \\ n_4 \end{array}\begin{array}{c} \begin{array}{cccc} n_1 & n_2 & n_3 & n_4 \end{array} \\ \begin{pmatrix} & 1+i & 1 & 1 \\ & & 1 & 0 \\ & & & i \\ & & & \end{pmatrix}\end{array} \tag{7}$$

4 Realization of Optimized Test Point Model Based on Improved Node Matrix

4.1 Realization Method

Optimal detection node set V can be established from improved node matrix. Let $b3$ and $b7$ direct short in figure 1, and follow the example (7), we get a $(n\text{-}1)\times(n\text{-}1)$ upper triangular matrix

$$G_{\Delta II}=\begin{array}{c} \\ n_1 \\ n_2 \\ n_3 \\ n_4 \end{array}\begin{array}{c} \begin{array}{cccc} n_1 & n_2 & n_3 & n_4 \end{array} \\ \begin{pmatrix} & 1+i & 1 & 1 \\ & & 1 & 1 \\ & & & i \\ & & & \end{pmatrix}\end{array} \tag{8}$$

If $b3$ branch is not associated with test points $n1$ and $n2$, then

$$G_{\Delta II}=\begin{array}{c} \\ n_1 \\ n_2 \\ n_3 \\ n_4 \end{array}\begin{array}{c} \begin{array}{cccc} n_1 & n_2 & n_3 & n_4 \end{array} \\ \begin{pmatrix} & 1+i & 1 & 1 \\ & & 1 & 1 \\ & & & \\ & & & \end{pmatrix}\end{array} \tag{9}$$

Let the different element between equation (8) and (9) is '0'. The fewer elements '0', the more node set V covers branches.

The more elements '1' in equation (9), the less repetition degree of branch associated with nodes in set V.

The smaller the variance $Var\{num[T(n_a)], num[T(n_b)], \ldots num[T(n_t)]\}$, the more average the number of branching-off associated with a node.

4.2 Essence of the Algorithm

Typical combinational optimization problems are set-covering, enchasement, binning and travelling salesman problem(TSP).

The set-covering problem can be described as following example [2].

$$c=\begin{bmatrix}10 & 15 & 11 & 10 & 8 & 2\end{bmatrix}$$

$$A=\begin{pmatrix}1 & 1 & 0 & 1 & 0 & 0\\ 0 & 0 & 1 & 1 & 0 & 0\\ 1 & 1 & 1 & 0 & 0 & 0\\ 0 & 0 & 0 & 0 & 1 & 1\\ 0 & 0 & 1 & 0 & 0 & 1\\ 0 & 0 & 0 & 1 & 1 & 1\end{pmatrix} \quad (10)$$

Its solution is $x=$ (1,0,1,0,1,0]) and cost is 29. Resource distribution truck dispatch [3], [4] and location of equipment [5] are also set covering problem.

5 Conclusion

Principles of building test set condition each other. A set does not exist meeting all condition. Stressing one principle does not accord with engineering practice. Neglecting branching-off coverage won't detect fault on circuit board. Neglecting repetition degree of branching-off associated with nodes will reduce precision of fault location. Too many branching-off associated with a node will damage status resolution of different branching-off. Few nodes will gather many faults on one node. With the electrical network spreading, too many nodes will make processing difficult and consume much resource. Node number relates to specific demand. You can set more nodes in production line and less node in detection. Optimal test set is realized by constructing adaptation degree function and each restriction condition.

Promoted node matrix can discover abnormality in circuit, reflect faulty status of electrical network, and is convenient for program.

References

1. Yuan, H., Chen, G.: Research on testability problem and fault diagnosis method in analog circuit. Journal of Electronic Measurement and Instrument 20(5), 17–20 (2006)
2. Balinski, M.: Integrate programming: method, uses and computation. Management Science 12(3), 253–313 (1965)

3. Valenta, J.: Capital equipment decisions: A model for optimal systems interfacing, pp. 30–33, 19. Massachusetts Institute of Technology Master thesis
4. Garey, M., Johnson, D.: Computers and Intractability: A Guide to the Theory of NP-Completeness, pp. 50–55. W H Freeman, New York (1979)
5. Walker, W.: Using the set-covering problem to assign fire companies to fire house. European Journal of Operational Research 22, 275–277 (1974)

Design of Monitoring and Alarming System for Urban Underground Gas Pipe Leakage Based on C8051F060

Shiwei Lin[1] and Yuwen Zhai[2]

[1] College of Information and Control Engineering, Jilin Institute of Chemical Technology, Jilin, China
13704406003@126.com

[2] Mechanical & Electrical Engineering College, Jiaxing University, Jiaxing, China
wanglei_new814@126.com

Abstract. The overall design of monitoring and alarming system for underground gas pipe leakage is introduced. Its block diagram is given, and the design principles of the data collector in the base station and the measuring subsystem in the field are described in detail, the design method of the system management software id presented. Each base station can be connected with 64 field measuring subsystem, and the data communication between base station and subsystems is implemented by signal carrier technology with DC power supply. The system can detect the real-time fuel gas content at the specified point in underground pipe net and transmit it to the monitor center in Natural Gas Supply Company. So the system can promote the centralized monitoring and control for urban underground gas pipe leakage.

Keywords: Gas pipe, Gas leakage, Monitoring and alarming, C8051F060.

1 Introduction

The supply of the gas brings people using easily, but safety and environmental pollution problems are brought because of its toxic and explosive. Much explosion that is caused of gas leak brings out personal injuries and economic losses significant each year in the world. At present, manual measurement is the only method that monitors urban underground gas leakage, the method is a heavy workload and not timely. Automatic monitoring and alarming system is no successful application [1]. The design of the system that monitors and alarms the urban underground pipe gas leak automatic and real-time is very necessary. The design uses a tree branch structure, the system consists of the main station, base station and sub-machine. The system can monitor the gas content data of underground gas pipe network in real time and upload the data to the gas Company Monitoring Center. Urban underground gas pipe leakage is monitored centralized by it.

2 The Overall Design

The system uses master station; base station and sub-machine triple tree brand structure. The system design diagram is shown in Figure 1. Taking into account the

D. Jin and S. Lin (Eds.): Advances in CSIE, Vol. 2, AISC 169, pp. 519–522.
springerlink.com

complete gas detection at the same time, the sub-machine also will collect data uploaded to the base station, while the combustible gas sensor current at 20mA or more, can not achieve long-term battery-powered remote power supply by 24VDC set by the base station collector. master Painter NET network (can also be designed to use GPRS and other wireless data transmission module) to access the base station up to can mount the 64 on-site detection of sub-machine each base station includes a base station PC and base station data acquisition in two parts, the base station data collector DC-powered carrier for each collection point on-site detection of sub-machine to provide a 24VDC power supply. site detection of sub machine will upload the gas concentration detected by the carrier collector to the base station, base station collected through the serial port will be collected from each site detection of sub-machine coming gas concentration data to the base station PC, the base station PC, network and The master control PC (gas company) to connect.

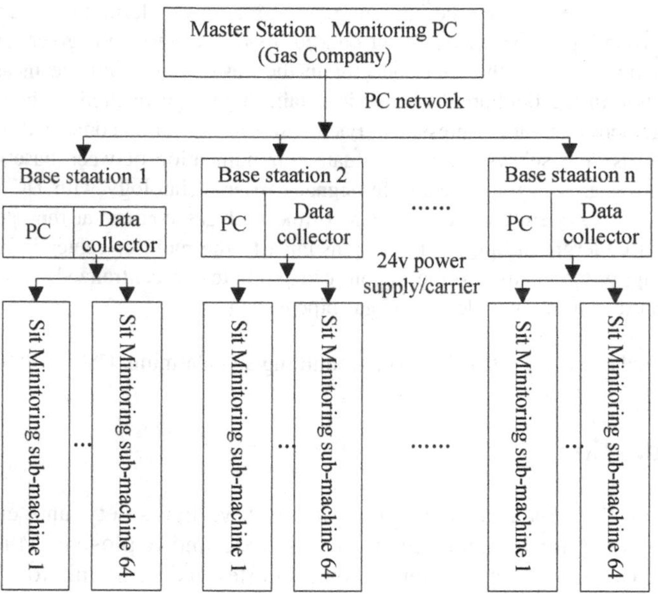

Fig. 1. The system overall design diagram

2 The Design of Site Monitoring Sub-machine

The function of the site monitoring sub-machine is monitor gas pipeline detection point of the site concentration of combustible gases, and the measured data to the base station data collector, its block diagram shown in Figure 2. The system-on-chip C8051F060 is the core of the site monitoring sub-machine core [2], the internal A / D converter to collect combustible gas sensor detection circuit output and the combustible gas concentration is proportional to the voltage signal, and the collected digital modulation circuit through the serial port to send superimposed to the supply line. transmitted through the supply line to the base station data collector selection with high sensitivity and low power consumption TP 1. 1A as a combustible gas sensor, a sampling of resistance and description control circuit detection circuit to real-time detection of combustible gas concentrations of methane to NE555 constitutes the variable duty cycle astable multivibrator constitute a serial port to send modulation circuit, the output duty cycle to 50% of intermittent oscillation signal. signal receiving demodulation circuit, including narrow-band amplification of two parts of the shaping circuit and serial demodulation circuit the shaping circuit narrowband amplification by the transistor and RC elements and serial demodulation circuit monostable trigger circuit, the output waveform to fully meet the requirements of serial receiver.

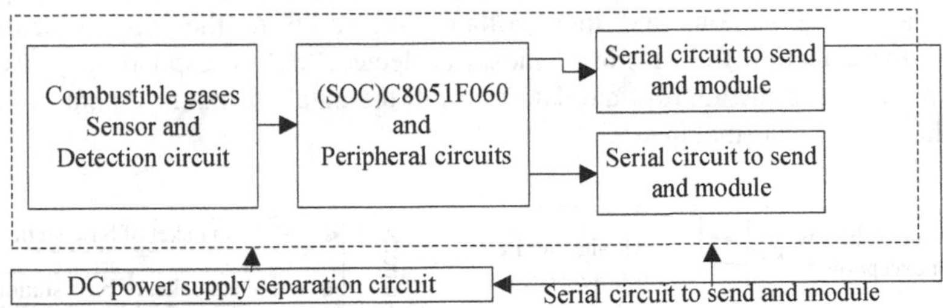

Fig. 2. Block diagram of field testing sub-machine

3 The Design of the Base Station Data Collector

The base station data acquisition for on-site detection of sub-machine provides DC power supply, field testing sub-machine detected the combustible gas content data collected by the carrier technology, the detection point data collected is uploaded to the base station PC, and then through the serial port its block diagram shown in Figure 3. C8051F060 chip system is the core of the base station data acquisition, C8051F060 serial port 1 (UART1) will be collected on-site detection of sub-machine data uploaded to the base station PC via the RS232 interface circuit, machine. Since each base station data acquisition to achieve sub-machine detection to combustible gas concentrations on the acquisition of 64 on-site

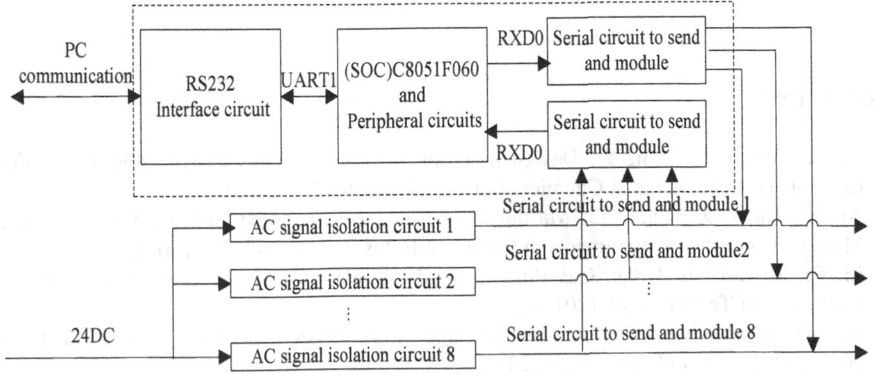

Fig. 3. Base station collector Schematic

detection, while 64 on-site detection of sub-machine power, but also DC-powered carrier to exchange data, consider to the line impedance, the base station data acquisition, including 8-way AC signal isolation circuit and serial modulation / demodulation circuit, the circuit structure and on-site detection of sub-machine.

4 The Design of the Management Software

Gas pipeline leak monitoring and alarming systems management software block diagram in Figure 4 below. Main station to monitor the PC receives the data packet sent to each base station, and then collate, according to the gas concentration values of the detection field testing handset collected, the corresponding processing and alarm; and save the appropriate data through the database management system to ensure that the post-normal query.

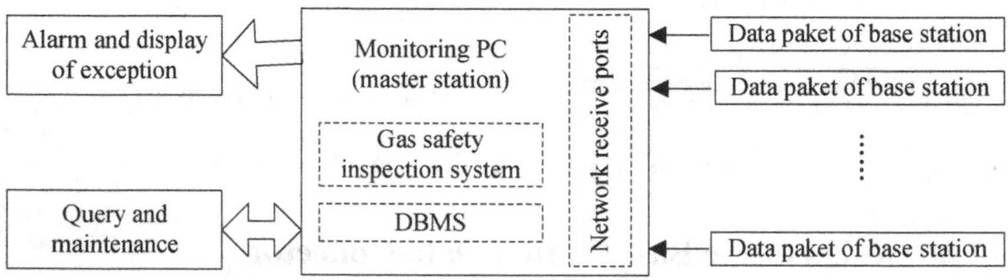

Fig. 4. Management software block diagram

5 Conclusions

The design uses the master station, base stations and sub-machine triple tree branch structure. The function of the underground gas pipeline leak monitoring and alarm in real-time is realized by the network communication technology and DC-powered carrier. By test, combustible gas (methane) detection range: (50 ~ 40 000) 10^{-6}, response time <10s, each base station to connect the handset number 64, the communication distance> 5 km.

References

1. Ai, X., Yang, X., Zhai, Y.: Design fo Fiedl-bus controller bansed on SOC C8051F606. Control and Instruments in Chemical Industry 35(4), 67–71 (2008)
2. Ai, X., Yang, X., Zhai, Y.: Master/Slave Telecommunication Scheme Based on MCU Modem. Control and Instruments in Chemical Industry 32(3), 47–49 (2005)
3. Li, C.: Research of Urban Gas Pipeline Leak Wireless Detection Technology. Shenyang University of Technology 8 (2011)
4. Ma, W.: Research for Oil Pipeline Leakage Alarm System Based on Vibration Theory. Northeast Petroleum University 5 (2011)
5. Yang, Q., Chen, K., Lv, Q., Bao, G.: Research on Simulation of Gas Pipeline Leak Detection and Location. Machine Tool & Hydraulics 38(9), 85–88 (2010)

The Research of Fast File Search Engine Based on NTFS and Its Application in Fast Electronic Document Destruction

Jun Huang and ShunXiang Wu

Department of Automation, Xiamen University, Xiamen, China, 361005
evolsin@sina.com, wsx1009@163.com

Abstract. NTFS has become the main file system of Windows. Based on NTFS, We propose an efficient method to build a fast file search engine by USN journal which is vividly called the secretary of NTFS. Using the engine, we can locate files or folders by name in all NTFS disks in a second. In addition, with the function of fast file location and the important MFT information stored in USN journal, the engine can apply to fast electronic document destruction. This research not only can be of great use to daily computer application, but also has significant value in computer privacy protection.

Keywords: NTFS, USN, fast file search, electronic document destruction.

1 Introduction

File search is a very common operation in computer application. Although Windows operating system provides the attached file search function and has a great improvement in the efficiency at new version, it still needs waiting for a long time, especially searching all disks. 'Everything' is the fastest filename search tool developed by voidtools. It can complete building hundreds of thousands of files' index within a minute and locate files or folders by name instantly, the search speed is very egregious, however, it is a pity that 'Everything' is not open source software, therefore, it is difficult for programmer to use its simple SDK to freely extend the application of their own. Besides, the general program commonly uses recursive or non-recursive method to iterate the files of disk, the efficiency is often not satisfactory. This paper, we research the USN journal which is vividly called the secretary of NTFS but lack of attention [1], we want to describe how to use Windows API of USN journal and SQLite database to build a fast file search engine, and its running process just like 'Everything' tool. What's more, with the function of fast file location and the important MFT information stored in USN journal, we can apply the engine to electronic document destruction which is one of the hot topics in information security field and find another way to implement destroying file fast, safely and thoroughly.

D. Jin and S. Lin (Eds.): Advances in CSIE, Vol. 2, AISC 169, pp. 523–528.
springerlink.com

2 Basic Concepts

2.1 USN Journal

USN (Update Service Number) journal is a dependable secretary which NTFS 5.0 brings in. USN journal can monitor every NTFS volume, while the event of adding, modifying or deleting file happened, USN journal stores the changed information [1]. Every USN record uses specific symbol to mark as a form of log, and it has a common structure called USN_RECORD. In the structure, there is a field 'USN' which is the unique number mark every record, there is field 'file reference number' which is the important MFT number we can use to locate file, there is field 'reason' which indicates that what change happened, it is an enumeration, here we only pay attention to the change of inserting, modifying or deleting file, there is a field 'time stamp' which records what time the change happened. In addition to this, the structure also records the filename and some other fields out of our attention. USN journal begins as an empty file, when NTFS volume has some change, all changed information will be added to this file immediately. Every NTFS disk saves only one USN journal. There is a fixed structure called USN_JOURNAL_DATA which records the basic information about the USN journal. In this structure, there is a field 'USN journal ID' which is unique to every journal, that is to say when we create USN journal each time, NTFS will return different USN journal ID. There is a field 'first USN' and a 'next USN', if we want to query all data of the USN journal, we need to know the start position called 'low USN' and end position called 'high USN', here the 'first USN' corresponds to 'low USN' and the 'next USN' corresponds to 'high USN', each of them is the unique number of a USN record. At last, noting that USN journal is only loyal to NTFS, not FAT or other file system.

2.2 MFT

Everything in NTFS is file. MFT (Master File Table) is the core of NTFS [3]. MFT consists of a series of MFT records. Each file has at least 1 MFT record. Each MFT record takes up 1024 bytes and owns a unique MFT number. The header of the record is a fixed structure to describe the basic information. Follow the header are attributes. In NTFS, all information which are relevant to data are called attribute [3]. Common attributes are the attribute of standard information, file name, data, index root, bitmap and so on. Each attribute has a feature code. All file operations will cause the change of the MFT record.

3 Building the Engine

3.1 Main Idea

The first time to use the engine, we should initialize USN journal of every NTFS disk and list all USN records, each USN record references to a file or folder, than we should insert information of all files to a database. Through optimizing performance, this procedure will take less than 2 minutes per million files. Now, we can input the key words of filename we want to find and get the output by querying database.

After using computer for some time, the files on the disk may have some change, it may have some new created files or folders, and some old files or folders may be renamed or deleted. Start the engine this time, we should check the USN journal to list all changed USN records, after analyzing them, we should do insert, update or delete operation to update the database to make sure the consistency between files of disk and file records in database. As the amount of the changed files is much fewer than the amount of files on disk, this procedure will completes in a short time, that is to say users can begin searching instantly after starting the engine.

3.2 The Selection and Design of Database

Before starting the engine, we should create a database to record the basic USN journal information and files' information of disk. Here we adopt SQLite as database because it's an in-memory database that can provides better read and write performance, besides, it's green software, and the core engine itself doesn't depend on the third party software and no need to install, so it decreases a lot of trouble in deployment. Furthermore, lightweight, portable, single file, cross platform are other features of it.

There are 2 tables we need to create at least:

USNJBasicInfo{id, diskName, usnJournalId, startUSN}

USNRecordInfo{id, mftNo, pMftNo, fileName, diskName}

As is mentioned above, every NTFS disk only has an USN journal, each USN journal has an unique id. If we replace a disk and create its USN journal, the returned new journal id must be different to the old, in this case, the relevant data about the old disk stored in USNRecordInfo are invalid, and we should delete the invalid data and insert all files' information of the new disk.

The 'startUSN' points the USN of the last record of journal in recent query and its initial value is 0. The next time to start the engine, we should not query journal from beginning to the end but from the value of 'startUSN' to get the changed records.

The 'mftNo' and 'pMftNo' are two important attributes. They are the unique identification number to a file's MFT record and its parent folder's MFT record, through them, we can get the file path and locate it, what's more, they are the entrance of electronic document destruction. Here we don't save the file path because get the file path needs some time cost, we can dynamically load the file path when user do searching.

3.3 Concrete Procedure

The first time to start the engine, the database is empty, we should fill the database with basic information of new created USN journal and all files' information of every NTFS disk. As single disk for example [2]:

First, we should check if the disk is NTFS format.

Second, get the handle of this NTFS disk.

Third, create or initialize the USN journal. Here we use the function 'DeviceIoControl' of Windows API and input the handle of this NTFS disk and the specific flag 'FSCTL_CREATE_USN_JOURNAL' and other parameters. NTFS will create an empty USN journal and put all files on the disk as USN records into the journal automatically.

Fourth, get the basic information of the new created USN journal above and store to database. Here we still use the function 'DeviceIoControl' with the specific flag 'FSCTL_QUERY_USN_JOURNAL'. This function will output a structure 'USN_JOURNAL_DATA', there are 3 important attributes mentioned above: 'UsnJournalId', 'FirstUSN', 'NextUSN. We should insert 'UsnJournalId' and "NextUSN' as 'startUSN' to table 'USNJBasicInfo' of database.

Fifth, list all USN records in the USN journal. Here we also use the function 'DeviceIoControl' with the specific flag 'FSCTL_ENUM_USN_DATA'. We should open a buffer to circularly receive the data of USN records and circularly analysis each USN record. After receiving all data, we have collected all files' information such as MFT number, the parent folder's MFT number, filename. We should insert them into the table 'USNRecordInfo' of database.

At this time, users can do searching by filename, the engine will query the database to get all similar filenames and dynamically get file path, and then return instantly.

Now, the question is how to get file path? As the filename and parent MFT number we know, we can use the function 'DeviceIoControl' with the specific flag 'FSCTL_GET_NTFS_FILE_RECORD' to find filename attribute to get name and the parent MFT number of parent folder. As this method to recur until reach the root of the disk (generally the MFT number is 5), and then combine all these filename, folder name and root name as reversed order to build the file path.

The next, when we start the engine next time, there is no need to initialize the USN journal, we only need to find the new added USN records which they record the changed files' information. Here, we also use the function 'DeviceIoControl' with the specific flag 'FSCTL_READ_USN_JOURNAL' and input the start position of the read range which is the value of the attribute 'StartUSN' stored in table 'USNJBasicInfo' of database, we also need to open a buffer to circularly receive the data of new added USN records and circularly each USN record especially the reason. According to the reason, we distribute the records into 3 assembles corresponding to 3 different operations. If the reason indicates that the file is new created, we should insert the information of this file to database. If the reason shows that the file is renamed, we should modify the filename of relevant record in database. If the reason means the file is deleted, we should delete the relevant record in database. After a short time to update the database, users can do searching.

3.4 Optimize Performance

First, in order to economy user's waiting time of initializing or updating database when start the engine, we adopt multi-thread programming to exert the multi-core process's performance sufficiently [5]. We can parallelly initialize USN journal of each disk and adopt thread-synchronized method to wait all USN records obtain completely and then bulk insert into database.

Second, In order to improve the efficiency of data query, we can consider establishing the clustered index on certain field of database [7].

4 Application in Electronic Document Destruction

4.1 Background

Electronic document destruction is one of the hot topics in information security field. Simply deleting file exists many security problems. While deleting a file, the file system only makes some marks but not erase the MFT record and data field, it leaves many traces on the disk. Through certain data recovery means, the deleted file still can be reconstructed [4]. Therefore, electronic document destruction is crucial importance to some secret files. At present, the published paper researching destroying electronic document is relatively few, we have proposed an efficient method before, now we have find another way by using this file search engine.

4.2 Resolving Method

In order to implement destroying electronic document thoroughly from file system, the most important thing is to get the MFT number of the file. Here, we can get the MFT number by file searching. We can create a query inputting disk name and filename, the MFT number is stored in table 'USNRecordInfo'. If there are some files of the same name return, we can get the file path by the method mentioned above to match the target file path to determine the unique MFT number.

Getting the MFT number of the file, we are ready for destruction [6]:

First, we should erase the file's data. Getting the handle of the file by file path, we can use WriteFile function of Windows API to cover the data section directly with 0x00. We can also get the data's sectors map which record sectors of data distribution by analyzing the data attribute (0x80) of MFT record and then erase the sectors.

Second, we should delete file using DeleteFile function of Windows API.

Finally, we should erase the MFT record. We can easily calculate the beginning sector of the MFT record by MFT number and then erase 2 continuous sectors (generally 1KB).

5 Summary

Upon the research, we have proposed an efficient method to build a fast file search engine by USN journal and make it to software [8]. In order to confirm the efficiency of our engine, we compare our software with 'Everything', a general file finder software which uses non-recursive method to iterate the files of disk and file search function which is Windows attached. We do test in the same condition which owns about 855,000 files on disk. The result of contrast in finding the same filename as the following table shows:

Table 1. Performance contrast (time measurement rounds to second)

	Everything	Our engine	File finder	Windows Search
Initialization time cost	70s	102s	287s	-
Query time cost	<=1s	2s	27s	205s
Update DB time cost	4s	5s	-	-
Best time cost	5s	7s	314s	205s
Worst time cost	71s	104s	314s	205s

From the contrast above, we can see that our engine runs a little slower than 'Everything' in performance of initialization but also can complete loading 1 million files about 2 minutes, it is much faster than others, so it can proved the validity of our proposed method and still has space to improve. Besides, in query performance and data update performance, they are similar, users can instantly get the list of relevant files by inputting the key words of the filename.

We will provide our source code later, it's convenient for extending and putting them into other application. Welcome readers to modify or propose improvement methods.

Acknowledgments. This project is supported by the Aeronautical Science Foundation (20115868009) and the Open Project Program of Key Laboratory of Intelligent Computing & Information Processing of Ministry of Education, Xiangtan University, China.

References

1. Meng, J.: The loyal secretary of NTFS and some things about USN journal. Computer Fan 1 (2009)
2. The research of reading USN journal on NTFS, http://univasity.iteye.com/blog/805234
3. Marin: Data reproduce. Tsinghua University Press, Beijing (2009)
4. Dai, S.: Technology of data recovery. Electronic Industry Press, Beijing (2005)
5. Wang, H.: Fast file search tool based on multi-thread. Computer Programming Skills and Maintenance 3, 16–19 (2010)
6. Huang, B.: Analysis of traces on storage media by file operation for NTFS file system. Computer Engineering 33, 281–283 (2007)
7. Silberschatz, A., Korth, H.F., Sudarshan, S.: Database system concepts. Electronic Industry Press, Beijing (2006)
8. Eckel, B., Allison, C.: Thinking in C++. Electronic Industry Press, Beijing (2011)

Improved Genetic Algorithm Based 2-D Maximum Entropy Image Segmentation Algorithm

Jianjun Wang

Engineering University of Chinese Armed Police Force,
Xi'an, China
dayu504@126.com

Abstract. Since simple genetic algorithm based 2-D maximum entropy image segmentation algorithm has the problem of premature, this paper proposes an improved genetic algorithm. Through using Fitness Extreme Distance (FED), the improved genetic algorithm proposed in this paper establishes fuzzy evaluation mechanism in the evolution procedure. Compared with the simple genetic algorithm, improved algorithm remarkably enhances the genetic algorithm's convergence and the overall search ability. Theoretical analysis and experiment results shows, compared with basic genetic algorithm, the proposed genetic algorithm based 2-D Maximum Entropy Segmentation algorithm's acquired segmentation effectiveness is better.

Keywords: Image segmentation, Genetic algorithm, 2-D maximum entropy.

1 Introduction

Image segmentation provides the preparative works for the high-level operations in image processing and computer vision. Since the variability of images and the presence of noise, image segmentation is still a challenging problem [1].

Among all kinds of segmentation methods, threshold based method is the most basic and the most widely applied technology. In many cases, it is often the first choice of image analysis, feature extraction and pattern recognition. But for an image, the pixels have strong consistency and correlation in position and gray level, the traditional methods ignore the pixels' detail distribution information, only considering the gray level histogram information, so when an image's optimal segmentation threshold dose not reflect in the valley of histogram, they are always helpless [2]. So people add pixel's neighborhood information as important information to the traditional segmentation strategy, and construct amount of 2-D histogram based image segmentation algorithm.

2-D maximum entropy image segmentation algorithm is a typical 2-D histogram based algorithm which perfectly uses the correlation of the gray level histogram information and the pixels' neighborhood information, effectively reducing the effect of noise. But theoretical analysis and experimental results show that its time complexity is extremely high, not suitable for real-time applications, therefore, in practical applications, it is often optimized by other intelligence algorithms [3].

D. Jin and S. Lin (Eds.): Advances in CSIE, Vol. 2, AISC 169, pp. 529–534.
springerlink.com

As an efficient method of parallel searching, genetic algorithm has incomparable advantages in optimizing the 2-D maximum entropy image segmentation algorithm [4]. However, in actual applications, genetic algorithm often faces the problem of premature. In order to solve this question, this paper proposes an improved genetic algorithm. The proposed algorithm introduces Fuzzy Penalty Function (FPF) in inner-generation and improves the algorithm's global search ability. When using the proposed improved genetic algorithm in the 2-D maximum entropy image segmentation algorithm's optimization, experimental results show, compared with simple genetic algorithm, the proposed method's segmentation effective is better.

2 Background Knowledge

2.1 Genetic Algorithm

A genetic search starts with a randomly generated initial population within which each individual is evaluated by means of a fitness function. Individual in this generation are duplicated or eliminated according to their fitness values [5].

There are usually three operators in a typical genetic algorithm. The first is the production operator which makes one or more copies of any individual that posses a high fitness value. Otherwise, the individual is eliminated from the solution pool. The second operator is the crossover operator. This operator selects two individuals within the generation and a crossover site and carries out a swapping operation of the string bits to the right hand side of the crossover site of both individuals. Crossover operations synthesize bits of knowledge gained from both parents exhibit better than average performance. Thus, the probability of a better performing offspring is greatly enhanced; the third operator is the 'mutation' operator. This operator acts as a background operator and is used to explore some of the invested points in the search space by randomly flipping a 'bit' in a population of strings. Since frequent application of this operator would lead to a completely random search, a very low probability is usually assigned to its activation.

2.2 Two-D Maximum Entropy Image Segmentation Algorithm

The thinking of the2-D maximum entropy image segmentation algorithm is using 2-D histogram entropies which are determined by using the gray value of the pixel and the local average gray value of the pixel [6].

Let the average gray-level values of each pixel's neighborhood are from 0 to $1-L$, as well as the gray level values of each pixel. At each pixel, both the average gray-level value of the neighborhood and its gray level are calculated. This forms a pair that belongs to a 2-dimensional bin: the pixel gray level and the average gray level of the neighborhood.

2-D maximum entropy image segmentation algorithm's realization procedure is as follows:

STEP1: Calculate each pixel's joint probability $p(i,j)$:

$$p(i,j)=\frac{N_{i,j}}{N_{image}} \quad (i=0,1,\cdots,255, \quad j=0,1,\cdots,255) \tag{1}$$

In formula (1), $N_{i,j}$ represents the number of pixels which value is i and average level is j, N_{image} represents the total number of the pixel in the whole image;

STEP2: Set an initial threshold $Th_{st}=Th_{st}(0)$ which divide the original image into two classes: C_1 and C_2;

STEP3: Calculate the 2-D histograms' relative average 2-D entropy E_1 and E_2 respectively:

$$\begin{cases} E_1=-\sum_{i=1}^{s}\sum_{j=1}^{t}(\frac{p_{ij}}{p_{st}})\ln(\frac{p_{ij}}{p_{st}}) \\ E_2=-\sum_{i=s+1}^{255}\sum_{j=t+1}^{255}(\frac{p_{ij}}{1-p_{st}})\ln(\frac{p_{ij}}{1-p_{st}}) \end{cases} \tag{2}$$

In formula 2 $p_{st}=\sum_{i=1}^{s}\sum_{j=1}^{t}p_{ij}$, $s=0,1,\cdots,255, \quad t=0,1,\cdots,255$;

STEP4: Choose the best value $Th_{st}=Th_{st}^{*}$, according to which the pixels are divided into class C_1 and class C_2 which satisfy the blow formula:

$$[E_1+E_2]|_{Th_{st}=Th_{st}^{*}}=\max\{E_1+E_2\} \tag{3}$$

STEP5: Segment image by Th_{st}.

3 Proposed Algorithm

3.1 Improved Genetic Algorithm

How to effectively describe and process the constraints and the constraint conditions to construct a suitable evaluation function is a key to solve the problem of optimization, penalty function is a common used method [7]. In order to construct penalty function, the proposed algorithm defines Fitness Extreme Distance (FED).

Sorting all individuals $\{f_1,f_2,...,f_N\}$ in current population set according to their fitness, and forming a ordered individual set $\{g_1,g_2,...,g_N\}$, in which N represents the size of population set, f_i represents individual i's fitness. $g_i\in\{g_1,g_2,...,g_N\}, 0<i,j\leq N$. Thus, in the new set, the first element has the highest fitness, and called Fitness Extreme value, accordingly, the individual's position represents the order of fitness, and called FED, represented as d_i.

By introducing the FED, based on considering the different evolution stages' features, improved genetic algorithm establishes a kind of Fuzzy Penalty Function (FPF) which punishes the Chromosomes according to their fitness with different ways. In this paper, FPF $u(d_i,x)$ defined as follows:

(1)In the prophase of evolution:

$$u(d_i,x)=\begin{cases}0.85, & d_i\le C_1N\\ 0.95, & C_1N<d_i<C_2N\\ 1, & di\ge C_2N\end{cases} \tag{4}$$

In formula 4, N defined as population size, C_1 and C_2 meet $C_1<C_2<1$, and defined as control parameters.
(2)In thc medium of evolution

$$u(d_i,x)=1-d_i/N \tag{5}$$

(3)In the later stage of evolution

$$u(d_i,x)=1-d_i/\exp(N-d_i) \tag{6}$$

In proposed improved genetic algorithm the new fitness function $eval(i,x)$ is defined as formula 7:

$$eval(i,x)=f(i,x)*u(d_i,x) \tag{7}$$

3.2 Proposed Segmentation Algorithm

The proposed improved genetic algorithm based 2-D maximum entropy image segmentation algorithm's realization thinking and steps are as follows:

Step1: Coding: there are 4 parameters whose values are from 0 to 255, so coding the chromosome with 32 bit binary strings;

Step2: Population generation and parameter setting: set 100 as the initial population size, set 45 as the maximum evolution generation, crossover probability $P_c=0.6$, mutation probability $P_m=0.6$, generate 100 chromosomes and initialize them with random values from 0 to 255;

Step3: Calculating fitness: calculate each chromosomes' fitness using formula 7,set $EvoGen$ as number of evolution times, when $0\le EvoGen\le 15$, the genetic algorithm in the prophase stage of evolution, when $16\le EvoGen\le 30$, the genetic algorithm in the medium stage of evolution, when $31\le EvoGen\le 45$, the genetic algorithm in the later stage of evolution;

Step4: Select: execute selection operation by using roulette wheel selection method;

Step5: Crossover: according to initialized crossover probability P_c to execute crossover operation;

Step6: Mutation: according to initialized mutation probability P_m to execute mutation operation;

Step7: Termination condition: if $EvoGen=45$, terminate the algorithm, else go to step 2;

Step8: Decoding and Image segmentation: decode the chromosome which with highest fitness as the best individual and using the decoded result as the optimal threshold to segment the image.

4 Experiment Results

Experiments are performed on 512×512 gray-scale image Lena which with stochastic noisy. Fig. 1 compares the segmentation results of different methods. Fig. 1 (3), Fig. 1 (4), Fig. 1 (5) shows three kinds of two-dimensional histogram block based algorithm's results, compare with Fig. 1. (2), they are affected by noisy lower. Among Fig. 1(3), Fig. 1(4), Fig. 1(5), Fig. 1(3) with the best quality, although Fig. 1(4) achieves good background and foreground segmentation, it just converges to a partial solution, the obtained threshold a little large. Fig. 1(5) achieves better segmentation effective, compares with Fig. 1(4), it not only suppresses the noisy, but also retains the image's necessary details, the segmentation results closer to Fig. 1(3).

Table 2 lists three kind of 2-D fuzzy entropy algorithm's operation time. Table 3 shows genetic algorithm and improved genetic algorithm can greatly shorten the segmentation process' running time. Compared to the basic genetic algorithm, due to the introduction of FPF, the proposed algorithm is more time consuming.

Fig. 1. Image segmentation results, (1) testing image with stochastic noisy, (2) reusult of One-dimensional fuzzy entropy,(3) result of two-dimensional histogram block, (4) reuslt of basic genetic algorithm , (5)the proposed method.

Table 1. Comparison of three kind of 2-D fuzzy entropy algorithm's operation time

Method	Operation time for one times	Average time of 10 times
Group list method	2.1 h	
Genetic algorithm		109.6s
Proposed method		136.7s

5 Conclusion

In this paper, we developed an improved genetic algorithm based 2-D maximum entropy image segmentation algorithm. This algorithm is able to improve the basic genetic algorithm because: (1) by using FED, the algorithm establishes the population' fuzzy penalty function, which makes the chromosome's evaluation more reasonable and fair; (2) the algorithm uses different fuzzy evaluation function in different stages of evolution, significantly improves the algorithm's convergence and global search ability. From theoretical analysis and experimental results, it clearly shown it can obtain better segmentation results. So we propose the improved genetic algorithm based 2-D maximum entropy image segmentation algorithm with high performance.

References

1. Lie, W.N.: An efficient threshold-evaluation algorithm for image segmentation based on spatial gray-level co-occurrences. Signal Processing 33(1), 121–126 (2004)
2. Robert, M.R., Teresa, E.M.: A strategy for reduction of noise in segmented images and its use in the study of angiogenesis. Intelligent and Robotic Systems 33(1), 1112–1115 (2009)
3. Davis, L.: Handbook of Genetic Algorithm. Van Nostrand, New York (1991)
4. Awadh, B.: A computer-aided process planning model based on genetic algorithm. Computer Operational Research 22, 841–856 (2005)
5. Turan, G.: Electric Power Distribution Systems Engineering, pp. 235–277. Mc.Graw-Hill International Edition (1986)
6. Chen, G.L., Cai, G.Q.: Fuzzy penalty algorithm for optimization of network system reliability. Fuzzy Math. 11, 501–509 (2003)
7. Isao, T.: An evolutionary optimization based on the immune system and its application to the VLSL floor-plan design problem. Electrical Engineering in Japan 124(4), 231–239 (2007)

A Design of the Urban Well Environment Monitoring System Based on M-BUS*

JianYun Ni[1,2], Jing Luo[3], ZeRen Yu[1], Lian Li[1], and Ming Zhao[1]

[1] Tianjin Key Laboratory for Control Theory & Applications in Complicated Systems, Tianjin 300384, China
nijianyun@tjut.edu.cn
[2] School of Electrical Engineering, Tianjin University of Technology, Tianjin 300384, China
[3] College of Electrical Engineering and Automation, Tianjin Polytechnic University, Tianjin 300160, China

Abstract. M-Bus is a special communication technology for transferring information of consumption measurement instrument and the counter. M-Bus is widely used in data collection in the fields of buildings and industrial energy consumption. City underground pipeline system is huge and the structure is complex. As a result, it is great inconvenient for a daily management and maintenance. With all kinds of sensors, the proposed monitoring system can monitor the temperature of the well environment, the water level, gas leaks, whether workers into the well, alarm information. Moreover, it can be applied in the long distance, multi-collection channel, inflammable and explosive application. Due to the new Bus communication, equipments are supplied with power by Bus. Consequently, the proposed system is manageable and suitable for site conditions which can't afford power supply. The system has high reliability, low cost, a broad prospect and practice value.

Keywords: Monitoring System, M-BUS, Communication, Remotely alarm.

1 Introduction

At present, urban underground pipeline system is huge and structure is complex, which results to greatly inconvenience for the daily management and maintenance. The underground pipeline system has long line, and complicated structure. Furthermore, it can't add any relay equipment, and it can't have any external power supply. Thus, the common fieldbus is not suitable for the city to well environment monitoring. Traditional M-Bus[1-3] communications Bus can meet the long distance communication and bus power supply, but slave computer can't realize active communication alarm. So, it is needed to design a new kind of fieldbus to monitor the well environment. All field equipment is supplied by bus power. The real-time monitoring and well environment alarm is realized. The system can realize the automatic monitoring and management, history data storage and management.

With all kinds of sensors, the proposed monitoring system can monitor the temperature of the well environment, the water level, gas leaks, whether workers into

* Supported by National Undergraduate Innovational Experimentation Program (091006011).

D. Jin and S. Lin (Eds.): Advances in CSIE, Vol. 2, AISC 169, pp. 535–540.
springerlink.com

the well, alarm information. Moreover, it can be applied in the long distance, multi-collection channel, inflammable and explosive application. Due to the new Bus communication, equipments are supplied with power by Bus. Consequently, the proposed system is manageable and suitable for site conditions which can't afford power supply. The system has high reliability, low cost, a broad prospect and practice value.

2 System Structure and Principle

The city environment monitoring system based on the improved M-BUS is mainly composed of three parts, the master machine, the slave machine and transmission lines. The monitoring software designed serial interface communication programs by VB[4] language and achieved urban underground pipeline monitoring system[5]. The master PC outputs RS-232 signal through the serial interface. Through a converting circuit, RS-232 signals are converted to TTL signal, and then TTL signals into the M-BUS, and through the M-BUS voltage signal are sent to the slave PC, the slave PC receives different instructions sent by master PC though the serial interface UART. Then, according to the different instructions, different operations can be realized, such as monitoring the temperature through the temperature sensor, monitoring the human body invasion, monitoring water level, etc.. Finally the monitoring information is converted into instructions and sent to the PC.

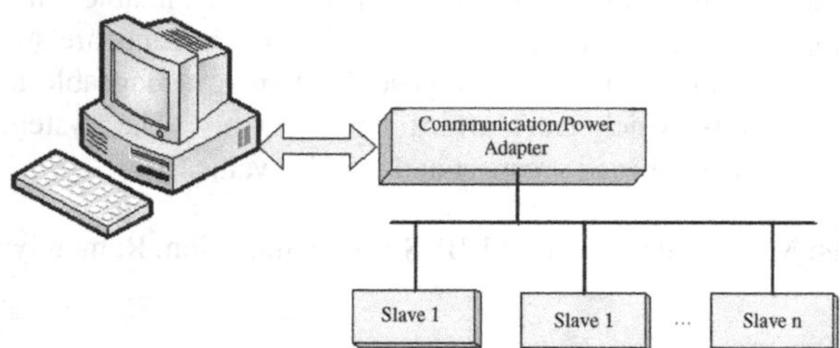

Fig. 1. System structure

3 Hardware Circuit Design

The improved M-BUS circuit mainly includes voltage sending unit, current receiving unit, power module and signal conversion module. Slave PC includes power supply circuit, communication interface, reset circuit and temperature sensor, human invasion and water level sensor and slave PC address settings.

In Fig.2, slave PC is based on STC12C2052AD and it includes the following six circuits. STC12C2052AD is the core of the system, and it mainly accomplishes calculation and controls surrounding circuit. LM35 series integrates analog integrated temperature sensor in a chip. It can complete temperature measurement and simulation signal output. It only measures temperature[6,7] and it has small temperature measurement error , low price, rapid response speed ,long transmission distance, small volume, low power consumption, which is suitable for long distance measuring temperature, controling temperature. The system does not need to undertake nonlinear calibration and its peripheral circuit is simple. Communication bus used 36 V power supply. Chip adopts 5 V power supply. The power supply module uses LM2576 chip to switch the 36 V to 5 V. BUS1 is 36 V, the other end of chip is 5 V power supply. Micro-consumption electronic infrared sensor is used for perception whether personnel are in well. Two water level sensors are used for percepting water levels of low and high.

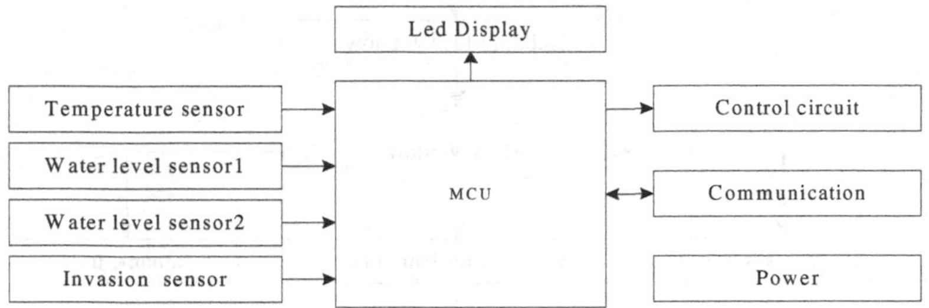

Fig. 2. System structure of slave PC

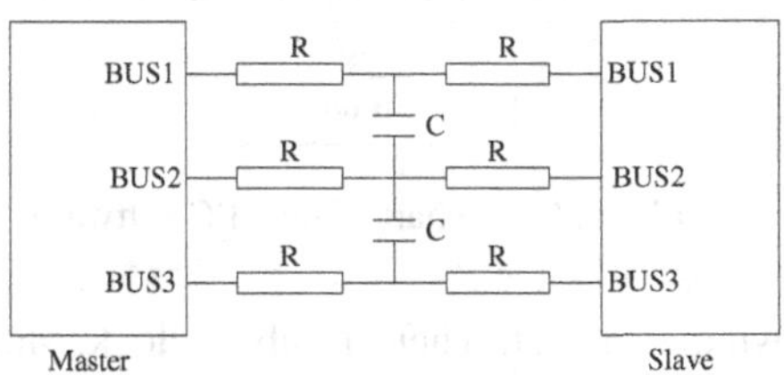

Fig. 3. Communication connection between host PC and slave PC

RS232 is connected to computer through serial interface by MAX232 . On the other hand, the RS232 signal is converted into TTL signal and is connected to the bus. Improved M-BUS[8,9] consists of three wires, i.e., BUS1, BUS2 and BUS3.

BUS1 is multi-use line and it is a power supply line and channel by which slave PC transfers information to host PC by current. BUS2 is common ground wire. BUS3 is a channel by which host PC transfers information to slave PC by voltage. BUS1 has the voltage limit of 36 V. The communication connection between host PC and slave PC is shown in Fig.3, R and C is circuit impedance and capacitance. The dashed box represents a slave PC.

4 Structure of System Software

4.1 Software Structure and Function of Host PC

This system applies VB6. 0 as a host PC software platform. It can easily make a friendly interface, and it has a powerful database which can be very good to deal with monitoring data, and is convenient for the user to understand the previous monitoring results, also can send and receive data and can accomplish real-time communication. As a result, it can meet the requirements of host PC application software platform of remote monitoring system.

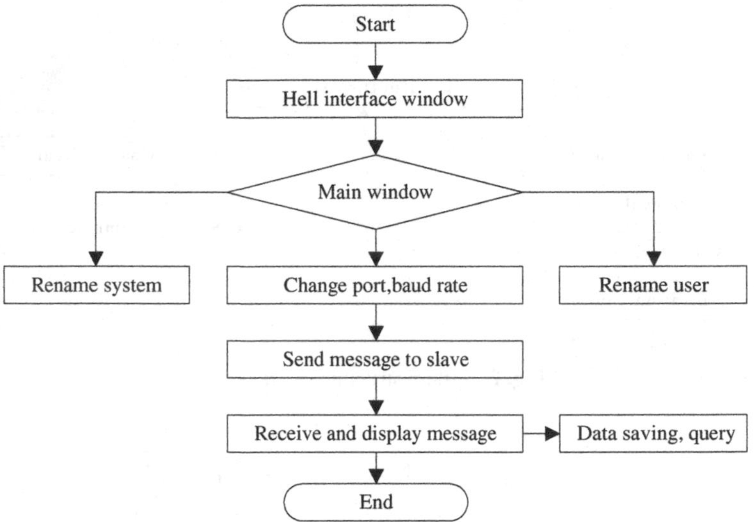

Fig. 4. Flow chart of host PC software

Firstly, the main window is set. Then, toolbar, clock, status bar, pictures, labels, text frame, button and tree structures can be added to the main window according to the requirements. Finally, their properties are set, respectively.Code is written for Form_Load events, which mainly completes the following several aspects: variables initialization; serial port initialization; tree structures initialization, etc..Form_Load still sets the size of main window; after operating, the status bar shows "welcome to use this system!! ". 10 site PC is shown on the operation interface. The code of serial communication is written, and MSComm1_OnComm is created. The function SEND is defined. SEND is applied when clicking a command button and the display system shows a "in communication......" .10 site PCS are established. The address of each site PC is assigned to the third bit of receiving data , and its name is saved for inquiring if needed. Due to receive window cannot save the information and no history records. If users want to know the previous monitoring results, such as want to understand whether alarm has happened during the past period of time, or want to compare different temperature during different time, the receive window does not meet these needs. Therefore, another database is built, which can be used to store the records of the site PC and temperature, and it is convenient for the user's query. The Timer1 control is programmed. it is used to time and set up the track of little red dot after the launch of the animation, which makes the system more vivid and directly. In the menu bar, it can achieve the following functions: changing the system name, software introduction, changing information and use instructions.

In the form welcome, the unit, the name, version number are filled. A timer1 clock control is added. The interval is 2000. It goes to the window after 2 seconds. In the engineering properties dialog box, the form of wellcome will be set to start object. Form a dialog form can change the name of the current system. After the form wellcome receive information, by clicking inquires buttons, users can inquires the record according to the machine name and operation type. The results will show the name and the operation of the machine, state information and the date and time of communication. From a dialog, users can select computer communication port com1 or com2, respectively. Different baud rate, 1200, 2400, 4800, 9600 can be chose too.

4.2 Site Machine Software

The machine is connected to PC through the two buses. The machine is monitored using the PC monitoring software. Due to the number of the machine is more than testing machine, so a lot of the machines are in parallel with the two buses. PC sent the instructions to the site machine through a serial port. The machine can feedback the invasion, and then output instructions to the PC through the temperature sensors, water level sensor and human body sensor. The instructions are translated into information we need through some agreement, such as the temperature, water level, human body invasion, whether to cancel the monitoring and so on. We can learn that the place of machine and make corresponding processing.

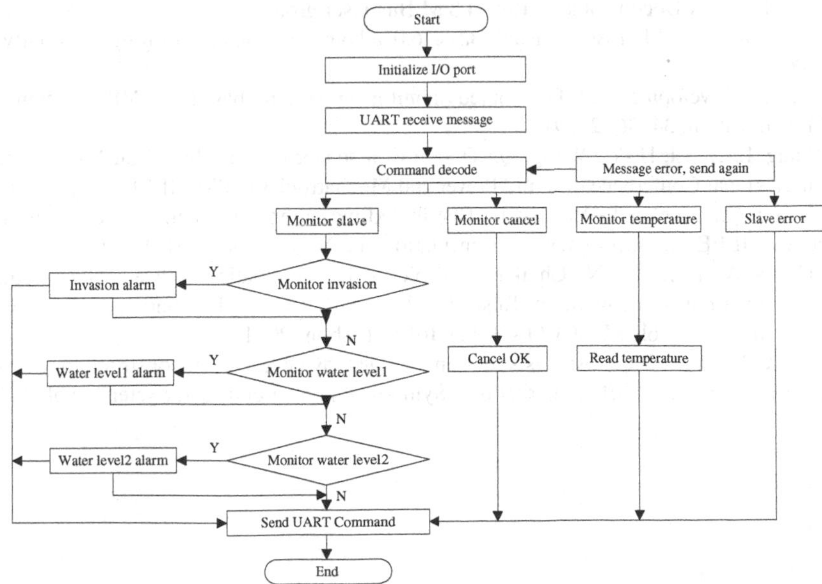

Fig. 5. Flow chart of slave software

5 Conclusions

This monitoring system uses a point-to-point master-slave response method and realizes the serial communication between PC and microcomputer. The scene applied the single-chip machine. In the integrated development environment, monitoring is accomplished through the C language. The experiments show that the system has high reliability and practicability, etc. Along with the continuous integration of serial communication equipment, this system can be better used in monitoring and control system. Communication applies the improved version of the M-BUS which makes further communication distance. The signal is more stable. So, the system has a good application value. This project used computer technology, fieldbus technology, the communication technology and microelectronics technology to develop environment monitoring system in the well. The system is mainly composed of the PC, communications bus, site PC. The system can monitor the water level, gas leaks, human body invasion and alarm etc, also can be applied in the fields of long distance, multi-collection, inflammable, explosive, bus power supply and so on. The system has a broad prospect of application and practice value. Furthermore, the system lay the foundation for the realization of digital city and information management of the underground pipes.

References

1. The M-BUS : A documentation version 4.8 (November 11, 1997)
2. TSS 721A METER-BUS TRANSCEIVER, Texas Instruments Incorporated (April 1997)
3. The M-Bus: A Documentation Rev. 4.8 M-Bus User group
4. TMS company, Microsoft Visual Basic 6.0 advanced program. Beijing university press (2000)
5. Yu, L.: Development of Distributed monitoring system based on VB. Computer and Modernization, 34–36 (2009)
6. Zhang, L.-L., Li, H.-Y., Wang, Q.: Design of a sensor network based on M-Bus. In: 2011 International Conference on Fluid Power and Mechatronics (FPM). IEEE, Beijing (2011)
7. Moczar, G., Csubak, T., Varady, P.: Distributed measurement system for heat metering and control. IEEE Transactions on Instrumentation and Measurement 51(4) (2002)
8. Zabasta, A., Kunicina, N., Chaiko, Y., Ribickis, L.: Automatic wireless meters reading for water distribution network in Talsi city. In: EUROCON - International Conference on Computer as a Tool (EUROCON 2011). IEEE, Lisbon (2011)
9. Vu, K.T., Liu, C.-C.: Analysis of tap-changer dynamics and construction of voltage stability regions. In: IEEE International Symposium on Circuits and Systems, vol. 2 (1988)

A Color Image Representation Method Based on Double-Rectangle NAM

Peng Wu[1], Kai Xie[1,2], Houquan Yu[1,2], Chuanbo Chen[3], and Chao Wu[1]

[1] School of Electronics and Information, Yangtze University, Jingzhou, China
[2] Key Laboratory of Oil and Gas Resources and Exploration Technology of Ministry of Education, Yangtze University, Jingzhou, China
[3] School of Computer Science and Technology, Huazhong University of Science and Technology, Wuhan, China
peng_wu@189.cn, pami2009@163.com, hq_yu@163.net

Abstract. In this paper, we propose a new color image representation method based on double-rectangle Non-symmetry Anti-packing pattern representation Model (NAM), which is used to reduce the number of subpatterns in image representation. The method adopts bit-plane optimization strategies at first to reduce the correlation between color image bit planes, then employ the correlation among pixels in bit planes and proceed with double rectangle NAM segmentation to decrease data storage space. Experimental results show that this method is an effective lossless color image coding method.

Keywords: image representation, image compression, Non-symmetry Anti-packing pattern representation Model, linear quadtree, bit-plane optimization.

1 Introduction

With the rapid development of computer network and communication technology, data compression has become a key technology in information storage and transmission, which is widely used in various businesses, such as seismic exploration, remote sensing and telemetry, network video transmission, multimedia communication, etc. Among numerous representation methods, quadtree [1-2] is the most studied hierarchical representation method. Early quadtree representation adopts pointer-based quadtree structure, in order to further reduce storage space, researchers have proposed various improvements [3-6] relating to quadtree. Although these improved quadtrees have different data structure, coding scheme and efficiency, they all have one thing in common, that is, they put too much emphasis on symmetry segmentation. This will cause original adjacent pixels to be separated, and increase the number of nodes for representation, so it needs to be further improved.

Non-symmetry Anti-packing pattern representation Model [7-8] is a new pattern representation method and abtain higher compression ratio owing to adopt predefined subpatterns and asymmetric segmentation. Related researchers have proposed various pattern representation methods in the light of NAM, but in general, on the flexibility and adaptability of different types of images, there is still in need of further improvement. Inspired by the idea of Parking Problem, rectangle subpattern is adopted in paper [7] and rectangle NAM(RNAM) representation is achieved for color

D. Jin and S. Lin (Eds.): Advances in CSIE, Vol. 2, AISC 169, pp. 541–546.
springerlink.com

image. Although the compression ratio has been improved a lot than the linear quadtree, this method is not obvious for non-block images. The reason is that the method adopts a single type of subpattern, which is not suitable for all kinds of images, especially the texture-rich images. In order to improve the adaptability, it is necessary to increase the types of predefined subpatterns. [8] put forward a bit-plane optimized NAM to represent gray images, the method make use of correlation between bit planes and change scanning scheme, finally a good result has been achieved for non-block images.

In this paper, we use double rectangle, i.e. regular rectangle and diagonal rectangle, as two types of subpatterns, which is refered to as double-rectangle NAM(DRNAM). Combines with bit-plane decomposition based optimization strategies[8], color images have been antipacked and the number of subpatterns has been significantly reduced, higher compression ratio is achieved as well.

2 DRNAM Principles

In the NAM representation method various predefined subpatterns are adopted to represent the information to be stored. This paper will put forward two rectangles as the predefined subpatterns and further improvements will be made in three aspects: 1, predefined subpatterns are classified into two types, normal rectangle and diagnal rectangle subpatterns. The former can be used to represent regular regions and the latter suits non-block regions; 2, allow two types of subpattern overlap to get more large blocks; 3, cancel various subpatterns' flag bits to save space, adopt the method of adding table header, that is, record the number of rectangle and diagnal rectangle firstly, then store subpatterns' data sequentially.

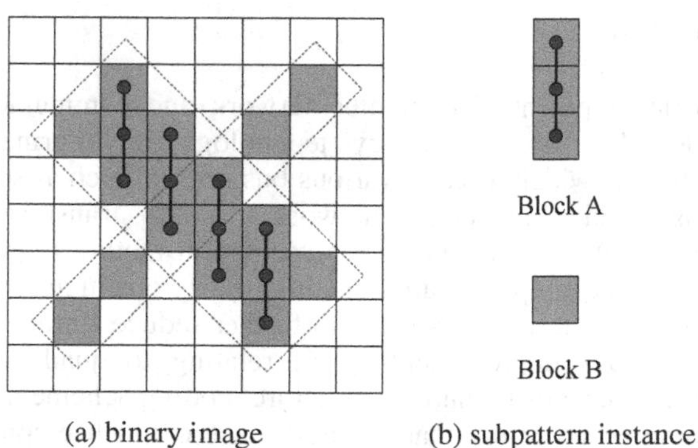

(a) binary image (b) subpattern instance

Fig. 1. binary image by using NAM segmentation

m	*NormalRect*	*DiagonalRect*
Grayscale	Number of normal rectangle	Number of diagnal rectangle

Fig. 2. DRNAM header structure

To illustrate the advantages of improved methods relative to rectangle NAM algorithm, there is an example in figure 1. According to raster scanning order to find the starting point of subpattern in the 8×8 binary image, it is obvious that the image can split up into two types of blocks. As is shown in figure 1(a), there are 10 subpattern instances in total, where the number of block A and B are 4 and 6 respectively.

When we employ DRNAM algorithm for segmentation, it is needed to scan two types of subpatterns and choose the largest one as the final type of subpattern. The coordinate of the starting point, the height, width and type of subpattern instances will be write down simultaneously. So the binary image can be divided into 2 overlapping diagonal rectangular blocks, which are enclosed by red dotted line in figure 1(a). The first block's height and width is 2 and 5, and the second block's height and width is 6 and 1. As a result the number of subpattern instances has been reduced sharply.

As for diagonal rectangle, two simplified strategies should be noted: 1, in order to simplify the problem, the diagonal rectangle is restricted to 45 degree angle; 2, all diagonal rectangles are assumed to be rotated 45 degree clockwise. Although the above strategies make diagonal rectangle subpattern a little special, it is proved that they're very effective and significantly reduce the number of subpattern instance.

In the above example, the number of subpattern instance is 10 and 2 respectively. In order to save storage space, DRNAM algorithm does not need additional identification information, the length of each subpattern is the same as rectangle NAM, just need to add a header node and store the number of double rectangle subpattern, where the space consumption is the same as two subpattern instances. Suppose that each subpattern instance occupies L bits, the given binary image will occupy $10L$ bits by using rectangle NAM and $4L$ bits by using DRNAM. It shows that DRNAM image representation method reduces storage space effectively.

3 DRNAM Algorithm Description

For a $2^n \times 2^n$ color image in DRNAM algorithm, it will be divided into three parts according to color components and produces three gray images. Suppose that the grayscale is m, it produces $3m$ binary images, which is divided by using two predefined subpatterns. In order to save space, we just record black pixels and finally get the DRNAM table for color images.

Input : a $2^n \times 2^n$ color image G with grayscale parameter m.

Output : DRNAM table Q for color image G.

Step 1: Regard color image G as three gray images according to color components, initiate i to 1 and represent the number of color components.

Step 2: Decompose gray image G_k into m binary images $BP_i(1 \le i \le m)$, set number of bit-planes i to 1, point to the bit-plane where exists the lowest bits of each pixel.

Step 3: Carry on exclusive-OR operations according to pixel value between the former m-1 binary images BP_i and their following binary image BP_{i+1}, the last bit plane BP_m remain constant;

Step 4: Set double-rectangle subpattern counter *rect_num* and *diag_rect_num* to 0, record the number of subpattern instances of rectangle and diagonal rectangle respectively, then define two arrays Q_{i1} and Q_{i2}, where the structure of data element is (*x*,*y*,*height*,*width*) which represent the starting point's coordinates, height and width of double-rectangle subpattern instances.

Steps 5: Establish an unmarked starting point (*x*,*y*) from the first entrance of the image BP_i, trace the double-rectangle subpatterns and find out the biggest subpattern in terms of the areas of the rectangles, save parameters to the corresponding array Q_{i1} or Q_{i2} according to the type of the biggest subpattern.

Step 6: In the binary image BP_i, mark the area of selected rectangle subpattern with 2, namely gray pixel, indicate that the pixels in the area cannot be used as the starting point, but can be reused and join subsequent subpattern instances.

Step 7: Scan the image array in raster order to find the next unmarked black pixel as a new starting point.

Step 8: Repeat Step 5 to Step 7 until there are no unmarked new starting points.

Step 9: Merge Q_{i1} and Q_{i2} sequentially and set table header according to the counter variable, in this way a bit plane decomposition result has been saved. Increase the number of bit plane by one, i.e., $i=i+1$. If $i \leq m$, then go to step 3, otherwise proceed to the next step.

Step 10: Merge the current bit plane decomposition results into Q_k, increase color component number k by one. If $k \leq 3$, return to step 3, dispose the next gray image, or go to the next step.

Step 11: Scan the subpattern instances one by one, employ K-Code transform of dimension reduction[7] and convert the coordinates to K-Code, that is, $sp \leftarrow K(x,y)$, then record the K-Code by using difference method.

Step 12: Output encoding result $Q=\{Q_1, Q_2, Q_3\}$.

4 Storage Structure and Data Amount Analyses of DRNAM

4.1 Storage Structure of DRNAM

For a color image, the output of DRNAM is a queue set $Q = \{Q_1, Q_2, Q_3\}$, each $Q_i(1 \leq i \leq 3)$ represents a color component of the image. As is shown in figure 3, storage structure of subpattern instances is made up of two types of elements. One is the starting point, the other is the parameters of subpattern instances. In general, the binary encoding lengths of x and y are both n. According to the definition of K-Code, the maximal lengths of both length and width are $n/2$. Since the storage structure of *sp* is represented by K-Code, storing *sp* need n bits. Therefore, storing a double-rectangle subpattern instance needs $2n$ bits altogether.

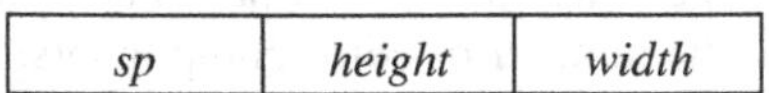

Fig. 3. Storage structure of subpattern instances

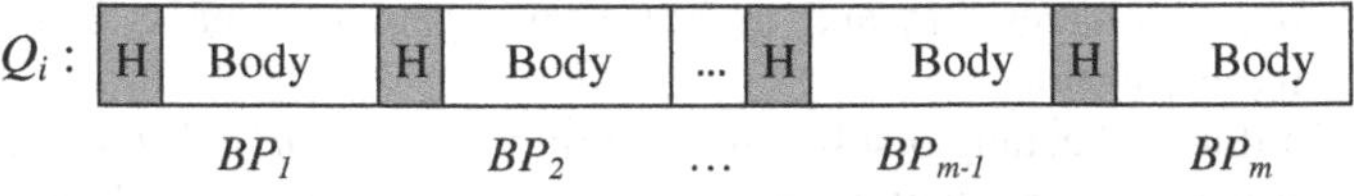

Fig. 4. Storage structure of a color component

Figure 4 shows the storage structure of a color component, which consists of the data of *m* bit planes, each bit plane is divided into two parts, table header and table body. The shaded area in figure 4 is table header, the white part is table body. The table header will record the number of double-rectangle and the table body will record the data of subpattern instances.

4.2 Data Amount Analyses of BPD-Based NAM

As for images with different complexity, the data amount represented by DRNAM are different. The following will make a comparative analysis between DRNAM and classic linear quadtree.

First of all, linear quadtree adopts symmetrical division, while DRNAM adopts asymmetrical and overlapping division, for the same test image, DRNAM will get less blocks than linear quadtree. Furthermore, each linear quadtree node needs $3n$-1 bits, while each node represented by DRNAM needs only $2n$ bits. Consequently for the same image, the storage requirement of linear quadtree representation will be 1.5 times or so that of the DRNAM representation.

5 Experimental Results

In this section, some representative color images of size $2^9 \times 2^9$ are analyzed to prove the obtained theoretical result. The gray level of these images is 8, i.e., $m = 8$. We implement the algorithm for the DRNAM and make a comparison with that of the classical linear quadtree and RNAM.

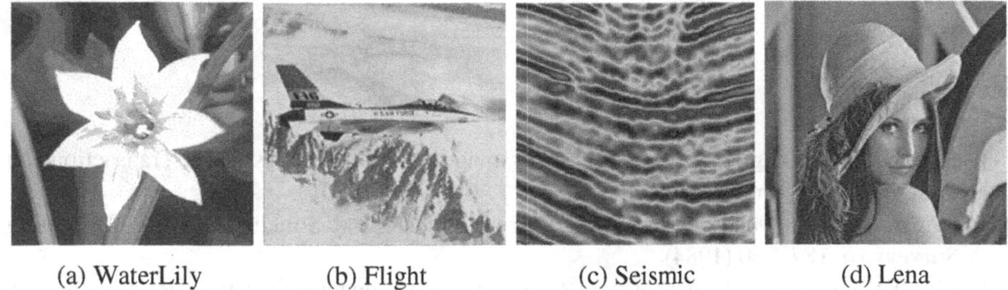

(a) WaterLily (b) Flight (c) Seismic (d) Lena

Fig. 5. Four test images of size $2^9 \times 2^9$

Table 1. Comparison of the performance with Quadtree and RNAM

Image	N			η	
	Quadtree	RNAM	DRNAM	Q/D	R/D
WaterLily	153621	54197	42268	3.636	1.282
Flight	729531	608628	385292	1.893	1.580
Seismic	771420	805883	530386	1.455	1.519
Lena	777006	746274	484751	1.603	1.540

Note: N: Number of subpatterns or nodes; η: compression ratio;
Q/D: Quadtree to DRNAM; R/D: RNAM to DRNAM.

Table 1 shows the comparison of the performance between Quadtree, RNAM and DRNAM. From the table, it can be easily seen from the value of *N* that the subpattern instances' number of the DRNAM is much less than the nodes' number of the Quadtree and RNAM. In particular, the total nodes of the linear quadtree is 1.455 to 3.636 times that of DRNAM, the actual amount of storage will be more greater. Compared with RNAM, DRNAM has the advantage as well, the compression ration range from 1.282 to 1.580.

The experimental results show that our algorithm for color images is much more effective than that of the popular linear quadtree and the late RNAM, so it is a better method to represent color images.

6 Conclusions and Future Work

In this paper, we present a Double-Rectangle Non-symmetry and Anti-packing pattern representation Model (DRNAM) and propose a novel algorithm for color images. By comparing our algorithm with that of the popular linear quadtree and RNAM, it is proved that the former is much more effective than the latters. In future work, we will consider adopting multi-subpattern methods, instead of one or two subpatterns, and we strongly believe that it will obtain better results.

Acknowledgements. The authors wish to acknowledge the support of the National Natural Science Foundation of China under Grant No.60973085, CNPC Innovation Foundation (2010D-5006-0304), Specialized Research Fund for the Doctoral Program of Higher Education of China (20070532077), Natural Science Foundation of Hubei Province of China (2009CDB308), Educational Fund of Hubei Province of China (Q20091211, B20111307).

References

[1] Hunter, G.M., Steiglitz, K.: Operations on images using quad trees. IEEE Transactions on Pattern Analysis and Machine Intelligence 1, 145–153 (1979)

[2] Samet, H.: The Quadtree and Related Hierarchical Data Structures. ACM Computing Surveys 16, 187–260 (1984)

[3] Lin, T.: Set operations on constant bit-length linear quadtrees. Pattern Recognition 30, 1239–1249 (1997)

[4] Yuh-horng, Y., Kuo-liang, C., Yao-hong, T.: A compact improved quadtree representation with image manipulations. Image and Vision Computing 18, 223–231 (2000)

[5] Vörös, J.: Quadtree-based representations of grid-oriented data. Image and Vision Computing 24, 263–270 (2006)

[6] Pješivac-Grbović, J., Bosilca, G., Fagg, G.E., Angskun, T., Dongarra, J.J.: MPI collective algorithm selection and quadtree encoding. In: Proceedings 13th European PVMMPI Users Group Meeting, vol. 33, pp. 613–623 (2006)

[7] Yun-ping, Z., Chuan-bo, C.: A Color Image Representation Method Based on Non-Symmetry and Anti-Packing Model. Journal of Software 18, 2932–2941 (2007)

[8] Peng, W., Chuan-bo, C., Yun-ping, Z.: Optimization Strategy for Non-symmetry Anti-packing Pattern Representation Model Based on Bit Plane Decomposition. Computer Engineering 36, 221–223 (2010)

Method for Image Shape Recognition with Neural Network

Wenpeng Lu

School of Science,
Shandong Polytechnic University
Jinan, Shandong, China, 250353
lwp@spu.edu.cn

Abstract. Shape recognition is important for image retrieval. The selection of shape features and recognition model would directly affect the effectiveness of shape recognition. In the paper, seven invariant moments, circularity degree, rectangle degree, sphericity degree, concavity degree and flat degree are selected as description features. With the shape features, image shape is recognized with BP neural network. Evaluation is performed over a manual dataset. Experimental result show that the method is a preferred strategy to recognize image shape.

Keywords: shape recognition, neural network, BP, shape feature.

1 Introduction

Content-based image retrieval (CBIR) is one of the hotspots of the current multimedia retrieval technology. CBIR directly analyze image content, such as color, shape, texture, et al, and describe images with reasonable features, which makes the retrieval more efficiently and more adaptively with human[1]. Image shape is one of most important visual features. Therefore, shape-based image retrieval is an important aspect in content-based image retrieval, which extracts shape feature of the image and retrieve relevant images by computing the similarity of shape feature[2]. The paper has proposed a method for image shape recognition with neural network, which extracts a series of shape features, train a BP neural network on a sample dataset and recognize image shape with it.

In the paper, the method for image shape recognition with neural network is described in detail. The rest of the paper is organized as follow. Section 2 introduces the model of BP neural network. The extracted features of the image are described in Section 3. Experiments are introduced in Section 4. As last, we give the conclusion and future work.

2 Model of BP Neural Network

2.1 Structure of BP Neural Network

The structure of three-layer BP Neural Network is shown as Fig.1[3]. Its input vector is $X = \{x_1, x_2, ..., x_i, ..., x_n\}^T$, $x_0 = -1$ is used to import threshold of hidden layer;

D. Jin and S. Lin (Eds.): Advances in CSIE, Vol. 2, AISC 169, pp. 547–551.
springerlink.com

output vector of hidden layer is $Y=\{y_1, y_2,..., y_i,..., y_n\}^T$, $y_0=-1$ is used to import threshold of output layer; output vector of output layer is $O=\{o_1,o_2,...,o_k,...,o_l\}^T$; expected output vector is $d=\{d_1,d_2,...,d_k,...,d_l\}^T$.

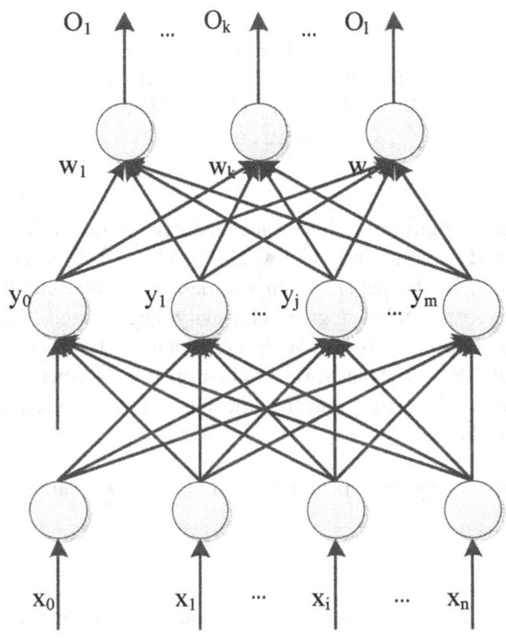

Fig. 1. Structure of BP Neural Network

The weight matrix from input layer to hidden layer is represented with V. $V=\{V_1,V_2,...,V_j,...,V_m\}$, in which V_j is weight vector of *j-th* neuron. The weight matrix from hidden layer to output layer is represented with W. $W=\{W_1,W_2,...,W_k,...,W_i\}$, in which W_k is weight vector of *k-th* neuron. The relation among different layers is as follow.

For output layer, the relations are shown as Eq.(1) and Eq.(2).

$$o_k = f(net_k), k=1,2,...,l \ . \tag{1}$$

$$net_k = \sum_{j=0}^{m} w_{jk} y_i, k=1,2,...,l \ . \tag{2}$$

For hidden layer, the relation are shown as Eq.(3) and Eq.(4).

$$o_k = f(net_k), k = 1,2,...,l \ . \tag{3}$$

$$net_k = \sum_{j=0}^{m} w_{jk} y_i, k = 1,2,...,l \ . \tag{4}$$

In Eq.(1) and Eq.(3), the transfer function $f(x)$ is single polarity Sigmoid function as Eq.(5).

$$f(x) = \frac{1}{1+e^{-x}} \ . \tag{5}$$

2.2 Basic Idea of BP Neural Network

The learning process of BP neural network algorithm is composed with forward propagation of input signal and backward propagation of error. In forward propagation, input samples are introduced from input layer and are transferred to output layer after being processed by each hidden layer. If the actual output is inconsistent with expected output, the back propagation of error would begin. In backward propagation, error would be transferred to input layer through hidden layers. The error would be apportioned to all of neurons, which would correct weight of each neuron.

The forward propagation and backward propagation would be repeated, which would adjust the weight of each neuron. This is the process of learning and training of BP neural network. The process would loop until that output error is reduced to an acceptable level or the preset number of learning is achieved.

3 Features of Image Shape

In the paper, twelve features of image shape is extracted, which include seven invariant moments, circularity degree, rectangle degree, sphericity degree, concavity degree and flat degree [4].

3.1 Invariant Moments

Invariant moments describe geometrical characteristic of a shape, which would not alter for size, rotation and translation of images. According to the research of Hu.m.K, we can obtain seven invariant moments[5-7].

3.2 Circularity Degree

Circularity degree also is referred as shape factor, which is defined as the ratio of square of the perimeter and the area.

3.3 Rectangle Degree

Rectangle degree also is referred as rectangle fitting factor, which is defined as the ratio of the area of the shape and the area of minimum enclosing rectangle of the shape.

3.4 Sphericity Degree

Sphericity degree is defined as the ratio of the radius of internally tangent circle of the shape and the radius of circumcircle of the shape.

3.5 Concavity Degree

Concavity degree also is defined with Eq.(6).

$$y = 1 - \frac{s_0}{s_r} . \qquad (6)$$

In which, s_0 is the area of approximation polygon of the shape and s_r is the area of minimum external convex.

3.6 Flat Degree

Flat degree is defined as the ratio of long axis and short axis, which reflects the degree of narrow and flat. For example, flat degree of square is 1, flat degree of rectangle is its aspect ratio.

4 Experiment

In the paper, based on the twelve shape features, we propose to utilize BP neural network to recognize image shape, which is provided in Matlab. Seven invariant moments, circularity degree, rectangle degree, sphericity degree, concavity degree and flat degree are sent to neural network as its input vector. The transfer function of hidden layer is logarithm Sigmoid function. The transfer function of output layer is Purelin function. Trainlm function is selected to train neural network on sample dataset[8, 9].

We have manually built an experimental dataset respectively for triangle, parallelogram, ellipse and arch. For each kind of shape, there are 80 images as training samples and 60 image for testing samples.

Precision and Recall is used to evaluate the effectiveness of the method.

Precision is computed as the percentage of correct answers given by the neural network, as Eq.(7).

$$P = \frac{\#\text{ correct answers provided}}{\#\text{ answers provided}} . \qquad (7)$$

Recall is computed as the ratio of correct answers given by neural network over the total number of answers to be given, as Eq.(8).

$$R = \frac{\#\ \text{correct answers provided}}{\#\ \text{total answers to provide}} \quad . \tag{8}$$

The detailed experimental result is shown in Table.1.

Table 1. Detailed experimental result

	Precision	Recall
Triangle	85%	83.33%
Parallelogram	81.67%	78.33%
Ellipse	80%	85%
Arch	76.67%	75%

5 Conclusions and Future Work

The paper proposes a method for image shape recognition with neural network, in which seven invariant moments, circularity degree, rectangle degree, sphericity degree, concavity degree and flat degree are selected as shape features and BP neural network are utilized to recognize image shape.

In the paper, the quantity and quality of training samples have great impact on neural network. In order to get better effectiveness, it is necessary to strengthen the construction of sample database, which is a time-consuming work. In the field of pattern recognition, neural network has shown some advantages. But, with the increase of pattern categories, the effectiveness of neural network would decline sharply. The recognition of complex shape needs the further development of neural network.

References

1. Ying, L., Dengsheng, Z., Guojun, L., Wei-Ying, M.: A survey of content-based image retrieval with high-level semantics. Pattern Recognition 40, 262–282 (2007)
2. Bayramoǧlu, N., Aydin Alatan, A.: Shape Index SIFT: Range Image Recognition Using Local Features. In: International Conference on Pattern Recognition, ICPR 2010, pp. 352–355. IEEE Press, New York (2010)
3. Haykin, S. (ed.): Neural Networks and Learning Machines. Prentice Hall, New Jersey (2009)
4. Gonzalez, R.C., Woods, R.E., Eddins, S.L. (eds.): Digital Image Processing Using MATLAB. Press of Electrical Industry, Beijing (2009)
5. Hu, M.K.: Visual Pattern Recognition by Moment Invariants. IRE Transaction of Information Theory IT-8 (1962)
6. Flusser, J.: On the Independence of Rotation Moment Invariants. Pattern Recognition 33, 1405–1410 (2000)
7. Flusser, J., Suk, T.: Rotation Moment Invariants for Recognition of Symmetric Objects. IEEE Transactions on Image Processing 15, 3784–3790 (2006)
8. Zhou, K., Kang, Y. (eds.): Neural Network Model and Simulation Program Design in Matlab. Press of Tsinghua University, Beijing (2005)
9. Zhu, K., Wang, Z. (eds.): Master Neural Networks in Matlab. Publishing House of Electronics Industry, Beijing (2010)

Brief Introduction of Back Propagation (BP) Neural Network Algorithm and Its Improvement

Jing Li, Ji-hang Cheng, Jing-yuan Shi, and Fei Huang

Aviation University of Air Force
130022, Changchun China
568026388@qq.com

Abstract. The back propagation (BP) neural network algorithm is a multi-layer feedforward network trained according to error back propagation algorithm and is one of the most widely applied neural network models. BP network can be used to learn and store a great deal of mapping relations of input-output model, and no need to disclose in advance the mathematical equation that describes these mapping relations. Its learning rule is to adopt the steepest descent method in which the back propagation is used to regulate the weight value and threshold value of the network to achieve the minimum error sum of square. This paper focuses on the analysis of the characteristics and mathematical theory of BP neural network and also points out the shortcomings of BP algorithm as well as several methods for improvement.

Keywords: Neural network, BP, algorithm.

1 Back Propagation Algorithm (BP Algorithm)

1.1 BP Algorithm

In fact, BP algorithm is a method to monitor learning. It utilizes the methods of mean square error and gradient descent to realize the modification to the connection weight of network. The modification to the connection weight of network is aimed at achieving the minimum error sum of squares. In this algorithm, a little value is given to the connection value of network first, and then, a training sample is selected to calculate gradient of error relative to this sample [1].

1.2 BP Learning Algorithm

The BP learning process can be described as follows:

(1) Forward propagation of operating signal: the input signal is propagated from the input layer, via the hide layer, to the output layer. During the forward propagation of operating signal, the weight value and offset value of the network are maintained constant and the status of each layer of neuron will only exert an effect on that of next layer of neuron. In case that the expected output can not be achieved in the output layer, it can be switched into the back propagation of error signal.

D. Jin and S. Lin (Eds.): Advances in CSIE, Vol. 2, AISC 169, pp. 553–558.
springerlink.com

(2) Back propagation of error signal: the difference between the real output and expect output of the network is defined as the error signal; in the back propagation of error signal, the error signal is propagated from the output end to the input layer in a layer-by-layer manner. During the back propagation of error signal, the weight value of network is regulated by the error feedback. The continuous modification of weight value and offset value is applied to make the real output of network more closer to the expected one [2].

2 Description of BP Algorithm in Mathematics

The ideology guiding the learning rules of BP network is: the modification to the weight value and threshold value of network shall be done along the negative gradient direction reflecting the fastest declining of function.

$$x_{k+1} = x_k - \eta_k g_k$$

In the formula mentioned above, xk represents the matrix of current weight value and threshold value; gk represents the gradient of current function; ηk represents the learning rate. Here, the three-layer BP network is taken as an example to describe the BP algorithm [3] in details.

As for the three-layer BP network, suppose its input node is xi, the node of hide layer is yj, and the node of output layer is zl. The weight value of network between the input node and node of hide layer is wji, and the weight value of network between the nodes of hide layer and output layer is vlj. When the expected value of the output node is tl, f(·) is the active function. The computational formula of the model is expressed as follows:

Forward propagation: output of computer network

Output of the node of hide layer

$$y_j = f(\sum_i w_{ji} x_i - \theta_j) = f(net_j)$$

including

$$net_j = \sum_i w_{ji} x_i - \theta_j$$

Computational output of the output node

$$z_l = f(\sum_j v_{lj} y_j - \theta_l) = f(net_l)$$

$$net_l = \sum_j v_{lj} y_j - \theta_l$$

Including

Error of the output node

$$E = \frac{1}{2}\sum_l (t_l - z_l)^2 = \frac{1}{2}\sum_l (t_l - f(\sum_j v_{lj} y_j - \theta_l))^2$$

$$= \frac{1}{2}\sum_l (t_l - f(\sum_j v_{lj} f(\sum_i w_{ji} x_i - \theta_j) - \theta_l))^2$$

Back propagation: the gradient descent method is adopted to regulate the weight value of all layers, and the learning algorithm of weight value is expressed as follows:

2.1 Modification of Weight Value

1. Derivation of output node by means of error function

$$\frac{\partial E}{\partial v_{lj}}=\sum_{k=1}^{n}\frac{\partial E}{\partial z_k}\cdot\frac{\partial z_k}{\partial v_{lj}}=\frac{\partial E}{\partial z_l}\cdot\frac{\partial z_l}{\partial v_{lj}}$$

E is a function containing several z_k, but only one z_l is related with v_{lj} and all the z_k are independent from each other, in this formula,

$$\frac{\partial E}{\partial z_l}=\frac{1}{2}\sum_k[-2(t_k-z_k)\cdot\frac{\partial z_k}{\partial z_l}]=-(t_l-z_l)$$

$$\frac{\partial z_l}{\partial v_{lj}}=\frac{\partial z_l}{\partial net_l}\cdot\frac{\partial net_l}{\partial v_{lj}}=f'(net_l)\cdot y_j$$

In this way,

$$\frac{\partial E}{\partial v_{lj}}=-(t_l-z_l)\cdot f'(net_l)\cdot y_j$$

Suppose the error of input node is

$$\delta_l=(t_l-z_l)\cdot f'(net_l)$$

In this way

$$\frac{\partial E}{\partial v_{lj}}=-\delta_l\cdot y_j$$

2. Deviation of the node of hide layer by error function

$$\frac{\partial E}{\partial w_{ji}}=\sum_l\sum_j\frac{\partial E}{\partial z_l}\cdot\frac{\partial z_l}{\partial y_j}\cdot\frac{\partial y_j}{\partial w_{ji}}$$

E is a function containing several z_l; it is targeted at certain w_{ji}, corresponding to one y_j, and related to all z_l, in this formula,

$$\frac{\partial E}{\partial z_l}=\frac{1}{2}\sum_k[-2(t_k-z_k)\cdot\frac{\partial z_k}{\partial z_l}]=-(t_l-z_l)$$

In this way,

$$\frac{\partial E}{\partial w_{ji}}=-\sum_l(t_l-z_l)\cdot f'(net_l)\cdot v_{lj}\cdot f'(net_j)\cdot x_i=-\sum_l\delta_l v_{lj}f'(net_j)\cdot x_i$$

Suppose the error of node of hide layer is

$$\delta'_j=f'(net_j)\cdot\sum_l\delta_l v_{lj}$$

In this way

$$\frac{\partial E}{\partial w_{ji}}=-\delta'_j x_i$$

As the modification of weight Δv_{lj} and Δw_{ji} is in proportion to the error functions and descends along the gradient, the formula showing the modification of weight of hide layer and output layer is expressed as follows:

$$\Delta v_{lj}=-\eta\frac{\partial E}{\partial v_{lj}}=\eta\delta_l y_j$$

In this formula, η represents the learning rate. The formula showing the modification between the input layer and hide layer is expressed as follows:

$$\Delta w_{ji} = -\eta' \frac{\partial E}{\partial w_{ji}} = \eta' \delta'_j x_i$$

$$\delta'_j = f'(net_j) \cdot \sum_l \delta_l v_{lj}$$

In this formula, η' represents the learning rate; $\sum_l \delta_l v_{li}$ in the node error of hide layer δ'_j expresses that the error δ_l of output node z_l is back propagated through the weight value v_{lj} to the node y_j to become the error of node of the hide layer.

2.2 Modification of Threshold Value

The threshold value θ is also a variation value and it also needs to be modified while the weight value is modified; the theory applied is the same as that used in the modification of weight value.

(1) Derivation of the threshold of output node by error function

$$\frac{\partial E}{\partial \theta_l} = \frac{\partial E}{\partial z_l} \cdot \frac{\partial z_l}{\partial \theta_l}$$

In this formula

$$\frac{\partial z_l}{\partial \theta_l} = \frac{\partial z_l}{\partial net_l} \cdot \frac{\partial net_l}{\partial \theta_l} = f'(net_l) \cdot (-1)$$

In this way

The formula expressing the modification of threshold value is

$$\Delta \theta_l = \eta \frac{\partial E}{\partial \theta_l} = \eta \delta_l$$

Namely $$\theta_l(k+1) = \theta_l(k) + \Delta \theta_l = \theta_l(k) + \eta \delta_l$$

(2) Derivation of the threshold of node of hide layer by error function

$$\frac{\partial E}{\partial \theta_j} = \sum_l \frac{\partial E}{\partial z_l} \cdot \frac{\partial z_l}{\partial y_j} \cdot \frac{\partial y_j}{\partial \theta_j}$$

In this formula

$$\frac{\partial y_j}{\partial \theta_j} = \frac{\partial y_j}{\partial net_j} \cdot \frac{\partial net_j}{\partial \theta_j} = f'(net_j) \cdot (-1) = -f'(net_j)$$

In this way

$$\frac{\partial E}{\partial \theta_j} = \sum_l (t_l - z_l) \cdot f'(net_l) \cdot v_{lj} \cdot f'(net_j) = \sum_l \delta_l v_{lj} f'(net_j) = \delta'_j$$

The formula expressing the modification of threshold value is

$$\Delta \theta_j = \eta' \frac{\partial E}{\partial \theta_j} = \eta' \delta'_j$$

Namely $$\theta_j(k+1) = \theta_j(k) + \Delta \theta_j = \theta_j(k) + \eta' \delta'_j$$

3 Improvement of BP Algorithm

In the application of artificial neural network, the BP network and its varied pattern are adopted in most of the neural network models; however, this does not mean that BP network is perfect and there are still inevitable defects in its algorithm, for example, falling into the minimum part of local part in the process of training, the convergence rate being rather slow, the network tending to have more redundancy and the samples newly added those may affect the samples learned, or others. The researcher has put forward many improved algorithms [4] to solve these defects. Its improved methods can be generally classified as three categories: one is to improve the speed of neural network training; the second is to improve the accuracy of training; and the third is to avoid dropping into the minimum point of local part. Among these methods, the rather typical ones are the additional momentum method and the variable learning rate method [5].

3.1 Additional Momentum Method

The additional momentum method is formed by introducing the momentum coefficient α based on the gradient descent algorithm. The formula that shows the adjustment of weight value and includes the additional momentum coefficient is expressed as follows:

$$\Delta w(t+1) = \alpha \Delta w(t) + \eta(1-\alpha)\frac{\partial E}{\partial w}$$

In this formula, $\Delta w(t+1)$ and $\Delta w(t)$ represent the weight corrections after the $(t+1)^{th}$ and t^{th} iteration; the value of the momentum coefficient α must be selected between 0 and 1, and generally 0.9 is selected. $\frac{\partial E}{\partial w}$ represents the negative gradient of the error sum of squares to weight in the BP algorithm.

The process of networked learning is that of weight modification; in this algorithm, the correction result of last time is used to affect the corrections of this time; when the correction of last time is oversized, the symbol of the second item in this formula will be contrary to that of correction of last time, in order to reduce the correction of this time and lower the oscillation. When the correction of last time is undersized, the symbol of the second item in this formula will be the same with that of correction of last time, in order to amplify the correction of this time and speed up the correction. It is thus clear that the application of additional momentum method always tries to increase the corrections those are in the same direction of gradient. This method has accelerated the convergence rate and reduced to a certain extent the probability of falling into minimal local part.

3.2 Variable Learning Rate Method

The error surface of network varies dramatically according to the variable parameter; the larger learning rate shall be selected for the areas whose error surfaces are very smooth; the smaller learning rate shall be selected for the areas whose error surfaces are very precipitous. The variable learning rate method is used for the self-adaptive adjustment of learning rate according to the change of error. The formula below shows the adjustment of learning rate:

$$\eta(t+1)=\begin{cases} k_{inc}\eta(t) & E(t+1)<E(t) \\ k_{dec}\eta(t) & E(t+1)>E(t) \\ \eta(t) & \end{cases}$$

In this formula, the incremental factor of learning rate $k_{inc}>1$, often 1.05; the reduction factor of learning rate $0<k_{dec}<1$, often 0.7; $E(t+1)$ and $E(t)$ represent the total error sum of squares after the $(t+1)^{th}$ and the tth iteration respectively; η represents the learning rate, and in standard BP algorithm, it is always a constant value between 0 and 1. If $E(t+1)<E(t)$, it represents that the tth iteration is effective, then multiplying the incremental factor to increase the learning rate; if $E(t+1)>E(t)$, it represents the tth iteration is ineffective, then multiplying the reduction factor to reduce the learning rate in order to reduce the ineffective iteration and accelerate the learning rate of network.

3.3 Optimization of Initial Weight

On one side, several local minimal points exist in the error surface of BP network; on the other side, the error gradient descent algorithm is adopted to adjust the weight of network; these two sides have caused the results of training of network easily to fall into the minimal point. Hence, the initial weight of network has exerted an enormous effect on the final result of training of network and it is one of the important factors those affect the possibility of network to achieve certain acceptable accuracy.

The initial weight of network is generally generated at random in certain interval; the training starts with an initial point and reaches gradually to a minimum of error along the slope of error function; that is to say, once the initial value is defined, the convergence direction of network is determined. In case that a bad initial weight is selected, the convergence direction of network can be towards the direction of divergence and this causes the non-convergence of concussion in network. In addition, the convergence rate of network training is also related to the selection of initial weight. As a result of that, it seems to be of great significance in selecting an appropriate initial weight, so as to accelerate the convergence rate of network training and avoid the concussion occurred in the process of studying. The genetic algorithm can be used to optimize the initial weight of neural network and enables the initial weight to jump out of the local extremum, speed up the convergence of BP network and improve the convergence accuracy of network.

References

1. Yan, P., Huang, R.: Artificial Neural Network — Model, Analysis and Application. Anhui Educational Publishing House, Hefei
2. Zhou, K., Kang, Y.: Neural Network Models and MATLAB Simulation Program Design. Tsinghua University Press, Beijing
3. Gao, J.: Artificial Neural Network Theory and Simulation Examples. China Machine Press, Beijing
4. Liu, J.: Intelligent Control. Electronic Industry Press, Beijing
5. Wang, X., Cao, L.: Genetic Algorithm – Theory, Application and Software Implementation. Xi'an Jiaotong University Press

Research of Colorde Image Digital Watermark

Cun Shang and Peng Yang

Service Computer Technology and Graphics Lab
Department of Computer Science and Technology
Xinyang Agricultural College, Henan, 464000, China
shang10278@163.com

Abstract. Digital watermarking is one of the effective technology which can protect the copyright of digital product and data security. For the encoding technology of color image, this paper proposed a method for embedding the watermarking using HSI color model. this paper propose an algorithm based on wavelet analysis, a colored watermark is embedded in a carrier, It embedded the image, which is disordered through ARNOLD transformation into the I space of the carrier image transformed. The results indicate that the watermarking technology proposed by this paper is secure and robust at resistance to compression and resistance to filtering.

Keywords: Digital watermarking, HSI color space, wavelet analysis, Brightness space.

1 Introduction

With the rapid development of digital technique and Internet, Digital watermarking technique becomes more and more important in protecting the intellectual property rights of digital products. This paper puts forward the digital watermarking algorithm based on wavelet analysis, a colored watermark is embedded in a colored carrier. First, It embedded the color image, which is disordered through ARNOLD transformation, In order to ensure high security and a strong stability. then the wavelet transformation into the I space of the carrier image transformed by small wave package. In the end, embedded the image into the I space, which is disordered through ARNOLD transformation. By experiment, this algorithm has a good robustness on common image processing and malicious attacks.

2 Selection of Color Space and the Formula of Transform

HSI color space describes the color from the human visual system, which presents color by the Hue, Saturation and Intensity. HSI color space, which presents color more naturally and directivity, is more suitable for the characteristics of human being's vision, and can handle the luminance component in the image in the separation of color and brightness [1]. HSI transform using the cylinder transform, the transformation is in accordance with formula 1:

D. Jin and S. Lin (Eds.): Advances in CSIE, Vol. 2, AISC 169, pp. 559–563.
springerlink.com

3 Pretreatment of Watermark Image

Watermark system should be pretreated before watermark embedding. This not only plays a key role in system robustness, but also has very important significance in the security of digital watermarking system. It increases the difficulty of guessing attack for attacker. In this paper, Arnold transform is used in the process of pretreatment because it is simple, easy to use, and cyclical.

$$I=\frac{1}{\sqrt{3}}(R+G+B)\;;\quad S=1-\frac{\sqrt{3}}{I}\min(R,G,B)\;;\quad H=\begin{cases}\theta,\text{当}G\geq B\\ 2\pi-\theta\,,\ \text{当}G<B\end{cases}\;;$$

$$\theta=\cos^{-1}\left[\frac{\frac{1}{2}[(R-G)+(R-B)]}{\sqrt{(R-G)^2+(R-B)(G-B)}}\right] \tag{1}$$

$$\begin{cases}\text{当}0^{\circ}\leq H<120^{\circ}\text{时},\ R=\frac{I}{\sqrt{3}}\left[1+\frac{S\cos(H)}{\cos(60^{\circ}-H)}\right],B=\frac{I}{\sqrt{3}}(1-S),G=\sqrt{3}I-R-B\\ \text{当}120^{\circ}\leq H<240^{\circ}\text{时},\ G=\frac{I}{\sqrt{3}}\left[1+\frac{S\cos(H-120^{\circ})}{\cos(180^{\circ}-H)}\right],R=\frac{I}{\sqrt{3}}(1-S),B=\sqrt{3}I-R-G\\ \text{当}240^{\circ}\leq H<360^{\circ}\text{时},\ B=\frac{I}{\sqrt{3}}\left[1+\frac{S\cos(H-240^{\circ})}{\cos(300^{\circ}-H)}\right],G=\frac{I}{\sqrt{3}}(1-S),R=\sqrt{3}I-G-B\\ \text{当}240^{\circ}\leq H<360^{\circ}\text{时},\ B=\frac{I}{\sqrt{3}}\left[1+\frac{S\cos(H-240^{\circ})}{\cos(300^{\circ}-H)}\right],G=\frac{I}{\sqrt{3}}(1-S),R=\sqrt{3}I-G-B\end{cases} \tag{2}$$

4 Based on Small Wave Package Transformation

The wavelet transform also has good local properties in time domain and frequency domain, and a variety of signal characteristics multi-resolution analysis of a great deal of adaptability, and it has been widely used in signal and image processing, speech recognition, synthesis technology areas. Through the use of small wave package transformation technique, The carrier image transformed by small wave package transformation. Four components of the watermark image is embedded in the four regions of the LL3, CV, CH, CD. LL3, CV, CH, CD is the low frequency components of small wave package transformation Respectively. The choice of lossy compression usually removes the high-frequency component of the image Which is very important for guaranteeing robustness of the system. The wavelet transformation into the I space of the carrier image transformed by small wave package as shown in Fig.1.

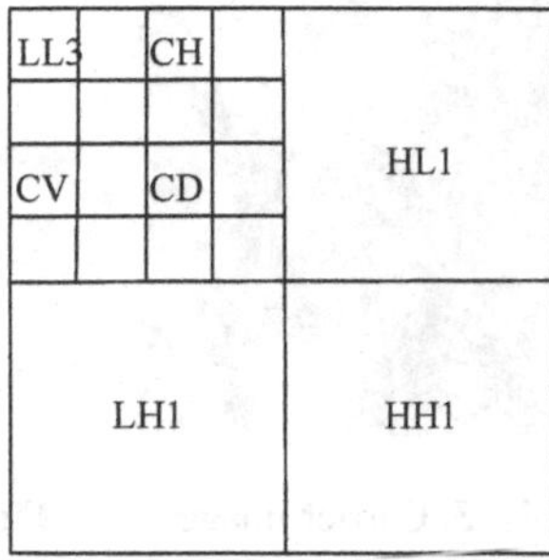

Fig. 1. Small wave package transformation

5 Implementation of Digital Watermarking

5.1 Watermark Embedding

Suppose original image is RGB true color image, in the size of 256×256. The watermark embedded is the gray image, in the size of 64×64. It embedded the gray image, which is disordered through ARNOLD transformation and one time wavelet transformation, four components of the watermark image is embedded in the four regions of the LL3, CV, CH, CD. the I space of the carrier image transformed by small wave package. then, original color image is converted from HSI mode RGB mode.

5.2 Performance Evaluation

Performance evaluations of image watermarking systems is used for quantitative evaluation to the carriers.

5.3 Test of Invisibility of the Watemark

The test of this algorithm is processed on Matlab 7.0 platform. According to the algorithm, take the best quantization coefficient, the image after embedding watermarking on I space is shown in Fig.2- Fig.4. By comparison, the image with embedded watermark has not degraded significantly through observation, and the visual system is not aware of any differences. It suggests that the digital watermarking technique proposed in this paper has good invisibility. Not having been attacked, the robust watermark is extracted from the image that has an embedded watermark, as shown in Fig.6.

5.4 Watermark Extraction Effect

The watermark is effectively extracted after image after embedding watermarking image. The process of watermark extraction as shown in Fig.5- Fig.7.The clarity of watermarking extracted is visible.

Fig. 2. Watermarking

Fig. 3. Carrier image

Fig. 4. The image after embedding watermarking on I space

Fig. 5. Watermarking image

Fig. 6. Image after embedding watermarking image

Fig. 7. The watermarking image extracted

6 Anti-attack Experiment

The effect of watermark extraction from the image extracted after the filter and compression. The results indicate that the watermarking technology proposed by this paper has good Anti-attack.

6.1 Filter Experiment

Fig. 8 the effect of filer with coefficient = 0.2

Fig. 9 the watermarking image extracted

6.2 Lossy Compression Experiment

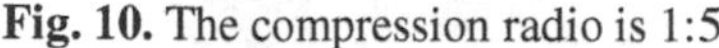

Fig. 10. The compression radio is 1:5

Fig. 11. The watermarking extracted

7 The References Section

In this paper, watermark scrambling algorithm, It embedded the image into the I space of the HSI color space, According to the characteristics of small wave package transformation,this paper propose an algorithm of watermark embedding based on wavelet transform. The results indicate that the watermarking technology proposed by this paper r is secure and a strong stability.

References

1. Du, Z.Y., Zhou, Y., Lu, P.Z.: Anoptimized spatial data hiding scheme combined with convolutional code sand hilbert scan. In: Proceedings of the 3rd IEEE Pacific Rim Conference on Multimedia, Taiwan, pp. 97–104 (2002)
2. Podilchuk, C.I., Zeng, W.: Image-adaptive watermarking gusing visual models. IEEE J. Select. Areas Commun. 16(4), 525–539 (1998)
3. Cox, I.J., Miller, M.L.: The First 50 years of Electronic watermarking. EURASIP J. of Applied Signal Proceessing 2, 126–132 (2002)
4. Neilson, G.M.: On Marching Cubes. IEEE Transactions on Visualization and Computer Graphics 9(3), 283–297 (2003)
5. Gao, J.G., Fowler, J.E., Younan, N.H.: An Image Adaptive Watermark Based on Redundant Wavelet Transform. In: Processing of the IEEE International Conference On Image Processing, Thessaloniki (2001)
6. Hsu, C.-T., Wu, J.-L.: Hidden Signature in Image processing 8(1), 58-68 (1999)

The Design of Web Services in e-Commerce Security Based on XML Technology

Wei Xu* and DaPeng Zhang

Henan Occupation Technical College, Zhengzhou, 450046, China
zhangzhiqiang70@sina.com

Abstract. XML is an extensible markup language, with its powerful descriptive, scalable, structured, platform-independent features are widely used in the presence of a large number of e-commerce data exchange. This article firstly analyses in-depth of XML technology. Then, the XML security technology in-depth is studied and analyses of existing XML security standards such as XML Encryption, XML Signature, and XML Key Management. The paper presents the web services in e-commerce security based on XML technology. The compared experimental results indicate that this method has great promise.

Keywords: web services, XML, e-commerce.

1 Introduction

Traditional information security technologies such as Secure Sockets (SSL) and IP layer security standards to a certain extent to meet the security requirements of XML, XML language for its characteristics of structured data information security technology proposed new requirements. Under the premise of the increasingly strong demand for XML security, the international community, the W3C, OASIS, the IETF and several other groups to participate in XML security standards development work, a series of new XML security standards for the application of XML as a data exchange carriers to provide security protection. The combination of XML and the security itself is a new area, it and the application of the "traditional" security has its own significant features [1]. XML security standards will be the traditional cryptographic techniques and XML technology to together to solve the XML data storage and transmission of data confidentiality (to prevent unauthorized users from stealing data), data integrity (to prevent data without authorized tampering, insert, delete, or retransmission), authentication (which requires data exchange of the identity of both can be identified), access control (different users can control data access), non-repudiation services (send to ensure data, the two sides can not afterwards deny the operation made by their own behavior) and other security issues.

The traditional information security technologies can not fully take into account the structure of XML, descriptive, and can not achieve the new demands of the

* Author Introduce: Wei Xu (1979), male, lecturer, Master, Hehan Occupation Technical College, Research area: web mining, data mining.

D. Jin and S. Lin (Eds.): Advances in CSIE, Vol. 2, AISC 169, pp. 565–570.
springerlink.com

particle size optional encryption, signature and multiple signatures. In order to solve the security problems in the storage and exchange of XML, Web-based e-commerce orders, in-depth study of XML security technology. Proposed an integrated XML security technology for data exchange, it is XML encryption / decryption and signature / verification. Solved in the storage and transmission of XML, Web-based e-commerce orders, data confidentiality, data integrity, authentication, access control, non-repudiation and so on security issues. The paper presents the web services in e-commerce security based on XML technology.

2 The Research of XML Security Technology

XML is an extensible markup language, with its powerful descriptive, scalable, structured, platform-independent features are widely used in the presence of a large number of e-commerce data exchange. This chapter describes the characteristics of the XML document, format, display, conversion and processing.

Data encryption is in order to achieve confidentiality of the confidentiality of the password algorithm and key in two ways. Encryption transform algorithm based on the confidentiality of the algorithm is kept secret [2]. Confidentiality of key encryption algorithm can be made public, but the algorithm uses the key is kept secret. Modern cryptography is generally based on the confidentiality of the keys. There are two types: symmetric encryption and asymmetric encryption key-based encryption technology, as is shown by equation 1.

$$W_j^{\lambda} f(x,y) = f \times \Psi_j^{\lambda}(x,y) = \iint_{R^2} f(u,v)\Psi_j^{\lambda}(x-u, y-v)dudv \qquad (1)$$

According to the relative position between the signature elements and the signature object, the XML Signature There are several types, its structure is shown in Figure 1.

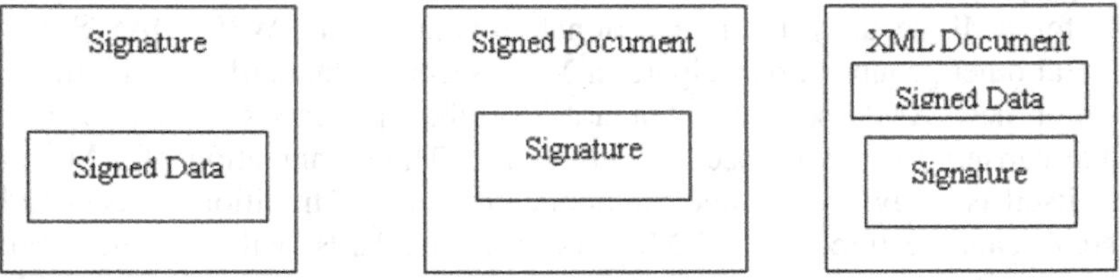

Fig. 1. Three types of XML Signature

(1) Packaged signature (Enveloping the Signatures): signature data encapsulated in XML signature element of the internal Signature element is similar to an envelope.

(2) Packaged signature (Enveloped the Signatures): the Signature element itself is embedded in the data is signed, the signature data to act as the envelope containing the signature.

(3) Detached signature (Detached the Signatures): the Signature element, and the signature data is separated from each other, does not exist to contain and inclusion relations between the two.

XML digital signature technology is jointly developed in the IETF and the W3C XML Digital Signature specification based on. W3C Recommendation defines the XML signature syntax and processing rules to it. XML signature can serve any data type of document. This document is a remote or local, or even the contents of an element or elements in the XML document. XML signature used to authenticate users to ensure data integrity and non-repudiation of data manipulation, as is shown by equation 2.

$$E_{jA}^{\xi}(m,n) = \sum_{m'\in J,n'\in K} w_{A}^{\xi}(m',n')[D_{jA}^{\xi}(m+m',n+n')]^2 \tag{2}$$

Have characteristics that we can see by the XML Digital Signature, XML digital signature has a broad application prospects in the transmission of network documentation, especially in e-commerce, e-government. It can better support the key features of e-commerce, making e-commerce can secure information exchange between different systems on the Internet [3]. Currently, XML digital signature technology has been gradually applied to electronic funds transfer, electronic data interchange and on-line contracts and signatures e-commerce activities, as is shown by equation 3.

$$W_{2j}f(n) = 2^{-j/2}\int_{-\infty}^{\infty} f(x)\overline{\psi(2^{-j}x-k)}dx \tag{3}$$

Reference to generate the digest value is calculating <Reference> elements. <Reference> Element's URI attribute specified to calculate the digest value of the data resources, data resources prior to calculate the digest value through <Transforms> said to transform its cascading transformation, and then the optional transform results to calculate the digest value data resources of the 8 byte stream. Last use <Digest Method> expressed digest algorithm to calculate the digest value of the 8 byte stream. The digest value stored in a Base-64 encoded <Digest Value> elements. The entire reference to the steps generated as follows.

1. The use of reference the <Reference> generate the formation of elements to produce the specified signature algorithm and standardized methods <Signed Info> elements. 2. Right <Signed Info> elements using standardized methods and signature algorithm. Use newly generated <Signature Value> and previously generated <Signed Info> and all optional elements and attributes to generate <Signature> parent element. XML signature generation process is shown in Figure 2.

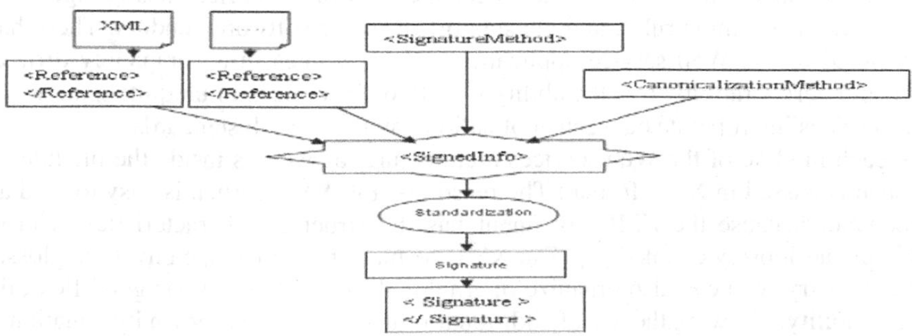

Fig. 2. XML signature generation process

XML Signature Recommendation regulations only reference to both validation and signature verification, signature to be valid. However, in practical applications, it is such provisions too binding. Some applications will make more relaxed requirements, reference validation fails or signature verification fails, the signature can also be seen as effective.

XML Encryption specification (XML Encryption Syntax and Processing) is the basis of XML encryption technology. Development by the W3C in September 2002 announced a recommended standard. The goal of the XML Encryption is a data encryption to use XML to describe Web resources, which can be HTML files, XML files, JPG files and any file can be any element in the XML file and content. In essence, XML encryption does not define a new encryption algorithm, but the XML technology and a combination of existing encryption algorithms, as is shown by equation4.

$$\widetilde{W}_{j,k} = \sum_{l=0}^{L_j-1} \widetilde{h}_{j,l} X_{k-l \bmod N} \tag{4}$$

XML Encryption can encrypt the contents of elements in the non-XML documents, XML documents, XML documents and XML document elements. So also with the generated <Encrypted Data> elements to the original encrypted plaintext replace the plaintext and cipher text to coexist with the transmission in the XML document.

XML encryption is generally composed of three entities: applications, encryption, and decryption device. The decryption parses the XML document in the package <EncryptedType> elements and decrypts it. Verify the results of the decryption operation and the synthesis of an XML document is a standardized and effective, are generally left to the application processing.<Encrypted Type> Element is the XML Encryption specification defines an abstract type, it provides the basic functionality for <Encrypted Key> elements and <Encrypted Data> of elements.

3 The Web Services and Security Technology

XML makes the transmission of information to be freed from the restrictions of the platform and programming language, provides a common standard for the communication of the various systems on the network. Service messaging, SOAP protocol defines simple rules, and the support of major software vendors. These have contributed to the Web service application. Web services in a completely different platform interoperability, it is the ability to call Web services in ubiquitous networks. Web services interoperate between applications by using Web standards.

In each module of the Web services architecture, as well as inside the module, the message is passed in XML format. The reason is: the XML format is easy to read and understand, because the XML document has the structural characteristics of cross-platform and loosely coupled [4]. The XML format of the message envelope glossary in the industry, you can also organize internal and external use, it has good flexibility and scalability, allowing the use of additional information, as is shown by equation5.

$$W_f(a,b) = \frac{1}{\sqrt{|a|}} \int_{-\infty}^{\infty} f(x) \overline{\psi\left(\frac{x-b}{a}\right)} dx \quad (5)$$

Symmetric encryption is divided into two types: block ciphers and stream ciphers. Block cipher to encrypt the plaintext fixed packet length. Stream ciphers use keys to tear down the function to generate a key stream, and then XOR operation between the plaintext byte and each byte of the key stream to generate the cipher text. Stream ciphers to implement than block ciphers faster, easier, but there are security risks reuse the same key stream, easy to attack. RSA Data Security's RC4 algorithm is a commonly used stream ciphers. Commonly used block cipher algorithm is DES (Data Encryption Standard, Data Encryption Standard), a variant of DES, 3DES, IDEA (International Data Encryption Standard, International Data Encryption Algorithm) and AES (Advanced Encryption Standard Advanced Encryption Standard).

4 The Development of Web Services in e-Commerce Security Based on XML Technology

XML as a markup language to describe the data that is used to its powerful description of features, scalability, structured semantics and platform-independent features, the Internet and distributed and heterogeneous environment as the main data transmission and exchange carrier has been widely used in e-commerce.

XML data exchange, in an open environment to ensure that information security is a primary condition for the smooth implementation of XML applications. XML data exchange security related to authentication, access control, data confidentiality, data integrity, non-repudiation, etc. By analysis of the actual application requirements for the safety of e-commerce data exchange, combined with the technical standards for XML security, an integrated XML security technologies used in e-commerce data exchange program, and implemented some of the features. The combination of XML security technology standards proposed an integrated XML security technology for data exchange, and it uses the NET platform, C # language to achieve one of the XML encryption / decryption, XML digital signature / verification module, as is shown by figure3.

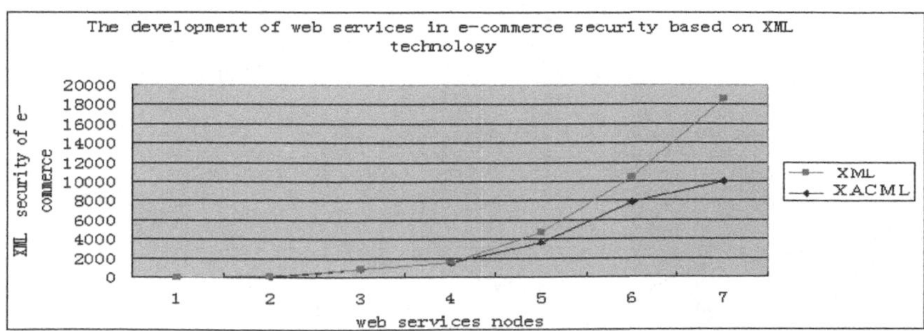

Fig. 3. The compare result of web services in e-commerce security based on XML and XACML

By the results of the safety analysis of the functional modules of the integrated XML security technology for data exchange program to achieve the desired functionality, provides data confidentiality, data integrity, authentication and non-repudiation of network security services. It can do a guarantee the security requirements in the store and exchange process the xml web-based e-commerce orders. XML Encryption can achieve different granularity of data confidentiality, XML digital signatures to achieve the multisignature traditional security function can not be achieved, has a broad application prospects.

5 Summary

The XML description, scalability, structured and platform-independent features fully meet the needs of the Internet and distributed heterogeneous environments, network data exchange of the main carrier, a strong impetus to the development of e-commerce network applications, the security has also been widespread concern. The paper presents the web services in e-commerce security based on XML technology.

Acknowledgement. This paper is supported by Education Department of Henan Province, 2011 Natural Science Research Program (2011C520019).

References

1. Kudo, M., Hada, S.: XML document security based on provisional authorization. In: Proc. of the 7th ACM Conf. on Computer and Communications Security (2000)
2. Takase, T., Uramoto, N., Baba, K.: XML Digital Signature System Independent of Existing Applications. In: Proceedings Applications and the Internet (SAINT) Workshops, pp. 150–157 (2002)
3. Verma, M.: XML Security: Control information access with XACML. IBM developer Works (2004)
4. Lu, J., Zhu, X., Peng, D., Huo, H.: Active XML for Service Discovery in Mobile Environment. JCIT 6(6), 47–53 (2011)

The Research of Agent Software Engineering Technology

Jinhao Lu[1], Chi Dong[2], and Fei Zhan[3]

[1] Department of Computer Science Northeast Petroleum University at Qin Huangdao, Qin Huangdao, China
[2] Petroleum Engineering Institute Northeast Petroleum University, Da Qing, China
[3] Sinopec ShengLi Oil Field Binnan Oil Production Plant, Bin Zhou China
{lujhteacher,dongchidongchi,zhanfei_ql}@163.com

Abstract. Compared with traditional development methods, Agent software engineering is a new method in the field of software engineering, which has a strong ability to complete higher level abstractions of active entities in a complex software system. This paper mainly researched engineering characteristics, analytical methods, and advantages of dealing with complex systems for agent software agent software engineering method was described.

Keywords: Agent, software engineering, object-oriented.

1 The Development of Software Engineering Technology

Software engineering has been a greater improvement through a few decades of rapid development. The development of software engineering was divided into four stages generally [1]. The first stage was the process-oriented software design methods. The second stage was the module-oriented software design methods. The third was the object-oriented software design methods. The last stage was the Agent-oriented software design method engineering. Various stages of software engineering methods were shown in table 1. In recent years, accompanied by the rapid development of computer network technology, software engineering development gradually changed to the [1]direction of the complication, intellectualization and large-scale. And the traditional software engineering methods (such as O-O) had more prominent shortcomings and defects, and became difficult to meet the needs of the future development of software engineering [2]. Therefore, based on the Agent software engineering research has important practical significance.

[1] The State Social Science Fund "11th five-year" plan task: "Study on employment-oriented integrated solution for IT mayor teaching of advanced vocational education". Number: BJA060049-ZKT030.

D. Jin and S. Lin (Eds.): Advances in CSIE, Vol. 2, AISC 169, pp. 571–575.
springerlink.com

Table 1. The compare of a variety of software engineering methods

Four stages	Analysis methods	Main features	Main languages	Applications range
process-oriented method	process-oriented program flowchart and language	reusability, versatility, scalability	FORTRAN, COBOL, ALO2GOL	the development of professional small-scale package
module-oriented method	The system is divided into functional modules, and achieved one by one, and then connected	A certain degree of reusability, versatility, scalability	PASCAL, C	structurization, modular programs, the development of large-scale system
object-oriented method	The system is abstracted as a collection of objects	better reusability, versatility, scalability	C++, Java, SmallTalk	The development of a wide range of applications of large-scale system
Agent-oriented method	System abstract is Agent collection with personification	better reusability, versatility, scalability and intelligence	Placa, AGENTO, GOLOG	the development of a variety of large and complex department

2 The Concept and Characteristics of the Agent

The concept of Agent is the key issues to study Agent-Based software engineering. This method not only has the characteristics of firmness and intelligence but also has strong flexibility. Besides it has strong adaptive capacity when it deals with complex problems which are difficult to predict and collaborate. So-called the concept of Agent, the international community recognized that the Agent can automatically and flexibly percept and self-adapt to the environment in the specified environment, and be able to act on behalf of users or designers to complete a certain task. It can achieve a computer entity with high autonomy ability [3]. The main goal of Agent is to accept the request and consignment of the other any entity (system, Agent, etc.), and provide appropriate services or assistance. On the basis of the goal-driven, the system consciously applies various types of necessary means, including learning, social and other methods. Agent has the following characteristics [4].

The first is initiative character. Agent can take the necessary behavior to show a goal-oriented way, and to achieve promise.

The second is social character. This feature is the collaboration between Agents. Specifically for the Agent in a certain social environment, it is not isolated entities, usually it collaborates between agents and the environment.

The third is reactivity character. Agent can consciously perceive the surrounding environment, which also can make reaction according to environment combined with relevant knowledge.

The fourth is the autonomy character. Agent should be self-controlled entity and has its own computing power. It also can judge and decide its own behavior combined with its perceived information and the internal situation in no direct manipulation of the outside and dynamic changes in environment and non-selected mode of the surrounding environment.

It not only has these main features, but also has the characteristics of compliance, mobility, honest and sensible.

3 Agent Software Analysis and the Advantages of Dealing with Complex Systems

3.1 Oriented Agent Software Analysis

Unlike traditional software engineering methods, the analysis of Agent-oriented regards the behavior as the core. The method is modeling analysis which is independent of the computational field and has nothing to do with the existing software. The purpose of traditional software engineering needs analysis is to obtain integral and a certain software implementation specification. But the analytical methods of Agent-oriented analysis are to create a type of problem solving mode. Therefore, needs analysis results of Agent-oriented have more reusability and stability, especially for the analysis of complex system.

Agent as a actor, in certain circumstances, can respond to external events and engage in certain activities to generate new events or implement state transformation in Agent-oriented software analysis. We depict Agent from three points: The first is the internal reasoning way. The second is the interface with the outside world. The third is its dynamic behavior way, the response to the outside world.

3.2 The Advantage of Dealing with Complex Systems

Understanding the concept of complex systems is a very intuitive concept. How to judge the complexity of the system did not give a clearly defined, usually we determine the complexity of the system by three ways [5].

The first is the feature of collaboration between system and its own environment. System accepts input information from the environment and gives the corresponding output, this is the most basic form of the interaction between system and environment, and then it terminates. But the interaction of complex system needs to maintain a never-ending, ongoing relationship between system and environment. This interaction apparently causes some difficulties in the development of complex systems.

The second is the characteristic of the system environment: system environment is certain or uncertain, not accessible or accessible, static or dynamic, these factors have an impact on the complexity of software development.

The third is the complex structure of system, which is the feature of the system itself. In software development, the complex structure makes abstract modeling become more difficult. Therefore, we carry out necessary decomposition layer by layer for complex systems, making it become an interrogational, smaller and relatively easy subsystem. When analyzing this complex system, the general method used in software engineering includes abstraction, decomposition and organization.

4 Application of Agent-Based Software Engineering Method

This paper mainly introduces the application of the Agent-Based Software Engineering Method in Remote Data Synchronization System. The main effect of RDSS is to complete the allopathic heterogeneous or allopathic homogeneous database synchronously through the Internet/Intranet network environment. That is, based on the change of the one part database content and combining with the user's task, the system can generate an E-mail from the change of database manually or automatically, and then make packaging and encryption, which will be sent to the specified destination in the task by Mail. According to the accepted E-mail data, the system would update the corresponding contents of local database automatically. It supports many kinds of database system driven by BDE or ODBC, such as SQL, Sybase and Oracle.

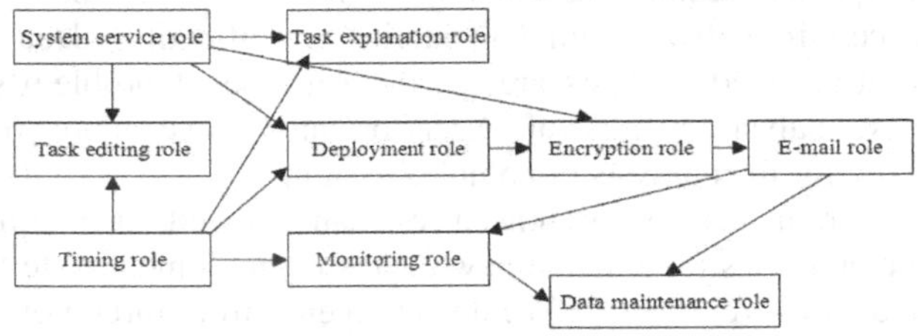

Fig. 1. The design model of the system

Every role in the model has its own agreement, authority and responsibility. Among them, system service role is the key part of The RDSS mechanism, which is responsible for the overall management and planning, used for main setting of the local system, and can monitor all the roles, so that it can make response timely for various requests. Task explanation role is used to explain the generated task documents from the compiled task script and implement corresponding works. Task editing role is mainly responsible for compiling and edit the task script, and generating corresponding task documents, waiting for execution. Timing role is to carry out missions according to time, which can carry out the task files in the task one by one automatically, including the generated outgoing relevant documents, that is we can activate other roles at set time or on time. Deployment role is responsible for distributing and Setting the Remote Data Synchronization System and database related to data transmission, which automate the generation of storage process, database trigger, buffer table, email address tracking allopathic data system, timer-triggered scheduling, scripts, key, etc. Through active monitoring, the monitoring role can analyze emails in all mailboxes of Remote Data Synchronization System, collect and analyze the information about the mails' retention time and number, the availability status of mailbox, the content and theme of retention mail, and report to the system service role promptly. Meanwhile, the system service role can make corresponding dynamic adjustment according to the working situation of sampling information monitoring system. Encryption role is mainly responsible for infilling the changed files in allopathic related database. Data maintenance role is responsible for the analysis and maintenance of data transmission in all databases, including analysis of sent data, optimization or deletion of data in the buffer table, retransmission of special data, etc, but also check the system completing transmission, and produce inspection record facilitate to analysis.

E-mail role is responsible for obtaining the recipient's e-mail address in the receiving department in the task file, then generate E-mail from encrypted allopathic files, send it regularly, and receive E-mail automatically when having reception instruction.

The roles above make up the whole system, combining with role models and the existing interactions, Agent model corresponding to each role have been established. Communication interaction model has been created in accordance with the agreement between the roles, and the service model has been created according to responsibility of the roles.

5 Conclusion

Agent idea has put forward the new method of programming, system analysis and requirement analysis, changes the process method of the entire software engineering, and improves software's self-learning ability, adaptive ability, and dynamic levels, which has laid a strong foundation for creating the next more intelligent software.

References

1. Huang, Z., Gu, Y.: Software Analysis model based on agent thinking. Computer Engineering (1) (2009)
2. Liu, D.: Development trends of Agent Research. Journal of Software (3) (2009)
3. Luck, M., Mcburney, P., Shehory, O., et al.: Agent Technology: Computing as Interaction-A Roadmap for Agent Based Computing. AgentLink (2005)
4. Mao, X., Chang, Z., Wang, J., et al.: Agent Oriented Software Engineering: Status and Challenges. Computer Research and Development (10) (2006)
5. Wooldridge, M., Jennings, N.R., Kinny, D.: The gaia methodology for agent—oriented analysis and design. International Journal of Autonomous Agents and Multi—agent System (3) (2008)

Prediction of the Busy Traffic in Holidays Based on GA-SVR

MeiLi Guo[1], DianJun Li[1], ChaoBen Du[1], ZhenHong Jia[1,*], XiZhong Qin[1], Li Chen[2], Lei Sheng[2], and Hong Li[2]

[1] College of Information Science and Engineering, Xinjiang University, Urumqi 830046, P.R. China
[2] China Mobile Group Xinjiang Company Limited, Urumqi, Xinjiang 830091, P.R. China
guomeili314@126.com, jzhh@xju.edu.cn

Abstract. The prediction of holiday's traffic has the characteristics of small historical sample size and strong nonlinear, which result in low prediction accuracy. Genetic algorithm (GA) is adopted in this paper to optimize the support vector regression machine (SVR) to forecast the busy traffic of Xinjiang in holidays and compared with the traditional SVR and the BP neural network. The result shows that the GA-SVR has a higher forecast precision and a less time-consuming, which is an effective method of busy traffic prediction.

Keywords: prediction of holiday's traffic, Support vector regression machine, Genetic algorithm.

1 Introduction

The mobile communication network facing the impact of high traffic during the major holidays such as the Spring Festival, the Mid-Autumn Festival .It is easy to cause the exchange system overload, the circuit congested and cause voice connection rate drop, which could cause an irreparable damage to the mobile operators and mobile users. Therefore, the traffic prediction is the basis of the communication security work, it is also the fundamental of network planning and construction for mobile operators, the prediction accuracy determines the rationality and the scientificity of the whole plan.

Many methods of traffic forecasting have been proposed, such as the ARIMA method [1], the support vector machine method [2], etc. The practical application of the above method achieved better results for the mean monthly traffic or busy monthly traffic. But a certain Time of the festival often has a traffic surge and a large fluctuation. In response to these problems, it is difficult to achieve the accurate prediction in traditional forecasting methods. In this paper, support vector machine [3] optimized by genetic algorithm is adopted to build the prediction model. The predicted performance of support vector machine is very sensitive for the selection of parameters. In the practical application, the experience or the test algorithm are often adopted to determine the parameters, resulted the inaccurate selection of parameter, which made the final prediction accuracy lower than the target accuracy. In this paper, the parameters of SVR are automatically searched and determined based on GA, the results show that GA-SVR has a better forecast effect.

[*] Corresponding author.

D. Jin and S. Lin (Eds.): Advances in CSIE, Vol. 2, AISC 169, pp. 577–582.
springerlink.com

2 Support Vector Regression

For a given set of data $T=\{(x_1,y_1),\ldots(x_i, y_i)\}\subset R^d\times R$, where the x_i is the size of input data R^d is the input feature space, y_i is the corresponding size of the output data. The regression problem [4] is estimated the relationship between x_i and y_i:

$$y=f(x)=<\omega,\Phi(x)>+b \qquad x\in R^d, y,b\in R \tag{1}$$

Where<·,·> corresponds to the inner product of Rd space. Φ(·) is a non-linear transformation, which can map the training data to the high-dimensional space F,so solving nonlinear problems in the original space is equivalent to solving the linear regression problems in a new high-dimensional space. To get the solution of the SVR is equivalent to seek the optimal solution of the following questions:

$$\begin{aligned} \max \quad W &= -\frac{1}{2}\sum_{i=1}^{n}\sum_{j=1}^{n}(\alpha_i-\alpha_i^*)(\alpha_j-\alpha_j^*)k(x_i,x_j) \\ &\quad -\varepsilon\sum_{i=1}^{n}(\alpha_i+\alpha_i^*)+\sum_{i=1}^{n}y_i(\alpha_i-\alpha_i^*) \end{aligned} \tag{2}$$

$$s.t.\begin{cases} \sum_{i=1}^{n}(\alpha_i-\alpha_i^*)=0 \\ 0\le\alpha_i,\alpha_i^*\le C,\quad i=1,2,\ldots n \end{cases} \tag{3}$$

Where ε is defined by the insensitive loss function $L\ (x, y, f)$, which determine the flat degree of the regression curve. C is the penalty factor, which means the penalty for right or wrong sub-sample. Thus, the support vector regression machine function can make (1) rewritten as:

$$f(x)=\sum_{i=1}^{SV}(\alpha_i-\alpha_i^*)k(x_i,x_j)+b \tag{4}$$

Kernel function is very important for support vector machine. The literature [5] suggests that the radial basis function is usually superior to other kernel functions, Therefore, RBF is used in the SVR in this paper.

3 The Traffic Forecasting Model Based on GA-SVR

The GA-SVR method [6] is adopted in this paper to build the prediction modeling and analysis for the holiday's traffic, the algorithm does not need put forward empirical hypothesis conditions for complex, nonlinear and uncertainty holiday traffic, which use the traffic data before holiday as input / output data for learning and predicting directly.

The process of GA search the best parameters of SVR and forecast the traffic are in Fig.1.

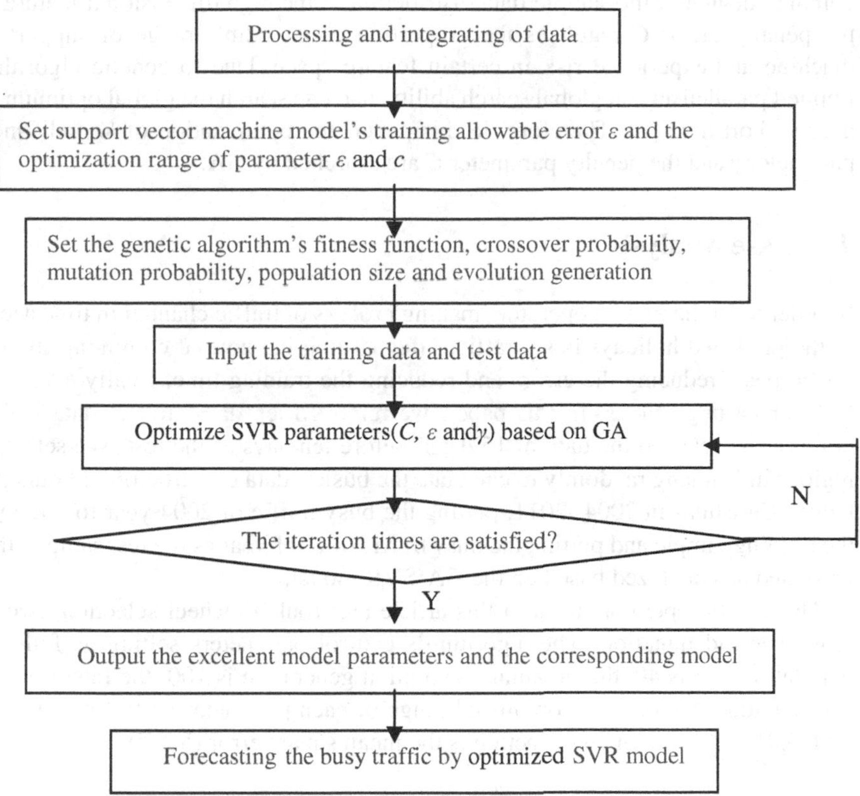

Fig. 1. The process of optimizing the SVR parameters with genetic algorithm

3.1 The Basic Principle of GA

Genetic algorithm [7] is developed based on the theory of evolution of Darwin's and the genetic theory of Mendel, it is simple and general, it also has strong robust, and it is suitable for parallel processing. GA produces a new generation of chromosomes by selection, crossover and mutation genetic operator. According to some convergence conditions, evolving from generation to generation, and finally converging to the individual that adapted to the environment, with the result that seeking the optimal solution.As a global optimization search method, GA has the advantages of simple, the strong generality, practical and parallel processing, widely used and so on.

3.2 The Realization of the Traffic Forecasting Model Based on GA-SVR

The main reason affecting the prediction accuracy of the traffic is the RBF kernel function's parameter γ and the penalty factor C in SVR, the parameter γ main affects the complex degree of the sample data distribution in the high dimensional feature space, the penalty factor C regulates the proportion of incredible range of support vector machine and experience risk in certain feature space. Due to genetic algorithm has implied parallelism and global search ability, and can search the global optimum within a very short time [8]. Therefore, in this paper, the optimization of kernel function's parameter γ and the penalty parameter C are searched by GA.

4 Case Analysis

In order to let the mobile operators making process of traffic channel in time according to the predicted holidays busy traffic, safeguarding the normal communication in the traffic peak, reducing the errors and reducing the training time, finally achieving the best forecasting effects. In this paper, we removed ten days' traffic data before the holiday and selected the data of 15 days' before ten days as the test. We selected five regions in Xinjiang randomly to calculate the busiest data of traffic of 24 hours per day before Christmas in 2004 -2011, putting the busy traffic of 2004 year to 2010 year as the training sample and putting the busy traffic of 2011 year as the test sample, then we predicted and analyzed based on the GA-SVR model.

The genetic operators used in this article are: roulette wheel selection, two points crossover and mutation. The algorithm's control parameters setting as follows: the population size is 40, the maximum evolution generation is 100, the rate of crossover and mutation 0.8, 0.1. The optimized range of each parameter is: $0 \leq C \leq 100$, $0 \leq \gamma \leq 100$, $0.01 \leq \varepsilon \leq 1$, the fitness function is the mean square error (MSE).

$$MSE = \sum_{i-0}^{n} (\hat{y}_i - y_i)^2 / n \tag{5}$$

Where y_i is the actual data value and $\hat{y}_i$ is the predicted data value in the i-group. Mean relative error (MRE) is the final assessment criteria, MRE closer to zero, the better the results.

The experimental data in this paper is the real-time mobile communications traffic data, including the traffic of per hour of 16 regions in Xinjiang from 2004 to 2011. The genetic algorithm's source code is written to find the optimal parameters in MATLAB, then the forecasting model based on SVR is established, which predicted the busy traffic on Christmas Day in 5 regions of Xinjiang ,take one region of Xinjiang for example,the prediction effect is shown in Figure 2.

In order to illustrate the efficiency of this method, this paper selected the other two methods -- the basic SVR and BP neural network compared with the GA-SVR. Due to the randomness of the algorithm, every time the experimental results will be different, therefore the error and time-consuming of the three kinds of methods are the average of program run 10 separate times. The comparison results are shown in Table 1.

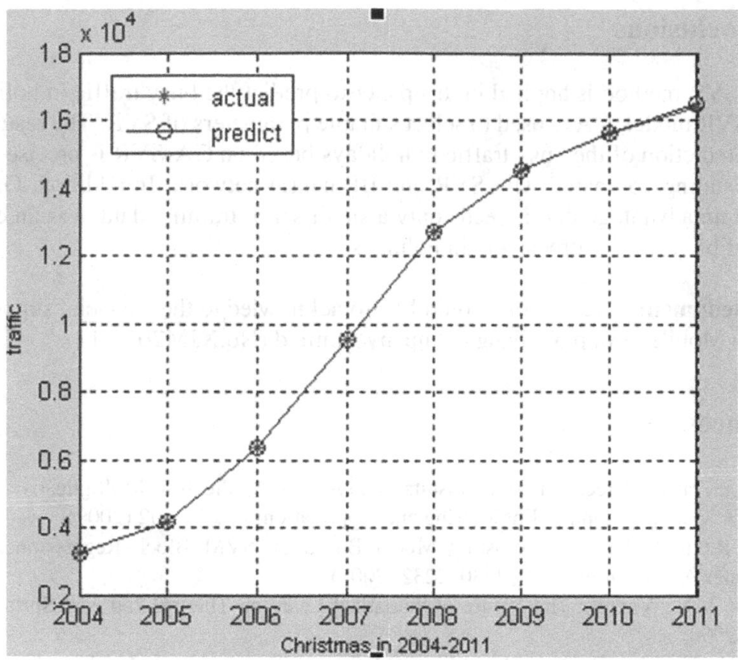

Fig. 2. Train and Test Set Regression Predict by GA-SVR

Table 1. comparison of the forecasting results in three ways

	GA-SVR		SVR		BP	
	MRE /%	Times/s	MRE /%	Times/s	MRE /%	Times/s
Region 1	-1.036	3.00	11.18	0.62	1.234	3.41
Region 2	1.888	3.14	6.9409	0.74	0.377	3.26
Region3	-3.098	3.26	11.54	0.64	11.262	3.72
Region 4	-1.703	2.97	25.5327	0.66	6.777	3.55
Region 5	-2.313	3.08	2.2823	0.64	-3.093	3.71

The Table 1 shows that the forecasting results is satisfactory based on GA-SVR, the MRE are all less than 5%, and it have small fluctuations, the run-time are about three seconds, fully meet the needs of practical prediction. Although the SVR model run fast, it is difficult to find the best model parameters, and need someone to test manually many times, and only one area error is less than 5%; BP network has three areas' MRE less than 5%, the effect is slightly better than SVR, but because the great randomness and fluctuation of BP network, the predicted results are fluctuating, so it is difficult to do an accurate prediction for each region. Therefore, whether the prediction accuracy or time-consuming, the GA-SVR are better than SVR and BP neural network model.

5 Conclusions

The GA-SVR method is applied in this paper to predict the busy traffic in holidays. In the GA-SVR model, GA is used to select suitable parameters of SVR. The result shows that the prediction of the busy traffic in holidays based on GA-SVR is precise and less time-consuming compared with SVR and BP neural network. In addition, GA- SVR has a certain advantage that it needs only a small set of training data. it is an effective method of busy traffic forecasting in holidays.

Acknowledgment. The authors would like to acknowledge the financial support from the China Mobile Group Xinjiang Company Limited (No.XJM2011-11).

References

1. Yu, Y.: Traffic Forecasting and Results Analysis of Applicating Multiplicative Seasonal ARIMA Model. Computer Engineering and Applications 45, 99–102 (2009)
2. Chen, R.: Risk Traffic Forecasting Model Based on SVM Block Regression Analysis. Computer Applications 28(9), 2230–2232 (2008)
3. Vapnik, V.N.: Vapnik: The Nature of Statistical Learning Theory, 2nd edn. Springer, New York (2000)
4. Deng, N., Tian, Y.: The New Data Mining Methods - Support Vector Machine. Science press (2004) (in Chinese)
5. Keerthi, K., Lin, C.: Asymptotic Behaviors of Support Vector Machines with Gaussian kernel. Neu. Com. 153, 1667–1689 (2003)
6. Saini, L.M.: Parameter Optimisation Using Genetic Algorithm for Support Vector Machine-based Price-forecasting Model in National Electricity Market. IET Gen. Tran. & Dis., 36–49 (2009)
7. Goldberg, D.: Genetic Algorithms in Search. Addison Wesley (1989)
8. Cheng, S., Arnold, D.P.: Optimization of Permanent Magnet Assemblies Using Genetic Algorithms. IEEE. Tran. on Mac. 47, 4104–4107 (2011)

Research on Software Simulation Technique for Qingha Oil Pipeline Production Run Process

Lixin Wei, Hang Dong, and Yu Wang

Northeast Petroleum University, Daqing, P.R. China
dh.123@163.com

Abstract. This paper established Qingha oil pipeline simulation technique mathematical model, corrected the pipeline flow and thermodynamic calculation models by software built on least square method and overall of coefficient of heat transfer calculated interpolation method. And on that basis, the summer and winter's temperature and pressure drop curve in different flow rate have developed by this model. In addition, Pipephase software has been used in the paper to provide the flow rate critical point special to the pipeline thermal running in summer, and make the point as the condition of whether switch on heating furnace on Zhongyi Station or not. This application of Pipephase software can save energy consumption efficiently, and guide the pipeline practical production running in theory.

Keywords: overall heat transfer coefficient, flow rate critical point, software, Pipephase software.

Qingha oil pipeline is one of the Daqing oil field pipelines outside. The pipeline is 182.8 kilometers long in total, ϕ377×6.4mm in diameter, four transfer stations along the line, and 2~2.8 million tons flowrate per year[1]. At present, due to lacking of the research on hydraulic characteristics and thermodynamic property for the pipeline practical running process, the scheme for pipeline running is short of guidance in theory and the running consumption is very high. On the other side, the policy in the "twelfth five-years plan" has pointed that low carbon economy has been the important developing goal of our country. So it is meaningful for researching on the oil pipeline production running scheme.

1 The Mathematical Model of Oil Pipeline

1.1 Flow Calculation Model

The oil in Qingha pipeline is high wax content oil. Usually the flow regime of oil is hydraulic smooth. The pressure drop can be calculated by the equation below:

$$\Delta P = P_1 - P_2 = \rho g(z_2 - z_1) + \rho g h_f \tag{1}$$

D. Jin and S. Lin (Eds.): Advances in CSIE, Vol. 2, AISC 169, pp. 583–588.
springerlink.com

Among,

$$h_f = \lambda \frac{L}{d}\frac{\upsilon^2}{2g} = 0.0246 \frac{Q^{1.75} \cdot v^{0.25}}{d^{4.75}} L \tag{2}$$

1.2 Correction of Flow Calculation

In practical application, the error between hydraulic analysis theory calculation result and practical running data is usually very big. To solve the problem, use the theory of least square method, take the lowest fitting degree as aim, correct the parameter in flow calculation model by optimization theory[2]. Because of the parameter which influenced the friction loss are all constant, such as flow velocity v, rate of flow Q, inside diameter d, and pipe length L. So it is necessary to correct hydraulic friction coefficient λ.

Bring in the correction factor k, and correct hydraulic friction coefficient by the formula below:

$$\lambda' = k\lambda \tag{3}$$

Establish unconstrained optimization model when solve k and n:

$$\text{Find k} \tag{4}$$

$$\text{Min } F(k) = \sum_{i=1}^{m} (\Delta P' - \Delta P(k))^2 \tag{5}$$

The result after solving is in Table1:

Table 1. Qingha oil pipeline flow friction correction factor

Name	Pipe length[km]	Diameter[mm]	*k*	Average fitting error[%]
Initial station to Zhongyi station	62.8	377	0.889	1.1
Zhongyi station to Zhonger station	62.1	377	0.924	1.8
Zhonger station to Terminal station	58.9	377	0.937	2.1

1.3 Thermodynamic Calculation Model

To calculate the temperate drop of Qingha oil pipeline, using Sudhoff temperature drop formula:

$$\Delta T = T_L - T_0 = (T_R - T_0)e^{-\frac{K\pi D}{Gc}L} \tag{6}$$

For the basic structure parameters and flow rate are constant, we can get the pipeline overall coefficient of heat transfer curves in condition of year-round operation by anti-inference method to improve the accuracy of pipeline thermodynamic calculation. The curves show from Figure1 to Figure3.

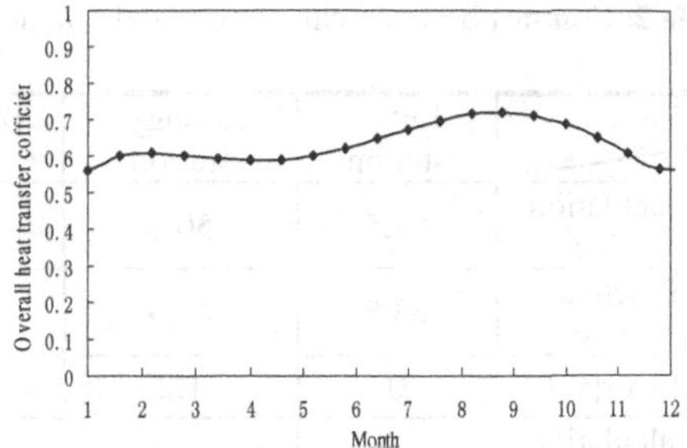

Fig. 1. Pipeline overall heat transfer coefficient curve from Initial to Zhongyi station

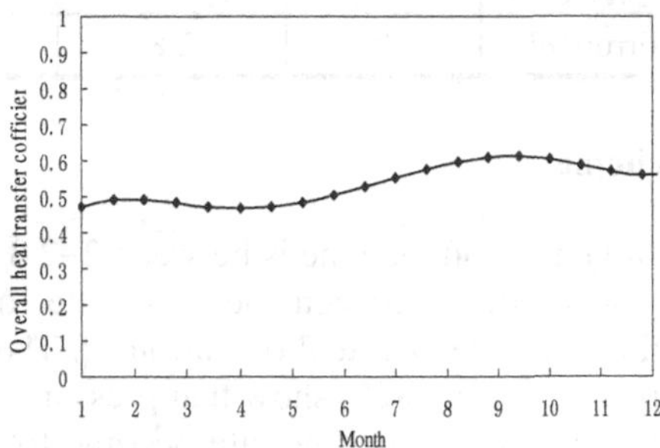

Fig. 2. Pipeline overall heat transfer coefficient curve from Zhongyi to Zhonger station

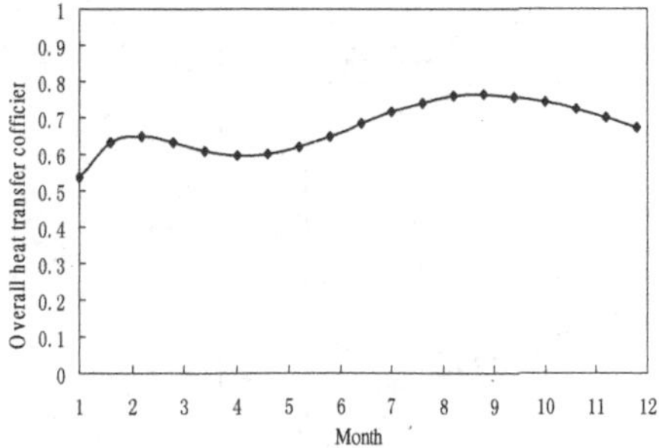

Fig. 3. Pipeline overall heat transfer coefficient curve for Zhonger to Terminal station

2 Calculation of Qingha Oil Pipeline Running Simulation Process

2.1 Example Calculation

The result of calculation for the operational states in July 15th, 2007(Figure2) shows that, the error between calculation and practical value is in 3%, which proves that the model is calculable, truthfulness and can be the guide for field application in theory.

Table 2. Contrast between calculation and practical value

		Initial station	Zhongyi station	Zhonger station	Terminal station
Inlet temperature[℃]	Calculation value	63.5	50.5	49.7	46.2
	Practical value	63.5	49.9	49.2	45.9
	Error[%]	0	1.2	1	0.6
Inlet pressure[MPa]	Calculation value	0.1	3.6	2.2	0.5
	Practical value	0.1	3.5	2.21	0.49
	Error[%]	0	2.8	0.5	2

2.2 Operational Scheme

The overall flow rate of Qingha oil pipeline is between 2~2.8 million tons per year, so choose the different rate of flow between the flow rate to calculate pressure and temperature drop for the reference as actual operation[3]. The pressure drop curves in winter and summer(Figure4 and Figure5) show that pressure for initial station become from 2.9MPa to 5MPa, when the year flow rate become from 2 to 2.8 million tons. The friction loss can be seen from the curves for the impact of heat furnaces in Zhongyi station and Zhonger station.

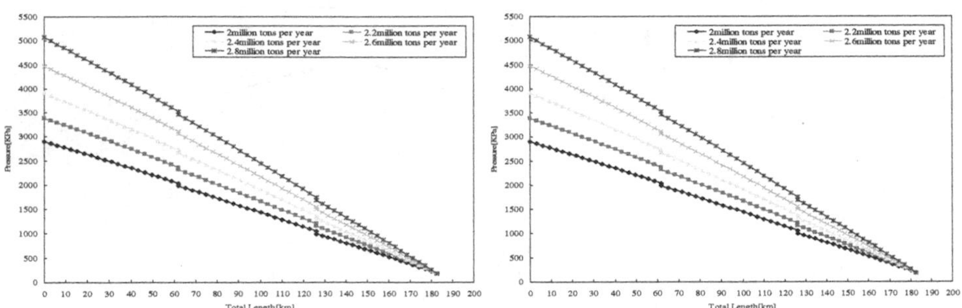

Fig. 4. Pressure drop curve in winter

Fig. 5. Pressure drop curve in summer

The initial station of Qingha oil pipeline gets the oil from Pubei oil depot. The outlet temperature of initial station is usually 62℃ for the high outlet temperature from the Pubei oil depot. To ensure the principle that smooth flow of crude oil, make each inlet temperature is maintained above 40℃(above condensation point 3℃). Calculation result by Pipephase software shows that to ensure the oil inlet temperature is above 40°C in winter, both the heat furnaces in Zhongyi station and in Zhonger station should be turned on, Figure6 is the temperature drop curve under different flow rate in winter. In summer, there is a flow rate critical point of pipeline operation which is 2.3 million tons per year. When the flow rate is higher than the critical point, the heat furnace has not to be turned on to make the oil flow into the next station, but when the flow rate is lower than it, both the heat furnaces should be switched on. Figure7 presents when the flow rate is above the critical point, the temperature drop curve under different flow rate in summer. Figure8 is that when the flow rate is under the critical point, the temperature drop curve under different flow rate in summer.

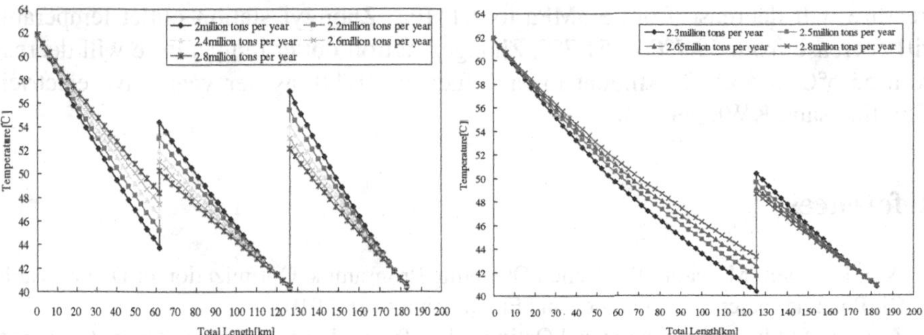

Fig. 6. Temperature drop curve under different flow rate in winter

Fig. 7. Above critical point, temperature under different flow rate in summer

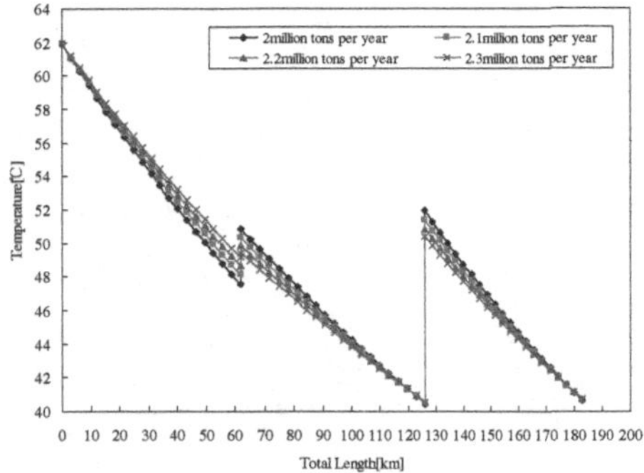

Fig. 8. Under critical point, temp drop curve under different flow rate in summer

The result shows that the heating station outlet temperature is lower than practical value, and the result can be proved to save energy for production and running. When the flow rate is 2.4million t/y, initial station outlet pressure will decrease from 5.3MPa to 4.1MPa, Zhongyi station outlet temperature will decrease from 59.7°C to 51.7°C, Zhonger station outlet temperature will decrease from 55.6°C to 53.8°C. Amount to save fuel oil 1024t/y, save electricity 1350 thousand KWh/y.

3 Conclusion

(1) The paper established the mathematical model of steady state for Qingha oil pipeline and presents the flow rate critical point as the condition for deciding whether to turn on the heat furnace in Zhongyi station in summer, which is 2.3million tons per

year. When the flow rate is higher than 2.3million tons per year, only the heat furnace in Zhonger station should be switched on. When flow rate is lower than 2.3million tons per year, both the heat furnaces in Zhongyi station and Zhonger station should be turned on. And on that basis, temperature drop curves in summer are established.

(2) The calculation result can be proved to save energy for production and running. For example, when the flow rate is 2.4million tons per year, initial station outlet pressure will decrease from 5.3MPa to 4.1MPa, Zhongyi station outlet temperature will decrease from 59.7°C to 51.7°C, Zhonger station outlet temperature will decrease from 55.6°C to 53.8°C. Amount to save fuel oil 1024tons per year, save electricity 1350 thousand KWh per year.

References

1. Wei, L., Chen, M., et al.: Production Operating Programme Optimization of Daqing-Harbin Oil Pipeline. Science Technology and Engineering (29) (2010)
2. Zhao, J.: Study on Simulation and Optimization Technology for Crude Storage Operational Process. Northesat Petroluem University (2010)
3. Liu, Y., Zhao, J., et al.: The Research on Optimizing the Operating Scheme of the Oil Depot. Science Technology and Engineering (35) (2010)

Study on Improved Collision Detection Algorithm

LiMei Fan*

Department of Information Engineering, Jilin Business and Technology College
130062, Changchun, China
flmflm123@163.com

Abstract. In order to improve the accuracy and real-time of collision detection algorithm, This paper study the oriented bounding box(OBB) algorithm and random collision detection algorithm based on improved particle swarm Optimization and hybrid collision detection algorithm.. The introduction of intelligent optimization algorithms, the complex three-dimensional model collision problem into simple two-dimensional discrete space optimization problem, to improve the real-time of collision detection algorithm; Using the OBB surrounded box surrounded by basic geometric elements as random collision detection algorithm feature sampling point method, Compensate for the random collision detection algorithm is easy to miss some of the interference element defects, thereby improving the accuracy of collision detection. Finally, through authentication for the hybrid collision detection algorithm, and compared with the basic OBB and based on improved PSO random collision detection algorithm, Verify the real-time and efficiency of this algorithm.

Keywords: OBB, collision detection algorithm, PSO, hybrid collision detection algorithm.

1 Introduction

In recent decades, the fields of collision detection, researchers at home and abroad have done quite a lot of meaningful work, made a number of efficient collision detection algorithm, and many algorithms have been applied to the virtual reality system. Their works play a significant role in promoting the development of virtual reality technology. However, the collision detection technology is still in the stage of development, many researchers in theoretical research and practical application of collision detection techniques. This paper is focus on improving the real-time and accuracy of collision detection algorithm. Collision detection algorithm combined with intelligent optimization algorithms form an efficient, real-time hybrid collision detection algorithm.

2 Collision Detection Algorithm

The collision detection problem is that the computer simulation, physical simulation, robot motion planning, and many other research areas of difficult problems, Real-time

* Corresponding author.

D. Jin and S. Lin (Eds.): Advances in CSIE, Vol. 2, AISC 169, pp. 589–594.
springerlink.com

and accurate collision detection algorithm can enhance the authenticity of the virtual environment and immersive. The collision detection is divided into two-dimensional and three-dimensional space collision detection, in two-dimensional plane; Objects can be used polygons to represent. Collision detection problem is polygon intersection; so detection algorithm is relatively simple. This paper is from two perspectives to Classification of the collision detection algorithm. First, it from the perspective of time domain; second it from the perspective of the spatial domain [1].

From the perspective of time domain points, The collision detection algorithm can be divided into the following three categories: static collision detection algorithms, discrete collision detection algorithm and continuous collision detection algorithm. Classification shown in Figure 1:

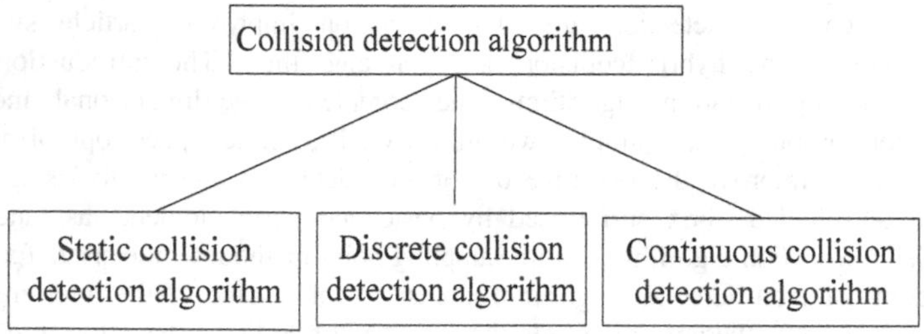

Fig. 1. The classification of collision detection algorithm based on the time domain

From the perspective of the spatial domain points, collision detection algorithm can be divided into two categories: First, the collision detection algorithm based on object space; second, collision detection algorithm based on image space.

3 Oriented Bounding Box Technology

The basic idea of hierarchical bounding box is use volume slightly larger and the geometrical properties simple of the bounding box approximation to describe complex geometric objects, Construct the tree hierarchy closer and closer to the geometry of the object model, until it is completely object geometry. In test, just part of the bounding box overlap exact test. root node of the hierarchy Surrounded the entire model , each parent node surrounded by geometric objects are geometric objects surrounded by all of its child nodes, the nodes from top to bottom and gradually approaching the geometric objects are surrounded. Collision detection Based on hierarchical bounding box is essentially a recursive overlap test based on bounding volume hierarchies. If the overlap, the pair node further overlap test; If you do not overlap, the end of this process. If overlap of the leaf nodes in the testing process. The will have the basic geometric elements intersect test , the intersection of the surface structure, and attach it to intersect in the structure of the table, back to the system, in order to further test of basic geometric elements intersect.

Build the OBB bounding box there are three ways: bottom-up approach, gradual insertion method, top-down method.

4 Random Collision Detection Algorithm Based Improved PSO

With the development of collision detection technology, it is emergence of random collision detection algorithm. It is based on accuracy for speed of collision detection algorithms, it can be used in relatively high real-time interactive systems, it become collision detection field a new of research in recent years [2].

4.1 Particle Swarm Optimization Algorithm

Particle Swarm Optimization(PSO) algorithm [3] is Propose by James Kennedy and Russell Eberhart in 1995; it is an evolutionary computing technology, The algorithm is simulated foraging behavior of birds flying, through the collaboration between individuals to search for the optimal solution. Set, in N dimensional search space, there are m particles, in which the i particle position vector is $X_i = (x_{i1}, x_{i2}, ..., x_{iN})(i = 1,2,...,m)$, Particle velocity vector is $V_i = (v_{i1}, v_{i2}, ..., v_{iN})$. the i-th particle best position (optimal solution) is P_{best}. PSO the best position denoted by G_{best},PSO algorithm formula:

$$V_{id}(t+1) = v_{id}(t) + c_1 * Rand_1 * (P_{best} - x_{id}(t)) + c_2 * Rand_2 * (G_{best} - x_{id}(t)) \quad (1)$$

$$X_{id}(t+1) = x_{id}(t) + v_{id}(t+1) \quad (2)$$

Which, $Rand_1, Rand_2$ is [0, 1] random number; c_1, c_2 is acceleration factor, the particle in each dimension has a maximum speed limit $V_{\max}(V_{\max} > 0)$, if one dimension exceeds set speed $V_{\max}$, the speed is limited to $V_{\max}$. The PSO basic algorithm schematic as figure 2.

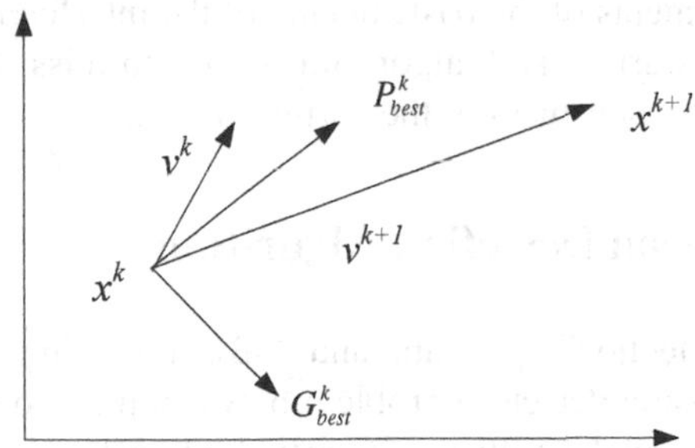

Fig. 2. The basic principle schematic drawing of PSO Algorithm

4.2 Random Collision Detection Algorithm

Random collision detection algorithm currently is divided into two categories: the first Average-Case algorithm is propose by Klein and Zachmann. Algorithm idea is based on the region bounding the intersection of two objects number of features as the

the test standard, and each node of the feature vectors do not do precise collision detection, to achieve the purpose of controlling the rate of detection [3].

The second algorithm is proposed by Guy and Debunne [4], Raghupathi [5] a surface such as random sampling, assuming random sampling of the collision, and then step through the sampling of the local minimum to reach the accurate detection algorithm. The movement of objects is very small time difference between time slots, typically a few hundredths of a second, so the location of the object only minor differences, the relative displacement between the characteristics of teams is small, so the movement of objects with spatial and temporal correlation [6, 7].f

4.3 The Improved Algorithm

In this paper use random for collision detection algorithm based on feature sampling, due to random collision detection algorithm random collision of two objects in question into the search feature on the distance between the issues. Therefore, the characteristics of the distance between should be able to be calculated. Algorithm real is to calculate the distance between the basic geometric elements; the algorithm requires the user to set the distance threshold. If the threshold is set large, easy to miss has occurred the geometric elements to interfere with and affect the algorithm accuracy. In contrast, a smaller set of values will increase the geometric elements on without interference, increasing the complexity of the algorithm.

Point-based particle swarm algorithm robustness is poor; to determine the spatial relationship between adjacent points is more complex. To overcome these shortcomings, this paper, the basic geometric elements surrounded by the bounding box as a sampling of feature objects. This paper selects the OBB bounding box. OBB bounding box making the model more elements intersection test simple. Before making accurate collision detection, first determine whether the OBB bounding box intersection, excluding most of the disjoint interference Yes, there are likely to intersect the basic elements of the basic details of the interference detection. Solve the point-based particle swarm search algorithm is easy to miss elements of interference problems. Enhance the robustness of the particle search.

5 Hybrid Collision Detection Algorithm

Random collision detection algorithm and OBB bounding box algorithm is only applicable to the collision detection problem between the two models, however, there is often more than one model collision in virtual interactive systems. To do this add a quick collision exclude test prior to the collision detection, Exclude far away from the model, thereby reducing the need for further collision detection of objects on the number. This paper presents a hybrid collision detection algorithm which is combining algorithm with OBB bounding box and random collision detection algorithm based on improved PSO.

The basic idea of the algorithm is: First for all objects in the virtual environment pretreatment, Spherical bounding box to exclude is not possible collisions model.

Then for the may intersect the model carry out OBB bounding box traversal, using random collision detection algorithm Based on Improved PSO to Collision Detection. Finally, for the basic geometric elements intersect the test and obtained the results of collision detection. This played their respective advantages of OBB hierarchy bounding box algorithm, and random collision detection algorithm based on improved PSO, to improve the efficiency of collision detection between complex objects.

6 Algorithm Experiments Analysis

In Order to test the hybrid collision detection algorithm performance and features, Do multiple sets of experiments and take the average of the data..Using the model of the formats 3DS, using C + + and OpenGL programming algorithm in VC + + 6.0 platform.

Compared with random collision detection algorithm Based on improved PSO and OBB bounding box algorithm and hybrid collision detection algorithm. Model in the scene is randomly generated and sports, there is certain randomness, In this study Use several tests and Find averaging method. This experiment test 10 times and take average. The performance of the algorithm shown in Figure 3.

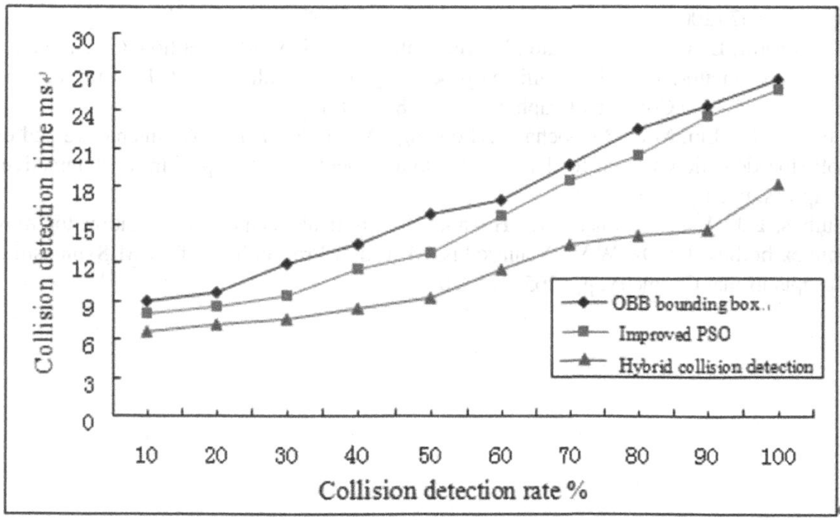

Fig. 3. The comparison of algorithm performance

From the above Figure shows, OBB bounding volume hierarchies' algorithm and the PSO-based random collision detector performance is almost the same, the performance of random collision detection algorithm based on improved PSO higher than the OBB bounding box algorithm, the hybrid collision detection algorithm performance is higher than them.

7 Conclusion

This paper describes a variety of collision detection algorithm, Introduced basic knowledge of Hierarchical bounding box, random collision detection algorithm and the OBB bounding box algorithm, And through experiment to simulation Verification, Compared with random collision detection algorithm Based on improved PSO and OBB bounding box algorithm and hybrid collision detection algorithm. The performance of random collision detection algorithm based on improved PSO higher than the OBB bounding box algorithm, the hybrid collision detection algorithm performance is higher than them

References

1. Wei, F.Z.: Real-time collision detection technology. Zhejiang University, Zhejiang (2003)
2. Kimmerle, S.: Collison detection and post-processing for physical cloth simulation, pp. 28–31. Tübingen (2003)
3. Klen, J., Zachmann, G.: Adb-trees: Controlling the error of time-cntical collision detection. In: 8th International Fall Workshop Vision, Modeling, and Visualization (VMV), Germany, pp. 19–21 (2003)
4. Guy, S., Debunne, G.: Monte-carlo collision detection. Technical Report RR-5136, INRIA, pp. 564–567 (2004)
5. Raghupathi, L., Grisoni, L., Faure, F., Marchal, D., Cani, M.-P., Chaillou, C.: An intestione surgery simulator: Real-time collision processing and visualization. IEEE Transactions on Visualization and Computer Graphics, 708–718 (2004)
6. Cohen, J.D., Lin, M.C., Manocha, D., Ponamgi, M.: I-COLLIDE: An interactive and exact collision detection system for large-scale environments. In: Symposium on Interactive 3D Graphics, pp. 189–196 (2005)
7. Guibas, L.J., Hsu, D., Zhang, L.: H-walk: Hierarchical distance computation for moving convex bodies. In: Oz, W.V., Yannakakis, M. (eds.) Proceedings of ACM Symposium on Computational Geometry, pp. 265–273 (2009)

Coefficient Statistic Based Modified SPIHT Image Compression Algorithm

Jianjun Wang and Ying Cui

Engineering University of Chinese Armed Police Force,
Xi'an, China
dayu504@126.com, cy_1224eve@yahoo.com

Abstract. Among all wavelet transform and zero-tree quantization based image coding algorithms, set partitioning in hierarchical trees (SPIHT) is well known for its simplicity and efficiency. But theoretical analysis and experimental results have shown there are still some key points need to be further improved. This paper proposes a coefficient Statistic based Modified SPIHT Lossless Image Compression Algorithm (MSPIHT-ICA). Through adding a new judgment to type A sets, MSPIHT-ICA effectively optimized the compression algorithm' outputted bit stream. Experiment results further show, compared with SPIHT algorithm, with the same bite rates, MSPIHT-ICA gains higher Peak Signal to Noise Ration (PSNR).

Keywords: Image compression, Wavelet transform, SPIHT, Modified SPIHT.

1 Introduction

Recently the massive use of digital images generates increasingly significant volumes of data. Compressing these digital images is thus necessary in order to store them and simplify their transmission. Wavelet-based coding provides substantial improvements in image quality at higher compression rations. Over the past few years, several very competitive wavelet based image compression algorithms with an embedded bit stream have been developed, such as Shapiro's embedded zerotree wavelet compression (EZW) algorithm[1], Said and Pearlman's set partitioning in hierarchical trees (SPIHT) algorithm[2], and Taubman's embedded block coding with optimized truncation (EBCOT) algorithm[3]. With an embedded bit stream, the reception of code bits can be stopped at any point, and the image can then be reconstructed.

SPIHT is an improved version of EZW. It improves the coding performance by exploiting the self-similarity of the coefficients across sub-bands more efficiently than EZW. Although it is less efficient in coding performance than EBCOT, Which forms the basis of JPEG2000 standard in image coding [4], it has much lower computational complexity than EBCOT. It is in part due to the fact that there is non explicit rate-distortion optimization in the SPIHT coder [5].

Although, SPIHT is well known for its simplicity and efficiency, theoretical analysis and experimental results show that, the algorithm still has some defects. In the encoding process of SPIHT algorithm, when a significant Type A Set with four

D. Jin and S. Lin (Eds.): Advances in CSIE, Vol. 2, AISC 169, pp. 595–600.
springerlink.com

un-significant direct child nodes which less than the threshold, still need to code the four coefficients respectively, and output 4 bit 0 repetitively, therefore ,the outputted bit stream are redundant, which affects the performance of the algorithm. In order to solve the problem, this paper proposes a coefficient Statistic based on MSPIHT-ICA. Based on the statistic analysis of wavelet coefficients, through add judge on type A Set, proposed MSPIHT-ICA divides the type A Set two types: S_1 and S_2, then , uses different stream output strategy coding for S_1 and S_2. Theoretical analysis and experimental results show, using this way, MSPIHT-ICA effectively optimized the compression algorithm' outputted bit stream.

2 SPIHT Coder

2.1 Wavelet Decomposition

Wavelet transform provides a compact multi-resolution representation of the image [6]. It has excellent energy compaction property which suitable for exploiting redundancy in an image to achieve compression. DWT can be implemented using two-channel wavelet filter bank in a recursive fashion [7]. For an image, 2D-DWT is calculated using a separable approach. Input image is first scanned in the horizontal direction and passed through low pass and high pass decomposition filters producing low-frequency and high-frequency data in horizontal direction. The outputted data then scanned in a vertical direction and again these filters are applied separately to generate different frequency sub-bands. The transform generates sub-bands LL, LH, HL and HH each with one-forth the size of the original image. Most of the energy is concentrated in low-frequency sub-band LL, whereas higher frequency sub-bands LH, HL and HH contain detailed information of the image in vertical, horizontal and diagonal directions, respectively. For high-level decomposition, DWT can be applied again to the LL sub-band recursively in a similar way to further compact energy into fewer low-frequency coefficients. After DWT, the decomposed data resembles a tree structure with different scales. A coefficient at a coarser scale in wavelet tree is called a parent, whereas coefficients of the next finer scale at the same orientation are termed as children. When children of all the finer scales at the same orientation grouped together, they are known as descendants. A parent contains four children at its next level, with the exception at the highest level where a parent only contains three children and at the lowest level sub-bands has no children.

2.2 Original SPIHT Coder

After transformation, all coefficients are quantized.

The SPIHT coder uses three types of sets: D(i,j) denoting the set of all descendants of a node (i,j), O(i,j) representing the set of all offspring of the node (i,j), and L(i,j) representing the set of all descendants excluding the immediate four offspring of the node (i,j), that is, L(i,j) = D(i,j) – O(i,j). The represented as C(i,j). C(i,j) is called significant with respect to given threshold T (T=2n , the initial $n = \lfloor |\log_2 C_{\max}| \rfloor$) if C(i,j)≥T, otherwise, it is called insignificant. To indicate the significance of a set U, we use the following notation: $S_n(U) = 1$, if $\max\{C(i, j)\} \geq 2n$, $(i, j) \in U$, otherwise, $S_n(U) = 0$.

In practical implementation, the important information is stored in three ordered lists: a list of insignificance sets (LIS), a list of insignificance pixels (LSP), and a list of significant pixels (LSP). In all lists, each entry is identified by a coordinate (i, j), which represents individual pixels in the LIP and LSP, and indicates either the set D(i,j) (a type A set) or L(i,j)(a type B set) in the LIS. At the initialization step, the coefficients in the highest level are added to LIP, and only those with descendants are added to LIS as type A entries. The LSP is set as an empty list. The SPIHT coder starts with the most significant bit plane. At every bit planc, it tests the three lists in order, starting with LIP, followed by LIS and LSP. The coefficients in the LIP are coded firstly, when a coefficient in LIP becomes significant it is moved to the end of the LSP and their signs are coded. Similarly, sets D(i,j) and L(i,j) are sequentially coded following the LIS order, and those that become significant are partitioned into subsets. The set partitioning rules are as follows: (1) if D(i,j) is significant, it is partitioned into L(i, j) plus four single-elements sets with $(k,l) \in O(i,j)$. (2) if L(i,j) is significant, it is partitioned in to four sets D(k, l) with $(k, l) \in O(i,j)$. Newly formed set L(i,j) and D(k,l) are ended to the end of LIS to be coded again before the same sorting pass ends. Finally, each coefficient in LSP except the ones added in the last sorting pass is refined in each refinement pass. The algorithm then repeats the above procedure for the next resolution, and then stopped at desired bit rates. The decoding algorithm can be obtained by duplicating the encoder's execution path.

3 Proposed Algorithm

3.1 Type A Set's Statistical Characteristics

As described in chapter 2, in the sorting phase of SPIHT algorithm, if a Type A Set is larger than the judgment threshold, it is call significant set, when corresponding sets of four direct child nodes' wavelet coefficient is smaller than the judgment threshold, called un-significant coefficients. In this case, SPIHT algorithm still needs to output 4 bit zero continuously. In a word, although knowing $S_n(O\ (i,j)) = 0$, SPIHT algorithm still need to coding 4 known un-significant wavelet coefficient sets. Therefore, when coding the Type A Sets' un-significant direct child nodes' coefficients sets, SPIHT algorithm consumes large amounts of bit stream.

Therefore, if a Type A Set is significant, this paper adds judgment to the corresponding wavelet coefficients sets of it's 4 direct child nodes. Then, according to the significance of 4 direct child nodes' wavelet coefficients set, divides the significant Type A Sets in to two following two categories:

(1)S0: Type A Set is significant with direct child nodes' correspondent coefficients Set is un-significant, in SPIHT algorithm represents as $Sn(O\ (i,j)) = 0$;

(2)S1: Type A Set is significant with direct child nodes' correspondent coefficients Set is significant, in SPIHT algorithm represents as $Sn(O\ (i,j)) = 1$.

Table 1 shows the ratio of S0 and S1 in different phase of sorting pass for ISO testing image Lena.

Table 1. Ratio of S0 and S1 in different phase of sorting pass for ISO testing image Lena

	prophase of sorting	Medium of sorting	Later of sorting
S_0	41.48%	45.64%	42.81%
S_1	58.52%	54.36%	47.19%

3.2 Proposed Coefficient Statistic Based MSPIHT-ICA

In SPIHT algorithm, when processing the significant Type A Set, presence the problem of repetitive output problem. In order to solve this question, when processing the significant Type A Set, the proposed MSPIHT-ICA adds type judgment to Type A Set, when it belongs to S1 or S1 processed differently. The other processes same with the original SPIHT algorithm.

Compared with SPIHT, in the sorting pass, MSPIHT-ICA uses the following modified way to process Type A Set:

(1) When a Type A Set belongs to type S0, output one bit with value 1 to represent the set is significant, then output one bit with value 0 to represent the set's four direct child nodes' coefficient sets are un-significant, and add them to the LIP respectively;

(2) When a Type A Set belongs to type S1, output one bit with value 1 to represent the set is significant, and output one bit with value 1 to indicate the set is belongs to type S1, then similar to SPIHT algorithm, output 4 direct child nodes' significance judging results respectively, if a child node is significant, add it to LSP, and out put one bit to indicate the coefficient's sign; if a child node is un-significant, add it to LIP.

Although add one bit to judge significant Type A Set belongs to type S_0 or type S_1 in proposed MSPIHT-ICA, when the set belongs to type S0, just output one bit with value 0 to indicate the 4 child nodes' corresponding coefficient sets are un-significant, while in SPIHT, requires to output 4 bits, therefore, in this case, MSPIHT-ICS's outputted bit stream saves 3 bit data. When the set belongs to type S1, compared with SPIHT, although the MSPIHT-ICA consumes extra 1 bit to indicate the set is belongs to type S1, from table 1, for ISO testing image Lena, in different sorting phase totally meet $3S0 > S1$, therefore, the outputted bit stream of proposed modified algorithm saved is larger than the type judgment increased extra bit consume, obviously the compression performance is improved. Experimental results further proved, in the different sorting pass of other ISO testing image Barbara and Baboon's also meet the condition of $3S0 > S1$.

4 Experiment Results

Experiments are performed on 512×512 gray-scale image Lena, using a six-level wavelet decomposition bases on the 9/7 Daubechies filters [6]. We compare our MSPIHT-ICA with the original SPIHT algorithm in the following criteria: PSNR values of the reconstructed image (for simplicity, the entropy coding is not used). The reconstructed images of M-SPIHT-IMA in various bit rates are shown in Fig. 1(2), Fig. 1(3) and Fig. 1(4). As shown in Fig. 1(2), Fig. 1(3) and Fig. 1(4), the reconstructed images show good visual quality in different bit rates. Table 2 lists the experiment results of encoding the image by the MSPIHT-ICA which compared with the results of SPIHT algorithm based coder. As show in table 1, the MSPIHT coder's PSNR is increased.

Fig. 1. Reconstructed coding results: (1) Original ISO testing image Lena, (2) Reconstructed Lena by using MSPIHT-ICA in the bitrate of 0.25bpp, (3) Reconstructed Lena by using MSPIHT-ICA in the bitrate of 0.50bpp, (4) Reconstructed Lena by using MSPIHT-ICA in the bitrate of 0.25bpp

Table 2. Comparison of three kind of 2-D fuzzy entropy algorithm's operation time

bit per pix / bpp	Algorithm	PSNR/ db
0.25	SPIHT	37.09
	MSPIHT-ICA	33.75
0.50	SPIHT	36.88
	MSPIHT-ICA	37.09
0.75	SPIHT	38.53
	MSPIHT-ICA	38.76

5 Conclusion

In this paper, we developed a coefficient Statistic based modified SPIHT image compression algorithm (MSPIHT-ICA) based on the same principle as Said and Pearlman's SPIHT algorithm. This algorithm is able to improve the performance of original SPIHT algorithm because: based on the statistical analysis of wavelet coefficients, though classifying the significant Type A Set into two different type and processing this two type sets in different way, MSPIHT-ICA effectively optimizes the outputted bit stream. Form Theoretical analysis and experimental results, it clearly shown that the reconstructed images' PSNR is increased in various bit rates in the proposed MSPIHT-ICA. So the proposed MSPIHT-ICA has good performance.

References

1. Shapiro, J.M.: Embedded image coding using zerotrees of wavelet coefficients. IEEE Trans. Signal Process. 41(12), 2445–2462 (1993)
2. Said, A., Pearman, W.A.: A new, fast, and efficient image codec based on set partitioning in hierarchical trees. EEE Trans. Circuit Syst & vedio Technol. 36(6), 243–250 (1993)
3. Taubman, D.: High Performance scalable image compression with EBCOT. IEEE Trans. Image Processing 44(9), 1158–1170 (1993)
4. Adams, M.D.: The JPEG, still image compression standardization. ISO/IEC JTCI/SC29/WG1 N2412, Geneva (2001)
5. Wheeler, F.W., Pearman, W.: SPIHT image compression without lists. In: IEEE Int. Conf. Acoustics, Speech and Signal Processing (2000)
6. Antonini, M., Barlaud, M., Mathiew, P., Daubechies, I.: Image coding using wavelt transform. In: IEEE Int. Conf. Image Process., vol. 2(10), pp. 1065–1068 (2004)
7. Daubechies, I.: Orthogonal bases of compactly supported wavelets transform. J. On Pure and Applied Mathematics 42(7), 909–996 (1998)

Holidays Busy Traffic Forecasting Based on MPSO-SVR Algorithm

Jiao Lan[1], DianJun Li[1], XiZhong Qin[1,*], ZhenHong Jia[1], Li Chen[2], Lei Sheng[2], and Hong Li[2]

[1] School of Information Science and Engineering, Xinjiang University, 830046 Urumqi, China
[2] China Mobile Group Xinjiang Company Limited, 830091 Urumqi, China
{lan0816206,qmqqxz}@163.com

Abstract. Holiday traffic prediction is the foundation of the whole communication network planning. In order to predict the busy traffic accurately and ensure the stability of the network, a support vector regression machine (SVR) combined with the improved particle swarm optimization algorithm (MPSO) is proposed, an inertia weight and shrinkage factor is introduced in the algorithm. The proposed algorithm is used to predict the busy traffic in Mid-autumn day. Simulation result shows that, compared with SVR algorithm and the basic particle swarm optimization optimize SVR (PSO-SVR) method, MPSO-SVR algorithm has a higher prediction precision.

Keywords: busy traffic forecasting, support vector regression machine, improved particle swarm optimization algorithm.

1 Introduction

When busy traffic forecast of holidays as a basis for holiday network planning of mobile operator, the predictive accuracy of the whole planning decided the reliability and scientific. The traditional time series prediction methods such as linear ARIMA model, BP neural network all has shortcomings such as prediction accuracy is not high, predict a long time.

SVR is widely used because of its good generalization ability, high precision accuracy [1]. But the learning performance of SVR have very strong link with the parameter selection of kernel function, super parameter selection of accurate or not directly affect the prediction precision of the traffic. When select SVR parameters, the algorithm have fast convergence speed, small amount of calculation and the global search ability request [2], this article puts forward an improved particle swarm algorithm to choose the super parameter of SVR. The algorithm introduced inertia weight and shrinkage factors on the basis of basic particles, inertia weight is used to expand the search space, and maintain the balance of global search and local search, shrinkage factors is used to improve convergence rate. The experimental results show

* Corresponding author.

D. Jin and S. Lin (Eds.): Advances in CSIE, Vol. 2, AISC 169, pp. 601–605.
springerlink.com

that the improved algorithm can not only improve the prediction accuracy, and also to speed up the convergence speed, has some of the feasibility and superiority.

2 The Theories of SVR

Support vector machine is a kind of pattern recognition method based on statistical learning theory. The method based on statistical learning theory of VC dimension and structural risk minimization principle, and by solving a quadratic programming problem better solved the small sample, nonlinear, high dimension and local minimum point practice problems [3]. With non-sensitive loss function ε, the support vector machine started to be used in nonlinear regression estimation problem. The main ideas of the SVR is map the input vector through the nonlinear mapping to high dimension in the feature space, then linear regression in the high dimension feature space [4].

3 Basic Particle Swarm Optimization Algorithm

Particle swarm optimization (PSO) algorithm is a global optimization technology based on the swarm intelligence [5]. Particle swarm algorithm takes each individual as a particle with no weight and volume in N d space, and moving these particles around in the search space according to simple mathematical formula over the particle's position and velocity [6]. This flight speed is dynamically adjusted by individual flying experience and group flight experience, which is updated as better position are found by other particles. This is expected to move the swarm toward the best solutions.

4 Improved Particle Swarm Optimization Algorithm

Particle swarm algorithm have advantages such as dependent less on experience parameters, fast convergence rate, but there are also some shortcomings such as search precision is not high, easy to fall into the local optimal solution[7]. A kind of improved PSO (MPSO) algorithm is adopted in this paper; inertial weight and shrinkage factors are introduced in the speed evolution equation in this algorithm.

Inertial weight is introduced in speed evolution equation in order to improve convergence performance of basic PSO algorithm, namely:

$$v_{ij}(t+1) = wv_{ij}(t) + c_1 r_{1j}(t)\left(p_{ij}(t) - x_{ij}(t)\right) + c_2 r_{2j}(t)\left(p_{gj}(t) - x_{ij}(t)\right) \tag{1}$$

Where ω called the inertial weight, it makes the particles keep inertia moving, make it have the trend of expanded the search space, and have the ability to explore new areas. Introduce of inertial weight can remove the need of v_{max} in basic PSO algorithm; because ω itself has the ability of maintain global and local search ability. Biggish ω have good global search capability, lesser ω have good local convergence ability. Therefore, with the increase of number of iterations, inertial weight should be declining, thus make the particle swarm algorithm is strong in the early stages of the global search ability, and later with strong local convergence ability. The self-adapting inertia meets the following formula:

$$w = w_{\max} - \frac{w_{\max} - w_{\min}}{k_{\max}} \times k \tag{2}$$

Where w_{max} is the initial weights, w_{min} is final weight, k_{max} is maximum iterating time, and k is current iteration times.

Shrinking factor λ can effectively control and restrict the flight speed of particles, and at the same time improve the local search capability of the algorithm. The evolution equation of the improved particle swarm algorithm is described as:

$$v_{ij}(t+1) = \lambda\left(wv_{ij}(t) + c_1 r_{1j}(t)\left(p_{ij}(t) - x_{ij}(t)\right) + c_2 r_{2j}(t)\left(p_{gj}(t) - x_{ij}(t)\right)\right) \tag{3}$$

$$x_{ij}(t+1) = x_{ij}(t) + v_{ij}(t+1) \tag{4}$$

$$\lambda = \frac{2}{\left|2 - \mu - \sqrt{\mu^2 - 4\mu}\right|} \qquad \mu = c_1 + c_2 > 4 \tag{5}$$

The process of MPSO is as follows:

Step1. Initialize the model parameters, such as set the initial position and speed, population size, the number of iteration of particle swarm;

Step2. Calculate the adaptive value of each particle;

Step3. For each particle, compared the adaptive value with the individual best position and the global best position and then take the optimal value as the current best position;

Step4. Update the speed and position according to the equation (3) (4);

Step5. If $k \neq k_{max}$, return to the Step2.

5 The Simulation Results and Analysis

5.1 Experimental Data and Initial Value Choice of the Improved Particle Swarm Algorithm

The experimental data of this paper is consists of two part, one is the training data and the other is test data. Seeing the experimental data from the lateral aspect, there is a 15×8 data matrix A, it composed by every day's busy traffic of 15 days before the Mid-Autumn festival from the year 2004 to 2011; Seeing the experimental data from the vertical aspect, there is a 1×8 data matrix B, it composed by the busy traffic on the day of the festival from the year 2004 to 2011. The front 7 column of matrix A are taken as the training sample, the front 7 data of matrix B are taken as training objectives; The last column of matrix A are taken as test sample, the last data of matrix B is taken as test goal. The test error criterion adopts the Mean Absolute Percentage Error function, the Mean Absolute Percentage Error is:

$$MAPE = \frac{1}{p}\sum_{i=1}^{p} \frac{|y_i - z_i|}{|y_i|} \tag{6}$$

Where p is prediction steps, y_i is the real value of the data; z_i is the SVR predictive value.

The initial values of improved Particle swarm algorithm in this paper are taken as below: maximum iterating time k_{max}=100, the number of particles N= 20, dimension W= 2, acceleration constant c_1,c_2= 0.7293, initial weight w_{max} = 1.4, eventually weight w_{min} = 0.67.

5.2 Busy Traffic Prediction Results and Analysis

In order to assess the prediction effect of MPSO-SVR model, the SVR model, PSO-SVR model and MPSO-SVR model are used to forecast the busy traffic separately and then compared their prediction results. The prediction effects of the busy traffic on the Mid-Autumn festival in two areas of Xinjiang province are shown in Fig.1 and Fig. 2.

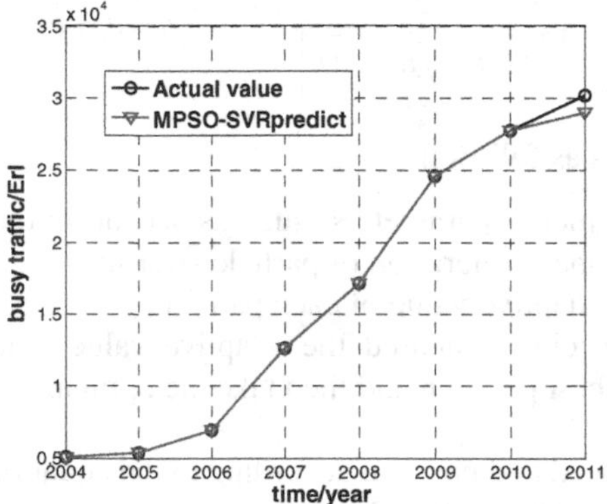

Fig. 1. Forecasting results chart of area A

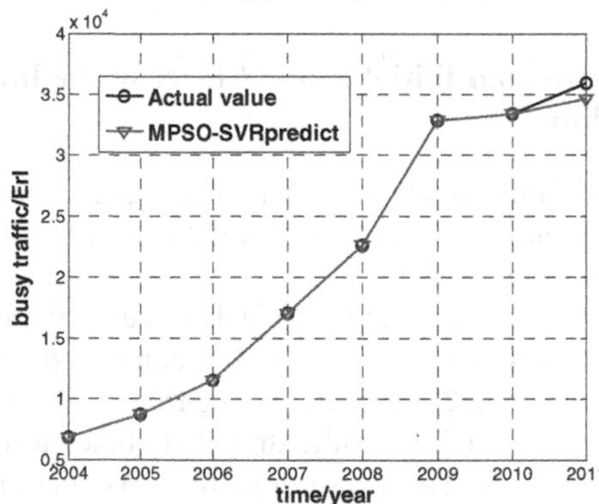

Fig. 2. Forecasting results chart of area B

From the prediction results, we know that the prediction effect of MPSO-SVR algorithm is better than PSO-SVR algorithm and SVR algorithm. Each method's Mean Absolute Percentage Error and time-consuming are given in table 1, because of the randomness of the initial population of particle swarm, so each of the experimental results will be slightly different. The average relative error of every method is the average of the program running 10 times separately.

Table 1. Prediction result of two areas in Xinjiang province

Prediction model	MAPE(%)	time(s)	Prediction model	MAPE(%)	time(s)
Area A			Area B		
SVR	9.972	15.431	SVR	10.483	17.926
PSO-SVR	7.396	14.047	PSO-SVR	7.224	15.012
MPSO-SVR	3.928	7.281	MPSO-SVR	2.619	7.992

6 Summary

A MPSO-SVR algorithm is proposed in this paper to predict the holiday busy traffic in mobile communication, and this model is used to predict the busy traffic of the Mid-Autumn festival of two areas in Xinjiang. The experimental result shows that, the introduction of the particle swarm algorithm of the SVR model is better than SVR model in prediction effect, but this model is easy to fall into local extreme value thus can't find the global optimal solution. The inertia weight and shrinkage factors introduced in the MPSO-SVR model offset the shortcoming of easy to fall into local extreme value in the PSO-SVR model. The MPSO-SVR model has higher forecast accuracy and faster convergence rate.

Acknowledgments. The authors would like to acknowledge the financial support from the China Mobile Group Xinjiang Company Limited (No.XJM2011-11).

References

1. Elattar, E.E., Goulermas, J.(Y.), Wu, Q.H.: Electric Load Forecasting Based on Locally Weighted Support Vector Regression. IEEE Transaction on Systems 40, 438–447 (2010)
2. Wang, J., Zhao, H.: ε-SVRM parameters optimization based on the difference of evolution. Computer Application 28, 2074–2076 (2008)
3. Camps-Valls, G., Muñoz-Marí, J., Gómez-Chova, L.: Biophysical Parameter Estimation With a Semisupervised Support Vector Machine. IEEE Geosciences and Remote Sensing Letters 6, 248–252 (2009)
4. Zhao, L., Jing, S., Butts, K.: Linear Programming SVM-ARMA With Application in Engine System Identification. IEEE Transaction on Automation Science and Engineering 8, 846–854 (2011)
5. Zhong, L., Pan, H.: Pattern recognition. Wuhan University Press, Wuhan (2006)
6. Ali, F.A., Selvan, K.T.: A Study of PSO and its Variants in respect of Microstrip Antenna Feed Point optimization, vol. 8, pp. 1817–1820 (2009)
7. Fernandez-Martinez, J.L., Garcia-Gonzalo, E.: Stochastic Stability Analysis of the Linear Continuous and Discrete PSO Models. IEEE Transactions on Evolutionary Computation 15, 405–423 (2011)

Busy Hour Traffic of Wireless Mobile Communication Forecasting Based on Hidden Markov Model

DongLing Zhang[1], ZhenHong Jia[1,*], XiZhong Qin[1], DianJun Li[1], ChaoBen Du[1], Li Chen[2], Lei Sheng[2], and Hong Li[2]

[1] College of Information Science & Engineering, Xinjiang University, Urumqi, Xinjiang 830046, P.R. China
[2] China Mobile Group Xinjiang Company Limited, Urumqi, Xinjiang 830063, P.R. China
zdl3555086@126.com, jzhh@xju.edu.cn

Abstract. In order to forecast mobile communication traffic quickly and accurately, Hidden Markov Models (HMM) is used to forecast busy hour traffic of wireless mobile communication forecasting in this paper. Because of the model having rigorous mathematical structure, reliable computing performance and the characteristics of describing the event, HMM becomes the ideal model for describing the traffic sequence. The experimental results shows that HMM model have higher precision and better stability compared with the method of SVM with DE-strategy in the area of forecasting mobile communication traffic.

Keywords: Hidden Markov Models, busy traffic, forecasting.

1 Introduction

As the rapid development in communications industry technology, mobile communication traffic forecasting work is more and more important for communication operators in recent years. And traffic load sizes will directly affect the economic interests of the company. However the traffic size in mobile network design, planning performance evaluation has a large influence. In particular, we are confronting with a new era of data in recent years, popularity of 3G, increasing the number of users, change of tariff levels, unexpected events occur, and so will affect the traffic load, carriers must also actively considering cost optimization strategy. Therefore, traffic load forecasting is very necessary. At present, intelligent algorithms of neural networks, support vector machines, is also widely used in traffic time series forecasting, and achieved certain results [1,2].

HMM is widely used in speech signal recognition[3], DNA sequence analysis[4], electrical signal prediction and image processing, economic forecasting[5,6],etc in the past few years. HMM model is a widely used as statistical model which is one of the most important statistical models. It has a strong mathematical structure and reliable computing performance. Because of the Markov chain can be used to describe hidden

* Corresponding author.

D. Jin and S. Lin (Eds.): Advances in CSIE, Vol. 2, AISC 169, pp. 607–612.
springerlink.com

in the sequence of random observations of time-varying characteristics, thus it has unique advantages in dealing with non-stationary random series. Hassan and Nath had experimentally noted that HMM has the similar performance with Artificial Neural Network ANN [7].His method are better than that of past approaches. The basic idea there is to combine HMM's data pattern identification method to partition the data-space with the generation of fuzzy logic for the prediction of multivariate financial time.

In this paper, we first make use of the well established Hidden Markov Model (HMM) technique in forecasting wireless mobile communication traffic. To verify its preciseness and rightness, several experiments are conducted on it. The experimental results show that it is a feasible model. And then we will provide a precise introduction in the following sections.

2 Hidden Markov Model

Hidden Markov Model (HMM) is a statistical model, which is used to describe an implied the unknown parameters of the Markov process. It was originally introduced as far back as 1957 and they have become increasingly popular in the last several years due to its strong mathematical structure and theoretical basis [8,9]. Its difficulty is determined from the observable parameters of hidden parameters of the procedure. Then use these parameters for further analysis, in normal Markov model the status is directly visible to the observation. The state transition probabilities are all parameters, but in HMM the state is not directly visible and affected by the state of some variables are visible. Every state in an output symbol there is a probability distribution. Output sequences of symbols to reveal status sequence number information.

A Hidden Markov model (HMM) is a triple (A,B,Π):

$A=\{a_{ij}\}$; A is an $n \times n$ state-transition matrix, a_{ij} is the state-transition probability from states, s_i to s_j i.e. where

$$a_{ij} = Pr(s_{j,t} \mid s_{i,t-1}) = Pr(s_{i,t-1} \longrightarrow s_{j,t}) \tag{1}$$

$B=\{b_{ij}\}$;B is an $n \times m$ confusion matrix, where b_{ij} is the probability of observing a state O_j, given the hidden state s_i, i.e. where

$$b_{ij} = Pr\left(o_{j,t} \middle| s_{i,t}\right) \tag{2}$$

$\Pi=\{\pi_i\}$; the vector of the initial state probabilities, where π_i is the probability of each state .Each probability in the state transition matrix and in the confusion matrix is time independent ,that is, the matrices do not change in time as the system evolves. In practice, this is one of the most unrealistic assumptions of Markov models about real processes.

3 Traffic Forecasting Based on HMM

Under the realistic circumstances, there are usually multiple related factors that impact on the traffic load. Such as the number of users, change of tariff levels. However, in this paper, we limit ourselves to use the first four days of historical data to predict the next day's. In this section we build a continuous HMM model for forecasting busy hour traffic of wireless mobile communication, While using the HMM, choice of the number of states and observation symbol becomes a tedious task.

As the Continuous Hidden Markov Model (CHMM) to model the traffic data as a time series, so the model formula can be written as:

$$\lambda = (A, B, \Pi) \tag{3}$$

Where *A* is the transition matrix whose elements give thc probability of transitioning from one state to another, *B* is the emission matrix giving $b_j(O)$ the probability of observing *O* when in state *j*, and gives the initial probabilities of the states at *t=1*. Further, The dataset being continuous, the probability of emitting symbols from a state cannot be calculated .We assume that the probability density function of the initial subject to three-dimensional Gaussian distribution. Its probability density function is usually represented as:

$$b_j(O) = \sum c_{jm} P[O, \mu_{jm}, U_{jm}] \qquad (1 \le j \le N) \tag{4}$$

Where *O* is vector of observations being modeled, c_j m is the weight of the m-th mixture component in state *j*. where

$$\sum_{m=1}^{M} c_{jm} = 1 \tag{5}$$

μ_{jm} is the mean vector for the m-th mixture component in state *j*. U_{jm} is the covariance matrix for the m-th mixture component in state *j*. *P* is a multi-dimensional Gaussian distribution.

Each observation is generated by the probability density function associated with one state at one time step. In the experiment, our task is using HMM to forecast the next day's busy hour traffic for the wireless mobile communication. We put the dataset into two sets, training dataset and test dataset. The trained HMM is used to identify and locate similar patterns in the historical data. First of all, we are using the training data to estimate parameters set $\lambda=(A,B,\Pi)$ of the HMM by EM algorithm; And then under parameters λ using *Viterbi* algorithm to solve this problem efficiently. Then find some similar data in the training data (that is identified the data which in the training data and test data have a similar output probabilities), With these similar figures calculated separately each test data's forecasting values.

4 Simulation Analysis of Experiment

4.1 Experimental Data

In order to verify the accuracy of HMM model, we divide the data into two parts, one part as training data, other part as test data. Determine model's relevant parameters on the training data, and then verify the accuracy of parameters and the validity of HMM model on the test data. *Table 1* shows the details of the experimental data.

As the *Table 1* shows that we use *2834* data for training and *89* data for testing. Because in our experimental data we use the first four days of historical data to predict the next day's, for instance we have used left-right HMM with 4 states.

Table 1. Experimental data

Region name	Training data	Test data
region A	2004-1-1~~2010-10-3	2010-10-4~2011-1-1
region B	2004-1-1~2010-10-3	2010-10-4~2011-1-1
region C	2004-1-1~2010-10-3	2010-10-4~2011-1-1

4.2 Performance Indicator

Here we take Mean Absolute Percentage Error (MAPE) criteria as the evaluation of prediction error. It is calculated by first taking the absolute deviation between the actual value and the forecast value. Then the total of the ratio of deviation value with its actual value is calculated. The equation shows as the following:

$$MAPE = \frac{\sum_{i=1}^{k} \{abs(y_i - c_i)/y_i\}}{k} \times 100\% \tag{6}$$

Where k is the number of testing data; y_i is the real traffic load in the day i; c_i is the prediction of traffic in the day i.

4.3 Experimental Results

We have trained four HMM for busy hour traffic of wireless mobile communication in three different regions. The *Fig.1, Fig.2, Fig.3* shows the prediction accuracy of the model in the three regions.

Fig.1~3 shows the comparison of predicted and actual traffic load. The red line represents actual busy hour traffic while the blue represents predicted traffic. For the aforementioned experiment the *MAPE=3.8086* in region A; *MAPE=3.2966* in region B; *MAPE=2.1419* in region C.

In order to test the superiority of the proposed model, we compare its performance with Han Rui 's model- SVM with DE-strategy[10], which is reported as having a good effect on predicting mobile telephone traffic. To make the comparison fair, the same data are used in Han Rui's model. And the *Table 2* shows the results. Clearly, the MAPEs in HMM method are far less than the SVM with DE-strategy method.

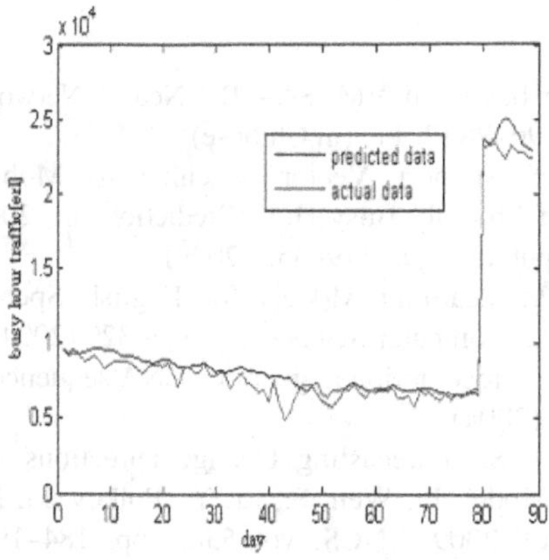

Fig. 1. Actual VS predicted data in region A

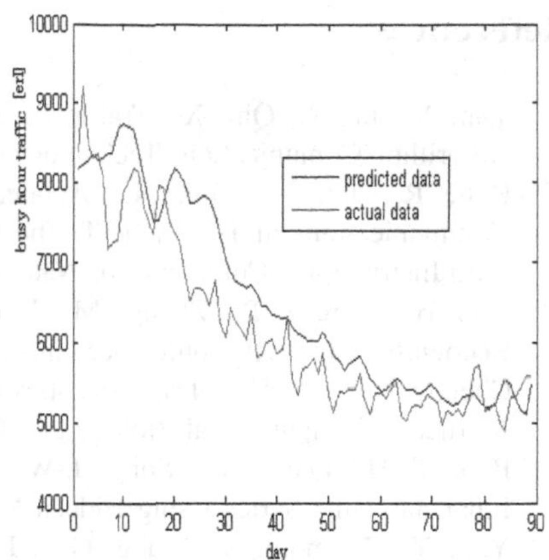

Fig. 2. Actual VS predicted data in region B

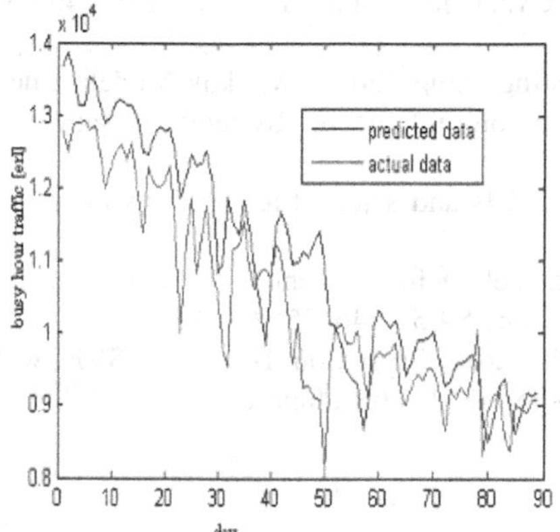

Fig. 3. Actual VS predicted data in region C

Table 2. Prediction accuracy of HMM and the SVM With DE-strategy method

method	region	MAPE
HMM	A	3.8086
	B	3.2966
	C	2.1419
SVM with DE-strategy	A	4.9650
	B	31.9783
	C	53.4416

5 Conclusion and Prospect

In this paper, a new method is proposed using HMM, which is used to forecast busy hour traffic of wireless mobile communication. And compare its performance with SVM with DE-strategy method which is reported as having a good effect on traffic forecasting. It is clear from *Table 2* that HMM is more accurate than SVM with DE-strategy method. Experimental results show that HMM in terms of traffic forecasting also has a better effect .The method to the analysis of network capacity and preventing network congestion problem has important significance. In the future we will do more in-depth research in HMM, combine it with other algorithms to improve its performance.

Acknowledgments. The authors would like to acknowledge the financial support from the China Mobile Group Xinjiang Company Limited Company (No.XJM2011-11).

References

1. Yan, X., Jia, Z., Qin, X.: Traffic Forecasting based on VMPSO—BP Neural Network Algorithm. Comunications Technology 44(01), 96–98 (2011) (in Chinese)
2. Han, R., Jia, Z., Qin, X.: Application of Support Vector Machine to Mobile Communications in Telephone Traffic Load of Monthly Busy Hour Prediction. In: 2009 Fifth International Conference on Natural Computation, pp. 349–352 (2009)
3. Xie, H., Anreae, P., Zhang, M., Warren, P.: Learning Models for English Speech Recognition. In: 27th Conference on Australasian Computer Science, pp. 323–329 (2004)
4. Cheung, L.W.-K.: Use of runs statistics for pattern recognition in genomic DNA sequences. Journal of Computational Biology 11, 107–124 (2004)
5. Park, S.-H., Lee, J.-H., Song, J.-W., Park, T.-S.: Forecasting Change Directions for Financial Time Series Using Hidden Markov Model. In: Wen, P., Li, Y., Polkowski, L., Yao, Y., Tsumoto, S., Wang, G. (eds.) RSKT 2009. LNCS, vol. 5589, pp. 184–191. Springer, Heidelberg (2009)
6. Li, S.T., Cheng, Y.C.: A Stochastic HMM-Based Forecasting Model for Fuzzy Time Series. IEEE Transactions on Systems, MAN, and Cybernetics—Part B: Cybernetics 40(05), 1255–1266 (2010)
7. Hassan, M.R., Nath, B.: Stock market forecasting using Hidden Markov Model: a new approach. In: 05th International Conference on Intelligent Systems Design and Applications, pp. 192–196 (2005)
8. Rabiner, L.R.: A tutorial on hidden Markov models and selected applications in speech recognition. Proc. IEEE 77(2), 257–286 (1989)
9. Longa, Z., Younan, N.H.: Statistical image modeling in the contour let domain using contextual hidden Markov models. Signal Processing 89(5), 946–951 (2009)
10. Han, R., Jia, Z., Qin, X.: Telephone Traffic Load Prediction Based on SVR with DE-strategy. Computer Engineering 37(2), 178–182 (2011) (in Chinese)

The Development of Web Log Mining Based on Improve-K-Means Clustering Analysis

TingZhong Wang*

College of Information Technology, Luoyang Normal University, Luoyang, 471022, China
wangtingzhong2@sina.cn

Abstract. The main objective of Web log mining is to extract interesting patterns from the Web access to records. Web log mining has been successfully applied to a personalized recommendation system improvement and business intelligence. This paper presents the development of Web log mining based on improve-K-Means clustering analysis. K-Means clustering algorithm is analyzed and the paper proposes effective index of the K-Means clustering algorithm and verified by experiment, and proposes automatically selected based on the initial cluster centers that this selection method can reduce the outlier and improve the clustering results.

Keywords: web log mining, K-Means clustering, isolated point clustering.

1 Introduction

Web log mining, also known as Web usage mining, namely the use of the data set to analyze the mining, data mining technology on the site use a lot of data (user access) and other relevant data to obtain valuable website access mode of knowledge the main objective of Web log mining is to extract interesting patterns from the Web access to records. Web log mining is mainly used in e-commerce, through the analysis and explores Web log records law, to identify potential customers, and enhance the quality of Internet information services to end-users, to improve the performance and structure of the Web server system. Currently studied in the Web Usage Mining techniques and tools can be divided into two categories: pattern discovery and pattern analysis.

Web log mining has two main research directions: user access pattern tracking and personalized use of the recorded track. Track user access patterns are to understand the user's access patterns and tendencies in order to improve the organizational structure of the site by analyzing the use of records [1]. Therefore, these data were analyzed to help understand user behavior, to improve the site structure and to provide users with personalized service.

Web access to the most common applications in the mining Web log mining, mining server's log files, draw the user access patterns, the article is based on Web

* Author Introduce: TingZhong Wang(1973.7-), Male, Han, Master of Henan University of Science and Technology, Research area: web mining, data mining, clustering.

D. Jin and S. Lin (Eds.): Advances in CSIE, Vol. 2, AISC 169, pp. 613–618.
springerlink.com

Log Mining for Personalized Recommendation. This paper presents the development of Web log mining based on improve-K-Means clustering analysis. In this paper, the K-Means clustering algorithm to cluster the user, therefore, described in detail below the K-Means clustering algorithm.

2 K-Means Clustering Algorithm and Improve

The cluster analysis used to discover the data distribution and patterns, is an important research direction in data mining. The clustering problem can be described as follows: collection of data points are divided into classes (called clusters, cluster), makes the greatest extent possible between each cluster of data points is similar to the data points in different clusters to maximize the cluster.

Web log clustering in two ways: user clustering and page clustering. User clustering user sessions, according to the user access to the action, looking for patterns of behavior similar to the user. User clustering results can be used as a library of smart Web site mode recommended mode, such as: Web Server analysis, judgment, user A and user B belong to the same group, assuming that the user profile in the group {a.html, b. html, c.html}, user A has access page contains a.html page, the smart Web site real-time recommendation module will be recommended to the user A - b.html and c.html two pages [2]. If user B has access to the page b.html page recommendation module is real-time user B should be recommended to a.html page and c.html page, that is equation 1:

$$\beta_n = -\kappa R\cos(\phi - \phi_n)\sin\theta \tag{1}$$

K-means clustering method is a common division-based clustering method, also known as K-means method, is a widely used algorithm. A form of clustering will make an objective criteria for the classification (often referred to as the similarity function, such as: distance, similarity coefficient) optimization. In this article we use the distance between the relatively simple and commonly used data to describe the similarity, the greater the distance, the smaller the similarity, on the contrary is the greater.

Its core idea is the data objects through an iterative clustering, in order to target function is minimized, so that the generated cluster as compact as possible and independent. This iterative relocation process is repeated until the objective function (generally used to mean square error as the standard measure function) to minimize so far, that is, until each cluster is no longer changes until. The objective function (error function) is generally used to mean square error as the standard measure function such as Equation 2.

$$E = \sum_{j=1}^{k} \sum_{i_l \in c_j} \left| i_l - w_j \right|^2 \tag{2}$$

In general, pre-determined the value of the clustering parameter k is very difficult, therefore, should be based on the data sets and clustering criteria to obtain the clustering parameter k. Ray and Turi, the measure of an effective index of the cluster distance and the distance between the clusters, and applied to image processing, the effective index such as (3),(4),(5) as shown.

$$Validity(k) = \frac{Intra(k)}{Inter(k)} \tag{3}$$

$$Intra(k) = \frac{1}{N}\sum_{i=1}^{k}\sum_{x \in Ci} \|x - Z_i\|^2 \tag{4}$$

$$Inter(k) = \min_{i,j}(\| Z_i - Z_j \|)^2 \tag{5}$$

This article will effectively index and K-Means clustering algorithm is proposed, which combines the K-Means clustering algorithm based on the effective index. The algorithm does not require the user to determine in advance the clustering parameter k, can be automatically determined, but required Kmax limit the number of clusters. Under normal circumstances, the cluster parameters is much smaller than the number of objects (k << n).The algorithm is described as follows.

Algorithmic thinking: the algorithm will be effective index and the K-Means clustering algorithm, the combination of effective index based on the average of the objects in the cluster and clustering;

Input: a data set of n objects where each object m attributes;

Output: the number of clusters k and the set of k clusters, which minimize the effective index of the clustering.

i, the While (k = 2 to of Kmax Step by a variable value);

ii, random selection of k objects as initial cluster centers: c1 (1), c2 (1), ..., ck (1);

iii, to re-allocate each object to the clustering of the object and the center of the cluster closest to;

iv, update cluster mean, using the following formula to calculate the object in each cluster mean as equation 6:

$$C_j(k+1) = \frac{1}{Nj}\sum_{x \in Cj(k)} X \tag{6}$$

Of which: j = 1,2, ..., k, the number of Properties Nj as to Cj (k) of the object;

v, repeat steps iii, iv until the cluster centers no longer change, for all j = 1,2, ..., k

$$C_j(k+1)=C_j(k)$$

Such as cluster centers no longer change, switch to the next step; validity of the effective index of vi, in accordance with the formula (1) - (3) to calculate the number of clusters is k (k);

vii, compare the effective index of validity (k) and the previous index of validity (k-1) to retain the make validity value smaller k;

viii, the end of the algorithm, the output of the most effective number of clusters k and k the center of a cluster and cluster C1, C2, C3, ..., Ck.

3 Web Log Mining Technology

Web log mining has been successfully applied to a personalized recommendation system improvement and business intelligence. Accumulation, especially in the business site with a large number of users access to log data, businesses can use these data to provide users with personalized services to improve customer trust and service quality [3]. Web access to the most common applications in the mining Web log mining, mining server's log files, draw the user access patterns, the article is based on Web Log Mining for Personalized Recommendation, as is shown by equation7.

$$u^{(0)} = \sum_{i=1}^{N} \alpha_i e_i + \sum_{i=N+1}^{M} \alpha_i e_i \tag{7}$$

Web log mining can be divided into three phases: data preprocessing, pattern mining and dig out the pattern analysis. Web server access log (Access Log) generally include: IP address, request time, the method (eg GET, POST), the URL of the requested file, the HTTP version number, the return code, transmit the number of bytes. Table 1 lists several Web server http://lpqf.haust.edu.cn access log. In Table 1 of the first log that user from the IP address 192.168.2.174 to a GET request transmission / comment/list4js.asp, this request is successfully transferred 93 bytes of data, 200 for return code, indicating that the response successfully.

Table 1. Content of Web-server's Access Log

IP Address	Time	Method/url	Status	Size
192.168.2.174	2006-10-16 00:23:40	GET /comment/list4js.asp	200	93
192.168.2.222	2006-10-16 00:25:02	GET /include/PageCount.asp	200	188
192.168.2.174	2006-10-16 00:25:48	POST /comment/comment.asp	200	242
192.168.2.233	2006-10-16 00:27:21	GET /include/functionhit.asp	188	231

Mainly based on the idea of the automatic evaluation methods: if the user is a long time or high frequency access to a site or a page, indicating their interest in the site or page high, therefore, you can access time and frequency as a hobby measure the weight, the algorithm is as follows: i calculate the user to access a url of the frequency obtained by the statistics of the url is the number of users to access. Taking into account the data cleaning stage to remove the occasional visits of the page, you can set the number of users to access a url in the fixed time period should be greater than or equal to a set value.

4 Experimental Results and Analysis

In order to verify the validity of the algorithm test, the log data after data cleaning, user identification, page recognition and other steps, the two sets of data: The first set of data consists of 201 users and 81 links; the two sets of data, including 792 different users and 1644 links. Effective index values, as shown in Figure 1.

It can be seen from the above two sets of test results to algorithm clustering k = 62 the minimum effective index, the clustering meet close and maximum reparability between the clusters, the largest cluster experiments to achieve the desired results. Using the clustering algorithm based on the size and number of data objects to be clustering to select the appropriate step. When the amount of data is small, choose smaller step length can improve the precision of clustering; when the large amount of data, increasing the step size reduces the computation for a large amount of data, a step increase in the accuracy of the algorithm the impact is negligible.

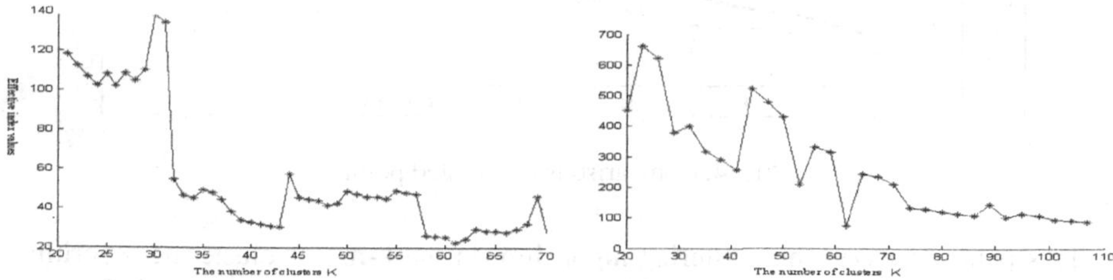

Fig. 1. Validity versus Cluster Number

Test using the mean of the conventional k-means method to cluster, the initial point selected were random and before the automatic cluster center selection algorithm, test users in 2658 (nine clusters) clustering, test results such as Figure 2 to Figure 3, the results show that the initial cluster centers automatically selects the algorithm is better than randomly selected. It can be seen from Figures 2, 3; automatic initial point selection method is superior to the random initial point selection method.

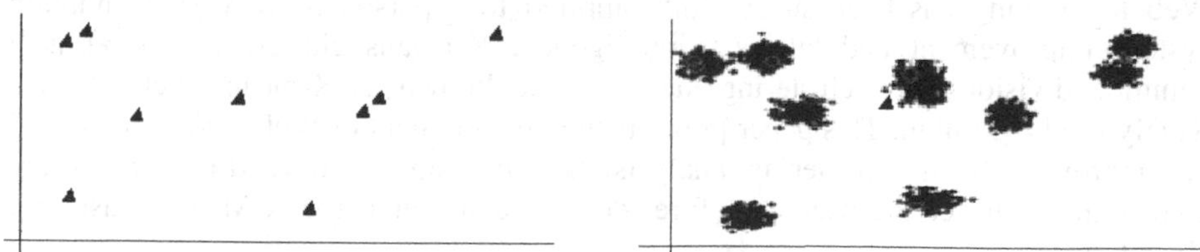

Fig. 2. Stochastic selection clusting initialization point and result of clusting

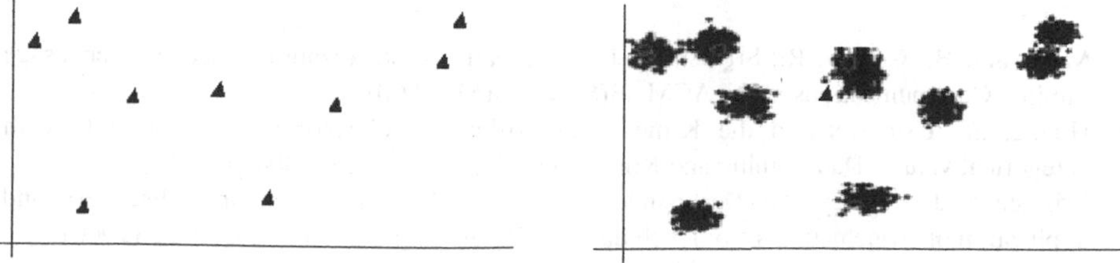

Fig. 3. Automatic selection clusting initialization point and result of clusting

Experiment cluster analysis to 2658 users, the results show that the initial cluster centers automatically selected a better solution to the problem of isolated points, the comparison shown in Figure 4. Is obvious from the figure can be seen: before the cluster center automatically selects the algorithm, reducing the initial cluster centers randomly selected to result in isolated points more.

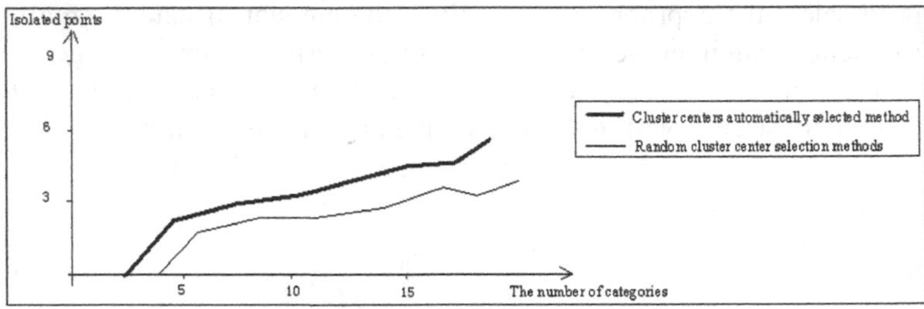

Fig. 4. Comparisons of isolated point

This paper analyzes the clustering algorithm on the k-means clustering algorithm, the initial value problem for a traditional clustering algorithm, improved K-Means clustering algorithm proposed effective index of the K-Means clustering algorithm and validated through experiments. Isolated points are more randomly selected from the initial point of clustering to reduce the outlier, automatically selected based on the initial cluster centers, the experiment found that this selection method can reduce the outlier and improve the clustering effect.

5 Summary

Web log mining has been successfully applied to a personalized recommendation system improvement and business intelligence. K-means clustering method is a common division-based clustering method, also known as K-means method, is a widely used algorithm. This paper presents the development of Web log mining based on improve-K-Means clustering analysis. In this paper, the K-Means clustering algorithm to cluster the user, therefore, described in detail the K-Means clustering algorithm.

References

1. Mobasher, B., Cooley, R., Srivastava, J.: Automatic personalization based on web usage mining. Communications of the ACM 43(8), 142–151 (2000)
2. Huang, Z.: Extensions to the K-means algorithm for clustering large data sets with categorical values. Data Mining and Knowledge Discovery 2, 283–304 (1998)
3. Srivastava, J., Cooley, R., Deshpande, M., et al.: Web usage mining: discovery and application of usage patterns from web data. SIGKDD Explorations 1(2), 12–23 (2000)

Automatically-Controlled System for Detecting Quartz Crystal Based on PLC

Wei Li[1,2], Guanying Zhu[1,2], and Bin Lin[1,2]

[1] State Key Lab of Modern Optical Instrumentation, Zhejiang University, Hangzhou, China
[2] NERC for Optical Instrument, Zhejiang University, Hangzhou 310027, China
lee919@qq.com

Abstract. Since the size of quartz crystals is tiny, the precision of the system needs to be high, the process of the system needs to be reliable and every component in the system needs to cooperate with each other. By the method of using manipulator of stepper motor, the class of quartz crystal is put into the special small chamfer which allows quartz crystal to fall onto the positioning platform. The fallen crystal is detected by machine vision. Finally the detected quartz crystals are classified into several groups under the manipulation of Programmable Logic Controller (PLC). The current system is of importance since it has demonstrated the possibility of getting the fine quartz crystal at the speed of 4600 pieces per hour and this result can meet most of the requirements in practical production.

Keywords: Automatically Detecting System, PLC, Stepper Motor.

1 Introduction

The quartz crystal is the core component of quartz crystal resonator, which is used to create an electrical signal with precise frequency. Quartz crystals are manufactured for frequencies from tens of kilohertz to tens of megahertz and are widely used in consumer devices such as clocks, radios, computers and cell phones.

Automatic visual inspection plays a key role in manufacturing processing of the semiconductor industry. The technology in detecting the defect of quartz crystal has relatively been mature and some commercial inspection devices can supply some kinds of different services [1-3]. With the development of the new technology and the improvement of the manufacturing, the quantity and the diversity demand become more and more critical, and miniaturization is the trend in the development of quartz crystal resonator. The methods are changing with fast speed; however, there exists some limitations. To speed up production, the author finds out that not only the methods of multi-point positioning and Hough transform but also a new control system based on PLC is of necessity and should be employed. Tested by thousands of trials, the detected results are analyzed and presented as follows.

D. Jin and S. Lin (Eds.): Advances in CSIE, Vol. 2, AISC 169, pp. 619–624.
springerlink.com

2 System Configuration

A block diagram of the system to detect flaws of quartz crystals is shown in Fig.1. The detecting station is completely controlled by PLC which receives data from the computer. The system is mainly composed of four units: delivery unit, image analysis unit, transfer unit and control unit.

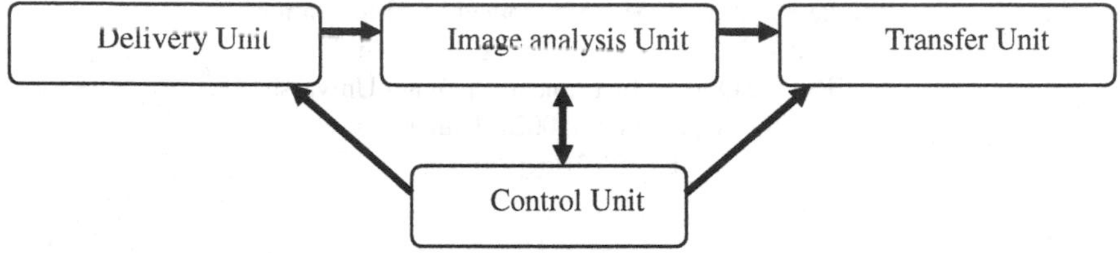

Fig. 1. The configuration of the auto-detecting system

2.1 Control Unit

The function of the control unit is to regulate and coordinate all devices in the detecting station. In order to achieve stronger capacity of resisting disturbance and keep high reliability, PLC has been adopted in the system. It needs to control alternating-current motor, stepper motors, linear electric motors, cameras, LED light sources, warning devices, external relays and other electronic components. Since there is not enough IO interface available, extension modules has been employed in this unit. The extension modules have 16 input pins and 16 output pins. In order to control stepper motor conveniently, a special module has been designed which can generate pulses directly in accordance with the received data from the computer. For the purpose of testing and operating with ease, a humanized type of PLC is introduced. Its human-computer interface is available and attractive to users.

2.2 Delivery Unit

The function of the delivery unit is to take the undetected quartz crystals out of hollow cylinder when the computer detects the quartz crystals on positioning platform is less than threshold or empty. The cylinder is driven by AC motor whose status is under control of an internal relay of PLC. The speed and the time duration of the motor can be set by computer. Then the quartz crystals fall uniformly on positioning platform with the platform twirls simultaneously.

2.3 Image Analysis Unit

The function of image analysis unit is to analyze the quality of quartz crystals. The image analysis unit includes three cameras with different lens and light sources. The first camera, which is set up above of the positioning platform, is used for crystal positioning, and the field of view is 100mmX80mm. The second and the third cameras are set up above of the small chamfer respectively, and the fields of view of them are both 5mmX4mm. All of them produce a 1280(H) X 1024(V) digital image. A ring setup with LEDs which emits blue lights is used as the lighting source.

2.4 Transfer Unit

The function of the transfer unit is to transfer undetected quartz crystals from positioning platform to detecting platform, and to transfer detected quartz crystals into different classified boxes which are placed in precalculated position.

The whole transfer unit includes two steps:

2.4.1 First Step

In this step, the undetected quartz crystals from hollow cylinder are transferred from positioning platform to detecting platform one by one. It consists of one linear electric motor, one stepper motor and one vacuum absorber. With the rotation of positioning platform, computer calculates the polar coordinates according to the image which is photographed by the positioning camera. These polar coordinates are sent to PLC, which drives the stepper motor and moves the vacuum absorber on the linear electric motor to the calculated point which is the center of the quartz crystal. Then the quartz crystal is transferred onto detecting platform through absorbing.

2.4.2 Second Step

In this step, the quartz crystals which have been detected are transferred from detecting platform to classified boxes. It also consists of one linear electric motor, one stepper motor and one vacuum absorber, additionally; several boxes with holes drilled on its top have been employed. The quartz crystals which have been analyzed for two procedures will be classified into three or more categories. The linear electric motor and the stepper motor work the same as which in the first step.

3 Operating Principle of System

This chapter investigates the operating principle of the quartz crystal detecting system. The positioning camera is placed on the positioning platform for the purpose of obtaining enough fields. One detecting camera lies upside and the other lies downside of the detecting platform. In order to detect front and back of quartz crystals, 18 stations have been made, the two detecting cameras are placed in the 6th and 10th separately. The operating sequence chart of the control system is shown in Fig.2.

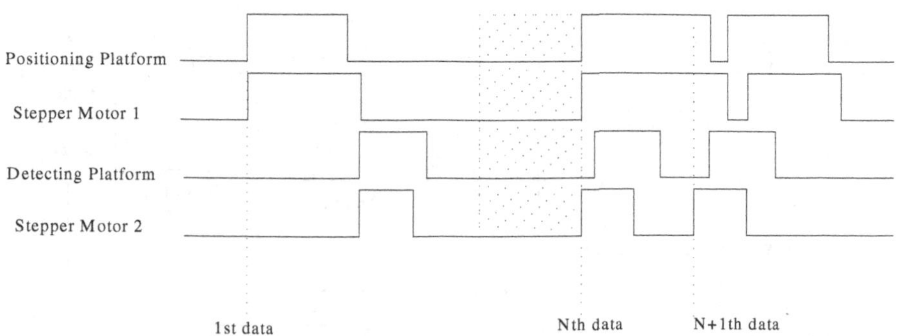

Fig. 2. The New Operating Sequence Chart

This system achieves high-coordinated steps and shows high stability. Applied with PLC high speed buffer and new operating sequence, quartz crystals are transferred in millisecond timescale on hypogynous computer section.

3.1 Initialization

When the system is powered on, PLC scans a cycle. The red indicator on top is lightened. Then PLC sets a flag which starts initializing. This flag can be reset after powering on or altering the internal register by software. Initialization is mainly concerning two platforms and two linear electric motors. The positioning platform rotates a circle and stops when the stop sensor gets the signal and feeds back to PLC. The detecting platform rotates as well, but when it obtains the stop signal, PLC takes this position as station one. Two linear electric motors return to the original point in a specific way. Afterwards, they move for a certain offset. If the entire process is not finished in 12 seconds, PLC considers that the initialization is unsuccessful and lightens the warning light.

3.2 Operational Scheme

After initialization succeeds, the related registers in PLC are given initial values which control the moving devices. The initial data which controls positioning platform is used for starting. The data packet received from computer through serial port includes three data: the first controls the positioning platform, the second controls linear electric motor one and the third controls the motor two.

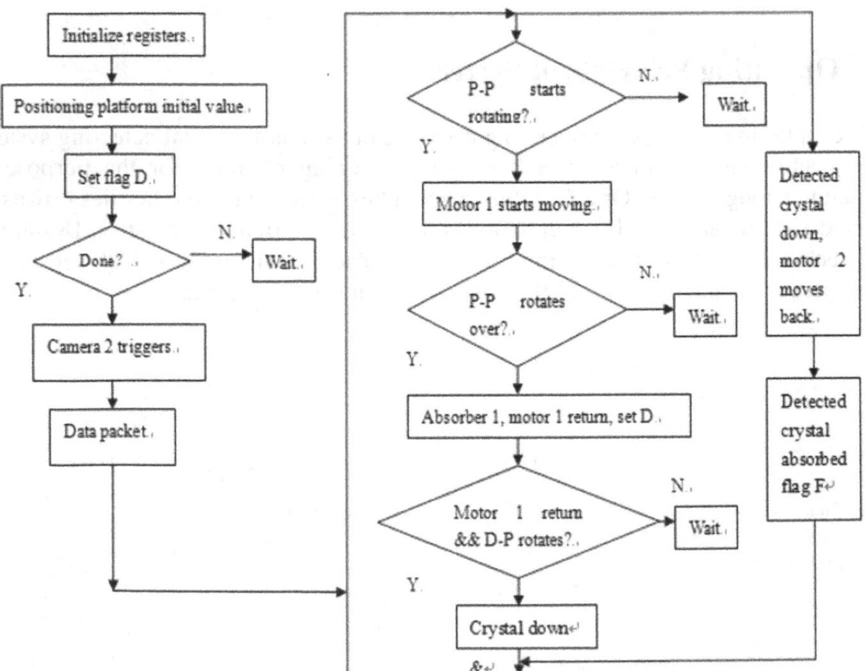

Fig. 3. The Flow Chart of the Auto-Detecting System

The data is processed by PLC. The positioning platform starts to rotate and sets flag A. The motor 1 gets the flag A and starts to move to the designated location, and sets flag B. The positioning platform rotates over, PLC sets flag C. Once B&C is true, PLC turns on the vacuum absorber to absorb one crystal and sets flag D. Then the motor 1 returns to the original point which is also the first station on detecting platform and sets flag E. Once the detecting platform gets flag E, it rotates for one station and the motor 2 absorb the detected crystal on the 16th station to the classified box. The image processing of crystal positioning and defect detection is not discussed in detail.

4 Results

A company has made an automatic sorting system which achieves a speed of 4000 pieces/hour [4].Through thousands times of our tests, the average speeds of detecting different quartz crystals are listed in table 1. It shows the capability of 4600 pieces/h plus. Smaller crystals will be detected in a shortened period.

Table 1. The Average Speed of Detecting Different Quartz Crystals

Sample	Size(mm)	Speed(pieces/h)
SMD	1. 5X1. 25	5200
	3. 0X1. 45	5300
	3. 225X1. 45	5500
49S	7. 9X1. 732	4700
	8X1. 732	4800

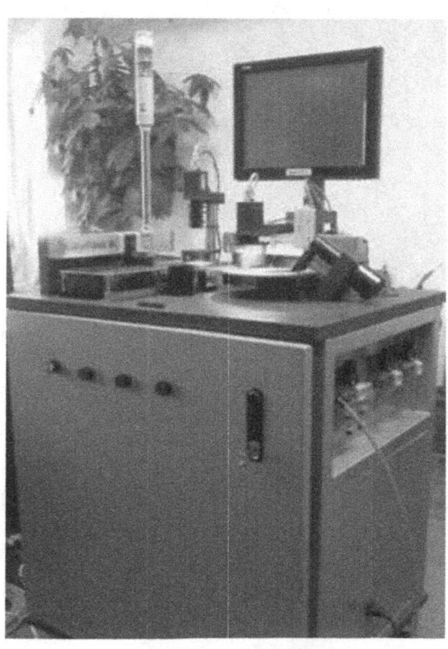

Fig. 3. The Photo of Automatically Control System of Detecting Quartz Crystals

The photo of our automatically-controlled system of detecting quartz crystals is shown in Fig.3.

References

1. Batchelor, B.G., Hill, D.A., Hodgson, D.C.: Automated visual inspection. Elsevier Science Pub., New York (1985)
2. Newman, T.S., Jain, A.K.: A survey of automated visual inspection. Computer Vision and Image Understanding 61(2), 231–262 (1995)
3. Thomas, A.D.H., Rodd, M.G., Holt, J.D., Neill, C.J.: Real-time Industrial Visual Inspection: A Review. Real-Time Imaging 1(2), 139–158 (1995)
4. Nanking Hanna Science CO. LTD., http://www.njhnkj.eb80.com

The Design of the Weeds Classification System Based on BP Neural Network

DongMing Li[1], Bo Du[2], and Li Zhang[1]

[1] Mechanical and Electrical College Agricultural University of Hebei, Baoding, 071001, China
[2] Foreign Language College Agricultural University of Hcbei, Baoding, 071001, China
Ldmmail@163.com

Abstract. This design analyzed the shape parameters of weeds. Through the data comparison and experimental verification, it comprehensively utilized the six effective shape parameters as a neural network input feature vector, which contained the broad and long edge ratio, the leafage and circumcircle area ratio, the leafage and girth's square ratio ,the leafage and circum-rectangle area ratio, the framework and area ratio, the framework and girth ratio. In this article, neural network was trained and improved. The frequency, sample rate and the recognition correction rate for the test samples should be considered in the network structure. The experimental results showed that Neural Network Classifier was able to identify crops and weeds well. For specific plants, the recognition rate could be improved using of a specific shape features.

Keywords: weeds, shape features, BP neural network, recognition rate.

1 Introduction

Weeds, the same as plant insect pests, were a major crop disaster. Weeds contended for soil, sunshine and water with crops, and they easily spread and caused incalculable economic loss. At present, the main method of weeding was the use of herbicide, which could not only remove the weeds in crops, but also improved the crop production. However, because of the excessive use of herbicides, on the one hand, this method increased the production input costs; on the other hand, it resulted in environmental pollution. Therefore, people felt an urgent need to restrict the use of herbicides. How to reduce the amount of herbicides and fully control the weeds growth has been a problem of concern. The best solution was to spray the herbicide on weeds gathered region, namely to realize variable spray implementation. To achieve variable herbicide spraying, first of all, we needed to solve the problem of weeds identification through computer vision techniques to determine the weeds region, and then the weeds region would be separated. With the development of computer technology, machine vision technology had been widely applied and developed quickly in the field of agriculture.

In this paper, wheat and associated weeds were treated as the research objects, MATLAB7.0 as analytical tools. At the basis of preliminary studies, the differences of

D. Jin and S. Lin (Eds.): Advances in CSIE, Vol. 2, AISC 169, pp. 625–630.
springerlink.com

the shape existed on the weeds and wheat was analyzed, and the BP neural networks were designed, which could implement the classification of weeds.

2 To Obtain the Shape Characteristic Parameters

Many shape characters were non-dimensional in the weed identification. They chiefly comprised the ratio of the width and the length, complete degrees, the roundness, the rectangle and so on. In addition to the use of common characteristics of parameters, this paper also proposed to use the ratio of the framework proportion, frame perimeter as characteristics parameter to identify weeds.

2.1 Tag Connected Domain

Partition could only separate images with different characteristics of gray or organization. Feature extraction was the further separation of the regional feature which had been extracted. For a number of image feature extraction in the region could be divided into two steps to complete: marking and feature extraction. First of all, we should tag different regions of the image, and then extracted the characteristics from different regions to complete the quantitative description.

2.2 Shape Feature Extraction

After calibrating connective regions, we could extract certain characteristics of all the connected regions. In order to reduce the number of pixels of different pictures on the impact of statistics, in addition to the area and perimeter, we mainly extracted the non-dimensional characteristics. After the extraction of the figure1, the data was shown in Table1.

After obtained the shape feature parameters of the image, in accordance with its typical characteristics, the next step was to extract the weeds or crops.

Table 1. The data of shape feature parameters extraction from the figure

Serial number	Girth	Area	Leafage area and circum-rectangle area ratio	…	Framework and girth ratio
1	0.8460e+003	1.3935e+004	8.1034e-001	…	5.1138e-001
2	8.4200e+002	1.5908e+004	5.7730e-001	…	6.0377e-001

3 Comparison and Analysis of Shape Characteristic Parameters

Through the study we could find that, for different weeds, different parameters of the shape could be found, which was largely different from the wheat. The ratio of width and length of large leafage category weeds could be shown as an example. So we could use the differences of shape characteristics parameters between the wheat and weeds on to distinguish them. Fig.1 was showing several kinds of common weeds with a typical representative in the wheat field in spring, which was the study object of this experiment.

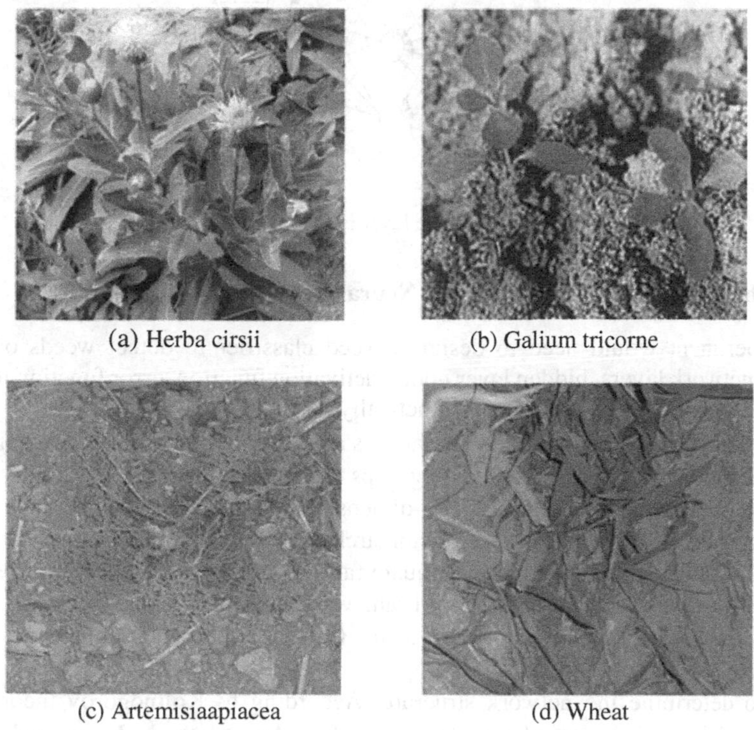

(a) Herba cirsii (b) Galium tricorne

(c) Artemisiaapiacea (d) Wheat

Fig. 1. Wheat and weeds image compare

3.1 Shape Features Selection

We selected 10 typical laminas separately in each type of weeds, and then coded them. Using the methods described above, the shape characteristics could be extracted separately.

As was shown from the figure, broad and long edge ratio and framework and area ratio of the wheat were significantly different from the other three kinds of weeds. So these two eigenvalues could be used to identify wheat.

3.2 BP Neural Network Classifier Design

As was shown in Fig.2, there was a Neural Network Image Classifier: network was divided into 3 layers, and interconnection-wide mode was used between the layer and layer. The connection between the same floors units was non-existent. Each input node of the network denoted a component data (gray value) of image feature vector. Output nodes denoted classification number, and the maximum output method could be used in classification judgments.

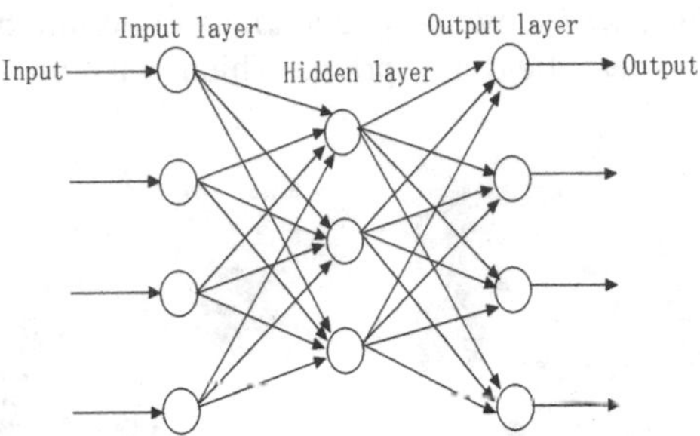

Fig. 2. Three layer BPNN frame

3.3 The Design and Training of BP Neural Network

This experiment would need to design a weed classifier to detect weeds or crops. Implicit network layers, hidden layer nodes, activation function, error function forms as well as the learning rate should be synthetically considered.

① First, the network input and the target sample should be required. According to the presentation above, we obtained 40 groups of shape characteristics samples, which respectively contained the 6 groups non-dimensional numerical value: Broad edge and long edge ratio, Leafage area and circumcircle area ratio, Leafage area and girth's square ratio, Leafage area and girth's square ratio, Leafage area and circum- rectangle area ratio, Framework and area ratio, Framework and girth ratio. Corresponding to different data sets, there were Herba cirsii, Galium tricorne, artemisiaapiacea and wheat.

② To determine the network structure. According to Kolmogorov theorem, we adopted a 3-layer BP network as weeds classifier. Among them, N denoted the heft number of the input feature vector, and M denoted the total number of output state categories. For this experiment, N equaled 6.In order to simplify the network structure, (0, 0) was used to express Herba cirsii, (0, 1) Galium tricorne, (1, 0) Artemisiaapiacea, (1, 1) Wheat expressed Galium. In this way, we could express these 4 kinds of plants with only designing two input neurons elements.

③ We used P to express input sample vector of the network, T objective vector of the network. Among them, the rows P and T were required to maintain the same number; otherwise the process would be running error.

④ MATLAB7.0 toolbox function was used to create a BP network to comply with these requirements. The training function of the network was "trainlm".

⑤ Network training. Network could not be directly put into use after it was created, which could be used as a weeds classifier after given the network configuration error and then trained to meet the requirement.

The experimental results showed that: the smaller the settings of the Network error, the longer the network training time. But this did not mean that the smaller the network error was, the higher the recognition rate of the sample was. Table2 was experimental effect under different studying error.

Table 2. The influence of identify result by learning error

Learning error	Training time	Swatch learning ratio (%)	Testing swatch right identify ratio (%)
0.001	1260	100	91
0.002	895	100	94
0.003	565	98	89
0.004	231	95	87

3.4 Weeds Test

We used neural network classifiers for weed identification experiment. In accordance with previous experimental results, the error for the study of neural network 0.002 was selected. The purpose of network testing was to determine whether the network met the requirements of practical applications. It should be noted that test data and training sample data should be different. Otherwise, the results of the test were always satisfied.

Weed identifiable experiment had two cases: the leaves covering were low (less than 10% coverage), the leaves covering were relatively high (greater than 30% coverage). Through 20 plants for each type of experiment, results showed in Table3 and Table4.

Table 3. The dentify result of bestrow ratio was low

Plant name	number	Identify right	Identify error	Right ratio (%)	Error ratio (%)
Herba cirsii	50	45	5	90	10
Galium tricorne	50	47	3	94	6
Artemisiaapiacea	50	43	7	86	14
Wheat	50	47	3	94	6

Table 4. The dentify result of bestrow ratio was high

Plant name	number	Identify right	Identify error	Right ratio (%)	Error ratio (%)
Herba cirsii	50	35	15	70	30
Galium tricorne	50	38	12	76	24
Artemisiaapiacea	50	34	16	68	32
Wheat	50	40	10	80	20

As could be seen from the experimental results: when the coverage was low, the overall average correct recognition rate was 91%, thereinto crop 94% correct recognition rate and weed average correct recognition rate 90%; when the coverage was high, the overall average correct recognition rate was 73.5%, thereinto crop correct recognition rate 80%, weeds average correct recognition rate 71.3%.

Obviously, when leaf coverage was low, the correct recognition rate was relatively high. But when leaves were blocked, recognition rate dropped significantly, this was the main problems that needed to be resolved in the future. In addition, this experiment also tested aiming at single feature neural network and found that, when the leaves were seriously blocked, the recognition rate of the wheat using framework and area ratio, which could reach 89%. So this feature may be considered for region of the leaves of serious coverage.

4 Conclusion

The paper analyzed how to identify of crops and weeds with different shape, studied the setting up and training of the neural network. For the average image, it had relatively better recognition rate, and achieved more satisfactory results.

Although this study has made some achievements, the deficiencies still remain to be improved further. Recommendations of study in the following areas in the near future:

(1) At different days, the differences of leaf shape of crops are significant, and the object of study of this subject is limited. The next step should focus on the features of different crops at different times. A database of the shape features should be set up, and the professional weed identifier should be developed.

(2)The recognition rate of leaves with serious coverage should be further improved.

Acknowledgement. This paper was supported by the Department of Science and Technology of Hebei Province (10220925) and Non-living Science Foundation of AUH (Fs20101303).

References

1. Marchant, J.A., Andersen, H.J., Onyango, C.M.: Evaluation of an imaging sensor for detecting vegetation using different waveband combinations. Computers and Electronics in Agriculture 32(2), 101–117 (2001)
2. Goel, P.K., Prasher, S.O., Patel, R.M., et al.: Use of airborne multi-spectral imagery for weed detection in field crops. Transactions of the ASAE 45(2), 443–449 (2002)
3. Alchanatis, V., Ridel, L., Hetzroni, A., et al.: Weed detection in multi-spectral images of cotton fields. Computers and Electronics in Agriculture 47(3), 243–260 (2005)
4. Karimi, Y., Prasher, S.O., Patel, R.M., et al.: Application of support vector machine technology for weed and nitrogen stress detection in corn. Computers and Electronics in Agriculture 51(2), 99–109 (2006)
5. Neto, J.-C., Meyer, G.E., Jones, D.D., et al.: Plant species identification using Elliptic Fourier leaf shape analysis. Computers and Electronics in Agriculture 50(2), 121–134 (2006)

The Research of Ontology Merging Based on Rough Concept Lattice Isomorphic Merging

HongSheng Xu* and MingLi Song

College of Information Technology, Luoyang Normal University, Luoyang, 471022, China
xhs_ls@sina.com

Abstract. Ontology merging is the process of merging two or more source ontology to goal ontology. Ontology merging means that getting some ontology existed in the domain together, eliminating the overlap and not harmonious part. The paper aims at different scale ontology, takes different approaches. For the lightweight ontology uses the ontology merging technology of RFCA-based for ontology merges; for the heavyweight ontology, the paper proposes the ontology merging methods based on rough concept lattice isomorphism merging.

Keywords: ontology merging, concept lattice, rough concept lattice.

1 Introduction

Ontology merging is two or more source ontology merging process of target ontology. Using conventional editing tools for ontology merging, it is manual merging time-consuming and error prone. There are scholars made a number of systems and frameworks to help the knowledge engineer to carry out ontology merging, they are dependent on the body taken by the engineers in the merged ontology syntax and semantics that matches the heuristic method.

Ontology merging is generally used only for the ontology construction process will be more than one ontology merging the ontology or the ontology maintenance phase, a "mini-ontology (mini-of ontology)" (a small part of the concepts and relationships) incorporated into the original body to body to be updated. Ontology merging method is to get the mapping ontology mapping, depending on the application need to perform the merge, and after the implementation of processing [1].

The concept lattice and rough set are two powerful tools of data analysis and dispose in the data mining; the project uses the reduction ideas of β-upper and lower distribution of VPRS in the reduction of form background, and brings forward the concept lattice structure model based on the VPRS. Its core idea is bring the strong capacity of variable precision rough set attributes reduction in the concept lattice reduction to greatly reduce the nodes number of concept lattice and improve the robustness and noise immunity of the system.

* Author Introduce: Xu HongSheng(1979-), Male, lecturer, Master, College of Information Technology, Luoyang Normal University, Research area: Data mining, web mining.

D. Jin and S. Lin (Eds.): Advances in CSIE, Vol. 2, AISC 169, pp. 631–636.
springerlink.com

In this paper, a different order of magnitude of the body, take a different approach. Lightweight ontology, ontology merging based on the RFCA ontology merging; heavyweight ontology, the paper proposed ontology merging method based on rough concept lattice structure generated. This paper presents ontology merging based on rough concept lattice isomorphic merging.

2 The Research of Ontology Merging Methods

Ontology merging is two or more source ontology merging process of target ontology. Using conventional editing tools for ontology merging, it is manual merging time-consuming and error prone. There are scholars made a number of systems and frameworks to help the knowledge engineer to carry out ontology merging, they are dependent on the body taken by the engineers in the merged ontology syntax and semantics that matches the heuristic method. Following is a brief introduction to these methods.

The main purpose of the development of ontology is a knowledge sharing and reuse. To achieve knowledge sharing requires people to use common knowledge representation forms, such as the common ontology. But this is clearly unrealistic, domain ontology, for example, even for the same area, different ontology developers also possible to set up the domain ontology, let all developers to follow the same methods and rules to build ontology is difficult. Therefore, the domain ontology knowledge reuse a key issue, then the solution is to integrate and consolidate data from multiple ontologies in the same area or different areas of knowledge base to achieve, that is, ontology merging.

Hovy the first ontology merging method [2], which describes the heuristic method is by identifying the corresponding concepts in different ontologies, such as name and the definition of natural language analysis comparing the two concepts, and check in the concept of class two concept of similarity relations, that is equation 1.

$$K_E(x) = \begin{cases} \frac{1}{2} c_d^{-1} (d+2)(1-\|x\|^2), & if\ \|x\| < 1 \\ 0, & if\ \|x\| \geq 1 \end{cases} \tag{1}$$

The Chalupsky system OntoMorph conversion and consolidation of the ontology presented two mechanisms. A: the rewriting mechanism based on syntactic support In two different knowledge representation language conversions; another: the semantic rewriting mechanism for inference-based conversion solution. The system is clearly more flexible conversion mechanism allows the conversion; you can violate existing semantic rules, as is shown by equation2.

$$p(s(k)) = \frac{1}{\left(2\pi\sigma_{s(k)}^2\right)^{1/2}} \exp\left[-\frac{\left(s(k)-s_0(k)\right)^2}{2\sigma_{s(k)}^2}\right] \tag{2}$$

The vocabularies of terms are from different source ontologies to merge to provide a method. In addition, it can not only help the coverage and correctness of the test body, but would also facilitate maintenance of the ontology. Although the Chimaera system provides a number of functions, but the underlying assumptions about the properties of the body structure is not yet made a clear explanation: equation 3.

$$\|x(k_0)\| < \delta(k_0,\varepsilon) \Rightarrow \|x(k)\| < \varepsilon \quad \forall\, k \geq k_0 \tag{3}$$

Stumme and Macdche2001 mention FCA ontology merging the FCA-MergeE tame. The method uses a bottom-up approach given the global process description of the ontology merging. The concept lattice is in fact the concept of source ontology clustering, Tong became a level of conceptual clustering concept lattice generated by bulk T lemma was to convert the results of ontology. This method is the ontology concepts and the FCA concept with their properties. The merging process is shown in Figure 1.

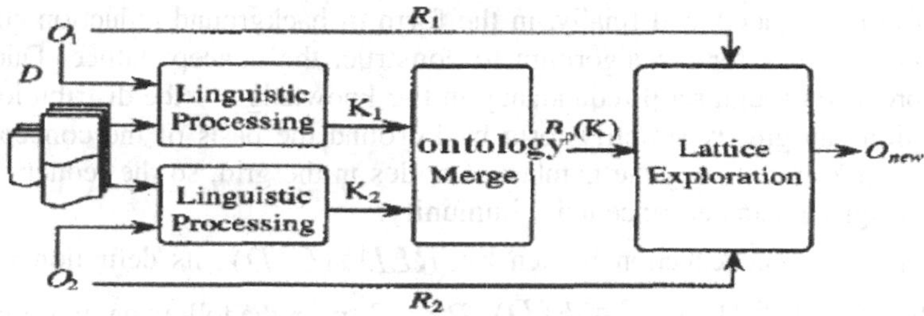

Fig. 1. The process of ontology merging

Noy and Musen, Protégé, 2000 ontology merging algorithm. The algorithm is identified that matches the class name is a starting point, to be able to automatically update on this basis to discover and resolve conflicts when merging the repeated iteration of the implementation of programs.

These tools, OntoMorph Chimaera system uses description logic-based methods, partial description of the merged ontology, for example: only describes the inclusion relationship between the test terminologies. These methods do not provide a global structure for the merged body of to describe the program, as is follows by equation 4.

$$L(x(k),u(k)) = (R(k) - x(k))^T (R(k) - x(k)) \tag{4}$$

3 Rough Concept Lattice Isomorphic Merging Model

The concept lattice (Concept of Lattice) is a concept hierarchy R.Wille such as binary relations, is an effective tool for data analysis and rule extraction. Extraction method based on concept lattice rules has been widely used; particularly the use of knowledge discovery in databases is quite mature. But the traditional concept lattice based on data pre-processing and construction methods used to make redundancy rules extraction based on concept lattice through the large, too sensitive to noise.

While proposed by Z.Pawlak, rough set theory to deal with vague, inaccurate or incomplete, a new tool in maintaining the same premise of the classification capability of information systems, knowledge reduction, export issues decision-making or classification rules [3]. Pawlak rough set theory can the data set reduction, but the over-reduction, resulting in the loss of necessary information, so that the generalization ability of the system weakened. Reduction based on variable precision rough set theory knowledge can effectively prevent the loss of necessary information, set the parameter β to increase the necessary redundancy, so that the generalization capability, fault tolerance, noise suppression have a very good performance.

Variable precision rough set of β-reduction ideas used in the form of the background distribution under the reduction is proposed based on variable precision rough set concept lattice structure. Its core idea is to first preprocess for the formal context, and then using the improved based on variable precision of β, the distribution of attribute reduction algorithm to the raw data to form the background attribute reduction of concept lattice, redundancy in the form of background knowledge as well as the noise is excluded, and finally, in the form of background reduction combined with improved rules for the algorithm to construct the concept lattice. Due to the variable precision rough set β-redundancy in the knowledge of the distribution under the reduction can greatly simplify form background the basis of the concept lattice can be effective in reducing the number of nodes in the grid, so the reduction in the form of background and enhance noise immunity.

The approximation reduction written by $RED^{\beta}(C,D)$, its definition is: to the given value $\beta \in (0.5,1]$, set $Z \in E(D)$, $P \subseteq C$ meet the following two conditions.

$$\gamma^{\beta}(C,D) = \gamma^{\beta}(RED^{\beta}(C,D),D) \tag{5}$$

General approximation algorithm is to gradually examine and delete each condition attribute a_i in decision-making according to the definition of approximation reduction, if the total number of objects with decision-making unchanges after deleting, that is:

Firstly, according to the different classification quality $\gamma^{\beta}(C,D)$ based on the condition attribute set C; define its β interval β^{C}, denote as equation 6.

$$\gamma^{\beta}(C,D) = \gamma^{\beta}(RED^{\beta}(C,D),D) \tag{6}$$

Iteration idea of the algorithm: first set reduction attribute set is empty; then add attribute $\{c_j\}$ $(j = 1,2,\ldots,|C|)$ to the attribute set, ie $R_1 = \{c_j\}$, $\{c_j\}$ has a minimum $\Delta\gamma^{\beta}_{P\cup C}(\{c_j\},D)$ value, and then add attribute continually, denote as R_2 , then as is shown by equation 7.

$$R_i = R_{i-1} \cup \{c_j\} \rightarrow \min(\Delta\gamma^{\beta}_{P\cup C}(R_{i-1} \cup \{c_j\},D)), c_j \in C - R_{i-1} \tag{7}$$

If L (K_1) and L (K_2) are two extension independent domain concept lattices, define L (K_1) $\cup$L (K_2) is concept lattice L, if L meets:

(1) one lattice node C_l of $L(K_l)$ and one lattice node C_2 of $L(K_2)$, make $C_3=C_1+C_2$, if any lattice node C'_1 bigger than C_1, has no C'_1 equals to or less than C_3; Likewise, any lattice node C'_2 more than C_2, are not has C'_2 equals to or less than C_3, then $C_3 \in L$.

(2) for any lattice node C_1 of L (K_1), if there is no lattice node equal to or less than C_1 in $L(K_2)$, then $C_1 \in L$.

(3) for any lattice node C_2 of L (K_2), if there is no lattice node equal to or less than C_2 in $L(K_1)$, then $C_2 \in L$.

For the formal contexts $K_1 = (G, M_1, I_1)$ and $K_2 = (G, M_2, I_2)$ of the same object domain, if $M_1 \subseteq M$, $M_2 \subseteq M$, $M_1 \cap M_2 = \varnothing$, then says K_1 and K_2, L (K_1) and L (K_2) were connotation independent; If $M_1 \subseteq M$, $M_2 \subseteq M$, $M_1 \cap M_2 \neq \varnothing$, for any $g \in G$ and arbitrary $m \in M_1 \cap M_2$ meet $gI_1m = gI_2m$, it says K_1 and K_2, L (K_1) and L (K_2) are respectively connotation consistent.

A given background (G, M, R), for objects and attributes can be defined according to weight, the following two equivalence relations were equation8 and equation9.

$$\text{value}(g) = \text{BinToDec}(a_1 a_2 \cdots a_l) \tag{8}$$

$$\text{ValueSequence}(A) = \left\langle \text{value}(g_1), \cdots, \text{value}(g_i), \text{value}(g_{i+1}), \cdots, \text{value}(g_s) \right\rangle \tag{9}$$

The combined form of background, after the variable precision rough set β, under the distribution and the corresponding discernibility matrix to form the background of table attributes reduction; Reduction of the formal context, making the grid algorithm concept lattice, and in accordance with the combination of ontology and the concept lattice, the concept lattice conversion cost of the body, that is, the merged ontology.

4 Ontology Merging Based on RFCA Isomorphic Merging

The paper aims at different scale ontology, takes different approaches. For the lightweight ontology uses the ontology merging technology of RFCA-based for ontology merges; for the heavyweight ontology, the project proposes the ontology merging methods based on the rough concept lattice isomorphism generated.

Merger process: first of all, for any ontology context O, it can be decomposed into the context O with less order of attributes; Next, in order to get any context K1, system check whether there are some isomorphism context with it in the context library, if so, set to K2, and deposit the mapping between them into the mapping library, otherwise fill K1 into the context library, build the concept lattice B(K1)and fill it into the concept lattice library; Then, form ontology according to the user needs, and use library mapping and isomorphism lattice B(K2), isomorphism generate B(K1), then, use the attribute joint distribution mapping method of multi-strategy merge all the sub-ontology B(Oi), in the end, get the final ontology B(O) by the ontology merging methods based on FCA.

Carry on theβ- upper and lower distribution attribute reduction of VPRS to the sub-ontology context. Each limited context can be transformed into a reduction form without changing its corresponding concept lattice structure, and this reduction form is exclusive (in the premise of isomorphism), that is the standard context.

Located in the original concept lattice inf(L) elements C1, C1 and add special handling properties of the object x * set f (x *) the intersection of computing f(x*)∩Intent(C1); then the lattice L the content of the remaining elements of the set content from small to large number of elements to sort, followed by removing elements and C_j, for computing Intent(C_j)∩f(x*), to determine the appropriate type C_j, for appropriate action. Sorting and use of database error handling, reducing the number of search and determine the process to improve the performance of the algorithm, thus forming the information search result as shown in figure 2.

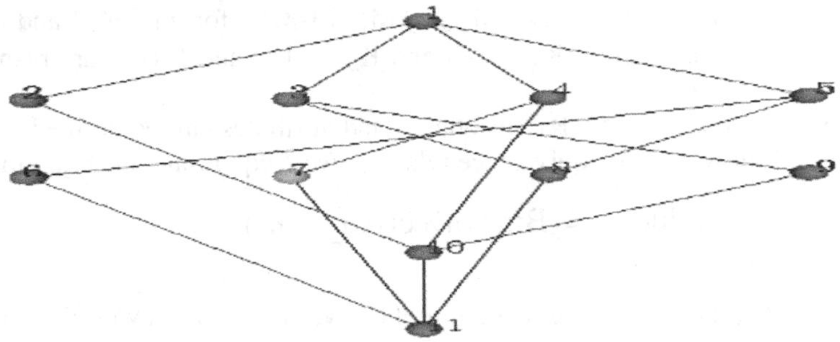

Fig. 2. The result of Ontology merging based on RFCA isomorphic merging

According to concept lattice of Hj derived the ontology structure of Oi, and in order to get its ontology B (Oi). For the sub-ontology pairs in the two systems which is to be merged, the project proposes attribute joint distribution mapping method of multi-strategy and combines with the RFCA-Merge method ti merge all the sub-ontology B (Oi) to get the final ontology B(O).

5 Summary

Ontology merging is to be mapped, depending on the application need to perform mapping combined ontology, and after the implementation process. This paper presents ontology merging based on rough concept lattice isomorphic merging.

References

1. Shamsfard, M., Barforoush, A.A.: Learning ontologies from natural language texts. Int'l Journal Human-Computer Studies 60(l), 17–63 (2004)
2. Missikoff, M., Navigli, R., Velardi, P.: Integrated Approach for Web Ontology Learning and Engineering. IEEE Computer 35(11), 60–63 (2002)
3. Choi, N., Song, I.Y., Han, Y.: A survey on ontology mapping. SIGMOD Record 35(3), 34–41 (2006)

An Improved Electric Vehicle Regenerative Braking Strategy Research

Qingsheng Shi[1,*], Chenghui Zhang[2], and Naxin Cui[2]

[1] College of electrical Engineering, Henan University of Technology, 450001, Zhengzhou, China
[2] School of Control Science and Engineering, Shandong University, 250061, Jinan, China
sdustone@gmail.com

Abstract. How to protect the battery from overcharging and recover regenerative energy effectively are the key issues during the braking control process for electric vehicle. Among the existed braking control strategies, most of them didn't consider the battery overcharging problem. according to vehicle's braking force safety requirements on wheels and battery state of charge(SOC), both of a braking force distribution controller (BFDC) and a braking force regulator (BFR) are designed respectively, followed by the proposal of a practical electric vehicle regenerative braking strategy that considering battery SOC. Finally, experiments are carried on ADVISOR simulation platform, and results show that the vehicle obtains a good anti-overcharging ability.

Keywords: Electric vehicle, state of charge, regenerative braking, anti-overcharging, optimal control.

1 Introduction

Electric vehicles, as the green and non-polluting transport means, have shown an accelerating development trend throughout the world [1-2]. However, short driving range is still the main bottleneck that hinders their development. Thanks to the specific configuration of energy storage system, electric vehicles can recover kinetic and potential energy in the process of regenerative braking, which provide a positive solution for improving the vehicle driving range [3]. Therefore, it is of great significance to improve energy efficiency and increase continuing driving mileage by designing effective regenerative braking strategy.

At present, the regenerative braking control strategies of electric vehicles can be categorized into three groups. The first is ADVISOR distribution strategy based on the vehicle speed [4]. The second is based on the ideal wheels brake-force distribution [3]. The third is based on optimal regenerative energy [5]. For the first group, the brake force proportions of different wheels in the total required brake force are adjusted according to vehicle speed, which highly relies on experience and easily

* Corresponding author.

D. Jin and S. Lin (Eds.): Advances in CSIE, Vol. 2, AISC 169, pp. 637–642.
springerlink.com

brings big control error. For the second group, short braking distance and good direction stability could be achieved, however, the regenerative energy is relatively small. For the third group, the regenerative energy can be recovered as many as possible while meeting the basic braking requirements, which has gradually become the mainstream design method. Recently, the performance simulation for a hybrid electric vehicle equiped with an EMB system is conducted considering battery overcharging [6]. When the SOC is above 80%, it does not perform regeneration. But, as to electric vehicle, battery is the main energy storage system and larger in size, so, the regenerative energy need not prohibit but just limit would be better if the SOC is above 80%. Therefore, when the battery state-of-charge is higher than 0.8, it is necessary to design appropriate regenerative braking strategy in order to effectively protect the battery from overcharging during the braking process.

Based on the analysis of vehicle braking force safety requirements and the state-of-charge constraint of the battery, a practical regenerative braking force distribution strategy considering the SOC of battery is proposed. Finally, simulation experiments are carried out to validate the performance of the proposed strategy on energy consumption rate and the anti-overcharging ability..

2 Novel Regenerative Braking Strategy Considering SOC

According to the braking strength, the electric vehicle braking mode can be classified into three different types:1) regenerative braking mode, which means the regenerative braking energy can be totally recycled if the brake strength is less than 0.1; 2) friction braking mode, which means the emergency braking occurs and a quick braking is required if the braking strength is greater than 0.7; 3) compound braking mode, which includes the regenerative braking and friction braking, if the brake strength is between 0.1 and 0.7. Therefore, it is crucial to properly achieve the braking force distribution in compound braking mode.

Electric vehicle regenerative braking strategy considering battery SOC can be designed as follows: first, a braking force distribution controller is designed according to the braking safety requirement, braking strength z and required braking force F_{req} are used to determine the distribution ratio of braking forces; then, braking force regulator is designed to adjust the braking forces output from braking force distribution controller.

(1) Braking force distribution controller design

According to the analysis in section III, the range of braking force distribution can be depicted curve *OABGFO* in Fig.1.

At present, hydraulic proportional valve distribution line (line *OAB* shown in Fig.1) is usually adopted to achieve braking force distribution in modern fuel vehicles. It is used as friction braking force distribution line in this paper and its equation is presented below:

$$OA:\ F_{u2} = \frac{1-\beta}{\beta} F_{u1} = 0.498 F_{u1},\ (0<z<0.35) \tag{1}$$

$$AB:\ F_{u2} = 0.247 F_{u1} + 667,\ (z>0.35) \tag{2}$$

where, β is 0.667 when $0<z<0.35$. The β in eq.(1) is just corresponding to certain hydraulic proportional valve. So, it's not a constant value, different hydraulic proportional valve have their own values.

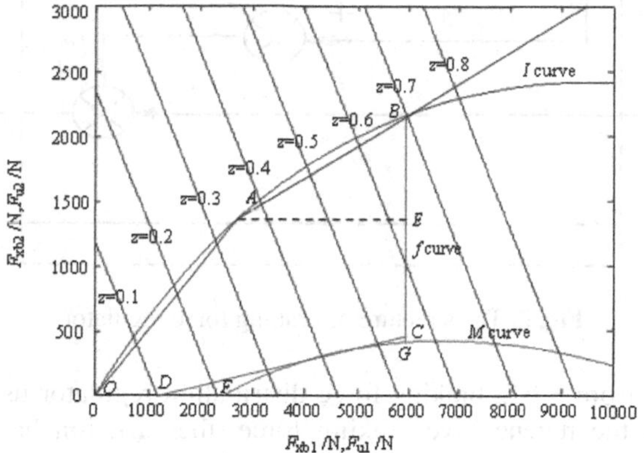

Fig. 1. Sketch of braking force distribution of wheels

In order to effectively avoid front wheels' lock, *M* curve's tangent line *CD* is used to replace *M* curve itself. It forms the composite braking force distribution line with *f* curve. Thus, sketch of braking force distribution of wheels can be achieved.

From Fig.3, the *f* curve and line *CD* are taken as the braking force distribution curve between the front wheels and rear wheels, and the line *OAB* as the braking force distribution curve between the front friction braking force and the rear friction braking force.

Taking certain front drive electric vehicles on a dry asphalt road with the adhesion coefficient of 0.7 as an example, the distribution ratio of the braking forces under different brake strengths can easily be derived. Thus, if the braking strength and total required braking force at time *t* is known, the different braking forces can be easily calculated.

(2) braking force distribution regulator design

To ensure the safety of battery, the braking forces output from the braking force distribution controller should be adjusted by a braking force distribution regulator. Fig.2 shows the structure of braking force distribution regulator. The inputs of braking force regulator are the outputs of braking force controller, and the outputs of braking force distribution regulator are defined as followings: F'_{re} is the adjusted regenerative braking force, F'_{u1} is the adjusted front wheels friction braking force, F'_{u2} is the rear wheels friction braking force, the adjustment function can be expressed as:

$$w = f(SOC) = \begin{cases} 0, & SOC \geq 0.9 \\ -10 \times SOC + 9, & 0.8 \leq SOC < 0.9 \\ 1, & SOC < 0.8 \end{cases} \tag{3}$$

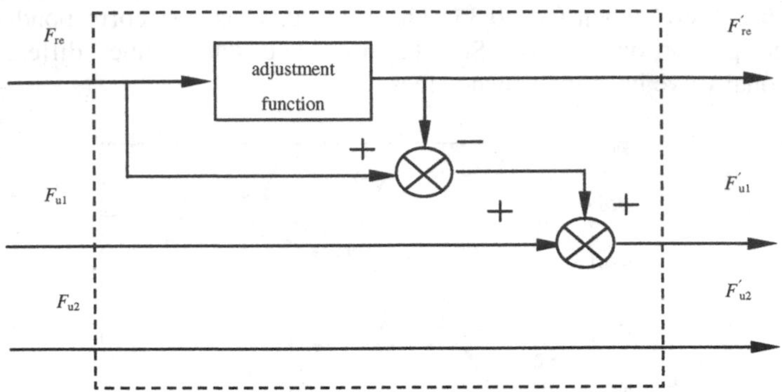

Fig. 2. The structure of braking force regulator

As can be seen from Fig.2, braking force distribution regulator use an adjustment function to adjust the regenerative braking force, then friction braking forces are adjusted through the difference between the unadjusted regenerative braking force and adjusted regenerative braking force.

3 Simulation and Analysis

The advanced vehicle simulator (ADVISOR) is used for simulation study. ADVISOR employs a combined forward/backward facing approach for the vehicle performance simulation. Five typical drive cycles are taken as test cycles: US06, UDDS, JA1015, FTP and HWFET. The main simulation parameters of electric vehicles are shown in Table 1.

Table 1. Simulation parameters of Electric Vehicle

Whole-car parameters	mass (kg)	1159
	g (m/s^2)	9.8
	L (m)	2.6
	b (m)	1.56
	h_g (m)	0.5
Motor parameters	type	Induction motor
	Peak power (kW)	75
	Maxim speed (r/min)	10000
	Rated torque (Nm)	271

Table 2 and Table 3 show the electric vehicles energy consumption rate and anti-overcharging ability under different cycles, respectively. The increase can be calculated by:

$$\Delta = \frac{x_2 - x_1}{x_1} \times 100\% \tag{4}$$

where, Δ is the increase after using the novel strategy, x_2 is the performance parameter using Strategy II, and x_1 is the performance parameter using Strategy I.

In Table 2, energy consumption rate means electric vehicle's consumed energy in per 100km. Generally, it can be calculated using the following formula:

$$ECR = 1.1 \times 10^{-7} \frac{E}{S} \tag{5}$$

where, S is the travel distance, km; E is the consumed energy during the vehicle's driving cycle, J; and 1.1×10^{-7} is the unit conversion coefficient.

It can be seen from Table 2, using the strategy proposed in this paper, the vehicle's energy consumption rate incerases 0.45%, 1.22%, 6.83%, 0.61% and 0.27% compared to the regenerative braking strategy that not considering the SOC. The energy consumption rate in drive cycle JA1015 increases relatively significant. This is because vehicle's frequent start-stop operation and short drive distance in drive cycle JA1015, which make battery SOC relatively high and restrain the braking energy recover. But overall, the average improvement on energy consumption rate is not great.

In Table 3, anti-overcharging ability is defined as the deviation between the terminal SOC and the initial SOC of battery, as shown in Eq.(6):

$$\eta = \sqrt{\frac{SOC_0 - SOC_t}{SOC_0}} \times 100\% \tag{6}$$

where, η is the anti-overcharging coefficient, the bigger its value is, the stronger the anti-overcharging ability is in the same condition; SOC_0 stands for the initial SOC of battery, here it is set to 0.9; SOC_t is the terminal SOC when a certain drive cycle is over.

It can be seen from Table 3, using the proposed strategy, the improvement is 2.49%, 5.48%, 10.55%, 3.71% and 0.9%, respectively. Among them, the vehicle energy consumption rate in drive cycle UDDS, JA1015 and FTP improve relatively big. This is because these cycles all belong to city cycles, in which vehicles start and stop frequently. When the battery is to be overcharged, the braking force regulator will timely adjust the regenerative braking force, thereby enhance the vehicle's anti-overcharging ability.

In short, after using the proposed strategy, vehicle's anti-overcharging ability improves significantly despite the energy consumption rate increase a bit.

Table 2. Energy consumption rate comparison

Drive cycle	energy consumption rate (kWh/100km)		Increase
	Strategy I	Strategy II	(%)
US06	22.2	22.3	0.45
UDDS	16.4	16.6	1.22
JA1015	16.1	17.2	6.83
FTP	16.5	16.6	0.61
HWFET	14.8	14.84	0.27

Table 3. Anti-overcharging ability comparison

Drive cycle	anti-overcharging ability (%)		Improvement
	Strategy I	Strategy II	(%)
US06	3.45	3.54	2.49
UDDS	2.83	2.98	5.48
JA1015	1.64	1.82	10.55
FTP	3.45	3.58	3.71
HWFET	3.33	3.36	0.9

Acknowledgments. This work was supported and funded by the Natural Science Research Program of the Education department of Henan province (2011A480001), the development funding of Zhengzhou City (20110312) and the doctoral funding of Henan University of Technology (2009BS061).

4 Conclusion

In this paper, a novel electric vehicle regenerative braking strategy is designed while considering battery SOC. In the strategy, a braking force distribution controller and a braking force regulator are designed respectively. The simulation results show that the novel strategy further complements the existing electric vehicle regenerative braking distribution strategy not considering battery SOC, and makes vehicle obtain a better anti-overcharging ability.

References

1. Chan, C.C., Chau, K.T.: Modern Electric Vehicle Technology. Oxford University Press, USA (2001)
2. Shi, Q., Zhang, C., Cui, N.: Estimation of battery state-of-charge using v-support vector regression algorithm. The International Journal of Automotive Technology 9(6), 759–764 (2008)
3. Gao, Y.: Electronic Braking System of EV And HEV–Integration of Regenerative Braking, Automatic Braking Force Control And ABS. SAE Technical Paper Series, 01–2478 (2001)
4. Wang, Z., Huang, M., Deng, Y.: Modelling and Simulation of the Regenerative Braking System for Electrical Vehicles. Journal of Chinese Wuhan University of Technology (Information & Management Engineering) 23(4), 102–105 (2001)
5. Shi, Q., Zhang, C., Cui, N.: Novel Electric Vehicle braking force distribution strategy in Regenerative braking process. Transactions of China Electrotechnical Society 22(s2), 125–129 (2007)
6. Ahn, J.K., Jung, K.H., et al.: Analysis of a regenerative braking system for hybrid electric vehicles using an electro-mechanical brake. IJAT 10(2), 229–234 (2009)
7. USABC Electric Vehicle Battery Test Procedure Manual, Revision 2. DOE/ID-10479. US Department of Energy, Washington, DC (1996)

An Improved Image Encryption Method Based on Total Shuffling Scheme

Yan Shen[1], Guoji Zhang[1], Xuan Li[2], and Qing Liu[2]

[1] School of Science, South China University of Technology, Guangzhou, 510640, China
[2] School of Computer Science and Engineering, South China University of Technology, China
shenyan_scut@foxmail.com, magjzh@scut.edu.cn

Abstract. In this paper, an improved image encryption method based on permutation-diffusion architecture and total shuffling scheme is proposed. In the permutation process, the *P-box*, which makes every image shuffle the position of pixels by its own *P-box*, depends on the plain-image. In the diffusion process, the keystream is related to the plain-image directly and a more secure feedback is employed to change the number of iterations of the chaotic map. Moreover, a reverse diffusion process is added to protect the final cipher-image. Our experimental results of statistical analysis, information entropy analysis and sensitivity analysis demonstrate that the improved algorithm is more secure and reliable than the original one and it can avoid the category of attacks similar to Ref. [5].

Keywords: Improvement, Security, Image encryption, Cryptanalysis.

1 Introduction

With the ever-increasing requirement of information transmission on Internet, the security of digital images transmission gains more and more attentions from both academia and enterprises [1~3]. Recently, a novel image encryption method was proposed by Zhang *et al.* [4]. This method holds a lot of advantages, like large key space, low encryption time, high key sensitivity and so on. However, there are still some flaws with it, because of these flaws, Wang *et al.* [5] broke this method via a successful chosen plaintext attack. In this paper, we propose an improved encryption cryptosystem on the basis of Zhang *et al.*'s scheme. The improved encryption method is able to address the flaws pointed out in Ref. [5] and to resist the similar attacks in Ref. [5].

2 The Improved Encryption Method

In this section, we propose an improved method base on the original one. The improved method can address the flaws mentioned above and resist a class of attack similar to [5], furthermore, the overall safety factor is also increased.

D. Jin and S. Lin (Eds.): Advances in CSIE, Vol. 2, AISC 169, pp. 643–650.
springerlink.com

A skew tent map [4]

$$F(x)=\begin{cases} x/p, & if\ x\in(0,p) \\ (1-x)/(1-p), & if\ x\in(p,1] \end{cases} \tag{1}$$

is adopted, where $x\in(0,1)$ is the state of the system and $p\in(0,1)$ is the control parameter. Transform an image of size $M\times N$ to a one-dimensional vector $P=\{p_1,p_2,\cdots,p_{MN}\}$, given $y_1\in(0,1)$, $y_2\in(0,1)$, $p\in(0,1)$ and $j\in[0\ ,\ MN-1]$, j is a position of P.

Step 1. Set the initial value $x_1=\dfrac{p_j\times y_1+0.01}{257}$, iterate the skew tent map $x_{i+1}=F(x_i)$ by using Eq. (1) for L times to get rid of transient effect, then continue to iterate the skew tent map for MN times, we can get a sequence X, where L is a constant;

Step 2. Remove the jth value from sequence X and sort X to get an index order sequence $T^*=\{t^*_1,t^*_2;\cdots,t^*_{MN-1}\}$. Obtain the P-box T by T^* with the following formula:

$$T=\begin{cases} t_n=t^*_n, & 1\le n\le j-1 \\ t_j=j, & n=j \\ t_n=t^*_{n-1}+1, & j+1\le n\le MN \end{cases} \tag{2}$$

After that, shuffle the value in P by the P-box to get $P'=\{p'_1,p'_2,\cdots,p'_{MN}\}$, where $p_{t_i}=p'_i\left(i=1,2,\cdots,M\times N\right)$;

Step 3. Let $i\leftarrow 1$, if i equals to j, go to step 4, otherwise, obtain an 8-bit random code d_i according to the following formula:

$$d_i=\mathrm{mod}(floor(x\times 2^{48}),256) \tag{3}$$

where x is the current state value of the skew tent chaos system. Then compute the corresponding pixel data:

$$c_i' = p_i' \oplus \mathrm{mod}(p_{i-1}' + d_i, 2^8) \tag{4}$$

Here $\oplus$ is bitwise XOR operator. We set the initial value $p'_0 = p_j'$, then go to step 5;

Step 4. Compute c_j' with the following formula:

$$c_j' = c_{j-1}' \oplus p_j' \oplus \mathrm{mod}(floor(y_2 \times 2^{48}), 2^8) \tag{5}$$

Step 5. Compute the rotation frequency k according to the following formula:

$$k = 1 + \mathrm{mod}(c_i', d_i) \tag{6}$$

Then, iterate the skew tent map $x \leftarrow F(x)$ for k times;

Step 6. Let $i \leftarrow i+1$, return to step 3 until i reaches MN;

Step 7. Iterate the $F(x)$ with initial value y_2 $L+MN$ times to produce a random value sequence Y, remove the first Lth value in Y to get rid of the transient effect and denote the remaining value as $Y=\{Y_1, Y_2, \cdots, Y_{MN}\}$, then execute the reverse diffusion process with given c_{MN+1}:

$$c_i = c_{i+1} \oplus c_i' \oplus \mathrm{mod}(floor(Y_i \times 2^{48}), 2^8), i = M \times N, \cdots, 2, 1 \tag{7}$$

$c=\{c_1, c_2, \cdots, c_{M\times N}\}$ is the final encrypted vector, c_{MN+1} should be provided to cipher out of sequence $c=\{c_1, c_2, \cdots, c_{M\times N}\}$.

3 Performance Test and Analysis for the Improved Cryptosystem

3.1 Key Space Analysis

In our improved vision, since the permutation process is irrelevant to the diffusion process, the key space consists of the cipher keys in both processes. The keys include $y_1 \in (0,1)$, $y_2 \in (0,1)$, $p \in (0,1)$ and the position $j \in [0, MN-1]$. For a $M \times N$ gray image, the key space is bigger than $M \times N \times (2^{52})^3 = M \times N \times 2^{156}$. According to the IEEE floating-point standard [6]. The computational precision of the 64-bit double-precision numbers is 2^{-52}.it is obviously that such a big key space can provide a sufficient security against brute-force attacks [7, 8].

3.2 Statical Analysis

3.2.1 Histograms of Cipher-Image

Figs. 1 and 2 depict the histograms of the plain-image "Lenna" and the corresponding cipher-image encrypted by the improved method with key:

$$\{y_1 = 0.123456789, y_2 = 0.246813579, p = 0.654321, j = 2222\} \tag{8}$$

The histograms of the encrypted images are nearly uniform, which can well protect the information of the image and withstand the statistical attack.

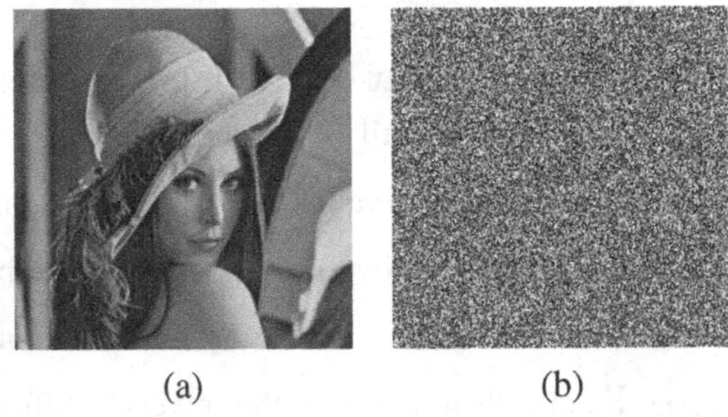

Fig. 1. (a) Plain-image "Lenna"; (b) Cipher-image using improved method.

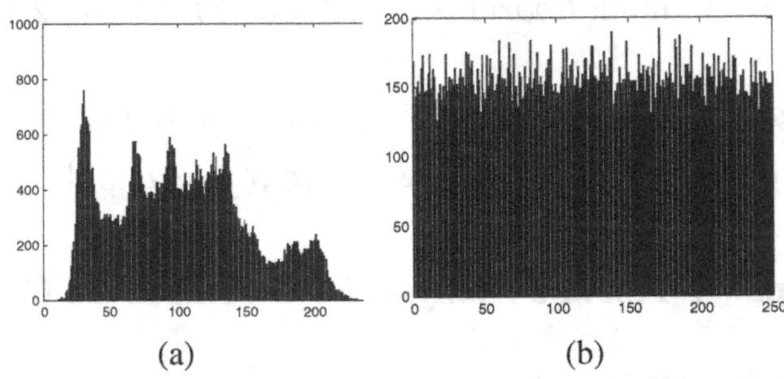

Fig. 2. Histogram (a) Plain-image "Lenna"; (b) Cipher-image using improved method.

3.2.2 Correlation of Two Adjacent Pixels

To test the correlation between two-adjacent pixels in plain-image and cipher-image, we randomly select 1000 pairs of two-adjacent pixels (in vertical, horizontal, and diagonal direction) from plain-image and cipher-image, and calculate the coefficient of each pair by using the following formulas as in [4]:

$$E(x)=\frac{1}{N}\sum_{i=1}^{N}x_i \tag{9}$$

$$D(x)=\frac{1}{N}\sum_{i=1}^{N}(x_i-E(x))^2 \tag{10}$$

$$\mathrm{cov}(x,y)=\frac{1}{N}\sum_{i=1}^{N}(x_i-E(x))(y_i-E(y)) \tag{11}$$

$$r_{xy}=\frac{\mathrm{cov}(x,y)}{\sqrt{D(x)}\sqrt{D(y)}} \tag{12}$$

where x and y are grey-scale values. The correlation of two horizontally adjacent pixels in the plain-image and cipher-image "Lenna" are shown in Fig. 3.

Table 1 proves that the our encryption scheme satisfies zero co-correlation, which is a private high-level security. Compared with the original algorithm, the improved one shows a better performance.

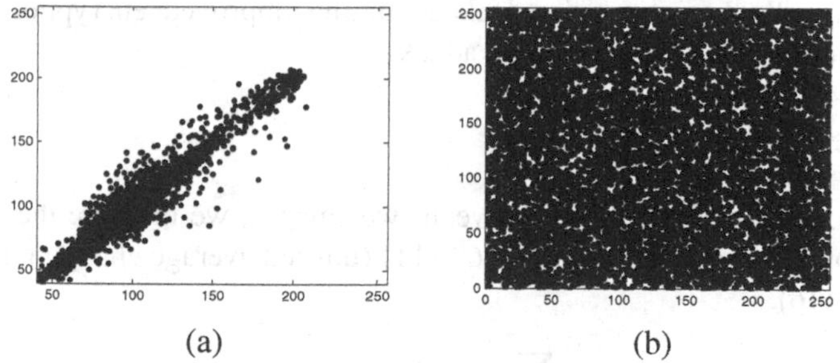

Fig. 3. Correlation of two horizontally adjacent pixels: (a) Correlation of the plain-image; (b) Correlation of the cipher-image using improved method.

Table 1. Correlation coefficients of two adjacent pixels in the plain-image and cipher-image

Direction	Plain-image	Original scheme	Improved scheme
Horizontal	0.9404491	0.000368341	0.00065647
Vertical	0.9532125	0.002605824	0.00013100
Diagnal	0.9022331	-0.000159639	-0.00052320

3.2.3 Information Entropy Analysis

The entropy is the most outstanding feature of the randomness [9]. The entropies of the cipher-image created by the original and improved method is shown in Table 2 and Table 3 shows that the entropies of improved algorithm compared with the existing algorithms mentioned in Ref. [4] is better.

Table 2. Entropy value for cipher-images

Cipher-image	Original scheme	Improved scheme
Lenna	7.9972	7.9973
Parrot	7.9973	7.9974
Couple	7.9966	7.9970

Table 3. Entropy value for cipher-image of "Lenna"

Methods	Entropy value
Baptista's	7.9260
Wong's	7.9691
Xiang's	7.9951
Sun's	7.9965
Original	7.9972
Improved	7.9973

The obtained value is very close to the theoretical value 8 and the entropy of the improved method is bigger than the original one, it means that the information entropy of our improved algorithm is better than the original one and many existing algorithms. In all, the information leakage in the improved encryption scheme is negligible and it is secure upon entropy attack.

3.3 Sensitivity Analysis

In order to test the different range between two images, we measure the *NPCR* [10] (number of pixels change rage) and *UACI* [11] (unified average changing intensity) in Eq. (15) and (16).

$$NPCR = \frac{\sum_{i,j} D(i,j)}{W \times H} \times 100\% \tag{13}$$

$$UACI = \frac{1}{W \times H} \sum_{i,j} \frac{|c_1(i,j) - c_2(i,j)|}{255} \times 100\% \tag{14}$$

Where C_1 and C_2 are two images with the same size $(W \times H)$. If $c_1(i,j) = c_2(i,j)$ $D(i,j)$=1, otherwise $D(i,j)$=0.

We encrypted a gray image "Couple" with size 256×256 using *key(15)*, and then encrypted the same plain-image with a slightly different keys as below:

$$\{y_1' = y_1 + 10^{-10}, y_2, p\} \quad (16)$$

$$\{y_1, y_1' = y_2 + 10^{-10}, p\} \quad (17)$$

$$\{y_1, y_2, p' = p + 10^{-10}\} \quad (18)$$

Table 4. *NPCR* and *UACI* among cipher-images with *key (19)* and with *key (16)~(18)*

key	*NPCR(%)*	*UACI (%)*
key(16)	99.6292	33.5481
key(17)	99.6078	33.5631
key(18)	99.6414	33.3653

We can find a slight change in key makes the cipher-image totally different and more than 99% pixels in cipher-image change their gray rate when the key change 10^{-10}. The improved algorithm provides high key sensitivity.

4 Conclusion

In this paper, we proposed an improved image encryption method based on total shuffling scheme. Experiment results and extensive security and performance analysis on the proposed image encryption scheme are carried out. The experimental results corroborate that the proposed improved scheme is suitable for the secure image storing and transmission.

References

1. Matthews, R.: On the derivation of a chaotic encryption algorithm. Cryptologia 13, 29 (1989)
2. Fridrich, J.: Image encryption based on chaotic maps. Systems, Man, and Cybernetics, Computational Cybernetics and Simulation 2(1), 1105–1110 (1997)
3. Shannon, C.E.: Communication Theory of Secrecy Systems. Bell Telephone Laboratories, Murray Hill (1949)
4. Zhang, G.J., Liu, Q.: A novel image encryption method based on total shuffling scheme. Optics Communications 284, 2775–2780 (2011)

5. Wang, X.Y., He, G.X.: Cryptanalysis on a novel image encryption method based on a skew tent map. Optics Communications 284, 5804–5807 (2011)
6. IEEE Computer Society. IEEE standard for binary floating-point arithmetic, ANSI/IEEE Std (754) (1985)
7. Stinson, D.R.: Cryptography: Theory and Practice. CRC (2007)
8. Sun, F., Lu, Z., Liu, S.: Optics Communications 283, 2066 (2010)
9. De Santis, A.F., Lisaand Masucci, B.A.: Discrete Applied Mathematics Coding and Cryptography 154, 234 (2006)
10. Bluman, A.G.: Elementary Statistics: a Step by Step Approach. McGraw-Hill, Boston (1997)
11. Mao, Y., Chen, G., Lian, S.: International Journal of Bifurcation and Chaos 14, 3613 (2004)

The Study and Analysis of Fast Iterative Algorithms in Industrial Computerized Tomography Image Reconstruction

RuiFang Zhang

College of Mechanical and Control Engineering, Guilin University of Technology, 541004, Guilin, China
tdrf@sina.com

Abstract. Iterative Algorithms can obtain good reconstruction image for enough Iterative times in the technique ct image reconstruction. The order method of the Weighted Distance Orthogonal (WDO) is introduced in this paper to accelerate the Iterative reconstruction, which improves greatly the access mode of ART to projection data, based on the crucial effect that the access order of projection date has on reconstruction quality and speed in algebraic reconstruction technique. The method speeded up image reconstruction and improved the quality as well in large scale by using information relation of radial traversing pixels in consecutive orthogonal projection angles.

Keywords: image reconstruction, iterative algorithm, projection data, the order of the weighted distance orthogonal.

1 Introduction

Industrial Computerized Tomography (ICT) technology can obtain the object's internal fault information of the image in nondestructive condition; with the advantages of other nondestructive testing methods can not be replaced. The core technology of CT is in more than one observation angle to obtain the target object a series of projection data, through the technology of image reconstruction to obtain the target image [1]. At present, in the practical application of CT image reconstruction, image reconstruction method there are basically two kinds: analysis method Based on the Radon transform and series expansion method based on iteration theory [2-4]. This paper based on the Expectation Maximization (EM) algorithm for fast reconstruction as the goal, according to the projection data sorting method, proposed one kind ordered subsets reconstruction method based on the weighted distance orthogonal, The method is according to the projection data orthogonal, The convergence of required the iteration number minimum, fastest convergence speed, effectively improve the reconstruction speed of EM algorithm.

2 Realization of the OSEM Algorithm

The basic idea of the iterative method reconstruction is: Assuming an initial image, according to some optimization criteria will be in accordance with the hypothesis of

D. Jin and S. Lin (Eds.): Advances in CSIE, Vol. 2, AISC 169, pp. 651–657.
springerlink.com

images obtained by the projection information and the actual detected projection information to comparison, then the correction to a new image, so to gradually approach the true image., then correction to obtain a new image, so to gradually approach the true image.

OSEM is referred to the Ordered Subsets Expectation Maximization [5]. It gradually applied in the clinical image reconstruction process. OSEM iterative model are as follows:

$$x_j^{(n+1)} = x_j^{(n)} + \lambda_n \sum_{i \in s_t} \frac{r_{ij}}{\sum_{i \in s_i} r_{ij}} (p^i - R^i x^{(n)}) \tag{1}$$

3 Realization of the WDO-OSEM Algorithm

3.1 The Access Mode of Projection Data

The access mode of projection data is consider how to arrange the hyper plane of discharge order, made during projection adjacent to each hyper plane satisfy certain relations, thus speeding up the convergence rate. Many scholars have put forward different projection data access mode [6]. As shown in figure 1, The iterative process is a continuous projection process. Obviously, the two line included angle is large, then the convergence required of iteration is small; the last picture show, when the angle between two lines for 90°, the convergence speed is fastest.

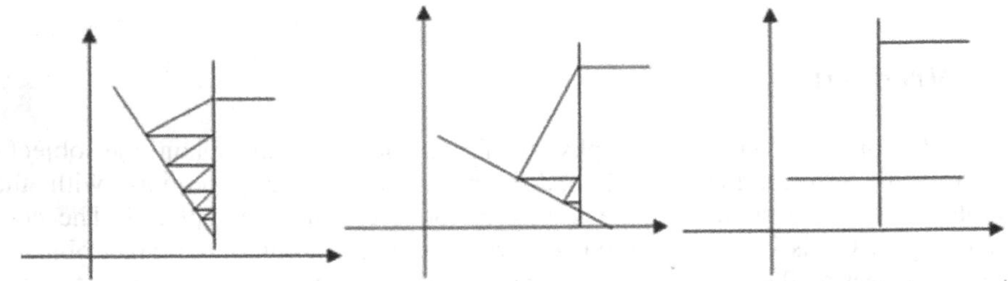

Fig. 1. The process of iterative

3.2 Weighted Distance Orthogonal (WDO) Sort Method

OSEM can well satisfy the above (formula 1) important condition, but ignore the importance of front and rear projection angle of orthogonal. As shown in Figure 1, the when the angle is 90°, the convergence speed is the quickest. Weighted distance orthogonal method in addition to meet the above conditions, also given before and after the choice of projection angles, introduce as follows:

Put $0 - 90^\circ$ range projection out into the array Λ, first projection Λ are sorted and then into the array Θ , Θ with Λ length is equal. From Λ Choose next projection, that is calculated in Λ each one p_p and "repulsion" between

p_q $(0 \le q \le Q-1)$ in Θ, Q means that the current Q projection angle sum. First Calculation the between distance of the projection, set p_p, p_q projection angle, respectively is p, n, the two projection p_p, p_q, can be calculated the between distance as

$$d_{pq} = \min(|p-n|, 90-|p-n|) \tag{2}$$

Then according to the distance size calculation p_q $(0 \le q \le Q-1)$ Applied p_q to the "repulsion" μ_p

$$\mu_p = \sum_{q=0}^{Q-1} w_q (45 - d_{pq}) / \sum_{q=0}^{Q-1} w_q \tag{3}$$

Power factor $w_q = (q+1)/Q$ is used to ensure that the newly accessible projection than the previously accessed projection has greater repulsive force, In addition the two projection distance d_{pq} is smaller, as during the repulsion is bigger, otherwise is smaller. Give priority to the repulsion small projection p_p. Then on the 0 - 360° range projection angle are arranged in groups as follows:

$\{p_1, p_1+90, p_1+180, p_1+270\}$, $\{p_2, p_2+90, p_2+180, p_2+270\}$, …… $\{p_n, p_n+90, p_n+180, p_n+270\}$

This projection selection sequence to ensure that within each group continuous selection projection orthogonal, while ensuring that the group between arranged are as uniform as possible.

3.3 Improvement of Reconstruction Speed

Weighted distance orthogonal each within groups beforc and after the projection angle difference is 90°. Ray rotated about center of rotation 90° and before and after through between the pixels has some relationship, in an angle of ray through the coordinates (x, y), the coordinates can be expressed as: $x = r\cos(a)$, $y = r\sin(a)$,

Rotary 90° the ray through the corresponding coordinates is $x' = r\cos(a+90) = -y$, $y' = r\sin(a+90) = x$, Rotary 180°, the ray through the corresponding coordinates $x' = r\cos(a+180) = -x$, $y' = r\sin(a+180) = -y$, Rotary 270°, the ray through the corresponding coordinates $x' = r\cos(a+270) = y$, $y' = r\sin(a+270) = -x$.

Set the image size is $s \times s$, Then the coordinate's points (x, y) corresponding to the pixel as the: $i = x + \frac{s}{2}$, $j = y + \frac{s}{2}$. When rotated 90°, corresponding to the pixel point: $i' = -y + \frac{s}{2} = s - j$, $j' = x + \frac{s}{2} = i$, When rotated 180° corresponding to the pixel point: $i' = -x + \frac{s}{2} = s - i$, $j' = -y + \frac{s}{2} = s - j$, When rotated 270° corresponding to the pixel point: $i' = y + \frac{s}{2} = j$, $j' = -x + \frac{s}{2} = s - i$, And these 4 angles corresponding pixel point weight r_{ij} are equal, So the sequence subsets reconstruction needs to know only one angle ray through the pixel information which can use this relationship directly drawn from other angle ray through the pixel information, thereby improving the speed of reconstruction. In addition to the 0-180° direction is uniformly distributed on the projection can be used for this method, but each group only taking the first two projections.

4 the Results Analysis of Simulation

Considering the particularity of the image processing, we selected which can conveniently realize the data visualization software MATLAB to simulation experiment, with Shepp-Logan head model (as shown in Figure 2) digital section as an example, There are many studies is use of the model for the image reconstruction simulation experiments and algorithm performance evaluation [7]. Model of section size is 128 x 128, gray level is 0 - 255.

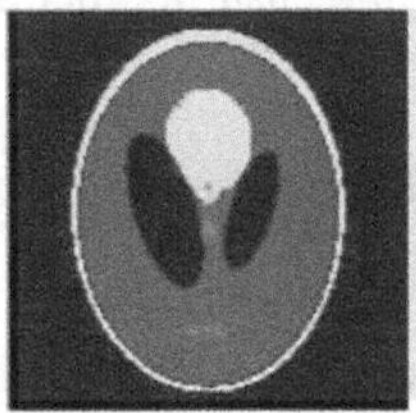

Fig. 2. Shepp-Logan head model

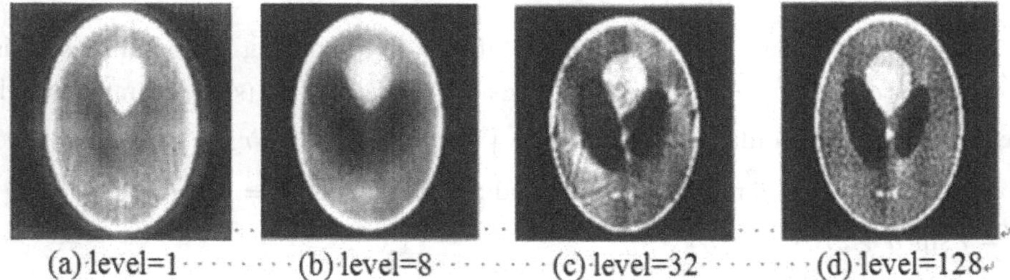

(a) level=1 (b) level=8 (c) level=32 (d) level=128

Fig. 3. The reconstruction of the OSEM first iteration

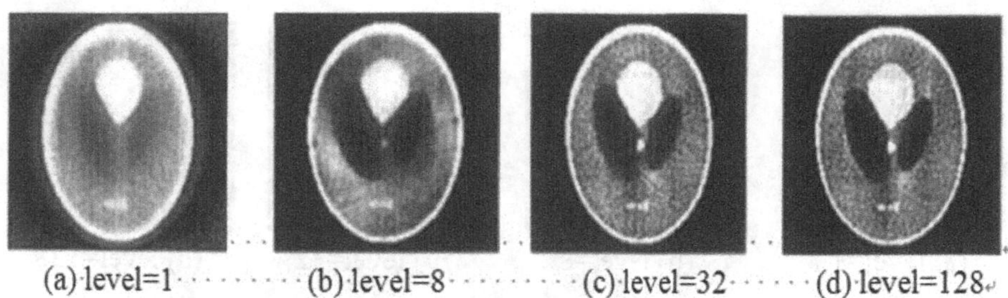

(a) level=1 (b) level=8 (c) level=32 (d) level=128

Fig. 4. The reconstruction of the WDO-OSEM first iteration

In this paper, using the projective geometry is the parallel beam projector; simulation experiment of projection data is within 180° to 180 directions of projection of the spacer, that projection data is 185×180 projection matrix.

Figure 3 and Figure 4 compares OSEM and WDO-OSEM algorithm the reconstruction in different subsets level after iteration. As can be seen in the low subset level, whether it is OSEM or WDO-OSEM iteration reconstruction image quality is not good, especially the intermediate (high frequency) is fuzzy; Relative to the former, in a subset of high level, two methods of reconstruction image quality are improved a lot, that there is a iterations, the subset number is greater, image reconstruction, high-frequency signal recovery is quicker. In the subset levels less than or equal to 8, the second projection direction seems to be slightly better than the first one, and when the subsets level greater than or equal to 32, second projection direction shows obvious advantages, Image quality has been greatly improved. By special order processing projection is beneficial image reconstruction, when the iterative to a certain extent, the two order of the projection image reconstruction is a very quit.

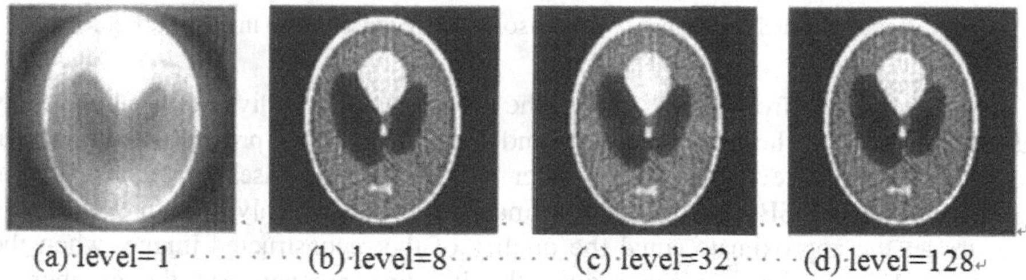

(a) level=1 (b) level=8 (c) level=32 (d) level=128

Fig. 5. The reconstruction of the OSEM 20 iteration

Figure 5 and Figure 6 compares OSEM and WDO-OSEM algorithm the reconstruction in different subsets level 20 iterations. As can be seen, the special order for low subsets level effect is small , in figure 5, subsets level reconstruction artifacts in images than in Figure 6 are slightly noticeable. But for the high subsets level, special order can get better image quality in a few iterations.

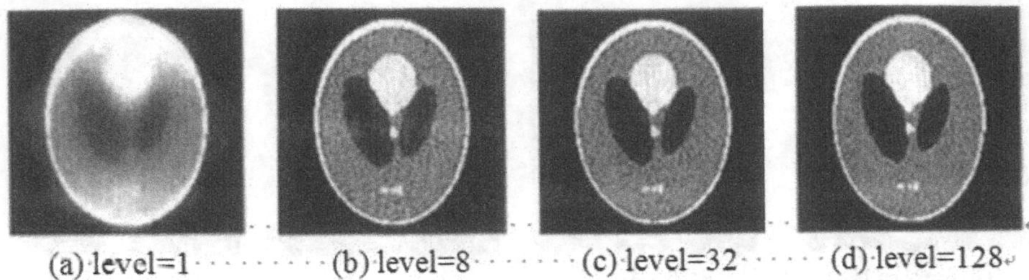

Fig. 6. The reconstruction of the WDO-OSEM 20 iteration

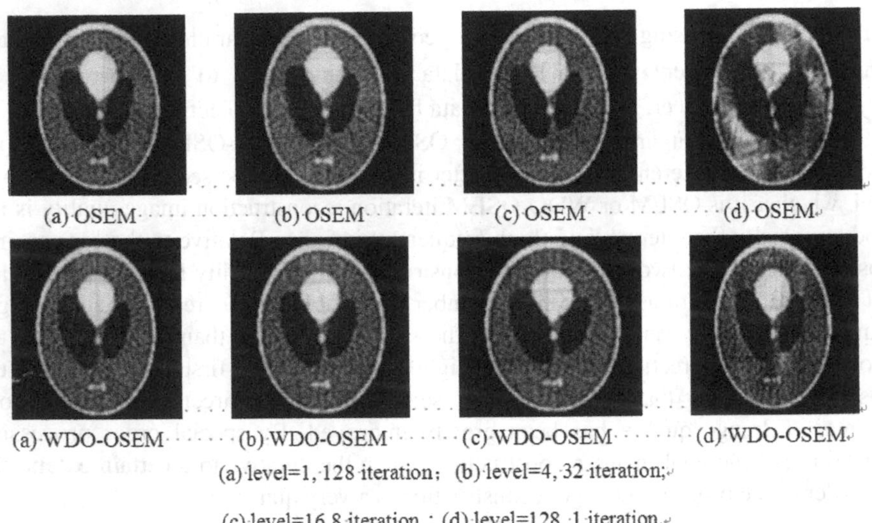

Fig. 7. Iterative matching number comparison after reconstruction in each subset level

As can been seen from Figure 7, both the projection data is divided into how many subsets, as long as the subset number and iteration number product determine, in matching (subsets level iteration number = 128) in the cases of reconstruction, OSEM and WDO-OSEM reconstruction image quality is roughly the same as. If you want to get the approximate equal the quality of the reconstructed image, when the subset number increases, it can reduce the iteration number. But the number of subsets is too large (L = 128), fewer iteration times matched to cases, the reconstructed image quality becomes bad. At the same time, the experimental results it can be seen from the above high subset level in the WDO-OSEM projection of early iterative reconstruction quality better than the OSEM.

6 Conclusion

CT image reconstruction accelerated has been a project and researchers active exploration target, from the quality of the reconstructed image can be seen, the intervention of weighted distance orthogonal sorting method, the iterative reconstruction algorithm in the reconstructed image quality and the speed of reconstruction have been greatly improved.

References

1. Yang, J.S., Guo, X.H., Kong, Q., et al.: Parallel implementation of Katsevichs FBP algorithm. International Journal of Biomedical Imaging, Article ID 17463 (2006)
2. Basu, S., Bresler, Y.: O (N2 log2 N): filtered back projection reconstruction algorithm f or tomography. IEEE Transactions on Image Processing 9(10), 1760–1773 (2000)
3. Sun, X.N., Liang, X.Z., Liu, B.: Fast image reconstruction with Hakopian interpolation. Journal of Computer Aided Design & Computer Graphics 18(3), 451–455 (2006)
4. Wang, J., Sun, H., Zhang, S., et al.: Two dimensional image reconstruction from small amount of projection paths by using joint two grade neural network. Journal of Computer Aided Design & Computer Graphics 16(9), 1284–1288 (2004)
5. Yin, Y., Li, L.: Fan-beam image reconstruction algorithm OSEM and subsets. Research on CT Theory and Application 12(3), 1–8 (2003)
6. Zhang, S., Zhang, D., Li, S., et al.: Research on ART algorithm of fast image reconstruction. Computer Engineering and Applications 24(1), 1–3 (2006)
7. Hu, J.-C., Johnson Roger, H.: A helical cone beam algorithm for large cone angles with minimal over scan. In: IEEE Nuclear Science Symposium Conference Record, vol. 2, pp. 1013–1017. IEEE Inc., Norfolk (2002)

Design and Implementation of FPGA Network Simulation and Verification Platform

Keke Wang, Bo Yang*, Zhenxiang Chen, and Tao Sun

Shandong Province Key Laboratory for Network Based Intelligent Computing,
School of Information Science and Engineering, University of Jinan, Jinan, Shandong, China
hnwangkeke@163.com, yangbo@ujn.edu.cn

Abstract. Network development platform with certain data transmission format provides the bidirectional simulation incentive, which transmits from PCI interfaces to network interfaces and inversely from network interfaces to PCI interfaces. The platform defines the data format and user expansion function interfaces, and then makes it come true. The platform provides a high speed data accessing interface at level of Gbps rates for developers. Developers embed several modules on their own without concerning how to achieve network protocols at hardware level. In this way, they can concentrate their energy and time on data processing. At the same time, the platform gives plenty of support modules to accelerate simulation and verify.

Keywords: FPGA, network verify environment, development platform.

1 Introduction

With the computer network popularization and development, it is very important for the application of network products to guarantee the stability, reliability and fast listed on the electronic marker. It is one of the directions for logic circuit design development basing on the design of the platform [1-2].

FPGA network development platform is created a logical design core framework which bases on practical simulation results of the hardware circuit. This frame is easy to customize and expand, provide the expansion of the interface and developers can start second development network data processing module on it quickly but not concern about complex interface logic [3]. Logical modules which is developed in this platform can be quickly implemented in hardware circuit and shorten the logical circuit design cycle greatly.

2 Hardware Module of the Platform

The platform function of logical parts provided some interfaces which is easy to design and expand. So, researchers can expand functions basing on the existing resources

* Corresponding author.

D. Jin and S. Lin (Eds.): Advances in CSIE, Vol. 2, AISC 169, pp. 659–664.
springerlink.com

quickly. The researchers of network developing simulation and verification platform design the network system in a high start point using NetFPGA board and shorten the time of system design basing on the system prototype of this platform. Designers have implemented the network traffic collector, network flow classifier and security network card and so on. Used FPGA characteristic of parallel and pipelining processing function, network development platform greatly enhance the ability of data processing. Therefore, it is greatly suitable for high speed network flow to the researchers.

The platform uses hardware for the figure 1 shown. the core is a Xilinx Virtex-5 FPGA chip, 4 giga PHY chips and FLASH[4] etc. Virtex-5 FPGA logic resources are very rich, and suit for a complex logic design.

3 Functions of the Platform

Recently, the main application of network is used in flow characteristics analysis, traffic data sets analysis, flow characteristics, Internet traffic classification algorithm theory research and Theory or model establish for real-time Internet traffic etc. For the above application, network simulation and verification platform provides high-speed network data flow and easy to realize expansion user interfaces in data link layer [5]. The platform also provides data incentive used by protocol interfaces and user expansion interfaces.

As a verification support environment, the platform not only provides network data incentive simulating of the practical hardware platform, but also has a check to data of input and output. It also provides incentives data for the PCI protocol interfaces and Ethernet protocol interfaces, and give the function of protocol check. So the verification support environment is comprised by two parts: synthesis part and no synthesis part. Synthesis part can divide into function modules and requirements analysis of bandwidth.

The platform is suited for researcher to use. Researchers cannot be care about the button protocol of the concrete realization, and analysis Gpbs network traffic using the user programmable expand interfaces the platform provided. Especially it is used the research of traffic collect and flow classification or analysis new network protocol and network security etc. System design standard protocol interfaces have source and detailed realization process, and developers can verify their design easily according to the corresponding protocol modules.

4 Composition of the Platform

Synthesis parts are divided by the transmission process of network data and physical interface source hardware platform provided. For the figure 2, synthesis parts are constituted by five modules: interface module of Ethernet protocol, cache control module of SRAM, PCI protocol interface module, add or remove header module and asynchronous signal processing module. Some large module is divided into several small modules according to the method of top to down. These small modules are easy to design and local some errors during the design.

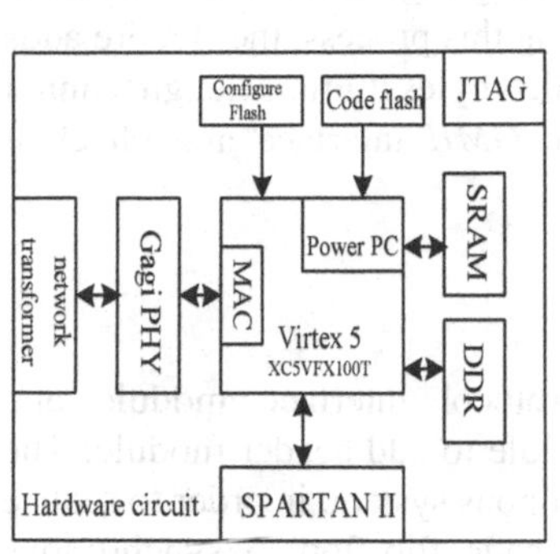

Fig. 1. Hardware platform

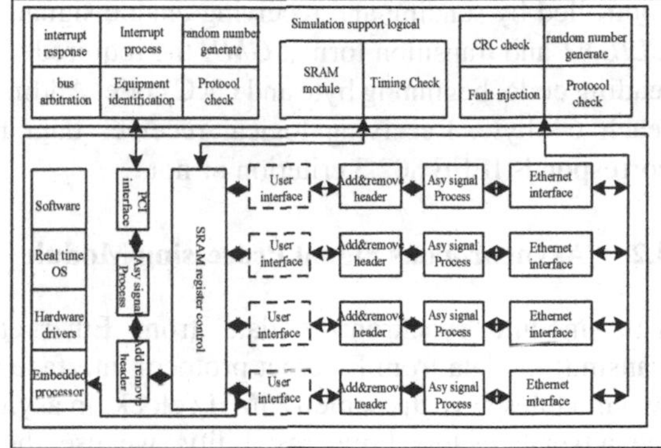

Fig. 2. System block diagram

The platform provides expand user interfaces, and gives network simulate at a Gpbs rate in the data link layer. Researchers can process these data according their requirements. Different modules added at user interface can finish different task and function. In the figure 2, the dotted line box is standard for the function of user extension logical interface module. PCI interface used by the platform is applied widely, so designer can design their products with PCI interface.

4.1 Ethernet Protocol Interface Module

Ethernet protocol interface module processes data from other modules based on IEEE802.3 giga MAC protocol. The flows are followings: remove frame synchronization code, CRC check, and providing frame good instructions signal, adding asynchronous code and checking code, add filled bytes, if frame length cannot reach the minimum bytes. We design gigabit Ethernet IP core for the figure 3.

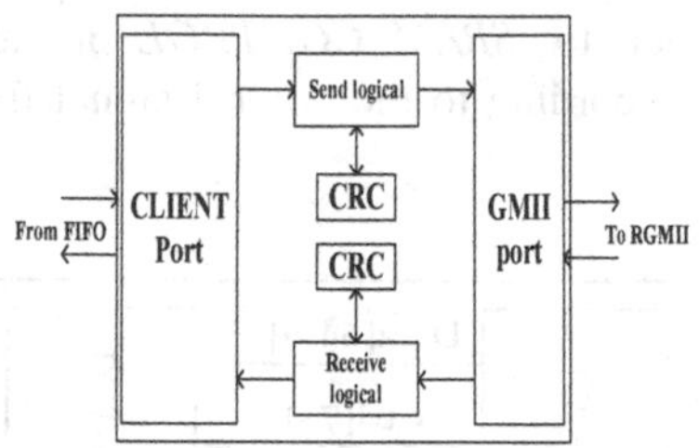

Fig. 3. Gigabit Ethernet IP core

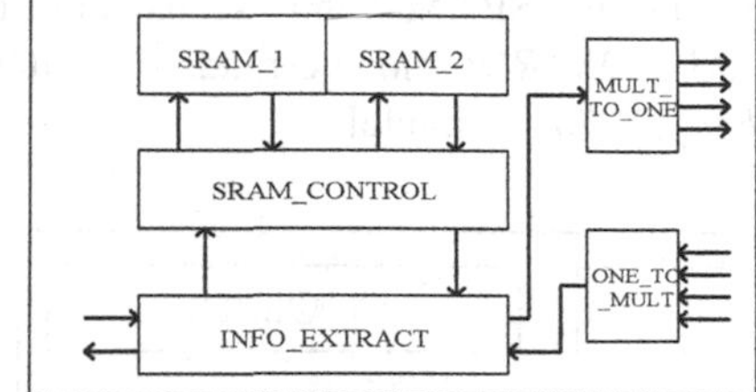

Fig. 4. SRAM control module

Design of *CLIENT* interface is considered to the maximum flexibility to meet the requirements of switch and network processing interface. The sending and receiving data width are eight. Sending data is synchronized by *gmiitxclk* clock signal and receiving data is synchronized by *gmiirxclk* clock signal. Input and output of data is controlled by sending and receiving enable signal. Sending logical receives data from *CLIENT* and transition format *GMII* the required. During this process, the IP core adds leading code, beginning byte and CRC code. Adding filled bytes if Frame length cannot reach 64 byte. Receiving logical receives data from *GMII* interface and check it corresponds IEEE 802.3 criterion or not.

4.2 Asynchronous Signal Processing Module

This module synchronizes data from Ethernet protocol interface module and transmitting data from Ethernet protocol interface module to add header module. The two modules have the same 125MHz clock. In asynchronous system, in order to reduce the adverse factors bring instability, we use the cascade flip-flop to synchronous asynchronous signals. As experience proves that the asynchronous processing method of the first fully complies with the design requirements and circuits work well.

4.3 Add and Remove Header Module

This module process data from asynchronous signal processing module and append a custom header to indicate a completion frame. The module also remove custom header and then send them to asynchronous signal processing module.

4.4 SRAM Buffer Control Module

SRAM buffer control module is constituted by 4 modules: *ONE_TO_MULT*, *MULT_TO_ONE*, *INFO_EXTRACT* and *SRAM_CONTROL*. The relationship between modules is shown in figure 4.

MULT_TO_ONE is choose one channel from four network interface and four host interfaces, and then transmit these data to *INFO_EXTRACT* module. the way of choose is by polling implementation, no priority. *ONE_TO_MULT* module solute to the one data channel from *INFO_EXTRACT* module to multiples. Solution of data use subordinate to the channel. *INFOR_EXTRACT* module extracts and records packet from or to SRAM, and sends control signal to *SRAM_CONTROL* module. *SRAM_CONTROL* module reads or writes data according to the control signal from *INFO_EXTRACT* module.

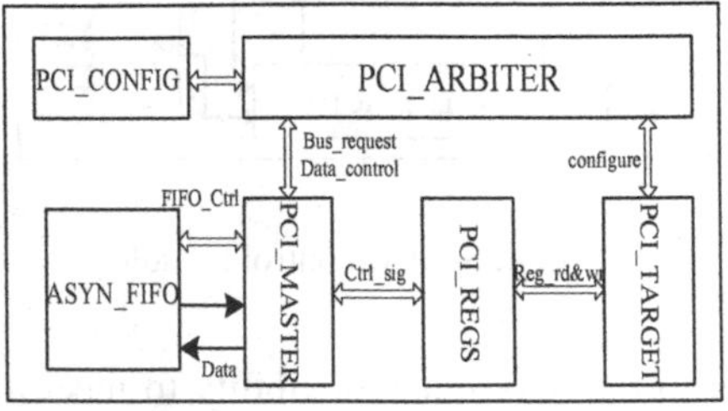

Fig. 5. PCI interface module

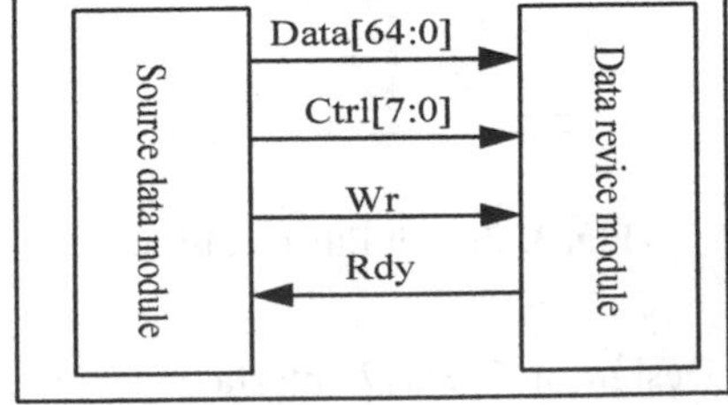

Fig. 6. Data interaction signal between modules

4.5 PCI Interface Module

PCI interface modules finishes packet exchange with host and implement the DMA control, PCI interruption and some other functions. PCI interface module is constituted to 6 sub modules: *PCI_TARGET*, *PCI_CONFIG*, *PCI_REGS*, *PCI_MASTE*, *PCI_ARBITER* and *ASYN_FIFO* module. The figure 5 is show the relation of them.

4.6 Data Format of User Interface

User interface is data interactive and data format between modules. User interface defines the way of data transmute in system and make data format equal in data processing. The platform use the NetFPGA style in data format and signal interactive.

For the figure 6 shown, the signals between modules are data, *ctrl*, *wr*, and *rdy*. The *rdy* indicates that for-module can send packet to post module when the source module make the *rdy* sinal high [6]. *Wr* express that the for-module sending message is effective. Data and ctrl show the packet' begin and end etc.

It means the begin of a packet when the *ctrl [7:0]* equal to 8'FF. *data [40:32]* expresses that how much words (width 64 bit) a packet includes. It means a packet data when *ctrl* is 8'00. *ctrl* only one 1 indicates that a frame is ending.

5 The Application of the Platform

Software of the platform operates in Windows OS, easy to use compare with Linux because it is hard to debug and check errors in Linux. It is needed more than half hour for synthesizing in Linux but only several minutes in Windows, so the platform save lots of time and energy when designers use it.

In our lab, we design a traffic collector to collect packets on the Internet using this platform [7]. Add the modules to the platform and begin to simulate. The waves are shown in Modelsim and packets flow the traffic collector module. And then port sends packets. A large number of project development practices certificate that FPGA network development and verification platform Improve the efficiency of development and save time and energy in it.

6 Conclusion

This paper describes the PCI protocol interface for adding registers and explains Ethernet interface and struts of SRAM according to the hardware logical of the platform in designing. It is also shown the state machine and state transition in detail. In addition, paper provides user extend interface and data format. Furthermore, because asynchronous signal may produce instability influence the stability of the system, this paper put forward the improvement cascade trigger edge detection methods and design the asynchronous signal processing module to Synchronous asynchronous clock signals in the same domain. In this way, the stability of the platform is improved greatly.

Acknowledgements. This research was supported by Natural Science Foundation of Shandong Province No. ZR2010FQ028, New Century Excellent Talents in University No. NCET-10-0863, Technology development Program of Shandong Province No. 2011GGX10116, Youth and mid-life scientist's award fund in Shandong province under Grant No. BS2009DX037.

References

1. Simpson, T.W., Marion, T., Hölttä-otto, K., et al.: Platform-based design and development, current trends and needs in industry. In: International Design Engineering Technical Conferences & Computers and Information in Engineering Conference. ASME, Philadelphia (2006)
2. Sangiovanni-Vincentelli, A., Martin, G.: Platform-based design and software design methodology for embedded systems. IEEE Design & Test of Computers, 23–33 (2002)
3. Lockwood, J.W., McKeown, N., Watson, G., et al.: NetFPGA–an open platform for gigabit-rate network switching and routing. In: Proceedings of MSE 2007. IEEE, Washington (2007)
4. Li, Z., Yang, B.: Design of development platform facing to network application based on NetFPGA. Shandong Science 23(5), 28–32 (2010)
5. IEEE STD 802 3. Part 3: Carrier sense multiple access with collision detection (CSMA/CD) access method and physical layer specifications (2002)
6. NetFPGA network open source projects [EB/OL], `http://www.netfpga.org`
7. Jin, L., Bo, Y.: Network traffic collector based on NetFPGA. University of Jinan Journal (01) (2011)

Application of Simulated Annealing Algorithm to Optimization Deployment of Mobile Wireless Base Stations

Zih-Ping Ho[1] and Chien-Sheng Wu[2]

[1] Information Center, National Taiwan Normal University,
No.162, Sec. 1, Heping East Road, Taipei City, Republic of China
zpho@ntnu.edu.tw
[2] D-Link Corporation,
3F-2, No.67, Tzyou Rd., Hsinchu City, Republic of China
jasonw.tw@gmail.com

Abstract. Deployment of mobile wireless base (transceiver) stations (MBTS, vehicles) is expensive, with the wireless provider often offering a basic coverage of BTS in a normal communication data flow. However, during a special festival celebration or a popular outdoor concert in a big city, the quality of the wireless connection would be insufficient. In this situation, the wireless service providers always increase the number of MBTS to improve the density of nets and speed up the data flow of communication. This research intended to construct an integer programming (IP) model to minimize the density gap between wireless request and supply. The solver used was an SA algorithm. In order to validate, the proposed approach was compared to other famous heuristics, such as Random with Tabu and Ransom Search; and it was found that SA outperformed RS by an average of 18%. This result suggested a reduction of the density gap between wireless request and supply by 18% if an MBTS company's allocation of transceiver stations is optimal, and an SA algorithm is used.

Keywords: Communication Technology, Simulated Annealing Algorithm, Wireless Deployment, Wireless Layout, Mobile Base Transceiver Stations.

1 Introduction

In the modern world, mobile wireless base (transceiver) stations (or called mobile base transceiver stations) are more important in a big city. Deployment of these wireless transceiver stations is expensive, therefore the wireless provider often offers a basic coverage of such stations in a normal communication data flow. However, during a special festival celebration, or a popular outdoor concert, the quality of the wireless connection would be insufficient. In this situation, the wireless service providers always increase the number of mobile base transceiver stations (trucks) to improve the density of nets and speed up the data flow of communication. Nowadays, a failure to connect to a wireless network is normally due to insufficient wireless transceiver

D. Jin and S. Lin (Eds.): Advances in CSIE, Vol. 2, AISC 169, pp. 665–670.
springerlink.com

stations or inefficient deployment of such stations. Thus, deployment of mobile base transceiver stations is a difficult decision making problem for a wireless provider company.

Thus the main purpose of this research was to construct an integer programming (IP) model to minimize the density gap between wireless request and supply; while maximizing wireless nets density for users and reducing the use of mobile base transceiver stations where possible. Others coverage related technology industries, such as mobile commerce, traffic tracking, weather prediction, pollution monitoring, critical infrastructure surveillance, disaster management, military battle communication, etc. [1, 2] could be also improved due to this study.

2 Literature Review

A coverage problem is one of the most important practical subjects relating to our daily life. In literatures, it is often classified into three catalogs. One is continuous vs. discrete coverage sub problems. Two is constrained vs. unconstrained. The last is weighted vs. unweighted space [3]. This study adds a fourth sub problem, that is mobile vs. static providers / sensors, which belongs to a discrete, constrained, weighted space, mobile providers and mobile requests mixed type coverage problem. What follows is a review of the related literature of the last decade.

Nesargi and Prakash [4] stated that in traditional cellular networks, the cell layout is dynamic due to the mobility of base stations, the cluster of cells within co-channel interference range changes with time. None-of the existing dynamic channel allocation algorithms for cellular, networks works in such a system. Chu and Lin [5] investigated the survivability of mobile wireless communication networks in the event of base transceiver station (BTS) failure. Zhao et al. [6] stated that a novel mobile position location procedure, based on the relaying capability of the mobile stations in a hybrid cellular or peer-to-peer ad hoc network to improve positioning coverage and accuracy, was proposed. They simulated the proposed positioning scheme and found coverage and density were improved. Esnaashari and Meybodi [7] proposed a scheduling algorithm to deal with the dynamic point coverage problem. Qi et al. [8] stated that a technique for using a receive signal strength indicator to measure the total isotropic sensitivity of a wireless handset is proposed in their paper. Hayashi et al. [9] applied potential game theory to a mobile sensor coverage problem. Tynan et al. [10] they proposed a new approach to wireless sensor network coverage problem. Luo et al. [11] proposed an interesting application of coverage problem. They measured the sensors which were located on the sea surface. Their sensors cannot be controllable. In this study, the mobile base transceiver stations can be controllable.

Although numerous literatures were related to coverage problems, there is only a limited amount of literature available on mobile providers and mobile requests mixed type coverage problem. Solving the complicated problem was the objective of this study.

3 Problem Formulation

This research adapts an IP (integer programming) model to minimize density gap between wireless request and supply. The distribution of users is showing discrete and irregular. Thus the model takes each point of a map to examine. It shows as follows:

A : Space of a map
$R_{(x,y)}$: request density at coordinate (*x*,*y*) by users, $R_{(x,y)} \in Non-Negative\ Integer$
$S_{(x,y)}$: supply density at coordinate (*x*,*y*) by mobile transceiver stations
$G_{(x,y)}$: density gap at coordinate (*x*,*y*) between users and mobile transceiver stations

$$G_{(x,y)} = \left\{ \begin{array}{ll} R_{(x,y)} - S_{(x,y)}, & \text{if } R_{(x,y)} > S_{(x,y)} \\ 0, & otherwise \end{array} \right\}$$

The IP model to minimize density gap between wireless request and supply is as follows:

$$\textbf{Min}\ \ G_{(x,y)} \tag{1}$$

s.t.

$$S_{(x,y)} \in A \tag{2}$$

$$S_{(x,y)} \in Non-Negative\ Integer \tag{3}$$

The objective (1) is to find a minimal density gap between wireless request and supply. In Eq. (2), the mobile transceiver stations have to locate on a predefined map. Eq. (3) declares non-negative integer constraints.

4 Simulated Annealing Algorithm

Simulated Annealing resembles the cooling process of molten metal through annealing[12]. It was introduced by Kirkpatrick et al. [13] and Cerny [14]. In SA algorithm [15], as new objective value is worse than the current one, it can still be accepted depending on a so-called probability of accepting worse solution [16]. The probability is obtained from the Boltzman distribution.

$$\mathrm{P}(\Delta \mathrm{E}) = \mathrm{e}^{-\frac{\Delta E}{kT}} \tag{4}$$

where $\mathrm{P}(\Delta \mathrm{E})$ is the probability of the optimization process to keep a modification that incurs an increase $\Delta \mathrm{E}$ of the objective function. *k* is a parameter of the process, and T is the instantaneous "temperature" of the process [17]. A cooling rate α is used to reduce the temperature of the annealing process [18]. In SA algorithm [15], Initialization, Objective function evaluation, Neighbor solution generation, Objective function evaluation, Comparison, Cooling, and Stopping were the standard procedures. The different points were as follows:

(1) Supply density at coordinate (*x*,*y*), which is a cumulative counter of $S_{(x,y)}$

(2) The solution represents the coverage of a mobile base station. It's a cycle. Besides, $S_{(x,y)} := S_{(x,y)} + 1$

5 Computation Results and Discussion

In this section, the algorithm coded in Visual Basic computer language with Celeron (R) Dual-Core CPU at 2.10GHz and 1GB of RAM memory. This IP model is a strong NP-hard problem [19]. Since the proposed model is a new problem, no published solution method is available for comparison. Thus, this research conducted an experimental design, at different mobile base transceiver stations, for comparing the performance of heuristics. The parameters of SA were listed in table 1. The parameters of RHT were listed in table 2.

Table 1. Set of parameters for *SA*

Parameter	Value
Cooling Rate α	0.90
T_0	100℃
T_n	0℃
Iterations at the T_i	10
Threshold Rate to P (ΔE)	0.50
Tabu	0
Termination time (sec)	10

Table 2. Set of parameters for *RHT*

Parameter	Value
Repeat times	100
Tabu	1
Termination time (sec)	10

In a problem size, mobile base transceiver stations is equal to 10, 15, or 20, the results are listed in table 3, and shown in figure 1.

Table 3. The comparison results of Heuristics

Problem (*n*)	Best solution (objective value) obtained		
	SA	*RHT*	*RS*
10	5640	5850	6789
15	5600	5780	6616
20	4900	4990	6375

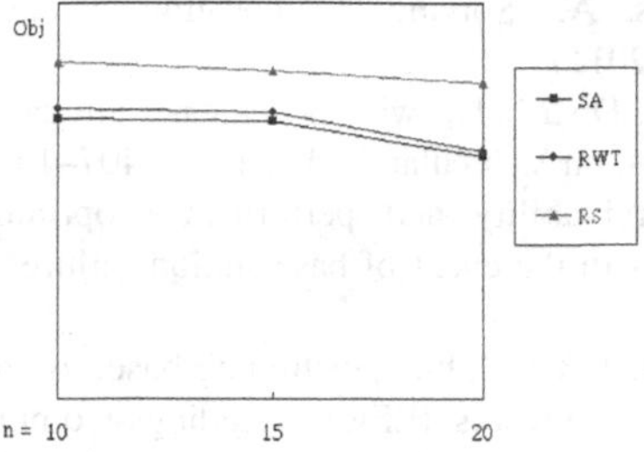

Fig. 1. Comparison results of objective values in different approach

The heuristics of SA and RHT both found an acceptable solution in reasonable time in this study. The best solution (objective value) is obtained by SA, followed by RHT. RS generated the worst solution (objective value). Furthermore, in table 3, we found that SA outperformed RS by an average of 18%. The results of RHT were close to SA. It represented that it would minimize density gap between wireless request and supply by 18% if a wireless mobile company allocated base transceiver stations optimally, and used SA. In real world observations, managers always over deploy mobile base transceiver stations.

6 Conclusion and Suggestion

Deployment of mobile base transceiver stations is a difficult decision making problem for a wireless provider company. Over deployment of mobile base transceiver stations often occurs. Numerous literatures were related to coverage problems, but there is only a limited amount of literature available on mobile providers and mobile requests mixed type coverage problem. This research is to construct an integer programming (IP) model to minimize density gap between wireless request and supply. The study proposed an SA algorithm to solve the model. From validation, the proposed approach was compared to other famous heuristics, such as Random with Tabu and Ransom Search. It found SA outperformed RS by an average of 18%. This represented a minimization of the density gap between wireless request and supply of 18% if a wireless mobile company allocated base transceiver stations optimally, and used SA. Besides, the results of RHT were close to SA, therefore, the heuristics of GAWT and RHT both found an acceptable solution in a reasonable time. However, it may result in a different density if some assumptions are revised. Future research suggests expanding this research to more discussion when some assumptions are changed.

References

1. Lambrou, T.P., Panayiotou, C.G.: A testbed for coverage control using mixed wireless sensor networks. J. Net. and Comp. App. 35(2), 527–537 (2012)
2. Cheng, P., Cao, X., Bai, J., Sun, Y.: On optimizing sensing quality with guaranteed coverage in autonomous mobile sensor network. Comp. Comm. (2011) (in press)

3. Davoodi, M., Mohades, A.: Solving the constrained coverage problem. App. Soft Comp. 11(1), 963–969 (2011)
4. Nesargi, S., Prakash, R.: Distributed wireless channel allocation in networks with mobile base stations. IEEE Trans. on Vehicular Tech. 51(6), 1407–1421 (2002)
5. Chu, K., Lin, F.: Survivability and performance optimization of mobile wireless communication networks in the event of base station failure. Comps. & Ele. Engi. 32(1), 50–64 (2006)
6. Zhao, L., Yao, G., Mark, J.W.: Mobile positioning based on relaying capability of mobile stations in hybrid wireless networks. IEEE Proceedings-Commu. 153(5), 762–770 (2006)
7. Esnaashari, M., Meybodi, M.R.: A learning automata based scheduling solution to the dynamic point coverage problem in wireless sensor networks. Comp. Net. 54(14), 2410–2438 (2010)
8. Qi, Y., Jarmuszewski, P., Zhou, Q.M., Certain, M., Chen, C.J.: An efficient TIS measurement technique based on RSSI for wireless mobile stations. IEEE Trans. on Instr. and Measure. 59(9), 2414–2419 (2010)
9. Hayashi, N., Ushio, T., Kanazawa, T.: Potential game theoretic approach to power-aware mobile sensor coverage problem. IEICE Trans. on Fundamentals of Ele. Comm. and Comp. Sci. E94A(3), 929–936 (2011)
10. Tynan, R., Muldoon, C., Hare, O., Grady, M.: Coordinated intelligent power management and the heterogeneous sensing coverage problem. Comp. J. 54(3), 490–502 (2011)
11. Luo, J., Wang, D., Zhang, Q.: On the double mobility problem for water surface coverage with mobile sensor networks. IEEE Trans. on Para. and Distri. Sys. 23(1), 146–159 (2012)
12. Vasan, A., Raju, K.S.: Comparative analysis of Simulated Annealing, Simulated Quenching and Genetic Algorithms for optimal reservoir operation. App. Soft Comp. 9(1), 274–281 (2009)
13. Kirkpatrick, S., Gelatt, C.D.: Optimization by simulated annealing. Science 220, 671–680 (1983)
14. Cerny, V.: Thermodynamical approach to the travelling salesman problem: An efficient simulation algorithm. J. Opti. Theory and App. 45, 41–51 (1985)
15. Bouleimen, K., Lecocq, H.: A new efficient simulated annealing algorithm for the resource-constrained project scheduling problem and its multiple mode version. E. J. Oper. Res. 149(2), 268–281 (2003)
16. Saraiva, J.T., Pereira, M.L., Mendes, V.T., Sousa, J.C.: A Simulated Annealing based approach to solve the generator maintenance scheduling problem. Electric Power Sys. Res. 81(7), 1283–1291 (2011)
17. Tavares, R.S., Martins, T.C., Tsuzuki, M.S.G.: Simulated annealing with adaptive neighborhood: A case study in off-line robot path planning. Exp. Sys. W. App. 38(4), 2951–2965 (2011)
18. Damodaran, P.: A simulated annealing algorithm to minimize makespan of parallel batch processing machines with unequal job ready times. Exp. Sys. W. App. 39(1), 1451–1458 (2012)
19. Rosen, K.H.: Discrete math. and its App. McGraw Hill Press (2006)

Low Computation Resource Allocation for Adaptive OFDM Power Line Communication

Lei Wang, Jun Lu, XianQing Ling and Qian Huang

School of Electrical and Electronic Engineering,
North China Electric Power University, Beijing 102206, China
ncepuleiwang@sina.com, lujun@ncepu.edu.cn,
{275386096,511597555}@qq.com

Abstract. This paper proposed a power and bit allocation algorithm for broadband power line communication system. A concept of Virtual Bit was introduced in the algorithm and the water line could be determined easily using the concept. After the water line was determined, the power and bits that exceeded upper limit were adjusted to achieve the optimal allocation target. This algorithm allocated bits in batches, and did not need iteration. The simulation result shows that the computational complexity decreased significantly compared with traditional bit adding algorithm, while the system throughput is close to the maximum.

Keywords: Power Line Communication, Power and Bit Allocation, OFDM.

1 Introduction

The broadband power line communication technology of low-voltage grid is a preferred solution for broadband access issues of the last mile. The application of adaptive orthogonal frequency division multiplexing and dynamic resource allocation optimization can effectively decrease the power line channel multipath, attenuation, and frequency-selective time-varying effects [1]. The allocation of resources in OFDM system involves the joint optimization of sub-carrier, power, adaptive modulation and bit resources [2].Through the rational allocation of these resources, the transmission speed and service quality of power line communications can be effectively improved.

Literature [3] proposed a forced convergence algorithm, which gave an effective solution to the convergence of the algorithm and the initial value problem, and achieved the rapid allocation of power and bits. Literature [4] proposed a non-iterative adaptive resource allocation algorithm based on MIMO-OFDM system that can approximate the maximum system throughput, but these algorithms are not for the power line channel. Literature [5] proposed an optimal allocation and adjustment algorithm based on the optimal criteria for the power line system, but the algorithm is too complex in the adjustment phase. In literature [6], the algorithm shows that bits were allocated even for every subcarrier firstly, then adjusted bit by bit using the power incremental table, which can get the optimal performance, but the computational complexity is relatively large.

D. Jin and S. Lin (Eds.): Advances in CSIE, Vol. 2, AISC 169, pp. 671–676.
springerlink.com

In this paper, we present a Virtual Bit allocation algorithm combined with constraints of power and bit upper limit due to power line channel characteristics. And two rules are introduced in the loading process based on the concept of Virtual Bit. Through simulation and analyses in typical power line environment, we can find that the new method has a relatively small computation and an optimal performance.

2 The Adaptive Power Resource Allocation Model under Single-User

In this paper, we study the resource allocation problem of the minimum power consumption in the case of single-user transmission rate is at a given value (MA model) in power line communication adaptive OFDM system. Let the set of subcarriers of the system is Ω, the total number of subcarriers is N. The mathematical model of this problem is:

$$\min \sum_{i=1}^{N} p_i, st: \sum_{i=1}^{N} r_i = R, 0 \le p_i \le \overline{p}_i, 0 \le r_i \le b \tag{1}$$

Which R is the total bit number of system required, b is the maximum number of bits on subcarriers, r_i, p_i are bits, power allocated on the subcarrier i and $\overline{p_i}$ is its upper limit of power.

And by the Shannon formula in the information theory, we can find the relationship between p_i and r_i.

$$p_i = (2^{r_i} - 1)\Gamma / g_i \tag{2}$$

Which g_i is per unit power signal to noise ratio of the subcarrier i. Γ is the difference between the signal to noise ratio, reflecting the signal to noise ratio difference between theoretical and actual. For quadrature amplitude modulation (QAM) systems of which the target bit error rate is P_e, the difference of signal to noise ratio can be expressed as:

$$\Gamma = [Q^{-1}(P_e / 4)]^2 / 3 \tag{3}$$

3 Virtual-Bit Algorithm Concepts and Criteria

From the formula (1), we can get the power increment formula:

$$\Delta p_i(r_i) = p_i(r_i + 1) - p_i(r_i) = \Gamma 2^{r_i} / g_j = 2\Delta p_i(r_i - 1) \tag{4}$$

$\Delta p_i(r_i)$ is the power increment in the case of subcarrier i with r_i bits.

Relevant Concepts and Optimal Criteria

1) For subcarrier i, its Virtual Bits v_i can be defined as:

$$v_i = \log_2(\Delta p_i(0) g_{best} / \Gamma) = \log_2(g_{best} / g_i) \tag{5}$$

Which g_{best} represents the largest per unit power signal to noise ratio.

2) Let subcarrier i has been allocated r_i bits, and its Virtual Bit is v_i, we define its Whole Bit W_i is:

$$W_i = r_i + m_i \tag{6}$$

3) If the system satisfied formula (7), the system has the best performance of allocation of bits[5], so we call it the optimal criteria.

$$\max_{i \in \Omega}(\Delta p_i(r_i)) < 2 \min_{j \in \Omega}(\Delta p_j(r_j)) \tag{7}$$

Rules

Based on the above concepts and the optimal criteria, we proposed rules (1) and (2).The bit allocation which satisfies the rules (1) and (2) must meets the optimal criteria.

1) The optimal subcarrier allocates a bit first and in the bit allocation process later, the difference of the Whole Bit among all subcarriers is not greater than 1.

Based on this rule, the bit allocation process can be seen as by rounds. The nth round makes the Whole Bit of all subcarriers be n.

2) The sequence of allocation in every round is according to the ascending order of m_i which equals to ceil(v_i)- v_i, ceil means to round the elements to the nearest integers greater than or equal to the elements.

4 The Implementation of Virtual-Bit Algorithm

The algorithm is divided into three steps: computation of the Virtual Bit number; allocation bits in batches; adjustment and allocation of the rest bits. The first two stages is to calculate the "water line", the last step is to ensure the optimal performance in the case of the system meets the constraints.

Calculate the Virtual Bit Number

According to the formula (5), the Virtual Bit number of every subcarrier can be calculated.

Allocation Bits in Batches

Firstly, we can determine the "water line" using the following formulas.

$$\sum_{i \in \Omega} W_i = R + \sum_{i \in \Omega} v_i \tag{8}$$

v_i, W_i are Virtual Bits and Whole Bits defined as above, let N be the number of the subcarriers, the water line Av describes as:

$$Av = foor(\sum_{i \in \Omega} W_i / N) \tag{9}$$

Then, we allocate bits for every subcarrier, and make sure the number of the Whole Bits on the subcarriers whose Virtual Bits less than Av is Av.

Adjustment and Allocation of the Rest Bits

According to the following formula ,the upper limit of bits of subcarrier i can be determined by power signal to noise ratio per unit r_i^g [7], the power limit r_i^p from formula (2),the maximum bit number b.

$$\overline{r_i} = \min(r_i^g, r_i^p, b) \tag{10}$$

The subcarriers that exceed the upper limit are removed the excess number of bits, make bits on them are the upper limit. Let the sum of excess bits is e_0.

The all remaining *e* bits can be calculated by:

$$e = e_0 + \sum_{i \in \Omega} W_i - Av \times N \tag{11}$$

The subcarriers are ordered in the ascending sequence of values of m_i mentioned above. Then the subcarriers whose bits less than the upper limit are allocated one bit in the arranged order circularly until there are no rest bits.

5 Simulations and Analysis

The comparison algorithm in this part is the traditional add-to-bit algorithm that has the optimal performance. And the simulation of the two algorithms is carried out in the power line channel environment; the parameters are shown in the following table.

Table 1. The power line channel parameters

Subcarrier Number	Frequency Range	Power Spectrum Upper Limit	Error Rate	Maximum Bit Number
128	0~20MHz	-50-0.8f (dBm/Hz)	10(-4)	6

The unit of f is MHz, the noise is colored noise and channel coding has not been used. Figure 1 shows the bit allocation results in the case of the total bit is 200

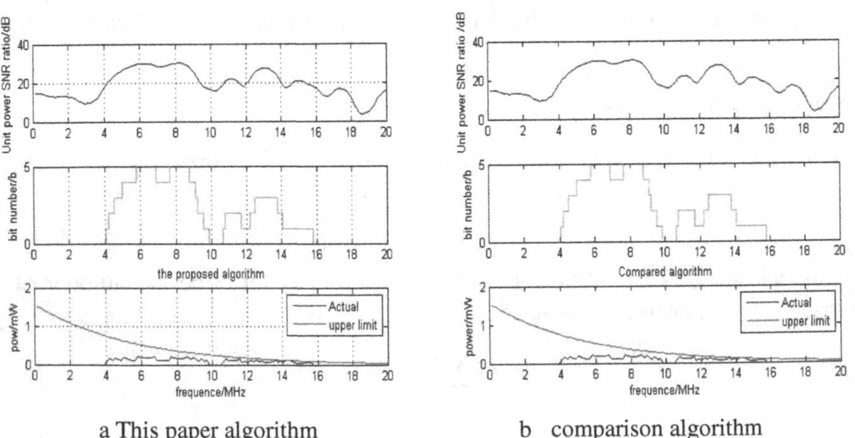

a This paper algorithm b comparison algorithm

Fig. 1. The bit allocation at a total number of 200

Table 2. The allocation of resources at a total number of 200

Type \ Parameters	Required Number	Actual Number	Power Consumption	Subcarrier Number
This paper algorithm	200	200	8.4669	70
comparison algorithm	200	200	8.4669	70

Figure 1 and Table 2 show this article algorithm and compared algorithms have completed the bit allocation that system required. And the allocation of power did not exceed the power limit. The use of subcarriers and the system rate of two algorithms are the same.

In order to verify the rules (1) and (2), also to compare the performance of two algorithms further, another eight groups of experiments were carried out. Each experiment takes different total required bits and the experimental results of are recorded as follows. We declare that this paper data were on the left side of '/' while comparison data on the right side (this paper/comparison) in the following table.

Table 3 shows the allocation results of this paper algorithm and add-to-bit algorithm are the same, both can achieve optimal performance. The up limit of bits that each OFDM symbol of the system can transmit was about 270. Also, we can find that this paper algorithm can exhibit stable performance even in the vicinity of the maximum transmission rate.

Table 3. The allocation results at different rate request

Required Number \ Parameters	Actual Number	Power Consumption	Subcarrier Number
50	50/50	0.5058/0.5058	34/34
100	100/100	1.8480/1.8480	44/44
150	150/150	4.3797/4.3797	59/59
250	250/250	14.7798/14.7798	84/84
260	260/260	23.6414/23.6414	95/95
270	268/268	29.4949/29.4949	99/99
280	265/265	30.6181/30.6181	99/99
300	277/277	32.0432/32.0432	99/99
350	268/268	32.2780/32.2780	99/99
500	273/273	32.4941/32.4941	99/99

6 Summary

Based on simulation and analysis in the power line environment, the computational complexity of the proposed algorithm is lower than the improved add-to-bit algorithm, far less than the add-to-bit algorithm. And the proposed Virtual Bit algorithm can achieve the optimal performance. Therefore, the proposed algorithm in this paper has a certain practicality in broadband power line OFDM system.

Acknowledgments. This work was supported by National Natural Science Foundation of China (No.51007021) and National Science and Technology Major Project of the Ministry of Science and Technology of China (No.2010ZX03006-005-01).

References

1. Xu, Z.: Study of Resource Allocation in Broadband over Power-line Systems North China Electic Power University 6 (2010)
2. Zhang, D.-M., Xu, Y.-Y., Cai, Y.-M.: High efficiency algorithm of power and bit allocation for OFDMA system. Journal on Communications 29(4) (April 2008)
3. Ding, L., Yin, Q.-Y., Dengke: A Computationally Efficient Transmit Power and Bit Allocations Algorithm for Wireless OFDM Systems. Journal of Electronics & Information Technology 29(7) (July 2007)
4. Xu, L., Xu, D.: Adaptive Non-Iterative Resource Allocation Algorithm for MIMO-OFDM System Based on Channel Coding. Acta Electronica Sinica 38(4) (April 2010)
5. Fang, Y., Xu, Z., Zhai, M.-Y.: Resource allocation for single user in power line communication adaptive OFDM systems. Power System Protection and Control 38(4) (February 2010)
6. Zhao, Y.-M., Guo, J.-B., Wang, Z.-J., et al.: Bit Swap and Power Reallocation Algorithm for Power-line High-speed Communication Systems. Journal of Tsinghua University: Natural Science Edition 46(10), 1645–1648 (2006)
7. Andrea, J.G., Soon, G.C.: Variable-rate Variable-power MQAM for Fading Channels. IEEE Trans. on Communications 45(10), 1218–1230 (1997)

The Study of Apodization of Imaging Fourier Transform Spectrometer in the Spectrum Reconstruction

XueJun Zhao and Bin Zhao

Department of Computer Science and Technology,
China University of Mining and Technology(Beijing), Beijing 100083, China

Abstract. During the process of imaging in interference spectrum, apodization is an important part of the spectrum reconstruction in imaging Fourier transform spectrometer (IFTS), and it has a powerful effect on the accuracy of reconstructed spectra. This paper analyzes the principle of apodization, and uses six common kinds of apodization functions to process the simulated interferogram. Combining with the principle of selecting apodization functions, analyzing experimental results by approaches of comparison and drawing conclusions. The conclusions show that Happ-Genzel function is best for simulated interferogram, and the experiment shows that stand or fall of selecting apodization functions impacts directly on the accuracy of reconstructed spectra.

Keywords: imaging Fourier transform spectrometer (IFTS), apodization, spectrum reconstruction.

1 Introduction

The original data which is get in imaging Fourier transform spectrometer (IFTS) is the interference data, and it must be changed into the spectroscopic data which can be used through the spectrum reconstruction. So the spectrum reconstruction is an important part of the data processing in IFTS, and it has a direct influence on the accuracy of the spectroscopic data[8].

The spectrum reconstruction is the process which change the original interference data into the spectroscopic data. The spectrum reconstruction process includes the relative radiometric correction, remove baseline, apodization filter, phase correction and Fourier transform[9].

During the process of imaging in interference spectrum, apodization is an important part of the spectrum reconstruction in imaging Fourier transform spectrometer (IFTS), and it has a powerful effect on the accuracy of reconstructed spectra[1].

In this paper, apodization which is part of the spectrum reconstruction in imaging Fourier transform spectrometer (IFTS) is researched.

2 The Principle of Apodization

During the process of imaging in interference spectrum, using formula 1 to obtain the reconstructed spectrum through Fourier transform.

D. Jin and S. Lin (Eds.): Advances in CSIE, Vol. 2, AISC 169, pp. 677–682.
springerlink.com

$$B(\sigma) = \int_{-\infty}^{+\infty} I(x) e^{-i2\pi\sigma x} dx = FT^{-1}(I(x)) \tag{1}$$

Fourier transform integral limit is infinite in theory, but the actual instruments optical path difference is limited. Assume that the maximum optical path difference of the interferogram obtained from the actual instrument is L, the actual reconstructed spectrum worked out is shown as formula 2:

$$B'(\sigma) = \int_{-L}^{+L} I(x) e^{-i2\pi\sigma x} dx = \int_{-\infty}^{+\infty} I(x) T(x) e^{-i2\pi\sigma x} dx = B(\sigma) * T(\sigma) \tag{2}$$

Rectangular truncation function is shown as formula 3:

$$T(x) = \begin{cases} 1, |x| < L \\ 0, |x| > L \end{cases} \tag{3}$$

The Fourier transform of Rectangular truncation function is shown as formula 4:

$$T(\sigma) = FT^{-1}[T(x)] = 2L \sin c(2\pi\sigma L) \tag{4}$$

Formula 4 is the instrument line shape function. Because the sinc function has obvious side-lobe effect, it causes obvious distortion between reconstructed spectrum and real spectrum. The side lobe is the main source of spectrum signal interference near the center wave number, and it covers weak spectrum information near the center wave number and seriously impacts on adjacent spectrum, especially the accurate determination of the weak spectrum[2]. In order to correct the disturbance of reconstructed spectrum due to interference pattern' appearing sharp discontinuous changes owing to interference pattern' suddenly truncating in the biggest optical path difference, we usually use the gradual change weight functions to multiply by interference patterns in the purpose of easing discontinuous degree of the interference patterns in the biggest optical path difference and called the gradual change weight functions apodization functions. The value of Apodization functions in the zero optical path difference is one, decreases with the increase of the optical path difference, and is zero in the maximum optical path difference L. The appropriate use of apodization functions obviously weakened the side lobe of the instrument line shape functions, but the cost is to enlarge the half height line width of spectral lines and reduce the resolution of the instrument.

There are more than 20 kinds of apodization functions, which include Rectangular function, Triangle function, Hanning function, Hamming function, Blackman function, KaiserBessel function, Happ-Genzel function and Norton-Beer functions, etc[3].

3 Principles of Selecting Apodization Functions

The selection of apodization functions should follow the following principles[4]:

(1) Functional form is simple and easy to calculate the amount of computation;
(2) To make the main lobe of the instrument line shape function as narrow as possible;

(3) To make side lobes as low as possible, in order to avoid substantial shock in the maximum optical path truncation;
(4) There is no change in the nature for spectrum calculation after the introduction of the apodization function;

In the actual selections, these four principles are difficult to satisfy at the same time. The lower side lobes, the greater the effect on the system resolution. And the width of the main lobe becomes larger. Because side-lobe suppression is at the cost of the loss of the resolution of the system[5]. The narrower the width of the main lobe, the less the impact on the system resolution. But it leads to larger side-lobe shocks become higher. So only from the theoretical point of view,it is difficult to explain what kind of function is optimal apodization function. Specific application must be determined with the specific realities. According to experimental results, it is determined whether the apodization or what kind of apodization function is selected.

4 Effects of Apodization Functions on Reconstructed Spectrum

In this paper six common kinds of apodization functions which include Rectangular function, Triangle function, Hanning function, Hamming function, Blackman function and Happ-Genzel function are used to process the simulated interferogram. By comparing the reconstructed spectral curve after different apodization function processing, based on the principles of selection of the apodization function, it is to identify which apodization function is optimum and analyse effects of different apodization functions on the accuracy of reconstructed spectra. Set maximum optical path difference as L, and the expressions of the above six kinds of apodization functions are shown in Table 1 [6], [7]:

Table 1. Six common kinds of apodization functions and their expressions

Apodization functions	**Expressions**
Rectangular apodization	$w(\delta)=\begin{cases}1, -L\le\delta\le L\\ 0, other\end{cases}$
Triangle apodization	$w(\delta)=1-(\lvert\delta\rvert/L)$
Hanning apodization	$w(\delta)=0.5+0.5\cos(\pi\frac{\delta}{L})$
Blackman apodization	$w(\delta)=0.42+0.5\cos(\pi\frac{\delta}{L})+0.08\cos(2\pi\frac{\delta}{L})$
Hamming apodization	$w(\delta)=0.54+0.46\cos(\pi\frac{\delta}{L})$
Happ-Genzel function	$w(\delta)=\begin{cases}a+b\cos\frac{\delta\pi}{2L}, -L\le\delta\le L\\ 0, other\end{cases}$

4.1 Apodization Processing on the Simulated Interferogram

In accordance with the formula of interferogram function

$$I_D(\Delta) = \int dI_D(\Delta,\sigma) = \int_{\sigma_{min}}^{\sigma_{max}} B_0(\sigma)[1+\cos(2\pi\sigma\Delta)]d\sigma \quad (5)$$

programming to generate simulated interferogram.

The original simulated interferogram curve is shown in Figure 1, and it is an ideal symmetrical cosine curve. The reconstructed spectral curve after different apodization function processing is shown as Figure 2. In Figure 2, the y axis is relative intensity.

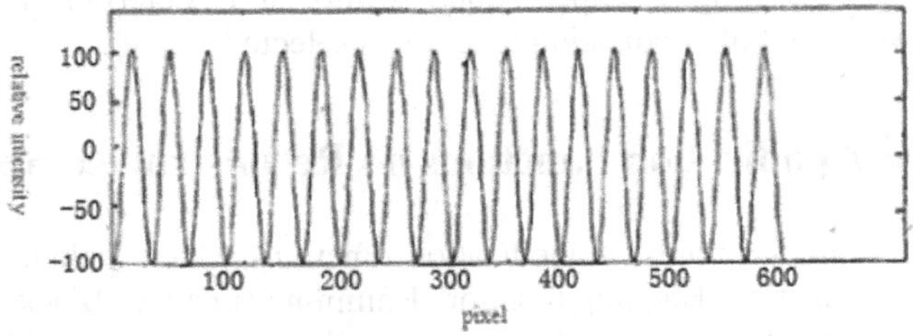

Fig. 1. The original simulated interferogram curve

It can be seen from Figure 2(a) and Figure 2(g), The effect of the reconstructed spectrum after Rectangular apodization function processing is even worse than the effect of the reconstructed spectrum after no apodization processing. Because the simulated interferogram is actually multiplied by Rectangular function, after another Rectangular apodization processing the side lobes of the reconstructed spectrum continue to expand and fluctuation is still very obvious in the region far from the main lobe and the width of the main lobe is not reduced compared with the reconstructed spectrum after no apodization processing, the use of Rectangular apodization is unnecessary. As shown in Figure 2(e) and Figure 2(g), after Triangle apodization processing the width of the main lobe is much narrower than after no apodization processing, but there are still many side lobes near the main lobe. If there is relatively weak spectral information near the main lobe, it may be concealed by the side lobes. So the effect after Triangle apodization function processing is not good enough. In Figure 2(c) and Figure 2(g), after Hanning apodization processing there is basically no side lobe and only the width of the main lobe is broadening compared with the original spectral line. The effect of apodization is relatively good. As shown in Figure 2(d) and Figure 2(g), in general the effect of apodization is also relatively good after Hamming apodization processing. As shown in Figure 2(b) and Figure 2(g), after Blackman apodization processing there is no side lobe and only the width of the main lobe is broadening compared with the original spectral line. The effect of apodization is also relatively good. In Figure 2(f) and Figure 2(g), the effect after Happ-Genzel apodization processing is also relatively good.

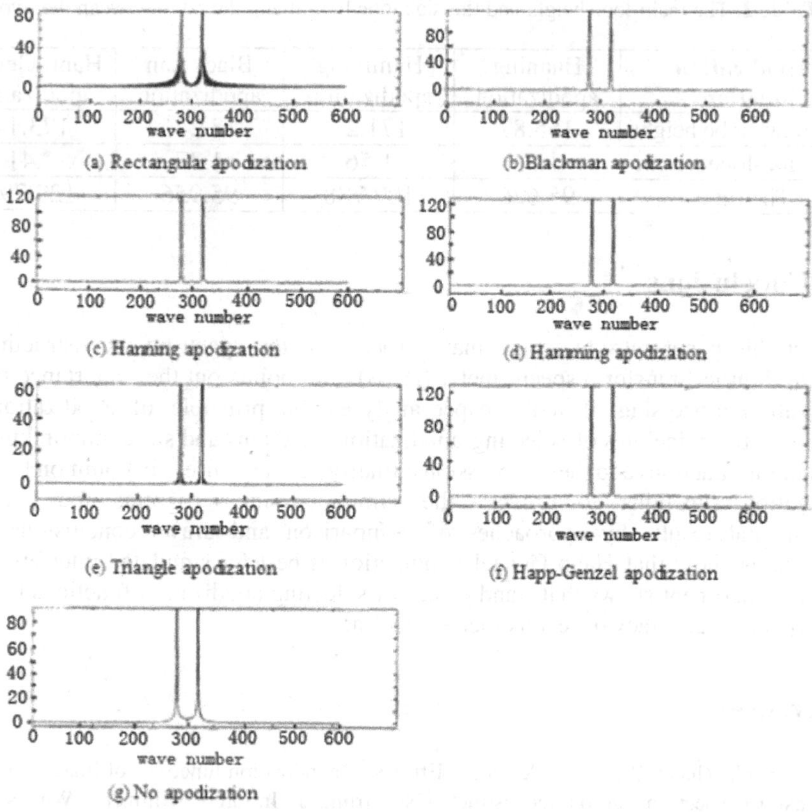

Fig. 2. The reconstructed spectral curve after different apodization function processing

In summary, considered from these aspects which include that the main lobe width is as narrow as possible and that the effect of the side-lobe suppression is as good as possible, the effect of four kinds of apodization functions which include Hanning apodization function and Blackman apodization function and Hamming apodization function and Happ-Genzel apodization function is relatively good. The reconstructed spectral curve is uncalibrated, so it is determined that the effect of which kind of apodization is the best through the ratio of the maximum intensity of the main lobe and the maximum intensity of the side lobe of these four kinds of apodization functions. The main-lobe height and the side-lobe height and the ratio between the two of the reconstructed spectral curve after these four apodization function processing on the simulated interferogram are shown in Table 2. It can be seen through the ratio from Table 2 that the effect of the side-lobe suppression after Happ-Genzel apodization processing is the best. On the whole, the effect of Happ-Genzel apodization is best for the simulated interferogram used in this paper.

Table 2. The main-lobe height and the-side lobe height and the ratio between the two

Apodization functions	Hanning apodization	Hamming apodization	Blackman apodization	Happ-Genzel apodization
The main-lobe height	155.87	171.27	131.46	173.14
The side-lobe height	1.63	1.56	1.37	1.41
The ratio	95.626	109.788	95.956	122.794

5 Conclusions

At first this paper introduces the main process of the spectrum reconstruction in imaging Fourier transform spectrometer (IFTS), and points out the importance of the apodization processing; then the paper analyses the principle of apodization,and introduces the principles of selecting apodization functions and six common kinds of apodization functions and their expressions; finally the paper uses six common kinds of apodization functions to process the simulated interferogram, and analyses experimental results by approaches of comparison and draws conclusions. The conclusions show that Happ-Genzel apodization is best for simulated interferogram, and the experiment shows that stand or fall of selecting apodization functions impacts directly on the accuracy of reconstructed spectra.

References

1. Zhang, W., Zhang, B., Zhang, X., et al.: Effects of apodization functions of imaging Fourier transform spectrometer on reconstructed spectrum. J. Infrared Millimetre Waves 27(3), 228–232 (2008) (in Chinese)
2. Bell, R.J.: Introductory Fourier Transform Spectroscopy. Academic Press, New York (1972)
3. Norton, R.H., Beer, R.: New apodizing functions for fourier spectrometry. J. Opt. Soc. Am. 66, 259–264 (1976)
4. Chen, K., Jiao, Q., Gao, X.: The choice of apodization function of the peak method. China Agricultural University Learned Journal 2(4), 21–27 (1997) (in Chinese)
5. Fan, S., Lin, C.: Effects of apodization functions on the phase correction in Fourier transform spectroscopy. Analytical Insrument 4(2), 70–73 (1990) (in Chinese)
6. Chen, Y., Pan, W., Guo, J.: Depth analysis of fiber and grating apodization. Optoelectronic Technology and Information 9(4), 38–41 (2005) (in Chinese)
7. Wang, K.: The study of Narrowband FBG filter apodization function. Fiber and Cable and Their Applications 6(2), 22–26 (2006) (in Chinese)
8. Tahic, M.K., Naylor, D.A.: Apodizing functions for Fourier Transform Spectroscopy. In: Fourier Transform Spectroscopy Topical Meeting of the Optical Society of America in Alexandria, Virginina (2005)
9. Lv, Q.-B.: Interference spectral imaging data processing technology. Institute of Optics and Fine Mechanics of Chinese Academy of Sciences, Xi'an (2007) (in Chinese)

The Application of Semi-supervised Clustering in Web Services Composition

QiPing Zheng* and Yi Wang

Guangdong Textile Polytechnic, Foshan, 528041, China
wangyi02023026@sina.com

Abstract. Semi-supervised classification algorithm using a large number of unlabeled data supporting the supervised learning process to improve the classification results. Web service composition is a complex interaction between the execution order to determine the different Web services and Web services. This article focuses on the semi-supervised clustering algorithm based on tags and constraints. The paper presents the application of Semi-supervised clustering in web services composition. The compared experimental results indicate that this method has great promise.

Keywords: web services, semi-supervised clustering, web application.

1 Introduction

XML is a uniform standard in all services information and news. SOAP (Simple Object Access the Protocol, Simple Object Access Protocol) XML-based information exchange agreements and request services. WSDL (Web Service Description Language, Web Services Description Language) is an XML-based language used to describe the service operation. The UDDI (Universal Description, Discovery and Integration Universal Description, Discovery and Integration) provides a publish and discover services registry. A variety of Web application development platform, it is in order to web services such as the Microsoft. NET, so the IBM Web Sphere on Web services standards with varying degrees of support.

Semi-supervised learning is a new research hotspot in the field of machine learning, the joint probability distribution of the tag data and unlabeled data to improve the performance of the classifier [1]. Semi-supervised learning can be divided into semi-supervised classification and unsupervised clustering.

In each module of the Web services architecture, as well as inside the module, the message is passed in XML format. The reason is: the XML format is easy to read and understand, because the XML document has the structural characteristics of cross-platform and loosely coupled. The XML format of the message envelope glossary in the industry, you can also organize internal and external use; it has good flexibility and scalability, allowing the use of additional information. This article focuses on the semi-supervised clustering algorithm based on tags and constraints. Semi-supervised

* Author Introduce: Zheng QiPing, Female, Associate Professor, Master, Department of Economics and Management, Guangdong Textile Polytechnic, Research area: Data mining.

D. Jin and S. Lin (Eds.): Advances in CSIE, Vol. 2, AISC 169, pp. 683–688.
springerlink.com

clustering is some data on the type of mark or constraints to assist non-supervised clustering process. As many cases, the category information of the data is incomplete, semi-supervised clustering can be used to tag data of the type of information or constraints on the data gathered, extended and modified the original category tags. The paper presents the application of Semi-supervised clustering in web services composition. Web service composition refers to a set of Web services to support the business process logic, which itself can be either the final application, can also be a new Web service. Web service composition is a complex interaction between the execution order to determine the different Web services and Web services. On Web service composition, there is currently no unified solution.

2 The Research of Semi-supervised Clustering

Semi-supervised classification algorithm using a large number of unlabeled data supporting the supervised learning process to improve the classification results.

If the monitoring data is a category labeled, and these markers data on behalf of all the relevant categories, the semi-supervised clustering and semi-supervised classification algorithm can be used for data classification. However, in many areas related to the type of tag information is incomplete, and the semi-supervised classification, semi-supervised clustering can gather data as the category with the initial labeled data, and expand, modify the existing class mark reflect the data of other law.

(1) Algorithms based on tag data and constraints. The algorithm uses the tag data and constraints to meet the objective function of the clustering results. The tag data indicate an instance of the category; the two instances should belong to the same cluster must-link or belong to different clusters can not-link constraints specified. Specific algorithm: modify the objective function to meet the specified constraints; artificially added constraints in the clustering process.

(2) Distance-based algorithm. The algorithm uses a distance measure function to meet the mark or constraints to achieve the clustering process. Specific algorithms: the Mahalanobis distance based on convex optimization algorithm; the Euclidean distance based on shortest path algorithm; discrete gradient descent Jensen-the Shannon [2]; EM algorithm improvements edit distance (string-edit distance).

K-means algorithm can only be monitoring information of unlabeled data modeling, but can not use the tag data. Explained below in semi-supervised clustering algorithm, the tag data of the monitoring information is incorporated into the K-means algorithm. First seed clusters generated by the tag data to initialize the clustering algorithm, and then use the constraints of the tag data, guidance on the unlabeled data clustering process. Select the appropriate seeds to avoid local minima, and produce and markings similar to the cluster, as is shown by equation 1.

$$\tilde{r}(n) = \sum_{i=1}^{N} \gamma^{n-i} d^*(i) x^H(i) \tag{1}$$

Given a data set, the K-means algorithm can produce about division when the objective function is a local optimum. Assume that seed set through the following steps to get the seed set, as is shown by equation 2.

$$X_{i+1,V}(m) = \overline{H}_i^* X_{i,V}(m) + \sum_{r=1}^{q-1} \overline{G}_{r,i}^* X_{i,r,D}(m) \tag{2}$$

Most Seeded K-means algorithm is initialized with the seed cluster centers, where center is not like K-means algorithm randomly mean, choosing instead to seed set by the center as the first cluster center. Most Seeded K-means algorithm is described below.

Input: data set $X = \{x_i\}_{i=1}^n, x_i \in \mathrm{R}^d$, the number of clusters, the initial seed $S = \bigcup_{h=1}^k S_h$ set.

Output: disjoint partition data sets $\{X_h\}_{h=1}^k$, making the best of the Most Seeded K-means objective function.

1) Seed clusters to initialize the centers $\mu_h^{(0)} \leftarrow \frac{1}{|S_h|}\sum_{x\in S_h} x, h=1,\ldots,k; t \leftarrow 0;$

2) To calculate the cluster $x \in S$ data the mean time, if the data $x \in S_h$ is assigned a class (set $X_h^{(t+1)}$); when the $x \notin S$ data $X_{h^*}^{(t+1)}$ will be assigned to class (collection $h^* = \arg\min_{h\in\{1,\ldots,k\}} \left\|x - \mu_h^{(t)}\right\|^2$);

3) Web page collection P = {p1, p2, ..., pn}, of xij (i = 1,2, ..., n; j = 1,2, ..., m), said the Web pages within a given time pi and pjin the same user session, the number of visits. Fuzzy clustering steps available to the original Web data matrix R0 = (xij) n × n, data standardization, calibration, etc. Web to get the order n Web fuzzy similarity matrix R '= (rij) n × n, then use the web fuzzy clustering approach, the fuzzy classification.

Constrained K-means algorithm implementation process, the seed cluster category tag is to remain unchanged, the only non-seed data to recalculate the mean, also the mark of non-seed data may change, no noise or need to change on seeds marking. Most Seeded K-means algorithm for noise on seeds, seed tag can be changed in the clustering process, in the initialization of the seed center; you can get rid of noise seed, as is shown by equation3 [3].

$$W_X X(m+1) = W_X \Phi(m) x(m,M) + W_X \overline{w}(m) \tag{3}$$

Gaussian mixture model (the Gaussian mixture model to describe the model of mixed density distribution) is the most classic, most complete statistical learning modeling algorithms. This model assumes that data from different data sources, each data source can be used to determine the form of mathematical modeling. Assume that the number of mixture components for the hybrid model can be expressed as Equation 4.

$$p(x \mid \mu, \Sigma) = \frac{1}{(2\pi)^{d/2} |\Sigma|^{1/2}} e^{-\frac{1}{2}(x-\mu)^T \Sigma^{-1} (x-\mu)} \tag{4}$$

Unlabeled data categories as this article is incomplete data. Assumption represents the observational data, said the incomplete data as "incomplete data", and said the unknown parameters. In the subsequent sections of this section, assume that the value of said parameters, assumptions and modified in each iteration of the EM algorithm, as is shown by equation5.

$$Y_m = X_m - \sum_{i=1}^{m-1} (X_m, Z_i) Z_i \tag{5}$$

Most Seeded K-means algorithm and Constrained K-means algorithm is essentially in the case of certain assumptions, the EM algorithm, Gaussian mixture model. Distribution in the semi-supervised K-means algorithm, assuming that all the data in all categories of obedience is a "larger" Gaussian mixture distribution, the Gaussian number is the number of categories, each a Gaussian distribution, it this "small" Gaussian distribution of hybrid formation of the distribution is Gaussian mixture distribution for each category can be broken down into "smaller" Gaussian distribution.

3 The Research of Web Service Composition

Currently, it is the Web service composition from the business sector (business domain) and semantic boundary (semantic domain). Business sector are in accordance with the syntax, functional description of a service-oriented, such as bell, Model Driven Service Composition. The Semantic Web community in accordance with the semantics, the main concern by the explicit declaration of the body where the pre-conditions, affect the reasoning of Web resources, such as Ai is planning-based Composition. According to the Web service composition generated by the program, the Web service composition can be divided into two categories: static and dynamic combination.

In order to build a global Web services market, in order to strengthen the large-scale reuse of services, more and more people make great efforts to study a combination of Web services.

Web service composition method in two directions: (1) XML-based workflow description language and workflow technology-based, representative of BPEL4WS, WSCL. They are able to describe the Web service process models can also be used as an executable language for the Web service process execution parsing engine; (2) Construct Web service semantic description of the model ontology based on a representative is the DAML-S [4]. It Web services as artificial intelligence (based Artificial Intelligence, AI) in the behavior, to describe the service parameters, the premise and results, and can map the formal description of the behavior, the combination of services can use the AI the method to solve.

$$W = \left(I - \frac{Y_1^* Y_1^T}{Y_1^T Y_1^*} - \frac{Y_2^* Y_2^T}{Y_2^T Y_2^*} - \cdots - \frac{Y_M^* Y_M^T}{Y_M^T Y_M^*} \right) W_0 \tag{6}$$

DAML-S ontology consists of three parts: (1) profile (Service Profile), describes the service auto-discovery is necessary properties, such as input and output and a prerequisite. (2) Service model (Service Model), describe the process model used to implement automatic service composition and run. (3) Service basis points (Service the Grounding), the associated process model describes the communication protocol and WSDL message description.

$$\sum_{h=1}^{k}\sum_{i=1}^{n}\|x_i - \mu_h\|^2 p(z_h \mid x_i, \mu_h) \tag{7}$$

Web service composition can be reused. Moap atomic Web services and Web service composition system are described as a series of behavioral sequence and state transition such composite services can be used as an available service for more complex systems.

4 The Application of Semi-supervised Clustering in Web Services Composition

The standard K-means algorithm does not provide any monitoring information, the mean of the E-step is chosen at random, and subsequent M-step is divided to the nearest mean. Each data set in the semi-supervised clustering data corresponding is in order to a mean value and the corresponding conditional distribution. E-step, data is randomly assigned to a cluster, which is equivalent to a conditional distribution of selected data from a conditional distribution.

The presence of incomplete data, the EM algorithm can be resolved through an iterative fashion model parameters is maximum likelihood estimation problem. "Incomplete data" generally refers to two situations: one is due to the limit or error of the observation process itself, resulting in the observed data error "incomplete data"; the other is the direct optimization of the parameters of the likelihood function is very difficult, the introduction of additional parameters (implicit or loss) after relatively easy to optimize, so the definition of the original observations plus additional parameters to form a "complete data". In fact, in the latter case is more common in machine learning and related fields, as is shown by equation8.

$$\overline{P}^{(\eta)}(m,s) = \left[I - \overline{K}^{(\eta)}(m,s)\Psi_w(m,s)\right]\overline{P}^{(\eta)}(m,s-1) \tag{8}$$

K-means algorithm to solve the problem of incomplete data, incomplete data category of the conditional distribution is, in theory, can solve this problem, but in fact this is incalculable. In the semi-supervised clustering, the user provides the conditional distribution of some categories of data and tag data constraints. For example, two data and the constraints of the must-Link, and the distribution is the same. In fact, all the data in the transitive closure of a group of related collection of must-link constraints are subject to the same distribution, monitoring information provided by the semi-supervised clustering is the conditional distribution of the data categories, as is shown by figure 1.

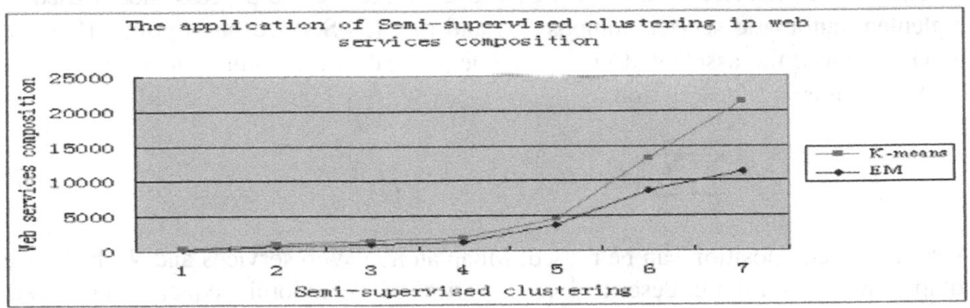

Fig. 1. The compare result of Sem-supervised clustering in web services K-means algorithm and EM algorithm

5 Summary

Web service composition refers to a set of Web services to support the business process logic, which itself can be either the final application, can also be a new Web service. Web service composition is a complex interaction between the execution order to determine the different Web services and Web services. The paper presents the application of Semi-supervised clustering in web services composition.

References

1. Zhu, X.J.: Semi-Supervised Learning Literature Survey. Computer Sciences TR 1530. University of Wisconsin–Madison (last modified on December 9, 2006)
2. Zhang, B.: Generalized K-harmonic means-boosting in unsupervised learning. Hewlett-Packard Laboratories (2000)
3. Wagstaff, K., Cardie, C., Rogers, S., Schroedl, S.: Constrained K-Means clustering with background knowledge. In: Proceedings of 18th International Conference on Machine Learning (ICML 2001), pp. 577–584 (2001)
4. Demiriz, A., Bennett, K.P., Embrechts, M.J.: Semi-supervised clustering using genetic algorithms. In: Artificial Neural Networks in Engineering (ANNIE 1999), pp. 809–814 (1999)

A Fast and Effective Algorithm for Image Co-segmentation

Ji Kai Chen[1], Shunxiang Wu[2,*], and YanYun Qu[3]

Department of Automation, Xiamen University
Xiamen, China
tenchu2009@sina.com, wsx1009@163.com

Abstract. The target of co-segmentation is to recognize the same object among many images and segment the object [1]. Generally, the model of co-segmentation is a Markov random field (MRF) model [1][2]. The complexity of optimizing the problem is NP-hard [2]. Actually, there is a high-speed approximate algorithm that can solve the problem well, this paper proposed an approximate algorithm which combines the color and shape information and image segmentation preprocessing to simplify the problem. Experiment results show that the algorithm has a high successful rate and its low error rate is also acceptable under the circumstance of its high speed.

Keywords: Co-segmentation, Computer Vision, Image segmentation, fast algorithm.

1 Introduction

Image co-segmentation is a fundamental problem in computer vision, it is used to recognize the same object in many images and segment the object. In general, the problem can be transformed into an optimization problem, then it can be solved in an optimization way. [6,7]. In totally un-supervised algorithm, automatically segmentation may produce many different errors, [6] but if we use interactive feedback method to reduce the errors, the cost seems to be large and unworthy for such an easy problem [7]. So it becomes a huge challenge to use automatically un-supervised segmentation. Our paper is focused on finding such an algorithm, which is effective with a low cost and can reduce the error rate as much as possible. The identification of similar objects in more than one image is a fundamental problem in computer vision and has relied on user annotation or construction of models [3,4, 5].

Different from the other algorithms, a direct, natural and easy algorithm is proposed in this paper.

* Corresponding author.

D. Jin and S. Lin (Eds.): Advances in CSIE, Vol. 2, AISC 169, pp. 689–695.
springerlink.com

2 Image Segmentation

In this paper, we scan from left to right and up to down, when there is a point that has not been assigned, build a super-pixel which includes the unassigned point, the maximum R,G,B around the point and R,G,B have the proximal ratios in R+G+B. In order to avoid diffusion, after every point is added, the mean values of these data of the whole super-pixel must be recalculated, the points which have the similar values as the mean values should also be added.

The formulations used in this process are:

$$Col\max = \max(R,G,B) \tag{1}$$

$$\mathrm{Re}\,dradio = \frac{R}{R+G+B} \tag{2}$$

$$Greenradio = \frac{G}{R+G+B} \tag{3}$$

$$Blueradio = \frac{B}{R+G+B} \tag{4}$$

3 Searching for the Similar Super Pixels

It is easy to recognize the similar colors, but picking the super-pixel with similar shape have a bit difficult. As to the magnitude of the pixels, a fast and effective algorithm is needed, in this paper, every pixel's distance from the barycentric of the super-pixel is sorted in ascending order, take distance samplings at an interval of 20 pixels as the features to determine the similarity of shape. Suppose that there are n pixels in a super-pixel, xi is the distance form ith pixel to the center of gravity of the super-pixel. We sort these n values by ascending order. Assume that the final result is x1,x2,x3…xn. Then we took points: x(n/20), x(n/20*2)…..x(n/20*20). We write them as , samx1, samx2….samx20. If n/20 is not an integer, then use interpolation method to take points. The interpolation formula is:

x(n/20)=x((int)(n/20))+(x((int)n/20+1)-x((int)n/20))*(n/20-(int)(n/20)) (5)

The other 19 points are also got in this way. If there are two similar blocks, Then the value Shapedis can be calculated as:

$$shapedis = \sum_{i=1}^{20} fabs(samxi - samyi) \tag{6}$$

To the two blocks X and Y, the value of Shapedis must be very small and a small Shapedis means a high probability of the shape similarity. Obviously, this criterion is rotationally invariant.

Another parameter that represents the average distance from every pixel to the center of gravity is needed. The average value is calculated by:

$$avg = \frac{\sum_{i=1}^{n} xi}{n} \tag{7}$$

4 A Fast Way to Recognize Foreground

When a human is focusing on the target, the background is ignored and becomes blurry, so we can also just care about the foreground and ignore the background. This process can be simulated by a normal distribution function.
Suppose this two super-pixels are x and y, the definition of similar shape and similar color are:
Similar shape:

$$Shapedis \leq \text{threshhold1} \tag{8}$$

$$Fabs(avgx - avgy) \leq \text{threshold2} \tag{9}$$

Where,threshhold1 and threshhold2 is a parameter you Specified.
Similar color:

$$\begin{aligned} Coldis = {} & fabs(redradiox - redradioy) + fabs(blueradiox - blueradioy) \\ & + fabs(greenradiox - greenradioy) \end{aligned} \tag{10}$$

$$Coldis \leq \text{threshlod3} \tag{11}$$

Where, threshhold3 is a parameter you Specified.

Assume that the first image A have n super-pixels, the second image B with m super pixels. Enumerate all the possible super-pixels x and y to find the pair of x and y that makes the following formula minimal:

$$\text{Argmin}\left\{ \begin{array}{l} \text{fabs}\left(\begin{array}{l} \sum_{\text{a}=1}^{\text{n}} \text{colorfun}(\text{a}) * \text{cnta} * \exp(-dis(x,a)/O) - \\ \sum_{b=1}^{m} colorfun(b) * cntb * \exp(-dis(y,b)/O) \end{array} \right) : \\ \text{x is super - pixel in A, and y is super - pixel in B, and} \\ \text{their shape and color is similar.} \end{array} \right\} \tag{12}$$

Where, Colorfun is a function about the color of the super-pixel, we use $sqrt(blueredio) + sqrt(redeadio)$ to enlarge the differences between different super-pixels. If we use blueradio+redradio, then actually, the same greenradio can mean the same colorfun., any colourfun can be used if appropriate, O decide the definition of the background.

5 Expanding From the Center to Find the Common Parts

Thinking about the expanding way, firstly, no matter how the image rotates and zooms, any two Super-pixel's standardized distance is invariable. Secondly, scan along the clockwise or the counterclockwise direction, the order that similar color appears is the same. Based on these two characteristics, we can quickly recognize the foreground.

The BFS algorithm seems to be easy, but it is hard to evaluate its complexity. The algorithm in this paper labels all the super-pixels with similar colors and shapes of image A and B as foreground, then expands from the best matching super-pixels x and y separately to find the interconnected branches including x and y. Next, the algorithm updates the foreground of two images by using each other, which can delete parts of background that be determined as foreground.

We also need to know the angle relationship between the two images, an easy method is scan along the distance order, there is no need to use too many distances here. Suppose the farthest super-pixel from x in A is farestx , the farthest super-pixel from y in B is faresty, the distances from farestx to x is disdis1, the distances from faresty to y is disdis2.

$$disdis = \min(disdis1, disdis2) \tag{13}$$

Do equal distance sampling at disdis*1/5,disdis*2/5,disdis*3/5,disdis4/5, every sampling point differs for 0.5 degree. We do two circles of sampling in image A and one circle for image B.

When the enumerate direction of A is counterclockwise:

$$sumcoldis = \sum_{i=0}^{359.5} coldis(x+i,i), \text{where x belongs to [0,360]} \\ \text{and i increase 0.5 each time} \tag{14}$$

When the enumerate direction of A is clockwise:

$$sumcoldis = \sum_{i=0}^{359.5} coldis(x-i,i), \text{where x belongs to [0,360]} \\ \text{and i increase 0.5 each time} \tag{15}$$

$Coldis(c,d)$ represents the color distance between two sampling points.

6 Experiment Results

In this paper, we discussed the image segmentation problem of two images. Here, we will show experiment results and make a comparison between our algorithm and other algorithms.

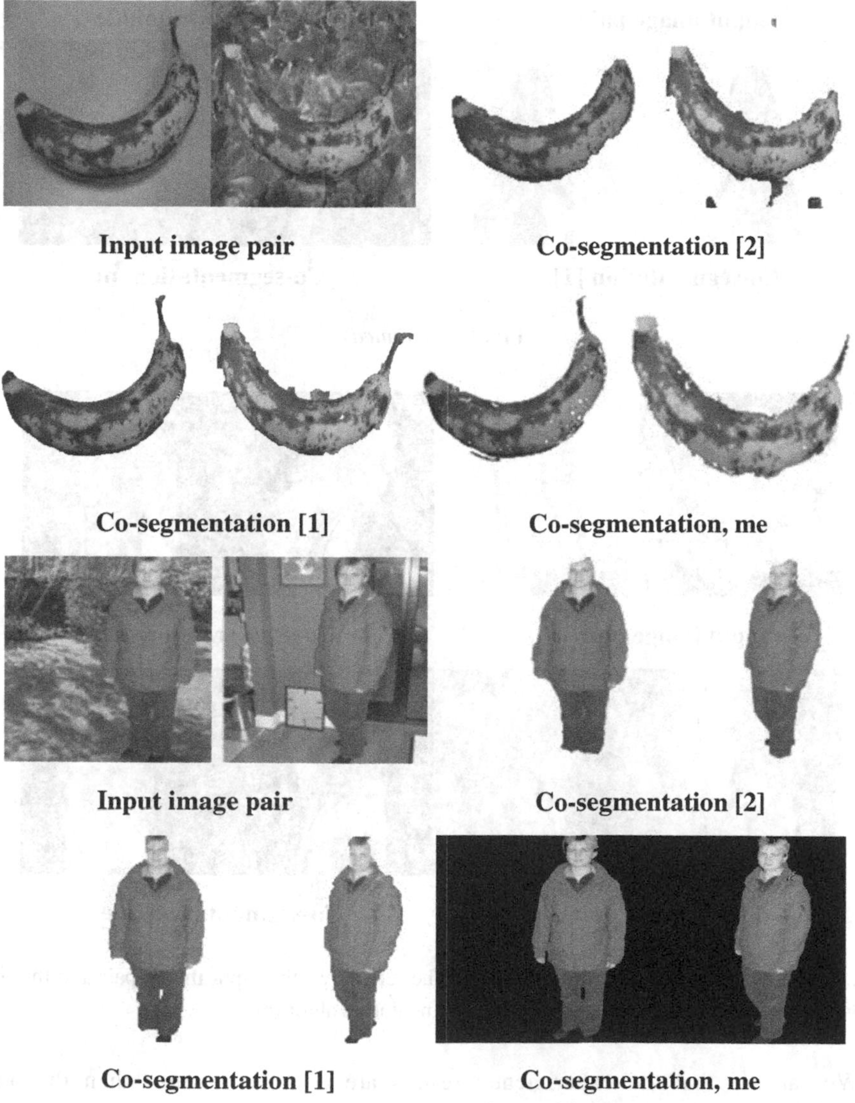

Fig. 1. The first row shows the input image pair and segmentations obtained using co-segmentation algorithms:[2]. The second row shows segmentations obtained using the two co-segmentation algorithms:[1] (left) and my solution (right).

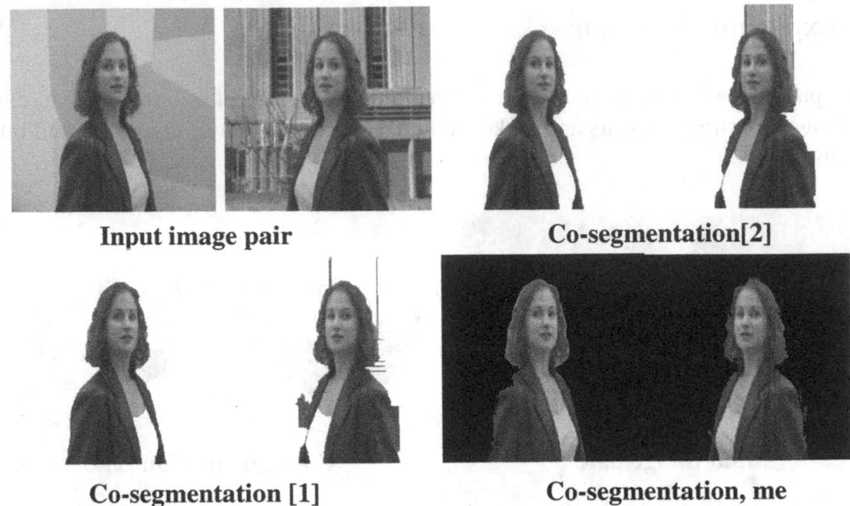

Fig. 1. *(continued)*

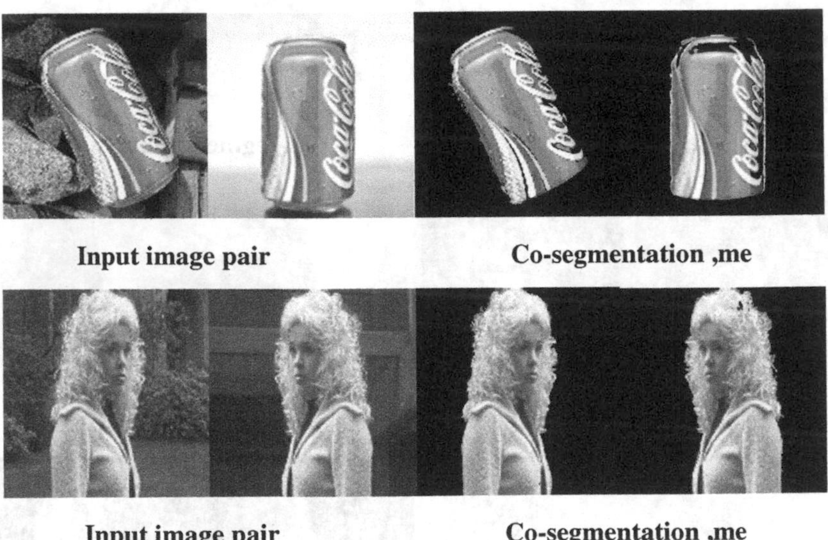

Fig. 2. The results of some other image pairs.The left shows the input image pair and the right shows segmentations obtained using my co-segmentation algorithms.

We can see that the experimental results are not much worse than the other algorithms, however, its speed is much faster than other algorithms.

7 Summary

The paper discussed a fast image co-segmentation algorithm, through its effect is a little worse than the best result, but its fast speed can compensate for this disadvantage. In this article, the image segmentation algorithm is not good, so the results is dependent on the parameter choice.The focus of the later work will be improving the effects of segmentation.

Acknowledgement. This project is supported by the Aeronautical Science Foundation (20115868009) and the Open Project Program of Key Laboratory of Intelligent Computing & Information Processing of Ministry of Education, Xiangtan University, China (No.2011ICIP04) and the Program of 211 Innovation Engineering on Information in Xiamen University (2009-2011) and "College Students Innovation Training Plan" of Xiamen University.

References

1. Hochbaum, D.S., Singh, V.: An efficient algorithm for Co-segmentation. In: 2009 IEEE 12th International Conference on Computer Vision (ICCV) (2009)
2. Rother, C., Minka, T., Blake, A., Kolmogorov, V.: Cosegmentation of image pairs by histogram matching – incorporating a global constraint into MRFs. In: Proc. of Conf. on Computer Vision and Pattern Recognition (2006)
3. Mukherjee, L., Singh, V., Dyer, C.R.: Half-integrality based algorithms for cosegmentation of images. In: Proc. of Conf. on Computer Vision and Pattern Recognition (2009)
4. Cootes, T., Taylor, C., Cooper, D., Graham, J.: Active shape models – their training and applications. Computer Vision and Image Understanding 61(1), 38–59 (1995)
5. Yuille, A.L., Hallinan, P.W., Cohen, D.S.: Feature extraction from faces using deformable templates. International Journal of Computer Vision 8(2), 99–111 (1992)
6. Shi, J., Malik, J.: Normalized cuts and image segmentation. PAMI 22(8), 888–905 (2000)
7. Boykov, Y., Jollie, M.-P.: Interactive graph cuts for optimal boundary and region segmentation of objects in N-D images. In: ICCV (2001)

Crowd Simulation Using Heat Conduction Model

Nan Xiang and Ling Lin

State Key Lab of CAD&CG, Zhejiang University, 310058 Hangzhou, China
xiangnan@zjucadcg.cn, linling156@163.com

Abstract. In this paper, we introduce a novel method for crowd simulation using heat conduction model. In our algorithm, individuals actions were affected by their neighbours in addition to the outside events and these affections were calculated grounded on head conduction theories. In order to improve the calculation speed, we adopted ADI method to generate the results. At last, we mapped these affections into their actions. Using this algorithm, we can simulate more realistic crowd behaviours.

Keywords: crowd simulation, heat conduction, spectator behaviors.

1 Introduction

Over the last decades, crowd simulation has become an important research field in computer graphics, virtual reality and social simulation. Many researchers have made impressive results [1-13].Computer games also adopted crowd simulation to improve the reality of virtual scene.

However, these crowd researches are mainly focus on the route or moving trajectories of individuals. Microscopic models were only designed to calculate motion and position of individuals but behaviour affections. Typically there are two kinds of model for crowd simulation, the microscopic model and macroscopic model[14]. As the spectators' positions would not be changed, we focus on microscopic models. Social force model, cellular automata (CA) model and rule-based model are the common local behavior models. Yang [4] used a two-dimensional CA model to simulate the kin behavior of evacuation. Kirehner [15] integrated the conception of friction into CA model to simulate the competition behaviors in evacuation. Yu [5] presented a cellular automaton model without step back to simulate the pedestrian counter flow in a channel. Parisi [12] studied the room evacuation problem using the social force model introduced by Helbing [11]. This model allows exploring different degree of panic. Berg [7] exploited a velocity obstacle to avoid collisions with the extrapolated trajectories of individuals. Besides individuals' behavior, group behaviors are generated in many researches, too. Musse and Thalmann [8] used group to generate crowd behaviors. Each group had its general behavior specified by the user and the individual behaviors were created by a random process through the group behavior. Gu [9] presented an intuitive yet efficient approach to generate arbitrary and precise group formations by sketching formation boundaries. CA models and social force models simulated the situation when

D. Jin and S. Lin (Eds.): Advances in CSIE, Vol. 2, AISC 169, pp. 697–702.
springerlink.com

people moved in a familiar environment and realized their goals with avoiding obstacles and other individuals. Such kinds of models are lack of physical and psychological elements embedded into the autonomy agents to provide humanlike decision-making [16].

In this paper we address the problem of simulating crowd behaviour based on individuals' affections. Crowd is affected by some certain individuals as they can affect the actions of the other individuals. This phenomenon is with many of the attributes of heat. For example, these affections are from nearby to distant and transmits faster when the crowd is dense or the outside stimulus are stronger. Furthermore, it keeps on spreading no matter the crowd is moving or not. This is just like heat conduction in gas or metal. As a result, we utilize two-dimensional heat conduction model to simulate the affections within crowd and adopt Alternate Direction Implicit (ADI) method to calculate the results. And then map these affections into actions.

2 Generation Loops

In this section, we will introduce how to construct individual personality model which would affect the way individual get other's influences. The generation pipeline was shown in figure 1.

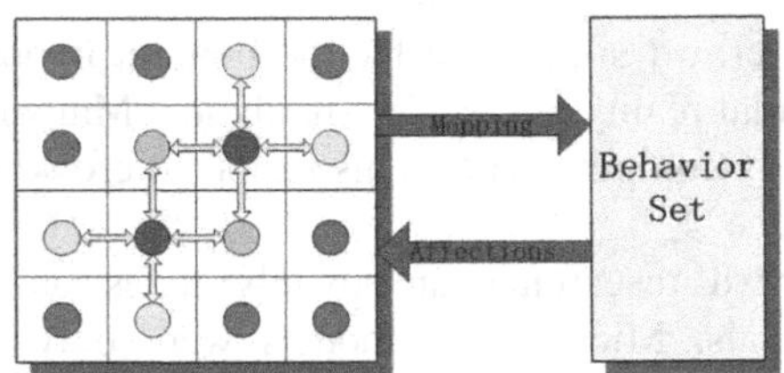

Fig. 1. Behavior generation circle

The questions before we found our crowd simulation model are: How can behaviors be spread? How do individuals supporting the same team affect each other? How do they do when they stand in the opposite? In our work, we believe the reason behaviors could be spread is because the individuals percept others' actions and have confidence or sympathy to the meaning of these actions. To which degree the individual is affect by the other people is corresponding to a parameter $\alpha \in [0,1]$ and it can also be treated as a behavior refresh rate. Assuming X_a^t, X_b^t represent the movement state of individuals a, b in time t, then movement state of a in time $t+1$ is:

$$X_a^{t+1} = \begin{cases} \alpha_a X_a^t + (1-\alpha_a) X_b^t, X_a^t X_b^t > 0 \\ \alpha_a X_a^t - (1-\alpha_a) X_b^t, X_a^t X_b^t < 0 \end{cases} \tag{1}$$

where $X_a^t X_b^t > 0$ means spectators are fans of the same team and $X_a^t X_b^t < 0$ means they are belong to opposite side.

Figure 2 shows that individuals can be affected by their neighbours and those in front of the person affect him much more than those behind him.

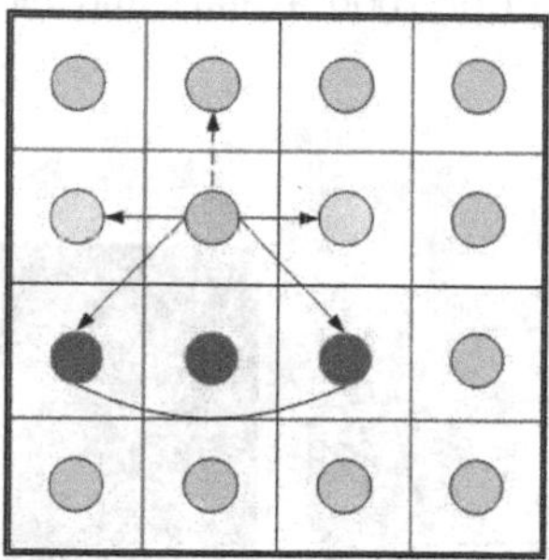

Fig. 2. Probabilities of affections from neighbours

Assuming individuals' positions are not changed, according to two-dimensional heat conduction equation, we construct the behaviour affection equation like:

$$\frac{\partial X}{\partial t} = K \frac{\partial^2 X}{\partial^2 U} \tag{2}$$

where K denotes the degree of difficulty of behaviour transmission and depends on individuals' personalities and position. U is two-dimensional space and has u, v axis. Besides the actions triggered by neighbours, people also react to the environment, which generates action A_t, then the equation is:

$$\frac{\partial X}{\partial t} = K \frac{\partial^2 X}{\partial^2 U} + A_t \tag{3}$$

We can use finite difference methods such as ADI method to solve the partial derivatives. Using ADI method, to square grid, equation (3) can be rewritten in two steps as

$$\begin{aligned} &-\lambda X_{u,v-1}^{t+1/2} + 2(1+\lambda) X_{u,v}^{t+1/2} - \lambda X_{u,v+1}^{t+1/2} \\ &= \lambda X_{u-1,v}^{t} + 2(1-\lambda) X_{u,v}^{t} + \lambda X_{u+1,v}^{t} \end{aligned} \tag{4-1}$$

$$\begin{aligned} &-\lambda X_{u-1,v}^{t+1} + 2(1+\lambda) X_{u,v}^{t+1} - \lambda X_{u+1,v}^{t+1} \\ &= \lambda X_{u,v-1}^{t+1/2} + 2(1-\lambda) X_{u,v}^{t+1/2} + \lambda X_{u,v+1}^{t+1/2} \end{aligned} \tag{4-2}$$

3 Implement

We exploited CUDA and DXUT framework to implement our algorithm on a computer with i5 CPU, 4G RAM and GTS 250 graphics card. Four cases were applied to simulate crowd behaviors of spectators. The models of the avatar are from Internet. We exploited a crowd of 1000 avatars and the output video run 15 frames per second.

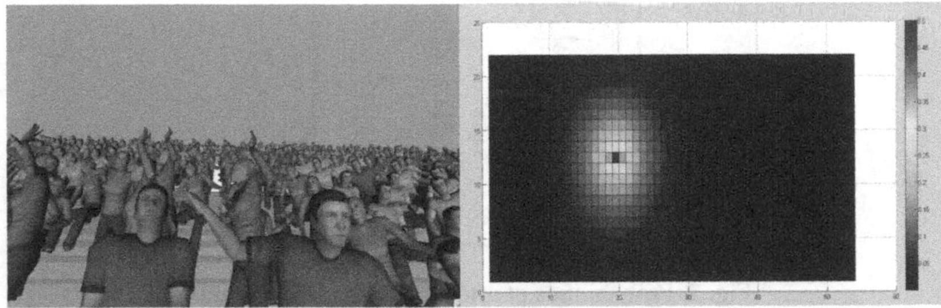

Fig. 3. Actions aroused by group leaders

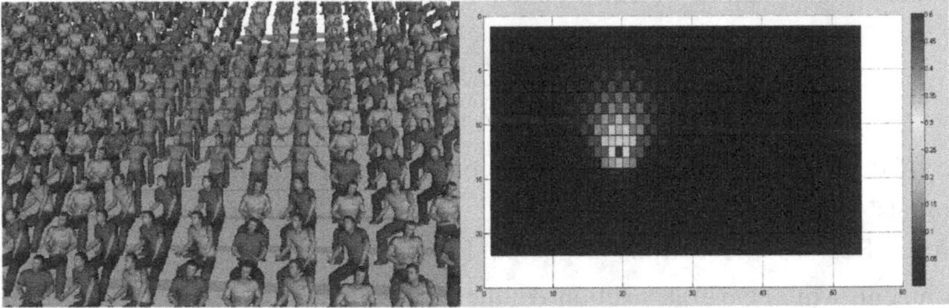

Fig. 4. Group behaviors

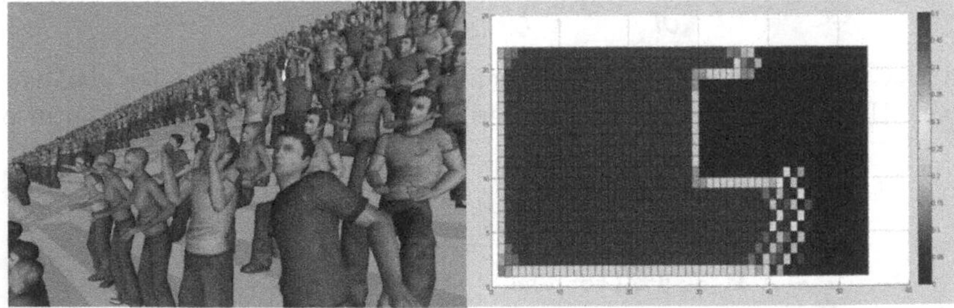

Fig. 5. Actions in sparse crowds

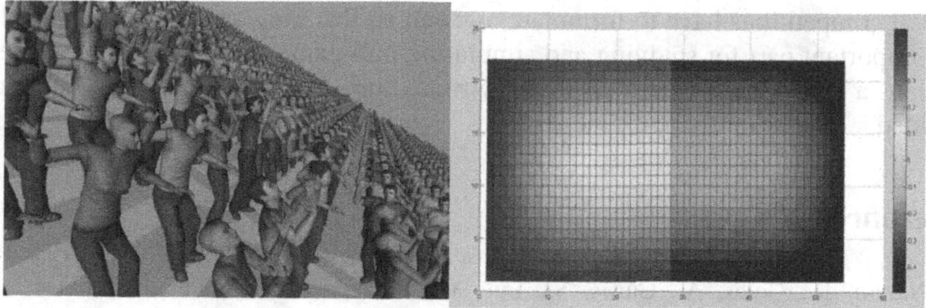

Fig. 6. Riot during competition

The left image of figure 3 shown the situation assuming the girl in pink was the group leader and the girls in green were her members and they were apt to accept and clone her actions. The other male individuals were not her group members. When the competition was in a stalemate, the leader was more active than other individuals and encouraging her team by dancing. Her actions had affected her neighbours, her group members were dancing in her way compared to the other individuals. Though the male individuals' emotions were aroused by her, they did not follow her behaviour but stand up and clapped. The right image of figure 3 shows their active state. The x and y axis are represent their positions and the colours represent their value of happiness, the redder the happier.

Figure 4 shown the situation that the individuals were likely be affected by the individuals in front her. When the competition was in a stalemate, the group behaviour formed according to the leader. As the same in figure 3, girls in green were in the same group. They were more active than the boys according to their personality and then chose to dance with their leaders when the competition was in a stalemate. Although some of the boys around the girls were affected by them too, they did not follow their actions. Figure 5 represented the situation that the crowd was sparse. The affections were slower than normal. Figure 6 simulated situation of riot during competition.

4 Conclusion

In this paper we lay our efforts on simulating spectator crowd behaviors. Different from the current works, we focus on individuals' affections inside the crowd and their transmission. Two- dimensional heat conduction model was used to calculate the results. In our simulation, emergence behaviours came out when group leader's movements was strong and his or her actions were proper to the competition situation, this could be seen in figure 6. Then the group leader became the global leader, as a result we treated the global behaviour a kind of special group behaviour. Although most of the current game simulation could formulate global spectator behaviour, yet we generated it spontaneously from bottom to up. Every person within the crowd was represented by an agent who could choose their own behaviours. Using our method, the spectator crowd expressed more diversified behaviours, especially the group behaviours. In real competitions, spectators must be affected by the atmosphere which created by both the players' performances and fans' encouraging. The fans of the football competitions have to formulate a group so that they can be noticed. And this is an important part for studying and simulating ball game competitions. However, to simulate a large scale crowd, a faster virtual human rendering method should be provided.

References

1. Narain, R., Golas, A., Curtis, S., Lin, M.C.: Aggregate dynamics for dense crowd simulation. ACM Transactions on Graphics 28(5), 122:1–122:8 (2009)
2. Treuille, A., Cooper, S., Popović, Z.: Continuum crowds. ACM Transactions on Graphics 25(3), 1160–1168 (2006)
3. Shao, W., Terzopoulos, D.: Autonomous pedestrians. Graphical Models 69(5-6), 246–274 (2007)
4. Yang, L., Zhao, D., Li, J., Fang, T.: Simulation of the kin behavior in building occupant evacuation based on Cellular Automaton. Building and Environment 40(3), 411–415 (2005)
5. Yu, Y., Song, W.: Cellular automaton simulation of pedestrian counter flow considering the surrounding environment. Physical Review E 75(4), 046112 (2007)
6. Jiang, H., Xu, W., Mao, T., Li, C., Xia, S., Wang, Z.: Continuum crowd simulation in complex environments. Computers & Graphics 34(5), 537–544 (2010)
7. van den Berg, J., Patil, S., Sewall, J., Manocha, D., Lin, M.: Interactive navigation of multiple agents in crowded environments. In: I3D 2008 Proceedings of the 2008 Symposium on Interactive 3D Graphics and Games, pp. 139–147 (2008)
8. Thalmann, D.: Crowd simulation. Wiley Online Library (2007)
9. Gu, Q., Deng, Z.: Formation sketching: an approach to stylize groups in crowd simulation, pp. 1–8. Canadian Human-Computer Communications Society
10. Isobe, M., Helbing, D., Nagatani, T.: Experiment, theory, and simulation of the evacuation of a room without visibility. Physical Review E 69(6), 066132 (2004)
11. Helbing, D., Farkas, I., Vicsek, T.: Simulating dynamical features of escape panic. Nature 407, 487–490 (2000)
12. Parisi, D., Dorso, C.: Microscopic dynamics of pedestrian evacuation. Physica A: Statistical Mechanics and its Applications 354, 606–618 (2005)
13. Braun, A., Bodmann, B.E.J., Musse, S.R.: Simulating virtual crowds in emergency situations. In: VRST 2005 Proceedings of the ACM Symposium on Virtual Reality Software and Technology, pp. 244–252 (2005)
14. Pelechano, N., Allbeck, J.M., Badler, N.I.: Virtual crowds: Methods, simulation, and control. Synthesis Lectures on Computer Graphics and Animation 3(1), 1–176 (2008)
15. Kirchner, A., Klüpfel, H., Nishinari, K., Schadschneider, A., Schreckenberg, M.: Simulation of competitive egress behavior: comparison with aircraft evacuation data. Physica A: Statistical Mechanics and its Applications 324(3), 689–697 (2003)
16. Pelechano, N., Malkawi, A.: Evacuation simulation models: Challenges in modeling high rise building evacuation with cellular automata approaches. Automation in Construction 17(4), 377–385 (2008)

Design of Semantic Web Retrieval Model Based on Ontology

Chaoyang Ji and Lingli Zhang

College of Computer Science and Technology
Xuchang University
461000 Xuchang, China
jcy@xcu.edu.cn

Abstract. The paper put forward a semantic web retrieval model based on ontology according to the defects of lack of knowledge representation and semantic processing capability in the traditional keyword-based information retrieval, and describes the main function modules and the key technologies.

Keywords: ontology, semantic web, information retrieval, model system.

1 Introduction

With the rapid development of the Information technology, there are more and more information on the network. How to obtain Information those users really need conveniently, fast and effectively on the network, it has been a research hotspot of computer application domain. Information retrieval is the process that finds the user's required information in the information collection. Keyword-based retrieval methods are the strictly mechanical matching of the keywords expressing user query request and the index term expressing information content in practice. Owing to one word having two or more meanings, one meaning with two or more words, there exist multiple expression differences between the keywords expressing user query request and the user's real needs, between keywords and index terms. Thus it cause that the precision of query result is low and the false detecting rate is high. The essential deficiency of the retrieval technology based on keyword is the lack of knowledge representation and semantic processing capabilities. The retrieval technology based on keyword is only syntactic matching of words and rarely able to come the word's semantic matching through knowledge inference, and it is difficult to get relevant results with the user's real needs. Therefore, it is the Key Problem of improving the efficiency of search systems that adding semantic information and enhancing knowledge processing capabilities of the retrieval system.

It can not be separated from knowledge representation and processing to add semantics for information retrieval. As a new method of knowledge representation, ontology extends knowledge representation to the semantic level and has ontology has a good concept hierarchy and logical reasoning support. It overcomes the limitations of keywords representation and can realize the semantic information representation, and it is the most important and difficult problems in research of the network

D. Jin and S. Lin (Eds.): Advances in CSIE, Vol. 2, AISC 169, pp. 703–708.
springerlink.com

information retrieval in recent years. Thus an ontology-based semantic retrieval model is presented in the paper.

2 A Summary of Semantic Web and Ontology

2.1 Semantic Web

Semantic Web is a concept proposed by Tim Berners-Lee who is the founder of the Internet in 1998. The core of the Semantic Web is that the document on the World Wide Web is added semantics which can be understood by the computer, so that the entire Internet becomes a common medium of information exchange. Architecture of the Semantic Web consists of seven layers, and bottom-up its layers function increases gradually.

The first layer is Unicode and URI. In the architecture of Semantic Web, this layer is the basis of the Semantic Web, which Unicode is responsible for resources encoding and the URI is responsible for the identification of resources. The second layer is XML+NS+XML Schema which are used to represent content and structure of data. The 3th Layer is RDF+RDF Schema which are used to describe resources and their types on the Web. The 4th Layer is Ontology vocabulary used to describe resources and their link with each other. The 5th Layer is logic, conducting logical reasoning operations on the basis of the following four levels. The 6th Layer is Proof, conducting verification in order to draw conclusions based on logical statements. The 7th Layer is trust, which is used to establish a trust relationship between users.

As the semantic layer, Ontology vocabulary is used to describe the heterogeneous information content in the web and is the core of the semantic retrieval.Different areas need to build different domain ontology and computers exchange the fields' information each other through the understanding to ontology. After the construction of domain ontology library, the query ontology and application ontology of information resources can be built using the domain ontology. It only needs to dynamically generate application ontology, and match with the query ontology.

2.2 The Concept of Ontology

The concept of ontology initially originated in the field of philosophy, and it is the systematic explanation or description of the objective existence. The research of ontology is the abstract nature of objective reality. Studer and other scholars believe that the ontology is the explicit formal specification of the shared conceptual model, and it includes conceptualization, explicit, formal and share four layers meaning. The conceptual model refers to the model obtained by abstracting some phenomena related concepts in the objective world, and the meaning of the conceptual model shown is independent of the specific state of the environment. Explicit refers to the concept used and the constraints using this concept have a clear definition. Formal refers to a precise mathematical description, and the extent of the description can reach the computer-readable level. Sharing refers to the common recognition knowledge embodied in the ontology, and it reflects the recognized concept set in the relevant fields, namely ontology for the consensus of the group rather than individual.

As a conceptual modeling tool, ontology can describe the information in the level of semantic and knowledge, provide semantic foundation of subject s exchange in the field and the share and common understanding to the meaning of terms, retrieve the documents pages which have different manifestations but semantic similarity. It is the key technologies of the Semantic Web and the research focus of intelligent information retrieval.

3 Semantic Retrieval Model Based on Ontology

3.1 The Design Idea of System Model

With the help of experts in the field, domain ontology library based on domain concept is established using ontology edit tool. According to established domain ontology, ontology metadata tagging of PDF, Web pages and other information resources is conducted. Using the ontology language RDF or OWL, information resources are represented and descripted in the semantic level. In the light of established domain ontology, collected data resource objects are stored in the application ontology library of metadata according to the prescribed format. Through the search interface the user submits a concept, selects a suitable keyword and domain for the need according to the keywords set and the domain set provided by the domain ontology, expresses the query needs in a maximum extent under the guidance of ontology, and constructs the query ontology. According to the constructed application ontology and query ontology, under the guidance of the semantic rules of the Knowledge Base, concept analysis and semantic reasoning are conducted through the semantics logical reasoning module. Information resource library which has been done by ontology metadata annotation is retrieved through the semantic search engine, and a matching list of ontology is got from the metadata database. Finally, retrieval results return to the user through the user search interface after the customized processing of removing duplicated information, merging and sorting, and users can also provide feedback to help update the relevant information of the knowledge library. The structure of ontology-based semantic retrieval mode is shown in Figure 1.

3.2 The Description of System Function

The ontology-based semantic retrieval model mainly consists of user interface module, semantic processing module, semantic retrieval module, document processing module and the output of search results.

(1) User interface module
It provides the user interface to interact with the system, and accepts the user's queries. After preprocessing, the query request will be submitted to the query semantic analyzer. The query request will be regulated using ontology, and then the knowledge base of domain ontology will be queried. All the fields which include the query request will be identified from the domain ontology knowledge base, and then the fields and the meaning of a query request in a field are listed to the user. According to his intentions, the user determines the field and the meaning he want to find through the interface.

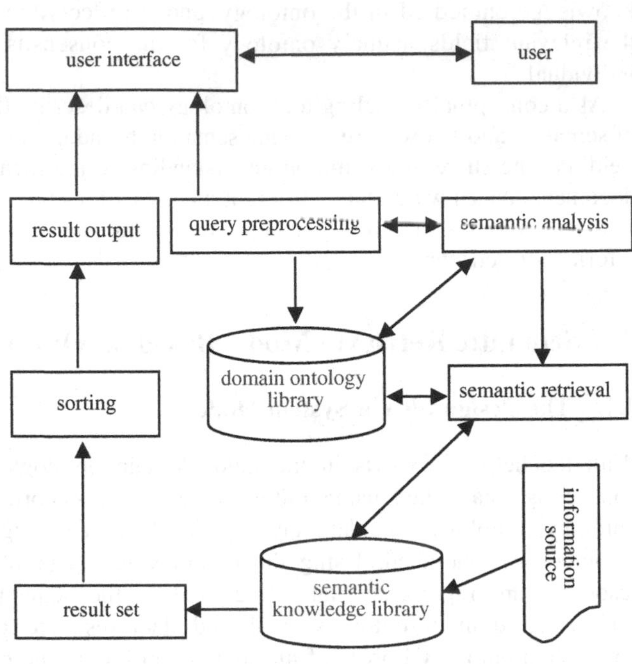

Fig. 1. Ontology-based semantic retrieval mode

(2) Semantic processing module
The query request which belong among the user's selecting query fields is analyzed. According to the similarity between the domain ontology, this module executes semantic query expansion to the query words or phrases. The user's query concept and relational map into the concept and relation in the ontology, and this as the basis of semantic expansion, this module will generate a new search expression. At the same time, this search expression will be submitted to the semantic retrieval module. According existing axioms and theorems, semantic retrieval module refers domain ontology, reasons the metadata in the semantic metadata database and adds new search words or phrases.

(3) Semantic retrieval module
Using lucene as indexing and retrieval tool, it can satisfy the retrieval requirements of strong extensibility and high performance, and can be easily integrated into applications to increase the index and search functions. Semantic knowledge library can be retrieved conveniently by search function after it is indexed by lucene.

(4) Document processing module
The module is mainly responsible for the indexing of information resources on the web, and gets the semantic knowledge library finally. That is extracting the useful information from the unstructured web documents described using HTML, XML, realizes semantic annotation, extracts the metadata, and then the metadata encoded by

the RDF triples is stored in the semantic metadata database. Documents can obtain using the Web crawler, and those also are the txt format documents which are various treated documents of the ontology.

(5) Output of search results

The vocabulary set of the domain ontology reasoning is as the basis for the retrieval of semantic knowledge database, and it match with the metadata stored in the semantic database. The results are prioritized according to the sorting algorithm, and return to the user through the user interface.

4 The Key Technique of Model Realization

Construction of ontology library, query preprocessing and similarity matching reasoning are the key technologies to achieve the model.

4.1 Construction of Ontology Library

The domain ontology plays an important role in the process of semantic query. Query disambiguation and query expansion need the support of domain ontology, and the collection, extraction, query and service of information sources are based on domain ontology. In the help of experts in the field building the domain ontology need to obtain the knowledge of related fields, provide a common understanding to the domain knowledge, determine the common vocabulary in the field, give a clear definition of the relationship between these terms from the formalization model of different levels, and then the domain ontology is built using some tools such as Protégé. The establishment of the domain ontology mostly uses manual mode, and also has semi-automated mode based on the original function words conversion. In the web environment, ontologies are stored in the form of RDFS files or OWL files.

4.2 Query Preprocessing

Information retrieval tools mainly provide the user with keyword-based search interface, but it is difficult to clearly express users' really search intention using a few key words in many cases, and this is one of the main reasons that existing retrieval systems can not meet users' needs. On the other hand, the user is not necessarily more familiar with the area he retrieved during the information retrieval, and it is possible that the user's retrieved content can not be expressed in accordance with the specifications in the field. Therefore, in order to allow the user to be able to better express his search intention, the retrieval tools should provide a search interface using natural language expression mode to the user. When the user submits questions to the system in the form of natural language, the module executes query preprocessing on the user's requests using the domain ontology knowledge and some simple natural language understanding technology, gets really user's search intention, and then submits pretreated search request to the query module.

4.3 Similarity Matching Reasoning

There are many methods for calculating the ontology similarity, and they are basically paired comparisons between entities that belong to different ontology. After Ontology is defined according to the areas, demand and grain size, the similarity between the ontology is calculated using ontology similarity matching theory. The similarity calculation mainly considers the similarity of concepts, attributes and relationships between the ontology, and it can use the method of fuzzy mathematics or threshold to define. Specific methods can use the restricted ontology similar technology to match similarity and it also can use joint probability distribution based on the concept of cluster to calculate the semantic similarity in the support of WordNet. On the development platform provided by Jena, using Racer or Pellet reasons the semantic relationship of user queries through calling predefined rules in the knowledge library.

5 Conclusion

According to the defects of lack of knowledge representation and semantic processing capability in the traditional keyword-based information retrieval, this paper put forward an ontology-based semantic web retrieval model. Compared with the traditional keyword-based retrieval model, through the application of ontology, semantic web technology, knowledge library, etc., this model solves the problem which has ambiguous relationships between words and semantic ambiguity in the simple keyword search. It can improve the completeness and accuracy of information retrieval through checking out the same or similar ontology and ontology instance in the retrieval. By means of the user feedback, it can also update the ontology library and amend the search results.

References

1. Wang, H., Sun, R.: Research of Semantic Retrieval System Based on Domain-ontology and Lucene. Journal of Computer Applications 06, 1655–1657, 1660 (2010)
2. Liu, L., Qin, J.: Research on Ontology-based Semantic Retrieval Model. Computer & Digital Engineering 12, 60–63 (2009)
3. Tian, X., Li, D.: Probability Estimation for Semantic Association on Domain Ontology. Computer Engineering and Applications 27, 136–140, 143 (2011)
4. Huang, H.: Concept Architecture of Semantic Search Engine Based on Ontology. Modern Electronics Technique 24, 90–92, 98 (2011)
5. Zhang, H., Zhang, M., Li, X.: E-learning resource library model based on domain ontology. Journal of Computer Applications 01, 191–195 (2012)

Integrated Scheduling Algorithm Based on Dynamic Essential Short Path

Zhiqiang Xie[1,2], Peng Wang[1], Zhongyan Gui[1], and Jing Yang[2]

[1] College of Computer Science and Technology, Harbin University of Science and Technology, Harbin 150080, China
[2] College of Computer Science and Technology, Harbin Engineering University, Harbin 150001, China
xzq0111@tom.com, {wangpengatc,hrbustguizhongyan}@126.com, yangjing@hrbeu.edu.cn

Abstract. In appearance, the total processing time of procedures on the longest path of the product processing tree is the lower bound of product completion time. In fact, essential paths which have great effects on scheduling result are formed due to serial processing of procedures on the same device. In order to reveal applications of essential path and avoid the occurrence of delaying in product processing as result of longer essential path being formed, an integrated scheduling algorithm based on dynamic essential short path has been proposed. Example validates the proposed algorithm can shorten the length of essential path formed by scheduling algorithm of long-path strategy and makes the product processing time less without increasing algorithm's time complexity.

Keywords: Integrated scheduling, Scheduling algorithm, Essential path, Dynamic essential short path, Serial processing.

1 Introduction

When production products belong to many varieties of small batch, especially belong to single piece of product with tree structure, processing and assembling respectively will inevitably affect internal relationship of parallel processing between them and affect scheduling results. Along with the increasing demand of product diversity, integrated scheduling algorithm of complex single product processing and assembling are getting more and more attention by people, and representative research results including ACPM algorithm [1] and algorithm considering long path strategy [2].

The ACPM algorithm [1] considers great effects of procedures on the long path on scheduling results and priority scheduling procedures on long path. Based on reference [1], the algorithm in reference [2] scheduling procedures with layer priority, short time strategy and long-path strategy, and realizes optimization in vertical and horizontal direction based mainly on horizontal direction and obtains better scheduling results. Above two representative algorithms all dynamically priority schedule procedures on the dynamic long path under certain conditions.

In the process of scheduling, the length of the longest path is changing as the already scheduled procedure being deleted from the processing tree. Factors which

D. Jin and S. Lin (Eds.): Advances in CSIE, Vol. 2, AISC 169, pp. 709–715.
springerlink.com

truly affect the lower bound of product completion time is not the long path of the static processing tree, but the essential long path proposed in this paper of which the longest process path is formed due to serial processing of procedures on the same device. In order to avoid the formation of much longer essential path in dynamic scheduling, the paper proposes an integrated scheduling algorithm based on dynamic essential short path and validates its advantages through examples.

2 Problem Description

2.1 Definition of Related Conceptions

Definition 1: Plan schedule procedure: the schedulable procedure on the longest path in the dynamic standby procedure set

Definition 2: Essential path: as serial processing of procedures on the same device, priority schedule some procedures will lead the start time of other procedures on the same device with them to move backward and form processing path.

Definition 3: Essential short path: the processing time of both the plan schedule procedure and other procedures on the same device with it in standby procedure sets are added to the front of the other's path length respectively to form essential path, compare these essential path length being formed and choose one which is shortest of all as the essential short path.

Definition 4: Essential schedule procedure: the schedulable procedure on essential short path.

2.2 Mathematical Description of the Problem

The paper constructs integrated scheduling model for complex single product processing and assembling on the basis of tree structure model proposed by reference [2].Though adopting essential short path method to control the length of the essential long path to make each procedure begin as early as possible under conditions that sequence order and essential path length taken short are being met, the mathematical model of the problem can be described as:

$$T=\min\{\max\{T_i\}\} . \tag{1}$$

$$s.t.\ \min\{t_y \mid H_{xy}=L_x+g_y=\min\{L_k+g_j, L_j+g_k\}, \text{thereinto, } L_k=\max\{L_i\}\} . \tag{2}$$

$$t_k \geq t_j+g_j . \tag{3}$$

$$t_r \geq \max(t_j+g_j) . \tag{4}$$

Thereinto, T_i denotes finishing time of procedure i (i=1, 2 ,… , n), T is the target of scheduling optimization. In equation (2), procedures k and j are schedulable procedures on the same device, L_k denotes the path length of procedure k, g_j denotes the processing time of procedure j, L_k+g_j denotes the essential path length which procedure j is on, H_{xy} denotes the length of essential short path when scheduling the plan schedule procedure k, t_y denotes start processing time of procedure y on essential short path H_{xy}. Equation (3) denotes procedures on the same device must process serially. Equation (4) denotes parent node procedures r can start processing after all its child node procedures j finish processing.

3 Scheduling Strategy Analysis

3.1 Analysis of Scheduling Strategy

During product processing process, the schedulable procedures are all leaf node procedures, though parts of these leaf node procedures has already be scheduled, the rest of the product are still with tree structure. As the equipment is unique, it needs to determine which procedure should be priority scheduled when there are many procedures needed to be processed on the same machine.

Adopting ACPM algorithms [1], algorithm in reference [2] and other scheduling rules [3-5] to scheduling procedure on the same device while considering constraints of processing order, as these algorithms all based on calculation of original data and do not take consideration of serial processing of procedures processing on the same device, thus make the essential path so long that it may prolong the product processing time. In order to reduce the essential long path, if the plan schedule procedure is unique, schedules it directly; if there exists schedulable procedures on the same device with the plan schedule procedure, adopts dynamic essential short path strategy and long-path strategy to confirm and schedule procedures; and first fit method also be adopted in order to make the product process as early as possible and reduce idle time on the machine.

3.2 Design of Scheduling Strategy

Dynamic Essential Short Path Strategy. provided that P_i and P_j denote procedures on the same device, T_i and T_j denote processing time of them respectively, L_i and L_j denote path length of them respectively, and the essential path length gained from cross addition is L_i+T_j and L_j + T_i. Choose the smallest one as the essential short path.

Long Processing Time Strategy. If essential path length are equal, namely $L_i + T_j = L_j + T_i$, if $T_i > T_j$ then $L_i > L_j$, that is the path length of the procedure P_i which has more processing time is longer, then the procedure P_i should be priority scheduled according to long-path strategy.

First Fit Scheduling Method. When scheduling a procedure, searches for the first idle time on the machine which is suit for the scheduling procedure and insert the scheduling procedure into the idle time under conditions that sequence constraints are being met [6].

4 Algorithm Design

Step1: Construct attribute $P_i/N_i/T_i/E_i/C_i/L_i$ for procedure P_i ($P_i \in P$, P denote set of all procedures) according to procedure information table.

Thereinto, attribute N_i denotes the only immediate successor procedure of P_i, when N_i is equal to 0 which denotes the procedure P_i does not have any successor procedure, namely P_i is the root node procedure; Attribute T_i denotes the working hours of procedure P_i; Attribute E_i denotes the processing machine of procedure P_i; Attribute C_i denotes the number of immediate predecessor procedures of procedure P_i; Attribute L_i denotes the path length to the root node procedure of procedure P_i.

Step 2: Calculate the number of immediate predecessor procedures and the path length L_i for each procedure.

Step 2.1: Confirm the attribute of immediate predecessor procedures C_i.

```
Begin C_i=0;    1≤i≤n
for j=1 to n {if (N_j=P_i)  C_i=C_i+1; }.
```

Step 2.2: Calculate path length of all procedures.

First, calculate path length of all leaf node procedures: provided that P_i denotes leaf node procedure, then looping execute $P_j= N_i$; $L_i= L_i +T_j$; $P_i =N_i$; until $N_i =0$.

Then, as path length of leaf node procedures P_i has been calculated, path length of procedure P_j which denotes immediate successor procedure of procedure P_i can be got through equation $L_j= L_i -T_i$; so all path length of non-leaf node procedures can be calculated through iteration of their immediate predecessor procedures' path length subtracting their immediate predecessor procedures' processing time.

Step 3: According to procedures' attribute, dynamic add schedulable procedures into standby procedure set S, confirm the plan schedule procedure, essential short path and essential schedule procedure in standby procedure set S.

Step 3.1: Confirm the plan schedule procedure.

Step 3.1.1: As procedures in standby procedure set S are dynamic and have no immediate predecessor procedures, so when C_i=0, then $P_i \in S$; At initial moment, all leaf node procedures are belong to S.

Step 3.1.2: In a moment, choose the procedure with the longest path length in standby procedure set S as the plan schedule procedure, if the procedure with the longest path length is not unique, choose one among them which has the shortest processing time as the plan schedule procedure according to short time strategy.

Suppose L_j=max{L_1, L_2, ..., L_m}; thereinto, m denotes the number of elements in standby procedure set S.

If there exist $L_i = L_j$ and $T_i<T_j$, then confirms P_i as the plan schedule procedure; or else confirms P_j as the plan schedule procedure.

Step 3.2: Confirm essential short path.

Step 3.2.1: Judge whether there are procedures on the same device with the plan schedule procedure P_j in the standby procedure set.

If there exist $E_i=E_j$; suppose H_{ij}, H_{ji} are essential path length being formed by cross addition of P_j and P_i, namely $H_{ij}=L_i+T_j$; $H_{ji}=L_j+T_i$. Otherwise, confirm the plan scheduling procedure P_j as essential scheduling procedure; turn to step 4.

Step 3.2.2: If H_{xi}=min{H_{j1}, H_{1j}, .. H_{ij}, H_{ji}.., H_{kj}, H_{jk}}, the essential path which P_i is on is essential short path.

Step 3.3: Confirm the schedulable procedure on essential short path be the essential schedule procedure, if there exist many essential short paths, choose the schedulable procedure with the longest processing time to be the essential schedule procedure.

Step 4: Scheduling the essential schedule procedure P_k to corresponding machine to process, and delete P_k respectively from set P and set S, namely P-{P_k}and S-{P_k}, for the immediate successor procedure Pi of procedure P_k, the attribute Ci of procedure P_i decreases 1, namely $C_i = C_i -1$, and judge whether there is a new schedulable procedure comes into being, if $C_i =0$, then procedure P_i can be schedulable and add it into standby procedure set *S*.

Step 5: If the standby procedure set *S* is empty, namely when *S*=Ø, the scheduling is finished, otherwise turn to step 3.

5 Example Comparision

Provided that there is a complex single product B with 18 procedures and processing on 4 machines respectively, and its product processing tree is shown as Fig.1. In the figure, the rectangle denotes procedure and the information in the rectangle is: procedure name/machine name/processing time.

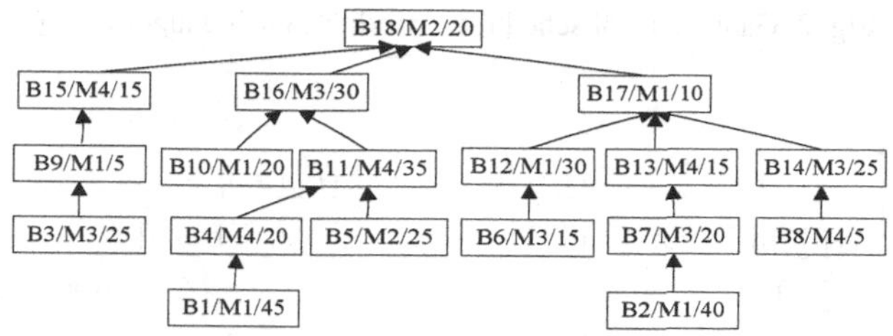

Fig. 1. Processing tree of product B

The following are scheduling the product B using①ACPM algorithm [1], ② algorithm in reference [2] and ③the essential short path proposed by this paper, advantages of the algorithm can be illustrated through contrasting Gantt chart.

Algorithm ① first calculates path length of leaf node procedures, and then confirms the critical path as (B1, B4, B11, B16, B18). The scheduling sequence by using ACPM algorithm is B1 B4 B5 B11 B10 B16 B2 B7 B13 B6 B12 B8 B14 B17 B3 B9 B15 B18. The Gantt chart is shown as Fig.2.

Algorithm ② mainly includes set priority for layers, short time priority strategy and long-path strategy, namely priority schedule procedures on the highest layer. The scheduling sequence of product B is: B2 B1 B8 B6 B4 B7 B5 B3 B9 B13 B10 B14 B12 B11 B17 B15 B16 B18. The Gantt chart is shown as Fig.3.

Algorithm ③ proposed in this paper scheduling procedures through essential short path strategy, long time strategy and first fit method. First, add all leaf node procedures into standby procedure set S, then confirm the procedure B1 which has the longest path length in set S as the plan schedule procedure, procedure B2 and B10 are on the same device with procedure B1, add the processing time of procedure B2 to the path length of procedure B1 to obtain an essential path length $H_{2,1}$, add the processing time of procedure B1 to the path length of procedure B2 to obtain essential path length $H_{1,2}$, then a set ($H_{2,1}$:150, $H_{1,2}$:190) can be got, in the same way, the set ($H_{10,1}$:115, $H_{1,10}$:170) of procedure B10 and the plan schedule procedure B1 can be obtained, in these two sets the path length of B1($H_{10,1}$:115) is shortest, so confirms procedure B1 as the plan schedule procedure. Delete B1 from standby procedure set S, and add the newly generated procedure B4 into the standby procedure set S. Repeat above operations. The scheduling sequence of the proposed algorithm is B1 B5 B4 B2 B8 B11 B6 B10 B7 B3 B9 B12 B14 B16 B13 B15 B17 B18, and the Gantt chart is shown as Fig.4.

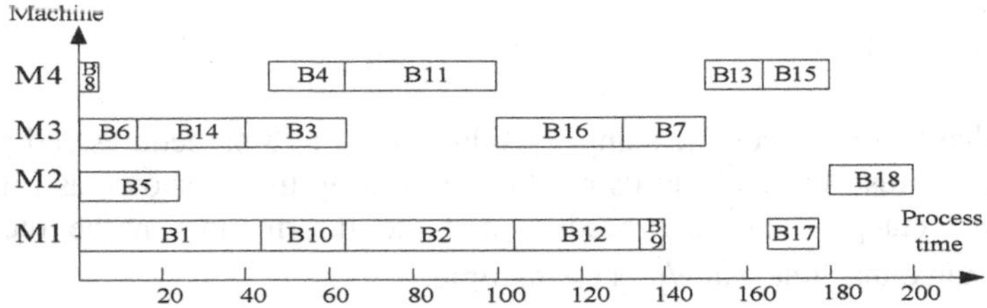

Fig. 2. Gantt chart of scheduling product B using algorithm①

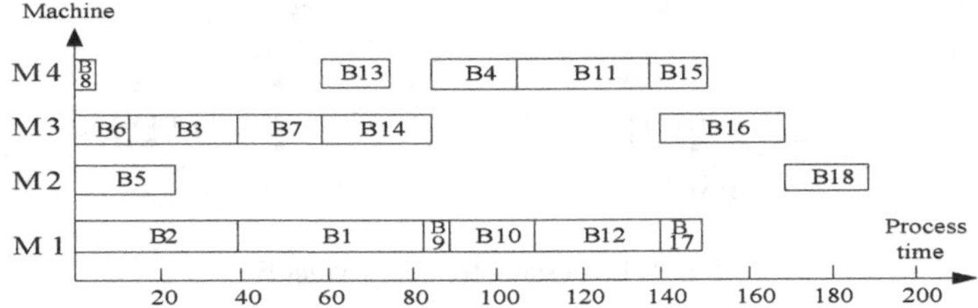

Fig. 3. Gantt chart of scheduling product B using algorithm②

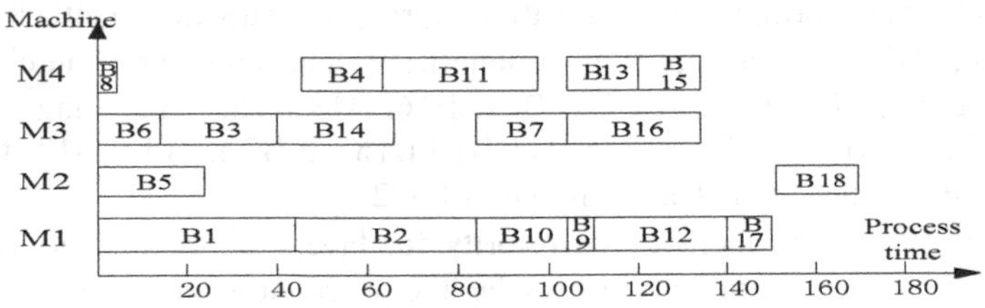

Fig. 4. Gantt chart of scheduling product B using algorithm③

Through contrasting Gantt chart it can be inferred that, under the premise without increasing time complexity of the algorithm, scheduling product procedures using algorithm ① the time consumed is 200 working hours, using algorithm ② the time consumed is 190 working hours and using algorithm ③ the time consumed is 170 working hours. So algorithm ③ is obviously better than algorithm ① and ②.

From above analysis, algorithm ③ absorbs advantages of long-path priority strategy of algorithm① and ②, meanwhile, it also considers the effect of serial processing of procedures on the same device on the scheduling result, through priority schedule procedures on dynamic essential short path to shorten the essential path length. So the scheduling result of algorithm ③ is better than algorithm ① and ②.

6 Conclusions

Aiming at the integrated scheduling problem of single complex product, the paper proposes and implements a method to control essential path length through choosing dynamic essential short path, further to make the product complete processing as early as possible. The conclusion as follows:

There existed essential paths in the process of integrated scheduling; adopts essential short path strategy can optimize the scheduling results of long-path strategy under conditions without increasing the algorithm's time complexity.

Acknowledgments. This project is supported by the National Natural Science Foundation of China (No. 60873019 and No.61073043), the Natural Science Foundation of Heilongjiang Province of China (No.F200901 and No.201101) and Harbin Outstanding Academic Leader Foundation (No.2010RFXXG054 and No.2011RFXXG015) and 2011 Graduate Student Innovation Research Foundation of Heilongjiang Province (No. YJSCX2011-035HLJ).

References

1. Xie, Z.Q., Liu, S.H., Qiao, P.L.: Dynamic Job-Shop Scheduling Algorithm Based on ACPM and BFSM. Journal of Computer Research and Development 40, 977–983 (2003)
2. Xie, Z.Q., Yang, J., Yang, G., Tan, G.Y.: Dynamic Job-Shop Scheduling Algorithm With Dynamic Set of Operation Having Priority. Chinese Journal of Computers 31, 502–508 (2008)
3. Gudehus, T., Kotzab, H.: Order Scheduling and Operating Strategies. Comprehensive Logistics, 247–270 (2009)
4. Jin, F.H., Kong, F.S., Kim, D.W.: Scheduling Rules for Assembly Job Shop Based on Machine Available Time. Computer Integrated Manufacturing Systems 14, 1727–1732 (2008)
5. Xie, Z.Q., Hao, S.Z., Ye, G.J., Tan, G.Y.: A new Algorithm for Complex Product Flexible Scheduling with Constraint between Jobs. Computers & Industrial Engineering 35, 983–989 (2009)
6. Wang, L.P., Jia, Z.Y., Wang, F.J., Meng, F.B.: Multi-product Complete Job-shop Scheduling Problem and its Solution. Systems Engineering—Theory & Practice 29(9), 73–77 (2009)

Author Index